30th International Symposium on Superconductivity (ISS2017)

Journal of Physics: Conference Series
Volume 1054

Tokyo, Japan
13 - 15 December 2017

Editors:

Atsutaka Maeda
Hiraku Ogino
Yoshiyuki Yoshida
Yasuhiro Iijima
Mutsuo Hidaka
Tsunehiro Hato
Naoyuki Amemiya
Tsuyoshi Wakuda

ISBN: 978-1-5108-6825-0
ISSN: 1742-6588

Printed from e-media with permission by:

Curran Associates, Inc.
57 Morehouse Lane
Red Hook, NY 12571

Some format issues inherent in the e-media version may also appear in this print version.

Printed by Curran Associates, Inc. (2018)

For permission requests, please contact the Institute of Physics at the address below.

Institute of Physics
Dirac House, Temple Back
Bristol BS1 6BE UK

Phone: 44 1 17 929 7481
Fax: 44 1 17 920 0979

techtracking@iop.org

Additional copies of this publication are available from:

Curran Associates, Inc.
57 Morehouse Lane
Red Hook, NY 12571 USA
Phone: 845-758-0400
Fax: 845-758-2633
Email: curran@proceedings.com
Web: www.proceedings.com

Table of contents

Volume 1054

30th International Symposium on Superconductivity (ISS2017)

13–15 December 2017, Tokyo, Japan

Accepted papers received: 28 June 2018
Published online: 26 July 2018

Preface

1. Physics and Chemistry

1.1 Novel materials

1.2 Cuprate superconductorsexperimental studies

1.3 Cuprate superconductorstheoretical studies

2.4 Iron-based superconducting materials and wires

2.5 Bulk materials and their applications

3. Electronic Devices

3.1 Detectors

4.2 Electric power devicessuperconducting cables

IOP Conf. Series: Journal of Physics: Conf. Series **1054** (2018) 011001 doi:10.1088/1742-6596/1054/1/011001

30th International Symposium on Superconductivity

I 超 S 電 S 導 2017

IINO HALL & CONFERENCE CENTER
TOKYO, JAPAN
December 13-15, 2017

30th International Symposium on Superconductivity (ISS2017)

Preface

It is our pleasure to report that the Proceedings of the commemorative 30th International Symposium on Superconductivity (ISS2017) held at Iino Hall and Conference Center, Tokyo, Japan, December 13–15, 2017, are now published in *Journal of Physics: Conference Series* (JPCS). ISS2017 was organized by the National Institute of Advanced Industrial Science and Technology (AIST), successively from the previous year. Total of 475 scientists, engineers, students, corporate executives and other participants (127 from abroad) gathered in ISS2017, and fruitful discussions were made to promote superconductor sciences and technologies. Total of 337 papers were presented, which included 6 plenary lectures, 88 invited talks, 29 contributed oral presentations and 214 poster presentations. The papers published in JPCS were categorized into relevant topics in the following four fields; (a) Physics and Chemistry, (b) Wires and Bulk, (c) Electronic Devices, (d) Large Scale System Applications.

To ensure the high publication standard mandated by JPCS, every paper was peer reviewed by a reviewer with expertise, and in some cases by two reviewers before it was accepted for publication. As editors of the Proceedings, we would like to express our sincere appreciation to all the reviewers involved in the evaluation of the papers for their invaluable contribution. Our special thanks go also to session chairs who recommended appropriate reviewers.

The Editors

Physics and Chemistry: Atsutaka Maeda (The University of Tokyo), Hiraku Ogino (AIST)

Wires and Bulk: Yoshiyuki Yoshida (AIST), Yasuhiro Iijima (Fujikura Ltd.)

Electronic Devices: Mutsuo Hidaka (AIST), Tsunehiro Hato (SUSTERA)

Large Scale System Applications: Naoyuki Amemiya (Kyoto University), Tsuyoshi Wakuda (Hitachi Ltd.)

IOP Conf. Series: Journal of Physics: Conf. Series **1054** (2018) 011001 doi:10.1088/1742-6596/1054/1/011001

Organization of ISS2017

General Chair:
Dr. Toshihiko Kanayama
Fellow, National Institute of Advanced Industrial Science and Technology (AIST)

Local Organizing Committee:
Secretary-General: Michiya Okada, TIA Central Office, AIST

Members:
Haruhiko Obara, Fumio Takemura, Teruhisa Horita, Mitsuho Furuse (Department of Energy and Environment, AIST)
Satoshi Haraichi, Masataka Ohkubo, Mutsuo Hidaka, Yoshiyuki Yoshida, Akira Iyo (Advisor), Shigeyuki Ishida (Department of Electronics and Manufacturing, AIST)
Susumu Ogawa, Hirofumi Yamasaki, Haruko Okazaki (TIA Central Office, AIST)

Supported by:
Applied Superconductivity Constellations of Tsukuba (ASCOT)
TIA—a platform for open innovation
Ministry of Economy, Trade and Industry (METI)
Ministry of Education, Culture, Sports, Science and Technology (MEXT)
New Energy and Industrial Technology Development Organization (NEDO)
Japan Science and Technology Agency (JST)
Cryogenics and Superconductivity Society of Japan (CSSJ)
The Japan Society of Applied Physics (JSAP)
The Institute of Electrical Engineers of Japan (IEEJ)

IOP Conf. Series: Journal of Physics: Conf. Series **1054** (2018) 011001 doi:10.1088/1742-6596/1054/1/011001

Committees of ISS2017

Program committee

Physics and Chemistry

Co-Chair: Atsutaka Maeda (The Univ. of Tokyo), Sub-Chair: Hiraku Ogino (AIST)

Members: Ryotaro Arita (RIKEN), Takekazu Ishida (Osaka Prefecture Univ.), Minoru Nohara (Okayama Univ.), Tsutomu Nojima (Tohoku Univ.), Takao Sasagawa (Tokyo Inst. of Technology), Takasada Shibauchi (The Univ. of Tokyo)

Advisors: Hiroshi Eisaki (AIST), Akira Iyo (AIST)

Wires and Bulk

Co-Chair: Yoshiyuki Yoshida (AIST), Sub-Chair: Yasuhiro Iijima (Fujikura Ltd.)

Members: Satoshi Awaji (Tohoku Univ.), Hiroshi Ikuta (Nagoya Univ.), Takeshi Kato (Sumitomo Electric Industries, Ltd.), Takanobu Kiss (Kyushu Univ.), Akiyoshi Matsumoto (National Inst. for Materials Science), Kaname Matsumoto (Kyushu Inst. of Technology), Jun-ichi Shimoyama (Aoyama Gakuin Univ.), Hideki Tanaka (Hitachi Ltd.)

Electronic Devices

Co-Chair: Mutsuo Hidaka (AIST), Sub-Chair: Tsunehiro Hato (SUSTERA)

Members: Yoshimi Hatsukade (Kindai Univ.), Masashi Ohno (The Univ. of Tokyo), Shigeo Sato (Tohoku Univ.), Naoto Sekiya (Yamanashi Univ.), Masamitsu Tanaka (Nagoya Univ.), Hirotake Yamamori (AIST)

Advisor: Masataka Ohkubo (AIST)

Large Scale System Application

Co-Chair: Naoyuki Amemiya (Kyoto Univ.), Sub-Chair: Tsuyoshi Wakuda (Hitachi Ltd.)

Members: Kazuhiro Kajikawa (Kyushu Univ.), Shinji Matsumoto (National Inst. for Materials Science), Shin-ichi Mukoyama (Furukawa Electric Co., Ltd.), Naoko Nakamura (MAYEKAWA MFG. Co., Ltd.), Taketsune Nakamura (Kyoto Univ.), Tetsuo Oka (Niigata Univ.), Tomonori Watanabe (Chubu Electric Power Co., Inc.), Shoichi Yokoyama (Mitsubishi Electric Corporation),

Advisor: Mitsuho Furuse (AIST)

International Advisory Committee

Physics and Chemistry

Hideo Aoki (The Univ. of Tokyo), Wai-Kwong Kwok (Argonne National Laboratory), Satoshi Okuma (Tokyo Inst. of Technology), Setsuko Tajima (Osaka Univ.), Hai-Hu Wen (Nanjing Univ.)

IOP Conf. Series: Journal of Physics: Conf. Series **1054** (2018) 011001 doi:10.1088/1742-6596/1054/1/011001

Wires and Bulk

David Cardwell (Univ. of Cambridge), Leonardo Civale (Los Alamos National Laboratory), Michael Eisterer (Vienna Univ. of Technology), Timothy J. Haugan (Air Force Research Laboratory), Bernhard Holzapfel (Karlsruhe Inst. of Technology), Teruo Izumi (AIST), Hiroaki Kumakura (National Inst. for Materials Science), David Larbalestier (Florida State Univ.), Qiang Li (Brookhaven National Laboratory), Xavier Obradors (Institut Ciencia de Materials de Barcelona), Venkat Selvamanickam (Univ. of Houston), Michael Sumption (Ohio State Univ.), Yutaka Yoshida (Nagoya Univ.)

Electronic Devices

Pascal Febvre (Univ. of Savoie Mont Blanc), Masataka Ohkubo (AIST), Shigetoshi Ohshima (Yamagata Univ.), Horst Rogalla (Univ. of Colorado at Boulder), Keiichi Tanabe (SUSTERA), Zhen Wang (SC2 & SIMIT, Chinese Academy of Sciences)

Large Scale System Application

Bob Buckley (Victoria Univ. of Wellington), Toru Fukushima (SuperPower Inc.), Mathias Noe (Karlsruhe Inst. of Technology), Hiroyuki Ohsaki (The Univ. of Tokyo), Minwon Park (Changwon National Univ.), Christopher M. Rey (Energy-to-Power Solutions), Pascal Tixador (Grenoble-INP), Liye Xiao (Chinese Academy of Sciences)

Peer review statement

All papers published in this volume of *Journal of Physics: Conference Series* have been peer reviewed through processes administered by the proceedings Editors. Reviews were conducted by expert referees to the professional and scientific standards expected of a proceedings journal published by IOP Publishing.

Structural Modulation in $LaO_{0.9}F_{0.1}BiSe_2$ Single Crystals Revealed by Scanning Tunneling Microscopy/Spectroscopy

N Ishida, S Demura, Y Fujisawa, S Ohta, K Miyata and H Sakata

Department of Physics, Tokyo University of Science, Shinjyuku-ku, Tokyo 162-8601, Japan

E-mail: 1215605@ed.tus.ac.jp

Abstract. We present scanning tunneling microscopy and spectroscopy measurements on a cleaved surface of the $LaO_{0.9}F_{0.1}BiSe_2$ single crystals. Tunneling spectra show a finite local density of states at E_F, which is consistent with metallic conductivity in bulk. In addition, the existence of the supermodulation running along the diagonal directions of Bi square lattice was revealed. The period of the supermodulation was about 3 to 5 times the length of the lattice constant. This period is close to that observed in $LaO_{0.5}F_{0.5}BiSe_2$.

1. Introduction

Enormous attentions have been paid to layered superconductors, such as cuprates and iron-based superconductors because of their high superconducting transition temperature T_c. Newly discovered BiS_2-based superconductors $LnO_{1-x}F_xBiCh_2$ (Ln = La, Pr, Ce, Nd, Yb and Bi, Ch = S, and Se) also have a layered structure [1]. A lot of BiS_2-based superconductors were synthesized with several kinds of Ln or Ch, where the carrier density is controlled by the amount of F.

Recently, scanning tunneling microscopy and spectroscopy (STM/STS) measurements on single crystals of $NdO_{0.7}F_{0.3}BiS_2$ revealed the existence of a novel electronic structure so-called "Checkerboard stripe" structure in BiS plane. However, spectroscopic gap was observed at the Fermi energy E_F despite of metallic bulk properties [2]. Furthermore, the unexpected supermodulation with the period of about 5 times the length of the lattice constant running along the diagonal direction of the Bi square lattice was revealed in $LaO_{0.5}F_{0.5}BiS_2$ (x = 0.5) single crystals by synchrotron X-ray measurements [3] and in $LaO_{0.5}F_{0.5}BiSe_2$ (x = 0.5) single crystals by STM/STS measurements [4]. Theoretically, in high F doping samples, such as $LaO_{0.5}F_{0.5}BiS_2$ (x = 0.5), the phonon instability spreading from near Γ point to M point is predicted while in low F doping samples, such as $LaOBiS_2$ (x = 0), the phonon instability around Γ point is predicted [5]. Thus, the measurements in different F doping samples give us the cue to understand the nature of the observed supermodulation. In this paper, we performed STM/STS measurements in $LaO_{0.9}F_{0.1}BiSe_2$ (x = 0.1) single crystals and observed the supermodulation. The results were compared with the supermoduation reported in $LaO_{0.5}F_{0.5}BiSe_2$ (x = 0.5).

2. Experimental

Single crystals of $LaO_{0.9}F_{0.1}BiSe_2$ were grown by CsCl flux method in vacuumed quartz tubes [6]. Mixtures of Bi (Mitsuwa Chemicals Co. Ltd., 99.9%), Se (Kojyundo Chemicals Co. Ltd., 99.99%), Bi_2O_3 (Kojyundo Chemicals Co. Ltd., 99.99%), and BiF_3 (Stella Chemifa Co. Ltd., 99.9%) were ground with nominal compositions of $LaO_{0.9}F_{0.1}BiSe_2$ except La. After grind into powder, La ribbons (Rare Metallic Co. Ltd., 99.9%) were mixed into the mixture. The mixture of 0.8 g was mixed with CsCl powder of 5 g, and sealed in an evacuated quartz tube. The thermal process is the followings. First, the tube was heated up to 800 °C at a rates of 100 °C/h, secondly, kept for 10 hours, finally, cooled down to 630 °C at a rates of 10 °C/h. After this thermal process, the heated material was washed by distilled water to remove the flux.

Superconducting transition temperatures T_c were determined from the temperature dependence of the resistivity. T_c of the sample of $x = 0.1$ and $x = 0.5$ were 3.2 K and 3.8 K, respectively. The difference in T_c indicates the difference in the actual amount of F between $x = 0.1$ and $x = 0.5$ samples [1].

STM/STS measurements were performed with laboratory-built STM in the He gas at 4.2 K. The clean surface of $LaO_{0.9}F_{0.1}BiSe_2$ single crystal was prepared by cleavage at 4.2 K *in situ*. A bias voltage was applied to the sample in all measurements. Tunneling spectra were obtained by numerically differentiation of the *I-V* curves.

3. Result

Figure 1 (a - c) show typical STM images and a tunneling spectrum in $LaO_{0.9}F_{0.1}BiSe_2$ ($x = 0.1$) single crystal. The STM image in Figure 1 (a), which was taken at a positive sample bias voltage, indicates a square lattice with a period of $a_0 = 4.7$ Å corresponding to the in-plane lattice constant. The observed bright and dark spots are Bi atoms and defects of Bi atoms, respectively [2, 4, and 7]. In addition, some Bi defects have dark streaks along one or two diagonal directions of the Bi square lattice. The density of defects was about 1% and was almost the same as that in $x = 0.5$ sample [4]. Thus, the Bi defect density is almost independent of the amount of F.

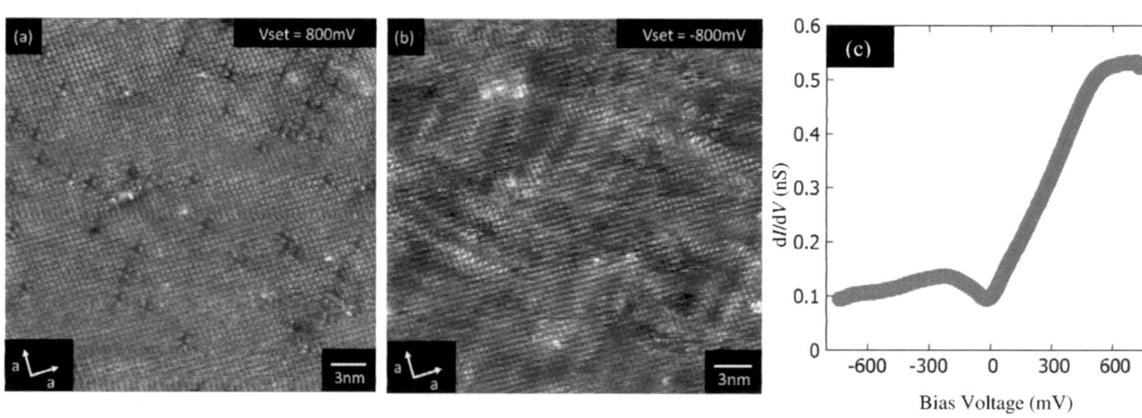

Figure 1 (Color online). (a) STM image of $LaO_{0.9}F_{0.1}BiSe_2$ ($x = 0.1$) single crystals taken at $V_{set} = 800$ mV and $I_{set} = 300$ pA. (b) STM image taken at $V_{set} = -800$ mV and $I_{set} = 300$ pA. (c) Typical tunneling spectrum taken at $V_{set} = 800$ mV and $I_{set} = 300$ pA.

Figure 1 (b) shows the STM image taken at a negative sample bias voltage for the same field of view as Figure 1 (a). In addition to the Bi square lattice, the supermodulation running along the diagonal directions of the Bi square lattice exists. Figure 2 (a) shows the low-pass-filtered image of Figure 1 (b) superimposed with the positions of Bi defects and dark streaks seen in Figure 1 (a). Bi defects and streaks are marked by gray circles and red dashed lines, respectively. The observed supermodulation is not two dimensional wave spreading over the surface uniformly, but consists of locally developed one dimensional waves. There seems to exist the relation between the supermodulation and Bi defects. There are two tendencies: (i) the supermodulation which is perpendicular to a dark streak terminates at Bi defects, (ii) dark streaks are located in the boundaries where the direction of supermodulation changes. This indicates that Bi defects affect the modulation seriously.

Next, we focus on the period of the supermodulation. Because the observed modulation is not regular, Fourier transformed images, shown in Figure 2 (b) and 2 (c), indicate that the signal corresponding to the supermodulation is no longer spots but broadened. Thus, the period was estimated from the real space image shown in Figure 1 (b).

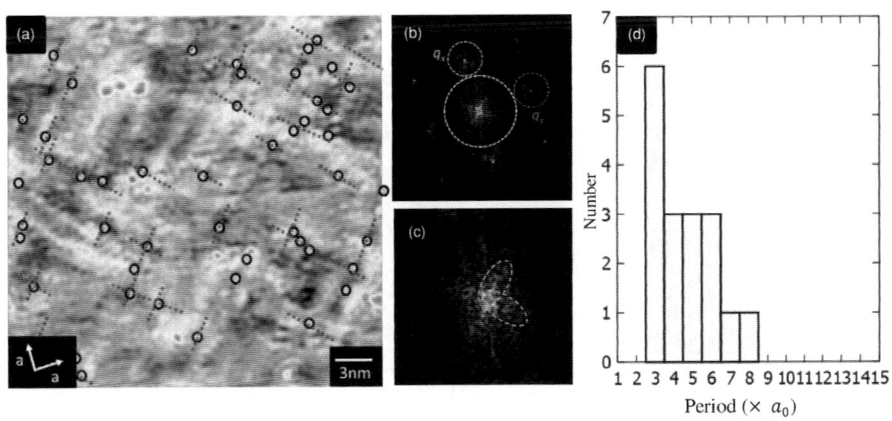

Figure 2 (Color online). (a) Low-pass-filtered image of Figure 1 (b). Cut off wave number is $|q| = 0.7 \times 2\pi/a_0$ marked by white dashed circle in Figure 2 (b). (b) Fourier transform image of Figure 1 (b). (c) Magnified image of Figure 2 (b) around Γ point. (d) Histogram of the period of the modulation.

To evaluate the period of the supermodulation, we took cross sections of the STM image and measured the distance between the crests of the modulation at several locations. Here, we call this distance as period. The histogram of the period is shown in Figure 2 (d). According to the Figure 2 (d), the estimated period was distributed between 1 and 2 nm, which corresponds to 3 to 5 times the length of the lattice constant. The period of 3 times the length of the lattice constant is the most frequent. Thus the observed period shows a little shorter value than that in $LaO_{0.5}F_{0.5}BiSe_2$ ($x = 0.5$).

4. Discussion

The STM observations revealed the existence of the supermodulation even in $LaO_{0.9}F_{0.1}BiSe_2$ ($x = 0.1$). Likewise $LaO_{0.5}F_{0.5}BiSe_2$ ($x = 0.5$), the observed supermodulation runs along the diagonal directions of Bi square lattice. Therefore, it is considered that the supermodulation in $LaO_{0.9}F_{0.1}BiSe_2$ ($x = 0.1$) is essentially the same as that in $LaO_{0.5}F_{0.5}BiSe_2$ ($x = 0.5$). Although the period of the observed supermodulation in $LaO_{0.9}F_{0.1}BiSe_2$ was a little shorter than that observed in $LaO_{0.5}F_{0.5}BiSe_2$ ($x = 0.5$), their features are almost same. Thus, it is difficult to say that the modulation is significantly modified with the amount of F. In theoretical calculation, doping dependence of the wave vector at which the phonon becomes instable in $LaO_{1-x}F_xBiS_2$ is predicted [5]. The observed results, however, does not consistent with the theoretical expectation. Thus, the origin of the formation of the supermodulation is still an open question.

It is noted, as shown in Figure 1 (c), the observed tunneling spectrum shows a finite density of states at the Fermi energy. This is consistent with the electric resistivity measurements [6, 8], indicating that the STM/STS measurements in this material represent the bulk property of the sample. Thus, the observed supermodulation is thought to also exist in the bulk. In fact, recent reports on synchrotron X-ray diffraction measurements in $LaO_{0.5}F_{0.5}BiS_2$ [3] claimed the existence of such supermodulation.

5. Summary

We succeeded in STM/STS observation in $LaO_{0.9}F_{0.1}BiSe_2$ single crystals and revealed the existence of the supermodulation running along the diagonal direction of the Bi square lattice at a negative bias voltage. Although there is fluctuation in the period of the supermodulation, the period was about from 3 to 5 times the length of the lattice constant, which is close to the value observed in $LaO_{0.5}F_{0.5}BiSe_2$ single crystals.

References

[1] Demura S 2015 *Novel Supercond. Mater.* **2** 1-15
[2] Machida T, Fujisawa Y, Nagao M, Demura S, Deguchi K, Mizuguchi Y, Takano Y and Sakata H 2014 *J. Phys. Soc. Jpn.* **83** 113701
[3] Sagayama R *et al* 2016 *The Phys. Soc. Jpn. 2016 Autumn Meeting*
[4] Demura S, Ishida N, Fujisawa Y and Sakata H 2017 *J. Phys. Soc. Jpn.* **86** 113701
[5] T Yildirim 2013 *Phys. Rev.* B **87** 020506(R)
[6] Nagao M, Tanaka M, Watauchi S, Tanaka I and Takano Y 2014 *J. Phys. Soc. Jpn.* **83** 114709
[7] Demura S, Fujisawa Y, Machida T, Nagao M, Takano Y and Sakata H 2016 *Phys. Proc.* **81** 49
[8] Ishida N *et al* 2016 *29th Int. Symp. Superconductivity*

Effect of non-magnetic rare earth substitution for Zr on mixed anion $Zr(P, Se)_2$ superconductors

K Iwakiri[1,2], T Nishio[2], K Kawashima[1,3], S Ishida[1], K Oka[1],

H Fujihisa[1], Y Gotoh[1], A Iyo[1], H Ogino[1], H Eisaki[1], Y Yoshida[1], and H Kitô[1]

[1]National Institute of Advanced Industrial Science and Technology (AIST), 1-1-1 Umezono, Tsukuba, Ibaraki 305-8568, Japan

[2]Department of Physics, Faculty of Science Division II, Tokyo University of Science, 1-3 Kagurazaka, Shinjuku-ku, Tokyo 162-8601, Japan

[3]IMRA Material R&D Co., Ltd, Kariya, Aichi 448-0032, Japan

E-mail: h.kito@aist.go.jp

Abstract. It is shown that the maximum superconducting transition temperature, T_c for $ZrP_{1.25}Se_{0.75}$ is increased from 5.8 K to 7.2 K by partially substituting a non-magnetic rare earth atom Lu for Zr and on the other hand is decreased by substituting other non-magnetic rare earth atoms Y and La. Obviously, T_c is raised when lattice constants in the *a*-axis and the *c*-axis increase with replacing Zr. Its implications for further increasing T_c are discussed.

1. Introduction

Recently, high temperature superconductivity has been discovered in layered compounds such as cuprates and iron-based superconductors [1-4]. $Zr(P_{2-x}Se_x)$ superconductors discovered in 2014 [5] have the PbFCl-type layered structure in which a double block layer [6] composed of Zr and $(P_{1-x}Se_x)$ is sandwiched between electrically conducting P-square lattice sheets, as shown in Fig. 1. Moreover, T_c in $Zr(P_{2-x}Se_x)$ exhibits a dome-like substitution dependence, as seen in cuprates [7]. According to the similarities, an effort to find new high-T_c materials having the PbFCl-type structure has been made until now [8]. Czulucki *et al.* [9] found $HfAs_{1.7}Se_{0.2}$ with T_c of 0.52 K. NbSiAs found by Ryu et al. [10] exhibits T_c of 8.2 K, which is relatively high in PbFCl-type compounds.

In this paper, we present T_c for $ZrP_{1.25}Se_{0.75}$ partially substituted by non-magnetic rare earth atoms

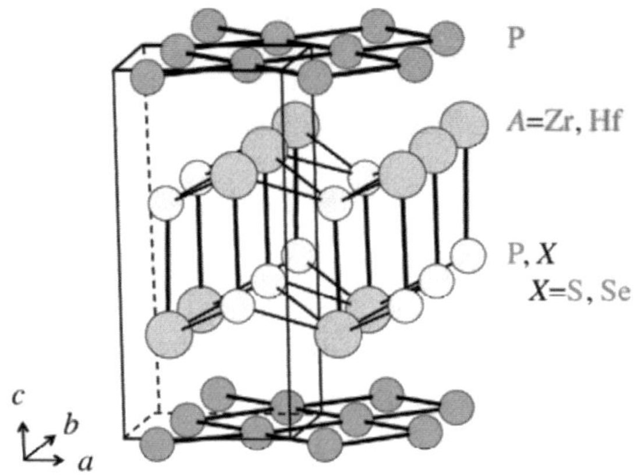

Figure 1. Schematic image of the crystal structure of $AP_{2-x}X_x$ (A = Zr, Hf; X = S, Se).

(Y, La and Lu) for Zr. It is shown that T_c is raised when lattice constants in the *a*-axis and *c*-axis increase with replacing Zr.

2. Experimental Details

Polycrystalline samples of $(Zr_{1-x}A_x)P_{1.25}Se_{0.75}$ (A = Y, La and Lu) were synthesized by a cubic-anvil high-pressure apparatus. First, powders of P, Se, Zr, and A at a molar ratio of 1.25: 0.75: 1-x : x were mixed in a N_2 atmosphere and then heated in a BN crucible at 1350 ~ 1450 °C under a pressure of 2 GPa for 2 hours. Powder X-ray diffraction (XRD, Rigaku Ultima-IV) measurements were made in a range of 5°~ 65° in 2θ at room temperature. Magnetization for samples as a function of temperature was measured at 20 Oe by a SQUID magnetometer (MPMS, Quantum design). Electrical resistivity was measured by a four-terminal method for rectangular samples. Scanning electron microscope (SEM, Hitachi High-Technologies TM3000) and energy dispersive X-ray spectrometry (Oxford Instruments SwiftED 3000) measurements were made for composition analysis.

3. Results and discussion

Figure 2 shows powder XRD patterns for $(Zr_{1-x}Lu_x)P_{1.25}Se_{0.75}$. Clearly, the PbFCl-type structure is preserved in a nominal composition range of $x < 0.5$, however, an impurity ($(Zr, Lu)Se_2$ and (Zr, Lu)Se) phase emerges prominently (2θ~ 14.5° and 32.0° *etc.*) at nominal composition range $x \gtrsim 0.25$. The PbFCl-type structure phase has not been obtained in a nominal composition range of $x > 0.5$.

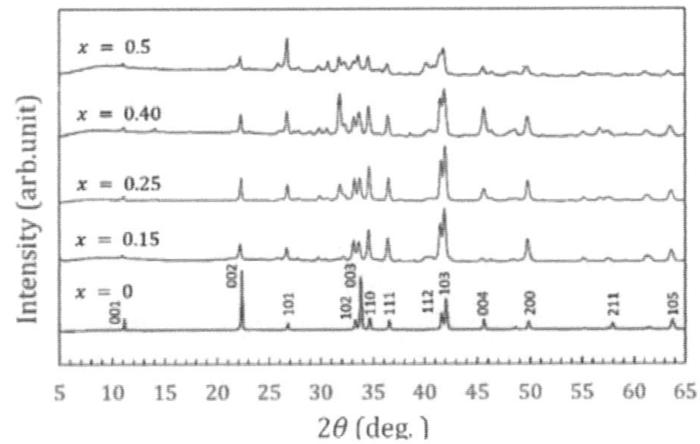

Figure 2. Powder XRD patterns of $(Zr_{1-x}Lu_x)P_{1.25}Se_{0.75}$ for nominal composition $x < 0.5$.

Figure 3 shows lattice constants in the *a*-axis and the *c*-axis for $(Zr_{1-x}A_x)P_{1.25}Se_{0.75}$ (*A*=Y, La and Lu) as a function of nominal composition *x*. Each lattice constant is approximately proportional to nominal composition *x*. Lattice constants do not increase with substituting Y and La but Lu, though these ionic radii are larger than it of Zr. We presume that there is a solubility limit of each *A* atom for Zr. In substituting La, a substantial amount of LaSe as an impurity has been observed in EDX measurements (not shown), so that in fact La ions do not occupy Zr sites at all but Zr ions react with Se ions to form LaSe. Se ions are removed out from $ZrP_{1.25}Se_{0.75}$, meanwhile, lattice constants decrease rapidly, as seen in Fig. 3.

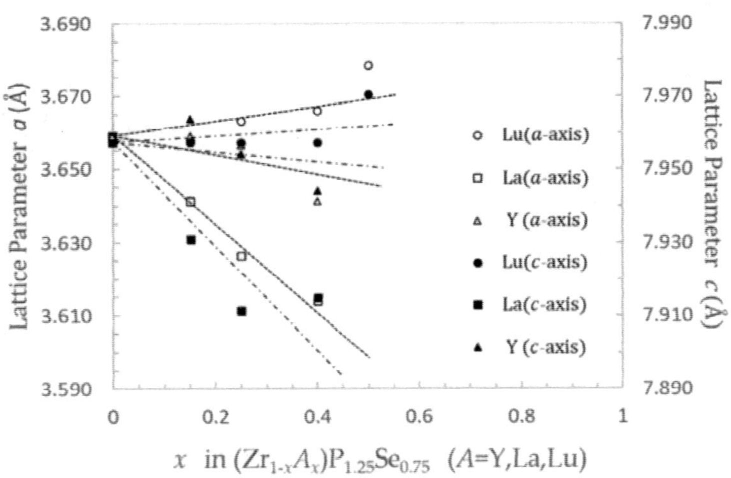

Figure 3. Lattice Parameters of the *a*-axis and the *c*-axis for $(Zr_{1-x}A_x)P_{1.25}Se_{0.75}$ (*A* = Y, La and Lu).

Figure 4(a) shows magnetization of $(Zr_{1-x}Lu_x)P_{1.25}Se_{0.75}$ as a function of temperature. T_c seems to increase with increasing nominal composition x, but on the other hand the superconducting volume fraction becomes lower. As shown in Fig. 4(b), in fact the nominal composition x dependence of T_c shows a dome-like behavior in $(Zr_{1-x}Y_x)P_{1.25}Se_{0.75}$ and $(Zr_{1-x}Lu_x)P_{1.25}Se_{0.75}$ but $(Zr_{1-x}La_x)P_{1.25}Se_{0.75}$. The maximum onset T_c is 6.4 K at x=0.15 in $(Zr_{1-x}Y_x)P_{1.25}Se_{0.75}$ and 7.2 K at x=0.4 in $(Zr_{1-x}Lu_x)P_{1.25}Se_{0.75}$, respectively. Using $AP_{2-x}X_x$ (A = Zr, X = S, x = 0.46) single crystal, the value of T_c = 3.5 K was reported [11, 12]. These maximum onset T_c values were higher than that of $AP_{2-x}X_x$ (A = Zr, X = S, x = 0.46). We consider these as follow. T_c increases as nominal composition x increases in a range of nominal composition x smaller than a solubility limit. When nominal composition x reaches a solubility limit, T_c becomes the maximum. T_c begins to drop with further increasing nominal composition x. Meanwhile, lattice constants decrease rapidly. T_c may rise further if Zr is substituted by trivalent ions, which dissolve well in $Zr(P_{1-x}Se_x)$.

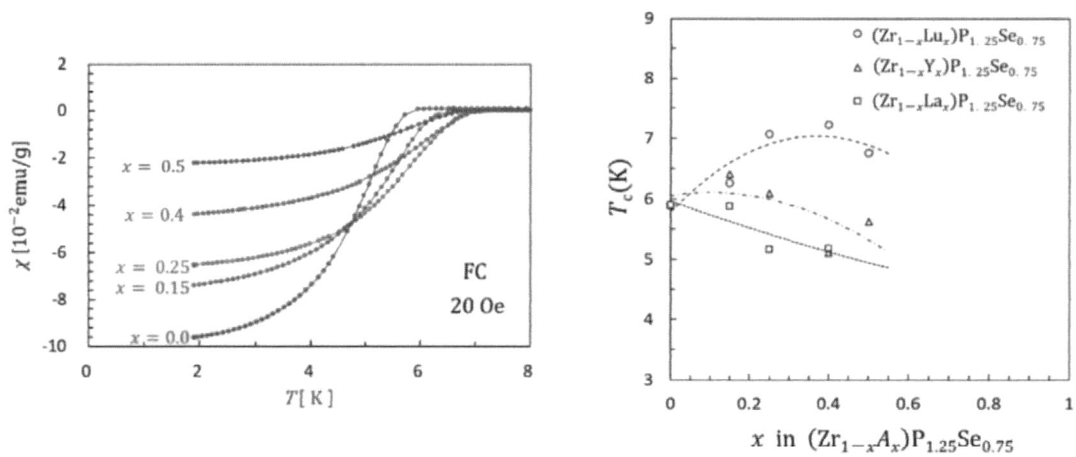

Figure 4. (a) T-dependence of magnetic susceptibility for $(Zr_{1-x}Lu_x)P_{1.25}Se_{0.75}$. (b) nominal composition x-T_c phase diagram of $(Zr_{1-x}A_x)P_{1.25}Se_{0.75}$ (A =Y,La and Lu)

Figure 5 shows electrical resistivity for $(Zr_{0.6}Lu_{0.4})P_{1.25}Se_{0.75}$ as a function of temperature. Resistance was confirmed to go to zero at 7.2 K. As shown in Fig. 4(a), superconducting volume fraction becomes lower with increasing nominal composition x. However, a bulk superconducting phase is indeed dominant at nominal composition x = 0.4. T_c observed in resistivity is in good agreement with that in magnetization.

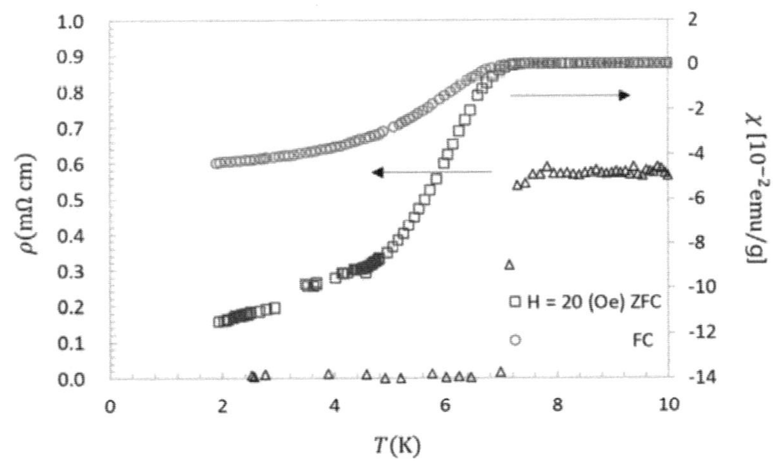

Figure 5. Electrical resistivity and magnetic susceptibility of $(Zr_{0.6}Lu_{0.4})\ P_{1.25}Se_{0.75}$.

4. Summary

In $(Zr_{1-x}A_x)P_{1.25}Se_{0.75}$ (A = Y, La and Lu), variations of T_c and the crystal structure indicate that T_c rises when Zr is adequately replaced by trivalent non-magnetic rare earth atoms. Because of a solubility limit, an increase of T_c is limited to 1 ~ 2 K as of now. In this research, it was clarified that T_c can be controlled by the nominal substitution concentration x of A site in A - (P, Se) frame units for $(Zr_{1-x}A_x)P_{1.25}Se_{0.75}$ [13].

Acknowledgement

This work partially supported by JSPS KAKENHI Grant-in-Aid for Scientific Research on Innovative Areas "Mixed anion" (Grant Number JP16H6439).

References

[1] Kamihara Y, Watanabe T, Hirano M, and Hosono H 2008 *J. Am. Chem. Soc.* **130** 3296-3297

[2] Rotter M, Tegel M, and Johrendt D 2008 *Phys. Rev. Lett.* **101** 107006

[3] Bednorz J G and Müller K A 1986 *Z. Phys. B* **64** 189-193

[4] Wu M K, Ashburn J R, Torng C J, Hor P H, Meng R L, Gao L, Huang Z J, Wang Y Q, and Chu C W 1987 *Phys. Rev. Lett.* **58** 908

[5] Kitô H, Yanagi Y, Ishida S, Oka K, Gotoh Y, Fujihisa H, Yoshida Y, Iyo A and Eisaki H 2014 *J. Phys. Soc. Japan* **83** 074713

[6] Ishida S, Fujihisa H, Hase I, Yanagi Y, Kawashima K, Oka K, Gotoh Y, Yoshida Y, Iyo A, Eisaki H and Kitô H 2016 *Supercond. Sci. Technol.* **29** 055004

[7] Takagi H, Ido T, Ishibashi S, Uota M, Uchida S and Tokura Y 1989 *Phys. B* **40** 2254

[8] Cichorek T, Bochenek L, Schmidt M, Czulucki A, Auffermann G, Kniep R, Niewa R, Steglich F and Kirchner S 2016 *Phys. Rev. Lett.* **117** 106601

[9] Czulucki A, Auffermann G, Bednarski M, Bochenek L, Böhme M, Cichorek T, Niewa R, Oeschler N, Schmidt M, Steglich F et al. 2010 *ChemPhysChem.* **11** 2639

[10] Ryu G, Kim S W, Mizoguchi H, Matsuishi S and Hosono H 2012 *Europhys. Lett.* **99** 27002

[11] Cichorek T, Bochenek L, Czulucki A, Schmidt M, Niewa R, Auffermann G, Prots Y, Wawryk R, Baenitz M, Steglich F and Kniep R, 2011 *Scientific Report MPG/CPfS* 103 (www.cpfs.mpg.de/2081022/B_2009-2010.pdf)

[12] Baenitza M, Lüders K, Kniep R, Steglich F and Schmidt M, 2016 *Physics Procedia* **81** 65

[13] Iwakiri K, 2016 Graduation Thesis, Tokyo Univ. of Science

Single Crystal growth of mixed anion $Zr(P, Se)_2$ superconductor and related materials

H Kitô[1], K Iwakiri[1,2], T Nishio[2], K Kawashima[1,3], S Ishida[1], K Oka[1], H Fujihisa[1], Y Gotoh[1], A Iyo[1], H Ogino[1], H Eisaki[1], and Y Yoshida[1]

[1]National Institute of Advanced Industrial Science and Technology (AIST), 1-1-1 Umezono, Tsukuba, Ibaraki 305-8568, Japan

[2]Department of Physics, Faculty of Science Division II, Tokyo University of Science, 1-3 Kagurazaka, Shinjuku-ku, Tokyo 162-8601, Japan

[3]IMRA Material R&D Co., Ltd, Kariya, Aichi 448-0032, Japan
E-mail: h.kito@aist.go.jp

Abstract. A new method to grow single crystals of PbFCl-type mixed anion $AP_{2-x}X_x$ (A = Zr, Hf, X = S, Se) superconductor is examined. Using a wedge-type, cubic-anvil, high-pressure apparatus, single crystals of $ZrP_{1.25}Se_{0.75}$ (A = Zr, X = Se) were grain grown from nominal composition melts under a pressure of 2.0 GPa. Obtained Plate-like single crystal with approximate edge sizes of $160 \times 160 \times 20\,\mu\,m^3$ was measured on electron probe microanalysis and X-ray diffraction indicate that the as-grown boules are a single phase $ZrP_{1.25}Se_{0.75}$. Superconducting transition temperature (T_c) is 6.31 K for $ZrP_{1.25}Se_{0.75}$ single crystal. It also succeeded in single crystal growth of substitution of nonmagnetic rare earth atoms for A (A = Zr) site, T_c = 6.91 K was reached in the $(Zr_{0.50}Lu_{0.50})PSe$ nominal composition. In this paper, the crystal growth of these compound are also reported.

1. Introduction

The mixed anion PbFCl-type intermetallic $AP_{2-x}X_x$ (A=Zr, Hf, X=S, Se) superconductor has a tetragonal layer-like structure consisting of alternately stacked A-(P, X) frame units and P square net layers along the c-axis. (see Fig. 1) Using polycrystalline sample study for $ZrP_{2-x}Se_x$, superconducting transition temperature (T_c) increases with increasing nominal composition x at first, reaches its maximum T_c value (6.31 K) at around nominal composition x=0.7, after which it decreases monotonically [1, 2]. T_c was controlled by the nominal substitution concentration x of (P, X) site in A-(P, X) frame units for $AP_{2-x}X_x$ (A = Zr, Hf, X = S, Se). The resultant phase diagrams are thus dome-shaped, mimicking the phase diagrams of high-T_c cuprates [3] or iron-based superconductors [4].

Non-magnetic rare earth (La, Y and Lu) substitution of A site in A-(P, X) frame units for $AP_{2-x}X_x$ (A = Zr, X = Se), nominal substitution concentration y-T_c phase diagram for $AP_{2-x}X_x$ (A= $Zr_{1-y}Ln_y$ (Ln= La, Y and Lu), X = Se) were also dome-shaped. In this polycrystalline samples research, T_c was also controlled by the nominal substitution concentration y of A site in A - (P, X) frame units for $AP_{2-x}X_x$ (A= $Zr_{1-y}Ln_y$ (Ln= La, Y and Lu), X = Se) [5, 6].

In this $AP_{2-x}X_x$ (A=Zr, Hf, X=S, Se) system, it is very important to understand for the expression mechanism of the superconductivity, so we try to growth the single crystal.

IOP Conf. Series: Journal of Physics: Conf. Series **1054** (2018) 012003 doi:10.1088/1742-6596/1054/1/012003

The high-pressure synthesis in geoscience research is very powerful technique for new material synthesis and single crystal growth. For example, MgB_2 superconductor [7], CoP_3 Skutterudite [8] had already reported. In this paper, grown the single crystal for $AP_{2-x}X_x$ (A = $Zr_{1-y}Ln_y$ (Ln= Y and Lu), X = Se) under the high-pressure were reported.

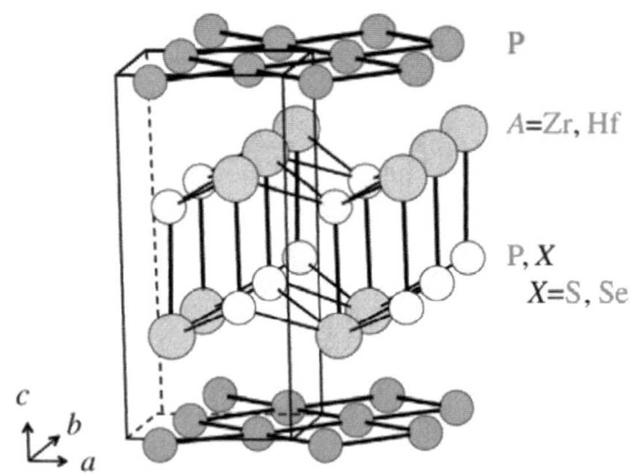

Figure 1. The crystal structure for mixed anion PbFCl-type intermetallic PbFCl-type $AP_{2-x}X_x$ (A = Zr, Hf; X = S, Se) superconductor. This PbFCl-type superconductor has a tetragonal layer-like structure consisting of alternately stacked A - (P, X) frame units and P square net layers along the c-axis.

2. Experiment and results

2.1. *Experiment*

$AP_{2-x}X_x$ single crystals were grown by a grain growth technique under the high pressure using a cubic anvil press. Zr (powder, 98 %), Lanthanide (powder, 99.9 %), P (grain, 99.9999%) and Se (grain, 99.999%) were mixed and pelletized in a nitrogen-filled glove box. The pellets (total mass: ~0.35 g) were put into BN crucibles and heated at ~2,050 ℃ under a pressure of approximately 4.0 GPa maintained for 3.0 h and then rapidly quenched to room temperature. Starting nominal composition were $ZrP_{1.25}Se_{0.75}$ (A = Zr, X = Se, x = 0.75), $(Zr_{0.5}Y_{0.5})PSe$ (A = $Zr_{0.5}Y_{0.5}$, X = Se, x = 1.0) and $(Zr_{0.5}Lu_{0.5})PSe$ (A = $Zr_{0.5}Lu_{0.5}$, X = Se, x = 1.0). The obtained single crystals had good cleavage, producing flat surfaces.

The compositions of the samples were analyzed by an energy dispersive X-ray spectrometry (Oxford Instruments SwiftED 3000) equipped in an electron microscope (Hitachi High-Technologies TM3000).

The crystal structure of the synthesized single-crystalline sample examined using the X-ray diffraction (XRD) technique. The intensity data were measured with a Rigaku Ultima-IV diffractometer using Cu Kα radiation and were collected over a 2θ range of 10~70° with a step width of 0.02° at 45 kV and 40 mA and a counting rate of 6° / minutes.

The temperature (T-) dependence of the DC magnetic susceptibility $\chi(T)$ was measured at 20 Oe using a superconducting quantum interference device (SQUID) magnetometer (Quantum Design MPMS). The data were collected during warming after zero-field cooling (ZFC) and then during field cooling (FC).

2.2. *Results*

Figure 2 shows the microstructure for obtained samples observed by scanning electron microscopy (SEM). For obtained samples, size of typical plate-like single crystals were ~160 × ~160 × ~20 μm^3 for $ZrP_{1.25}Se_{0.75}$ (A = Zr, X = Se, x = 0.75), ~100 × ~80 × ~20 μm^3 for $(Zr_{0.5}Y_{0.5})PSe$ (A =

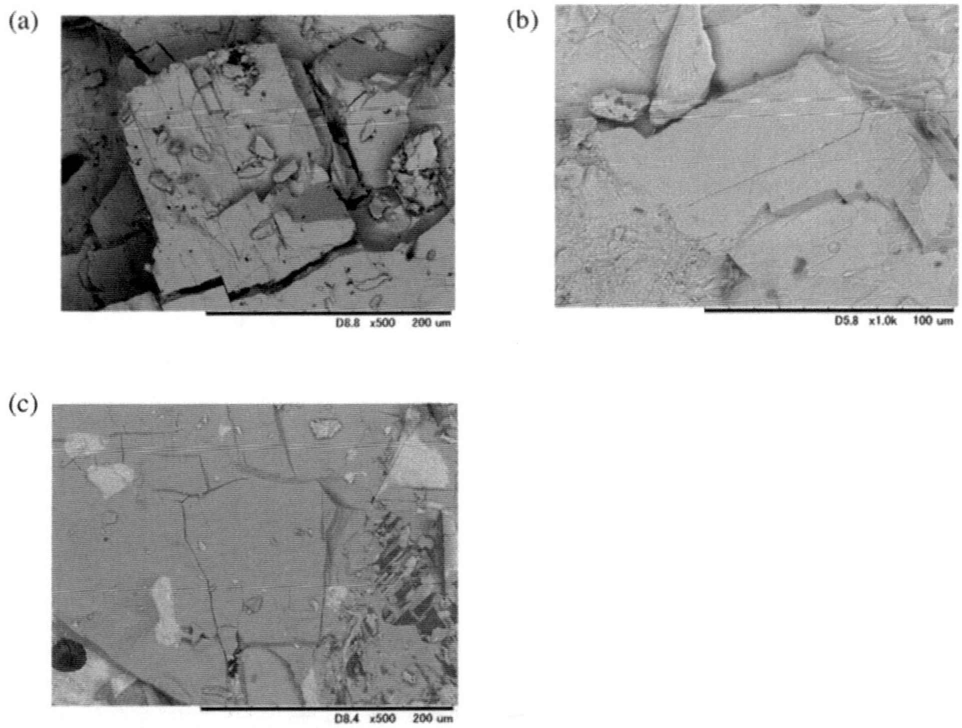

Figure 2. Morphology of the obtained typical plate-like $AP_{2-x}X_x$ crystal by SEM image; (a) $ZrP_{1.25}Se_{0.75}$ (A = Zr, X = Se, x = 0.75), (b) $(Zr_{0.5}Y_{0.5})PSe$ ($A = Zr_{0.5}Y_{0.5}$, X = Se, x = 1.0) and (c) $(Zr_{0.5}Lu_{0.5})PSe$ ($A = Zr_{0.5}Lu_{0.5}$, X = Se, x = 1.0).

$Zr_{0.5}Y_{0.5}$, X = Se, x = 1.0) and ~100 × ~80 × ~20 μm^3 for $(Zr_{0.5}Lu_{0.5})PSe$ ($A = Zr_{0.5}Lu_{0.5}$, X = Se, x = 1.0), respectively.

To investigate the chemical compositions for obtained $AP_{2-x}X_x$ single crystal, we performed elemental analysis by using energy dispersive x-ray spectrometry (EDS). The averaged Zr (or Zr : Ln) : P : Se ratios were determined to be Zr : P : Se = 1.0 : 1.04(4) : 0.642(13) for $ZrP_{1.25}Se_{0.75}$ (A = Zr, X = Se, x = 0.75), Zr : Y : P : Se = 0.82(9) : 0.054(3) : 1.159(5) : 0.841(5) for $(Zr_{0.5}Y_{0.5})PSe$ ($A = Zr_{0.5}Y_{0.5}$, X = Se, x = 1.0) and Zr : Lu : P : Se = 0.9013(12) : 0.0987(12) : 1.303(14) : 0.556(5) for

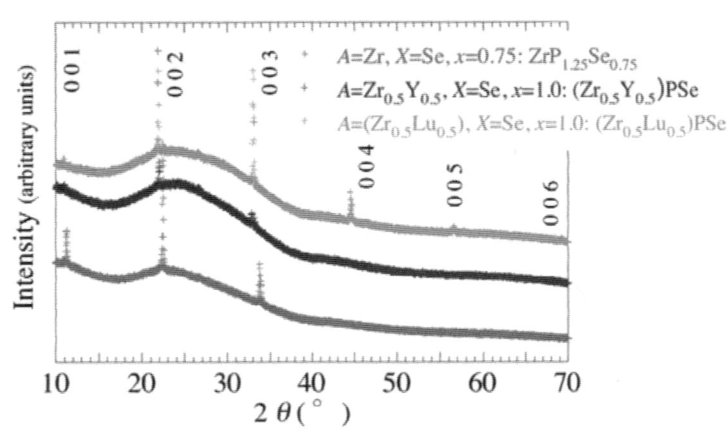

Figure 3. X-ray diffraction patterns of the well-developed plane of the obtained typical $AP_{2-x}X_x$ single crystal along c-axis at room temperature.

$(Zr_{0.5}Lu_{0.5})PSe$ (A = $Zr_{0.5}Lu_{0.5}$, X = Se, x = 1.0), respectively. The difference between the nominal composition and the obtained elemental analysis result is mainly thought to be due to the presence of the (Zr, *Ln*)P and (Zr, *Ln*)Se_2 impurity phase.

Figure 3 shows X-ray diffraction for obtained $AP_{2-x}X_x$ single crystals. These diffraction pattern only shows 00*l* diffraction peaks indicating that the crystals are highly oriented along the *c*-axis.

Figure 4 is the results for the magnetic susceptibility for obtained $AP_{2-x}X_x$ single crystals. The superconducting transition was drastically improved for the single crystal, showing a sharp transition. The large difference between the zero-field cooling and field cooling curves of the obtained sample indicates that the material has a fairly large flux pinning force resulting in the trapping of magnetic flux in the field cooling condition. T_c(onset) was ~6.31 K for $ZrP_{1.25}Se_{0.75}$ (A = Zr, X = Se, x = 0.75), T_c(onset) was ~6.71 K for $(Zr_{0.5}Y_{0.5})PSe$ (A = $Zr_{0.5}Y_{0.5}$, X = Se, x = 1.0) and T_c(onset) was~6.91 K for $(Zr_{0.5}Lu_{0.5})PSe$ (A = $Zr_{0.5}Lu_{0.5}$, X = Se, x = 1.0). Using $AP_{2-x}X_x$ (A = Zr, X = S, x = 0.46) single crystal, the value of T_c = 3.5 K was reported [9, 10]. These T_c(onset) values were higher than that of $AP_{2-x}X_x$ (A = Zr, X = S, x = 0.46).

Figure 5 shows nominal substitution concentration y-T_c phase diagram for $AP_{2-x}X_x$ (A = $(Zr_{1-y}Ln_y)$, X = Se, x = 1.0). T_c increases with increasing nominal compostion y at first, reaches its maximum value at around nominal composition y=0.5, after which it decreases monotonically. T_c was controlled by the

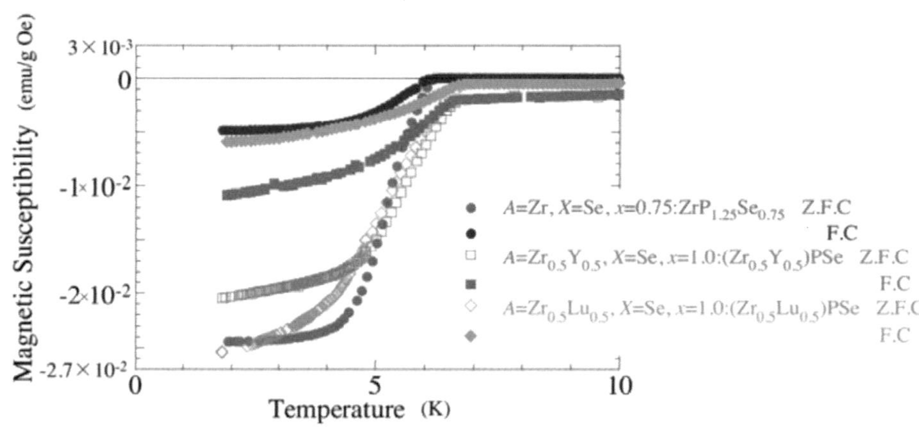

Figure 4. The temperature dependence of the magnetic susceptibility under H=20 Oe for the obtained piece of $AP_{2-x}X_x$ single crystal.

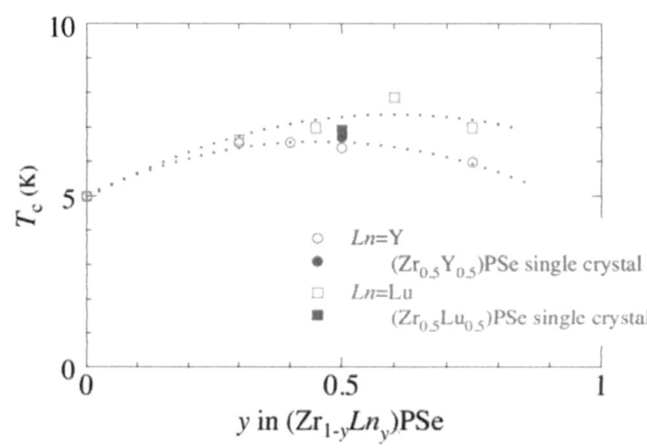

Figure 5. Nominal substitution concentration y-T_c phase diagram for $AP_{2-x}X_x$ (A = $(Zr_{1-y}Ln_y)$, X = Se, x=1.0). The filled squares were shown as results for single crystal samples and unfilled squares were shown as results for the polycrystalline samples.

substitution concentration y of A site in A-(P, X) frame unit for $AP_{2-x}X_x$ (A = $(Zr_{1-y}Ln_y)$, X = Se). This phase diagrams are also thus dome-shaped.

3. Summary

By the grain growth technique using the high-pressure apparatus, $AP_{2-x}X_x$ ($ZrP_{1.25}Se_{0.75}$ and $(Zr_{0.5}Y_{0.5})PSe$ and $(Zr_{0.5}Lu_{0.5})PSe$) single crystals were obtained for the first time. The magnetic susceptibility measurements shows T_c(onset) ~6.31 K for $ZrP_{1.25}Se_{0.75}$ (A = Zr, X = Se, x = 0.75), T_c(onset) ~6.71 K for $(Zr_{0.5}Y_{0.5})PSe$ (A = $Zr_{0.5}Y_{0.5}$, X = Se, x = 1.0) and T_c(onset) ~6.91 K for $(Zr_{0.5}Lu_{0.5})PSe$ (A = $Zr_{0.5}Lu_{0.5}$, X = Se, x = 1.0). Using the single crystal, the detail phase diagrams of the exactly substitution concentration y-T_c for $(Zr_{1-y}Ln_y)(P_{2-x}X_x)$ (X=Se, x = 1.0) will be obtained more clearly.

Acknowledgement

This work partially supported by JSPS KAKENHI Grant-in-Aid for Scientific Research on Innovative Areas "Mixed anion" (Grant Number JP16H6439).

References

[1] Kitô H, Yanagi Y, Ishida S, Oka K, Gotoh Y, Fujihisa H, Yoshida Y, Iyo A, and Eisaki H, 2014 *J. Phys. Soc. Jpn.* **83** 074713
[2] Ishida S, Fujihisa H, Hase I, Yanagi Y, Kawashima K, Oka K, Gotoh Y, Yoshida Y, Iyo A, Eisaki H, and Kito H, 2016 *Supercond. Sci. Technol.* **29** 055004
[3] Takagi H, Ido T, Ishibashi S, Uota M, Uchida S, and Tokura Y, 1989 *Phys. Rev.* B40 2254
[4] Rotter M, Pangerl M, and Johrendt D, 2008 *Angew. Chem. Int. Ed.* **47** 7949
[5] Iwakiri K, 2016 Graduation Thesis, Tokyo Univ. of Science
[6] Iwakiri K, Nishio T, Kawashima K, Ishida S, Oka K, Fujihisa H, Gotoh Y, Iyo A, Ogino H, Eisaki H, Yoshida Y, and H Kitô, Submitted to *J. Phys. Conf. Series*
[7] Tsuda S, Yokoya T, Takano Y, Kito H, Matsushita A, Yin F, Itoh J, Harima H, and Shin S, 2003 *Phys. Rev. Lett.* **91** 127001
[8] Lee C.H, Kito H, Ihara H, Akita K, Yanase N, Sekine C, and Shirotani I, 2004 *J. Cryst. Growth* **263** 358
[9] Cichorek T, Bochenek L, Czulucki A, Schmidt M, Niewa R, Auffermann G, Prots Y, Wawryk R, Baenitz M, Steglich F, and Kniep R, 2011 *Scientific Report MPG/CPfS* 103 (www.cpfs.mpg.de/2081022/B_2009-2010.pdf)
[10] Baenitza M, Lüders K, Kniep R, Steglich F, and Schmidt M, 2016 *Physics Procedia* **81** 65

IOP Conf. Series: Journal of Physics: Conf. Series **1054** (2018) 012004 doi:10.1088/1742-6596/1054/1/012004

Electronic Structure of Novel Binary Superconductor $SrGe_2$: A First-Principles Study

I Hase[1], T Yanagisawa[1], A Iyo[1], H Eisaki[1], Y Yoshida[1] and K Kawashima[2]

[1] National Institute of Advanced Industrial Science and Technology (AIST), Tsukuba, 305-8568, Japan
[2] IMRA Material R&D Co. Ltd.

E-mail: i.hase@aist.go.jp

Abstract. Due to the simplicity of its composition, many binary superconductors have been developed so far. Some are practically used like Nb_3Ge, and others have high T_c like MgB_2 or single layer FeSe. Recently we discovered that tetragonal $SrGe_{2\text{-d}}$ shows superconductivity by high pressure synthesis. This compound attracts our attention because it has the highest T_c among AGe_2 series (A: alkaline-earth or rare-earth element). In this paper we investigated the electronic structure of $SrGe_2$ from first principles. There are several polymorphs of $SrGe_2$, and the obtained density of states at the Fermi level (=$D(E_F)$) per $SrGe_2$ is 1.00, 1.20, 1.12 eV^{-1} for trigonal, hexagonal and tetragonal structure, respectively. $D(E_F)$ of tetragonal $SrGe_2$ is not more than the isostructural and isovalent compound $BaGe_2$, which has lower T_c than $SrGe_2$.

1. Introduction

Now that quantum computers are becoming a reality, the necessity of developing new superconductors is increasing more and more. Among them, binary superconductors are very important in terms of their applications because of their simple composition. For example, Nb_3Ge and MgB_2 are used as superconducting wires. Moreover, some binary superconductors have very high T_c, such as MgB_2, monolayer FeSe, and H_2S with applying very high pressure.

Recently we have discovered a new binary superconductor $SrGe_{2\text{-d}}$, which has $ThSi_2$-type tetragonal structure[1]. $SrGe_2$ has several polymorphs. It crystalizes into $BaSi_2$-type orthorhombic structure under ambient pressure, and transforms to $EuGe_2$-type trigonal structure under high pressure[2]. This experimental result is also supported by a theoretical study using first-principles calculation[3]. Therefore, our finding of the tetragonal phase of $SrGe_2$ (hereafter called *te*-$SrGe_2$) is rather unexpected. The band structure calculation of this *te*-$SrGe_2$ has not been reported yet. There are several binary superconductors MX_2 (M=Ba,Sr,La; X=Si,Ge) which have the same crystal structure as *te*-$SrGe_2$. Interestingly, *te*-$SrGe_2$ has the highest T_c among them.

The crystal structure of *te*-$SrGe_2$ has also a unique feature. Focusing on the network of Ge atoms, it forms the same network as what is called hyper-graphite or 3D-honeycomb lattice. There are intensive studies on hyper-graphite, and there are reported novel electronic structure such as Dirac point and Dirac loop [4-6]. It is interesting whether *te*-$SrGe_2$ has also the similar electronic state.

In this paper, we investigated the electronic structure of $SrGe_2$ by first-principles calculation. We found that the calculated density of states at the Fermi level (=$D(E_F)$) is much larger than that of the trigonal phase, and comparable with the AlB_2-type hexagonal phase. We also compare the electronic structure of tetragonal $BaGe_2$, $SrGe_2$ and $SrSi_2$ and found that they have pseudo-Dirac points, where three bands are entangled instead of two.

2. Methods

We have calculated the electronic structure of tetragonal $BaGe_2$, $SrGe_2$ and $SrSi_2$ from first-principles. As for $SrGe_2$, we also calculated that for three types of crystal structure: tetragonal, trigonal and hexagonal one. This calculation is based on the density functional theory (DFT) and the full-potential linearized augmented plane wave method (FLAPW), and the exchange-correlation potential is approximated by general gradient approximation (GGA) [7]. The whole calculation is implemented by the computer code WIEN2k [8]. We used the parameter RK_{max}=7.0, and the muffin-tin radii are set as r(Sr/Ba)=2.5 a.u. and r(Ge/Si)=2.19 a.u. We used the experimentally determined lattice parameters and space group (#141, $I4_1$/amd), and for $SrGe_2$ we optimized the internal parameter u of Ge(0, 1/4, u) within the above restrictions. We obtained u=0.2905, which agrees well with the experimental value for tetragonal $BaGe_2$ ($u_{exp,Ba}$=0.29178). As for the structure optimization and the self-consistent calculation we used the 1000 k-points mesh in the 1st Brillouin zone. After convergence of potential, we performed the DOS calculation using 8000 k-points mesh. Spin-orbit interaction is not included in this calculation.

3. Results and Discussions

First we show the band dispersion of tetragonal $BaGe_2$, $SrGe_2$ and $SrSi_2$ in figure 1. Since they are isovalent, their band structures are very similar. In these band structures, k-point similar to the Dirac point can be seen. That is, E_F is located near the point where the bands are crossed. However, unlike the usual Dirac point, three bands are crossing instead of two. In this case, perturbation such as spin-orbit interaction cannot open gap due to anti-crossing, as shown in figure 2. Therefore, they can neither be topological insulators, nor the point-nodal semimetals, predicted for the hypothetical 3D-honeycomb lattice model [4-6]. This kind of band crossing is similar to that in MoP [9], which is a topological semimetal and in which curious three-component fermion is observed. However, in MoP it differs in that two of the three bands are strictly degenerated. We speculate that this new type of band crossing in $BaGe_2$, $SrGe_2$ and $SrSi_2$ may emerge novel topological states.

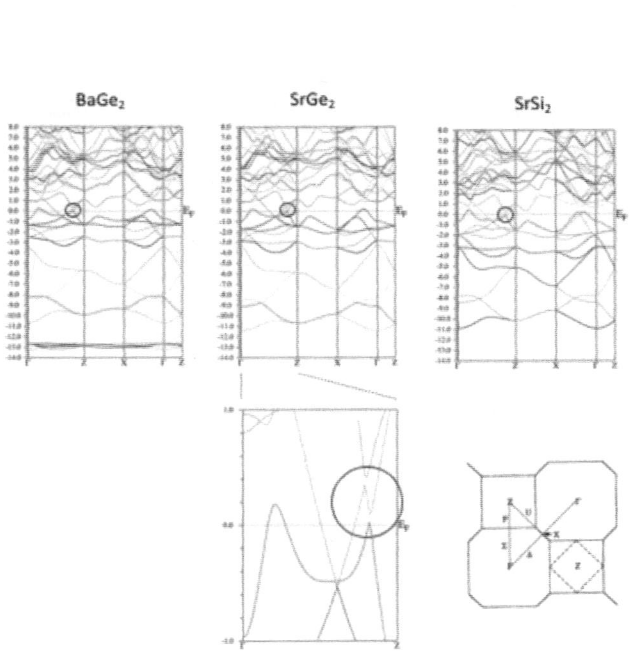

Figure 1. Energy dispersion curves of tetragonal $BaGe_2$, $SrGe_2$ and $SrSi_2$. The units of vertical axis are eV. The lower panel shows the close-up view near the pseudo-Dirac point (shown in circle) in $SrGe_2$, and the notation of the k-points.

IOP Conf. Series: Journal of Physics: Conf. Series **1054** (2018) 012004 doi:10.1088/1742-6596/1054/1/012004

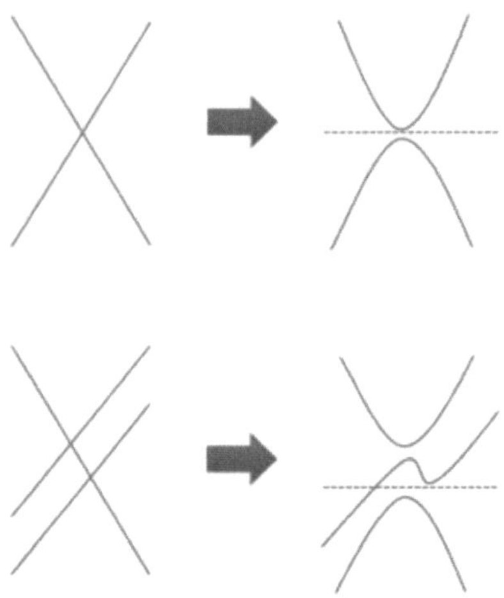

Figure 2. Schematic view of normal Dirac point and pseudo-Dirac point. Small perturbation can open a gap around normal Dirac point, while it cannot open a gap around pseudo-Dirac point.

Next we compare the density of states curves for tetragonal $BaGe_2$, $SrGe_2$ and $SrSi_2$, shown in figure 3. Overall peak structure is very similar since they are isovalent, while for $BaGe_2$ there are large peaks which are composed of Ba5p and Ba5d orbitals at the energy ~ -13eV and 9eV. We can see that Ba/Sr partial DOS is almost zero at E_F, and the shape of the Ba/Sr partial DOS is completely different with the Ge/Si partial DOS. This result means that the hybridization between Ba/Sr and Ge/Si is quite small, and these compounds can be regarded as forming the Zintl phase, $(Ba/Sr)^{2+}(Ge/Si)^{-}_2$. As for the unoccupied states, Ge/Si partial DOS is almost flat: i.e. it hardly depends on energy. This shape of partial DOS may suggest a 2D free electron system. However, these three compounds have 3D-honeycomb network of Ge/Si. And as we show later, the polymorphs with 2D-honeycomb network of Ge also have similar shape of Ge partial DOS for unoccupied states. Therefore, we cannot attribute this characteristic shape of the Ge/Si partial DOS to the 2D free electrons.

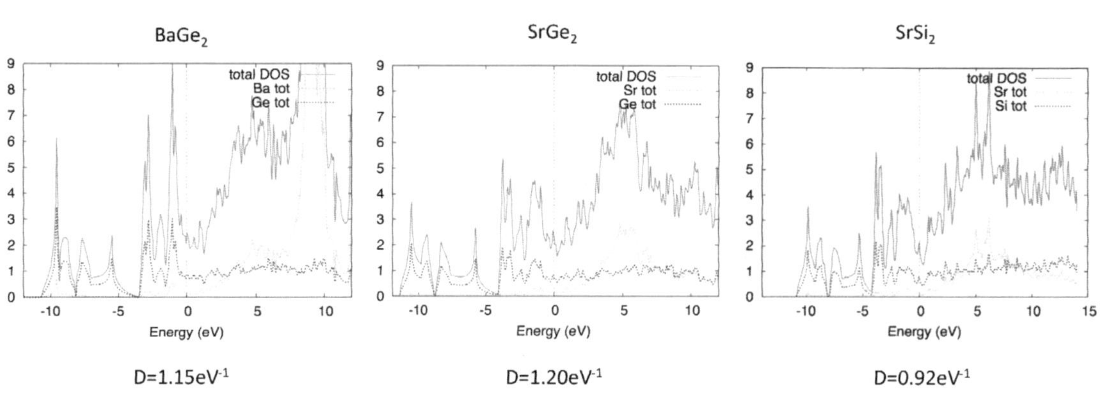

Figure 3. Density of states curves of tetragonal $BaGe_2$, $SrGe_2$ and $SrSi_2$. The units of vertical axis are eV^{-1} per unit cell.

Here we consider the material dependence of T_c for tetragonal $BaGe_2$, $SrGe_2$ and $SrSi_2$. In most mechanisms of superconductivity, most important quantity is the DOS at E_F (=D). We plot D vs T_c for these three compounds in figure 4, filled circle. Since there are no magnetic atoms, we assume that the mechanism of superconductivity is simple BCS type. In the standard BCS theory T_c is estimated as:

$$T_c = 1.14\theta_D \exp\left(-\frac{1}{DV}\right)$$

where V is the magnitude of the attracting interaction (usually caused by electron-phonon interaction) and θ_D is the Debye temperature. We can regard this equation as the fitting curve with two parameters θ_D and V. Then we obtained the fitting curve as the solid line in figure 4. The quality of fitting is not so good for the following reasons. First, θ_D and V can be varied with compound. Second, randomness can decrease T_c for the real sample. Finally, D can be also changed by off-stoichiometry of the sample. In fact, we found that *te*-$SrGe_2$ has the composition $SrGe_{1.66}$, which means that ~16% of Ge is deficient [1]. Since the Zintl phase-like description Sr^{2+} $(Ge_{1.66})^{2-}$ is valid as shown above, we can estimate the valence of Ge is $Ge^{-1.2}$. For stoichiometric sample the valence of Ge is $Ge^{-1.0}$, the Ge atom in the Ge-deficient sample has 0.2 excess electrons. This situation can be described by the rigid band model with shifting E_F so as to contain 0.2 excess electrons per Ge. We obtain D for $SrGe_{1.66}$ as D = 0.91 eV^{-1}. Since the amount of Ge/Si deficiency in $BaGe_2$ and $SrSi_2$ is not estimated, we tentatively assume they also have 0.2 excess electrons per Ge/Si. In this way we obtained D = 0.97 and 0.77 eV^{-1} for $BaGe_{1.66}$ and $SrSi_{1.66}$, respectively. We also plot D vs T_c for these three compounds in figure 4, filled triangle.

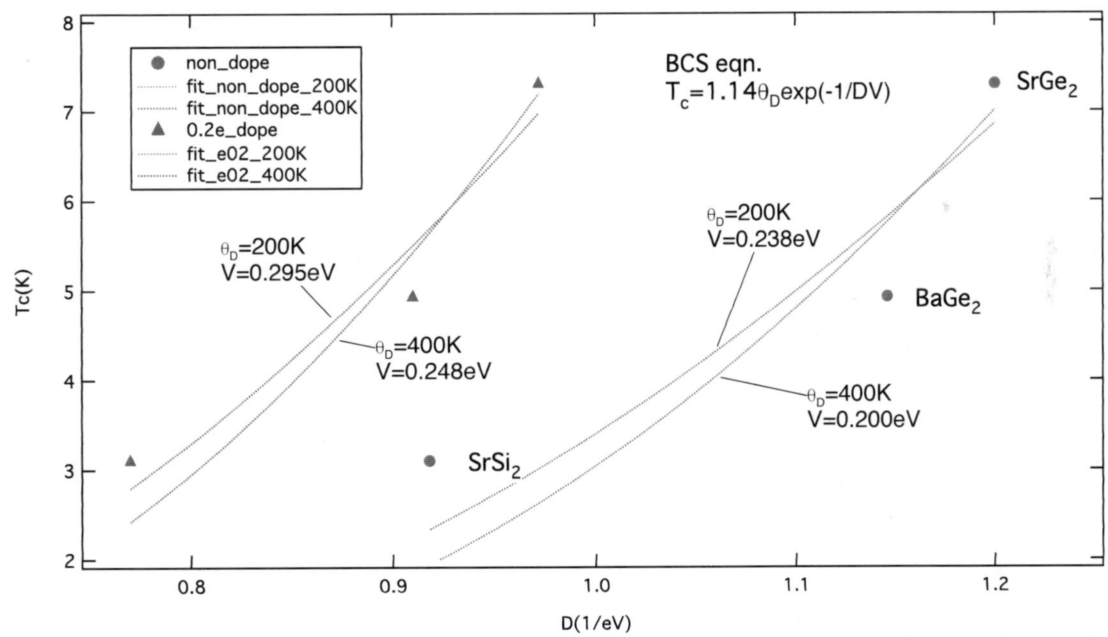

Figure 4. D(=DOS at E_F) vs T_c plot for tetragonal $BaGe_2$, $SrGe_2$ and $SrSi_2$. The filled circles are for stoichiometric case, and the filled triangles are for Ge/Si deficient case. We assumed that all of these three compounds have the same amount of Ge/Si deficiency, and treated within the rigid band model with 0.2 excess electrons by shifting E_F. Dotted lines and solid lines are the BCS curves for stoichiometric case and Ge/Si deficient case, with fixing the parameters θ_D and V.

Although the quality of fitting is not good, this result is still suggestive. First, when we increase θ_D by factor 2, only ~15% decrease of V can mostly reproduce the fitting curve. In other words, for realistic θ_D we can estimate V as 0.2~0.3eV for these compounds. This means that the electron-phonon coupling constant $\lambda = DV$ is estimated as 0.2~0.35, which indicates that these compounds are in the weak-coupling regime. We also note that the D vs T_c plot is almost similar for non-doped and 0.2 electron-doped case. This result may show that our rigid-band approach is essentially good, and Ge/Si deficiency increases D for all these three compounds almost in the same manner.

Next we show the DOS curves for the several polymorphs of $SrGe_2$ in figure 5. Since the local coordination for Ge atom is common between these three polymorphs, i.e. they all have (pseudo) planer triangle coordination of nearest-neighbor Ge atom, these three polymorphs have similar DOS curves. As for the trigonal structure, the buckling of the Ge atom along the c-axis splits the peak at E_F and opens a pseudogap. The total energy of the trigonal structure is lower than that of the hexagonal structure. This behavior is explained by the Peierls transition, i.e. the lattice modulation doubles the unit cell, opens a gap and the system gains kinetic energy. The total energy of the tetragonal structure is also higher than that of the trigonal structure, agreeing with the previous calculation[3]. However, it is difficult to transform from tetragonal to trigonal structure due to the steric barrier. As a result, once $SrGe_2$ forms the tetragonal structure, it does not easily transform to the more stable trigonal structure with less $D(E_F)$, and keeps the high $D(E_F)$ which leads to relatively high T_c.

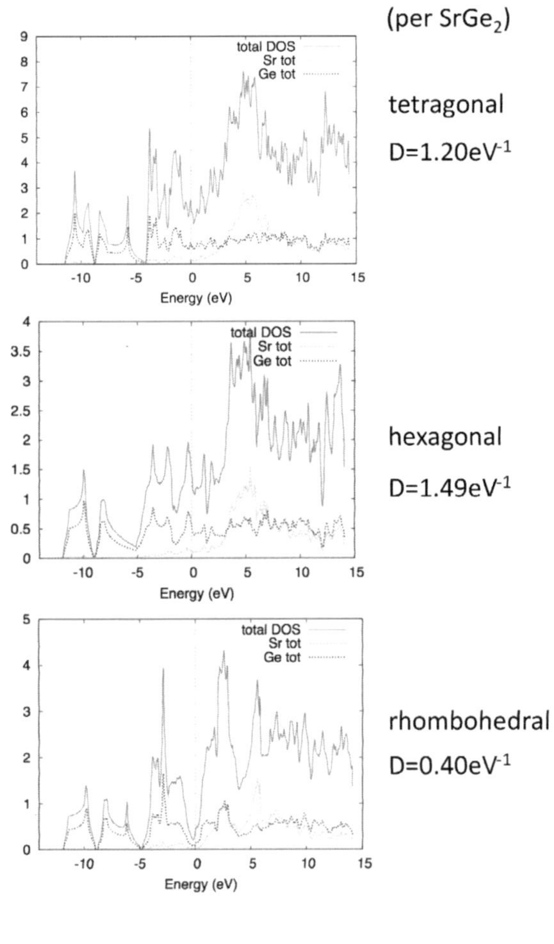

Figure 5. Density of states curves of tetragonal, hexagonal and rhombohedral polymorphs of $SrGe_2$. The units of vertical axis are eV^{-1} per unit cell.

4. Conclusion

IOP Conf. Series: Journal of Physics: Conf. Series **1054** (2018) 012004 doi:10.1088/1742-6596/1054/1/012004

We have calculated the band structure of tetragonal $SrGe_2$, $BaGe_2$ and $SrSi_2$ from first-principles. They have Dirac-point-like band structure, but are not topological insulator because three bands are entangled instead of two. BCS analysis suggests that these compounds are in the weak coupling regime. We also calculated the electronic structure for several polymorphs of $SrGe_2$, and the tetragonal phase has rather high $D(E_F)$ which is advantageous for high T_c.

Acknowledgments

We thank to Prof. H. Aoki for fruitful discussions. This work was partially supported by KAKENHI (Grant No. JP26400379) from Japan Society for the Promotion of Science (JSPS).

References

[1] Iyo A, Kawashima K, Hase I, Ishida S, Kito H, Takeshita N, Oka K, Fujihisa H, Goto Y, Yoshida Y and Eisaki H 2017 *Inorg. Chem.* **56** 8590
[2] Evers J, Oehleniger G and Weiss A 1980 *Z. Naturforsch. Growth B***35** 397
[3] Wang J, Chen C and Kawazoe Y, 2015 *Phys. Rev. B***91** 054107
[4] Mullen K, Uchoa B and Glatzhofer D T 2015 *Phys. Rev. Lett.* **115** 026403
[5] Ezawa M 2016 *Phys. Rev. Lett.* **116** 127202
[6] Nandkishore R 2016 *Phys. Rev. B***93** 020506
[7] Perdew J P, Burke K, Ernzerhof M 1996 *Phys. Rev. Lett.* **77** 3865
[8] Blaha P, Schwarz K, Madsen G K H, Kvasnicka D and Luitz J 2001 WIEN2k, *An Augmented Plane Wave + Local Orbbitals Program for Calculating Crystal Properties* (Wienna: Vienna Univ. of Technology)
[9] Lv B Q et al. *Nature* 2017 **546** 627

TEM and electron backscatter diffraction analysis (EBSD) on superconducting nanowires

A. Koblischka-Veneva, M. R. Koblischka, X. L. Zeng, J. Schmauch, U. Hartmann

Experimental Physics, Saarland University, Campus C 6 3, 66123 Saarbrücken, Germany

E-mail: a.koblischka-veneva@mx.uni-saarland.de

Abstract. Electrospun, superconducting nanowires are characterized concerning the grain orientation, their texture and the respective grain boundary misorientations by means of electron backscatter diffraction (EBSD) analysis. The individual nanowires in such electrospun, nonwoven nanowire networks of $Bi_2Sr_2CaCu_2O_x$ (Bi-2212) are polycrystalline, have average diameters up to 250 nm and their grains are in the 20-50 nm range. This requires a high spatial resolution for the analysis in the scanning electron microscope. However, the small diameter of the nanowires enables the application of the newly developed transmission EBSD (t-EBSD) technique without the preparation of TEM slices. Here, we present TEM images of individual nanowires and several EBSD mappings on Bi-2212 nanowires and compare their microstructure to those of filaments of the first generation tapes.

1. Introduction

Electrospinning is a versatile technique to fabricate long, polycrystalline nanowires of ceramic materials, including high-T_c superconductors [1]. In the literature, this enabled already the growth of (La,Sr)CuO_4 [2,3], $YBa_2Cu_3O_x$ [4,5] and $Bi_2Sr_2CaCu_2O_x$ (Bi-2212) [6–9] nanowires. The resulting nanowires are poycrystalline with grain sizes of about 40-60 nm, but are up to several hundreds of micrometers long. After the electrospinning process, a fabric-like material consisting of polymer nanowires containing the ceramic precursor is created. After the heat treatment, the polymer is burned off and grains of the superconducting material are formed. By means of focused ion-beam milling, individual nanowires can be cut out from the fabric-like sheets. These nanowires can now serve as objects for further studies of physical properties concerning superconductivity in reduced dimensions. Therefore, a detailed understanding of the resulting microstructure within an individual nanowire is strongly required. For this purpose, we employ here the electron backscatter diffraction (EBSD) analysis on such nanowires in order to obtain information on the crystallographic grain orientation and the resulting texture. Due to the small sizes of the Bi-2212 grains within the nanowires, we employ the recently developed transmission EBSD (t-EBSD) technique [10–12] on our samples. As only a small amout of nanowire pieces is transparent to the electron beam, we used a modified approach by using the sample holder as for t-EBSD, but working in reflection mode. This enables the study of nanowire pieces without any further surface preparation.

In the present contribution, we describe the transmission electron microscopy (TEM) investigation and the approach to perform EBSD analysis on pieces of Bi-2212 nanowires.

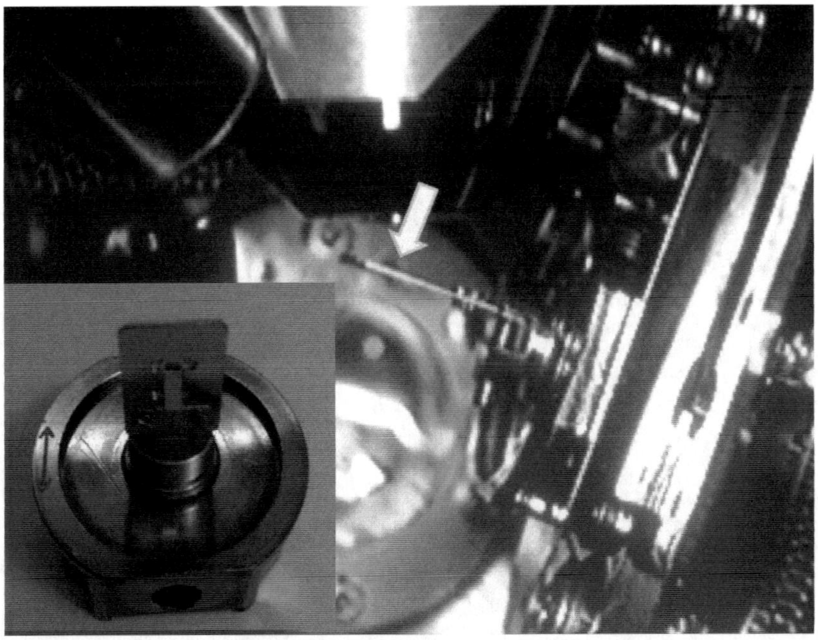

Figure 1. Arrangement of the TEM-slice sample holder in the SEM chamber, inclined by −20 degrees which enables the same detector position to be used as for standard EBSD. The inset shows the custom-built sample holder.

2. Experimental procedures

Bi-2212 nanowire networks were grown by the electrospinning technique, employing acetate powders of all consistuents and PVP (polyvinyl pyrrolidone, M.W.1,300,000) dissolved in propionic acid. After the electrospinning process, a heat treatment is required to remove all organic material from the as-prepared nanowire networks and to form the superconducting phase. Finally, an oxygenation step in pure O_2 is required. X-ray analysis confirmed that the samples are pure Bi-2212 phase with some residing carbon. Details of the nanowire fabrication procedure are given elsewhere [2,3]. Samples for TEM/t-EBSD were prepared by ultrasonication of nanowire network pieces in ethanol for 5 s. Then, a drop of this liquid was placed on a carbon-coated TEM grid and dried. Such a Bi-2212 piece can now serve for TEM and t-EBSD. An image is presented in Fig. 2.

The TEM analysis was performed by JEOL JSM-7000 F transmission electron microscope operating at 200 kV with a LaB_6 cathode. For TEM investigations, several nanowire pieces separated from the nanowire fabric by ultrasound treatment were placed on a carbon-coated TEM copper grid. This enabled to search nanowire regions which were thin enough to enable the transmission mode. No additional treatment of the sample surface was carried out.

For EBSD measurements, the TEM-grids were mounted in the scanning electron microscope (SEM) on a specially fabricated sample holder allowing for the correct 70° inclination of the sample required for EBSD. The stage with the sample holder is inclined to an angle of 20°, which enables, together with the sample mounting, the same detector position to be used for the EBSD detector as in the standard configuration. Here, the electron beam is passing through the sample (transmission mode) and the electron cone is formed on the backside of the sample. The holder further enables a second way of operation of EBSD like for standard EBSD in reflection mode together with the TEM-prepared sample. This second operation mode was chosen for the

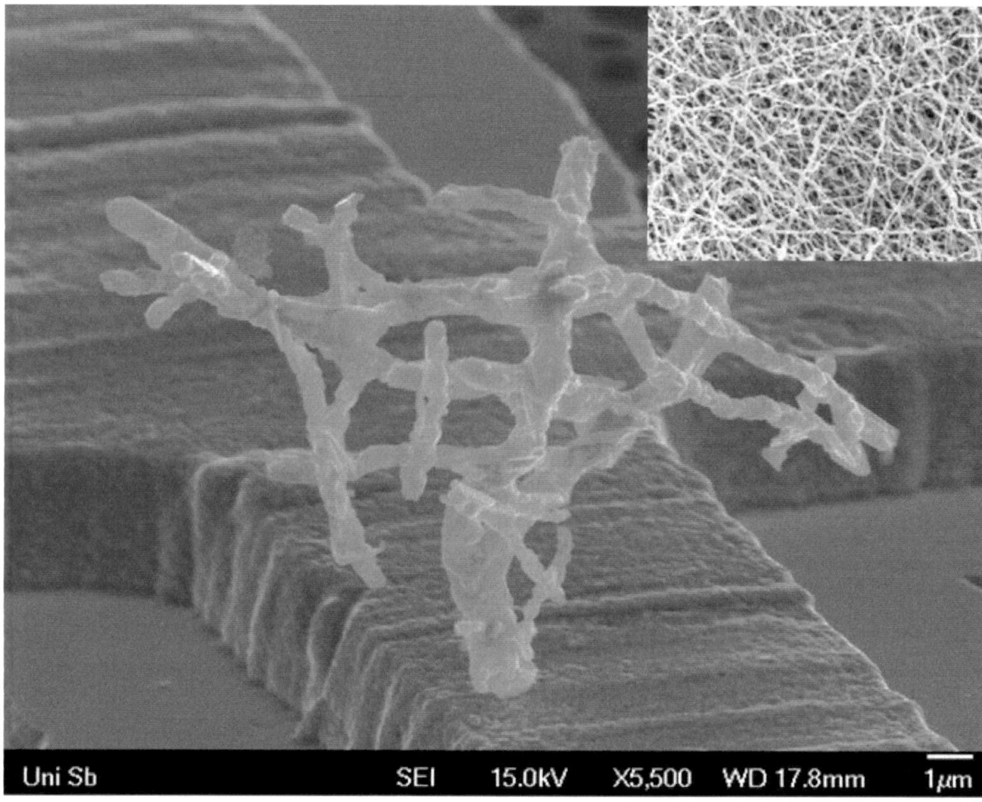

Figure 2. SEM image of a sectioned Bi-2212 nanowire fabric on the TEM grid. The inset shows a SEM image of a complete nanowire fabric.

present nanowire samples not to be limited to only electron-beam transparent sections. The electron beam operates at 30 kV, and the working distance is set to 5 mm. The EBSD stepsize was chosen to be 5 nm. An image of our sample holder and the entire arrangement within the SEM chamber is presented in Fig. 1.

This approach enables here for the first time a proper analysis of nanowire samples with nanometer-sized grains. However, as the nanowire samples were mostly too thick to operate EBSD in transmission, we operated EBSD also in reflection mode but using the custom-built sample holder for TEM grids.

3. Results and discussion

In previous publications, SEM images of the entire nanowire fabrics were presented [7, 8], see also the inset to Fig. 2. Here, we now focus on the microstructure of individual nanowire pieces. Firstly, Fig. 2 shows an SEM image of a nanowire fabric piece on the TEM grid, ready for further investigations. One can clearly see the Bi-2212 grains with their specific elongated shape. Some crystallites are even pointing away from the main nanowire. This image further demonstrates that for t-EBSD in either mode a high SEM magnification is required; in our case up to 70000×.

Figure 3 presents TEM images of Bi-2212 nanowires showing sections with interconnects between two nanowires (a,b). These interconnects of the nanowires, which are formed during the processing due to shrinking effects, are essential for the current flow through a nanowire fabric sample. This was discussed in detail in Ref. [13] together with a modelling approach. The

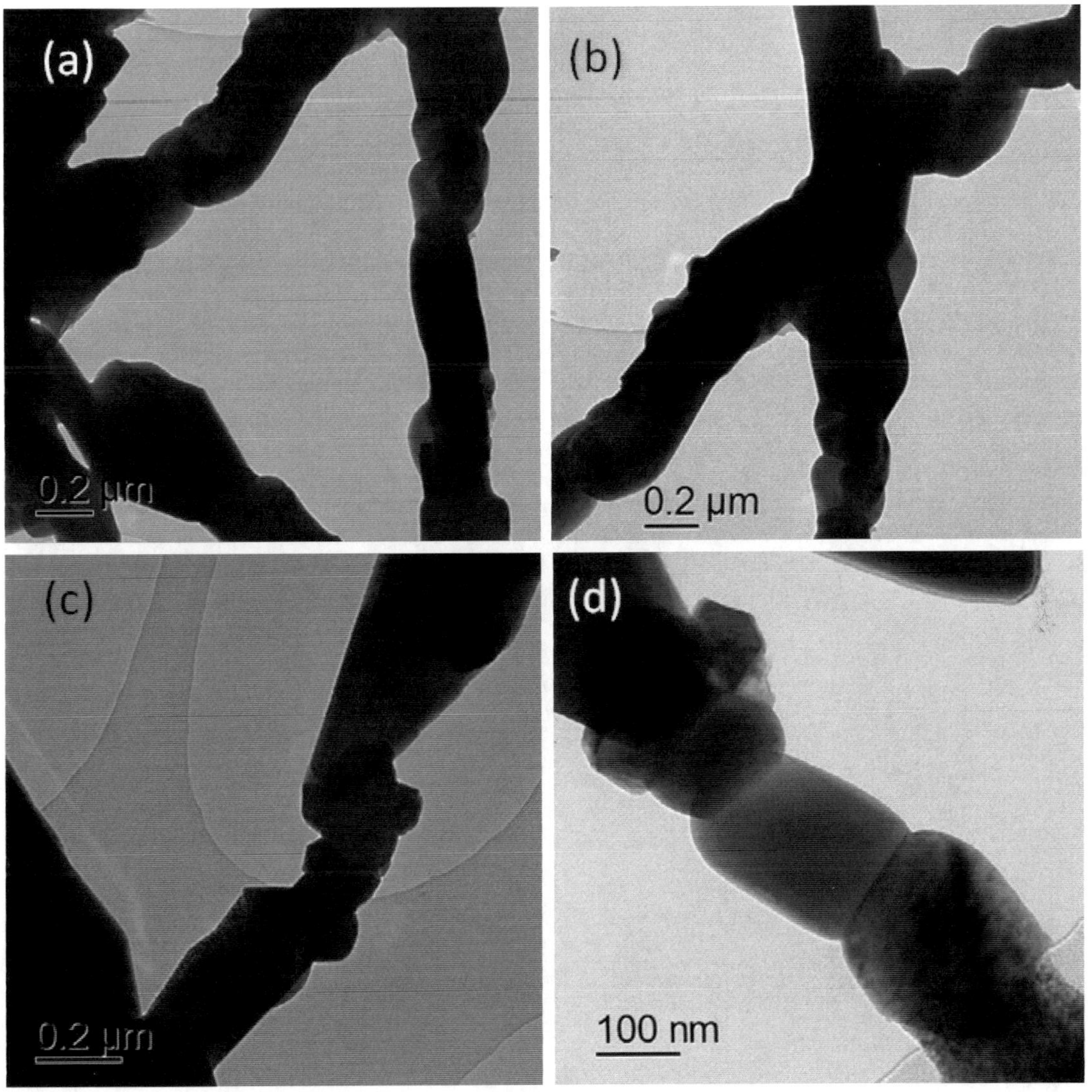

Figure 3. TEM images of Bi-2212 nanowires showing sections with an interconnect between two nanowires (a,b). Images (c) and (d) present sections of individual nanowires. In (d), one can see that the nanowire consists only of one single Bi-2212 grain.

magnetic properties of the interconnects determine the overall field dependence of the critical currents, so it is important to understand how these interconnects are built up. Images (c) and (d) present sections of individual, thin nanowires. In Fig. 3(d), one can see that the nanowire consists in the center only of one single Bi-2212 grain, which is about 200 nm long, but only $\sim$ 100 nm wide.

Figure 4 (a) – (d) give more details about the grain boundaries (GBs) between the grains within a nanowire. Here, it is important to note that a large number of the interconnects and many connections between the grains have survived the ultrasonic treatment. This indicates that the grains stick closely together, thus providing good electric contacts for the current flow. Transport measurements on such individual nanowire pieces will enable the study of the

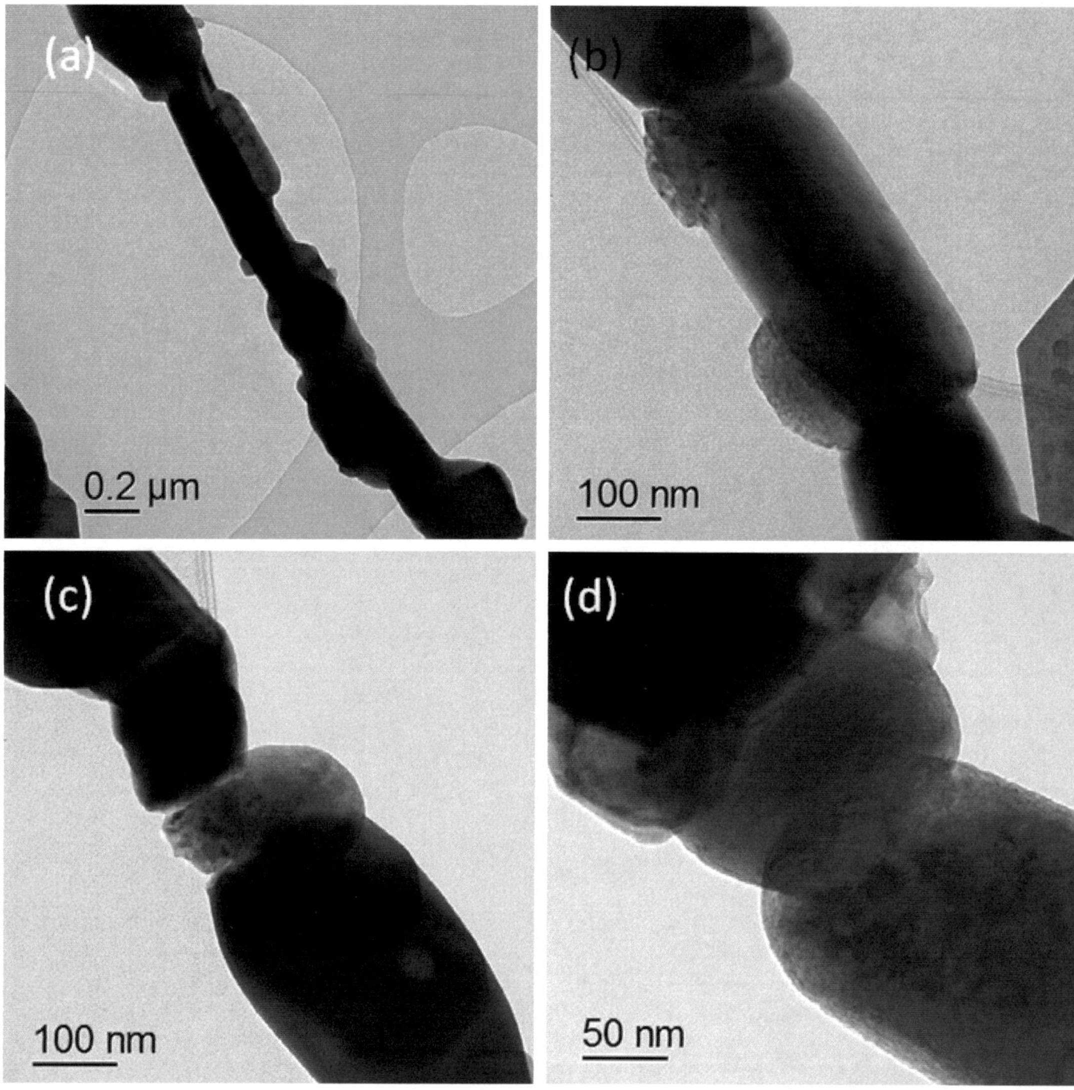

Figure 4. TEM images of individual Bi-2212 nanowire sections. Images (a), (b) and (c) present several thin nanowire sections, and (d) shows a grain boundary within the nanowire in high magnification. Note also that the edges of the grains are transparent to the electron beam.

GB influence on the current flow. It is further remarkable that in (a) the nanowire exhibits connections between elongated Bi-2212 grains, whereas images (c) and (d) show a small grain quasi as contact element between two bigger ones. The TEM image suggests that this small grain is misoriented by a high angle which can be up to 90°.

Now, we will turn to the EBSD measurements on such nanowire sections. As already stated before, the t-EBSD can be carried out in reflection mode in order to enable the investigation of samples which have a thickness larger than one single grain. EBSD measurements using the t-EBSD technique were already done in the literature on superconductors like MgB_2 [14–16], but now yet on ceramic high-T_c superconductors.

Figure 5 presents an EBSD-mapping on a nanowire section. Here, we selected a large piece

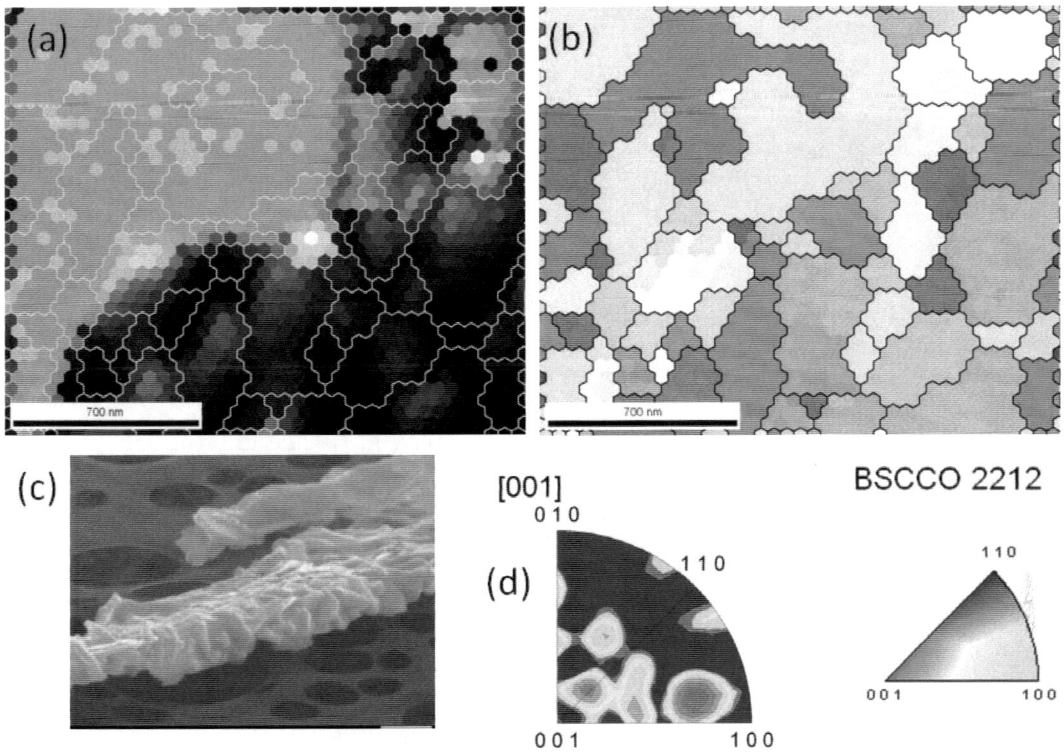

Figure 5. EBSD mapping on a nanowire section. The map (a) is an image quality (IQ) map resembling a SEM image. Additionally, the map is overlaid with the EBSD-detected grain boundaries (yellow). The map (b) shows an inverse pole figure map giving the crystallographic orientations. The color code is given in the stereographic triangle below the map. (c) is a SEM image of the investigated nanowire section. (d) shows the inverse pole figure in [001]-direction. Only some specific grain orientations are obtained.

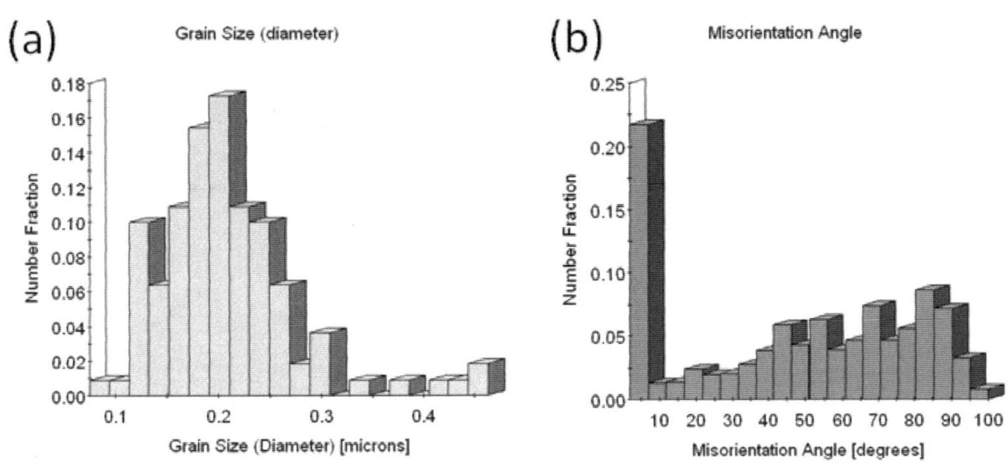

Figure 6. EBSD results for the grain size (a) and the misorientation angles (b).

allowing an EBSD mapping. The selected area spans 2000 × 1600 nm; the EBSD step size for this measurement was 40 nm. Fig. 5 (a) shows an image quality (IQ) map, together with the EBSD detected GBs marked in yellow. The resulting IQ of the Kikuchi patterns obtained was quite high despite that there was no surface preparation. The strong variation of the IQ values stems from the height differences in the nanowire piece, see the inset to Fig. 5 (a). Fig. 5 (b) is an inverse pole figure (IPF) map, giving the crystallographic orientation of the grains. The color code for this map is given in the stereographic triangle below the map. The distribution of the colors suggests directly that there are only some orientations; the grains in the nanowire are far from being randomly oriented. This is manifested in the inverse pole figure in [001]-direction presented in (c). Here only 8 spots are found. The highest peak is obtained close to the [100]-direction, which says that most of the grains have their *c*-axis in the nanowire plane. Here, we can state that the electrospun Bi-2212 nanowires are similar to tiny filaments extracted from multifilamentary wires or tapes [17–19]. As the as-spun nanowires are formed as polymer nanowires containing the precursor material, there is still the information of the main wire direction at the beginning of the Bi-2212 formation, so the growth direction of the Bi-2212 grains takes place in the direction of the former polymer nanowire. This explains why we find so many elongated Bi-2212 grains with only some orientations.

In Fig. 6, more results of the EBSD analysis are presented. (a) shows the EBSD-detected grain size. Grains can be as small as 40 nm, but also big grains with size up to 450 nm can be found. Image (b) shows the misorientation angles. The largest number fraction are orientations of up to 10°, but then there are many misorientations in the range from 20°C to 90° (c). As expected, the misorientation distribution does not correspond to the result which would be obtained in a fully random situation, which demonstrates again that the grains in the nanowire are oriented along the nanowire direction, however, also high misorientations of the Bi-2212 grains exist as seen in the TEM-images (Figs. 4 (c,d)).

Further EBSD analysis of different types of nanowire sections found in the TEM images will enable to construct a complete picture of the building principles of such nanowires. This knowledge will help to prepare nanowire networks with improved properties for certain applications in the future.

4. Conclusions

To conclude, we presented a TEM investigation on individual Bi-2212 nanowire pieces stemming from electrospun nanowire fabrics. We further have investigated the crystallographic orientation of Bi-2212 grains within individual nanowire pieces in detail using the t-EBSD technique in reflection mode. The analysis demonstrates that the Bi-2212 grains within a nanowire are mainly oriented with their *c*-axis parallel to the nanowire direction.

Acknowledgments

This work is subsidized by the DFG project Ko2323/8, which is gratefully acknowledged.

References

[1] Wu H, Pan W, Lin D and Li H 2012 *J. Adv. Ceramics* **1** 2
[2] Li J-M, Zeng X L, Moa A D and Xu Z-A 20111 *CrystEngComm* **13** 6964
[3] Zeng X L, Koblischka M R and Hartmann U 2015 *Mat. Res. Express* **2** 095022
[4] Duarte E A, Rudawski N G, Quintero P A, Meisel M W and Nino J C 2015 *Supercond. Sci. Technol.* **28** 015006
[5] Rotta M, Zadorosny L, Carvalho C L, Malmonge J A, Malmonge L F and Zadorosny R 2016 *Ceram. Int.* **42** 16230
[6] Duarte E A, Quintero P A, Meisel M W and Nino J C 2013 *Physica C* **496** 109
[7] Koblischka M R, Zeng X L, Karwoth T, Hauet T and Hartmann U 2016 *AIP Adv.* **6** 035115

[8] Koblischka M R, Zeng X L, Karwoth T, Hauet T and Hartmann U 2016 *IEEE Trans. Appl. Supercond.* **26** 1800605
[9] Zeng X L, Koblischka M R, Karwoth T, Hauet T and Hartmann U 2017 *Supercond. Sci. Technol.* **30** 035014
[10] Trimby P W 2012 *Ultramicroscopy* **120** 16
[11] Keller R and Geiss R 2012 *J. Microsc.* **245** 245
[12] Sneddon G C, Trimby P W and Cairney J M 2016 *Materials Science and Engineering* R **110** 1
[13] Zeng X L, Karwoth T, Koblischka M R, Hartmann U, Gokhfeld D, Chang C and Hauet T 2017 *Phys. Rev. Mater.* **1** 044802
[14] Yeoh W K, Cui X Y, Gault B, De Silva K S B, Xu X, Liu H W, Yen H-W, Wong D, Bao P, Larson D J, Martin I, Li W X, Zheng R K, Wang X L, Dou S X and Ringer S Pl 2014 *Nanoscale* **6** 6166
[15] Wong D C K, Yeoh WK, Trimby P W, De Silva K S B, Bao P, Li W X, Xu X, Dou S X, Ringer S P and Zheng R K 2015 *Scripta Mater.* **101** 36
[16] Koblischka-Veneva A, Koblischka M R, Schmauch J, Inoue K, Muralidhar M, Berger K and Noudem J 2016 *Supercond. Sci. Technol.* **29** 044007
[17] Kametani F, Jiang J, Matras M, Abraimov D, Hellstrom E E and Larbalestier D C 2015 *Sci. Rep.* **5** 8285
[18] Koblischka-Veneva A, Koblischka M R, Qu T, Han Z and Mücklich F 2008 *Physica* C **468** 174
[19] Skov-Hansen P, Koblischka M R, Vase P, Kovac P and Marti F 2001 *IEEE Trans. Appl. Supercond.* **11** 3740

IOP Conf. Series: Journal of Physics: Conf. Series **1054** (2018) 012006 doi:10.1088/1742-6596/1054/1/012006

Cu hyperfine coupling constants of $HgBa_2CaCu_2O_{6+\delta}$

Yutaka Itoh[1], Takato Machi[2] and Ayako Yamamoto[3]

[1]Department of Physics, Graduate School of Science, Kyoto Sangyo University, Kamigamo-Motoyama, Kika-ku, Kyoto 603-8555, Japan

[2]AIST Tsukuba East, Research Institute for Energy Conservation, 1-2-1 Namiki, Tsukuba, Ibaraki 305-8564, Japan

[3]Graduate School of Engineering and Science, Shibaura Institute of Technology, 3-7-5 Toyosu, Koto-ku, Tokyo 135-8548, Japan

E-mail: yitoh@cc.kyoto-su.ac.jp

Abstract. We estimated the ratios of ^{63}Cu hyperfine coupling constants in the double-layer high-T_c superconductor $HgBa_2CaCu_2O_{6+\delta}$ from the anisotropies in Cu nuclear spin-lattice relaxation rates and spin Knight shifts to study the nature of the ultraslow fluctuations causing the T_2 anomaly in the Cu nuclear spin-echo decay. The ultraslow fluctuations may come from uniform magnetic fluctuations spread around the wave vector $q = 0$, otherwise the electric origins.

1. Introduction

Spin polarized neutron scattering experiments indicate the emergence of an intra-unit-cell (IUC) $q = 0$ magnetic moments in the pseudogap states of the high-T_c cuprate superconductors, while no NMR and μSR experiment indicates any static ordering of local magnetic moments [1]. The IUC moments are associated with the loop current ordered state [2]. Recently discovered ultraslow fluctuations in the pseudogap states of $HgBa_2CaCu_2O_{6+\delta}$ (Hg1212) via ^{63}Cu nuclear spin-echo decay experiments [3] might reconcile an issue on the IUC moments. No wipeout effect on NMR spectra is characteristic of the ultraslow fluctuations of Hg1212, in contrast to the spin-charge stripe orderings [4, 5].

Knowledge of the hyperfine coupling constants helps us to clarify the nature of the local field fluctuations in NMR measurements [6]. In this paper, we report the estimation of the ^{63}Cu hyperfine coupling constants in the double-CuO_2-layer high-T_c superconductors Hg1212 from the anisotropies in ^{63}Cu nuclear spin-lattice relaxation rates and spin Knight shifts [7], and discuss the nature of the ultraslow fluctuations [3].

2. Estimations of ^{63}Cu hyperfine coupling constants

The ^{63}Cu hyperfine coupling parameters (A^{hf}_{cc} and A^{hf}_{ab}) consist of the anisotropic on-site A_{cc} (the c axis component) and A_{ab} (the ab plane component) due to the $3d$ electrons and the isotropic supertransferred component $B(> 0)$ [6]. The ratios of the individual components in the three coupling constants can be estimated from the anisotropy data [7] of the ^{63}Cu Knight shifts ($^{63}K_{cc}$ and $^{63}K_{ab}$) and the ^{63}Cu nuclear spin-lattice relaxation rates [$(1/T_1)_{cc}$ and $(1/T_1)_{ab}$] via the antiferromagnetic dynamical spin susceptibility [6, 8, 9, 10, 11]. The subscript indices

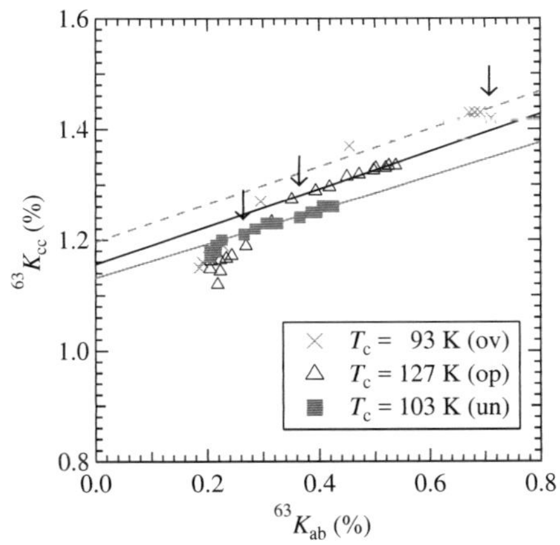

Figure 1. $^{63}K_{cc}$ versus $^{63}K_{ab}$ with temperature as an implicit parameter for Hg1212 from underdoped to overdoped [7]. The arrows indicate T_c.

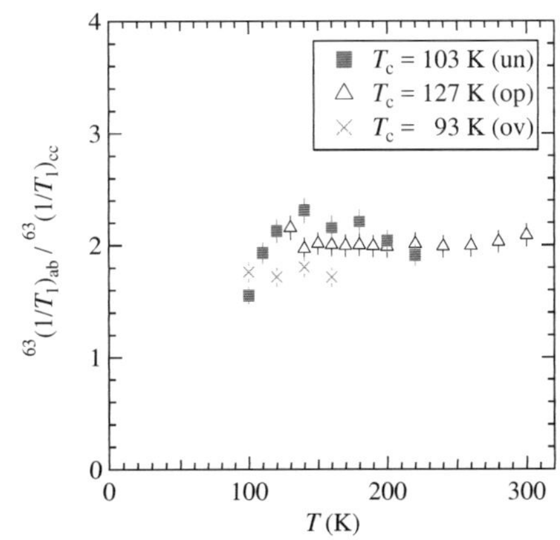

Figure 2. Anisotropy ratio $^{63}(1/T_1)_{ab}/^{63}(1/T_1)_{cc}$ against temperature for Hg1212 from underdoped to overdoped [7].

of cc or ab of ^{63}K and $1/T_1$ indicate the direction of a static magnetic field applied along the c axis or in the ab plane. The procedure to estimate the coupling constant ratios is shown below.

2.1. ^{63}Cu Knight shifts

The ^{63}Cu Knight shifts $^{63}K_{cc,ab}$ at a magnetic field along the c axis or in the ab plane are the sum of the spin shift K_{spin} and the orbital shift K_{orb} as $^{63}K_{cc,ab} = K_{spin}^{cc,ab}(T) + K_{orb}^{cc,ab}$. The spin shift is $K_{spin}^{cc,ab}(T) = A_{cc,ab}^{hf}\chi_s^{cc,ab}(T)$ with the hyperfine coupling parameters $A_{cc,ab}^{hf}$ and the uniform spin susceptibility $\chi_s^{cc,ab}(T)$. For a temperature-dependent isotropic spin susceptibility $\chi_s^{cc} = \chi_s^{ab}$, the ratio $\Delta K_{spin}^{cc}/\Delta K_{spin}^{ab} = (dK_{spin}^{cc}/dT)/(dK_{spin}^{ab}/dT)$ is equal to the ratio A_{cc}^{hf}/A_{ab}^{hf}.

Figure 1 shows $^{63}K_{cc}$ plotted against $^{63}K_{ab}$ with temperature as an implicit parameter for Hg1212 from underdoped to overdoped, which are adopted from [7]. The sold straight lines are the least-squares fitting results. The dashed straight line for overdoped Hg1212 is a visual guide with assuming the same slope as the optimally doped Hg1212. The straight lines show nearly parallel shift. Since the orbital shifts of $K_{orb}^{cc} = 1.14$–1.16 % and $K_{orb}^{ab} = 0.19$–0.20 % are estimated below T_c, then the parallel shift indicates a constant spin component above T_c. Similar parallel shift is found in the single crystal NMR for $HgBa_2CuO_{4+\delta}$ [12].

An easy plane magnetic anisotropy causes such a constant spin component in the paramagnetic spin susceptibility [13]. The anisotropic superexchange interaction in the $S = 1/2$ XXZ Heisenberg Hamiltonian yields the easy plane anisotropy in the paramagnetic state [14, 15]. The optimally hole doping makes the anisotropy weak [13, 16]. Although the multicomponents in the spin susceptibility are suggested from the anisotropic spin Knight shifts [12, 17], we believe that the doped superconductors with a single spin component can show a finite anisotropy and that the constant spin component does not impede a single spin component analysis to estimate the Cu hyperfine coupling constants.

The ^{63}Cu hyperfine coupling parameters A_{cc}^{hf} and A_{ab}^{hf} are expressed by A_{cc}, A_{ab}, and B as $A_{cc}^{hf} = A_{cc} + 4B$ and $A_{ab}^{hf} = A_{ab} + 4B$ [6, 8, 18]. Then, the anisotropy ratio r_u of the temperature-dependent K_s^{cc} and K_s^{ab} is given by

$$r_u \equiv \frac{\Delta K_s^{cc}}{\Delta K_s^{ab}} = \frac{A_{cc} + 4B}{A_{ab} + 4B}. \tag{1}$$

Figure 1 shows $r_u = 0.31$ for the underdoped and 0.34 for the optimally doped samples. The value of $r_u = 0.34$ is assumed for the overdoped sample.

2.2. ^{63}*Cu nuclear spin-lattice relaxation rate*

Figure 2 shows the ratio of $^{63}(1/T_1)_{ab}/^{63}(1/T_1)_{cc}$ plotted against temperature for Hg1212 from underdoped to overdoped (adopted from Ref. [7]). The anisotropy ratio r_{AF} of $(1/T_1)_{ab}$ and $(1/T_1)_{cc}$ is given by

$$r_{\mathrm{AF}} \equiv \frac{(1/T_1)_{ab}}{(1/T_1)_{cc}} \approx \frac{1}{2}\left\{1 + \left(\frac{A_{cc} - 4B}{A_{ab} - 4B}\right)^2\right\}, \tag{2}$$

for the leading term of the enhanced antiferromagnetic susceptibility [11]. For convenience, we introduce an alternative parameter of $r_A = \sqrt{2r_{\mathrm{AF}} - 1}$. We adopted the values of r_{AF} (r_A) = 2.3 (1.90), 2.0 (1.73), and 1.8 (1.61) from underdoped to overdoped (figure 2) to estimate the coupling constant ratios.

2.3. ^{63}*Cu hyperfine coupling constant ratios*

From the constraints of $A_{cc} < 0$ [18] and $A_{ab}/4B < 1$ on (1) and (2), we obtain the expressions of the ratios of the ^{63}Cu hyperfine coupling constants,

$$\frac{A_{cc}}{4B} \approx -\frac{r_A + r_u - 2r_A r_u}{r_A - r_u}, \tag{3}$$

$$\frac{A_{ab}}{4B} \approx \frac{r_A + r_u - 2}{r_A - r_u}, \tag{4}$$

and then

$$\frac{A_{ab}}{A_{cc}} \approx -\frac{r_A + r_u - 2}{r_A + r_u - 2r_A r_u}. \tag{5}$$

Thus, (3)-(5) with a set of r_u and r_A enable us to estimate the ratios of the ^{63}Cu hyperfine coupling constants.

Table 1 shows the estimated ratios of $A_{cc}/4B$, $A_{ab}/4B$ and A_{ab}/A_{cc} for Hg1212 from (3)-(5) with the experimental r_u and r_{AF} in figures 1 and 2. The on-site coupling ratio A_{ab}/A_{cc} depends on the hole concentration in Hg1212.

Table 1. Anisotropies (r_u and r_{AF}) of $^{63}K_s$ and $^{63}T_1$, and the ratios of ^{63}Cu hyperfine coupling constants (A_{cc}, A_{ab}, B) for underdoped (un), optimally doped (op) and overdoped (ov) Hg1212. T_c is in kelvin. The value of r_u for overdoped Hg1212 is assumed after the optimally doped value in figure 1.

	T_c	r_u	r_{AF}	$A_{cc}/4B$	$A_{ab}/4B$	A_{ab}/A_{cc}
un	103	0.31	2.3	− 0.65	+ 0.13	− 0.20
op	127	0.34	2.0	− 0.64	+ 0.05	− 0.078
ov	93	0.34	1.8	− 0.67	− 0.04	+ 0.058

The $3d(x^2 - y^2)$ orbital electron of Cu^{2+} in the tetragonal crystal field produces the on-site hyperfine fields. The ratio A_{ab}/A_{cc} is expressed as

$$\frac{A_{ab}}{A_{cc}} \approx -\frac{\kappa + \frac{2}{7} - \frac{11}{7}\gamma}{-\kappa - \frac{4}{7} - \frac{62}{7}\gamma}, \quad (6)$$

where $\kappa(> 0)$ is the core polarization parameter, $2/7$ and $-4/7$ are the spin-dipole field coefficients, and $\gamma(< 0)$ is the spin-orbit coupling parameter [9, 19, 20]. The empirical values of $\kappa = 0.25$ and 0.325 were estimated for Cu^{2+} ions in the dilute copper salts [20]. The first-principles cluster calculations give $\kappa = 0.289$ for the density functional $< 1/r^3 >$ and 0.455 for the Hartree-Fock $< 1/r^3 >$ in La_2CuO_4 [21]. The value of $\kappa = 0.41$ is found in $CuGeO_3$ [22]. For Hg1212, A_{ab}/A_{cc} in Table 1 through (6) leads to the core polarization parameter $\kappa = 0.265$ (un), 0.315 (op) and 0.387 (ov), assuming $\gamma = -0.044$ [9, 19].

2.4. ^{63}Cu hyperfine coupling constants of $HgBa_2CuO_{4+\delta}$ and Hg1212

Let us show the ^{63}Cu hyperfine coupling constants of the optimally doped single-CuO_2-layer superconductor $HgBa_2CuO_{4+\delta}$ ($T_c = 98$ K). From the uniform spin susceptibility $\chi_s = 1.47\times10^{-4}$ emu/mole-f.u. [23] and the in-plane ^{63}Cu $K^{ab}_{spin} = 0.48$ % [24], we estimated the in-plane ^{63}Cu hyperfine coupling parameter $A^{hf}_{ab} = A_{ab} + 4B = (N_A\mu_B/\chi_s)K^{ab}_{spin} = 182$ kOe/μ_B for $HgBa_2CuO_{4+\delta}$ (N_A is Avogardro's number and μ_B is the Bohr magneton). Substituting $r_u = 0.53$ and $r_{AF} = 1.8$ [24] into (3)-(5) and using $A_{ab} + 4B = 182$ kOe/μ_B, we obtained the values of

$$A_{cc} = -65,\ A_{ab} = 21,\ \text{and}\ B = 40\ \text{kOe}/\mu_B$$

for the optimally doped $HgBa_2CuO_{4+\delta}$.

By adopting $A_{ab} + 4B = 182$ kOe/μ_B for Hg1212 after $HgBa_2CuO_{4+\delta}$, we estimated the individual components of A_{cc}, A_{ab}, and B (Table 2). Figure 3 shows A_{cc}, A_{ab}, and B (Table 2) plotted against the hole concentration P_{sh} [7] for Hg1212. In Table 2 and figure 3, with increase in the hole concentration, the absolute value of the negative A_{cc} increases, A_{ab} shows a sign change, and the B term slightly increases.

The reported B term is in the range from 36 to 155 kOe/μ_B in the other cuprate superconductors [10, 25, 26, 27, 28], assuming *a priori* the fixed values of $A_{cc} = -170$ and $A_{ab} = 37$ kOe/μ_B [25, 27, 28]. The cation-cation supertransferred hyperfine field B between $3d$ and $4s$ orbitals depends on the strength of the p-d covalent bond parameter [29]. The doping dependent B term in Table 2 indicates the development of the covalency with the hole concentration in Hg1212.

Table 2. ^{63}Cu hyperfine coupling constants in units of kOe/μ_B for underdoped (un), optimally doped (op) and overdoped (ov) Hg1212, assuming $A_{ab} + 4B = 182$ kOe/μ_B.

	A_{cc}	A_{ab}	B
un	−105	+21	40
op	−111	+8.7	43
ov	−127	−7.4	47

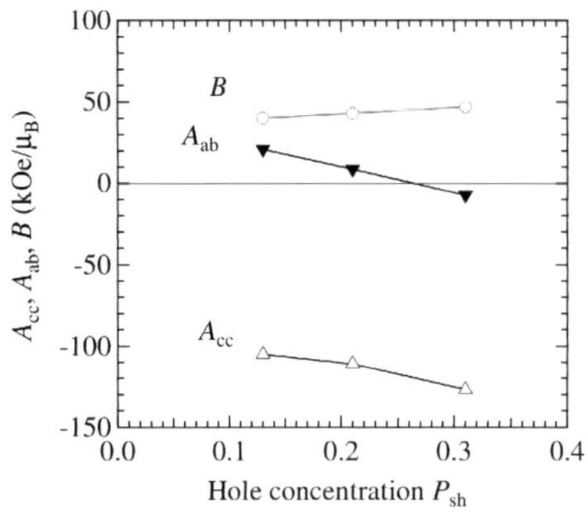

Figure 3. ^{63}Cu hyperfine coupling constants A_{cc}, A_{ab}, and B plotted against hole concentration P_{sh} for Hg1212. The solid curves are visual guides.

3. Local field fluctuations in ^{63}Cu nuclear spin-echo decay rate $1/T_{2L}$

3.1. ^{63}Cu nuclear spin-echo decay rate $1/T_{2L}$

Figures 4(a)-4(c) show the ^{63}Cu nuclear spin-echo decay rates $(1/T_{2L})_{ab,cc}$'s for Hg1212 from underdoped (a), optimally doped (b) and overdoped (c) [3]. The notations conform to those in [3]. The enhancements in $(1/T_{2L})_{ab,\ cc}$ at 220−240 K indicate the ultraslow fluctuations [3]. The peak temperature of $(1/T_{2L})_{cc}$ is nearly independent of the doping level, but the enhancement is suppressed by overdoping.

Figure 4(d) shows the anisotropy ratio of the local field fluctuations $\Delta J_{cc}/\Delta J_{ab} \equiv [(1/T_{2L})_{cc} - (1/T_{2R})_{cc}]/[(1/T_{2L})_{ab} - (1/T_{2R})_{ab}]$ derived from $1/T_{2L}$ and $1/T_1$ (Redfield's $1/T_{2R}$) [3]. $\Delta J_{\gamma\gamma}$ ($\gamma\gamma = cc$ and ab) expresses the additional fluctuations causing the enhancement in $1/T_{2L}$. One should note that $\Delta J_{cc}/\Delta J_{ab} < 1$ is characteristic of the ultraslow fluctuations.

Figure 4(e) shows the phase diagram of Hg1212, where the superconducting transition temperature T_c, the pseudo spin-gap temperature defined by the maximum temperature of

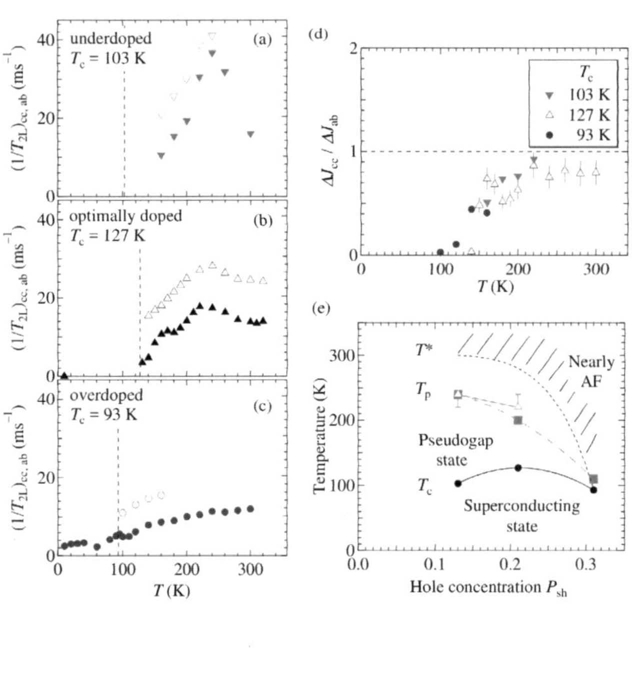

Figure 4. (a)−(c) T dependences of $(1/T_{2L})_{cc}$ (closed symbols) and $(1/T_{2L})_{ab}$ (open symbols) from underdoped to overdoped Hg1212 [3]. Each dashed line indicates T_c. (d) T dependences of $\Delta J_{cc}/\Delta J_{ab}$ [3]. (e) Phase diagram of Hg1212: T_c (closed circles), the pseudo spin-gap temperature defined by the maximum temperature of $1/T_1T$ [7] (closed squares), T_p defined by the peak temperature of $(1/T_{2L})_{cc}$ [3] (open triangles) against hole concentration P_{sh}. The dotted curve with a shaded region is a visual guide for the onset temperature T^* of decrease in ^{63}Cu Knight shift [7]. Nearly AF stands for the Curie-Weiss law in Cu $1/T_1T$ [7].

$1/T_1T$, T_p defined by the peak temperature of $(1/T_{2L})_{cc}$, and the onset temperature T^* of the decrease in the Cu Knight shift are plotted against the hole concentration P_{sh} in $Cu^{2+P_{sh}}$ [3, 7]. With hole doping, T^* decreases, while T_p is nearly independent of the hole concentration P_{sh}. The ultraslow fluctuations emerge in the underdoped regime and diminish in the overdoped regime.

3.2. Local field fluctuations

Local field fluctuations of J_{ab} ($B \perp$c axis) and J_{cc} ($B \parallel$c axis) causing the nuclear spin relaxations of T_1 and T_2 are defined by

$$J_{\gamma\gamma} = \sum_{\boldsymbol{q}} F_{\gamma\gamma}(\boldsymbol{q}) S(\boldsymbol{q}, \nu_n), \quad (7)$$

$$F_{\gamma\gamma}(\boldsymbol{q}) \equiv (4B)^2 f_{\gamma\gamma}(\boldsymbol{q}) = [A_{\gamma\gamma} + 2B\{\cos(q_x) + \cos(q_y)\}]^2, \quad (8)$$

where $\gamma\gamma = ab$ and cc, and ν_n is an NMR frequency [3]. The electron spin-spin correlation function $S(\boldsymbol{q}, \nu)$ (a frequency ν) is related to the dynamical spin susceptibility $\chi''(\boldsymbol{q}, \nu)$ through the fluctuation-dissipation theorem. $F_{\gamma\gamma}(\boldsymbol{q})$ is called the form factor of the wave vector $\boldsymbol{q}$ dependent hyperfine coupling constant, whose filtering effects in the $\boldsymbol{q}$ space play a significant role in the anisotropy and the site differentiation on NMR [8, 9, 10]. $\Delta J_{\gamma\gamma}$ expresses the additional fluctuations to $J_{\gamma\gamma}$ [3].

Figure 5 shows the q dependence of $f_{ab,\ cc}(\boldsymbol{q})$ for Hg1212 along the diagonal $\boldsymbol{q} = (q,\ q)$ in the first Brillouin zone, using the estimated coupling constant ratios in Table 1. Since $f_{cc}(\pi,\ \pi) > f_{ab}(\pi,\ \pi)$, the antiferromagnetic spin fluctuation $\chi'(\boldsymbol{q})$ localized around $\boldsymbol{q} = [\pi,\ \pi]$ leads to the anisotropy $J_{cc}/J_{ab} \sim 3$ in contrast to the experimental ratio $\Delta J_{cc}/\Delta J_{ab} < 1$ in figure 4(d) [3]. $\Delta J_{\gamma\gamma}$ expresses the development of the ultraslow fluctuations [3]. Thus, the antiferromagnetic fluctuations are excluded from the ultraslow fluctuations.

Let us assume a toy model of $\chi'(\boldsymbol{q},\ \nu_n) = \chi'_0\Theta(q_c - |q_x|)\Theta(q_c - |q_y|)$ ($\Theta(x)$ is the Heaviside step function). q_c is a cut-off wave number. $\chi'(\boldsymbol{q}) \propto S(\boldsymbol{q},\ \nu_n)$ takes a constant value χ'_0 over

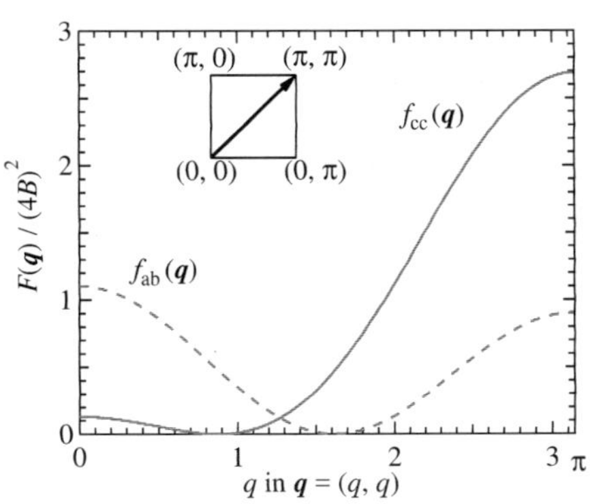

Figure 5. ^{63}Cu hyperfine coupling form factors $f_{cc}(\boldsymbol{q})$ and $f_{ab}(\boldsymbol{q})$ as functions of q in the wave vector $\boldsymbol{q} = (q,\ q)$ [$(0,\ 0) \to (\pi,\ \pi)$]. Inset shows the diagonal in the first Brillouin zone.

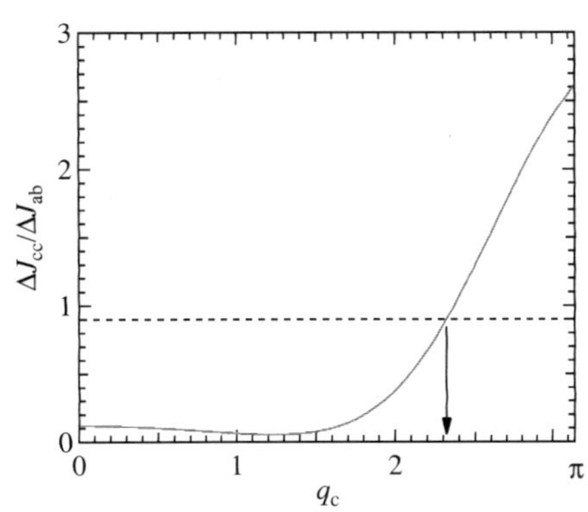

Figure 6. $\Delta J_{cc}/\Delta J_{ab}$ as a function of the cut off q_c in a toy model upon a static spin susceptibility $\chi'(\boldsymbol{q}) = \chi'_0$ ($|q_{x,y}| < q_c$) and 0 ($|q_{x,y}| > q_c$). Experimental constraint leads to $q_c < 2.3$.

$|q_{x,y}| < q_c$. For this toy model, the ratio $\Delta J_{cc}/\Delta J_{ab}$ is calculated as

$$\frac{\Delta J_{cc}}{\Delta J_{ab}} = \frac{\sum_{|q_{x,y}|<q_c} f_{cc}(\boldsymbol{q})}{\sum_{|q_{x,y}|<q_c} f_{ab}(\boldsymbol{q})}. \tag{9}$$

Figure 6 shows the numerical $\Delta J_{cc}/\Delta J_{ab}$ as a function of q_c. The experimental $\Delta J_{cc}/\Delta J_{ab} < 0.9$ in figure 4(d) imposes on the function in figure 6 and then leads to $q_c < 2.3$. The magnetic ultraslow fluctuations must be confined within $q_c < 2.3$. If the magnetic ultraslow fluctuations have the easy plane anisotropy, the upper limit of the cut-off value q_c will be smaller than 2.3. Thus, we obtained a model constraint on the magnetic ultraslow fluctuations, using the anisotropic hyperfine coupling constants.

Although the step function $\chi'(\boldsymbol{q})$ with $q_c < 2.3$ is not localized at $q = 0$, it is parallel to the IUC $q = 0$ magnetic moments observed by the spin polarized neutron scattering method [1]. The ultraslow fluctuations may be associated with the IUC $q = 0$ magnetic moments. However, if the enhancement in $1/T_{2L}$ is due to quadrupole fluctuations, one should explore the alternative fluctuations of charge or lattice for the electric ultraslow fluctuations.

4. Conculsions

The systematic hole doping dependences of the ^{63}Cu hyperfine coupling constants (A_{cc}, A_{ab} and B) were found for Hg1212 from underdoped to overdoped. A model constraint on the magnetic ultraslow fluctuations in Hg1212 was derived from the anisotropy ratios of the ^{63}Cu hyperfine coupling constants. The model expresses the magnetic fluctuations spread around $q = 0$. Possible electric ultraslow fluctuations causing the T_2 anomaly remain to be explored.

Acknowledgments

We thank Jun Kikuchi for fruitful discussions on the hyperfine coupling constants.

References

[1] Bourges P and Sidis Y 2011 *C. R. Physique* **12** 461
[2] Varma C M 2014 *J. Phys.: Condens. Matter* **26** 505701
[3] Itoh Y, Machi T and Yamamoto A 2017 *Phys. Rev.* B **95** 094501
[4] Singer P M, Hunt A W, Cederström A F and Imai T 1999 *Phys. Rev.* B **60** 15345
[5] Hunt A W, Singer P M, Cederström A F and Imai T 2001 *Phys. Rev.* B **64** 134525
[6] Mila F and Rice T M 1989 *Physica* C **157** 561
[7] Itoh Y, Tokiwa-Yamamoto A, Machi T and Tanabe K 1998 *J. Phys. Soc. Jpn.* **67** 2212
[8] Millis A J, Monien H and Pines D 1990 *Phys. Rev.* B **42** 167
[9] Monien H, Pines D and Slichter C P 1990 *Phys. Rev.* B **41** 11120
[10] Imai T 1990 *J. Phys. Soc. Jpn.* **59** 2508
[11] Itoh Y, Hayashi A and Ueda Y 1995 *J. Phys. Soc. Jpn.* **64** 3074
[12] Rybicki D, Kohlrautz J, Haase J, Greven M, Zhao X, Chan M K, Dorow C J and Veit M J 2015 *Phys. Rev.* B **92** 081115
[13] Shimizu T, Aoki H, Yasuoka H, Tsuda T, Ueda Y, Yoshimura K and Kosuge K 1993 *J. Phys. Soc. Jpn.* **62** 3710
[14] Hanzawa K 1994 *J. Phys. Soc. Jpn.* **63** 264
[15] Okabe Y and Kikuchi M 1988 *J. Phys. Soc. Jpn.* **57** 4751
[16] Terasaki I, Hase M, Maeda A, Uchinokura K, Kimura T, Kishio K, Tanaka I and Kojima H 1992 *Physica* C **193** 365
[17] Haase J, Jurkutat M, Kohlrautz J 2017 *Condense. Matter* **2** 16
[18] Takigawa M, Hammel P C, Heffner R H, Fisk Z, Smith J L and Schwarz R B 1989 *Phys. Rev.* B **39** 300
[19] Pennington C H, Durand D J, Slichter C P, Rice J P, Bukowski E D and Ginsberg D M 1989 *Phys. Rev.* B **39** 2902
[20] Bleaney B, Bowers K D and Pryce M H L 1955 *Proc. Roy. Soc. London*, Ser. A **228** 166
[21] Hüsser P, Suter H U, Stoll E P and Meier P F 2000 *Phys. Rev.* B **61** 1567

[22] Itoh M, Sugahara M, Yamauchi T and Ueda Y 1996 *Phys. Rev.* B **53** 11606
[23] Itoh Y, Machi T and Yamamoto A 2017 *Phys. Rev.* B **96** 235118
[24] Itoh Y, Machi T, Adachi S, Fukuoka A, Tanabe K and Yasuoka H 1998 *J. Phys. Soc. Jpn.* **67** 312
[25] Kitaoka Y, Fujiwara K, Ishida K, Asayama K, Shimakawa Y, Manako T and Kubo Y 1991 *Physica* C **179** 107
[26] Kambe S, Yasuoka H, Hayashi A and Ueda Y 1993 *Phys. Rev.* B **47** 2825
[27] Ishida K, Kitaoka Y, Asayama K, Kadowaki K and Mochiku T 1994 *J. Phys. Soc. Jpn.* **63** 1104
[28] Shimizu S, Iwai S, Tabata S-I, Mukuda H, Kitaoka Y, Shirage P M, Kito H and Iyo A 2011 *Phys. Rev.* B **83** 144523
[29] Huang N L, Orbach R, Šimánek E, Owen J and Taylor D R 1967 *Phys. Rev.* **156** 383

Effects of the Ba-substitution for Sr in the Bi-2201 phase of $BiPb(Sr,La)_2CuO_{6+\delta}$

D Hayakawa, T Watanabe, T Luo, T Kawamata, T Noji, M Kato and Y Koike

Department of Applied Physics, Graduate School of Engineering, Tohoku University, 6-6-05 Aoba, Aramaki, Aoba-ku, Sendai 980-8579, Japan

E-mail: kato@teion.apph.tohoku.ac.jp

Abstract. Polycrystalline samples of the Bi-2201 phase of $BiPbSr_{1.15-x}Ba_xLa_{0.85}CuO_{6+\delta}$ have been prepared by the conventional solid-state reaction method. Through the Ba-substitution for Sr, the lattice becomes large while T_c decreases gradually. With increasing x(Ba), the hole concentration in the CuO_2 plane decreases despite the increase in the number of excess oxygen. The suppression of the superconductivity may be due to the decrease in the hole concentration in the CuO_2 plane and the increase in the local lattice-distortion caused by excess oxygen.

1. Introduction

Among a lot of superconductors, the Bi-2223 cuprate superconductor is most expected for the practical application of superconductivity at the liquid nitrogen temperature because of its relatively high value of the superconducting transition temperature T_c ~110 K. However, it is 15-25 K lower than those of poisonous Tl-2223 and Hg-1223 cuprate superconductors, the crystal structures of which are analogous to that of the Bi-2223 cuprate superconductor. Elucidating why its T_c is low and increasing T_c up to ~135 K will promote the practical application of the Bi-2223 cuprate superconductor. There have been many efforts since the discovery of the Bi-2223 cuprate superconductor, but the problem has not yet been solved. In order to solve this problem, it is meaningful to investigate the Bi-2201 phase among three kinds of Bi-based cuprate superconductors of $Bi_2Sr_2Ca_{n-1}CunO_{2n+4+\delta}$ (n = 1, 2, 3), because the crystal structure is simple and the synthesis can readily be done, though its T_c is low.

The value of T_c in the Bi-2201 phase of $Bi_{2+x}Sr_{2-x}CuO_{6+\delta}$ at the time of discovery is ~ 8 K [1]. It increases up to ~ 15 K in $Bi_2Sr_2CuO_{6+\delta}$ in which Bi does not occupy the Sr site [2]. However, this value is much smaller than 90 - 95 K in the Tl-2201 phase of $Tl_2Ba_2CuO_{6+\delta}$ and the Hg-1201 phase of $HgBa_2CuO_{4+\delta}$ [3,4]. One of causes of the low T_c in $Bi_2Sr_2CuO_{6+\delta}$ is a large amount of excess oxygen introduced into the Bi_2O_2 layer which supplies the conductive CuO_2 plane with holes excessively to cause the local distortion. Therefore, T_c increases up to ~30 K by decreasing the hole concentration through the partial substitution of La^{3+} for Sr^{2+}[5]. Moreover, T_c is improved to ~40 K in $(Bi,Pb)_2(Sr,La)_2CuO_6$ by removing excess oxygen through the partial substitution of Pb^{2+} for Bi^{3+} with keeping the hole concentration at the optimum [6,7].

According to the empirical rule of high T_c in cuprate superconductors, T_c increases with increasing distance between Cu and O. The Cu-O distance in $Bi_2Sr_2CuO_{6+\delta}$ is shorter than that in $HgBa_2CuO_{4+\delta}$ and $Tl_2Ba_2CuO_{6+\delta}$, because smaller Sr^{2+} ions in $Bi_2Sr_2CuO_{6+\delta}$ are replaced by larger Ba^{2+} ions in $HgBa_2CuO_{4+\delta}$ and $Tl_2Ba_2CuO_{6+\delta}$. In this research, we have attempted to increase T_c further in the Bi-

2201 phase by extending the Cu-O distance through the substitution of Ba for Sr in $(Bi,Pb)_2(Sr,La)_2CuO_{6+\delta}$.

2. Experimental

Polycrystalline samples of $BiPbSr_{1.15-x}Ba_xLa_{0.85}CuO_{6+\delta}$ were prepared by the solid-state reaction method. Stoichiometric amounts of Bi_2O_3, PbO, $SrCO_3$, $BaCO_3$ and CuO powders were mixed, pressed into pellets and heated at 770°C for 24 h in flowing gas of Ar for 24 h. The products were then annealed at 500°C for 3 h in flowing gas of O_2. The structural analysis was carried out by the powder x-ray diffraction using CuKα radiation. The iodometric titration was carried out to estimate the oxygen content. Measurements of the thermoelectric power at room temperature were carried out to estimate the hole-concentration in the CuO_2 plane [8]. Magnetic susceptibility measurements were performed for the powdered samples using a SQUID magnetometer in a magnetic field of 10 Oe on warming after zero-field cooling in order to determine T_c.

3. Results

Figure 1 shows the powder x-ray diffraction patterns of $BiPbSr_{1.15-x}Ba_xLa_{0.85}CuO_{6+\delta}$. All the samples are almost of the single phase with the Bi-2201 structure, although a small amount of $Ba(Bi,Pb)O_3$ is included for $x \geq 0.8$. The crystal structure is orthorhombic for all the samples. As shown in Fig. 2, the lattice parameters increase linearly with increasing x(Ba) for x(Ba) $\leq$ 1.0, suggesting that the Cu-O distance increases through the Ba-substitution.

It is well known that Ba cannot be substituted for Sr at all in $Bi_2Sr_2CuO_6$. In contrast, Ba-substituted samples are successfully synthesized in the present system. This may be explained as follows. In $Bi_2Sr_2CuO_6$, the Ba-substitution is impossible owing to the enhancement of the size mismatch between the smaller Bi-O layer and the larger Sr-O layer. On the other hand, the size mismatch is reduced in $(Bi,Pb)_2(Sr,La)_2CuO_6$, because larger Pb is substituted for smaller Bi and smaller La is substituted for larger Sr, which allows for the Ba-substitution for Sr in the present system.

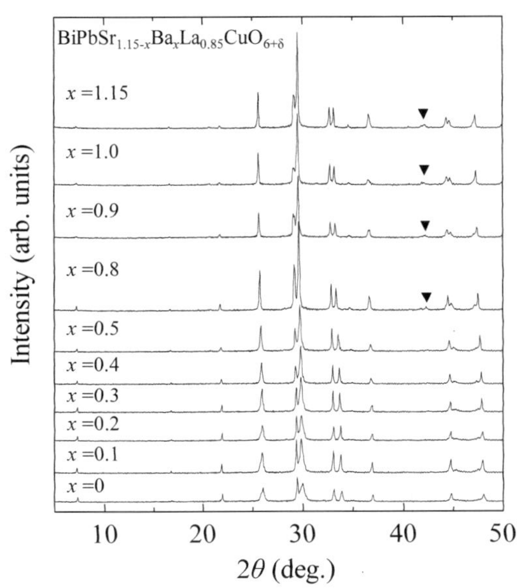

Figure 1. Powder x-ray diffraction patterns for $BiPbSr_{1.15-x}Ba_xLa_{0.85}CuO_{6+\delta}$. Peaks marked by ▼ are due to $Ba(Bi,Pb)O_3$.

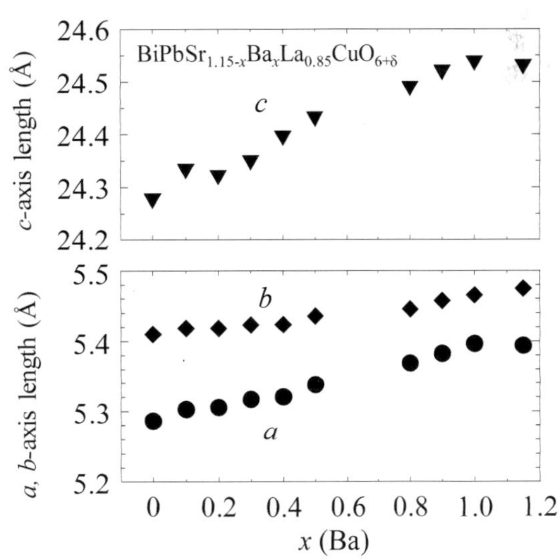

Figure 2. Dependence of lattice parameters on the Ba content x(Ba) for $BiPbSr_{1.15-x}Ba_xLa_{0.85}CuO_{6+\delta}$.

Figure 3 shows the temperature dependence of the magnetic susceptibility χ for $BiPbSr_{1.15-x}Ba_xLa_{0.85}CuO_{6+\delta}$. A diamagnetic signal due to the shielding effect is observed for all the samples. It is found that T_c tends to decrease and the superconducting volume fraction decreases with increasing x(Ba).

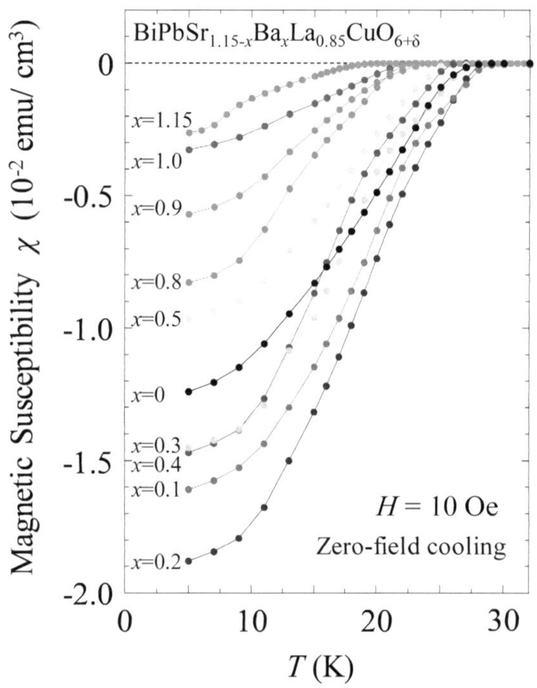

Figure 3. Temperature dependence of the magnetic susceptibility χ for $BiPbSr_{1.15-x}Ba_xLa_{0.85}CuO_{6+\delta}$.

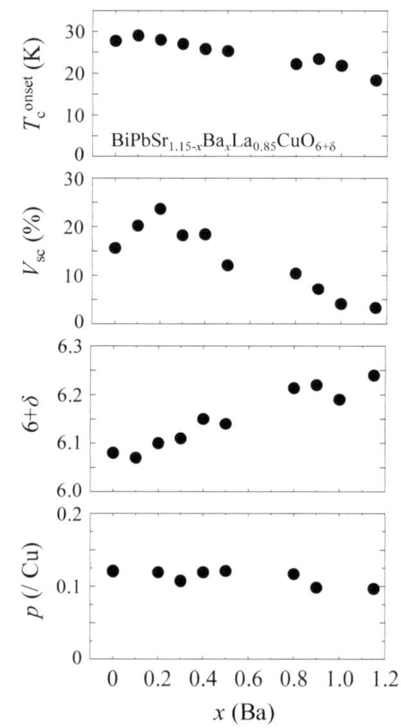

Figure 4. Dependences of T_c and the superconducting volume fraction at 5 K V_{sc}, the oxygen content $6+\delta$, the hole concentration in the CuO_2 plane p on the Ba content x(Ba) in $BiPbSr_{1.15-x}Ba_xLa_{0.85}CuO_{6+\delta}$.

Figure 4 displays the x(Ba)-dependences of T_c, defined at the temperature where χ changes by 0.1% of the value indicating the perfect-diamagnetism from the normal-state value, and the superconducting volume fraction V_{SC} estimated from the value of χ at 5 K. The x(Ba)-dependences of the hole concentration in the CuO_2 plane p and the oxygen content $6+\delta$ are also shown. Although the lattice becomes large monotonically through the Ba-substitution for Sr as shown in Fig. 2, both T_c and V_{SC} decreases gradually contrary to the empirical rule of high T_c in cuprate superconductors. The x(Ba)-dependence of T_c is quite similar to that of p. The p is as low as ~0.12 and tends to decrease slightly with increasing x(Ba), though the excess oxygen content increases with increasing x(Ba). This suggests that holes generated by excess oxygen are supplied to the $(Bi, Pb)_2O_2$ layer rather than to the CuO_2 plane. Form these results, it is concluded that the suppression of the superconductivity through the Ba-substitution is due to the decrease in p and the increase in the local lattice-distortion caused by excess oxygen. Similarly, the reason why the maximum value of T_c ~ 30 K in the present system is lower than the highest T_c ~40 K in the Bi-2201 phase reported [6,7] is concluded to be the low hole-concentration ~0.12 and the local lattice-distortion caused by excess oxygen.

4. Conclusions

Polycrystalline samples of the Bi-2201 phase of $BiPbSr_{1.15-x}Ba_xLa_{0.85}CuO_{6+\delta}$ have been prepared by the conventional solid-state reaction method. With increasing x(Ba), the lattice becomes large, while T_c decreases from ~30 K gradually. The unexpected suppression of the superconductivity may be due to the local lattice-distortion caused by excess oxygen, the number of which increases with increasing x(Ba). Despite the increase in excess oxygen, the hole concentration in the CuO_2 plane decreases slightly, indicating that holes generated by excess oxygen are supplied to the $(Bi,Pb)_2O_2$ layer rather than to the CuO_2 plane. This decrease in the hole concentration in the CuO_2 plane also suppresses the superconductivity. Therefore, the removal of excess oxygen through the reduction annealing and the increase in the hole concentration in the CuO_2 plane through the decrease in the La-content will allow for the investigation of effects of the enlargement of the Cu-O distance on the superconductivity in the present system.

Acknowledgments
This work was supported by JSPS KAKENHI Grant Number JP15K06433.

References

[1] Akimitsu J, Yamazaki A and Sawa H 1987 *Jpn. J. Appl. Phys.*, **26** L2080
[2] Niinae T, Ikeda Y, Bando Y, Takano M, Kusano Y and Takada J 1999 *Physica* C **313** 29
[3] Sheng Z Z and Hermann A M 1988 *Nature* **332** 55
[4] Putillin S N, Antipov E V, Chmaissem O and Marezzio M 1993 *Nature* **362** 226
[5] Schlögl A E, Neumeier J J, Diederichs J, Allgeier C, Schilling J S and Yelon W, *Physica* C **216** 417
[6] Kambe S, Kawai M and Kawai T 1990 *Solid State Commun.* **75** 435
[7] Arao Y, Tange M, Ikeda H, Koyano T and Yoshizaki R 2006 *Physica* C **445-448** 440
[8] Nakajima T, Tange M, Kizuka T, Ikeda H, Yoshizaki R and Sasaki S 2007 *Physica* C **463-465** 93
[9] Obertelli S D, Cooper J R and Tallon J L 1992 *Phys. Rev.* B **46** 14928

Superconductivity above 100 K in the Bi-2212 phase of $(Bi,Pb)_2Sr_2CaCu_2O_8$

K Sugawara, C Sugimoto, T Luo, T Kawamata, T Noji, M Kato and Y Koike

Department of Applied Physics, Graduate School of Engineering, Tohoku University, 6-6-05 Aoba, Aramaki, Aoba-ku, Sendai 980-8579, Japan

E-mail: kato@teion.apph.tohoku.ac.jp

Abstract. Polycrystalline samples of $Bi_{2-x}Pb_xSr_2CaCu_2O_8$ ($0.32 \leq x(Pb) \leq 0.42$) have been prepared by the conventional solid-state reaction method. It has been found that the maximum value of T_c, defined as the onset temperature of the shielding effect in the magnetic susceptibility measurements, is 102 K for $x(Pb)=0.36$. The present compound is the first superconductor with T_c above 100 K in the Bi-2212 phase.

1. Introduction

Among many superconducting materials, the Bi-based cuprate superconductor is one of the promising candidates for practical application of superconductivity at the liquid nitrogen temperature 77 K because of its relatively high value of the superconducting transition temperature T_c. The Bi-2223 phase of $Bi_2Sr_2Ca_2Cu_3O_{10+\delta}$, which is a member ($n$=3) of the homologous series of the Bi-based cuprate superconductor $Bi_2Sr_2Ca_{n-1}Cu_nO_{2n+4+\delta}$, has the highest T_c = 110 K among the Bi-based cuprate superconductors. However, it is known as a material whose sample of the single phase is extremely hard to be synthesized. On the other hand, the synthesis of the Bi-2212 phase of $Bi_2Sr_2CaCu_2O_{8+\delta}$ with $n = 2$ is relatively easy. However, the value of T_c of the Bi-2212 phase prepared in air is ~ 80 K, which is too low for the practical application at 77 K. Therefore, a further increase in T_c is desired.

T_c is well known to be sensitive to the hole concentration per Cu in the CuO_2 plane p. With the doping of hole carriers, superconductivity appears at $p \sim 0.05$. With increasing p, T_c increases in the underdoped region, takes the maximum at $p \sim 0.16$ and then decreases in the overdoped region. The Bi-2212 phase prepared in air is situated in the overdoped region, because excess oxygen atoms are easily introduced in the BiO plane. Therefore, its T_c increases up to about 95 K by the decrease in the hole concentration through the reduction annealing or the substitution of Y^{3+} for Ca^{2+} [1, 2]. In addition, the highest T_c of 98.5 K is observed in Pb-substituted $Bi_{1.6}Pb_{0.4}Sr_2CaCu_2O_8$ annealed in flowing gas of N_2 [3]. This increase in T_c is because excess oxygen atoms causing the local distortion are completely removed both through the Pb-substitution and through the annealing in flowing N_2 gas.

In this research, we have attempted to increase T_c further in Pb-substituted $Bi_{2-x}Pb_xSr_2CaCu_2O_8$ without excess oxygen by optimizing the Pb content, namely, the hole concentration in the CuO_2 plane.

2. Experimental

Polycrystalline samples of $Bi_{2-x}Pb_xSr_2CaCu_2O_8$ ($0.32 \leq x(Pb) \leq 0.42$) were prepared by the conventional solid-state reaction method as follows. Stoichiometric amounts of Bi_2O_3, PbO, $SrCO_3$, $CaCO_3$, CuO powders were mixed and calcined at 800°C for 24 h in air. The calcined mixture was reground and pressed into pellets, and sintered at 830°C for 40 h in air. The sintered pellets were annealed in an Ar gas flow of 50 ml/min at 775°C for 8 h. The structural analysis was carried out by the powder x-ray diffraction using Cu Kα radiation. The iodometric titration was carried out to estimate the oxygen content. Measurements of the thermoelectric power at room temperature were carried out to estimate the hole-concentration in the CuO_2 plane [4]. Magnetic susceptibility measurements were performed for the powdered samples using a SQUID magnetometer in a magnetic field of 10 Oe on warming after zero-field cooling in order to determine T_c. Electrical resistivity measurements were also performed by the standard dc four-probe method to determine T_c.

3. Results

Figure 1 shows the powder x-ray diffraction patterns of $Bi_{2-x}Pb_xSr_2CaCu_2O_8$ with $0.32 \leq x(\mathrm{Pb}) \leq 0.42$. It is found that all samples are of the single Bi-2212 phase. The peaks are indexed on the basis of the orthorhombic symmetry. As reported in the literature [3], the crystal structure of $Bi_{1.6}Pb_{0.4}Sr_2CaCu_2O_{8+\delta}$ with $\delta \sim 0$ is monoclinic. For our samples, however, the peak splitting due to the monoclinic distortion cannot be observed clearly within our experimental accuracy, but the broadening of (113), (115) and (117) peaks due to its distortion is observed. Figure 2 shows the dependence of the lattice parameters on the Pb-content x(Pb). The a-axis length tends to decrease slightly with increasing x(Pb) and both b-axis and c-axis lengths are almost constant.

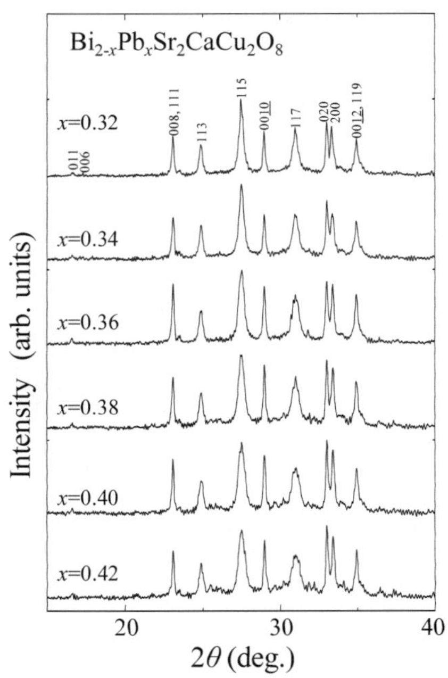

Figure 1. Powder x-ray diffraction patterns of $Bi_{2-x}Pb_xSr_2CaCu_2O_8$ ($0.32 \leq x(\mathrm{Pb}) \leq 0.42$).

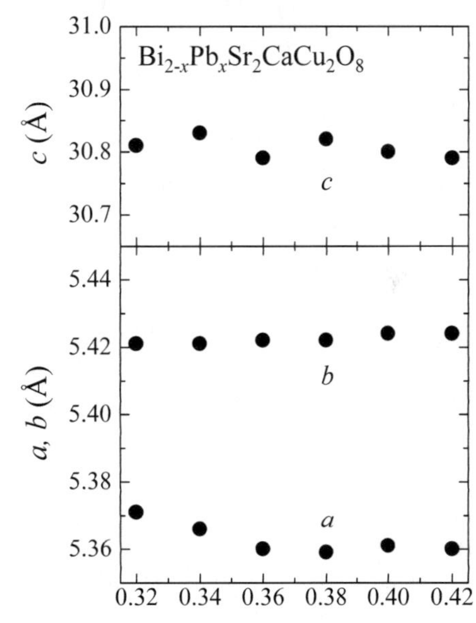

Figure 2. Variations of the lattice parameters a, b and c with the Pb-content x(Pb) in $Bi_{2-x}Pb_xSr_2CaCu_2O_8$ ($0.32 \leq x(\mathrm{Pb}) \leq 0.42$).

As shown in Fig. 3, values of the average oxygen-content $8+\delta$ are lower than the stoichiometric oxygen-content of 8 for all x(Pb), which may be due to the decomposition in the surface of crystalline grains through the reduction annealing at a temperature as high as 775°C rather than due to the oxygen deficiency in the Bi-2212 phase. Moreover, the average oxygen-content $8+\delta$ decreases with increasing x(Pb). This is because the melting point decreases with increasing x(Pb) so that the decomposition is more likely to occur.

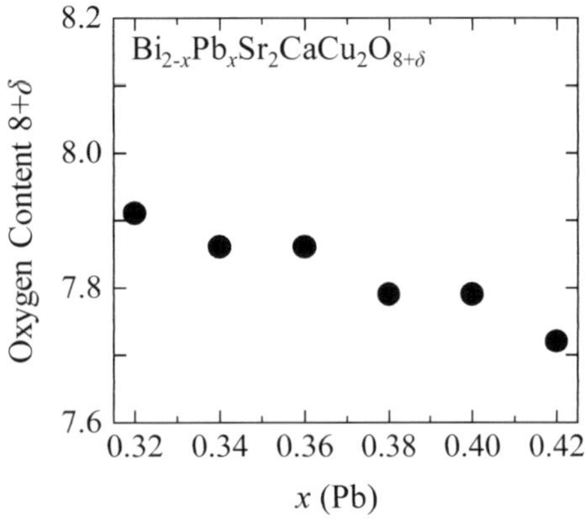

Figure 3. Dependence of the oxygen content $8+\delta$ on the Pb content x(Pb) in $Bi_{2-x}Pb_xSr_2CaCu_2O_8$.

Figure 4(a) shows the temperature dependence of the magnetic susceptibility χ for $Bi_{2-x}Pb_xSr_2CaCu_2O_8$. A single-step diamagnetic signal due to the shielding effect is observed for all samples. The superconducting volume fraction estimated from the shielding signal at 10 K is about 20-25%, indicating the appearance of bulk superconductivity. The value of T_c, defined as the onset temperature of the shielding effect, is about 100 K for all samples, as shown in Fig. 4(b). In particular, the maximum T_c of 102 K is observed for x(Pb)=0.36 and is larger than the maximum value of 98.5 K observed so far for x(Pb)=0.40 [3]. The present compound is the first superconductor with T_c above 100 K in the Bi-2212 phase.

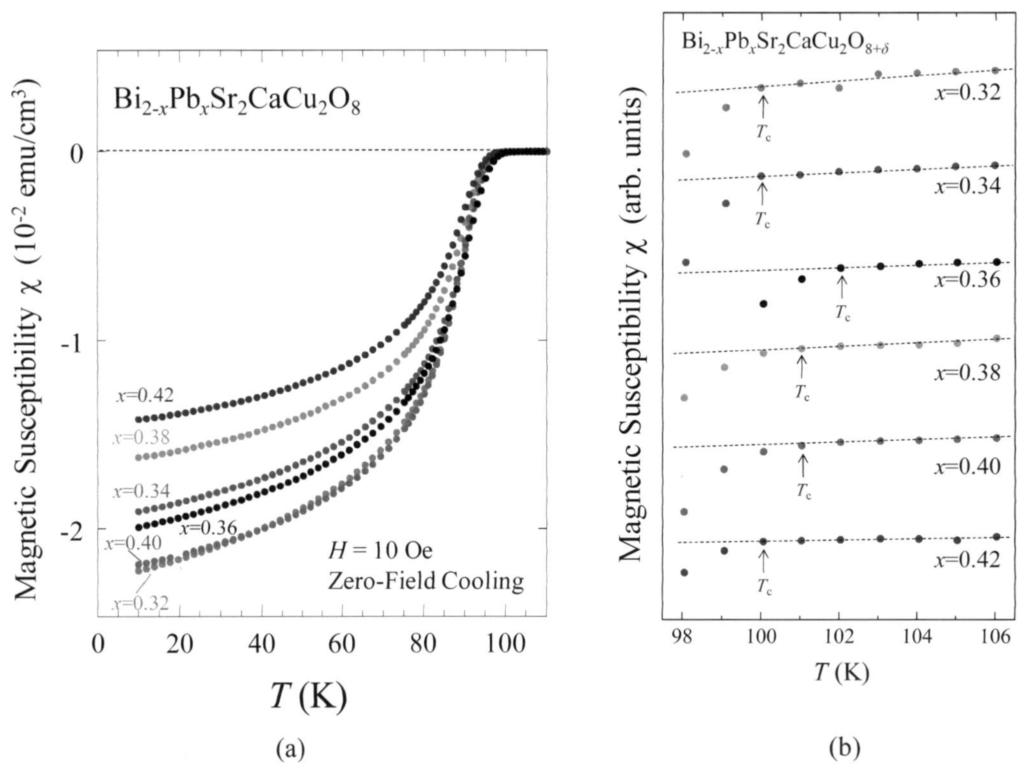

Figure 4. (a) Temperature dependence of the magnetic susceptibility χ for $Bi_{2-x}Pb_xSr_2CaCu_2O_8$, and (b) that around T_c.

Figure 5 shows the temperature dependence of the electrical resistivity ρ for $Bi_{2-x}Pb_xSr_2CaCu_2O_8$ with x(Pb)=0.36. With decreasing temperature, ρ starts to deviate from the extrapolated line in the normal state below ~110 K owing to the superconducting fluctuation and decreases sharply below ~102 K which is in good agreement with the value of T_c^{onset} determined from the magnetic susceptibility measurement. The value of T_c, defined at the midpoint of the transition curve, is 90.5 K. Then, a somewhat broad tail in the ρ-T curve is observed below ~85 K and then the zero-resistivity is realized at low temperatures below 72 K. The hole concentration in the CuO_2 plane is estimated to be about 0.17 from the value of the thermoelectric power at room temperature, indicating that holes are doped almost optimally. Therefore, the broad tail and semiconductive behaviour in the normal state may be due to the weak link between crystalline grains, the surface of which is decomposed through the reduction annealing at a high temperature as mentioned above.

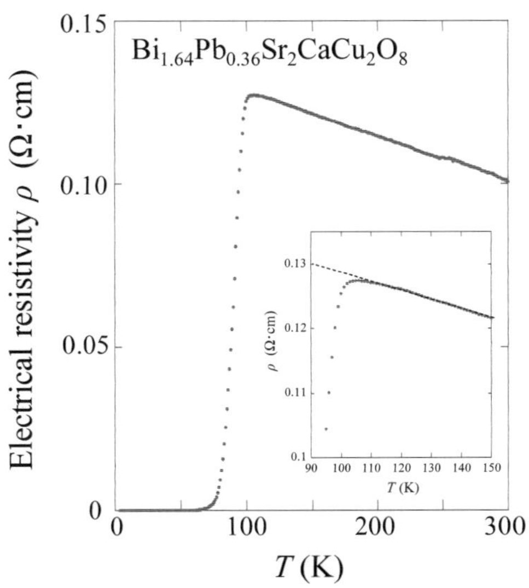

Figure 5. Temperature dependence of the electrical resistivity ρ for $Bi_{2-x}Pb_xSr_2CaCu_2O_8$ with x(Pb)=0.36.

4. Conclusions

We have successfully synthesized the Bi-2212 phase with T_c above 100 K in $Bi_{2-x}Pb_xSr_2CaCu_2O_8$ for the first time. It has been found that the maximum value of T_c^{onset} is 102 K for x(Pb)=0.36. Unfortunately, the zero resistivity is exhibited only below T_c^{zero} = 72 K due to the weak link between crystalline grains, the surface of which is decomposed by the reduction annealing at a high temperature. Nevertheless, the improvement of the annealing condition will allow for the increase in T_c^{zero}.

Acknowledgments

This work was supported by JSPS KAKENHI Grant Number JP15K06433.

References

[1] Triscone G, Genoud J -Y, Graf T, Junod A and Muller J 1991 *Physica* C **176** 247
[2] Eisaki H, Kaneko N, Feng D L, Damascelli A, Mang P K, Shen K M, Shen Z -X and Greven M 2004 *Phys. Rev.* B **69** 064512
[3] Kambe S, Matsuoka T, Takahashi M, Kawai M and Kawai T 1990 *Phys. Rev.* B **42** 2669
[4] Obertelli S D, Cooper J R and Tallon J L 1992 *Phys. Rev.* B **46** 14928

Preparation of (11n) oriented $Bi_2Sr_2CaCu_2O_{8+x}$ thin films by solution methods using $NdGaO_3$ (100) substrates

Y Yamada[1] and T Okamoto[2]

[1] Department of Innovative Electrical and Electronic Engineering, National Institute of Technology, Oyama College, 771 Nakakuki, Oyama, Tochigi 323-0806, Japan
[2] Electrical, Electronics and Information Engineering, Nagaoka University of Technology, 1603-1 Kamitomioka, Nagaoka, Niigata 940-2188, Japan

yyamada@oyama-ct.ac.jp

Abstract. We have prepared (11n) oriented $Bi_2Sr_2CaCu_2O_{8+x}$ (Bi2212) thin films by KCl flux method which is one kind of solution method. The (100) substrate of $NdGaO_3$ (NGO) was used. Heat treatment was performed with four different temperature profiles. Only the (11n) peak appeared in the θ-2θ X-ray diffraction pattern. From the SEM image, it was found that elongated plate-like crystal grains of Bi2212 were grown. From the viewpoint of lattice matching, this elongated crystal grain is considered to be (010) (or (100)) oriented crystal grains.

1. Introduction

$Bi_2Sr_2CaCu_2O_{8+x}$ (Bi2212) has a layered perovskite structure and forms an intrinsic Josephson junction (IJJ). When a voltage is applied in the c-axis direction, an alternating current with a frequency proportional to the voltage is generated. A vibration mode excited by matching the frequency of the alternating current with a cavity resonance frequency can be used for a terahertz oscillator, which is considered to be a promising device that fills a frequency domain called the "terahertz gap". It has already been proved that a μW class continuous coherent terahertz wave can be oscillated [1-4].

The major preparation methods thus far have been reported on c-axis oriented Bi2212 that requires complicated procedures and dry etching, such as a precisely controlled etching process in depth direction in order to form the c-axis current paths. If a non-c-axis oriented thin film of which the c-axis parallel or incline to the substrate surface can be prepared, planar type IJJ devices can be fabricate simply by forming current paths parallel to the substrate [5,6]. In preparing such non-c-axis oriented thin films, it is important that selection of substrates focusing on lattice matching. If such substrates are selected and solution or powder material is coated on that substrate by a printing method, it is considered that an IJJ devices can be fabricated only by performing a heat treatment.

In the case of preparing Bi2212 thin films, c-axis oriented thin films were formed by using $SrTiO_3$ (STO) (100) substrates [7,8], and (117) oriented thin film with c-axis twin structure ("c-twinning") are formed by using STO (110) substrates [9]. In order to fabricate Bi2212 IJJ devices, it is necessary to prepare thin films having no c-twinning in order to form c-axis current paths. The c-axis lattice constant of Bi2212 is $c = 30.6 \sim 30.9$ Å [10,11], which is relatively close to 8 times the lattice constant of STO ($3.095 \times 8 = 31.24$ Å). However, as mentioned above, actually (117) oriented thin film is formed. $NdGaO_3$ (NGO) has a lattice constant of $a = 5.427$ Å, $b = 5.497$ Å, $c = 7.707$ Å [12]. Four times the c-axis length of NGO is $7.707 \times 4 = 30.82$ Å, which is almost equal to that of Bi2212. This

IOP Conf. Series: Journal of Physics: Conf. Series **1054** (2018) 012009 doi:10.1088/1742-6596/1054/1/012009

value is closer than 8 times the *c*-axis length of STO. If an NGO (100) substrate is used, it is expected that the *c*-axis of Bi2212 grows lattice-matched to the *c*-axis of the NGO (100) substrate.

In this study, we report fabrication of (11*n*) oriented Bi2212 thin films using NGO (100) substrate by KCl flux method which is one kind of solution method. A typical flux method requires a large amount of solution and mechanical motion. On the other hand, in this report, a prototype was made by a simple method of simply coating the substrate and heat treating it in order to apply it to the printing method. Evaluation of the thin film was carried out through X-ray diffraction (XRD) pattern, observation of surface morphology by scanning electron microscope (SEM) image.

2. Experimental Procedure

Bi2212 thin films under study were prepared by the KCl flux method. Substrates used were NGO (100) substrates with the size of 10 × 10 × 0.5 mm. Preparation procedures were as follows:

(1) The raw material powder is weighed so that the molar ratio is Bi_2O_3 : $SrCO_3$: $CaCO_3$: CuO : KCl = 1:2:1:2:94.
(2) The weighed powder is mixed using an agate mortar and pestle in 2-propanol for 2 h.
(3) The mixed powder is dried using an IR lamp.
(4) Approximately 30 mg of the dried mixed powder is coated on the NGO (100) substrate.
(5) The dried mixed powder is coated on an NGO (100) substrate. The amount of powder is about 30 mg.
(6) Heat treatment is carried out using a box furnace with four temperature profiles.
(7) KCl flux is dissolved and removed using purified water.

The samples were taken into the box furnace at room temperature. Then, the samples were taken out after being cooled to room temperature. Figure 1(a) ~ (d) shows the temperature profile of a box furnace. The crystal structures of the samples were investigated by θ-2θ XRD patterns with CuKα radiation (λ = 1.54 Å). The surface morphologies were observed by SEM.

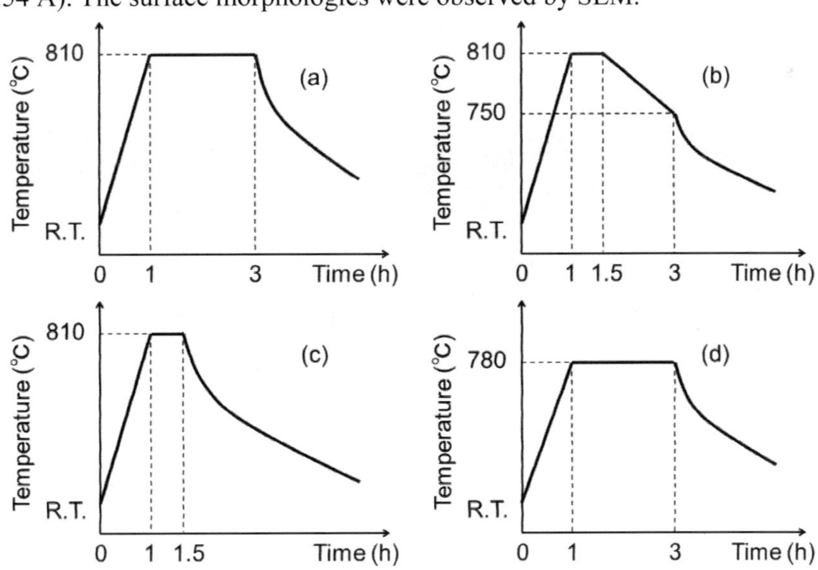

Figure 1. The temperature profile of a box furnace.

3. Result and Discussion

Figure 2 shows comparison of XRD patterns of samples with different temperature profiles. (a) ~ (d) of figure 2 correspond to (a) ~ (d) of figure 1, respectively. Only peaks of Bi2212 phase and NGO substrates appeared in these XRD patterns. As the time maintained at (or close to) 810 °C is longer, the (11*n*) peak of Bi2212 tends to become larger.

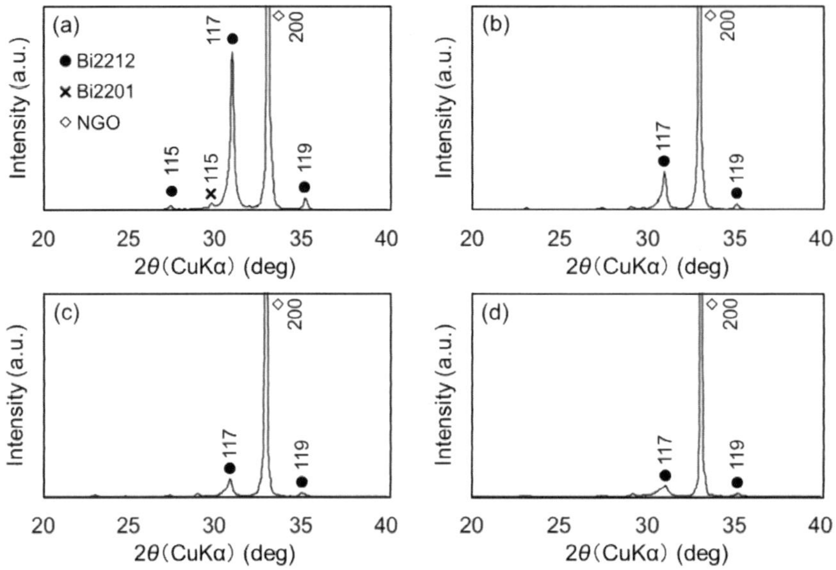

Figure 2. θ-2θ XRD patterns of samples with different temperature profiles using NGO (100) substrates.

Figure 3 shows a comparison of the observed surface morphologies by SEM with different temperature profiles. (a) ~ (d) of figure 3 correspond to (a) ~ (d) of figure 1, respectively. From the viewpoint of lattice matching, elongated plate-like crystal grains along the white broken arrow in the figure are considered to be (010) (or (100)) oriented crystal grains. The (010) (or (100)) oriented crystal grains should appear as (020) (or (200)) peaks on $2\theta = 33.16°$ on the XRD pattern. However, it is considered that it can not be observed in figure 2 because it is located at the position overlapping the peak of the NGO substrate.

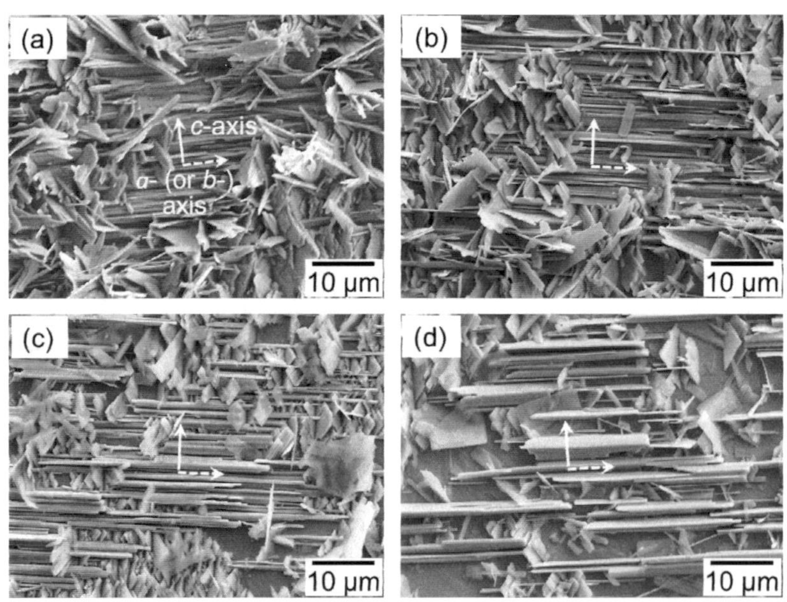

Figure 3. The observed surface morphologies by SEM. The white solid arrow in the figure indicates the c-axis direction of the Bi2212, and the white broken line arrow indicates the a-axis (or b-axis) direction.

Figure 4 show a top schematic view (imaginary view) of lattice matching. The lengths of the *a*-axis and the *b*-axis of Bi2212 were set to the same average length (= 5.41 Å) in the figure. Since the crystal grains of Bi2212 grow to extend in the *ab* plane direction, it is considered that the crystal grains oriented in the (010) (or (100)) orientation become elongated in the *a*-axis (or *b*-axis) direction.

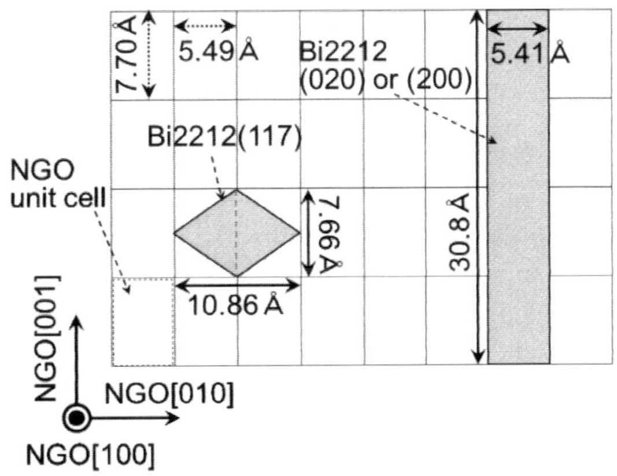

Figure 4. A top schematic view (imaginary view) of lattice matching for Bi2212 and NGO substrate.

4. Conclusion

In order to realize fabrication of non-*c*-axis oriented thin film that can be used for Bi2212 IJJ device by printing method, we attempted to fabricate non-*c*-axis oriented thin film by KCl flux method. For that purpose, focusing on lattice matching, NGO (100) substrates was used. The crystallinity and orientation were evaluated by the θ-2θ pattern of XRD. As a result, the followings were clarified.

(1) In comparison of XRD patterns of samples with different temperature profiles, as the time maintained at (or close to) 810 °C is longer, the (11*n*) peak of Bi2212 tends to become larger.

(2) From the viewpoint of lattice matching, elongated plate-like crystal grains in SEM images are considered to be (010) (or (100)) oriented crystal grains.

In this report, we succeeded in growing (010) (or (100)) oriented Bi2212 crystal grains. However, (11*n*) oriented crystal grains with *c*-twinning are also formed, which is a problem to be solved in the future.

Acknowledgments

This work was supported by JSPS KAKENHI Grant Number JP17K06377.

References

[1] Ozyuzer L, Koshelev A E, Kurter C, Gopalsami N, Li Q, Tachiki M, Kadowaki K, Yamamoto T, Minami H, Yamaguchi H, Tachiki T, Gray K E, Kwok W -K, and Welp U 2007 *Science* **318** 1291

[2] Benseman T M, Gray K E, Koshelev A E, Kwok W -K, Welp U, Minami H, Kadowaki K and Yamamoto T 2013 *Appl. Phys. Lett.* **103** 022602

[3] Kashiwagi T, Yamamoto T, Kitamura T, Asanuma K, Watanabe C, Nakade K, Yasui T, Saiwai Y, Shibano Y, Kubo H, Sakamoto K, Katsuragawa T, Tsujimoto M, Delfanazari K, Yoshizaki R, Minami H, Klemm R A and Kadowaki K 2015 *Appl. Phys. Lett.* **106** 092601

[4] Kashiwagi T, Sakamoto K, Kubo H, Shibano Y, Enomoto T, Kitamura T, Asanuma K, Yasui T, Watanabe C, Nakade K, Saiwai Y, Katsuragawa T, Tsujimoto M, Yoshizaki R, Yamamoto T, Minami H, Klemm R A and Kadowaki K 2015 *Appl. Phys. Lett.* **107** 082601

[5] Tsai J S, Fujita J and Kupriyanov M Yu 1995 *Phys. Rev.* B **51** 16267

[6] Kawahara T, Ishibashi T, Kaneko H and Sato K 1999 *IEEE Trans. Appl. Supercond.* **9** 4519
[7] Iwasaki H, Machikawa S, Yufune S, Ishibashi T and Sato K 2005 *IEEE Trans. Appl. Supercond.* **15** 3058
[8] Deguchi Y, Kikuchi H, Mori N, Yamada Y, Atsumi T, Yoshida K and Ishibashi T 2013 *Physics Procedia* **45** 193
[9] Yamada Y, Mori N, Atsumi T, Kato T and Ishibashi T 2015 *Physics Procedia* **65** 165
[10] Sunshine S A, Siegrist T, Schneemeyer L F, Murphy D W, Cava R J, Batlogg B, van Dover R B, Fleming R M, Glarum S H, Nakahara S, Farrow R, Krajewski J J, Zahurak S M, Waszczak J V, Marshall J H, Marsh P, Rupp L W, Jr and Peck W F 1988 *Phys. Rev.* B **38** 893
[11] Horiuchi S, Maeda H, Tanaka Y and Matsui Y 1988 *Jpn. J. Appl. Phys.* **27** L1172
[12] Vasylechko L, Akselrud L, Morgenroth W, Bismayer U, Matkovskii A and Savytskii D 2000 *J. Alloys Compd.* **297** 46

IOP Conf. Series: Journal of Physics: Conf. Series **1054** (2018) 012010 doi:10.1088/1742-6596/1054/1/012010

Intermediate Phase Evolution of YBCO Superconducting Films Fabricated by Fluorine Free MOD Method

J L Zuo[1], Y Zhao[1, 2], W Wu[1, 2], J Y Chu[1], X Y Wu[3], Z W Zhang[1, 2], Z Y Hong[1, 2], Z J Jin[1]

1 Shanghai Jiaotong University, 800th Dongchuan Road Shanghai, 200240, China

2 Shanghai Superconductor Co,.Ltd. 25# 1388th Zhangdong Road Pudong New District Shanghai, 201210, China

3 State Grid Shanghai Municipal Electric Power Company, Pudong Power Supply Company, Shanghai, 200120, China

E-mail: wei.wu@sjtu.edu.cn

Abstract. The second generation high temperature superconducting tapes, which use ReBCO film as superconducting layer, are considered to have enormous potential in practical applications because of its high operation temperature (up to 77 K), high superconducting performance under strong field and good mechanical properties relative to other practical superconducting wires / tapes. Among its preparation processes, fluorine free-metal organic deposition (FF-MOD) technique has been a research hotspot in recent years because of its unnecessity of vacuum environment, fast pyrolysis and crystallization growth rate, and environmental friendliness. In this paper, YBCO films were prepared on the CeO_2 / IBAD-MgO / Y2O_3 / Al_2O_3 / Hastelloy C276 technical substrate by FF-MOD technique. To systematically study the process of its phase evolution, the samples were quenched in different temperatures and followed by the characterization of XRD, ATR-IR and SEM. It is shown that the reaction between $BaCO_3$ and the oxides of Y and Cu elements does not produce YBCO directly, but experienced a more complicated intermediate process. Besides, the formed YBCO crystals are randomly oriented at first, and then converted into a biaxially textured structure.

1. Introduction

In recent years, superconducting technology has become a hot spot in many application scenario including power transmission cables, transformers, current limiters, magnetic levitation, magnetic resonance and particle accelerators. The second generation of high temperature superconducting (2G-HTS) tapes, which known as coated conductor, are considered to have a good prospect in practical applications. Because they have high critical temperature (T_c), high critical current density (J_c) and good mechanical properties (small critical bending radius [1], high critical tensile strength [2]).

By now, the main technical issue of preparing high-performance ReBCO coated conductors is to deposit biaxially textured ReBCO films on biaxially textured substrates [3]. In this process, three methods were wildly used: pulsed laser deposition (PLD), metal organic chemical vapour deposition (MOCVD) and metal organic deposition (MOD). Among them, the MOD method does not require high vacuum environment (which means low equipment cost), and has the unique advantage of making wide tapes [4]. However, in the YBCO films made by MOD method in earlier years, the $BaCO_3$ generated in pyrolysis process was still stable at the temperature of YBCO formation, and remained on the grain boundary, which is the obstacle of current transmission in the superconducting state [5]. Only by strict controlling of the temperature and oxygen partial pressure during heat treatment [6], YBCO films with J_c> 1MA / cm^2 (77K, 0T) could be prepared. On the other hand, Trifluoroacetic acid metal organic deposition (TFA-MOD) method was proposed by Gupta et al[7]., which replaced the $BaCO_3$ intermediate phase by more stable BaF_2 during the pyrolysis process by introducing F element into the precursor solution, and eliminated F element in the form of HF gas by introducing H_2O in the crystallization process[8]. After extensive development, YBCO films with J_c of more than 5 MA / cm^2 and commercial long tapes with critical currents of more than 200 A (77 K, 0 T) could been prepared by the TFA-MOD method [9] with wider process window.

However, in TFA-MOD method, the crystallization speed is severely limited by the removal speed of HF gas [10], and the HF gas itself is harmful on plants and animals. In contrast. The fluorine-free metal organic deposition (FF-MOD) method which has been studied before the invention of TFA-MOD method, is more environmentally friendly and more time efficient in heat treatment process. So in recent years, the FF-MOD technology has drawn the researchers' attention again [11]. As introduced above, during FF-MOD process, the Ba element exists in the form of $BaCO_3$ after the pyrolysis process. Therefore, a more in-depth study on the phase evolution of $BaCO_3$ and YBCO in the heat treatment process will be important for improving the performance of YBCO films made by FF-MOD method. Xu et al [12] removed $BaCO_3$ phase by using trimethylacetate as a precursor and introducing a wet atmosphere in the crystallization process in which $BaCO_3$ reacts with H_2O to form $Ba(OH)_2$. Then $Ba(OH)_2$ react with other oxides to form YBCO phase. Vermeir et al. [13] studied the phase evolution of YBCO films deposited on STO single crystal substrate by a water based FF-MOD solution. They found that $BaCO_3$ would experience a three-step reactions: first $BaCO_3$ reacts with CuO to form $BaCuO_2$ which will continue to react with CuO to form a liquid phase. At last, the liquid phase reacts with Y_2O_3 to form YBCO crystals.

In this paper, a solution based on yttrium acetate, barium acetate, copper acetate and propionic acid was used to prepare YBCO thin films on a technical substrate (CeO_2 / IBAD-MgO / Y_2O_3 / Al_2O_3 / Hastelloy C276). The phase evolution of $BaCO_3$ and YBCO during the crystallization process is studied.

2. Experiments

2.1. Solution preparation and coating

A precursor solution [11] with total cation concentration of 1.5mol/L was prepared by mixing Y $(CH_3COO)_3 \cdot 4H_2O$, Ba $(CH_3COO)_2$, Cu $(CH_3COO)_2 \cdot H_2O$ salts and propionic acid.
The wet films were uniformly coated on CeO_2 / IBAD-MgO / Y_2O_3 / Al_2O_3 / Hastelloy C276 technical substrates by dip coating method with SYDC-200 pulling coating machine, the pulling speed was 1.5mm/s.

2.2. Heat treatment

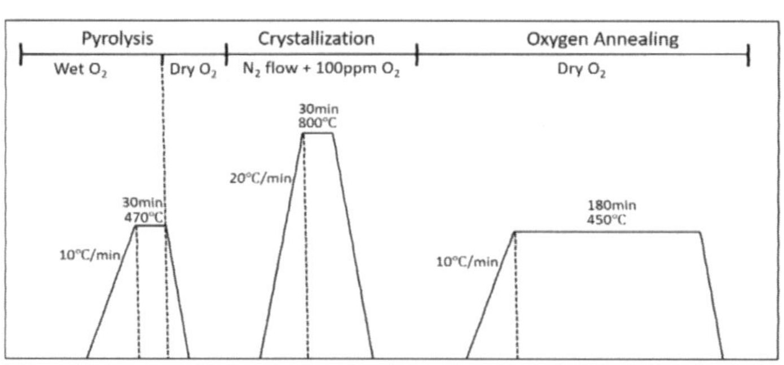

Figure 1. Diagram of temperature - atmosphere in heat treatment process

The heat treatment process, as shown in Figure 1, is divided into three steps: pyrolysis, crystallization and oxygen annealing. During the pyrolysis process, the temperature was raised to 470 ° C at a rate of 10 ° C / min and held for 30 minutes within an atmosphere of pure wet O_2 flow (500ml/min)[14].During the crystallization process, the temperature was raised to 800 ° C at a rate of 20 ° C / min and held for 30 minutes in an atmosphere of 100ppm O_2 flow (300ml/min, balanced with N_2), the atmosphere was kept until the end of the cooling (To study the phase evolution, samples were quenched at different temperatures in the crystallization process.) The oxygen annealing of the crystallized YBCO films was conducted at 450 ° C held for 180 minutes in an atmosphere of pure dry O_2 flow (50ml/min).

2.3. Characterization

The phase composition and texture of the samples were studied by a Bruker D8 Advance X-ray Diffraction (XRD) instrument using $CuK_{\alpha1}$ (K_β signal included) signal source with a Bruker VÅNTEC-500 area detector. The surface micro-morphology of the samples were studied by Zeiss Auriga scanning electron microscope (SEM) in Inlens mode. Using a Nicolet Model 6700 Infrared Spectrometer, Fourier transform infrared (FT-IR) spectra of the sample were obtained under attenuated total reflection (ATR) mode. $BaCO_3$ residues up to 2% by mass can be detected by this method. The hysteresis loop of the fully processed films were obtained by a Physical Property Measurement System (PPMS). The critical current density (J_c) was calculated using equation (1), which is based on the extended Bean critical state model [15]:

$$J_c = \frac{20\Delta M}{va\left(1 - \frac{a}{3b}\right)} \qquad (1)$$

Where Δm is the opening of the magnetization hysteresis loop; *a*, *b* are the length and width of the sample respectively ($a \leqslant b$); *v* is the volume of the YBCO film

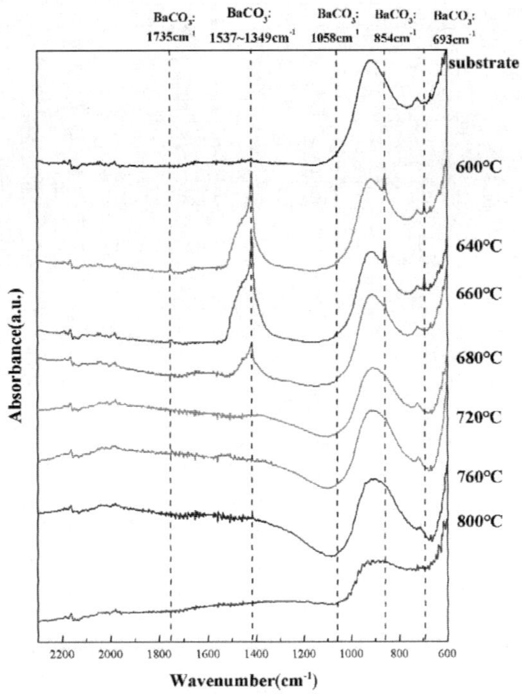

Figure 2. Infrared Absorption Spectrum on samples quenched in different temperature on artificial substrate

3. Results and Discussions

When samples were prepared on technical substrate, XRD and ATR-FT-IR characterization were performed on the samples quenched from different temperatures, the results is shown in Figure 2, Figure 3 and Figure 4. In order to give better expression to the change of $BaCO_3$ phase, larger scale is used in the χ direction in Figure 3 (a), 3 (b), and YBCO diffraction pattern is labeled with the abbreviation Y, and all patterns from the substrate(such CeO_2 at 32.7° and Y_2O_3 at 28.8°) were labeled with symbol "◊" in figure 3.

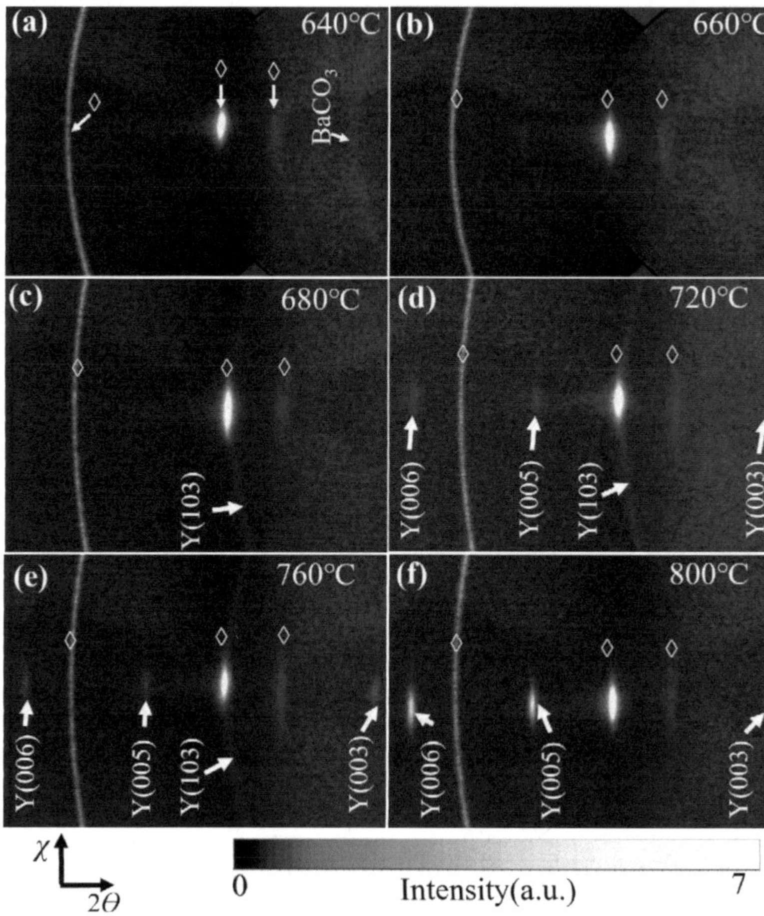

Figure 3. XRD results on samples quenched in different temperature on technical substrate
(a) 640°C (b) 660°C (c) 680°C (d) 720°C (e) 760°C (f) 800°C

As can be seen from the figure: 1) In the infrared spectrum, all peaks other than $BaCO_3$ phase are derived from the substrate itself; 2) the $BaCO_3$ phase in the film is always exists until the temperature reaches 660℃. When the temperature reaches 640 ℃, there is still obvious $BaCO_3$ phase on the XRD and ATR-FT-IR characterization results. At this time, the surface is mostly uniformly distributed Nano grains; 3) When the temperature reached 660 ℃, $BaCO_3$ phase was no longer visible in the XRD characterization, and YBCO(*103*) did not appear. In the infrared absorption spectrum, the corresponding absorption peak of $BaCO_3$ could still be observed, but with an obvious intensity attenuation, which demonstrated that $BaCO_3$ has been involved in the chemical reaction at this stage. No YBCO product was detected, and there was no significant change in the film surface morphology; 4) YBCO (*103*) began to appear when the temperature reached 680 ℃. At this stage, the $BaCO_3$ phase in the infrared absorption spectrum completely disappeared and crystalline morphology (lamellar grains) began to appear on the surface of the film; 5) With further increase of temperature, coexistence of YBCO (*00l*) and YBCO (*103*) begin to appear in XRD result, and as the temperature increases, the intensity of the YBCO (001) peaks and number and scale of lamellar crystals are also increasing. 6) When the temperature reached 800, the (*00l*) spots of YBCO phase were bright and sharp without the coexistence of (*103*) ring. It can be seen in the SEM result that at this time there are coexistence of well reacted area (the bright part on the left side) and Areas that have not yet fully reacted (non-glossy part on the right side) on the YBCO film surface. Meandering boundaries of these two areas is an

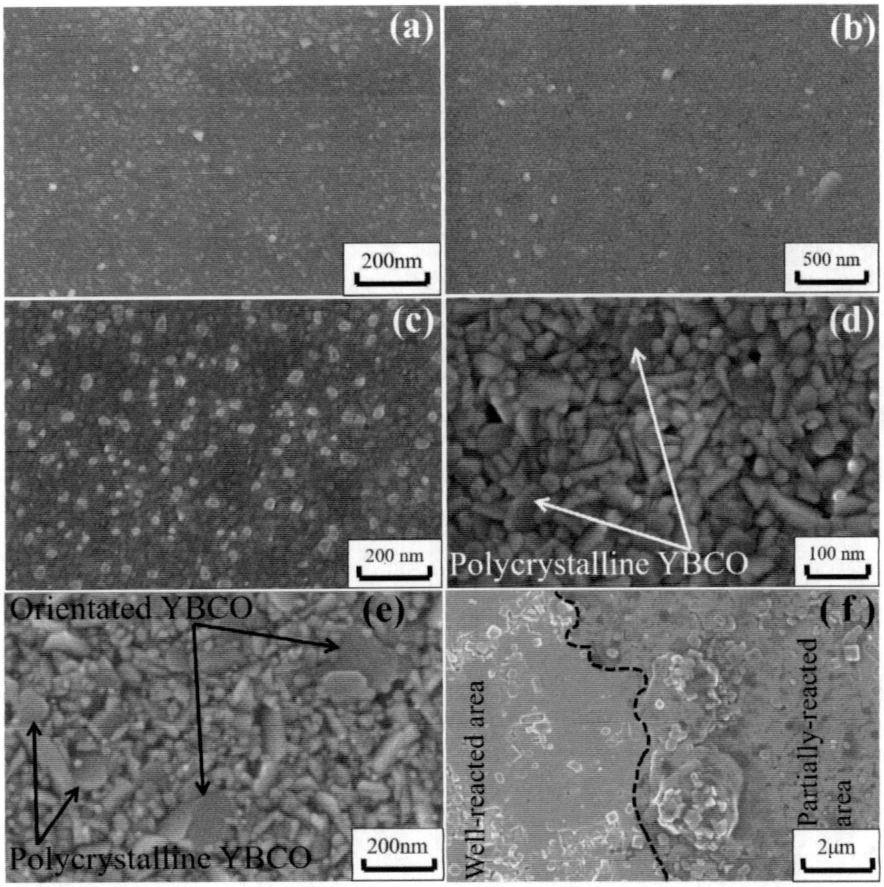

Figure 4. SEM results on samples quenched in different temperature
(a) 640℃ (b) 660℃ (c) 680℃ (d) 720℃ (e) 760℃ (f) 800℃

indication of the presence of transient liquid phase when the temperature is near 800℃.

According to the results above, it can be deduced that the reaction between $BaCO_3$ and other oxides does not produce YBCO directly, but to form intermediate products including liquid phase, which reacts with other reactants to form YBCO crystals in later process. This is consistent with the studies on phase evolution in FF-MOD method by Vermeir et al. [13].In their study $BaCO_3$ is regarded to involve in the following reactions:

$$BaCO_3 + CuO \rightarrow BaCuO_2 + CO_2 \uparrow \tag{2}$$

$$2BaCuO_2 + CuO \rightarrow liquid + xO_2 \uparrow \tag{3}$$

$$\frac{1}{2}Y_2O_3 + liquid \rightarrow YBa_2Cu_3O_{6-x} \tag{4}$$

The evolution YBCO phase are basically follow the rules below: 1) YBCO phase begins to form near 680 ℃; 2) YBCO phase experienced an evolution of " no YBCO phase - only the presence of YBCO(*103*) - the coexistence of YBCO(*103*) and YBCO(*00l*) - only YBCO (*00l*) " in crystallization process; 3) Corresponding to the phase evolution pattern of YBCO, the surface morphology also experienced an evolution of "irregular grains – only polycrystalline YBCO crystals – coexistence of polycrystalline and oriented YBCO crystals – only oriented YBCO crystals" in crystallization process. This indicates that, initially YBCO film appear as polycrystalline gains at 680°C. Afterwards, the polycrystalline YBCO grains reorganizes to orientated ones at higher temperature, and the liquid phase generated in intermediate process is involved in this reorganization process.

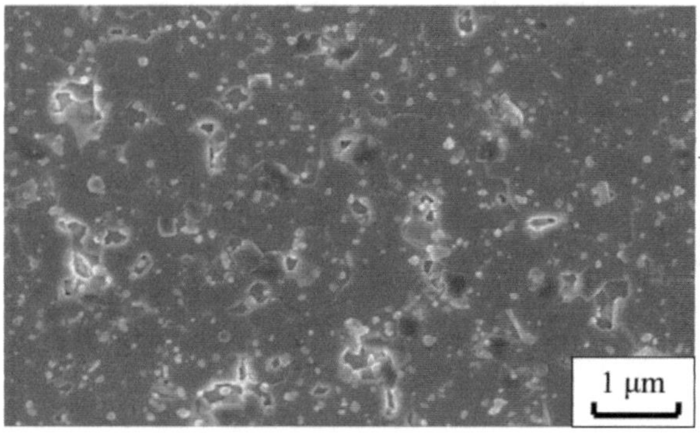

Figure 5. SEM images on the fully reacted YBCO films

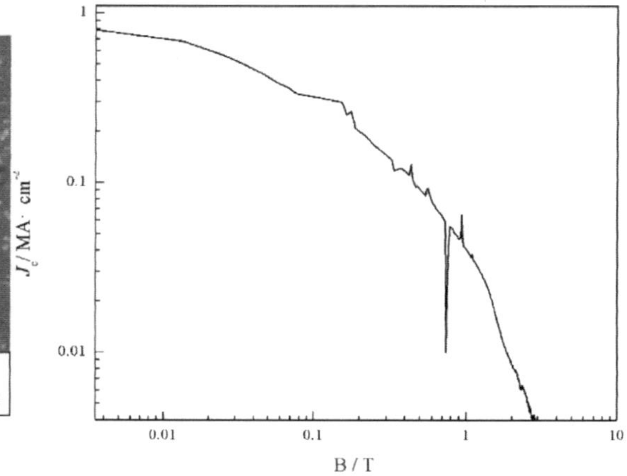

Figure 6. J_c-B curve of the fully reacted YBCO films

Samples prepared by the complete pyrolysis, crystallization and oxygen annealing process were measured by SEM and PPMS system, results was calculated from the equation (1). As shown in Figure 5, the surface of the film is basically flat and composed of plate-like c-axis oriented grains. There are also nanoscale pores on the sample surface, we think it is caused by the roughness of the substrate itself (R_a = 2.0nm, 5x5 μm scale) and the interface reaction at the CeO_2 / YBCO interface [16] since such morphology cannot be observed in the YBCO films deposited on LAO single crystal substrates. And as shown in figure 6,YBCO film prepared on technical substrates has J_c = 0.8 MA / cm^2 (77K, 0T, with 190nm thickness).

4. Conclusion

In this paper, YBCO thin films were prepared technical substrates by FF-MOD method. Phase evolution of $BaCO_3$ and YBCO phases during crystallization process was studied by quenching and characterizing samples at different temperatures.

The ATR-FT-IR, SEM and two-dimensional XRD characterization of samples quenched at different temperatures indicated that the reaction of $BaCO_3$ with oxides of Y and Cu elements will not directly generate YBCO, but experience some intermediate process. The formed YBCO crystal are randomly oriented at 680℃, and then gradually reorganize into textured crystals at higher temperature. Transient liquid phase generated in the intermediate process is involved in the YBCO reorganization process.

Acknowledge

This work supported by National Key Research and Development Program 2017YFB0902301.

IOP Conf. Series: Journal of Physics: Conf. Series **1054** (2018) 012010 doi:10.1088/1742-6596/1054/1/012010

References

[1] Vander L D C and Ekin J W 2008 *Superconductor Science and Technology* **21** 115002-6
[2] Hao C C, Tang Y J and Jiao S F 2013 *Cryogenics & Superconductivity* **41 (6)** 17-21
[3] Nasui M, Petrisor J T, Mos R B, Gabor MS, Mesaros A, Goga F, Ciontea L and Petrisor T 2015 *Ceramics International* **41** 4416-21
[4] Zhao Y, Tang X, Wu W and Grivel J C 2017 *IEEE Transactions on Applied Superconductivity* **27** 1-4
[5] Parmigiani F, Chiarello G, Ripamonti N, Goretzki H and Roll U 1987 *Physical Review B* **36** 7148-50
[6] Kumagai T, Manabe T, Yamaguchi I, Kondo W, Imai F, Murayama K and Mizuta S 2001 *Physica C: Superconductivity* **357-360** 1346-49
[7] Gupta A, Jagannathan R, Cooper E I, Giess E A, Landman J I and Hussey B W 1988 *Applied Physics Letters* **52** 2077-79
[8] Sathyamurthy C T A M W R S 2016 *Springer Series in Materials Science* **224** 36
[9] Shi D, Xu Y, Yao H, Han Z, Lian J, Wang L, Li A, Liu H K and Dou S X 2004 *Superconductor Science and Technology* **17** 1420-25
Wu W, Feng F, Shi K, Zhai W, Qu T, Huang R, Tang X, Wang X, Hu Q, Grivel J C and Han Z 2013 *Superconductor Science and Technology* **26** 055013(10)
Wu W, Feng F, Zhao Y, Tang X, Xue Y, Shi K, Huang R, Qu T, Wang X, Han Z and Grivel J C 2014 *Superconductor Science and Technology* **27** 055006(10)
Tang X, Suo HL, Ye S, Liu M, Wu Z P and Zhou M L 2010 *Journal of Inorganic Materials* **25** 971-4
[10] Llordes A, Zalamova K, Ricart S, Palau A, Pomar A, Puig T, Hardy A, Van Bael MK and Obradors X 2010 *Chemistry of Materials* **22** 1686-94
[11] Ma X L, Hirayama T, Yamagiwa K, Hirabayashi I and Ikuhara Y 1998 *Physica C: Superconductivity* **306** 8
Matsui H, Tsukada K, Tsuchiya T, Sohma M, Yamaguchi I, Kumagai T and Manabe T 2011 *Physica C: Superconductivity and its Applications* **471** 960-2
Tang X, Zhao Y, Wu W, Andersen N H and Grivel J C 2015 *Journal of the European Ceramic Society* **35** 1761-69
Zhao Y, Torres P, Tang X, Norby P and Grivel J C. *Inorganic Chemistry* **54** 10232-8
[12] Xu Y 2004 *High J c Epitaxial YBa 2 Cu 3 O 7-δ Films Through a Non-Fluorine Approach for Coated Conductor Applications* (Doctoral dissertation, University of Cincinnati)
[13] Vermeir P, Cardinael I, Schaubroeck J, Verbeken K, Bäcker M, Lommens P, Knaepen W, D'haen J, De Buysser K and Van Driessche I 2010 *Inorganic Chemistry* **49** 4471-7
[14] Eatough M O, Ginley D S, Morosin B and Venturini E L 1987 *Applied Physics Letters* **51** 367-8
[15] Bean C P 1962 *Physical Review Letters* **8(6)** 250
[16] Zhao Y, Tang X, Wu W and Grivel J C 2017 *IEEE Transactions on Applied Superconductivity* **27 No.4** 6600304(4)

IOP Conf. Series: Journal of Physics: Conf. Series **1054** (2018) 012011 doi:10.1088/1742-6596/1054/1/012011

Dependence of T_c on the *RE*-ion size in (*RE*,Ca)$Ba_2Cu_3O_6$

K Nakagawa, Y Sumino, H Chiba, K Kim, T Noji, M Kato and Y Koike

Department of Applied Physics, Graduate School of Engineering, Tohoku University, 6-6-05 Aoba, Aramaki, Aoba-ku, Sendai 980-8579, Japan

E-mail: nakagawak@teion.apph.tohoku.ac.jp

Abstract. Polycrystalline samples of $RE_{0.6}Ca_{0.4}Ba_{1.6}Sr_{0.4}Cu_3O_6$ (*RE* = Nd, Gd, Dy, Y, Yb, Lu) without oxygen in the Cu(1)O chain have been prepared by the conventional solid-state reaction method. No superconductivity appears for larger *RE* = Nd, Gd, Dy, while superconductivity with T_c ~50 K appears for smaller *RE* = Y, Yb, Lu. This is presumably because holes produced by the Ca-substitution for *RE* are more supplied to the conductive Cu(2)O_2 plane than to Cu(1) as the lattice becomes small. The maximum T_c = 53 K for *RE* = Lu is considerably lower than 90 K for $REBa_2Cu_3O_7$. This may be due to the low hole-concentration in the Cu(2)O_2 plane.

1. Introduction

Among numerous high-T_c superconducting cuprates, the so-called *RE*-123 system of $REBa_2Cu_3O_7$ (*RE* = rare-earth elements) is the most promising superconductor for practical application because of its high critical-current-density. However, its T_c ~ 90 K is too low for practical application at the liquid nitrogen temperature. Therefore, the development of superconductors with T_c ~ 100 K in the *RE*-123 system is desired.

It is known that T_c increases with an increase of the distance between Cu in the CuO_2 plane and the so-called apical-oxygen O_{ap}. The distance between Cu(2) in the CuO_2 plane and O_{ap} distance in $REBa_2Cu_3O_6$ (*RE* : rare-earth elements) is longer than that in $REBa_2Cu_3O_7$ with T_c ~90 K [1], so that its T_c is expected to increase up to more than 90 K. However, $REBa_2Cu_3O_6$ without the oxygen O_{chain} in the Cu(1)O chain is an insulator, because O_{chain} ions supply holes to the Cu(2)O_2 plane.

There have been several reports on the appearance of superconductivity by doping holes through the substitution of Ca^{2+} for RE^{3+} in $REBa_2Cu_3O_6$. In the case of *RE* = Y, the solubility limit of Ca is only x = 0.2 in $RE_{1-x}Ca_xBa_2Cu_3O_6$ and the doped holes are not sufficient. For this reason, the value of T_c is much lower than 90 K in $YBa_2Cu_3O_7$ and varies from 10-50 K depending on the reports, which may be due to the difference in the oxygen content or the definition of T_c [2-5]. On the other hand, no superconductivity appears for *RE* = Er, Gd and Eu in $RE_{0.8}Ca_{0.2}Ba_2Cu_3O_6$ [4]. As for small *RE* = Yb, moreover, Fujinami *et al.* [6] have reported superconductivity in $Yb_{1-x}Ca_xBa_{1.6}Sr_{0.4}Cu_3O_{6.0}$, in which the solubility limit of Ca is extended up to x(Ca) ≈ 0.40 through the partial substitution of Sr for Ba such that holes are doped sufficiently. In this system, superconductivity appears for x(Ca) ≥ 0.2 and T_c increases up to about 50 K as x(Ca) increases. Unfortunately, it does not reach 90 K.

In this paper, we report the dependence of T_c on *RE* ionic size in $RE_{1.6}Ca_{0.4}Ba_{1.6}Sr_{0.4}Cu_3O_6$ in which holes are expected to be doped sufficiently.

2. Experimental

Polycrystalline samples of $RE_{0.6}Ca_{0.4}Ba_{1.6}Sr_{0.4}Cu_3O_{6+\delta}$ (*RE* = Nd, Gd, Dy, Y, Yb and Lu) were prepared by the solid-state reaction method. The partial substitution of Sr^{2+} for Ba^{2+} was done in order to extend the solid solubility limit of Ca^{2+} for RE^{3+}. Stoichiometric amounts of starting materials were mixed and calcined at 890°C for 24 h in air. The calcined mixture was reground and then sintered at 890°C for 24 h in air twice. Then, the annealing with three steps was carried out as follows. In the first step (annealing 1), the sintered samples were annealed at a high temperature of 800°C in flowing gas of Ar for 24 h. Through this Ar-annealing, the solid solution of Ca is expected to be promoted through the removal of oxygen ions so as to keep the charge neutrality. In the second step (annealing 2), the samples obtained thus were annealed at a low temperature of 400°C in flowing gas of O_2 for 24 h in order to fill up oxygen vacancies induced in the previous annealing process with keeping the 123-type structure. In the third step (annealing 3), the products were annealed at 350-800°C in flowing gas of Ar for 3-6 h in order to remove O_{chain} ions.

The structural analysis was carried out by the powder x-ray diffraction using Cu Kα radiation. The iodometric titration was carried out to estimate the oxygen content. Magnetic susceptibility measurements were performed for the powdered samples using a SQUID magnetometer in a magnetic field of 10 Oe on warming after zero-field cooling in order to determine T_c.

3. Results

Figure 1 shows the powder x-ray diffraction pattern of $Lu_{0.6}Ca_{0.4}Ba_{1.6}Sr_{0.4}Cu_3O_{6+\delta}$ obtained through the annealing 3 at 350-800°C. All the samples are almost of the single phase with the *RE*-123 structure, though a tiny amount of $BaCuO_2$ or $BaCu_2O_2$ is included for the samples annealed above 700°C. With increasing annealing-temperature, the oxygen content has been found to decreases. The oxygen content of the sample annealed at 750°C had been estimated to be about 6.0.

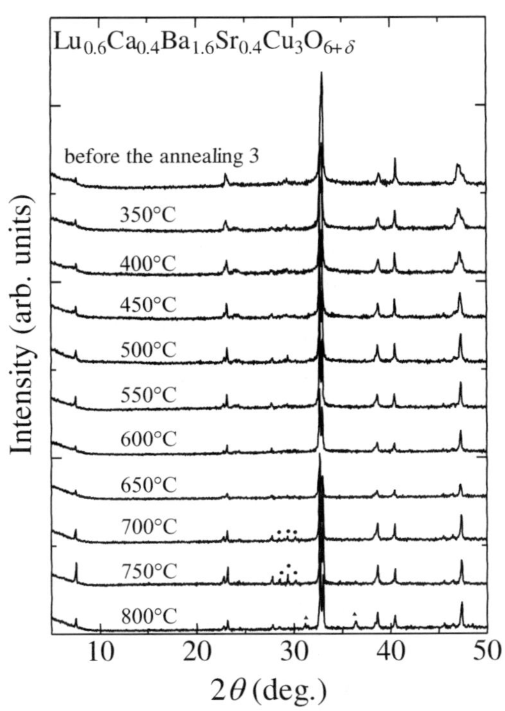

Figure 1. Powder x-ray diffraction patterns using Cu Kα radiation for the samples of $Lu_{0.6}Ca_{0.4}Ba_{1.6}Sr_{0.4}Cu_3O_{6+\delta}$ obtained through the annealing 3 at 350-800°C. For reference, the pattern for the sample before the annealing 3 is also shown. Peaks marked by ● and ▲ are due to $BaCuO_2$ and $BaCu_2O_2$, respectively

Figure 2 shows the temperature dependence of the magnetic susceptibility χ for $Lu_{0.6}Ca_{0.4}Ba_{1.6}Sr_{0.4}Cu_3O_{6+\delta}$ obtained through the annealing 3 at 350-800°C. A single-step diamagnetic signal due to the shielding effect is observed for all samples. The value of T_c, defined as the onset temperature of the shielding effect, is plotted against the oxygen content $6+\delta$ in Fig. 3. The value of T_c is 53 K for $6+\delta \sim 6.0$ at which O_{chain} is removed almost completely. For $6.0 < 6+\delta < 6.2$, T_c remains unchanged, indicating that holes doped by O_{chain} ions are not supplied not to the conductive Cu(2)O2 plane but to the Cu(1) site. With increasing oxygen-content, T_c increases up to 82 K at $6+\delta \sim 6.5$. This increase in T_c is because the holes are doped in the $Cu(2)O_2$ plane. That is to say, one of the reasons why T_c~50 K for $6+\delta \sim 6.0$ is much lower than 90 K is due to the low hole concentration in the $Cu(2)O_2$ plane. With further increasing the oxygen content, T_c decreases down to ~70 K for $6+\delta \sim 6.7$ because of the overdoping of holes in the $Cu(2)O_2$ plane.

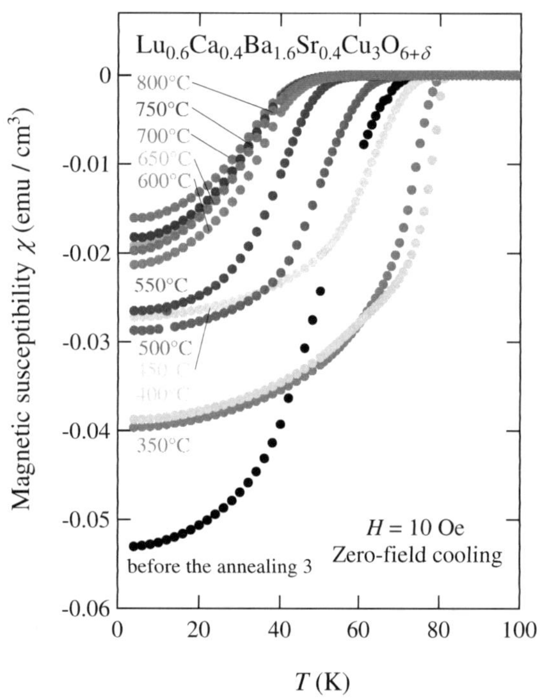

Figure 2. Temperature dependence of the magnetic susceptibility χ for $Lu_{0.6}Ca_{0.4}Ba_{1.6}Sr_{0.4}Cu_3O_{6+\delta}$ samples obtained through the annealing 3 at various temperatures.

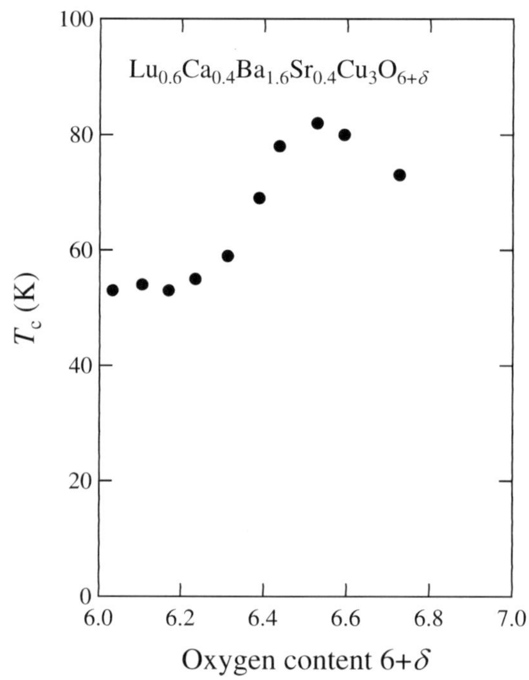

Figure 3. Dependence of T_c on the oxygen content $6+\delta$ for $Lu_{0.6}Ca_{0.4}Ba_{1.6}Sr_{0.4}Cu_3O_{6+\delta}$ samples obtained through the annealing 3.

Figure 4 shows the powder x-ray diffraction patterns of $RE_{0.6}Ca_{0.4}Ba_{1.6}Sr_{0.4}Cu_3O_{6+\delta}$ (*RE* = Nd, Gd, Dy, Y, Yb and Lu) before and after the annealing 3 at 750°C. All the samples before the annealing 3 at 750°C are almost of the single phase with the *RE*-123 structure, while the samples after the annealing 3 at 750°C include a tiny amount of $BaCuO_2$. The crystal structure is tetragonal for all the samples after the annealing 3 at 750°C. As listed in Table 1, their lattice parameters tend to decrease as the ionic radius of *RE* decreases. It is found that the samples of $RE_{0.6}Ca_{0.4}Ba_{1.6}Sr_{0.4}Cu_3O_{6+\delta}$ with $\delta \sim 0$ is obtained through the annealing 3 at 750°C.

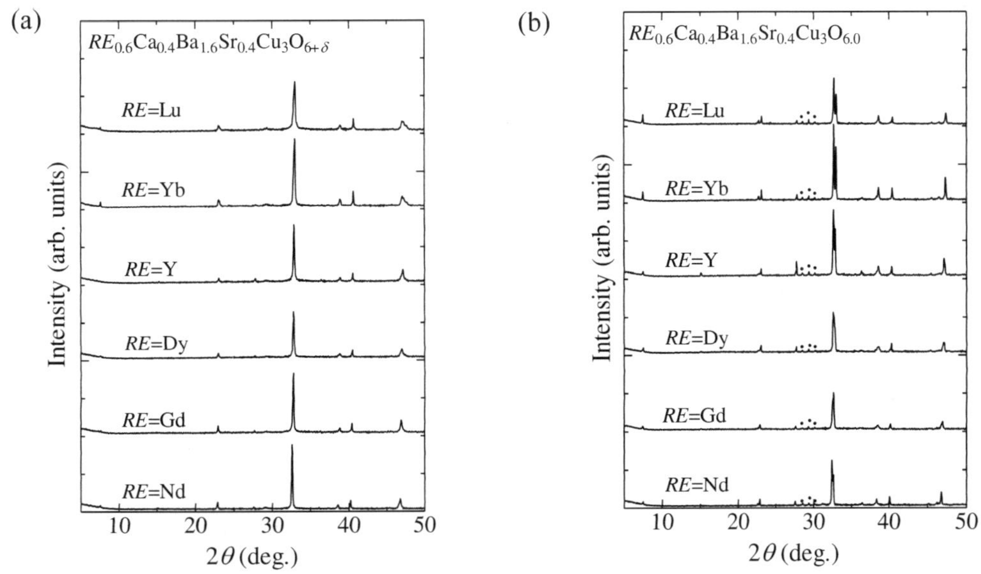

Figure 4. Powder x-ray diffraction patterns using Cu Kα radiation for $RE_{0.6}Ca_{0.4}Ba_{1.6}Sr_{0.4}Cu_3O_{6+\delta}$ (*RE* = Nd, Gd, Dy, Y, Yb, Lu) samples (a) before and (b) after the annealing 3 at 750°C. Peaks marked by • are due to $BaCuO_2$.

Table 1. Ionic radius of *RE* ion, lattice parameters, oxygen content and T_c for $RE_{0.6}Ca_{0.4}Ba_{1.6}Sr_{0.4}Cu_3O_{6+\delta}$ (*RE* = Nd, Gd, Dy, Y, Yb, Lu) samples obtained through the annealing 3 at 750°C.

RE	r_{RE} (Å)	*a* (Å)	*c* (Å)	6+δ	T_c (K)
Nd	1.11	3.882	11.775	6.06	-
Gd	1.05	3.872	11.699	6.07	-
Dy	1.03	3.859	11.705	6.05	-
Y	1.02	3.854	11.709	6.05	49
Yb	0.985	3.841	11.713	6.01	49
Lu	0.977	3.837	11.712	6.03	53

Figure 5 shows the temperature dependence of the magnetic susceptibility χ for $RE_{0.6}Ca_{0.4}Ba_{1.6}Sr_{0.4}Cu_3O_{6+\delta}$ (*RE* = Nd, Gd, Dy, Y, Yb, Lu) with δ ~ 0. No superconductivity appears for larger *RE* = Nd, Gd and Dy, whereas superconductivity with T_c ~50 K appears for smaller *RE* = Y, Yb and Lu. The values of T_c are listed in Table 1. In the present work, the maximum T_c = 53 K for *RE* = Lu is considerably lower than 90 K for $REBa_2Cu_3O_7$. This may be due to the low hole-concentration in the $Cu(2)O_2$ plane, because part of holes produced by the Ca-substitution are supplied to Cu(1) also, as mentioned above. In addtion, it is found that T_c tends to increase and the superconducting volume fraction increases as the ionic radius of *RE* decreases. This is presumably because holes are more supplied to the $Cu(2)O_2$ plane than to Cu(1) with decreasing *a*-axis-length.

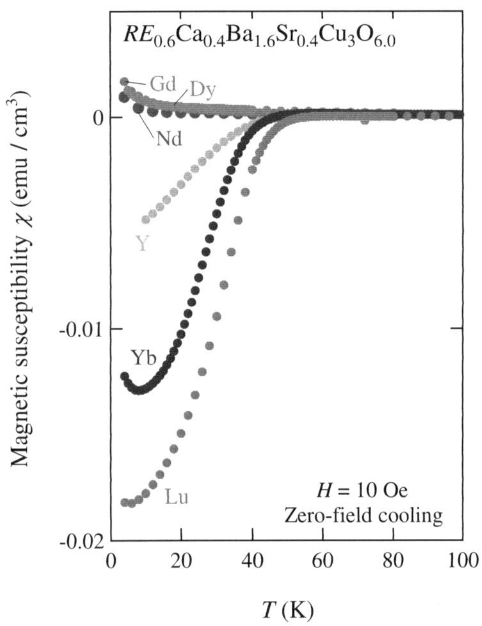

Figure 5. Temperature dependence of the magnetic susceptibility χ for $RE_{0.6}Ca_{0.4}Ba_{1.6}Sr_{0.4}Cu_3O_{6+\delta}$ (*RE* = Nd, Gd, Dy, Y, Yb, Lu) samples obtained through the annealing 3 at 750℃.

4. Conclusions

We have synthesized polycrystalline samples of $RE_{0.6}Ca_{0.4}Ba_{1.6}Sr_{0.4}Cu_3O_6$ (*RE* = Nd, Gd, Dy, Y, Yb, Lu) with the *RE*-123 structure. No superconductivity appears for larger *RE* = Nd, Gd, Dy, while superconductivity with T_c ~50 K appears for smaller *RE* = Y, Yb and Lu. In the present work, the maximum T_c = 53 K for *RE* = Lu is considerably lower than 90 K for $REBa_2Cu_3O_7$. This may be due to the low hole-concentration in the $Cu(2)O_2$ plane. It has been found that holes are more supplied to the $Cu(2)O_2$ plane than to Cu(1) with decreasing *a*-axis-length. The contraction of CuO(2) plane, for example, by the partial substitution of Sr for Ba may allow for the increase in T_c.

Acknowledgments

This work was supported by JSPS KAKENHI Grant Number JP15K06433.

References

[1] Guillaume M, Allenspach P, Henggeler W, Mesot J, Roessli B, Staub U, Fischer P, Furrer A and Trounov V 1994 *J. Phys.: Condens. Matter* **6** 7963.
[2] McCarron III E M, Crawford M K and Parise J B 1989 *J. Solid State Chem.* **78** 192.
[3] Tallon J L, Bernhard C, Shaked H, Hitterman R L and Jorgensen J D 1995 *Phys. Rev. B* **51** 12911.
[4] Hatada K and Shimizu H 1998 *Physica* C **304** 89.
[5] Berastegui P, Eriksson S -G, Johansson L G, Kakihana M, Osada M, Mazaki H and Tochihara S 1996 *J. Solid State Chem.* **127** 56.
[6] Fujinami K, Karppinen M and Yamauchi H 1998 *Physica* C **300** 17.

Frequency-dependent structure of superconducting gap function in two-dimensional Hubbard model

Shiro Sakai
Center for Emergent Matter Science, RIKEN, Wako, Saitama 351-0198, Japan
E-mail: shiro.sakai@riken.jp

Abstract. The frequency dependence of the superconducting gap function represents the dynamics of the electrons forming the Cooper pairs. Bearing the high-temperature superconducting cuprates in mind, we calculate the superconducting gap function in the two-dimensional Hubbard model by means of a cluster extension of the dynamical mean-field theory. We scrutinize the frequency dependence and its change with temperature, doping concentration and interaction strength. The calculated superconducting gap function shows various different behaviors from the anomalous self-energy, owing to a strong frequency dependence of the normal self-energy. The calculated results also suggest that the superconductivity is stronger when the interaction strength is as large as or slightly smaller than the bare band width and is suppressed for larger interactions.

1. Introduction

The Cooper pairing in a superconductor ricochets through the dynamics of quasiparticles. In conventional superconductors, the understanding of the quasiparticle dynamics served to establish the phonon-mediated mechanism of the pairing [1]: A frequency-dependent superconducting gap function, extracted from an experimental tunneling-conductance spectrum through the Migdal-Eliashberg theory [2], conformed to a measured phonon spectrum, giving a firm evidence of the phonon glue.

In cuprate high-temperature superconductors, however, no consensus has been achieved on the pairing mechanism. It is then of great interest to elucidate the quasiparticle dynamics in the cuprates. In particular, in analogy with the conventional superconductors, the frequency dependence of the superconducting gap function may hold a key to unveiling the pairing mechanism. However, because of the d-wave momentum dependence of the superconducting gap, it is not easy to extract the gap function from the tunneling data. Moreover, the applicability of the Migdal-Eliashberg theory itself has not been established for the cuprates [3]. These difficulties have prevented the cuprates from in-depth studies of quasiparticle dynamics.

A theoretical development in the past few decades, on the other hand, has made it possible to calculate the superconducting gap function in an unbiased nonperturbative way. In particular, cluster extensions [4, 5, 6] of the dynamical mean-field theory [7, 8] (CDMFT) have been developed and applied to the two-dimensional (2D) Hubbard model, a standard model for the cuprates. These studies have revealed the presence of a sharp low-energy peak in the superconducting gap function or the anomalous self-energy [9, 10, 11, 12, 13, 14]. This is reminiscent of the phonon peak in the conventional strong-coupling superconductors but is emergent from strong electronic correlation effects in the Hubbard model. A quantitative analysis of this peak's contribution [9] indeed showed that the peak gives a major contribution to the superconductivity and therefore is at the origin of high critical temperature T_c. In Ref. [13] we revealed that this peak originates from a hybridization of a low-energy electron with a hidden fermionic excitation emergent from the strong correlation effect. We then clarified in detail the changes of the self-energy structure with doping, temperature, and interaction strength [15]. In this paper, we focus on the changes of the superconducting gap function against these parameters, providing a complementary data to the self-energy presented in Ref.[15].

2. Model and method

We study the 2D single-band Hubbard model on a square lattice. The Hamiltonian reads

$$H = \sum_{\mathbf{k}\sigma} \epsilon(\mathbf{k}) c^\dagger_{\mathbf{k}\sigma} c_{\mathbf{k}\sigma} + U \sum_i n_{i\uparrow} n_{i\downarrow}, \tag{1}$$

where $c_{\mathbf{k}\sigma}$ ($c^\dagger_{\mathbf{k}\sigma}$) annihilates (creates) an electron with spin σ and momentum $\mathbf{k} = (k_x, k_y)$, $c_{i\sigma}$ ($c^\dagger_{i\sigma}$) is its Fourier component at site i, and $n_{i\sigma} \equiv c^\dagger_{i\sigma} c_{i\sigma}$. Here, U denotes the onsite Coulomb repulsion and

$$\epsilon(\mathbf{k}) \equiv -2t(\cos k_x + \cos k_y) - 4t' \cos k_x \cos k_y - \mu \tag{2}$$

is the bare dispersion with t (t') being the (next-)nearest-neighbor transfer integral and μ being the chemical potential. We take $t = 1$ as the unit of energy and $t' = -0.2$ throughout the paper. For real cuprates, t is estimated to be ~ 0.3eV.

CDMFT [5] maps in a self-consistent way the model (1) onto an effective impurity problem consisting of a small interacting cluster coupling to a dynamical mean field which is a function of Matsubara frequency $\mathrm{i}\omega_n = \mathrm{i}(2n+1)\pi T$. Here, T is the temperature. We solve the effective impurity problem with a finite-temperature version [16] of the exact diagonalization method [17] which is extended to a superconducting state [18, 13]. We use a 2×2 cluster and eight-site bath which represent the dynamical mean field. After a self-consistent loop converges, we calculate the real-frequency (ω) properties, by replacing $\mathrm{i}\omega_n$ with $\omega + i\eta$, where $\eta = \eta_0(1+\omega^2)$ with $\eta_0 = 0.05$ is a smearing factor (though in the following equations, we omit the factor $i\eta$ for simplicity). For more details, we refer the readers to Ref.[15].

A typical output of the CDMFT is the single-particle Green's function. In the superconducting state, it is written in the Nambu-matrix form as

$$\hat{G}(\mathbf{k}, \omega) = \begin{pmatrix} G(\mathbf{k}, \omega) & F(\mathbf{k}, \omega) \\ F(\mathbf{k}, \omega) & G(\mathbf{k}, -\omega)^* \end{pmatrix}, \tag{3}$$

where G (F) is the retarded normal (anomalous) Green's function. Here we have chosen the $U(1)$ gauge to make the two anomalous components be the same. The normal and anomalous self-energies, Σ^{nor} and Σ^{ano}, are defined by

$$\begin{pmatrix} \Sigma^{\mathrm{nor}}(\mathbf{k}, \omega) & \Sigma^{\mathrm{ano}}(\mathbf{k}, \omega) \\ \Sigma^{\mathrm{ano}}(\mathbf{k}, \omega) & \Sigma^{\mathrm{nor}}(\mathbf{k}, -\omega)^* \end{pmatrix} = \begin{pmatrix} \omega - \epsilon(\mathbf{k}) & 0 \\ 0 & \omega + \epsilon(\mathbf{k}) \end{pmatrix} - \left[\hat{G}(\mathbf{k}, \omega)\right]^{-1}. \tag{4}$$

With Σ^{nor} and Σ^{ano} obtained above, the superconducting gap function Δ is defined by [19]

$$\Delta(\mathbf{k}, \omega) = \frac{\Sigma^{\mathrm{ano}}(\mathbf{k}, \omega)}{1 - \frac{1}{2\omega}[\Sigma^{\mathrm{nor}}(\mathbf{k}, \omega) - \Sigma^{\mathrm{nor}}(\mathbf{k}, -\omega)^*]}. \tag{5}$$

With the 2×2 cluster, a direct calculation is possible only at four momentum points; (0,0), (π,0), (0,π) and (π,π), among which only the (π,0) and (0,π) points (both of which are equivalent by symmetry except for the sign of the anomalous component) possess the anomalous component in the d-wave superconducting state. Therefore, we focus only on $\mathbf{k} = \mathbf{k}_{\mathrm{AN}} \equiv (0, \pi)$ in the following.

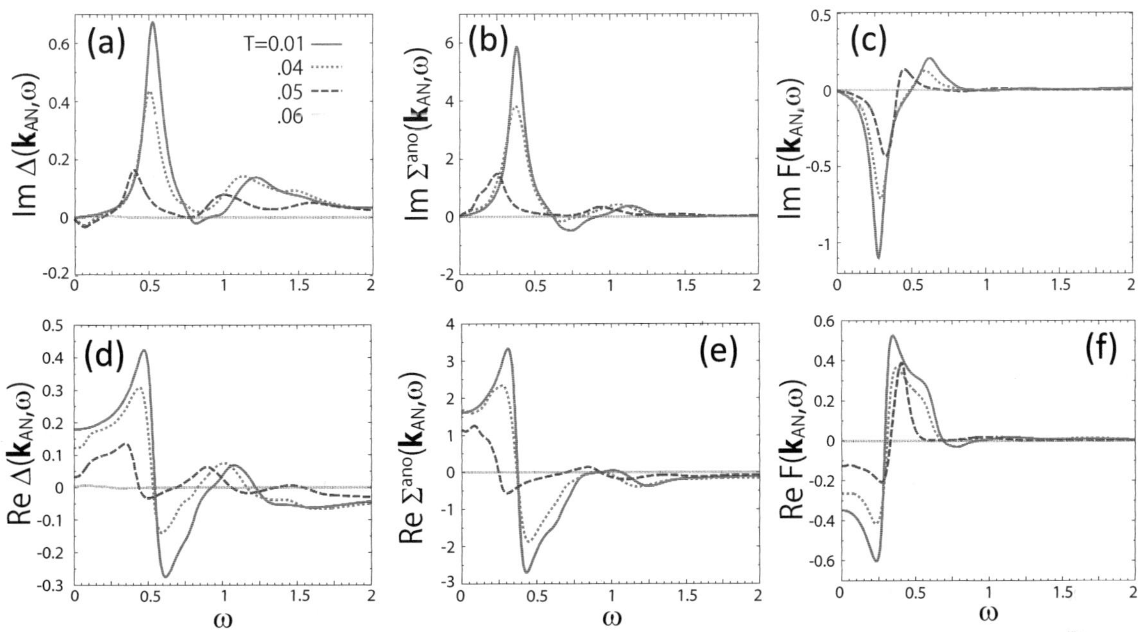

Figure 1. (Color online). Anomalous functions for various temperatures at $U = 8$ and $n \simeq 0.95$. $t = 1$ is the unit of energy. (a), (b), (c) Imaginary parts of Δ, $\Sigma^{\rm ano}$, and F, respectively. (d), (e), (f) The corresponding real parts. Note that the real (imaginary) part is (anti)symmetric with respect to $\omega = 0$ and we have shown only the positive-energy side.

3. Results and discussions

We first show in figure 1 the temperature evolution of the anomalous functions, Δ, $\Sigma^{\rm ano}$, and F at the electron density $n \simeq 0.95$. As we see that these anomalous functions vanish at $T = 0.06$ and become finite for $T \leq 0.05$, we find $0.05 < T_{\rm c} < 0.06$. As T decreases, the amplitudes of Δ, $\Sigma^{\rm ano}$ and F increase, with a peak developing at low energy in their imaginary parts. The scale of Δ is much smaller than that of $\Sigma^{\rm ano}$ because $\Sigma^{\rm nor}$, appearing in the denominator of Eq. (5), is substantial and suppresses Δ, as we discuss in more detail below. It would be interesting to point out here that these low-energy peaks of ImΔ and Im$\Sigma^{\rm ano}$ shift to higher energy with lowering T while the (negative) peak of ImF shifts to lower energy.

Figure 2 shows the doping evolution of the anomalous functions at $T = 0.01$ in the superconducting state. Note that in the present 2×2 CDMFT calculation, the optimal doping is located around 10%, i.e., $n \simeq 0.9$ (see figure 2(a) in Ref.[15]). As the doping increases, the peaks of ImΔ and Im$\Sigma^{\rm ano}$ shift to lower energies with decreasing amplitude. Around the peak energy of ImΔ, ReΔ changes its sign and is enhanced at low energy. This lifts the value of the "BCS" (static) gap, defined by

$$\mathrm{Re}\Delta(\mathbf{k}, \omega = 0) = \frac{2}{\pi} \int_0^\infty \frac{\mathrm{Im}\Delta(\mathbf{k}, \omega')}{\omega'} d\omega', \tag{6}$$

as was first pointed out in Ref.[9].

In order to quantify the contribution from the low-energy peak in ImΔ to the static part of the gap, we define the following function,

$$I(\mathbf{k}, \Omega) \equiv \frac{2}{\pi \mathrm{Re}\Delta(\mathbf{k}, \omega = 0)} \int_0^\Omega \frac{\mathrm{Im}\Delta(\mathbf{k}, \omega')}{\omega'} d\omega', \tag{7}$$

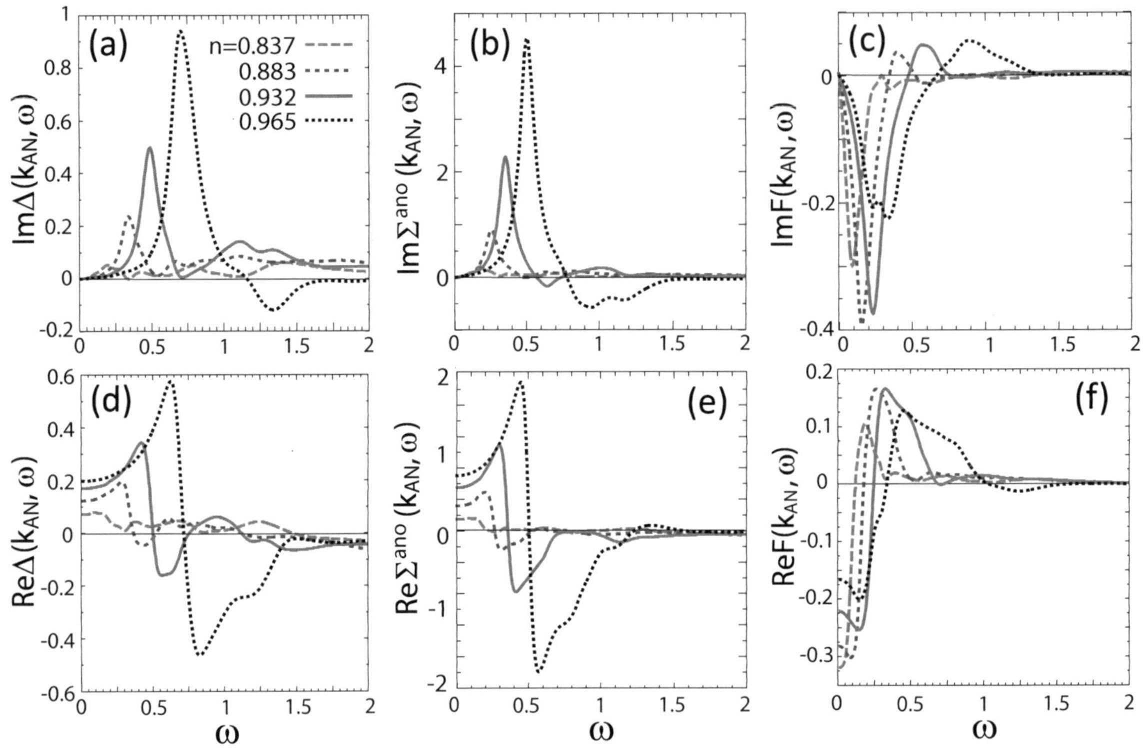

Figure 2. (Color online). Anomalous functions for various electron densities at $T = 0.01$ and $U = 8$. $t = 1$ is the unit of energy.

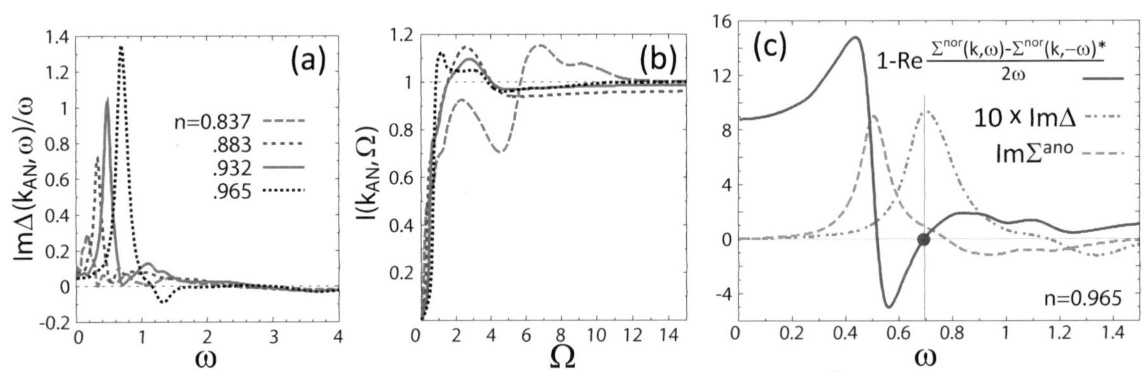

Figure 3. (Color online). (a) $\mathrm{Im}\Delta(\mathbf{k}_{\mathrm{AN}}, \omega)/\omega$ at various dopings for $T = 0.01$ and $U = 8$. $t = 1$ is the unit of energy. (b) $I(\mathbf{k}_{\mathrm{AN}}, \Omega)$ [defined by Eq. (7)] corresponding to the data in panel (a). (c) $\mathrm{Im}\Delta$, $\mathrm{Im}\Sigma^{\mathrm{ano}}$, and $1 - \mathrm{Re}[\frac{1}{2\omega}\{\Sigma^{\mathrm{nor}}(\mathbf{k}, \omega) - \Sigma^{\mathrm{nor}}(\mathbf{k}, -\omega)^*\}]$ [i.e., the real part of the denominator of Eq. (5)] at $\mathbf{k} = \mathbf{k}_{\mathrm{AN}}$ for $T = 0.01$, $U = 8$ and $n = 0.965$. The blue dot denotes the zero of $1 - \mathrm{Re}[\frac{1}{2\omega}\{\Sigma^{\mathrm{nor}}(\mathbf{k}, \omega) - \Sigma^{\mathrm{nor}}(\mathbf{k}, -\omega)^*\}]$, which gives the peak of $\mathrm{Im}\Delta$.

in a way similar to that used in Ref.[9] for Σ^{ano}. Figure 3(a) plots the integrand of Eq. (7) as a function of ω at $\mathbf{k} = \mathbf{k}_{\mathrm{AN}}$. It shows a low-energy peak for all the dopings studied here. As the doping increases, the peak shifts to a lower energy with decreasing the intensity. Figure 3(b) shows that $I(\mathbf{k}_{\mathrm{AN}}, \Omega)$ increases rapidly at around the peak energy. For example, for $n = 0.965$, the low-energy peak in figure 3(a) is located below $\omega \simeq 1$. Then, looking at figure 3(b), we find that $I(\mathbf{k}_{\mathrm{AN}}, \Omega = 1)$ exceeds 1: A small negative part of $\mathrm{Im}\Delta/\omega$ above $\omega \simeq 1$ remedies the overshoot eventually. For $n = 0.932$ and 0.883, too, $I(\mathbf{k}_{\mathrm{AN}}, \Omega = 1)$ reaches ~ 0.8, manifesting the significant contribution from the low-energy peak. For $n = 0.837$, higher-energy structures also gives a substantial contribution [11] while the low-energy peak still has a major contribution [$I(\mathbf{k}_{\mathrm{AN}}, \Omega = 0.3) \simeq 0.5$].

According to the change of the self-energies, the peak of ImF also shifts to lower energy with doping while the height of the (negative) peak shows a nonmonotonic behavior [figure 2(c)]: It seems to be maximized around the optimal doping. On the other hand, the width of the peak decreases monotonically with doping. As a consequence, $|\mathrm{Re}F(\mathbf{k}_{\mathrm{AN}}, \omega = 0)|$ increases monotonically with doping while the low-energy area of $\mathrm{Re}F < 0$ becomes narrower. Furthermore, above the energy of that structure, there remains a substantial area of $\mathrm{Re}F > 0$ even at the lowest density $n = 0.837$. These nontrivial behaviors give the domelike doping dependence of the superconducting order parameter [figure 2(a) in Ref. [15]], which corresponds to an ω integral of F.

Another interesting observation in figures 1 and 2 is that the peak of ImΔ is always located at an energy higher than that of $\mathrm{Im}\Sigma^{\mathrm{ano}}$. This is different from conventional superconductors, where $\Delta \propto \Sigma^{\mathrm{ano}}$ holds so that their peak positions coincide. According to Eq. (5), the difference between Δ and Σ^{ano} is ascribed to the denominator on the right hand side of the equation. In conventional superconductors, the frequency dependence of Σ^{nor} is so weak that this denominator is well approximated by an ω-independent renormalization factor at low frequencies. On the other hand, in the superconducting state of the 2D Hubbard model, Σ^{nor} has a strong frequency dependence; in Eq. (5) it eliminates the peak of $\mathrm{Im}\Sigma^{\mathrm{ano}}$ in the numerator and creates another peak at a higher energy. In fact, as we showed in Ref. [13, 15], $\mathrm{Im}\Sigma^{\mathrm{nor}}$ shows peaks at the same energy as $\mathrm{Im}\Sigma^{\mathrm{ano}}$, cancelling the peak of $\mathrm{Im}\Sigma^{\mathrm{ano}}$. The peak of Im$\Delta$ is then generated by the zero of $1 - \mathrm{Re}\frac{1}{2\omega}[\Sigma^{\mathrm{nor}}(\mathbf{k}, \omega) - \Sigma^{\mathrm{nor}}(\mathbf{k}, -\omega)^*]$, i.e., the real part of the denominator in Eq. (5). Figure 3(c) demonstrates this: At the energy where ImΔ shows a peak, the equation, $1 - \mathrm{Re}\frac{1}{2\omega}[\Sigma^{\mathrm{nor}}(\mathbf{k}, \omega) - \Sigma^{\mathrm{nor}}(\mathbf{k}, -\omega)^*] = 0$, is satisfied while the peak of $\mathrm{Im}\Sigma^{\mathrm{ano}}$ is located at a lower energy.

The strongly frequency-dependent structure of Σ^{nor} is produced by an underlying self-energy pole, which generates the pseudogap above T_{c} [20, 21, 22, 23, 24, 25, 26, 27, 28, 29]. The self-energy pole can emerge purely from the strong electronic correlation effect in the proximity to the Mott insulator, without breaking any symmetry. This self-energy pole can be interpreted as a fermionic excitation (hidden fermion) coupling to the electron. Then, by considering a Cooper pairing of these hidden fermions, we can explain at least qualitatively all the above nontrivial relations between Δ, Σ^{ano} and Σ^{nor} [13, 15].

Figure 4 compares the anomalous functions at several values of U. From $U = 6$ to 8, we see that the amplitude of Σ^{ano} increases. Nevertheless, Δ does not change significantly, which should again be attributed to the increase of Σ^{nor}. From $U = 8$ to 12, however, Δ shrinks in accordance with the suppression of Σ^{ano}. In contrast to these behaviors, ImF shows a strongest peak at $U = 6$, where $|\mathrm{Re}F(\mathbf{k}_{\mathrm{AN}}, \omega = 0)|$ is also largest, and shrinks monotonically as U increases. This is qualitatively the same as the behavior of the order parameter; $\langle c_{i\uparrow} c_{j\downarrow} \rangle = 0.025$ ($U = 6$), 0.019 ($U = 8$), and 0.014 ($U = 12$), where i and j denote the nearest neighboring sites. Thus, the anomalous functions behave differently from each other while they all indicate the strongest superconductivity for $U = 6 - 8$ and a suppression of it for $U = 12$.

4. Conclusion

We have calculated the frequency dependence of the anomalous functions, Δ, Σ^{ano} and F, in the superconducting state of the 2D Hubbard model, with systematically changing the temperature, electron density and interaction strength. In particular, we have focused on the behavior of the

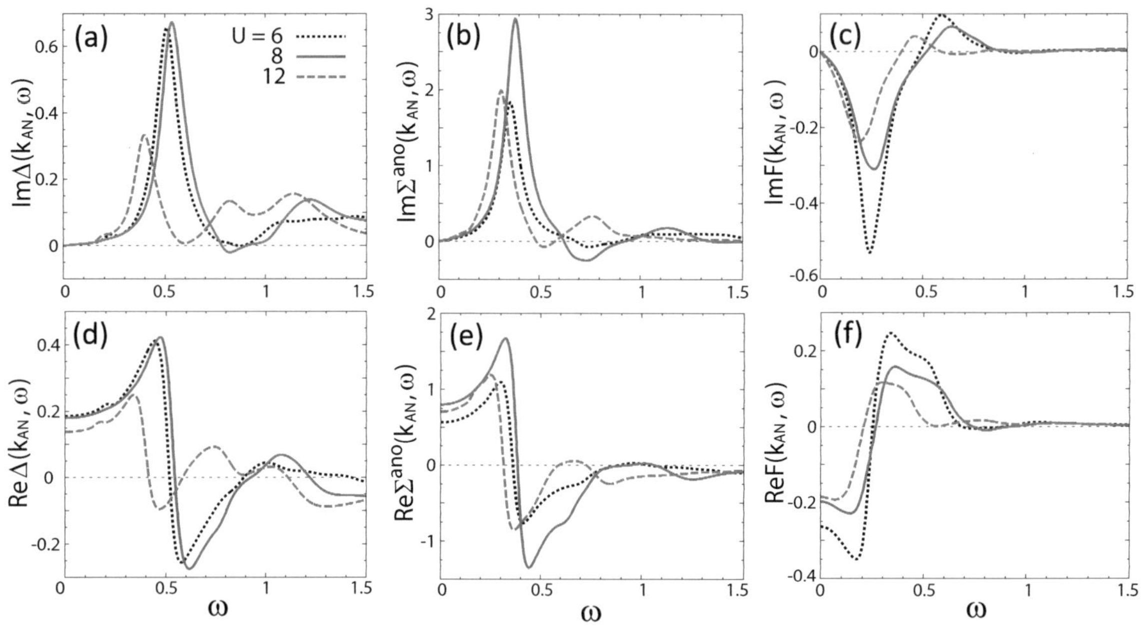

Figure 4. (Color online). Anomalous functions at several values of U for $\mathbf{k} = \mathbf{k}_{\rm AN}$, $T = 0.01$ and $n = 0.95$. $t = 1$ is the unit of energy.

low-energy peak seen in ImΔ, which is at the origin of high $T_{\rm c}$. We have shown that this peak is located at the zero of $1 - {\rm Re}\frac{1}{2\omega}[\Sigma^{\rm nor}(\mathbf{k},\omega) - \Sigma^{\rm nor}(\mathbf{k},-\omega)^*]$ rather than the peak position of ${\rm Im}\Sigma^{\rm ano}$, unlike the conventional strong-coupling superconductors. This highlights a crucial role of frequency-dependent $\Sigma^{\rm nor}$, which also manifests itself in the U dependence of Δ. This latter result suggests that $U = 6 - 8$ is an optimal interaction strength for superconductivity. The different behaviors between Δ and $\Sigma^{\rm ano}$, owing to the significant contribution from $\Sigma^{\rm nor}$, suggest that a special care (as indicated in Ref. [13, 15]) is needed to extract the superconducting gap function, as well as the self-energy, from the experimental data on cuprates.

Acknowledgments

The author acknowledges fruitful discussions with Masatoshi Imada and Marcello Civelli. This work is supported by JSPS KAKENHI (Grant No. JP17K14350 and JP16H06345).

References

[1] McMillan W L and Rowell J M 1965 *Phys. Rev. Lett.* **14**(4) 108–112 URL `http://link.aps.org/doi/10.1103/PhysRevLett.14.108`
[2] Eliashberg G M 1960 *J. Exp. Theor. Phys.* **11**(3) 696–702 URL `http://www.jetp.ac.ru/cgi-bin/e/index/e/11/3/p696?a=list`
[3] Anderson P W 2007 *Science* **316** 1705–1707 URL `http://science.sciencemag.org/content/316/5832/1705`
[4] Lichtenstein A I and Katsnelson M I 2000 *Phys. Rev. B* **62**(14) R9283–R9286 URL `http://link.aps.org/doi/10.1103/PhysRevB.62.R9283`
[5] Kotliar G, Savrasov S Y, Pálsson G and Biroli G 2001 *Phys. Rev. Lett.* **87**(18) 186401 URL `http://link.aps.org/doi/10.1103/PhysRevLett.87.186401`
[6] Maier T, Jarrell M, Pruschke T and Hettler M H 2005 *Rev. Mod. Phys.* **77**(3) 1027–1080 URL `http://link.aps.org/doi/10.1103/RevModPhys.77.1027`
[7] Metzner W and Vollhardt D 1989 *Phys. Rev. Lett.* **62**(3) 324–327 URL `http://link.aps.org/doi/10.1103/PhysRevLett.62.324`

[8] Georges A, Kotliar G, Krauth W and Rozenberg M J 1996 *Rev. Mod. Phys.* **68**(1) 13–125 URL `http://link.aps.org/doi/10.1103/RevModPhys.68.13`
[9] Maier T A, Poilblanc D and Scalapino D J 2008 *Phys. Rev. Lett.* **100**(23) 237001 URL `http://link.aps.org/doi/10.1103/PhysRevLett.100.237001`
[10] Kyung B, Sénéchal D and Tremblay A M S 2009 *Phys. Rev. B* **80**(20) 205109 URL `https://link.aps.org/doi/10.1103/PhysRevB.80.205109`
[11] Civelli M 2009 *Phys. Rev. Lett.* **103**(13) 136402 URL `https://link.aps.org/doi/10.1103/PhysRevLett.103.136402`
[12] Gull E and Millis A J 2015 *Phys. Rev. B* **91**(8) 085116 URL `http://link.aps.org/doi/10.1103/PhysRevB.91.085116`
[13] Sakai S, Civelli M and Imada M 2016 *Phys. Rev. Lett.* **116**(5) 057003 URL `http://link.aps.org/doi/10.1103/PhysRevLett.116.057003`
[14] Reymbaut A, Bergeron D and Tremblay A M S 2015 *Phys. Rev. B* **92**(6) 060509 URL `https://link.aps.org/doi/10.1103/PhysRevB.92.060509`
[15] Sakai S, Civelli M and Imada M 2016 *Phys. Rev. B* **94**(11) 115130 URL `http://link.aps.org/doi/10.1103/PhysRevB.94.115130`
[16] Liebsch A and Tong N H 2009 *Phys. Rev. B* **80**(16) 165126 URL `http://link.aps.org/doi/10.1103/PhysRevB.80.165126`
[17] Caffarel M and Krauth W 1994 *Phys. Rev. Lett.* **72**(10) 1545–1548 URL `http://link.aps.org/doi/10.1103/PhysRevLett.72.1545`
[18] Sakai S, Civelli M, Nomura Y and Imada M 2015 *Phys. Rev. B* **92**(18) 180503 URL `https://link.aps.org/doi/10.1103/PhysRevB.92.180503`
[19] Scalapino D J, Schrieffer J R and Wilkins J W 1966 *Phys. Rev.* **148**(1) 263–279 URL `https://link.aps.org/doi/10.1103/PhysRev.148.263`
[20] Dzyaloshinskii I 2003 *Phys. Rev. B* **68**(8) 085113 URL `https://link.aps.org/doi/10.1103/PhysRevB.68.085113`
[21] Stanescu T D and Kotliar G 2006 *Phys. Rev. B* **74**(12) 125110 URL `http://link.aps.org/doi/10.1103/PhysRevB.74.125110`
[22] Berthod C, Giamarchi T, Biermann S and Georges A 2006 *Phys. Rev. Lett.* **97**(13) 136401 URL `https://link.aps.org/doi/10.1103/PhysRevLett.97.136401`
[23] Yang K Y, Rice T M and Zhang F C 2006 *Phys. Rev. B* **73**(17) 174501 URL `https://link.aps.org/doi/10.1103/PhysRevB.73.174501`
[24] Stanescu T D, Phillips P and Choy T P 2007 *Phys. Rev. B* **75**(10) 104503 URL `https://link.aps.org/doi/10.1103/PhysRevB.75.104503`
[25] Sakai S, Motome Y and Imada M 2009 *Phys. Rev. Lett.* **102**(5) 056404 URL `http://link.aps.org/doi/10.1103/PhysRevLett.102.056404`
[26] Sakai S, Motome Y and Imada M 2010 *Phys. Rev. B* **82**(13) 134505 URL `http://link.aps.org/doi/10.1103/PhysRevB.82.134505`
[27] Yamaji Y and Imada M 2011 *Phys. Rev. Lett.* **106**(1) 016404 URL `https://link.aps.org/doi/10.1103/PhysRevLett.106.016404`
[28] Imada M, Yamaji Y, Sakai S and Motome Y 2011 *Annalen der Physik* **523** 629–637 ISSN 1521-3889 URL `http://dx.doi.org/10.1002/andp.201100028`
[29] Sakai S, Blanc S, Civelli M, Gallais Y, Cazayous M, Méasson M A, Wen J S, Xu Z J, Gu G D, Sangiovanni G, Motome Y, Held K, Sacuto A, Georges A and Imada M 2013 *Phys. Rev. Lett.* **111**(10) 107001 URL `http://link.aps.org/doi/10.1103/PhysRevLett.111.107001`

IOP Conf. Series: Journal of Physics: Conf. Series **1054** (2018) 012013 doi:10.1088/1742-6596/1054/1/012013

Phase diagram of high-temperature superconducting cuprates and pnictides

Kazuhisa Nishi

School of Engineering, University of Hyogo, Himeji, Hyogo 671-2201, Japan

tyus51kzhs@ares.eonet.ne.jp

Abstract. Phase diagram of high-temperature superconducting cuprates and iron pnictides is considered based on the theory emphasizing that the electronic state of superconductors can be described by the composed fermions. The theory is constructed with the Hamiltonian so modified by the unitary transformation as to apply the mean field approximation. It is indicated that the phase diagram of these superconductors can be explained from the viewpoint unifying the electronic states of superconducting cuprates and pnictides. The issues relevant to exotic phases are also discussed.

1. Introduction

Though there has been much study about high-temperature superconductors such as cuprates *etc*, its microscopic understanding still remains an insufficient stage [1-8]. Recently, the author proposed a unified theory to explain generally the properties of high-temperature superconducting cuprates and iron pnictides [9]. The theory is based on the extended *d-p* model emphasizing that the electronic state of superconductors can be described by the composed fermions constructed with *d-p* operators, and is indicated to provide a universal explanation about the superconducting and normal properties of these superconductors. In this paper the phase diagrams of high-temperature superconducting cuprates and iron pnictides will be investigated and then evaluated comparing with the experimental facts.

2. Basic equations

The Hamiltonian is assumed to be an extended *d-p* model for a single layer of square planar

$$H = \sum_{in\sigma} \varepsilon_{dn} d^+_{in\sigma} d_{in\sigma} + \sum_{jl\sigma} \varepsilon_{pl} p^+_{jl\sigma} p_{jl\sigma} + \sum_{inl\sigma} \sum_{j\in\{inl\}} (\varepsilon_{injl} d^+_{in\sigma} p_{jl\sigma} + \mathrm{H.c.}) + \sum_{inm\sigma\sigma'} U_{nm} d^+_{in\sigma} d_{im\sigma} d^+_{in\sigma'} d_{im\sigma'} + \sum_{injl\sigma\sigma'} V_{injl} d^+_{in\sigma} d_{in\sigma} p^+_{jl\sigma'} p_{jl\sigma'} \quad (1)$$

where the operator $d^+_{in\sigma}$ creates electrons or holes of *n-th* M (Cu or Fe) 3*d* orbital at site *i*, $p^+_{jl\sigma}$ creates *l-th* L (oxide or pnictide) *p* electrons or holes at site *j*, $\varepsilon_{dn} = 0$, and ε_{injl} is the nearest-neighbor hopping integral. U_{nn} and U_{nm} $(n \neq m)$ are the Coulomb repulsion and Hund coupling at M site, respectively, and V_{injl} is the interaction between neighboring M and L sites. In (1), the vacuum is defined as M d^x and L p^y states ($x = 10$, $y = 6$ in the cupurate, $x = 6$, $y = 6$ in the pnictide). Considering the *d-p* covalency effect, the operator combining the L states around M ion is defined as

$$\tilde{p}^+_{inl\sigma} = \sum_{j\in\{inl\}} \varepsilon_{injl} \varepsilon^{-1}_{nl} p^+_{jl\sigma}, \ \varepsilon_{nl} = \sqrt{\sum_{j\in\{inl\}} \varepsilon^2_{injl}} \quad (2)$$

where $j = i + \boldsymbol{x}_j$ and $\boldsymbol{x}_j$ presents the directions of M-L bonding. Since the relation of $\sum_j p^+_{jl\sigma} p_{jl\sigma} = N_v \sum_i \tilde{p}^+_{inl\sigma} \tilde{p}_{inl\sigma}$ (N_v is the valence ratio of d to p) is satisfied in the M-L compounds, Hamiltonian (1) is so changed as

$$H = N_v \sum_{inl\sigma} \varepsilon_{pl} \tilde{p}^+_{inl\sigma} \tilde{p}_{inl\sigma} + \sum_{inl\sigma} (\varepsilon_{nl} d^+_{in\sigma} \tilde{p}_{inl\sigma} + \text{H.c.}) + \sum_{inm\sigma\sigma'} U_{nm} d^+_{in\sigma} d_{im\sigma} d^+_{in\sigma'} d_{im\sigma'} + N_v \sum_{inl\sigma\sigma'} V_{inl} d^+_{in\sigma} d_{in\sigma} \tilde{p}^+_{inl\sigma'} \tilde{p}_{inl\sigma'} \tag{3}$$

Since the operator $\tilde{p}^+_{inl\sigma}$ is not orthogonal between the neighboring M sites, it does not exactly satisfy anti-commutation relations. However it is approximated here that $\tilde{p}^+_{inl\sigma}$ can be well-defined fermion operator.

To the second order in perturbation theory on the condition of $\varepsilon_{nl} < U_n (= U_{nn})$, let us find out the effective Hamiltonian of Eq. (3). For simplicity, Hund coupling and the interaction between different $3d$ orbitals are neglected. First in the second term of Eq. (3), the terms acting on the interaction U_n as a perturbation can be selected as follow as

$$H' = \sum_{inl\sigma} \{\varepsilon_{nl} [n_{d\sigma'} d^+_{in\sigma} \tilde{p}_{inl\sigma} (1 - n_{p\sigma'}) + (1 - n_{d\sigma'}) d^+_{in\sigma} \tilde{p}_{inl\sigma} n_{p\sigma'}] + \text{H.c.}\} \tag{4}$$

where H' indicates the terms in the presence of fermions with the opposite spin at either d or p sites because the ground state is not assumed here to include the double occupancy states. Further, let us average the occupancy factors such as $(1 - n_{d\sigma'})$ by using $\langle 1 - n_{d\sigma'} \rangle$. The Coulomb interaction U_n in the ground state can be effectively eliminated from the starting Hamiltonian because double occupancy states at d-sites are inhibited by Coulomb repulsion. Considering the second term except H', Hamiltonian (3) is thus modified by

$$\begin{aligned} H = {} & N_v \sum_{inl\sigma} \varepsilon_{pl} \tilde{p}^+_{inl\sigma} \tilde{p}_{inl\sigma} + \sum_{inl\sigma} \langle 1 - n_{p\sigma'} \rangle (\varepsilon_{nl} d^+_{in\sigma} \tilde{p}_{inl\sigma} + \text{H.c.}) + N_v \sum_{inl\sigma\sigma'} V_{inl} d^+_{in\sigma} d_{in\sigma} \tilde{p}^+_{inl\sigma'} \tilde{p}_{inl\sigma'} \\ & + \sum_{inl\sigma \neq \sigma'} \left(\langle 1 - n_{d\sigma'} \rangle \langle n_{p\sigma'} \rangle + \langle 1 - n_{p\sigma'} \rangle \langle n_{d\sigma'} \rangle \right)^2 (-J_1 d^+_{in\sigma} d_{in\sigma} \tilde{p}^+_{inl\sigma'} \tilde{p}_{inl\sigma'} - J_2 d^+_{in\sigma} \tilde{p}_{inl\sigma} \tilde{p}^+_{inl\sigma'} d_{in\sigma'}) \end{aligned} \tag{5}$$

where $J_1 = \varepsilon^2_{nl} (U_n - N_v \varepsilon_{pl} - N_v V_{inl})^{-1}$, $J_2 = \varepsilon^2_{nl} (N_v \varepsilon_{pl} - N_v V_{inl})^{-1}$. Though the former part of the forth term in (5) corresponds to the kinetic energy due to the hopping between sites, it is neglected here due to the second order kinetic energy. The latter part corresponds to the effective anti-ferromagnetic interaction between d-p fermions. Thus, the effective Hamiltonian is finally obtained as

$$H_{eff} = N_v \sum_{inl\sigma} \varepsilon_{pl} \tilde{p}^+_{inl\sigma} \tilde{p}_{inl\sigma} + \sum_{inl\sigma} \langle 1 - n_{p\sigma'} \rangle (\varepsilon_{nl} d^+_{in\sigma} \tilde{p}_{inl\sigma} + \text{H.c.}) + N_v \sum_{inl\sigma\sigma'} V d^+_{in\sigma} d_{in\sigma} \tilde{p}^+_{inl\sigma'} \tilde{p}_{inl\sigma'} \tag{6}$$

where $V = V_{inl} + N_v^{-1} V_{\sigma\sigma'}$ (V_{inl}: Coulomb interaction between the nearest neighboring sites in (1), $V_{\sigma\sigma'}$: anti-ferromagnetic interaction $V_{\sigma\sigma'} = -\left(\langle 1 - n_{d\sigma'} \rangle \langle n_{p\sigma'} \rangle + \langle 1 - n_{p\sigma'} \rangle \langle n_{d\sigma'} \rangle \right)^2 J_1 < 0 \quad \sigma \neq \sigma'$).

Now the ground state of effective Hamiltonian (6) will be generally considered in two states depending on the doping conditions. One of these is the states in the neighborhood of the insulator (so-called Mott insulator) and the other is the superconducting state based on the band picture. First let us consider the region in the neighborhood of the insulator. The effective Hamiltonian (6) is appropriate in this case because its representation in real space is very useful in the treatment of this region. In order to enable to apply the mean field approximation, the operators $d^+_{in\sigma}$, $\tilde{p}^+_{inl\sigma}$ are changed by unitary transformations of

$$b^+_{inl\sigma} = \alpha_{nl} d^+_{in\sigma} + \beta_{nl} \tilde{p}^+_{inl\sigma}, \quad c^+_{inl\sigma} = \beta_{nl} d^+_{in\sigma} - \alpha_{nl} \tilde{p}^+_{inl\sigma}, \quad \alpha^2_{nl} + \beta^2_{nl} = 1 \tag{7}$$

where b_i^+, c_i^+ are defined as the mixing operators of M and L fermions. For the condition of $\alpha_{nl} \beta_{nl} N_v \varepsilon_{pl} = \langle 1 - n_{p\sigma'} \rangle \varepsilon_{nl} (\beta^2_{nl} - \alpha^2_{nl})$, the Hamiltonian (6) is transformed into

$$H = N_v \sum_{inl\sigma} \varepsilon_{pl} (\beta^2 - \alpha^2)^{-1} (\beta^2 b_{i\sigma}^+ b_{i\sigma} - \alpha^2 c_{i\sigma}^+ c_{i\sigma}) + N_v \sum_{inl\sigma\sigma'} V (\alpha^2 \beta^2 b_{i\sigma}^+ b_{i\sigma} b_{i\sigma'}^+ b_{i\sigma'} + \alpha^4 b_{i\sigma}^+ b_{i\sigma} c_{i\sigma'}^+ c_{i\sigma'} + \beta^4 c_{i\sigma}^+ c_{i\sigma} b_{i\sigma'}^+ b_{i\sigma'} + \alpha^2 \beta^2 c_{i\sigma}^+ c_{i\sigma} c_{i\sigma'}^+ c_{i\sigma'} + \cdots) \tag{8}$$

Notice that the indexes n, l of $\alpha_{nl}, \beta_{nl}, b_{nl\sigma}^+, c_{nl\sigma}^+$ *etc* are omitted to avoid the complex representation. Since there appear many interaction terms in the transformed Hamiltonian (8), this transformation might make the problem somewhat difficult. However, the mixing of *d* and *p* fermions can be directly built in this Hamiltonian, which consists of two free fermions and the interaction terms between them. As a result this treatment has an advantage to enable to apply the mean field approximation because of the explicit representation of free fermions and their interaction terms. Next let us consider the case of the superconducting state based on the band picture. Since the representation in momentum space is appropriate in this region, the effective Hamiltonian of (6) is transformed into

$$H = \sum_{\mathbf{k}\,nl\sigma} \varepsilon_{pnl} p_{\mathbf{k}\,nl\sigma}^+ p_{\mathbf{k}\,nl\sigma} + \sum_{\mathbf{k}\,nl\sigma} \langle 1 - n_{p\sigma'} \rangle \varepsilon_{nl} s_{nl}(\mathbf{k}) (-i d_{\mathbf{k}\,n\sigma}^+ p_{\mathbf{k}\,l\sigma} + i p_{\mathbf{k}\,l\sigma}^+ d_{\mathbf{k}\,n\sigma}) + N^{-1} \sum_{\mathbf{k}\mathbf{k}'\mathbf{k}''nl\sigma\sigma'} V \theta_{nl}(\mathbf{k},\mathbf{k}') d_{\mathbf{k}\,n\sigma}^+ d_{\mathbf{k}'n\sigma} p_{\mathbf{k}''l\sigma'}^+ p_{\mathbf{k}-\mathbf{k}'+\mathbf{k}'',l\sigma'} \tag{9}$$

Here $s_{nl}(\mathbf{k}) = \sum_{j\in\{inl\}} \varepsilon_{injl} \varepsilon_{nl}^{-1} \exp(i\mathbf{k}\cdot\mathbf{x}_j)$, $\theta_{nl}(\mathbf{k},\mathbf{k}') = \sum_{j\in\{inl\}} \varepsilon_{injl}^2 \varepsilon_{nl}^{-2} \exp[i(\mathbf{k}-\mathbf{k}')\cdot\mathbf{x}_j]$, $\mathbf{x}_j(x, y)$ refers to the axis for the unit cell, and N is number of M sites in a single layer. The operators $p_{\mathbf{k}l\sigma}^+, d_{\mathbf{k}n\sigma}^+$, which are Fourier transformation of $p_{inl\sigma}^+, d_{in\sigma}^+$, are reconstructed with fermion operators defined by unitary transformation

$$b_{\mathbf{k}\,nl\sigma}^+ = \alpha_{\mathbf{k}\,nl} d_{\mathbf{k}\,n\sigma}^+ + i\beta_{\mathbf{k}\,nl} p_{\mathbf{k}\,l\sigma}^+, \quad c_{\mathbf{k}\,n\sigma}^+ = \beta_{\mathbf{k}\,nl} d_{\mathbf{k}\,n\sigma}^+ - i\alpha_{\mathbf{k}\,nl} p_{\mathbf{k}\,l\sigma}^+, \quad \alpha_{\mathbf{k}\,nl}^2 + \beta_{\mathbf{k}\,nl}^2 = 1, \tag{10}$$

where $b_{\mathbf{k}\,nl\sigma}^+, c_{\mathbf{k}\,nl\sigma}^+$ satisfy anti-commutation relations. Notice that the operators $p_{\mathbf{k}l\sigma}^+$ satisfy anti-commutation relations exactly, but those are approximated operators of $\tilde{p}_{inl\sigma}^+$ in momentum space. For the condition of $\alpha_{\mathbf{k}nl}\beta_{\mathbf{k}nl}\varepsilon_{pl} = \langle 1 - n_{p\sigma'} \rangle \varepsilon_{nl} s_{nl}(\mathbf{k})(\beta_{\mathbf{k}nl}^2 - \alpha_{\mathbf{k}nl}^2)$, the Hamiltonian (9) is given by

$$H = \sum_{nl} \sum_{\mathbf{k}\sigma} \varepsilon_{pl} (\beta_{\mathbf{k}}^2 - \alpha_{\mathbf{k}}^2)^{-1} (\beta_{\mathbf{k}}^2 b_{\mathbf{k}\sigma}^+ b_{\mathbf{k}\sigma} - \alpha_{\mathbf{k}}^2 c_{\mathbf{k}\sigma}^+ c_{\mathbf{k}\sigma}) + N^{-1} \sum_{nl} \sum_{\mathbf{k}\mathbf{k}'\mathbf{k}''\sigma\sigma'} V \theta_n(\mathbf{k},\mathbf{k}') (\alpha_{\mathbf{k}} \alpha_{\mathbf{k}'} \beta_{\mathbf{k}''} \beta_{\mathbf{k}-\mathbf{k}'+\mathbf{k}''} b_{\mathbf{k}\sigma}^+ b_{\mathbf{k}'\sigma} b_{\mathbf{k}''\sigma'}^+ b_{\mathbf{k}-\mathbf{k}'+\mathbf{k}'',\sigma'} + \alpha_{\mathbf{k}} \alpha_{\mathbf{k}'} \alpha_{\mathbf{k}''} \alpha_{\mathbf{k}-\mathbf{k}'+\mathbf{k}''} b_{\mathbf{k}\sigma}^+ b_{\mathbf{k}'\sigma} c_{\mathbf{k}''\sigma'}^+ c_{\mathbf{k}-\mathbf{k}'+\mathbf{k}'',\sigma'} + \beta_{\mathbf{k}''} \beta_{\mathbf{k}-\mathbf{k}'+\mathbf{k}''} \beta_{\mathbf{k}} \beta_{\mathbf{k}'} c_{\mathbf{k}\sigma}^+ c_{\mathbf{k}'\sigma} b_{\mathbf{k}''\sigma'}^+ b_{\mathbf{k}-\mathbf{k}'+\mathbf{k}'',\sigma'} + \beta_{\mathbf{k}} \beta_{\mathbf{k}'} \alpha_{\mathbf{k}''} \alpha_{\mathbf{k}-\mathbf{k}'+\mathbf{k}''} c_{\mathbf{k}\sigma}^+ c_{\mathbf{k}'\sigma} c_{\mathbf{k}''\sigma'}^+ c_{\mathbf{k}-\mathbf{k}'+\mathbf{k}'',\sigma'} + \cdots) \tag{11}$$

where the indexes n, l of $\alpha_{\mathbf{k}\,nl}, \beta_{\mathbf{k}\,nl}, b_{\mathbf{k}\,nl\sigma}^+, c_{\mathbf{k}\,nl\sigma}^+$ *etc* are omitted to avoid the complex representation. Based on these effective Hamiltonians, the phase diagram of cuprate and pnictide superconductors will be considered in next chapters.

3. Cuprates

3.1 Electronic state

The characteristics of cuprate superconductors is that it is a single *d*-band, *p*-band (n = 1, l =1), superconductive in hole doping, and insulator in non-doping state. Though *p*-band is multi-orbitals, a single band of l = 1 will be only considered here. First let us consider the case of the nearly non-doped region. Assuming that there is no *p*-holes in half-filling, the relations of $\langle 1 - n_{d\sigma'} \rangle \approx 0$, $\langle 1 - n_{p\sigma'} \rangle \approx (1 - 0.5\delta)$ are derived. For the condition of $2\alpha\beta\varepsilon_p = (1 - 0.5\delta)\varepsilon(\beta^2 - \alpha^2)$, the Hamiltonian (8) is represented as

$$H = 2\varepsilon_p (\beta^2 - \alpha^2)^{-1} \sum_{i\sigma} (\beta^2 b_{i\sigma}^+ b_{i\sigma} - \alpha^2 c_{i\sigma}^+ c_{i\sigma}) + 2\sum_{i\sigma\sigma'} V (\alpha^2 \beta^2 b_{i\sigma}^+ b_{i\sigma} b_{i\sigma'}^+ b_{i\sigma'} + \alpha^4 b_{i\sigma}^+ b_{in\sigma} c_{i\sigma'}^+ c_{i\sigma'} + \beta^4 c_{i\sigma}^+ c_{i\sigma} b_{i\sigma'}^+ b_{i\sigma'} + \alpha^2 \beta^2 c_{i\sigma}^+ c_{i\sigma} c_{i\sigma'}^+ c_{i\sigma'} + \cdots) \tag{12}$$

Here N_V (the valence ratio of *d* to *p*) = 2 and $V = V_c + V'(1 - 0.5\delta)^2$ (V_c is Coulomb interaction between the nearest neighboring sites, $V' = \varepsilon^2 V_0 / \varepsilon_p^2$). For the parameters of cuprates, it will be selected that $\varepsilon = 0.5 \sim 1\,\text{eV}$, $\varepsilon_p = 1\,\text{eV}$, $U = 5$ eV, $V_c = 1\,\text{eV}$, $V' = -1 \sim 2$ eV. It is well known that the ground state of superconducting cuprates shows antiferromagnets (so-called Mott insulator) in the non-doped region [10,11]. Using the composite operators defined here, the non-doped wave function corresponds to

$$|\Psi_{MI}\rangle = \prod_{i\in A}^{N/2}(\alpha b_{i\uparrow}^{+} + \beta c_{i\uparrow}^{+})\prod_{j\in B}^{N/2}(\alpha b_{j\downarrow}^{+} + \beta c_{j\downarrow}^{+})|0\rangle \tag{13}$$

where A or B shows the sub-space of antiferromagnetic lattice and N is number of Cu sites in a single layer. Since there exists no p-hole in half-filling Mott insulator, it will not be necessary to consider the interaction V in this case. However, in the existence of p-holes, since $V_{\sigma\sigma'}(\sigma \neq \sigma')$ is interpreted to be attractive due to d-p exchange antiferromagnetic interaction, the case of $V < 0$ will play an important role for determining the ground state. Considering the experimental facts that the doping can almost supply the p-holes, doping holes are expected to occupy the quantum state corresponding to this situation. For the condition $\alpha^2 < \beta^2$ which allows the reliable value of ε_p and ε, the term $b_{i\sigma}^{+}b_{i\sigma}c_{i\sigma'}^{+}c_{i\sigma'}$ ($c_{i\sigma}^{+}c_{i\sigma}b_{i\sigma'}^{+}b_{i\sigma'}$) can be the most dominant interaction. This means that b-c or c-b pair will mainly contribute to determine the ground state. Thus, for the equivalency of sites, the normalized wave function is assumed to be

$$|\Psi_A\rangle = \prod_{i}^{N}(s + t b_{i\uparrow}^{+}c_{i\downarrow}^{+})(s + t c_{i\uparrow}^{+}b_{i\downarrow}^{+})|\Psi_{MI}\rangle, \quad s^2(s^2+t^2)=1 \tag{14}$$

where the coefficient t indicates the probability of local-antiferro pair state. On the approximation using the relation of $\langle\Psi_A|\sum p_{i\sigma}^{+}p_{i\sigma}|\Psi_A\rangle \approx 2Ns^2t^2 = N\delta$ (δ: the doping ratio relative to half-filling), the ground-state energy is given by

$$E_p = \langle\Psi_A|H|\Psi_A\rangle = N\delta(\varepsilon_p + V), \quad V<0 \tag{15}$$

The energy of the excited states will be evaluated by using the presumed excited state wavefunction as follow as

$$|\Psi_{EX}\rangle = \prod_{i}^{N}(s + t b_{i\uparrow}^{+}c_{i\downarrow}^{+})(s + t c_{i\uparrow}^{+}b_{i\downarrow}^{+})(\beta b_{i'\uparrow}^{+} - \alpha c_{i'\uparrow}^{+})|\Psi_{MI}\rangle \tag{16}$$

This is orthogonal to the ground state function and corresponds to breaking up a b-c (c-b) pair in i', the spin-up member going to $b_{i'\uparrow}^{+}, c_{i'\uparrow}^{+}$ The excited energy is obtained as $N\delta(\varepsilon_p + V) - 2V$. This indicates that there is the energy gap between the ground and the excited states. Thus, the wave function (14) will be identified as the pseudogap (PG) state in the nearly non-doped region. The PG energy is then estimated to be $\Delta_p = |V|$ and the PG temperature T_p is approximated as the $\lambda\Delta_p$ ($\lambda < 1$). What the excited state (16) indicates the creation of $\tilde{p}_{i\sigma}$ fermions is consistent with the experimental result that the PG in momentum space is only in the antinodal regions of the Brillouin zone, because the location of O ions correspond to the same regions in Brillouin zone. Notice that the wavefunctions $|\Psi_A\rangle$, $|\Psi_{EX}\rangle$ are derived on the approximation that $b_{i\sigma}^{+}$ and $c_{i\sigma}^{+}$ satisfy anti-commutation relations. However, since these operators are not exactly orthogonal between neighboring Cu sites, Hamiltonian (12) has the matrix elements between the neighboring states. This suggests that the states (14) and (16) will form the narrow bands reflecting the wavefunction overlap of the neighboring sites.

Next let us consider the case in the neighborhood of optimally doped region. In this doping region the relation of $\langle 1 - n_{d\sigma'}\rangle \approx 0$, $\langle 1 - n_{p\sigma'}\rangle \approx 1 - 0.5\delta$ will be generally reasonable. On the condition of $\alpha_{\mathbf{k}}\beta_{\mathbf{k}}\varepsilon_p = (1-0.5\delta)\varepsilon s(\mathbf{k})(\beta_{\mathbf{k}}^2 - \alpha_{\mathbf{k}}^2)$, the Hamiltonian (9) is represented by

$$\begin{aligned} H = {} & \sum_{\mathbf{k}\sigma}\varepsilon_p(\beta_{\mathbf{k}}^2 - \alpha_{\mathbf{k}}^2)^{-1}(\beta_{\mathbf{k}}^2 b_{\mathbf{k}\sigma}^{+}b_{\mathbf{k}\sigma} - \alpha_{\mathbf{k}}^2 c_{\mathbf{k}\sigma}^{+}c_{\mathbf{k}\sigma}) + N^{-1}\sum_{\mathbf{k}\mathbf{k}'\mathbf{k}''\sigma\sigma'} V\theta(\mathbf{k},\mathbf{k}')(\alpha_{\mathbf{k}}\alpha_{\mathbf{k}'}\beta_{\mathbf{k}''}\beta_{\mathbf{k}-\mathbf{k}'+\mathbf{k}''} b_{\mathbf{k}\sigma}^{+}b_{\mathbf{k}'\sigma}b_{\mathbf{k}''\sigma'}^{+}b_{\mathbf{k}-\mathbf{k}'+\mathbf{k}'',\sigma'} \\ & + \alpha_{\mathbf{k}}\alpha_{\mathbf{k}'}\alpha_{\mathbf{k}''}\alpha_{\mathbf{k}-\mathbf{k}'+\mathbf{k}''} b_{\mathbf{k}\sigma}^{+}b_{\mathbf{k}'\sigma}c_{\mathbf{k}''\sigma'}^{+}c_{\mathbf{k}-\mathbf{k}'+\mathbf{k}'',\sigma'} + \beta_{\mathbf{k}''}\beta_{\mathbf{k}-\mathbf{k}'+\mathbf{k}''}\beta_{\mathbf{k}}\beta_{\mathbf{k}'} c_{\mathbf{k}\sigma}^{+}c_{\mathbf{k}'\sigma}b_{\mathbf{k}''\sigma'}^{+}b_{\mathbf{k}-\mathbf{k}'+\mathbf{k}'',\sigma'} + \beta_{\mathbf{k}}\beta_{\mathbf{k}'}\alpha_{\mathbf{k}''}\alpha_{\mathbf{k}-\mathbf{k}'+\mathbf{k}''} c_{\mathbf{k}\sigma}^{+}c_{\mathbf{k}'\sigma}c_{\mathbf{k}''\sigma'}^{+}c_{\mathbf{k}-\mathbf{k}'+\mathbf{k}'',\sigma'} + \cdots) \end{aligned} \tag{17}$$

where $s(\mathbf{k})=\sin k_x+\sin k_y$, $\theta(\mathbf{k},\mathbf{k}')=\cos(k_x-k_x')+\cos(k_y-k_y')$, and the Cu-O distance is used as the length unit. Here $b_{\mathbf{k}\sigma}^+$ and $c_{\mathbf{k}\sigma}^+$ satisfy anti-commutation relations. Note that $V=V_c+V_{\sigma\sigma'}\approx V'(1-0.5\delta)^2<0$ because in this doping region the Coulomb interaction V_c can be neglected due to the screening effect of carriers. In (17) $b_{\mathbf{k}\sigma}^+$ or $c_{\mathbf{k}\sigma}^+$ fermion has the possibility of creating Cooper pairs, but $b_{\mathbf{k}\sigma}^+$ will contribute to the Cooper formation because the Fermi surface exists in only $b_{\mathbf{k}}^+$ band. Thus the BCS-like wave function is given by

$$|\Psi_S\rangle=\prod_{\mathbf{k}\sigma}(u_{\mathbf{k}}+v_{\mathbf{k}}b_{\mathbf{k}\uparrow}^+b_{-\mathbf{k}\downarrow}^+)|0\rangle,\quad u_{\mathbf{k}}^2+v_{\mathbf{k}}^2=1 \tag{18}$$

Then, the gap equation is obtained as

$$\Delta_{\mathbf{k}}=-0.5\sum_{\mathbf{k}'}\Delta_{\mathbf{k}'}V_{\mathbf{k}\mathbf{k}'}(\Delta_{\mathbf{k}'}^2+\varepsilon_{\mathbf{k}'}^2)^{-1/2},\quad V_{\mathbf{k}\mathbf{k}'}=N^{-1}V\alpha_{\mathbf{k}}\alpha_{\mathbf{k}'}\beta_{\mathbf{k}}\beta_{\mathbf{k}'}\theta_{\mathbf{k}\mathbf{k}'} \tag{19}$$

where the relation $\varepsilon_{\mathbf{k}}=0.5\{\varepsilon_p+[\varepsilon_p^2+4(1-0.5\delta)^2\varepsilon^2 s^2(\mathbf{k})]^{1/2}\}-\varepsilon_F$ is defined, measuring the energy relative to the Fermi level ε_F. Replacing the sum in gap equation by an integral, the solution which is even in $\mathbf{k}$ is given by $\Delta_{\mathbf{k}}=\Delta_0\alpha_{\mathbf{k}}\beta_{\mathbf{k}}(\cos k_x\pm\cos k_y)$. The solution decreasing the Coulomb interaction is $\Delta_{\mathbf{k}}\propto(\cos k_x-\cos k_y)$ which agrees with the experimental fact about the anisotropic gap of high-temperature cuprate superconductors [12]. Thus Δ_0 is determined by the relation

$$1=-0.5\pi^{-2}V\int\alpha_{\mathbf{k}}^2\beta_{\mathbf{k}}^2(\cos k_x-\cos k_y)\cos k_x[\Delta_0^2\alpha_{\mathbf{k}}^2\beta_{\mathbf{k}}^2(\cos k_x-\cos k_y)^2+\varepsilon_{\mathbf{k}}^2]^{-1/2}d\mathbf{k} \tag{20}$$

Using several approximation on the band structure, the superconductive maximum gap energy and the ground-state energy are given by

$$\begin{aligned}\Delta_s&=2\varepsilon_p\delta[1+4(1-0.5\delta)^2\varepsilon^2/\varepsilon_p^2]^{1/2}\exp(1/N_FV),\\ E_s&=N\delta\varepsilon_p-N\varepsilon_p\delta^2[1+4(1-0.5\delta)^2\varepsilon^2/\varepsilon_p^2]^{1/2}\exp(2/N_FV)\end{aligned} \tag{21}$$

where N_F is the density of states at Fermi level and $V=V_c+V'(1-0.5\delta)^2$ $(V'=\varepsilon^2V_0/\varepsilon_p^2)$. Denote that the energy relative to Cu 3*d* level is measured and the contribution of the $c_{\mathbf{k}\sigma}$ fermions to the total system energy is considered.

3.2 Phase diagram
The antiferromagnetic long-range order between copper sites disappears by holes doping, but there still exists the short-range order interaction which plays an important role in the presence of holes. As shown in the preceding section, the ground states in the neighborhood of the insulator and the superconducting region can be then calculated. However, the doping dependency of these states is not clear. There may be possible intermediate cases constructed with these mixed states. It will be then needed to consider the over-all phase diagrams. Now, an overall wavefunction is assumed to be $|\Psi(\delta)\rangle=f^{1/2}|\Psi_p(\delta_p)\rangle+(1-f)^{1/2}|\Psi_s(\delta_s)\rangle$. Here δ_p, δ_s are the doping quantity of PG and superconductive states and the total doping quantity δ is defined to be $\delta=f\delta_p+(1-f)\delta_s$. The total free energy is then given by $G=fE_p(\delta_p)+(1-f)E_s(\delta_s)$, $(0<f<1)$ and the intermediate region is determined by minimizing G with respect to δ, δ_p, δ_s. Let us evaluate the doping quantity δ_{pm}, δ_{sm} by minimizing $E_p(\delta_p)$ and $E_s(\delta_s)$ and the minimum point of the total free energy in the condition of $\delta_{pm}<\delta_{sm}$. As the case of $\delta<\delta_{pm}$ means $\delta_p=\delta$, $\delta_s=0$, the system will show a pure PG state. At $\delta=\delta_{pm}$ superconductivity appears in the PG state and the mixed state will be maintained until the PG state disappear at the doping quantity δ_{sm} satisfying $\delta=\delta_s$. In more increasing of δ there exists a pure superconductor. Thus, the overall wave function will be represented as three regions of $f=1$ $(\delta<\delta_{pm})$, $f=(\delta_{sm}-\delta)(\delta_{sm}-\delta_{pm})^{-1}$ $(\delta_{pm}<\delta<\delta_{sm})$, and $f=0$ $(\delta>\delta_{sm})$, respectively. For both regions of $\delta<\delta_{pm}$ and

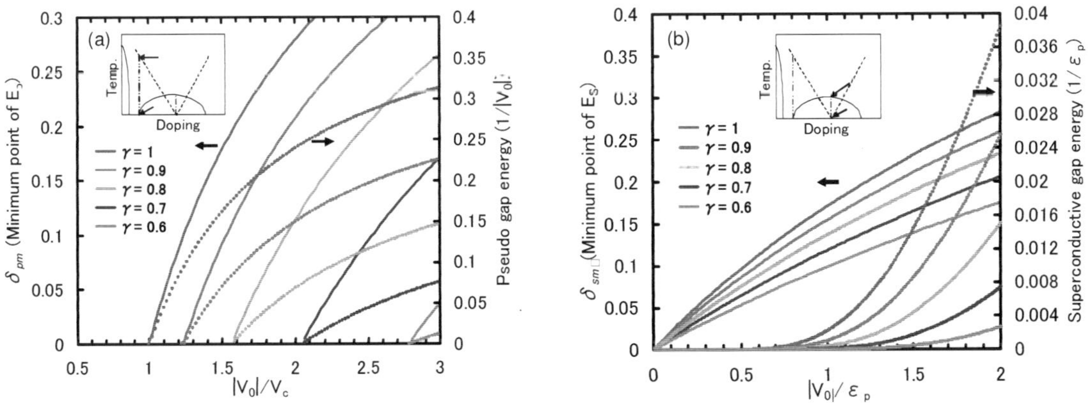

Figure 1 (a) *d-p* AF interaction dependence of δ_{pm} and pseudo gap energy. V_0 is the *d-p* antiferromagnetic-like attractive interaction and V_c is Coulomb interaction between the nearest neighboring sites. (b) *d-p* AF interaction dependence of δ_{sm} (optimal doping) and superconductive gap energy. $\gamma = \varepsilon/\varepsilon_p$. Insets show the schematic phase diagram and calculated points, respectively.

$\delta_{pm} < \delta < \delta_{sm}$, Δ_p and T_p will decrease with increase of δ and vanish at $\delta = \delta_{sm}$. Since the superconductive region appearing at $\delta = \delta_{pm}$ can be identified to be composed of very small superconductive region, the sufficient coherency to detect the net superconductivity will be less formed. Therefore, the superconductive state for $\delta_{pm} < \delta < \delta_{sm}$ could be given by the effective gap $\Delta_{eff} = (\delta - \delta_{pm})\,(\delta_s - \delta_{pm})^{-1}\Delta_s(\delta_s)$. This also indicates that the superfluid density δ_{sf} changes as $\delta - \delta_{pm}$ in the region of $\delta_{pm} < \delta < \delta_{sm}$. The superconductor will be the optimally doped state at the neighborhood of $\delta = \delta_{sm}$, and in the over-doped region Δ_s and T_c will decrease with δ according to doping dependency. Thus, an overall phase diagram is theoretically obtained by considering three types of electronic states.

To reveal the reality of phase diagram, let us estimate δ_{pm}, δ_{sm}, $\Delta_p(\delta_{pm})$ and $\Delta_s(\delta_{sm})$ under the realistic values of system parameters such as ε_p, ε, V_c, $V'(=\varepsilon^2 V_0/\varepsilon_p^2)$. Figure 1 (a),(b) show the dependency of δ_{pm} and pseudo gap energy, and δ_{sm} and superconductive gap energy on the *d-p* antiferromagnetic-like attractive interaction, respectively. Figure 1 (a) shows that for $\varepsilon/\varepsilon_p = 0.8$ and $|V_0|/V_c = 1.5 \sim 2.0$ ($V_c = 1$ eV), $\delta_{pm} = 0 \sim 0.1$ and $\Delta_p(\delta_{pm}) = 0 \sim 200$ meV. Similarly, Figure 1 (b) shows that for $\varepsilon/\varepsilon_p = 0.8$, $\varepsilon_p = 1$ eV, $|V_0| = 0.8 \sim 1.2$ eV, $\delta_{sm} = 0.15 \sim 0.2$, $\Delta_s(\delta_{sm}) = 1 \sim 10$ meV $T_s(\delta_{sm}) = 3 \sim 30$ K.

3.3 Exotic phases

Recently there has been much interest in various exotic phases of cuprates superconductors *etc.* [13]. Here let us discuss briefly the charge density wave (CDW), electron-nematic order and strange metal in normal state. First, it has been reported that CDW can coexist with superconductivity in underdoped region [14]. It has been confirmed that the observed CDW is static and short-range order. According to this paper, the state in underdoped region is constructed with PG and superconductive mixed phases, and the mixing state is evaluated by minimizing simply the free energy of these phases. However, more exactly, it is necessary to consider the space-like change of charge distribution due to the mixing of phases. Then, though PG ground state in mixing phase is not equivalent to CDW, it will possibly contain the charge order such as CDW. It has been also reported that the electron-nematic order can exist in the normal state of underdoped region [15]. Such orders have been interpreted as phases breaking the fourfold symmetry, which are translational invariant, but break rotational symmetry. The symmetry breaking suggests that the two oxygen ions in the copper-oxide unit are inequivalent. Since the electron-nematic order occurs in the PG phase, it may be related with the excited PG state. The reason is that there are the excitations of O holes in the excited PG state. Last let us discuss the strange metal. This has been discussed in the relation with quantum critical point (QCP) [13]. The strange metal region in phase diagram has been explained by interpretation that in temperature plane there is a quantum critical wedge opening up from the QCP. Here, since the optimally doped point is the same as minimum doping of superconductivity without PG state, this point will be interpreted as the QCP.

4. Pnictides

4.1 Electronic state

The characteristics of pnictide superconductors is that it is multi d band, p band (n >1, l >1) metal in non-doping, and superconductive in electron doping [16]. For the parameters of pnictides, it will be selected that $\varepsilon_n = \varepsilon = 0.1 \sim 0.3\,\mathrm{eV}$, $\varepsilon_{pl} = \varepsilon_p = 1\,\mathrm{eV}$, $U = 2.5$ eV, $V_c = 0 \sim 2\,\mathrm{eV}$, $V' = -1 \sim 2$ eV. Since the parent materials of pnictide are metal in non-doping, it seems to be appropriate to treat nearly non-doped region in the momentum representation. However, considering the antiferromagnetism (AFM) in non-doping, it is considered here to start from Eq. (8) in the real space presentation. The metallic state in nearly non-doped region can be interpreted to originate from the nonexistence of PG state due to the relation of $V = V_c + V' > 0$. This situation may correspond to the metallic state. In the momentum representation, V_c can be neglected due to the screening effect of carriers, then $V \approx V' < 0$. In similarity with the cuprates, the superconductive gap equation derived from Eq. (9) is given by

$$\Delta_{\mathbf{k}n} = -0.5\sum_{\mathbf{k}'}\Delta_{\mathbf{k}'n}V_{\mathbf{kk}'n}(\Delta^2_{\mathbf{k}'n} + \varepsilon^2_{\mathbf{k}'n})^{-1/2},\quad V_{\mathbf{kk}'n} = N^{-1}V\alpha_{\mathbf{k}n}\alpha_{\mathbf{k}'n}\beta_{\mathbf{k}n}\beta_{\mathbf{k}'n}\theta_n(\mathbf{k},\mathbf{k}') \tag{22}$$

Since Fe-3d electrons are multi orbitals, it is needed to select the dominant orbitals of Fe contributing to superconductivity [17]. The orbitals of d_{xz}, d_{yz} that have the strong hopping transition to the nearest Pn sites and intersect the Fermi surface are considered here. First let us consider the symmetry of superconductive paring wavefunctions [18]. The gap equation (22) is then considered in the case of two only orbitals d_{xz}, d_{yz}. For simplicity single p-orbital will be only considered. Considering that the square lattice of Fe^{2+} is tetrahedrally coordinated with Pn anions $s_{xz}(\mathbf{k}), s_{yz}(\mathbf{k}), \theta_{xz}(\mathbf{k},\mathbf{k}')$ and $\theta_{yz}(\mathbf{k},\mathbf{k}')$ are calculated as $s_{xz}(\mathbf{k}) = \cos k_x, s_{yz}(\mathbf{k}) = \cos k_y$, $\theta_{xz}(\mathbf{k},\mathbf{k}') = \cos(k_x - k'_x)$, $\theta_{yz}(\mathbf{k},\mathbf{k}') = \cos(k_y - k'_y)$ where the half of Fe-Fe distance is used as the length unit. In the case of d_{xz} (n = 1), replacing the sum in (24) by an integral, the solution which is even in $\mathbf{k}$ is given by $\Delta_{\mathbf{k}1} = \Delta_0\alpha_{\mathbf{k}1}\beta_{\mathbf{k}1}\cos k_x$. Then, Δ_0 is then determined by the relation similar with (19). The gap energy $\Delta_{\mathbf{k}1}$ is then obtained by solving the gap equation regarding Δ_0. In the case of d_{yz} (n = 2), using the transformations of $k_x \to k_y$, $V_1 \to V_2$, $\alpha_{\mathbf{k}1} \to \alpha_{\mathbf{k}2}$, $\beta_{\mathbf{k}1} \to \beta_{\mathbf{k}2}$, the similar gap equation is obtained. Since the band of d_{xz}, d_{yz} are almost the same configuration, the value of ε_1 will be equal to that of ε_2. These indicates that the superconductive paring of Fe-Pn is s-like state different from the d-state of cuprates.

4.2 Phase diagram

Though some Fe-Pn materials exhibit low-temperature (T_c~4K) superconductivity without doping, the high T_c iron-based superconductors appear by a partial replacement of F^- or H^- ion at the oxygen site [19,20]. In general these iron-based superconductors indicate the phase diagram that the T_c appears when the AFM disappear or weaken, and disappear with increasing of doping level. There are two phase diagrams of the 1111 and 122 system. For the 1111 system, the T_c appears when ferromagnetism disappears. On the other hand, the AFM and superconductivity coexist in the 122 system. These phase diagrams can be explained in a similar way to the case of cuprates. In non-doping state, Fe ions are the superexchange antiferromagnetic state through the sites of pnictides that are non-magnetic materials (closed shell). However, it is recognized that there is the change from the closed shell state related to the mechanism of superexchange AFM. This means that the doped electrons will be supplied in Fe-3d sites and the partial empty sites of pnictide. Therefore, in the similar situation with the case of cuprates, the doping will create the antiferromagnetic interaction contributing the superconductive state and reversely will weaken the superexchange antiferromagnetic interaction between Fe ions. Let us discuss the superconductive phases peculiar to iron pnictides. It has been reported that the doping by replacement of H^- ion generally leads to the dome structure in phase diagrams of $REFeAsO_{1-x}H_x$

IOP Conf. Series: Journal of Physics: Conf. Series **1054** (2018) 012013 doi:10.1088/1742-6596/1054/1/012013

(1111 system). The interesting thing is that the optimal doping level indicating the highest T_c is decreased with decreasing the size of rare earth ion [21]. According to the theory, $T_c \propto \exp[-A/N_F V(1-c\delta)^2]$. Since doping dependence of T_c is determined by $V(1-c\delta)$, it will be needed to consider this variance relating with the size of RE ion. On the condition of fixed V (*d-p* AF interaction), decreasing the ion size will weaken the correlation of fermions in the system. As a result, V will be inclined to increase, and at the same time Fe-Fe AF interaction will weaken by carrier doping. Since the value of V and c increase with decreasing the size of RE ion, the optimal doping level decreases with increasing of T_c. Considering the above relations, the fitting curves of doping dependency of T_c are shown in figure 2. Theoretical doping dependency of T_c is consistent with the data of the highest T_c in $REFeAsO_{1-x}H_x$ superconductors.

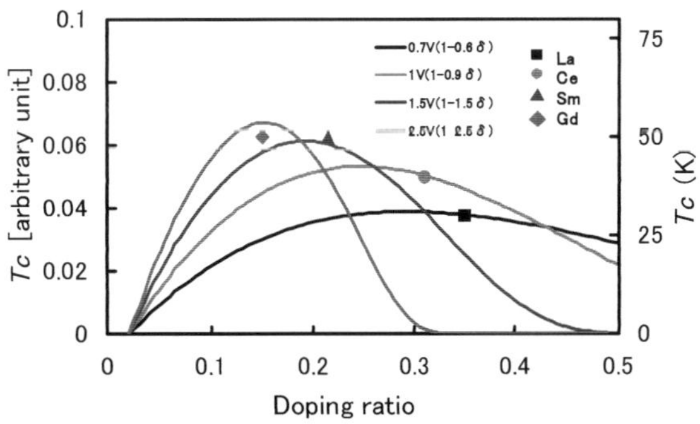

Figure 2 Calculated doping dependency of T_c.

5. Conclusions

Phase diagrams of high-temperature superconducting cuprates and iron pnictides are investigated by extended *d-p* theory. In cuprates, PG and superconductive gap energy are evaluated, emphasizing on the minimum and optimal doping points in the superconductive state which are indispensable for determining the phase diagrams. It is then found that the phase diagram can be constructed with pure pseudo gap and superconductive states, and their mixed state. In pnictides, the phases constructed with the anti-ferromagnetism and superconductor can be explained in similar situation with the case of cuprates. The theory can also explains how the superconductive phases of $REFeAsO_{1-x}H_x$ change with the size of RE ion.

Acknowledgments

The author thanks the Yukawa Institute for Theoretical Physics at Kyoto University, where this work was initiated during the YITP-W-17-02 on "Recent topics and developments on superconductors"

References

[1] Anderson P W 1987 *Science* **235** 1196
[2] Emery V J 1987 *Phys. Rev. Lett.* **58** 2794
[3] Zhang F C and Rice T M 1988 *Phys. Rev.* **B 37** 3759
[4] Massidda J Yu and Freeman A J 1988 *Physica* **C152** 273
[5] Varrma C M *et al.* 1989 *Phys. Rev. Lett.* **63** 1996
[6] Monthoux P, Balatsky A V and Pines D 1991 *Phys. Rev. Lett.* **67** 3448
[7] Alexandrov A S and Mott N F 1993 *Supercond. Sci. Technol.* **6** 215
[8] Leggett A J 2006 *Nat. Phys.* **2** 134
[9] Nishi K 2017 *J. Phys. Conf. Ser.* **871** 012033
[10] Vaknin D *et al.* 1987 *Phys. Rev. Lett.* **58** 2802
[11] Shirane G *et al.* 1987 *Phys. Rev. Lett.* **59** 1613
[12] Tsuei C C and Kirtley J R 2000 *Rev. Mod. Phys.***72** 969
[13] Keimer B *et al.* 2015 *Nature* **518** 179
[14] Wu T *et al.* 2011 *Nature* **477** 191
[15] Ando Y *et al.* 2002 *Phys. Rev. Lett.* **88** 137005
[16] Hosono H and Kuroki K 2015 *Physica* **C514** 399
[17] Singh D J *et al.* 2008 *Phys. Rev. Lett.* **100** 237003
[18] Mazin I I *et al.* 2008 *Phys. Rev. Lett.* **101** 057003
[19] Kamihara Y *et al.* 2006 *J. Am. Chem. Soc.* **128** 10012
[20] Kamihara Y *et al.* 2008 *J. Am. Chem. Soc.* **130** 3296
[21] Iimura S *et al.* 2012 *Nat. Commun.* **3** 943

IOP Conf. Series: Journal of Physics: Conf. Series **1054** (2018) 012014 doi:10.1088/1742-6596/1054/1/012014

Impurity-induced Mott transitions in partially filled antiferromagnetic states

Hisatoshi Yokoyama[1], **Ryo Sato**[1] **and Kenji Kobayashi**[2]

[1]Department of Physics, Tohoku University, Aoba-ku, Sendai, 980-8578, Japan

[2]Department of Natural Science, Chiba Institute of Technology, Narashino, 275-0023, Japan

E-mail: yoko@cmpt.phys.tohoku.ac.jp

Abstract. In view of cuprate superconductors, effects of point-type repulsive impurity potential (V) on partially filled antiferromagnetic (AF) states are studied for a strongly correlated square-lattice Hubbard model ($U/t = 12$) with a diagonal transfer ($t' = -0.3t$), using a variational Monte Carlo method. To simplify interpretation, here we focus on the case of $\delta_{\rm imp} = \delta$, where $\delta_{\rm imp}$ and δ are densities of impurity sites and doped holes, respectively, with $\delta = 0.04 - 0.20$. As a result, a filling-control-type Mott transition occurs at $V = V_{\rm M}$ ($\sim 2t$) irrespective of δ; a partially filled AF state becomes Mott insulating for $V > V_{\rm M}$. Such impurity-induced Mott transitions extensively appear beyond the above parameter settings.

1. Introduction

Recent numerical studies on uniform Hubbard-type models using refined techniques [1, 2] argued that antiferromagnetic (AF) long-range orders (or phase separation) prevail in wide ranges of doping rate δ and frustration t'/t. To cope with the inconsistency with experiments on cuprate superconductors, in which AF orders rapidly vanish on carrier doping, we have paid attention to the effects of impurity potential caused by the dopant ions inherent in cuprates. Another source of impurity effects is apical oxygen in the T'-structure cuprates (typical for electron-doped systems), which is considered to induce robust AF orders, especially, in as-grown samples [3].

In the preceding report [4], we studied effects of point-type impurity potential V on an AF state, mainly regarding U/t dependence and cases of attractive potential ($V < 0$). As a result, screening of V owing to U is so effective in a strongly correlated regime ($U \gtrsim W$, W: band width) with $-U/2 \lesssim V < 0$ that the effects of V thoroughly disappear, namely, the properties of uniform AF state ($V = 0$) are preserved in the above range of V [5]. A doped AF state remains metallic for $V < 0$.

As a continuation, in this report, we discuss the cases of repulsive impurity potential ($V > 0$), which actually correspond to the potential owing to the carrier dopants and apical oxygen atoms in cuprates, in a strongly correlated regime. It is found that, in contrast with the attractive cases, filling-control-type Mott transitions extensively arise at $V \sim 2t$, and partially filled AF states often become insulating.

2. Formalism

2.1. Model

In this study, we consider a Hubbard model ($U \geq 0$) with a point-type spin-independent impurity potential (V) on extended square lattices:

$$\mathcal{H} = \mathcal{H}_{\rm kin} + \mathcal{H}_U + \mathcal{H}_{\rm imp} = -\sum_{(i,j),\sigma} t_{ij}\left(c^\dagger_{i\sigma}c_{j\sigma} + {\rm H.c.}\right) + U\sum_j n_{j\uparrow}n_{j\downarrow} + V\sum_{\ell=1}^{N_{\rm imp}}\sum_\sigma n_{\ell\sigma}, \quad (1)$$

where $n_{j\sigma} = c^\dagger_{j\sigma}c_{j\sigma}$, (i,j) indicates the sum of pairs on sites i and j, and $N_{\rm imp}$ is the number of impurity sites, which we will explain shortly. Here, we set the hopping integral t_{ij} as t for nearest neighbors (≥ 0), t' for diagonal neighbors, and 0 otherwise ($\mathcal{H}_{\rm kin} = \mathcal{H}_t + \mathcal{H}_{t'}$). We use t and the lattice spacing as the units of energy and length, respectively. Here, we focus on parameters typical of hole-doped cuprates, namely, $t'/t = -0.3$ and $U/t = 12$. Electron-doped cases ($t'/t \sim 0.3$) have similar features, as far as the Mott transition discussed below is concerned. We assume that the impurity potential V comes from (i) a carrier dopant situated in an adjacent block layer or from (ii) an apical oxygen atom in a T'-structure system [3], and acts on a nearby single site ("impurity site") for simplicity. If we grant this simplification, the index ℓ in $\mathcal{H}_{\rm imp}$ runs over the impurity sites $\mathbf{r}_\ell$ ($\ell = 1, \cdots, N_{\rm imp}$). Here, we study a special but important case, in which the numbers of impurity site and of doped holes are equal: $N_{\rm imp} = N_{\rm s} - N$ or $\delta_{\rm imp} = \delta$, with $N_{\rm s}$ and N being the numbers of total sites and electrons, respectively, $\delta_{\rm imp} = N_{\rm imp}/N_{\rm s}$, and $\delta = 1 - N/N_{\rm s}$. This is a special case of (i), for which interpretation of the result becomes simple, as we will see. In general cases ($\delta_{\rm imp} \neq \delta$), which we will describe elsewhere, physics of Mott transitions becomes sensitive to the parameters and diversified. Furthermore, we assume that $\mathbf{r}_\ell$ are randomly distributed.

For both hole-doped and electron-doped cuprates, V is positive (repulsive) as far as we treat the model in a less-than-half-filled band, and can be estimated at $V \sim 2$eV ($\sim U$) as an electrostatic energy. Practically, the value of V should somewhat vary owing to quantum effects, etc. Anyway, the effect of such a large V should not be disregarded thoughtlessly. Therefore, we widely vary it (even for $V < 0$) to grasp overall features of impurity effects. Although the above setting of impurity potential is possibly simplistic, we believe that fundamentals of impurity effects with strong correlation can be captured.

2.2. Method

We apply a many-body variation theory to the above model. In computing expectation values, we employ a variational Monte Carlo (VMC) method [6, 7], which enables us to exactly treat many-body wave functions. As a trial wave function, we use a Jastrow type: $\Psi = \mathcal{P}\Phi$. Below, we use the data of the best wave function among what we checked (six types, not shown).

As a many-body factor, a form $\mathcal{P} = \mathcal{P}_{\rm H}\mathcal{P}^{\rm I}_Q\mathcal{P}_{\rm G}$ is used. In the onsite (Gutzwiller) factor $\mathcal{P}_{\rm G}$ [8] and the nearest-neighbor doublon-holon (D-H) binding factor $\mathcal{P}^{\rm I}_Q$, we distinguish impurity and ordinary (non-impurity) sites as,

$$\mathcal{P}_{\rm G} = \prod_{j(\neq\ell)}\left[1 - (1-g)n_{j\uparrow}n_{j\downarrow}\right]\prod_{\ell=1}^{N_{\rm imp}}\left[1 - (1-g_{\rm imp})n_{\ell\uparrow}n_{\ell\downarrow}\right], \quad (2)$$

where g and $g_{\rm imp}$ (≥ 0) are variational parameters, and ℓ (j) runs over impurity (ordinary) sites. If the optimized value of g (or $g_{\rm imp}$) is smaller [larger] than 1, the effective onsite interaction is repulsive [attractive]. As a D-H factor [9, 10], which is relevant for Mott physics, we use

$$\mathcal{P}^{\rm I}_Q = \prod_{j(\neq\ell)}(1 - Q_j)\prod_{\ell=1}^{N_{\rm imp}}\left\{\left(1 - \tilde{Q}_\ell\right)\prod_\tau\left[1 - (1-\eta_{\rm I})\,h_\ell d_{\ell+\tau}\right]\right\} \quad (3)$$

with

$$Q_J = \zeta_{\rm d}\ d_j \prod_{\tau'} (1 - h_{j+\tau'}) + \zeta_{\rm h}\ h_j \prod_{\tau'} (1 - d_{j+\tau'}) \tag{4}$$

$$\tilde{Q}_\ell = \zeta_{\rm I}\ d_\ell \prod_{\tau} \left(1 - h_{\ell+\tau}\right) \tag{5}$$

where $d_i = n_{i\uparrow}n_{i\downarrow}$, $h_i = (1 - n_{i\uparrow})(1 - n_{i\downarrow})$, and impurity sites are excluded in the product of τ' in Eq. (4). $\eta_{\rm I}$, $\zeta_{\rm d}$, $\zeta_{\rm h}$ and $\zeta_{\rm I}$ are variational parameters. We assume in constructing Eq. (3) that holons occupying impurity sites are isolated (doped) holes and do not compose D-H pairs. In this sense, a two-body Jastrow-type factor seems favorable for holons at impurity sites. Similarly, doublons in the ordinary sites are not assumed to combine with holons in the nearest-neighbor impurity sites. In addition, we take account of a nearest-neighbor repulsive correlation between holons (doublons), which are effective, especially, for high doping rates:

$$\mathcal{P}_{\rm H} = \prod_{\langle i,j\rangle} \left[1 - (1 - \eta)\left(d_i d_j + h_i h_j\right)\right], \tag{6}$$

where $\langle i, j\rangle$ indicates a nearest-neighbor pair of sites, and η $(0 \leq \eta \leq 1)$ is a variational parameter. In $\mathcal{P}_{\rm H}$, we do not distinguish impurity and ordinary sites for simplicity.

For obtaining the one-body part Φ, we diagonalize the following mean-field Hamiltonian,

$$\mathcal{H}_{\rm MF}^{(\sigma)} = \mathcal{H}_{\rm kin}^{\rm BR} + U\sum_j \langle n_{j\sigma}\rangle n_{j-\sigma} + \mathcal{H}_{\rm V}^{\rm scr}. \tag{7}$$

In the kinetic part $\mathcal{H}_{\rm kin}^{\rm BR}$, a band-renormalization effect is introduced by following Ref. [2], as

$$\mathcal{H}_{\rm kin}^{\rm BR} = -t\sum_{\langle i,j\rangle} \left(c_{i\sigma}^\dagger c_{j\sigma} + {\rm H.c.}\right) - \sum_{(i,j)} t_\nu \left(c_{i\sigma}^\dagger c_{j\sigma} + {\rm H.c.}\right), \tag{8}$$

where the first term is the nearest-neighbor hopping and the second term indicates virtual hoppings up to fifth neighbors (see Fig. 1(b) in Ref. [2]). Here, t_ν's (ν = 1–4) are variational parameters, which are crucial for stabilizing, especially, the AF state. In the second term of Eq. (7), we take account of an AF long-range order of the nesting vector $\mathbf{Q} = (\pi, \pi)$ as

$$\langle n_{j\sigma}\rangle = \frac{n}{2} + \delta_{\rm AF}\ {\rm sign}(\sigma) e^{i\mathbf{Q}\cdot\mathbf{r}_j}, \tag{9}$$

where $n = N/N_{\rm s}$, ${\rm sign}(\sigma) = 1$ or -1 according to $\sigma =\uparrow$ or $\downarrow$, and $\delta_{\rm AF}$ is a variational parameter that controls the magnitude of AF order. In the last term of Eq. (7), we consider screening effects of impurity potential V owing to electron correlation,

$$\mathcal{H}_{\rm V}^{\rm scr} = \sum_{\ell=1}^{N_{\rm imp}} \left(V_{\rm eff}\ n_{\ell,\sigma} + V_{\rm nn} \sum_\tau n_{\ell+\tau,\sigma}\right), \tag{10}$$

where τ runs over the nearest-neighbor sites of impurity site ℓ, and $V_{\rm eff}$ and $V_{\rm nn}$ are variational parameters. The first term $(0 \leq V_{\rm eff}/V \leq 1)$ represents a direct screening effect in the impurity sites; for $V_{\rm eff}/V = 0$, the effect of impurity potential V is considered to be completely screened out by the electron correlation. By $V_{\rm nn}$, the electron density in the adjacent sites of impurity sites can be optimized. Diagonalizing $\mathcal{H}_{\rm MF}^{(\sigma)}$ of $N_{\rm s} \times N_{\rm s}$ matrices, we obtain eigenenergies ε_n and eigenvectors $\phi_{n\sigma}$ with corresponding quasiparticle operators $\alpha^\dagger_{n\sigma}$, which are linear combinations of $c^\dagger_{j\sigma}$. Using $\alpha^\dagger_{n\sigma}$, one-body part is constructed as

$$\Phi = \prod_{n\ (\varepsilon_n \leq \varepsilon_{\rm F})} \prod_{\sigma} \alpha^\dagger_{n\sigma}|0\rangle, \tag{11}$$

where $\varepsilon_{\rm F}$ indicates the energy of the highest occupied level (HOMO). The correlated AF state $\Psi_{\rm AF} = \mathcal{P}\Phi_{\rm AF}$ is reduced to a correlated paramagnetic (PM) state $\Psi_{\rm PM} = \mathcal{P}\Phi_{\rm PM}$ for $\delta_{\rm AF} \to 0$.

We compute expectation values with respect to Ψ using a VMC method similar to that in Ref. [2]. We use systems of $N_{\rm s} = 10\times10$ sites with the periodic-antiperiodic boundary conditions, because the system-size dependence is small and only quantitative. A configuration of impurities $\{\mathbf{r}_\ell\}$ is randomly chosen and fixed through a single sweep of VMC calculation (optimization and measurement). $\{\mathbf{r}_\ell\}$ dependence exists to some extent and its degree depends on the model parameters. In a VMC sweep, we use 250,000 samples. In most cases, we will display data of 4 – 8 kinds of $\{\mathbf{r}_\ell\}$ simultaneously.

3. Results

In the preceding report [4], we showed that properties of the uniform cases ($V = 0$) are preserved for $V_{\rm L} \lesssim V \lesssim V_{\rm M}$ in a strongly correlated regime ($U \gtrsim W$, $W = 8t$: band width), because the effect of impurity potential is almost completely screened out, namely $V_{\rm eff}/V \sim 0$ and $V_{\rm nn}/V \sim 0$. Here, $V_{\rm L} \sim -U/2$ and $V_{\rm M} \sim t - 2t$. We concluded that the impurity potential is ineffective in weakly repulsive and moderately attractive cases for a large U/t.

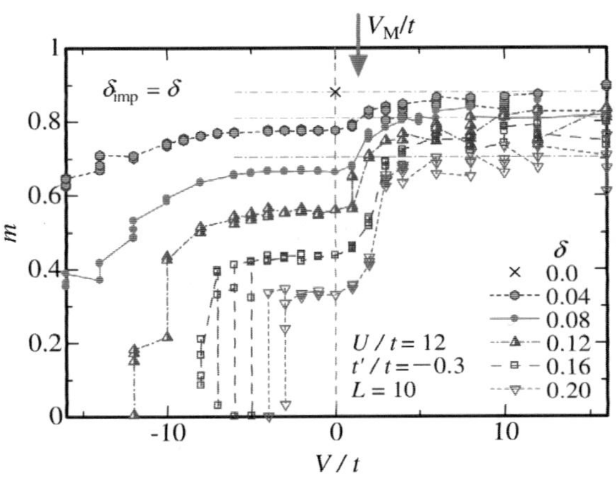

Figure 1. Staggered magnetization as function of impurity potential for several doping rates. For each parameter set (V, δ), data of four impurity configurations $\{\mathbf{r}_\ell\}$ are plotted at the same time. The value at half filling for $V = 0$ ($m_0 \sim 0.88$) [2] is indicated by a cross and extended with a gray dash-dotted line. We also show $m_0(1 - \delta_{\rm imp})$ for $\delta_{\rm imp} = 0.08$ and 0.20 with dash-dotted lines of coordinate colors. A Mott transition point is indicated by a thick gray arrow. Incidentally, an AF order survives up to $\delta \sim 0.27$ for $V = 0$ [2].

Since Ref. [4] mainly treated U/t dependence and cases of attractive potentials, here, we focus on cases of repulsive potential ($V > 0$) in the Mott-insulating regime of U/t ($U > U_{\rm c} \sim W$). In contrast with for $V_{\rm L} \lesssim V \leq 0$, the value of $V_{\rm eff}/V$ of the optimized AF state for $V > 0$ does not completely vanish (not shown). It is found that, when V surpasses $V_{\rm M}$ ($\sim t - 2t$), convergence in the VMC optimization suddenly becomes not easy and dependence on impurity configuration $\{\mathbf{r}_\ell\}$ becomes serious, especially, for the PM state. Consequently, fluctuations in expectation values become large (sometimes too large to be managed). It is natural to consider that some transition arises at $V = V_{\rm M}$ in the electronic state. Figure 1 shows V dependence of the staggered magnetization (an AF order parameter)

$$m = \frac{2}{N_{\rm s}} \left| \sum_j e^{i\mathbf{Q}\cdot\mathbf{r}_j} \langle S^z_j \rangle \right|, \tag{12}$$

IOP Conf. Series: Journal of Physics: Conf. Series **1054** (2018) 012014 doi:10.1088/1742-6596/1054/1/012014

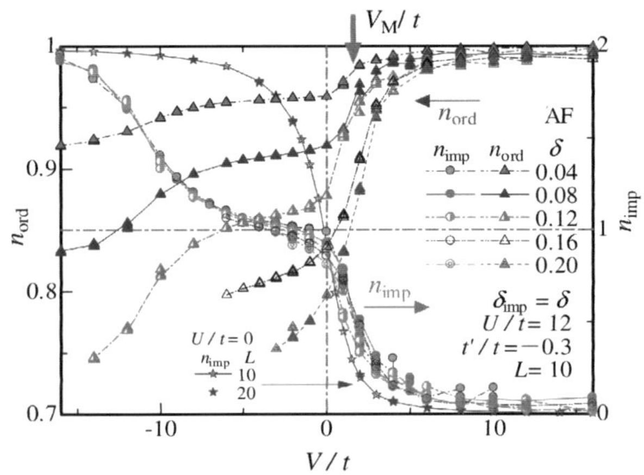

Figure 2. Average electron densities at impurity sites ($n_{\rm imp}$, right axis) and ordinary sites ($n_{\rm ord}$, left axis) as functions of impurity potential for several doping rates. As a reference, the data of $n_{\rm imp}$ for $\delta = 0.08$ and $U = 0$ ($L = 10$ and 20) are also plotted with stars. A Mott transition point is indicated by a thick gray arrow.

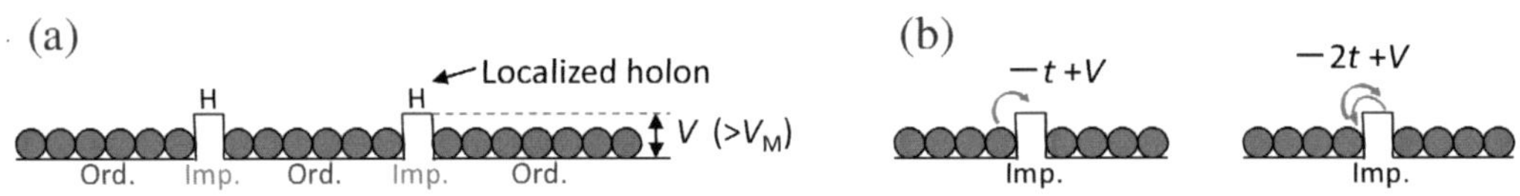

Figure 3. (a) Schematic figure of real-space electron configuration in Mott insulating state realized for $V > V_{\rm M}$. Steps at "Imp." sites represent the repulsive impurity potential ($V > 0$). A blue circle represents an electron, and "H" indicates a holon (empty site). (b) Schematic figure for rough estimation of Mott-transition point $V_{\rm M}/t$.

for several doping rates. We find, as V/t increases, m rapidly increases around $V = V_{\rm M}$ and approaches $m_0(1 - \delta_{\rm imp})$ for $V > V_{\rm M}$, where m_0 (~ 0.88) is the value of m in the uniform case at half filling [2]. It follows that the staggered moment of the magnitude at half filling arises in the ordinary sites. Figure 2 shows V/t dependence of the average electron densities on the impurity sites ($n_{\rm imp}$) and the ordinary sites ($n_{\rm ord}$) calculated as

$$n_{\rm imp} = \frac{1}{N_{\rm imp}} \sum_{\ell=1}^{N_{\rm imp}} \sum_{\sigma} \langle n_{\ell\sigma} \rangle, \qquad n_{\rm ord} = \frac{N - N_{\rm imp} n_{\rm imp}}{N_{\rm s} - N_{\rm imp}}, \tag{13}$$

respectively. The behavior of $n_{\rm imp}$ is almost independent of δ; as V/t increases from $V/t = 0$, $n_{\rm imp}$ rapidly decreases around $V = V_{\rm M}$ and almost vanishes for $V > V_{\rm M}$. Correspondingly, $n_{\rm ord}$ rapidly increases and approaches 1, the value at half filling. In fact, the behavior in Figs. 1 and 2 reflects the relation $\delta_{\rm imp} = \delta$, and can be interpreted as a filling-control-type Mott transition as we explain in the following.

The uniform AF state at half filling is insulating for any U/t (> 0) and its character changes from a Slater type to a Mott type as a crossover around $U = W$ [11]. The density of a doubly occupied site (doublon) is suppressed in the Mott-insulating regime, e.g. down to 0.027 at $U/t = 12$ [12]. In an impurity system of $V > V_{\rm M}$, almost all doped holes are trapped in the impurity sites to avoid raising energy by V. When the condition $\delta_{\rm imp} = \delta$ is imposed, empty sites (holons) have one-to-one correspondence to impurity sites, and (almost) all ordinary sites are occupied by a single electron as shown in Fig. 3(a), namely, a half-filled band is virtually

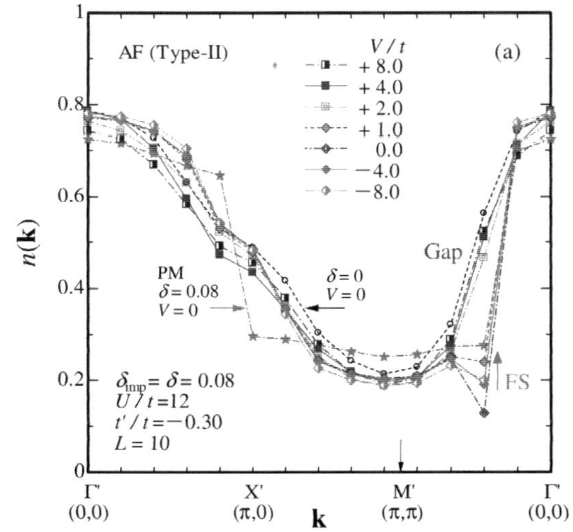

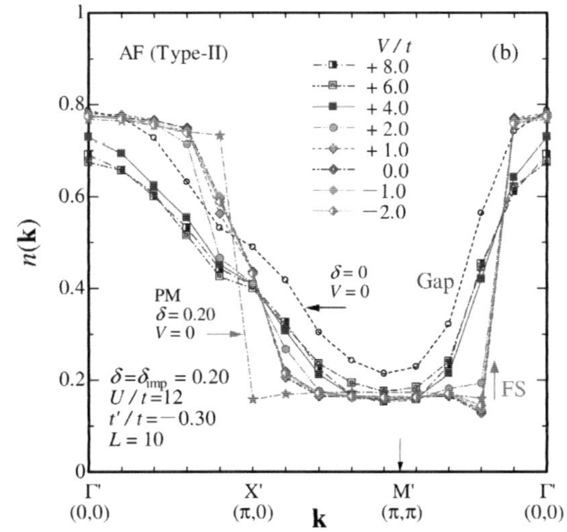

Figure 4. Momentum distribution function of type-II AF state for (a) $\delta = 0.08$ and (b) $\delta = 0.20$ along path $(0,0) \to (\pi,0) \to (\pi,\pi) \to (0,0)$. In each panel, evolution of $n(\mathbf{k})$ as V/t changes is to be noticed. For each value of V, the result of a single $\{\mathbf{r}_\ell\}$ is plotted. For comparison, the data for two typical reference values, a uniform AF Mott-insulating state at half filling (black open circles) and a uniform PM state (gray stars) are also drawn. The values of U/t, t'/t, and L are common to all cases.

formed within the ordinary sites. In other words, some sites of the uniform half-filled lattice are depleted by the impurities, but the residual $N_s(1-\delta_{\rm imp})$ sites preserve the properties of the uniform half-filled band. Thus, a Mott-insulating state is realized at a partially filling by impurity potential.

Next, let us discuss why the Mott transition point is situated at $V_{\rm M}/t \sim 1-2$ regardless of U/t and t'/t (data not shown). The variation of energy when an electron hops to an impurity site or hops to an impurity site and return to the original site, as shown in Fig. 3(b), is roughly estimated at $\Delta\varepsilon = V - t$ or $V - 2t$. For $\Delta\varepsilon < 0$ ($\Delta\varepsilon > 0$), an electron hops via an impurity site without (at) a loss of energy. In the case of $\delta_{\rm imp} = \delta$ with large U ($\gg t$), a global flow becomes possible only if electrons can hop to impurity sites. As a result, the state becomes conductive or insulating according as $\Delta\varepsilon$ is negative or positive, resulting in $V_{\rm M}/t = 1-2$.

Finally, we corroborate this Mott transition by the momentum distribution function,

$$n(\mathbf{k}) = \frac{1}{2}\sum_\sigma \langle c^\dagger_{\mathbf{k}\sigma} c_{\mathbf{k}\sigma}\rangle. \tag{14}$$

Showing in Figs. 4(a) and 4(b) is evolution of $n(\mathbf{k})$ as V/t changes for an underdoped ($\delta = 0.08$) and an overdoped ($\delta = 0.20$) cases, respectively. As discussed in Ref. [2], the uniform AF state for $t'/t = -0.3$ is classified as type II, which is insulating at half filling but becomes metallic for $\delta > 0$ with a pocket Fermi surface (FS) around $\mathbf{k} = (\pi/2, \pi/2)$ and a gap near antinodal $\mathbf{k} = (\pi, 0)$. Such metallic behavior—discontinuity with a dip near $(\pi/2, \pi/2)$ and smoothness near $(\pi, 0)$—can be confirmed by the curves for $V = 0$ in Fig. 4, and is preserved for $V \lesssim V_{\rm M}$ (drawn with warm colors). For $V \gtrsim V_{\rm M}$ (cold colors), however, the pocket FS near $(\pi/2, \pi/2)$ vanishes, and the behavior of $n(\mathbf{k})$ becomes similar to that of the uniform AF state at half filling. It indicates a Mott transition.

4. Conclusion

We studied effects of point-type repulsive impurity potential on a strongly correlated AF state, using a many-body variation theory. We showed that a filling-control-type Mott transition takes place in partially-filled AF states at $V = V_{\rm M} \sim 2t$, if the densities of impurity is equal to the doping rate. We have found that such Mott transitions are not restricted to the present special case ($\delta_{\rm imp} = \delta$) but manifest themselves widely ($\delta_{\rm imp} \geq \delta$ and in attractive-potential cases). We will soon report them elsewhere. It is important to consider how the Mott transition discussed here is related to the behavior of cuprates, or more generally to phenomena in other materials.

Acknowledgments

The authors thank Y. Koike, T. Adachi and H. Tsuchiura for useful discussions. This study is supported in part by Grants-in-Aid from the Ministry of Education, Culture, Sports, Science and Technology, Japan.

References

[1] Otsuki J, Hafermann H and Lichtenstein A I 2014 *Phys. Rev. B* **90** 235132; Misawa T and Imada M 2014 *Phys. Rev. B* **90** 115137; Zheng B-X and Chan G K-L 2016 *Phys. Rev. B* **93** 035126
[2] Sato R and Yokoyama H 2016 *J. Phys. Soc. Japan* **85** 074701
[3] Adachi T, Mori Y, Takahashi A, Kato M, Nishizaki T, Sasaki T, Kobayashi N and Koike Y 2013 *J. Phys. Soc. Jpn.* **82** 063713
[4] Yokoyama H, Sato R and Kobayashi K 2017 *J. Phys. Conf. Ser.* **871** 012032
[5] Such strong screening effects had been pointed out in similar contexts. For instance, Fukushima N, Chou C-P, Lee T K 2009 *Phys. Rev. B* **79** 184510; Farhoodfar A, Chen X, Gooding R J, Atkinson W A 2009 *Phys. Rev. B* **80** 045108
[6] Yokoyama H and Shiba H 1987 *J. Phys. Soc. Jpn.* **56** 1490.
[7] Umrigar C J, Wilson K G and Wilkins J W 1988 *Phys. Rev. Lett.* **60** 1719
[8] Gutzwiller M C 1963 *Phys. Rev. Lett.* **10** 159
[9] Kaplan T A, Horsch P and Fulde P 1982 *Phys. Rev. Lett.* **49** 889
[10] Yokoyama H and Shiba H 1990 *J. Phys. Soc. Jpn.* **59** 3669
[11] Tamura S and Yokoyama H 2015 *J. Phys. Soc. Jpn.* **84** 064707
[12] Yokoyama H, Ogata M, Tanaka Y, Kobayashi K and Tsuchiura H 2013 *J. Phys. Soc. Jpn.* **82** 014707

IOP Conf. Series: Journal of Physics: Conf. Series **1054** (2018) 012015 doi:10.1088/1742-6596/1054/1/012015

Interhole correlation and phase separation in t-J model

Ryo Sato and Hisatoshi Yokoyama

Depertment of Physics, Tohoku University, Sendai 980-8578, Japan

E-mail: satoryo@cmpt.phys.tohoku.ac.jp

Abstract. Interhole and interelectron correlations as well as a transition to phase separation in the t-J model near half filling are reconsidered using a variational Monte Carlo method. As a trial wave function, we use a state that can simultaneously represent a d-wave superconducting and antiferromagnetic (AF) orders and allows for nearest-neighbor hole-hole and spin-spin correlation factors and band renormalization effects. It is found that a transition occurs at $J = J_{\rm PS} \sim 2.5t$ for the doping rates $\delta = 0.08$ and 0.20 from a seemingly uniform state to phase separation ($J > J_{\rm PS}$) of singly occupied sites and empty sites. For $J < J_{\rm PS}$, where the values used for cuprate superconductors ($0.25 \lesssim J/t \lesssim 0.5$) are included, the interhole correlation is always repulsive. This result is negative to an intuitive picture for cuprates: Two holons coherently move in the background of AF orders.

1. Introduction

In early research on cuprate superconductors, it was often demonstrated that two holons (empty sites) can coherently move without destroying the background antiferromagntic (AF) order, but a single holon cannot. A similar notion that the interhole correlation is attractive even for a small J/t (plausible for cuprates) seems still alive. Furthermore, a problem as to the phase separation (PS) near half filling for small J/t is still not settled [1, 2]. In such contexts, it is significant in the square-lattice t-J model [3] to clarify at what values of J/t the interhole correlation is switched from repulsive to attractive ($J_{\rm I}/t$) and PS occurs ($J_{\rm PS}/t$) when we fix the doping rate δ ($= 1 - N_{\rm e}/N_{\rm s}$ with $N_{\rm e}$ and $N_{\rm s}$ being number of electrons and number of sites, respectively). It is certain for $J/t \to 0$ that total energy (E/t) is lowered only by the hopping (t) term, so that the correlation should be repulsive to keep other holes away and the state should be uniform. For $J/t \to \infty$, because E/t is reduced only by the superexchange (J) term, the correlation should be attractive and the state phase separates. For $\delta \to 1$, it is exactly known that $J_{\rm I}/t = 2$ [4, 5] and $J_{\rm PS}/t = 3.4367$ [4], whereas for a small δ of our interest, the problems are still unresolved; it remains unclear even whether $J_{\rm I}/t$ and $J_{\rm PS}/t$ differ or coincide.

In this study, we reconsider intersite correlation and PS in the t-J model using a many-body variation theory with recently developed wave functions and procedures. As a first step, in the Jastrow-type trial wave function, we consider nearest-neighbor interhole and interspin correlation factors, which is simple but can cope with the present problem. In this trial state, we allow for d-wave superconducting (d-SC) and antiferromagnetic (AF) orders simultaneously, for both of which band renormalization effects (BRE) [6] are introduced. Using this wave function, we found $J_{\rm I} = J_{\rm PS}$ for underdoped ($\delta = 0.08$) and overdoped ($\delta = 0.20$) regimes. Then, we analyze the properties of PS arising for $J > J_{\rm PS} \sim 2.5t$.

IOP Conf. Series: Journal of Physics: Conf. Series **1054** (2018) 012015 doi:10.1088/1742-6596/1054/1/012015

2. Model and Method

We consider the standard t-J model [3] ($J/t \geq 0$) on the square lattice,

$$\mathcal{H} = \mathcal{H}_t + \mathcal{H}_J = -t \sum_{\langle i,j \rangle \sigma} \left(\tilde{c}^\dagger_{i\sigma} \tilde{c}_{j\sigma} + \mathrm{H.c.} \right) + J \sum_{\langle i,j \rangle \sigma} \left(\mathbf{S}_i \cdot \mathbf{S}_j - \frac{1}{4} \tilde{n}_i \tilde{n}_j \right), \tag{1}$$

where $c_{j\sigma}$ is an operator which annihilates an electron of spin σ at site j, $\tilde{c}_{i\sigma} = c_{i\sigma}(1 - n_{i-\sigma})$ with $n_{i\sigma} = c^\dagger_{i\sigma} c_{i\sigma}$, $\mathbf{S}_j = \frac{1}{2} \sum_{\alpha,\beta} c^\dagger_{j\alpha} \sigma_{\alpha\beta} c_{j\beta}$ (σ: Pauli matrix), $\tilde{n}_j = \sum_\sigma \tilde{c}^\dagger_{j\sigma} \tilde{c}_{j\sigma}$. $\langle i,j \rangle$ in the summations indicates a pair of nearest-neighbor sites. We use t and the lattice spacing as the units of energy and length, respectively. This model was derived from a more realistic d-p model, but it is also connected to the Hubbard model near half filling through a strong-coupling expansion for $J/t \to 0$ and $t/U \to 0$ with $J = 4t^2/U$. The parameters of this model are the ratio of interaction strength J/t and doping rate δ. As values plausible to cuprates, $J/t = 0.25 - 0.5$ are often used in literatures; this model becomes unrelated to the Hubbard model for $J/t \gtrsim 1$.

To this model, we apply a VMC method, which enables us to exactly compute variational expectation values with respect to many-body trial wave functions. As a trial wave function, we adopt a Jastrow type: $|\Psi\rangle = \mathcal{P}_J \mathcal{P}_G |\Phi\rangle$. In the correlation factors, the complete exclusion of doubly occupied sites by $\mathcal{P}_G = \prod_j (1 - n_{j\uparrow} n_{j\downarrow})$ is inevitable for the t-J model. The choice of intersite correlation factor $\mathcal{P}_J$ is important to the present subject. As a first step, we restict it to between the nearest-neighbor sites, but take account of both charge density and spin correlations: $\mathcal{P}_J = \mathcal{P}_h \mathcal{P}_s$. The charge part is written as an interhole correlation:

$$\mathcal{P}_h = \prod_{\langle i,j \rangle} \left[1 - (1 - \xi_{h-h}) h_i h_j \right], \tag{2}$$

with $h_j = (1 - n_{j\uparrow})(1 - n_{j\downarrow})$, and the spin-dependent part is given by

$$\mathcal{P}_s = \prod_{\langle i,j \rangle} \left[1 - (1 - \xi_{\uparrow-\uparrow}) (n_{i\uparrow} n_{j\uparrow} + n_{i\downarrow} n_{j\downarrow}) \right] \left[1 - (1 - \xi_{\uparrow-\downarrow}) (n_{i\uparrow} n_{j\downarrow} + n_{i\downarrow} n_{j\uparrow}) \right], \tag{3}$$

where ξ_{h-h}, $\xi_{\uparrow-\uparrow}$ and $\xi_{\uparrow-\downarrow}$ are variational parameters. If the optimized value is $1/\xi_{h-h} > 1$ ($0 \leq 1/\xi_{h-h} < 1$), the interhole correlation is repulsive (attractive). This is also the case for $\xi_{\uparrow-\uparrow}$ and $\xi_{\uparrow-\downarrow}$. Using $\mathcal{P}_h$ and $\mathcal{P}_s$, we can assign distinct weights to all different nearest-neighbor bond configurations. $\mathcal{P}_s$ plays a corrective role for the (often overestimated) AF order in Φ_{AF} discussed below.

Now, we turn to the one-body state $|\Phi\rangle$ in $|\Psi\rangle$. To construct a mixed state of d-wave SC and AF orders, we first mention pure states of each order [6]. For a d-SC wave function, we use a BCS state with the electron number fixed:

$$|\Phi_{d\text{-SC}}\rangle = \left(\sum_{\mathbf{k}} A(\mathbf{k})\, c^\dagger_{\mathbf{k}\uparrow} c^\dagger_{-\mathbf{k}\downarrow} \right)^{\frac{N_e}{2}} |0\rangle, \tag{4}$$

$$A(\mathbf{k}) = \frac{\Delta_d}{\varepsilon^{SC}_{\mathbf{k}} - \zeta + \sqrt{(\varepsilon^{SC}_{\mathbf{k}} - \zeta)^2 + (\Delta_d)^2}}, \tag{5}$$

where we assume a $d_{x^2-y^2}$-symmetry SC gap $\Delta_d = \Delta_{SC}(\cos k_x - \cos k_y)$. Δ_{SC} and ζ are variational parameters, which correspond to the SC gap and chemical potential, respectively, for $U/t \to 0$. BRE explained below is introduced into $\varepsilon^{SC}_{\mathbf{k}}$. For an AF state, we employ a Hartree-Fock type at half filling,

$$|\Phi_{AF}\rangle = \prod_{\mathbf{k} \in \{\mathbf{k}_{occ}\}} a^\dagger_{\mathbf{k}\uparrow} a^\dagger_{\mathbf{k}\downarrow} |0\rangle \tag{6}$$

IOP Conf. Series: Journal of Physics: Conf. Series **1054** (2018) 012015 doi:10.1088/1742-6596/1054/1/012015

with AF quasi-particle operators

$$a^{\dagger}_{\mathbf{k},\sigma} = \alpha_{\mathbf{k}} c^{\dagger}_{\mathbf{k},\sigma} + \mathrm{sgn}(\sigma)\beta_{\mathbf{k}} c^{\dagger}_{\mathbf{k}+\mathbf{Q},\sigma} \quad (\mathbf{k} \in \text{magnetic BZ}), \tag{7}$$

$$a^{\dagger}_{\mathbf{k}+\mathbf{Q},\sigma} = -\mathrm{sgn}(\sigma)\beta_{\mathbf{k}} c^{\dagger}_{\mathbf{k},\sigma} + \alpha_{\mathbf{k}} c^{\dagger}_{\mathbf{k}+\mathbf{Q},\sigma} \quad (\mathbf{k}+\mathbf{Q} \notin \text{magnetic BZ}), \tag{8}$$

where $\mathbf{Q} = (\pi, \pi)$, $\mathrm{sgn}(\sigma) = 1$ (-1) for $\sigma =\uparrow$ $(\downarrow)$, and

$$\alpha_{\mathbf{k}}(\beta_{\mathbf{k}}) = \frac{1}{\sqrt{2}}\sqrt{1 - (+)\frac{\varepsilon^{\mathrm{AF}}_{\mathbf{k}}}{(\varepsilon^{\mathrm{AF}}_{\mathbf{k}})^2 + (\Delta_{\mathrm{AF}})^2}}. \tag{9}$$

BRE is also introduced into $\varepsilon^{\mathrm{AF}}_{\mathbf{k}}$. A mixed state actually used is obtained by replacing the bare electron operators $c^{\dagger}$ in Eq. (4) by the AF quasiparticle operators $a^{\dagger}$ in Eqs. (7) and (8) [7]:

$$|\Phi_{\mathrm{Mixed}}\rangle = \left(\sum_{\mathbf{k}} A(\mathbf{k})\; a^{\dagger}_{\mathbf{k}\uparrow} a^{\dagger}_{-\mathbf{k}\downarrow}\right)^{\frac{N_{\mathrm{e}}}{2}} |0\rangle\,. \tag{10}$$

It is vital, especially for the AF order, to take account of BRE [6]. To this end, we fit $\varepsilon^{\Lambda}_{\mathbf{k}}$ ($\Lambda =$ AF or d-SC) in $|\Phi_{\Lambda}\rangle$ to a form of tight-binding bands up to the three-step distance as,

$$\begin{aligned}\varepsilon^{\Lambda}_{\mathbf{k}} = &-2t(\cos k_x + \cos k_y) - 4t^{\Lambda}_1 \cos k_x \cos k_y - 2t^{\Lambda}_2(\cos 2k_x + \cos 2k_y) \\ &-4t^{\Lambda}_3(\cos 2k_x \cos k_y + \cos k_x \cos 2k_y) - 2t^{\Lambda}_4(\cos 3k_x + \cos 3k_y).\end{aligned} \tag{11}$$

Here, four hopping integrals t^{Λ}_j ($j = 1$–4) for each Λ work as band-adjusting parameters, which are simultaneously optimized with the other variational parameters. Note that it is important to optimize $t^{d\text{-SC}}_j$ and t^{AF}_j in $|\Phi_{\mathrm{Mixed}}\rangle$ independently.

In optimization, the stochastic reconfiguration method [8] is used. The VMC calculations were performed on the N_{s} $(= L \times L)$-site lattice with $L = 10$, and the periodic-antiperiodic boundary conditions are imposed. Typically, 2.5×10^5 samples are used for estimating physical quantities.

3. Results

We start with the character of intersite correlation. Figure 1 shows the evolution of the optimized inverse correlation parameters ($1/\xi$) as J/t increases. The interhole correlation preserves $1/\xi_{\mathrm{h-h}} > 0$ up to $J/t = J_{\mathrm{PS}}/t \sim 2.5$ and tends to rise for $J \sim J_{\mathrm{PS}}$, especially, in the underdoped regime in (b). That is, the interhole correlation is repulsive and becomes stronger as J/t increases. In contrast, the electron correlations both for parallel and antiparallel spins are attractive ($1/\xi_{\uparrow-\uparrow}, 1/\xi_{\uparrow-\downarrow} < 1$) and become strong as J/t increases. Because a repulsive (attractive) correlation between holes (electrons) encourages the hopping (exchange) term, the two terms compete with each other in this regime of J/t. In Fig. 1, all $1/\xi$ suddenly drop to small values ($1/\xi \ll 1$) at $J = J_{\mathrm{PS}}$; all intersite correlations become strongly attractive for $J > J_{\mathrm{PS}}$. As expected and as we will confirm below, this anomaly indicates a transition to PS. To sum up, the interelectron correlations are always attractive and become strong as J/t increases, whereas the interhole correlation is repulsive up to $J = J_{\mathrm{PS}}$ far above the plausible range of cuprates ($0.25 \lesssim J/t \lesssim 0.5$). The interhole correlation becomes attractive only when PS is arising ($J > J_{\mathrm{PS}}$). Incidentally, the value of J_{PS}/t is ~ 2.5 both for $\delta = 0.08$ and 0.20. This probably originates in the short-range Jastrow factor used. We would like to check such points using long-range factors in future.

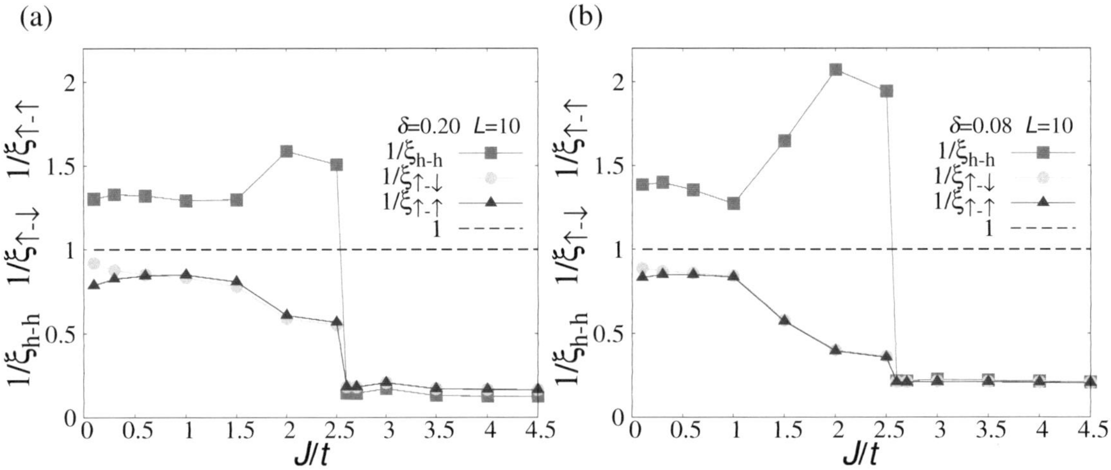

Figure 1. J/t dependence of optimized inverse intersite correlation parameters for (a) overdoped ($\delta = 0.20$) and (b) underdoped ($\delta = 0.08$) cases. The dashed lines indicate $\xi = 1$, which is the boundary of whether the intersite correlation is attractive ($1/\xi < 1$) or repulsive ($1/\xi > 1$).

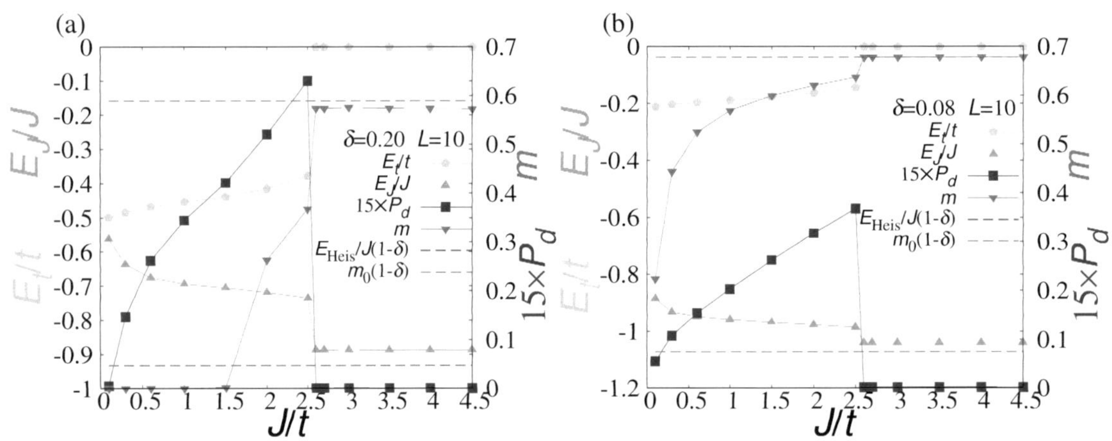

Figure 2. The two elements of energy per site (left axis) and the order parameters of d-SC (P_d) and AF (m) orders (right axis) are plotted as functions of J/t. The doping rates are (a) 0.20 and (b) 0.08. For ease of eyes, P_d is shown by 15 magnifications. The dashed lines indicate $(1-\delta)m_0$ [red] and $(1-\delta)E_{\rm Heis}/J$ [green] with m_0 and $E_{\rm Heis}$ being m and E_J at half filling, respectively.

Next, let us consider the d-SC and AF orders. In Fig. 2, J/t dependence of the order parameters for both orders are shown; m is the staggered magnetization defined as

$$m = \frac{2}{N_{\rm s}} \left| \sum_j e^{i\mathbf{Q}\cdot\mathbf{r}_j} \langle S_j^z \rangle \right|, \tag{12}$$

and P_d is the d-SC correlation function for nearest-neighbor singlet pairs given as

$$P_d(\mathbf{r}) = \frac{1}{N_{\mathrm{S}}} \sum_i \sum_{\tau,\tau'=\hat{\mathbf{x}},\hat{\mathbf{y}}} (-1)^{1-\delta_{\tau,\tau'}} \left\langle \Delta_\tau^\dagger(\mathbf{R}_i) \Delta_{\tau'}(\mathbf{R}_i + \mathbf{r}) \right\rangle , \tag{13}$$

where $\hat{\mathbf{x}}$ ($\hat{\mathbf{y}}$) denotes the lattice vector in x (y) direction and $\Delta_\tau^\dagger(\mathbf{R}_i)$ is the creation operator of a nearest-neighbor singlet pair around site $\mathbf{R}_i$: $\Delta_\tau^\dagger(\mathbf{R}_i) = (c_{i\uparrow}^\dagger c_{i+\tau\downarrow}^\dagger + c_{i+\tau\uparrow}^\dagger c_{i\downarrow}^\dagger)/\sqrt{2}$. For $P_d(\mathbf{r})$, we will plot the value at the farthermost point $\mathbf{r} = (L/2, L/2)$ as P_d. We find that the d-SC order appears for any J/t of $J < J_{\mathrm{PS}}$ in both doping rates, whereas the AF order vanishes for $J/t < 1.5$ in the overdoped case. This tendency is consistent with that for the Hubbard model with moderate values of U/t (e.g. $= 12$) [6]; the d-SC order is somewhat more favored in the t-J model. Both order parameters monotonically become large as J/t increases, indicating that the driving force of both orders is, in common, magnetic interaction. When the present-type PS occurs, P_d vanishes. In contrast, AF orders become more robust and almost constant $m = m_0(1-\delta)$ for $J > J_{\mathrm{PS}}$, where m_0 is the value of m at half filling 0.737 (Heisenberg model). The subtle difference between this value (dashed lines) and the corresponding VMC data for $J > J_{\mathrm{PS}}$ stem from finite-size effects.

Now, we consider the state for $J > J_{\mathrm{PS}}$. The above behavior of m suggests that the state phase separates into two domains: the domain of half-filled Heisenberg model and the domain of holons (empty sites). In Fig. 2, we also show the J/t dependence of energy elements ($E_t = \langle \mathcal{H}_t \rangle$, $E_J = \langle \mathcal{H}_J \rangle$) per site. For $J > J_{\mathrm{PS}}$, the hopping energy vanishes, whereas the exchange energy becomes almost constant $E_J = E_{\mathrm{Heis}}(1-\delta)$ with E_{Heis} ($= -1.165J$) being the energy of the Heisenberg model ($\delta = 0$). This behavior also supports the above PS. In Fig. 3, snapshots of typical electron configurations taken in VMC sweeps for $J/t = 0.3$ and 3.0 ($\delta = 0.20$) are shown to corroborate PS of the present type. In the former ($J < J_{\mathrm{PS}}$), electrons are uniformly distributed. In the latter ($J < J_{\mathrm{PS}}$), however, the electron configuration is composed of a holon island ($\delta = 0$) and a Heisenberg sea ($\delta = 1$).

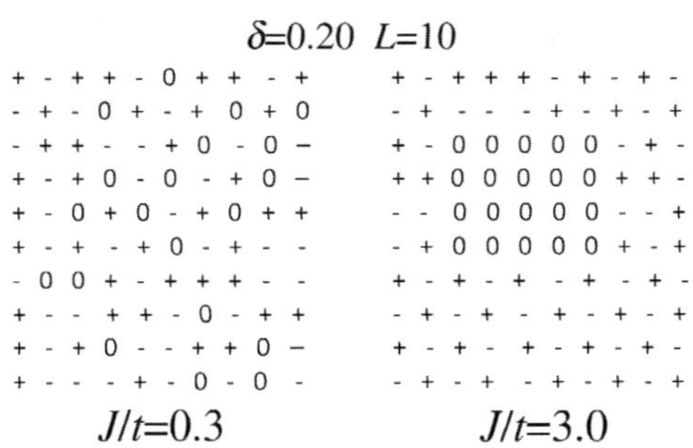

Figure 3. Snapshots of typical electron configurations in VMC sweeps with the optimized states are compared between $J/t = 0.3$ and 3.0. The characters '+', '−', and '0' indicate up- and down-spin sites, and empty site, respectively. For $J/t \gtrsim 2.5$, phase separation occurs with an AF order. The AF order vanishes for $J/t \lesssim 1.5$ at $\delta = 0.20$.

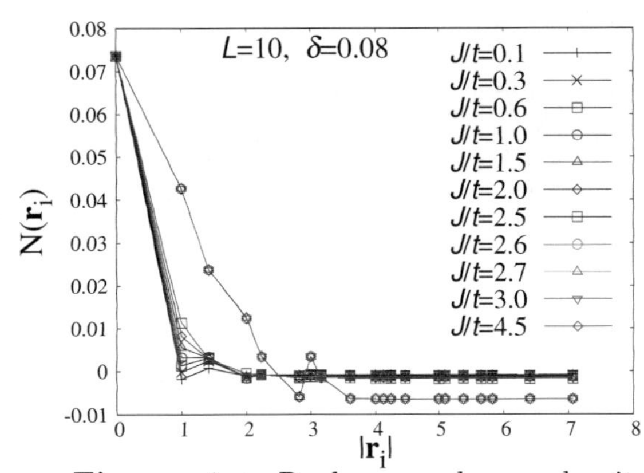

Figure 4. Real-space charge density correlation function for various values of J/t. Cold (warm) colors are used for $J < J_{\mathrm{PS}}$ ($J > J_{\mathrm{PS}}$).

In Fig. 4, we show the charge density correlation function,

$$N(\mathbf{r}_i) = \frac{1}{N_\mathrm{s}} \sum_j \langle n_j n_{j+i} \rangle - n^2 = \frac{1}{N_\mathrm{s}} \sum_j \langle h_j h_{j+i} \rangle - h^2. \tag{14}$$

where $n_j = \sum_\sigma n_{j\sigma}$, $n = N_\mathrm{e}/N_\mathrm{s}$, and $h = 1 - n$. For $J > J_\mathrm{PS}$, reflecting PS, $N(\mathbf{r}_i)$ exhibits appreciable values for $|\mathbf{r}_i| \le R$, compared to the value in the long-distance limit $N(\sqrt{2}L/2)$, where R is the longest intersite distance possible in the holon island: $R = \sqrt{10}$ (4×2 island) in the present case. R depends on the system size. On the other hand for $J < J_\mathrm{PS}$, $N(\mathbf{r}_i)$ converges to the long-distance value for $|\mathbf{r}_i| > D$, where D is determined by the length of correlation factors: $D = 1$ (nearest neighbors) in the present case. D will not be varied even if L is varied. It means that the system is uniform for $J < J_\mathrm{PS}$.

4. Summary and discussions

We studied interhole and interelectron correlations, as well as the phase separation, by applying a VMC method with a wave function that can exhibit AF and d-SC orders to the square-lattice t-J model. In this report, the intersite correlation factor is restricted to that between nearest-neighbor sites. It is found that (i) the interhole correlation remains repulsive up to the transition point to PS $J = J_\mathrm{PS} \sim 2.5t$ both for $\delta = 0.08$ and 0.20. This contrasts with an early intuitive arguement of SC in cuprates by coherent motion of hole pairs. In this regime, both AF and d-SC order become robust as J/t increases. (ii) A transition to PS is observed within a single wave function. For $J > J_\mathrm{PS}$, the state separates into the domains of $\delta = 0$ and 1 (Fig. 3).

In previous VMC studies [5, 9], the transition point to PS was estimated by comparing the variational energy of the uniform state with the exact estimation of the separated phase. Therefore, the transition point tends to be underestimated e.g. $J/t \sim 2$ for $\delta = 0.20$. There has been intensive debate as to whether the transition point to PS for $\delta \to 0$ is $J/t = 0$ or a finite small value, e.g. $J/t = 0.7$ [10]. In many related studies [11], charge compressibility becomes zero in appreciable ranges of δ near half filling, meaning that arbitrary electron densities in these ranges are allowed as separated phases. It is an important remaining problem whether or not these separated phases continue to what we treated in this report as J/t is increased.

References

[1] Emery V J, Kivelson S A and Lin H Q 1990 *Phys. Rev. Lett.* **64** 475
[2] Putikka W O, Luccini M U and Rice T M 1992 *Phys. Rev. Lett.* **68** 538
[3] Anderson P W 1987 *Science* **235** 1196; Zhang F C and Rice T M 1988 *Phys. Rev. B* **37** 3759(R)
[4] Hellberg C S and Manousakis E 1995 *Phys. Rev. B* **52** 4639
[5] Yokoyama H and Ogata M 1996 *J. Phys. Soc. Jpn.* **65** 3615
[6] Sato R and Yokoyama H 2016 *J. Phys. Soc. Jpn.* **85** 124707
[7] Giamarchi T and Lhuillier C 1991 *Phys. Rev. B* **43** 12943
[8] Sorella S 2001 *Phys. Rev. B* **64** 024512
[9] Valentí R and Gros C 1992 *Phys. Rev. Lett.* **68** 2402
[10] For instance, Lugas M, Spanu L. Becca F and Sorella S 2006 *Phys. Rev. B* **74** 165122
[11] As an early study, Himeda A and Ogata M 1999 Phys. Rev. B **60** R9935

Interplay between staggered flux and d-wave superconducting orders in t-t'-J model

Kenji Kobayashi[1], **Hisatoshi Yokoyama**[2] **and Yuta Toga**[3]

[1]Department of Natural Science, Chiba Institute of Technology, Shibazono, Narashino 275-0023, Japan
[2]Department of Physics, Tohoku University, Aoba-ku, Sendai, 980-8578, Japan
[3]ESICMM, National Institute for Materials Science, Tsukuba 305-0047, Japan

E-mail: kenji.kobayashi@it-chiba.ac.jp

Abstract. With cuprate superconductors in mind, we check whether the d-wave superconducting (dSC) state coexists with or excludes the staggered flux (SF) state, which was revealed to be a strong candidate for the pseudogap state. To this end, we use a variational Monte Carlo method for the square-lattice t-J model with diagonal transfer t'. We construct a trial wave function which can simultaneously represent dSC and SF orders for a continuous description of their interplay. Furthermore, a band renormalization effect is introduced into both orders independently. It is found that the SF state does not coexist with dSC and is unstable for a wide area in the parameter space. However, a strange state having both SF and antiferromagnetic orders is found to be stable for very low doping rate.

1. Introduction

In research for understanding superconductivity (SC) in high-T_c cuprates, what is the origin of pseudogap phenomena, which behave like a shadow of SC, has been a long-standing problem [1, 2]. A staggered flux (SF) state is one of the likely candidates for the pseudogap state proposed so far: Using a variational Monte Carlo (VMC) method, it was revealed that the SF state has properties consistent with the behavior of the pseudogap phase in cuprates and is moderately stable for the t-J model [3], and for the Hubbard model with diagonal transfer t' [4]. In the latter, it is shown that the SF state becomes more stable than the normal state (projected Fermi sea) for a strongly correlated and underdoped area, although the $d_{x^2-y^2}$-wave SC (dSC) state always has lower energy. Especially, for $t'/t \sim -0.3$, the area where the SF state is more stable than the normal state expands in a wide doping range, and the energy of SF state approaches to that of dSC. Meanwhile, coexistence or exclusivity of the dSC gap and the pseudogap has been debated, and some recent experiments reported coexistence of the two gaps [5].

Thus, it is important to check whether the SF and dSC orders coexist or are mutually exclusive. In the preceding study [6], we directly checked this problem using a mixed state of the SF and dSC orders, by applying a VMC method to the Hubbard (t-t'-U) model. It revealed that the SF order competes with the dSC order and is not stabilized. In Ref. [6], however, band renormalization effect (BRE) is introduced only into the SC order. If BRE is introduced into SC and SF orders independently, the result will possibly change. Another problem is that the Hubbard model is possibly lacking in some important elements for describing cuprates because the antiferromagnetic (AF) state (or phase separation) becomes stable against dSC for a wide area in the model-parameter space [7]. In this context, it is significant to check whether or not the t-J model exhibits essentially similar behavior as an effective model in a large-U/t regime.

In this work, we will check these points by applying a VMC method with a mixed state of SF and dSC orders, which includes individual BRE for both orders, to a t-t'-J model with three-site hopping $J^{(3)}$. On the basis of the data of the optimized states, we found that the SF and dSC orders are mutually exclusive in a wide area of the parameter space, and the SF order does not appear in the ground state. As an exception, for $J^{(3)}/J = -1$ near half-filling, a coexisting state of SF and AF becomes stable and dSC order vanishes. This extraordinary state has properties relevant to the present topics.

2. Method

We consider a t-J model on the square lattice with the diagonal transfer (t') and the three-site ($J^{(3)}$) terms:

$$\begin{aligned}\mathcal{H}_{\rm tot} &= \mathcal{H}_t + \mathcal{H}_{t'} + \mathcal{H}_J + \mathcal{H}_{J^{(3)}} \\ &= -t \sum_{\langle \mathbf{rr}' \rangle, \sigma} \left(\hat{c}^\dagger_{\mathbf{r},\sigma} \hat{c}_{\mathbf{r}',\sigma} + \mathrm{h.c.} \right) - t' \sum_{(\mathbf{rr}'), \sigma} \left(\hat{c}^\dagger_{\mathbf{r},\sigma} \hat{c}_{\mathbf{r}',\sigma} + \mathrm{h.c.} \right) + J \sum_{\langle \mathbf{rr}' \rangle} \left(\mathbf{S}_\mathbf{r} \cdot \mathbf{S}_{\mathbf{r}'} - \frac{1}{4} n_\mathbf{r} n_{\mathbf{r}'} \right) \\ &\quad - \frac{J^{(3)}}{4} \sum_{\mathbf{r},\sigma} \sum_{\boldsymbol{\tau} \neq \boldsymbol{\tau}'} \left(\hat{c}^\dagger_{\mathbf{r},-\sigma} \hat{c}_{\mathbf{r},-\sigma} \hat{c}^\dagger_{\mathbf{r}+\boldsymbol{\tau},\sigma} \hat{c}_{\mathbf{r}+\boldsymbol{\tau}',\sigma} + \hat{c}^\dagger_{\mathbf{r}+\boldsymbol{\tau},-\sigma} \hat{c}_{\mathbf{r},-\sigma} \hat{c}^\dagger_{\mathbf{r},\sigma} \hat{c}_{\mathbf{r}+\boldsymbol{\tau}',\sigma} \right), \end{aligned} \quad (1)$$

where $\hat{c}^\dagger_{\mathbf{r},\sigma} = c^\dagger_{\mathbf{r},\sigma}(1 - n_{\mathbf{r},-\sigma})$, $c^\dagger_{\mathbf{r},\sigma}$ creates an electron of spin σ at site $\mathbf{r}$, $n_\mathbf{r} = \sum_\sigma n_{\mathbf{r},\sigma} = \sum_\sigma c^\dagger_{\mathbf{r},\sigma} c_{\mathbf{r},\sigma}$, and $\mathbf{S}_\mathbf{r} = \frac{1}{2} \sum_{\sigma,\sigma'} c^\dagger_{\mathbf{r},\sigma} \boldsymbol{\sigma}_{\sigma,\sigma'} c_{\mathbf{r},\sigma'}$ with $\boldsymbol{\sigma} = (\sigma_x, \sigma_y, \sigma_z)$ being the vector of Pauli matrices. $\langle \mathbf{rr}' \rangle$ and $(\mathbf{rr}')$ denote indices of nearest-neighbor and diagonal-neighbor pairs, respectively. Note that $\mathcal{H}_{\rm tot}$ works in the space with no double occupancy. The three-site term $\mathcal{H}_{J^{(3)}}$, in which $\boldsymbol{\tau}, \boldsymbol{\tau}' = (\pm 1, 0), (0, \pm 1)$ with the unit of length being the lattice constant, is often disregarded but may be significant in comparing with the Hubbard model; $\mathcal{H}_{J^{(3)}}$ is the same order as $\mathcal{H}_J$ in the t/U expansion with $J = 4t^2/U$ and $J^{(3)} = J$, corresponding to electron-doped cases [8]. On the other hand, it was argued that $J^{(3)}/J$ could be negative for hole-doped cases [9, 10]. Therefore, in this work, we treat $J^{(3)}$ as a model parameter independent of J.

To this model, we apply a VMC method using the correlated measurement for efficient optimization [11]. We use a Jastrow-type trial wave function, $\Psi = \mathcal{P}^{\rm C} \mathcal{P}^{\rm S} \Phi$, where $\mathcal{P}^{\rm C}$ and $\mathcal{P}^{\rm S}$ are intersite correlation factors of Jastrow-type for charge and spin, respectively:

$$\mathcal{P}^{\rm C} = \prod_{\mathbf{rr}'} \prod_{\sigma\sigma'} \left[1 - \left\{ 1 - \eta^{\rm C} \left(|\mathbf{r} - \mathbf{r}'| \right) \right\} n_{\mathbf{r},\sigma} n_{\mathbf{r}',\sigma'} \right], \quad (2)$$

and

$$\mathcal{P}^{\rm S} = \prod_{\mathbf{rr}'} \left[1 - \left\{ 1 - \eta^{\rm S} \left(|\mathbf{r} - \mathbf{r}'| \right) \right\} S^z_\mathbf{r} S^z_{\mathbf{r}'} \right], \quad (3)$$

with $S^z_\mathbf{r} = \frac{1}{2} (n_{\mathbf{r},\uparrow} - n_{\mathbf{r},\downarrow})$. To exclude double occupancy, we impose the condition $\eta^{\rm C}(0) = 0$. For a short-range part of charge correlations, we consider up to the fourth-neighbor sites as illustrated in Fig. 1(b): $\eta^{\rm C}_\ell \equiv \eta^{\rm C} (|\mathbf{r} - \mathbf{r}'|)$ for $1 \leq \ell \leq 4$ is an independent variational parameter. For long-range correlations, we assume a Tomonaga-Luttinger-liquid form [8, 12],

$$\eta^{\rm C} \left(|\mathbf{r} - \mathbf{r}'| \right) = \left[\frac{L}{\pi} \sqrt{\sin^2 \left(\frac{\pi x}{L} \right) + \sin^2 \left(\frac{\pi y}{L} \right)} \right]^\nu \quad \text{for} \quad \sqrt{5} < |\mathbf{r} - \mathbf{r}'|, \quad (4)$$

where $\mathbf{r} - \mathbf{r}' = (x, y)$ and L is a linear dimension of the square lattice composed of $L \times L$ ($= N_{\rm s}$) sites. A variational parameter ν in Eq. (4) works as repulsive (attractive) correlation for $\nu > 0$

($\nu < 0$). For spin part $\mathcal{P}^{\mathrm{S}}$, correlations up to the tenth-neighbor sites are taken into account: $\eta_\ell^{\mathrm{S}} \equiv \eta^{\mathrm{S}}(|\mathbf{r}-\mathbf{r}'|)$ for $1 \le \ell \le 10$ is an independent variational parameter, and spin correlations for $11 \le \ell$ are disregarded.

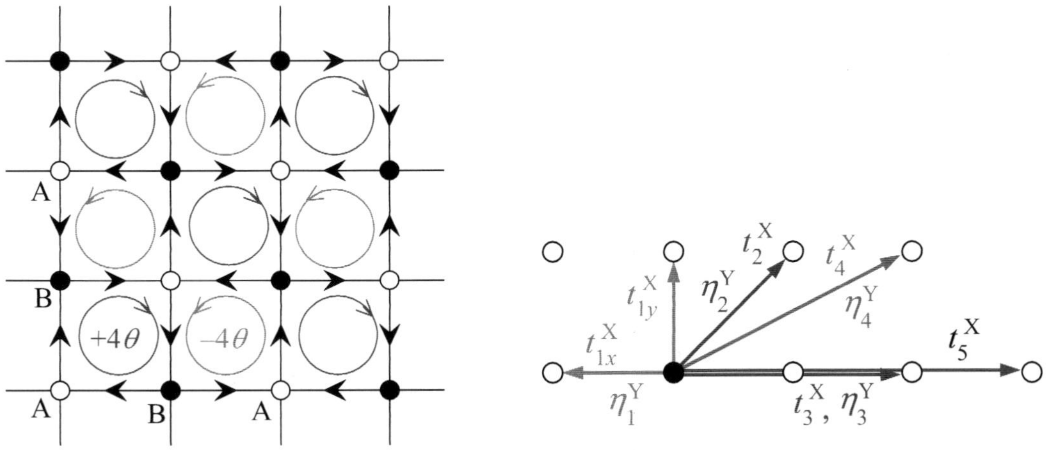

(a) Staggered Flux (b) Definition of variational parameters

Figure 1. (a) Schematic figure of the staggered flux state. Arrows denote the local circular current, which flows in alternate directions in each plaquette. (b) Definition of variational parameters, which are classified by intersite distance: Band parameters t_ℓ^{X} for optimizing $\varepsilon^{\mathrm{X}}(\mathbf{k})$ with X = SF or SC, and intersite correlation parameters η_ℓ^{Y} in $\mathcal{P}^{\mathrm{Y}}$ with Y = C or S.

For the one-body part Φ, we adopt a mixed state of SF and dSC for a continuous description of their interplay. For N_{e} electrons, Φ with a nearest-neighbor-pairing gap Δ_d is written as [13],

$$\Phi = \left(\sum_{\mathbf{k}} \sum_{\lambda=\pm} \varphi_{\mathbf{k}} \alpha_{\lambda,\mathbf{k},\uparrow}^{\dagger} \alpha_{\lambda,-\mathbf{k},\downarrow}^{\dagger} \right)^{\frac{N_{\mathrm{e}}}{2}} |0\rangle , \tag{5}$$

where

$$\varphi_{\mathbf{k}} = \frac{\Delta(\mathbf{k})}{\varepsilon^{\mathrm{SC}}(\mathbf{k}) - \zeta + \sqrt{(\varepsilon^{\mathrm{SC}}(\mathbf{k}) - \zeta)^2 + |\Delta(\mathbf{k})|^2}} , \tag{6}$$

with $\Delta(\mathbf{k}) = \Delta_d(\cos k_x - \cos k_y)$.

$\alpha_{\lambda,\mathbf{k},\sigma}^{\dagger}$ in Eq. (5) is a creation operator of a quasi-particle in SF state that is constructed by diagonalizing a SF Hamiltonian $\mathcal{H}^{\mathrm{SF}}$ as follows. The noninteracting SF Hamiltonian with transfers up to the fifth-neighbor sites in the sublattice (A, B) representation, shown in Fig. 1, is written as [4],

$$\mathcal{H}^{\mathrm{SF}}(\theta) = \sum_{\mathbf{k},\sigma} \begin{pmatrix} a_{\mathbf{k},\sigma}^{\dagger} & b_{\mathbf{k},\sigma}^{\dagger} \end{pmatrix} \begin{pmatrix} \varepsilon_2^{\mathrm{SF}}(\mathbf{k}) + \varepsilon_3^{\mathrm{SF}}(\mathbf{k}) & u(\mathbf{k},\theta) \\ u^*(\mathbf{k},\theta) & \varepsilon_2^{\mathrm{SF}}(\mathbf{k}) + \varepsilon_3^{\mathrm{SF}}(\mathbf{k}) \end{pmatrix} \begin{pmatrix} a_{\mathbf{k},\sigma} \\ b_{\mathbf{k},\sigma} \end{pmatrix} , \tag{7}$$

where θ is the phase of local circular current in SF state, $a_{\mathbf{k},\sigma}$ ($b_{\mathbf{k},\sigma}$) is a Fourier transformation of an annihilation operator on a sublattice A (B),

$$u(\mathbf{k},\theta) = -2t_{1x}^{\mathrm{SF}} e^{i\theta} \cos k_x - 2t_{1y}^{\mathrm{SF}} e^{-i\theta} \cos k_y + \varepsilon_4^{\mathrm{SF}}(\mathbf{k}) + \varepsilon_5^{\mathrm{SF}}(\mathbf{k}) , \tag{8}$$

and

$$\begin{aligned}
\varepsilon_2^{\rm X}(\mathbf{k}) &= -4t_2^{\rm X}\cos k_x \cos k_y \;, \\
\varepsilon_3^{\rm X}(\mathbf{k}) &= -2t_3^{\rm X}\left[\cos(2k_x)+\cos(2k_y)\right], \\
\varepsilon_4^{\rm X}(\mathbf{k}) &= -4t_4^{\rm X}\left[\cos(2k_x)\cos(k_y)+\cos(k_x)\cos(2k_y)\right], \\
\varepsilon_5^{\rm X}(\mathbf{k}) &= -2t_5^{\rm X}\left[\cos(3k_x)+\cos(3k_y)\right],
\end{aligned} \tag{9}$$

with X = SF. Note that the phase θ is accompanied by only the nearest-neighbor transfer $t_1^{\rm SF}$ but not by $t_2^{\rm SF}$ - $t_5^{\rm SF}$, and asymmetry in $t_1^{\rm SF}$ is also introduced between x and y directions ($t_{1x}^{\rm SF}$ and $t_{1y}^{\rm SF}$) to deal with the Pomeranchuk instability [14].

By applying a Bogoliubov transformation,

$$a_{\mathbf{k},\sigma} = \frac{\Gamma(\mathbf{k},\theta)}{\sqrt{2}}\left(-\alpha_{-,\mathbf{k},\sigma}+\alpha_{+,\mathbf{k},\sigma}\right), \qquad b_{\mathbf{k},\sigma} = \frac{1}{\sqrt{2}}\left(\alpha_{-,\mathbf{k},\sigma}+\alpha_{+,\mathbf{k},\sigma}\right), \tag{10}$$

with

$$\Gamma(\mathbf{k},\theta) = \frac{u(\mathbf{k},\theta)}{S(\mathbf{k},\theta)}, \qquad S(\mathbf{k},\theta) = \sqrt{u(\mathbf{k},\theta)\; u^*(\mathbf{k},\theta)}\;, \tag{11}$$

$$\begin{aligned}
u(\mathbf{k},\theta)\; u^*(\mathbf{k},\theta) = {}& 4\left[\left(t_{\rm 1x}^{\rm SF}\cos k_x\right)^2 + \left(t_{\rm 1y}^{\rm SF}\cos k_y\right)^2 + 2t_{\rm 1x}^{\rm SF}t_{\rm 1y}^{\rm SF}\cos(2\theta)\cos k_x\cos k_y\right] \\
& -4\cos\theta\left(t_{\rm 1x}^{\rm SF}\cos k_x + t_{\rm 1y}^{\rm SF}\cos k_y\right)\left[\varepsilon_4^{\rm SF}(\mathbf{k})+\varepsilon_5^{\rm SF}(\mathbf{k})\right] + \left[\varepsilon_4^{\rm SF}(\mathbf{k})+\varepsilon_5^{\rm SF}(\mathbf{k})\right]^2,
\end{aligned} \tag{12}$$

$\mathcal{H}^{\rm SF}(\theta)$ in Eq. (7) is diagonalized as

$$\mathcal{H}^{\rm SF}(\theta) = \sum_{\mathbf{k},\sigma}\sum_{\lambda=\pm}\varepsilon_\lambda^{\rm SF}(\mathbf{k},\theta)\;\alpha_{\lambda,\mathbf{k},\sigma}^\dagger\alpha_{\lambda,\mathbf{k},\sigma}\;, \tag{13}$$

where the upper and lower band dispersions are given by,

$$\varepsilon_\pm^{\rm SF}(\mathbf{k},\theta) = \pm S(\mathbf{k},\theta) + \varepsilon_2^{\rm SF}(\mathbf{k}) + \varepsilon_3^{\rm SF}(\mathbf{k}). \tag{14}$$

Corresponding creation operators of the quasi-particle for the upper and lower band, which constitute Cooper pairs in Eq. (5), are obtained by the inverse transformation of Eq. (10) as

$$\alpha_{\pm,\mathbf{k},\sigma}^\dagger = \frac{1}{\sqrt{2}}\left(\pm\Gamma(\mathbf{k},\theta)a_{\mathbf{k},\sigma}^\dagger + b_{\mathbf{k},\sigma}^\dagger\right), \tag{15}$$

where the phase θ, which is introduced into Eq. (5) through $\alpha_{\lambda,\mathbf{k},\sigma}^\dagger$, is a variational parameter to be optimized together with the other parameters.

In addition, BRE owing to electron correlation is separately introduced into the SC dispersion different from the SF one. To this end, we assume that $\varepsilon^{\rm SC}(\mathbf{k})$ in Eq. (6) is fitted by another tight-binding band composed of transfers up to fifth-neighbor sites as shown in Fig. 1(b),

$$\varepsilon^{\rm SC}(\mathbf{k}) = -2(t_{1x}^{\rm SC}\cos k_x + t_{1y}^{\rm SC}\cos k_y) + \sum_{i=2}^{5}\varepsilon_i^{\rm SC}(\mathbf{k}), \tag{16}$$

where $\varepsilon_i^{\rm SC}(\mathbf{k})$ is given in Eq. (9) with X = SC, and asymmetry in $t_1^{\rm SC}$ between x and y directions is introduced again for the Pomeranchuk instability.

Hopping parameters in $\varepsilon^{\rm SC}(\mathbf{k})$ and $\varepsilon_\pm^{\rm SF}(\mathbf{k},\theta)$ are optimized for BRE with some exceptions: $t_{1x}^{\rm X}$ (X = SF and SC) is fixed to t in Eq. (1) as the unit of energy, and $t_2^{\rm SF}$ and $t_3^{\rm SF}$ included in the diagonal elements in Eq. (7) are disregarded because of difficulties in the optimization procedure.

As a result, variational parameters to be optimized in Φ amount to Δ_d, ζ, θ, $t_{1y}^{\rm SC}$, $t_2^{\rm SC}$-$t_5^{\rm SC}$, $t_{1y}^{\rm SF}$, $t_4^{\rm SF}$, and $t_5^{\rm SF}$. Besides these eleven in Φ, variational parameters in the correlation factors are $\eta_\ell^{\rm C}$ for ℓ =1-4 and ν in $\mathcal{P}^{\rm C}$, and $\eta_\ell^{\rm S}$ for ℓ =1-10 in $\mathcal{P}^{\rm S}$. According to a standard procedure of VMC, we obtain the optimal set of the 26 parameters by minimizing the energy. Using the optimized wave functions, we calculate various quantities. Among them, we mainly discuss the following order parameters of respective orders. For dSC, we use the real-space $d_{x^2-y^2}$-wave SC correlation function for the nearest-neighbor pairing defined by,

$$P_d(\mathbf{r}) = \frac{1}{N_{\rm s}} \sum_{\mathbf{r}'} \sum_{\boldsymbol{\tau},\boldsymbol{\tau}'} \lambda_{xy}(\boldsymbol{\tau})\lambda_{xy}(\boldsymbol{\tau}')\langle \Delta^\dagger(\mathbf{r}'+\mathbf{r},\boldsymbol{\tau})\Delta(\mathbf{r}',\boldsymbol{\tau}')\rangle \ , \tag{17}$$

with

$$\Delta(\mathbf{r},\boldsymbol{\tau}) = \frac{1}{\sqrt{2}}\left(c_{\mathbf{r}\uparrow}c_{\mathbf{r}+\boldsymbol{\tau}\downarrow} - c_{\mathbf{r}\downarrow}c_{\mathbf{r}+\boldsymbol{\tau}\uparrow}\right) \ ,$$

$\lambda_{xy}(\boldsymbol{\tau}) = +1$ for $\boldsymbol{\tau} = (\pm 1, 0)$, and $\lambda_{xy}(\boldsymbol{\tau}) = -1$ for $\boldsymbol{\tau} = (0, \pm 1)$. Actually, we show $P_d(\mathbf{r})$ at the farthermost distance $\mathbf{r} = (L/2, L/2)$ as a measure of dSC order. For the SF phase, we employ the circular current defined as,

$$J_{\rm c}/t = \frac{1}{N_{\rm s}} \sum_{\mathbf{r}\in{\rm A},\sigma} \sum_{\boldsymbol{\tau}} \lambda_{xy}(\boldsymbol{\tau})\ {\rm Im}\langle c^\dagger_{\mathbf{r}+\boldsymbol{\tau},\sigma}c_{\mathbf{r},\sigma} - c^\dagger_{\mathbf{r},\sigma}c_{\mathbf{r}+\boldsymbol{\tau},\sigma}\rangle. \tag{18}$$

The staggered magnetization,

$$m_s = \frac{1}{N_{\rm s}}\left|\sum_{\mathbf{r}} e^{i\mathbf{Q}\cdot\mathbf{r}}\langle S_{\mathbf{r}}^z\rangle\right|, \tag{19}$$

with $\mathbf{Q} = (\pi, \pi)$ is also calculated as the order parameter of AF state.

As pilot calculations, J/t and t'/t are fixed at typical values of hole-doped cuprates, 0.3 and -0.3, respectively, and we compare the results of several values of $J^{(3)}/J$ for systems of $L = 10$ and $L = 12$ with periodic-antiperiodic boundary conditions.

3. Results and discussion

Let us begin with the expectation values of energy components (Fig. 2) and order parameters (Fig. 3), estimated by the optimized Ψ. In almost the whole range of the present calculations, the circular current $|J_{\rm c}|/t$ and staggered magnetization m_s vanish, but the pair correlation function $P_d(L/2, L/2)$ has finite values, and monotonically increases as δ increases within $\delta \le 0.2$. Thus the pure dSC is realized, and robust as compared to the strongly correlated Hubbard model. The SF order is totally suppressed in a wide range of model parameters. This is explained, by considering $E_{d\rm SC} < E_{\rm SF}$, as follows: The predominant dSC state opens a gap in the momentum space except for $k_x = k_y$, whereas the subordinate SF state needs to have a finite pocket Fermi surface (FS) around $\mathbf{k} = (\pi/2, \pi/2)$ [4]. Thus, the requirements of the two electronic states do not meet. Including the exceptional case discussed below, the dSC order and the SF and AF (type-II in Ref. [7]) orders are mutually exclusive. The overall feature is consistent with that in the Hubbard model of $U/t = 12$ [6].

An exception occurs for $J^{(3)}/J = -1$ and very low doping rate ($\delta \sim 0.03$), where extraordinary behavior appears both in energy components and order parameters, as seen in Figs. 2(c) and 3(c), respectively. A kink appear in each energy component, and $|J_{\rm c}|/t$ and m_s sharply rise in contrast with $P_d(L/2, L/2)$, which vanishes. This coexistent state of SF and AF orders is realized by combining reformation of the bands [Fig. 4(b)] with a variation in spin

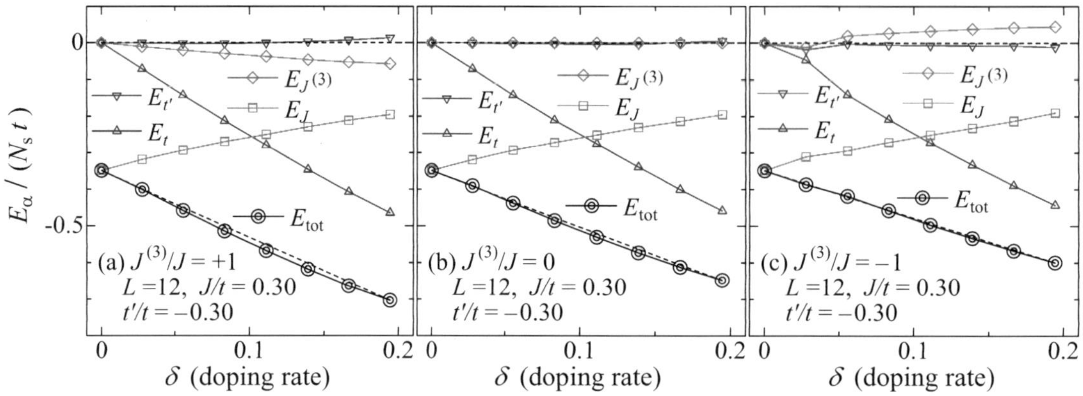

Figure 2. The total energy per site and its components, $E_\alpha/(N_{\rm s}t)$, are plotted as functions of doping rate δ for (a) $J^{(3)}/J = +1$, (b) $J^{(3)}/J = 0$, and (c) $J^{(3)}/J = -1$. Here, $E_\alpha = \langle\Psi|\mathcal{H}_\alpha|\Psi\rangle/\langle\Psi|\Psi\rangle$ with $\alpha =$ tot, t, t', J, $J^{(3)}$ as in Eq. (1). Symbols indicate raw VMC data for $L = 12$, and straight dashed lines connecting both ends of each $E_{\rm tot}$ curve are drawn as a guide to the eye.

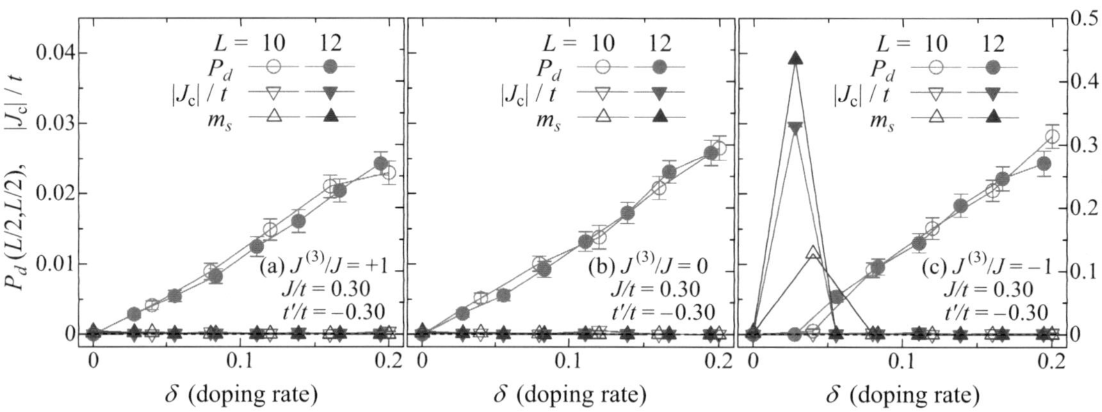

Figure 3. The expectation values of the order parameters, namely, $P_d(\mathbf{r})$ (circles) for dSC, $|J_{\rm c}|/t$ (inverted triangles) for SF, and m_s (upward triangles) for AF, are plotted as functions of δ for (a) $J^{(3)}/J = +1$, (b) $J^{(3)}/J = 0$, and (c) $J^{(3)}/J = -1$.

correlation, particularly, AF correlation of the nearest-neighbor factor $\eta_1^{\rm S}$ [Fig. 4(d)]. As a result, the sign of $E_{J^{(3)}}$ is reversed and $E_{t'}$ is reduced, stabilizing the coexistent state. Although the regime of this exception is narrow and special, we take notice of it because the realized state has features of our interest. We take up two points:

(1) In this state, the orders of SF and AF coexist. In studies for the Hubbard model with $t'/t = 0$ [15], the AF order excludes the SF order and never coexists. This is interpreted as follows: The pure metallic AF state for $t'/t = 0$ is classified as type-I, which shows a pocket FS around the antinodal $\mathbf{k} = (\pi, 0)$ [7]. In contrast, the pure metallic SF state has a pocket FS around $\mathbf{k} = (\pi/2, \pi/2)$ in the nodal direction [4]. Therefore, the FS position of the subordinate

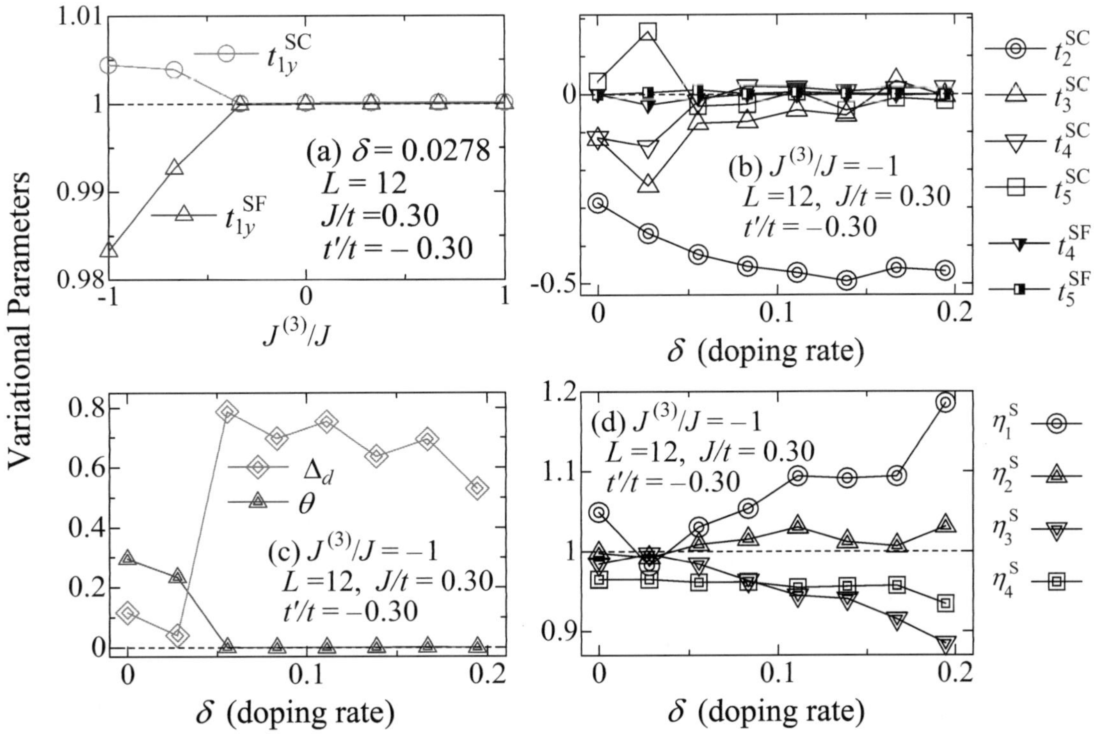

Figure 4. Optimized values of important variational parameters are shown. (a) $J^{(3)}/J$ dependence of $t_{1y}^{\rm SC}$ and $t_{1y}^{\rm SF}$ for $\delta = 0.0278$, where $t_{1x}^{\rm SC}/t = t_{1x}^{\rm SF}/t = 1$. In (b)–(d), we depict δ dependence of other parameters for $J^{(3)}/J = -1$: (b) Band parameters other than those in (a); (c) pairing gap Δ_d and SF phase θ; (d) spin correlation factors $\eta_\ell^{\rm S}$ in $\mathcal{P}^{\rm S}$ in a short-range part ($\ell = 1 - 4$).

SF state is incompatible with that of the predominant AF state. On the other hand in the present case of $t'/t = -0.3$, the pure metallic AF state is classified as type-II [7], which has a pocket FS around $\mathbf{k} = (\pi/2, \pi/2)$ in accordance with the SF state. The FS's in the two states probably overlap with each other, resulting in the coexistence of SF and type-II AF orders.

(2) As seen in Fig. 4(a), small Pomeranchuk instability or nematicity—asymmetry between x and y directions—appears in this coexistent state, in contrast with the SC state. It is already known that Pomeranchuk instability vanishes in the dSC state [16, 17] but appears in the AF state up to $|1 - t_{1y}^{\rm AF}/t_{1x}^{\rm AF}| \sim 0.1$ [18]. We have confirmed that this instability remains $|1 - t_{1y}^{\rm SF}/t_{1x}^{\rm SF}| \sim 0.02$ for $\delta < 0.17$ in metastable pure SF states found (not shown). Thus, the 4-fold rotational symmetry is subtly broken in the optimized SF state. If the SF state gives a sizable contribution to the pseudogap phase in finite temperatures, this symmetry breaking will be observed.

4. Summary

We have studied the coexistence or exclusivity of the SF and dSC orders in the t-t'-J model on the bases of VMC calculations. Using a mixed state which can simultaneously represent both orders, we found that, in almost all cases, the SF order is excluded in the more stable dSC state, and the two orders never coexist. This feature is consistent with that in the Hubbard model.

An exception is found for $J^{(3)}/J = -1$ near half filling, in which the optimized state has finite SF and AF orders but no dSC order. By analyzing this state, two interesting facts are found: (1) type-II AF and SF orders coexist, and (2) Pomeranchuk instability appears in the SF state, similarly to the AF state. Detailed description will be given elsewhere.

References

[1] Yoshida T, Hashimoto M, Vishik I M, Shen Z -X and Fujimori A 2012 *J. Phys. Soc. Jpn.* **81** 011006
[2] Fujita K, Schmidt A R, Kim E -A, Lawler M J, Lee D H, Davis J C, Eisaki H and Uchida S 2012 *J. Phys. Soc. Jpn.* **81** 011005
[3] Ivanov D A and Lee P A 2003 *Phys. Rev. B* **68** 132501
[4] Yokoyama H, Tamura S and Ogata M 2016 *J. Phys. Soc. Jpn.* **85** 124707
[5] For instance, Liu Y H, Toda Y, Shimatake K, Momono N, Oda M and Ido M 2008 *Phys. Rev. Lett.* **101** 137003; Uykur E, Tanaka K, Masui T, Miyasaka S and Tajima S 2013 *J. Phys. Soc. Jpn.* **82** 033701; Yoshida T *et al.* 2016 *Phys. Rev. B* **93** 014513
[6] Kobayashi K and Yokoyama H 2017 *J. Phys.: Conf. Series* **871** 012031
[7] Sato R and Yokoyama H 2016 *J. Phys. Soc. Jpn.* **85** 074701
[8] Yokoyama H and Ogata M 1996 *J. Phys. Soc. Jpn.* **65** 3615
[9] Batista C D and Aligia A A 1993 *Phys. Rev. B* **48** 4212; Erratum 1994 *Phys. Rev. B* **49** 6436
[10] Matsukawa H and Okamoto K 1996 *Czech. J. Phys. Suppl.* **46** 2647
[11] Umrigar C J, Wilson K G and Wilkins J W 1988 *Phys. Rev. Lett.* **60** 1719
[12] Valentí R and Gros C 1992 *Phys. Rev. Lett.* **68** 2402; Erratum 1992 *Phys. Rev. Lett.* **69** 996
[13] Giamarchi T and Lhuillier C 1991 *Phys. Rev. B* **43** 12943
[14] Yamase H, Oganesyan V and Metzner W 2005 *Phys. Rev. B* **72** 035114
[15] Toga Y and Yokoyama H 2016 *Phys. Procedia* **81** 13 and unpublished
[16] Yamase H and Kohno H 2000 *J. Phys. Soc. Jpn.* **69** 2151
[17] Edegger B, Muthukumar V N and Gros C 2006 *Phys. Rev. B* **74** 165109
[18] Kobayashi K and Yokoyama H 2014 *JPS Conf. Proc.* **1** 012120

IOP Conf. Series: Journal of Physics: Conf. Series **1054** (2018) 012017 doi:10.1088/1742-6596/1054/1/012017

Optimized wave function by kinetic renormalization effect in strongly correlated region of the three-band Hubbard model

[1]**Takashi Yanagisawa**, [1]**Izumi Hase**, [2]**Mitake Miyazaki**, [1]**Kunihiko Yamaji**

[1]Electronics and Photonics Research Institute, National Institute of Advanced Industrial Science and Technology (AIST), Tsukuba Central 2, 1-1-1 Umezono, Tsukuba 305-8568, Japan
[2]Hakodate National College of Technology, 14-1 Tokura, Hakodate 042-8501, Japan

E-mail: t-yanagisawa@aist.go.jp

Abstract. We investigate the ground state of the two-dimensional single-band Hubbard model and three-band Hubbard model (d-p model) by using a variational Monte Carlo method. We employ an optimization variational Monte Carlo method. In this method the many-body wave function is improved and optimized by introducing variational parameters in the form: $\psi = \exp(-S)\psi_0$ where ψ_0 is a wave function, for example, the Gutzwiller function, and S is an operator used to include the many-body effect.

The strong magnetic correlation is induced by the on-site Coulomb repulsive interaction U. The magnetic correlation plays a significant role in correlated electron systems as an origin of electron pairing as well as a main competitor of superconductivity. High-temperature superconductivity occurs in the strongly correlated region where U is greater than the bandwidth. We investigate the stability of antiferromagnetic state in the three-band d-p model.

1. Introduction

It is important to clarify the mechanism of high-temperature superconductivity. The physics of high-temperature superconductivity has been studied intensively since the discovery of cuprate high-temperature superconductors[1]. The electron correlation plays an important role for the appearance of high-temperature superconductivity, because the parent materials without carries are Mott insulators and it has been established that the Cooper pairs of cuprate superconductors have the *d*-wave symmetry.

The CuO_2 plane is commonly contained in high-temperature cuprates, where the CuO_2 plane consists of oxygen atoms and copper atoms. The electronic model for this plane is the called d-p model (or three-band Hubbard model)[2, 3, 4, 5, 6, 7, 8, 9, 10, 11, 12, 13, 14, 15, 16, 17, 18]. It appears very hard to elucidate the phase diagram of the d-p model because of strong correlation between electrons. We often examine simplified models such as the two-dimensional single-band Hubbard model[19, 20, 21, 22, 23, 24, 25, 26, 27, 28, 29, 30, 31, 32, 33, 34, 35, 36, 37] or ladder model[38, 39, 40, 41] to make clear the phase diagram of correlated electron systems.

A variational Monte Carlo method is a useful method to investigate electronic properties of strongly correlated electron systems by calculating the expectation values numerically. A variational wave function can be improved by introducing new variational parameters to control the electron correlation. We have proposed wave functions that are optimized by multiplying an initial wave function by $\exp(-S)$-type operators[37, 42], where S is a suitable correlation operator. The Gutzwiller function is also written in this form. An optimization process is performed in a systematic way by multiplying by the exponential-type operators repeatedly[42]. The ground-state energy is lowered considerably by using this type of wave functions[37].

2. Hamiltonian

The three-band model that explicitly includes oxygen p and copper d orbitals contains the parameters U_d, U_p, t_{dp}, t_{pp}, t'_d, ϵ_d and ϵ_p. Our study is within the hole picture and the Hamiltonian is written as

$$\begin{aligned} H_{dp} &= \epsilon_d \sum_{i\sigma} d^\dagger_{i\sigma} d_{i\sigma} + \epsilon_p \sum_{i\sigma} (p^\dagger_{i+\hat{x}/2\sigma} p_{i+\hat{x}/2\sigma} + p^\dagger_{i+\hat{y}/2\sigma} p_{i+\hat{y}/2\sigma}) \\ &+ t_{dp} \sum_{i\sigma} [d^\dagger_{i\sigma}(p_{i+\hat{x}/2\sigma} + p_{i+\hat{y}/2\sigma} - p_{i-\hat{x}/2\sigma} - p_{i-\hat{y}/2\sigma}) + \text{h.c.}] \\ &+ t_{pp} \sum_{i\sigma} [p^\dagger_{i+\hat{y}/2\sigma} p_{i+\hat{x}/2\sigma} - p^\dagger_{i+\hat{y}/2\sigma} p_{i-\hat{x}/2\sigma} \\ &- p^\dagger_{i-\hat{y}/2\sigma} p_{i+\hat{x}/2\sigma} + p^\dagger_{i-\hat{y}/2\sigma} p_{i-\hat{x}/2\sigma} + \text{h.c.}] \\ &+ t'_d \sum_{\langle\langle ij\rangle\rangle\sigma} \epsilon_{ij} (d^\dagger_{i\sigma} d_{j\sigma} + \text{h.c.}) + U_d \sum_i d^\dagger_{i\uparrow} d_{i\uparrow} d^\dagger_{i\downarrow} d_{i\downarrow}. \end{aligned} \quad (1)$$

$d_{i\sigma}$ and $d^\dagger_{i\sigma}$ represent the operators for the d hole. $p_{i\pm\hat{x}/2\sigma}$ and $p^\dagger_{i\pm\hat{x}/2\sigma}$ denote the operators for the p holes at the site $R_{i\pm\hat{x}/2}$, and in a similar way $p_{i\pm\hat{y}/2\sigma}$ and $p^\dagger_{i\pm\hat{y}/2\sigma}$ are defined. t_{dp} is the transfer integral between adjacent Cu and O orbitals and t_{pp} is that between nearest p orbitals. $\langle\langle ij\rangle\rangle$ denotes a next nearest-neighbor pair of copper sites. t'_d was introduced (as shown in Fig.1) to reproduce the Fermi surface[43] reported in several cuprate superconductors such as $Bi_2Sr_2CaCu_2O_{8+\delta}$[44] and $Tl_2ba_2CuO_{6+\delta}$[45]. ϵ_{ij} takes the values ± 1 according to the sign of the transfer integral between next nearest-neighbor d orbitals. U_d is the strength of the on-site Coulomb repulsion between d holes. In this paper we neglect U_p among p holes because U_p is small compared to U_d[46, 47, 48, 49, 50]. U_p is not important in the low-doping region where the p-hole correlation effect is small. The values of band parameters were estimated as, for example, $U_d = 10.5$, $U_p = 4.0$ and $U_{dp} = 1.2$ in eV[47] where U_{dp} is the nearest-neighbor Coulomb interaction between holes on adjacent Cu and O orbitals. We neglect U_{dp} because U_{dp} is small compared to U_d. We use the notation $\Delta_{dp} = \epsilon_p - \epsilon_d$. The number of sites is denoted as N, and the total number of atoms is $N_a = 3N$. The energy unit is given by t_{dp}.

The single-band Hubbard model can be regarded as a simplified version of the three-band d-p model. This model has been studied more intensively. The Hamiltonian is given by

$$H = \sum_{ij\sigma} t_{ij} c^\dagger_{i\sigma} c_{j\sigma} + U \sum_i n_{i\uparrow} n_{i\downarrow}, \quad (2)$$

where $\{t_{ij}\}$ are transfer integrals and U is the on-site Coulomb energy. The transfer integral t_{ij} for nearest-neighbor pairs $\langle ij\rangle$ is given as $t_{ij} = -t$ and that for next-nearest neighbor pair $\langle\langle ij\rangle\rangle$ is $t_{ij} = -t'$. Otherwise, t_{ij} vanishes. The number of sites is denoted as N and the number of electrons is as N_e. The energy unit is given by t. $n_{i\sigma}$ is the number operator: $n_{i\sigma} = c^\dagger_{i\sigma} c_{i\sigma}$. The second term in the Hamiltonian represents the on-site repulsive interaction between electrons.

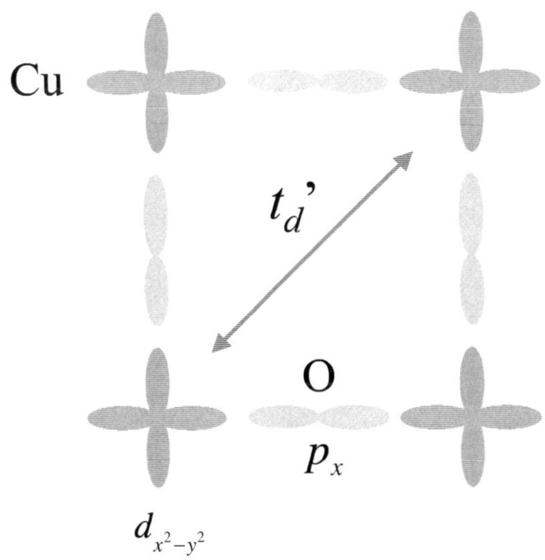

Figure 1. The transfer integral t'_d in the CuO_2 plane. t'_d is the transfer integral between next nearest-neighbor copper sites. We can also consider the transfer integral t'_{pp} between next nearest-neighbor oxygen atoms.

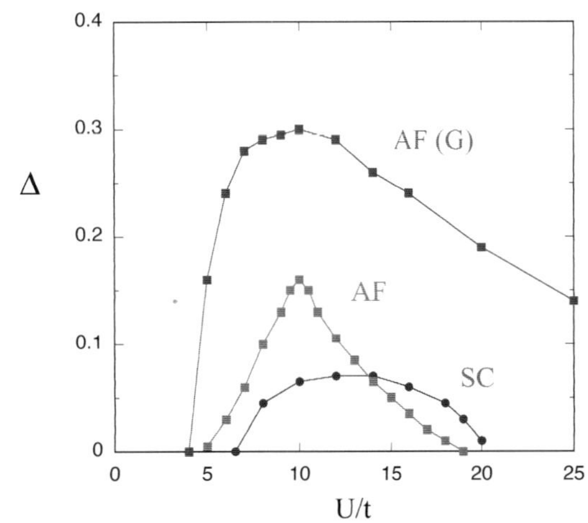

Figure 2. AF and SC order parameters as a function of U/t for the two-dimensional Hubbard model on a 10×10 lattice[37]. The upper curve AF(G) indicates the result obtained by the Gutzwiller function, and those for AF and SC are obtained using ψ_λ. We expect a similar phase diagram for the d-p model.

3. Optimized wave functions

The Gutzwiller-projected wave function is given as $\psi_G = P_G\psi_0$, where P_G is the Gutzwiller operator to control the double occupancy of d holes:

$$P_G = \prod_i [1-(1-g)n_{di\uparrow}n_{di\downarrow}] = \exp\left(-\alpha\sum_i n_{di\uparrow}n_{di\downarrow}\right), \tag{3}$$

where $\alpha = \ln(1/g)$. g is a variational parameter in the range from 0 to unity. $n_{di\sigma} = d^\dagger_{i\sigma}d_{i\sigma}$ is the number operator for d holes. ψ_0 is the Fermi sea where the lowest band is occupied up to the Fermi energy μ.

The Gutzwiller wave function is improved in the following way[37, 42, 51, 52, 53, 54, 55]:

$$\psi_\lambda = \exp(-\lambda K)\psi_G, \tag{4}$$

where λ is a variational parameter and K is the kinetic part of the Hamiltonian. The expectation values are evaluated by using the auxiliary field method[42, 56].

The correlated superconducting (SC) state is represented by the projected-BCS wave function. This is given as

$$\psi_S = P_{N_e}P_G\psi_{BCS}, \tag{5}$$

where ψ_{BCS} indicates the BCS wave function:

$$\psi_{BCS} = \prod_{\mathbf{k}}(u_{\mathbf{k}} + v_{\mathbf{k}}\alpha^\dagger_{\mathbf{k}\uparrow}\alpha^\dagger_{-\mathbf{k}\downarrow})|0\rangle. \tag{6}$$

$\alpha^\dagger_{\mathbf{k}\sigma}$ indicates the creation operator of the state in the lowest band with the momentum $\mathbf{k}$ which is represented by a linear combination of d and p electron operators $d^\dagger_{\mathbf{k}\sigma}$ and $p^\dagger_{\mathbf{k}\sigma}$. The BCS

IOP Conf. Series: Journal of Physics: Conf. Series **1054** (2018) 012017 doi:10.1088/1742-6596/1054/1/012017

parameters $u_{\mathbf{k}}$ and $v_{\mathbf{k}}$ are given by the ratio $v_{bfk}/u_{\mathbf{k}} = \Delta_{\mathbf{k}}/(\xi_{\mathbf{k}} + \sqrt{\xi_{\mathbf{k}}^2 + \Delta_{\mathbf{k}}^2})$ where $\xi_{\mathbf{k}}$ is the dispersion relation of the lowest band. We assume the d-wave symmetry $\Delta_{\mathbf{k}} = \Delta(\cos k_x - \cos k_y)$ and we regard Δ as the superconducting gap. Δ is a variational parameter that is optimized to give the lowest ground energy. P_{N_e} is a projection operator which extracts only the states with a fixed total hole number.

It is convenient to perform the particle-hole transformation for down-spin holes[37, 52] in multiplying by the operator $\exp(-\lambda K)$:

$$\psi_\lambda = \exp(-\lambda K)P_G\psi_{BCS} = \exp(-\lambda K)P_G\prod_{\mathbf{k}}(u_{\mathbf{k}}\beta_{\mathbf{k}}^\dagger + v_{\mathbf{k}}\alpha_{\mathbf{k}}^\dagger)|\tilde{0}\rangle, \tag{7}$$

where $\beta_{\mathbf{k}}^\dagger = \alpha_{-\mathbf{k}\downarrow}$ and $\alpha_{\mathbf{k}}^\dagger = \alpha_{\mathbf{k}\uparrow}^\dagger$. $|\tilde{0}\rangle$ denotes the vacuum for newly defined α and β particles satisfying $\alpha_{\mathbf{k}}|\tilde{0}\rangle = \beta_{\mathbf{k}}|\tilde{0}\rangle = 0$.

4. Antiferromagnetic state

We show the gap function Δ_{AF} (antiferromagnetic (AF) state) and Δ (superconducting state) as a function of U/t for the two-dimensional Hubbard model in Fig.2[37]. In the strongly correlated region where U is larger than the bandwidth, the antiferromagnetic correlation is suppressed and as a result superconducting state can be stable. We would expect a similar phase diagram for the d-p model.

It is important to have control of the antiferromagnetic state in strongly correlated electron systems. It appears however, that the AF state is considerably stable in the three-band d-p model, and is more stable than in the single-band Hubbard model. It is thus necessary to introduce a new parameter t_d'. The Fermi surface will have a large curvature as finite t_d' is introduced. This is shown in Fig.3. t_d' also plays a role concerning a stability of the AF state.

We show the ground-state energy E_g and the kinetic energy E_K as a function of λ in Fig.4 where E_K is the expectation value of the kinetic part K of the Hamiltonian. The antiferromagnetic state is suppressed by t_d'. This is shown in Fig.5 and Fig.6, where the level difference is chosen as $\Delta_{dp} = \epsilon_p - \epsilon_d = 2t_{dp}$ and $4t_{dp}$, respectively. When the level difference is large, we need large t_d' to reduce the antiferromagnetic correlation.

5. Discussion

We discuss how the wave function is improved compared to the Gutzwiller function. We are in a tendency to predict a large AF phase on the basis of the Gutzwiller function. It is necessary to improve and optimize the wave function to examine whether the AF region is indeed so large. The energy gain of the improved wave function ψ_λ mainly comes from the kinetic-energy gain as shown in Fig.4. The conduction electrons become rather localized due to the Gutzwiller projection, from which the electronic state favors magnetic correlation. This means that the kinetic-energy gain is relatively reduced compared with that of the Coulomb energy. The exponential kinetic operator $e^{-\lambda K}$ has an effect to move electrons to get the kinetic-energy gain. This will increase the double occupancy, and the variational parameters are determined to lower the ground-state energy by balancing the kinetic energy against the Coulomb repulsive energy. We can lower the ground-state energy further by multiplying the wave function by the Gutzwiller projection again.

The importance of multiplying by $e^{-\lambda K}$ lies in that the AF phase is reduced by suppressing AF correlation. In fact, in the 2D Hubbard model, the AF correlation is considerably suppressed when U is large being greater than the bandwidth where the kinetic-energy gain is dominant over the AF energy lowering. This is also the case for the d-p model. By using the Gutzwiller function, the optimized value of the AF order parameter is very large and is larger than that of

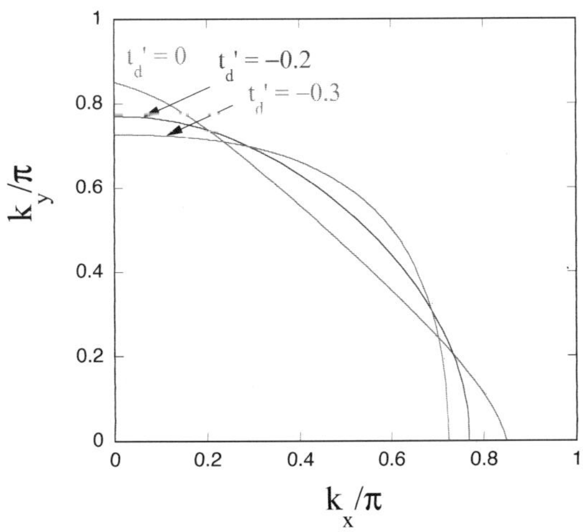

Figure 3. Fermi surface of the three-band d-p model in the non-interacting case. We used $t_{pp} = 0.4t_{dp}$ and $\epsilon_p - \epsilon_d = 2t_{dp}$ for $t'_d = 0$, $-0.2t_{dp}$ and $-0.3t_{dp}$. The hole-doping rate is about 0.13.

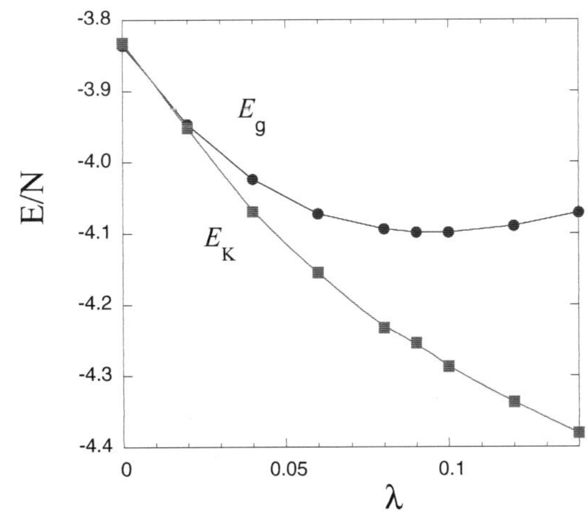

Figure 4. Ground-state energy E_g and the kinetic energy E_K as a function of λ on 8×8 lattice with 76 holes. The band parameters are $\epsilon_p - \epsilon_d = 2$, $t_{pp} = 0.4$, $t'_d = -0.2$, $U_d = 10$ and $U_p = 0$ in units of t_{dp}.

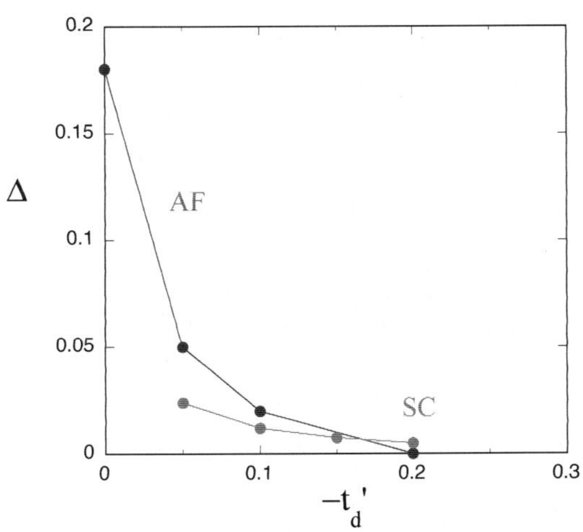

Figure 5. AF and SC order parameters as a function of t'_d in units of t_{dp} on 8×8 lattice with 76 holes. The band parameters are $\epsilon_p - \epsilon_d = 2$, $t_{pp} = 0.4$, $U_d = 10$ and $U_p = 0$. We used the Gutzwiller-BCS wave function for SC state.

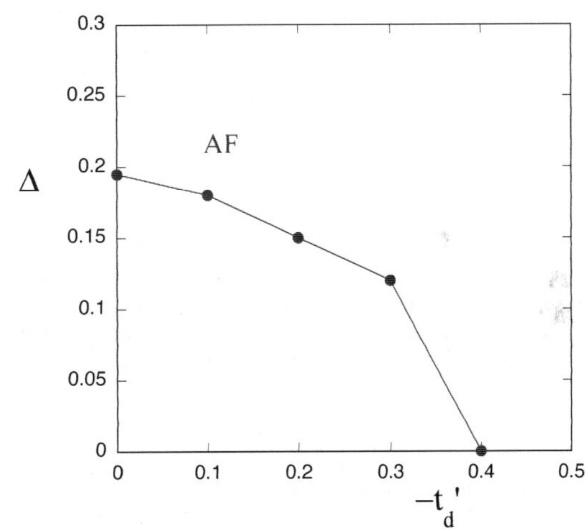

Figure 6. Antiferromagnetic order parameter as a function of t'_d on 8×8 lattice with 76 holes. The band parameters are $\epsilon_p - \epsilon_d = 4$, $t_{pp} = 0.4$, $U_d = 10$ and $U_p = 0$. When the level difference $\epsilon_p - \epsilon_d$ becomes large, the AF state is more stabilized.

the superconducting order parameter more than by one order of magnitude in the d-p model. The AF order parameter is reduced to be of the same order as the superconducting one based on our wave function. This would give a possibility of superconductivity.

6. Summary

We investigated the ground state of the two-dimensional three-band Hubbard model on the basis of the optimization variational Monte Carlo method. The antiferromagnetic correlation is very strong and the antiferromagnetic state is very stable in the d-p model. It is thus necessary to introduce long-range transfer intergrals such t'_d to conquer the antiferromagnetism in the d-p model. We expect that there also occurs a crossover between strongly correlated and weakly correlated regions by controlling the antiferromagnetic correlation in the d-p model.

There is a crossover in the two-dimensional Hubbard model[37]. A large fluctuation presumably exists in the crossover region. This indicates a possibility of high-temperature superconductivity. A crossover from weakly to strongly coupled systems is universal phenomenon that exists ubiquitously in the world. There may a class of phenomena that shows a crossover between weakly and strongly interacting regions. This class may include, for example, the Kondo effect[57, 58, 59, 60, 61, 62, 63], QCD[64], BCS-BEC crossover[65], Hubbard model, sine-Gordon model[66, 67, 68] and the Kosterlitz-Thouless transition.

References

[1] Bednorz J G and Müller K A 1986 Z. Phys. B64 189.
[2] Emery V J 1987 Phys. Rev. Lett. 58 2794.
[3] Hirsch J E, Loh E Y, Scalapino D J and Tang S 1989 Phys. Rev. B39 243.
[4] Scalettar R T, Scalapino D J, Sugar R L and White S R 1991 Phys. Rev. B44 770.
[5] Oguri A, Asahata T and Maekawa S 1994 Phys. Rev. B49 6880.
[6] Koikegami S and Yamada K 2000 J. Phys. Soc. Jpn. 69 768.
[7] Yanagisawa T, Koike S and Yamaji K 2001 Phys. Rev. B64 184509.
[8] Koikegami S and Yanagisawa T 2001 J. Phys. Soc. Jpn. 70 3499.
[9] Yanagisawa T, Koike S and Yamaji K 2003 Phys. Rev. B67 132408.
[10] Koikegami S and Yanagisawa T 2003 Phys. Rev. B67 134517.
[11] Koikegami S and Yanagisawa T 2006 J. Phys. Soc. Jpn. 75 034715.
[12] Yanagisawa T, Miyazaki M and Yamaji K 2009 J. Phys. Soc. 78 013706.
[13] Weber C, Lauchi A, Mila F and Giamarchi T 2009 Phys. Rev. Lett. 102 017005.
[14] Lau B, Berciu M and Sawatzky G A 2011 Phys. Rev. Lett. 106 036401.
[15] Weber C, Giamarchi T and Varma C M 2014 Phys. Rev. Lett. 112 117001.
[16] Avella A, Mancini F, Paolo F and Plekhanov E 2013 Euro. Phys. J. B86 265
[17] Ebrahimnejad H, Sawatzky G A and Berciu M 2016 J. Phys. Cond. Matter 28 105603.
[18] Tamura S and Yokoyama H 2016 Phys. Procedia 81 5.
[19] Hubbard J 1963 Proc. Roy. Soc. London 276 238.
[20] Hubbard J 1963 Proc. Roy. Soc. London 281 401.
[21] Gutzwiller M C 1963 Phys. Rev. Lett. 10 159.
[22] Zhang S, Carlson J and Gubernatis J E 1997 Phys. Rev. B55 7464.
[23] Zhang S, Carlson J and Gubernatis J E 1997 Phys. Rev. Lett. 78 4486.
[24] Yanagisawa T and Shimoi Y 1996 Int. J. Mod. Phys. B10 3383.
[25] Yanagisawa T and Shimoi Y 1995 Phys. Rev. Lett. 74 4939.
[26] Nakanishi N, Yamaji K and Yanagisawa T 1997 J. Phys. Soc. Jpn. 66 294.
[27] Yamaji K, Yanagisawa T, Nakanishi T and Koike S 1998 Physica C304 225.
[28] Yamaji K, Yanagisawa T and Koike S 2000 Physica B284-288 415.
[29] Bulut N 2002 Advances in Phys. 51 1587.
[30] Yokoyama H, Tanaka Y, Ogata M and Tsuchiura H 2004 J. Phys. Soc. Jpn. 73 1119.
[31] Yokoyama H, Ogata M and Tanaka Y 2006 J. Phys. Soc. Jpn. 75 114706.
[32] Aimi T and Imada M 2007 J. Phys. Soc. Jpn. 76 113708.
[33] Miyazaki M, Yanagisawa T and Yamaji K 2004 J. Phys. Soc. Jpn. 73 1643.
[34] Yanagisawa T 2008 New J. Phys. 10 023014.
[35] Yanagisawa T 2013 New J. Phys. 15 033012.
[36] Yanagisawa T, Miyazaki M and Yamaji K 2014 J. Mod. Phys. 4 33.
[37] Yanagisawa T 2016 J. Phys. Soc. Jpn. 85 114707.
[38] Noack R M, White S R and Scalapino D J 1995 EPL 30 163.
[39] Yamaji K, Shimoi Y and Yanagisawa T 1994 Physica C235 2221.
[40] Yanagisawa T, Shimoi Y and Yamaji K 1995 Phys. Rev. B52 R3860.

[41] Nakano T, Kuroki K and Onari S 2007 Phys. Rev. B76 014515.
[42] Yanagisawa T, Koike S and Yamaji K 1998 J. Phys. Soc. Jpn. 67 3867.
[43] Yanagisawa T and Miyazaki M 2014 EPL 107 27004.
[44] McElroy K et al. 2003 Nature 422 592.
[45] Hussey N E et al. 2003 Nature 425 814.
[46] Weber C, Haule K and Kotliar G 2008 Phys. Rev. B78 134519.
[47] Hybertsen M S, Schluter M and Christensen N E 1989 Phys. Rev. B39 9028.
[48] Eskes H, Sawatzky G A and Feiner L F 1090 Physica C160 424.
[49] McMahan A K, Annett J F and Martin R M 1990 Phys. Rev. B42 6268.
[50] Eskes H and Sawatzky G 1991 Phys. Rev. B43 119.
[51] Otsuka H 1992 J. Phys. Soc. Jpn. 61 1645.
[52] Yanagisawa T, Koike S and Yamaji K 1999 J. Phys. Soc. Jpn. 68 3608.
[53] Eichenberger D and Baeriswyl D, Phys. Rev. B76, 180504 (2007).
[54] Baeriswyl D, Eichenberger D and Menteshashvii M, New J. Phys. 11, 075010 (2009).
[55] Baeriswyl D, J. Supercond. Novel Magn. 24, 1157 (2011).
[56] Yanagisawa T 2007 Phys. Rev. B75 224503.
[57] Kondo J 2012 *The Physics of Dilute Magnetic Alloys* (Cambridge University Press, Cambridge).
[58] Wilson K G 1975 Rev. Mod. Phys. 47 773.
[59] Nagaoka Y 1965 Phys. Rev. 138 A1112.
[60] Zittartz J and Müller-Hartmann E 1968 Z. Phys. 212 380.
[61] Yanagisawa T 2012 J. Phys. Soc. Jpn. 81 094713.
[62] Yanagisawa T 2015 J. Phys. Soc. Jpn. 84 074705.
[63] Yanagisawa T 2015 J. Phys. Conf. Ser. 603 012014.
[64] Ellis R K, Stirling W J and Webber B R 1996 *QCD and Collider Physics* (Cambridge University Press, Cambridge).
[65] Nozieres P and Schmitt-Rink S 1985 J. Low Temp. Phys. 59 195.
[66] Rajaraman R 1989 *Solitons and Instantons* (North-Holland, Amsterdam).
[67] Solyom J 1979 Adv. Phys. 28 201.
[68] Yanagisawa T 2016 EPL 113 41001.

IOP Conf. Series: Journal of Physics: Conf. Series **1054** (2018) 012018 doi:10.1088/1742-6596/1054/1/012018

Electrotransport and magnetic measurements on bulk FeSe superconductors

T Karwoth[1], K Furutani[1,2], M R Koblischka[1,2], X L Zeng[1], A Wiederhold[1], M Muralidhar[2], M Murakami[2], and U Hartmann[1]

[1] Experimental Physics, Saarland University, Campus C 6 3, 66123 Saarbrücken, Germany
[2] Superconducting Materials Laboratory, Department of Materials Science and Engineering, Shibaura Institute of Technology, Tokyo 135-8548, Japan

E-mail: `m.koblischka@mx.uni-saarland.de`

Abstract. The superconducting properties of bulk, polycrystalline FeSe samples were characterized through magnetic and electric transport measurements. In order to improve the superconducting properties, the sintering temperature was varied up to 900 °C and to improve the grain connectivity, silver was added (0 to 7 wt.-%). Electric properties of the samples were investigated by the four point probe method (R-T measurements and V-I characteristics). Generally, the sample with 4 wt.-% Ag-addition showed the highest critical transition temperature among all the samples fabricated with the same preparation parameters. The critical currents were obtained from V-I measurements in various applied magnetic fields up to 6 T. Via Arrhenius plots, the pinning potential U_0 was determined for all samples studied. The magnetic properties (M-T and M-H) of the samples were measured using an extraction magnetometer in a Quantum Design PPMS with applied magnetic fields up to 7 T. The scaling of the normalized volume pinning force versus the reduced field indicated a peak position at ~0.2 for the Ag-added sample, which points to a pinning at the grain boundaries.

1. Introduction

The polycrystalline, bulk FeSe superconductor ('11'-compound) is considered as candidate for several types of applications like superconducting supermagnets despite the low superconducting transition temperature, T_c [1, 2]. FeSe offers a simple preparation route, is free of rare-earth elements and shows many similarities to the cuprate high-temperature superconductors like high critical fields, and no flux jumps as in the case of MgB_2. The transport and flux pinning properties of this compound, however, have yet to be explored in detail as only some reports concerning the electric transport and flux pinning of the 11-system can be found in the literature [3–6]. Furthermore, to improve the sample resistivity and the mechanical strength of the otherwise porous granular material, Ag was added to the samples. This was done in the literature already for conventional superconductors like Nb_3Sn [7] and MgB_2 [8, 9] and for $YBa_2Cu_3O_x$, where Ag was found to be located at the grain boundaries, thus improving the connectivity between the grains [10–13]. Recently, it was found that in FeSe some Ag enters into the crystal structure itself [14, 15], leading to an improvement of the superconducting transition temperature, T_c. This clearly demonstrates that more detailed investigations of the superconducting properties of the FeSe-Ag-system are required. Therefore, we investigate in the present contribution a series of FeSe samples with Ag content ranging up to 7 wt.-% by means

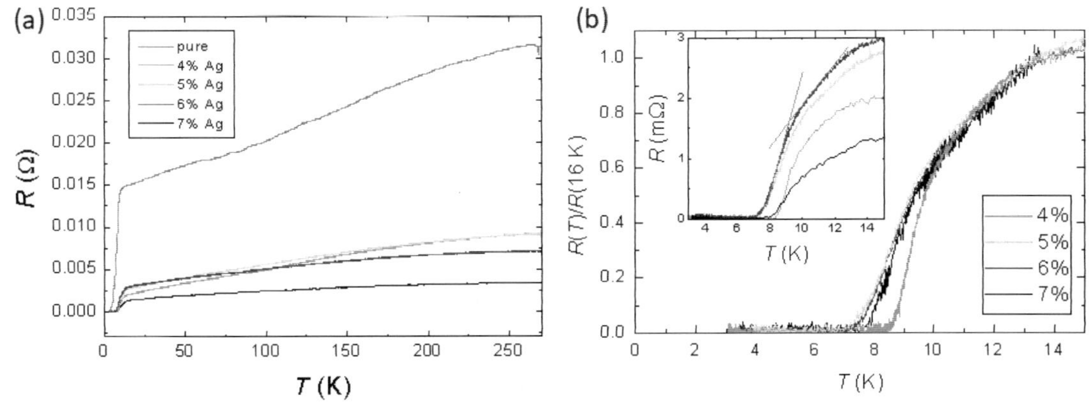

Figure 1. (a) Resistance vs. temperature in the range 3 K $< T <$ 300 K, zero applied field. (b) presents the normalized superconducting transistions of the Ag-containing samples; the inset shows the superconducting transitions in detail. The red lines indicate the determination of T_c.

of electric transport and magnetization measurements. Using the scaling approach of the flux pinning forces, we determine the acting flux pinning mechanism.

2. Experimental procedures

Table 1. Properties of the FeSe samples investigated in this contribution.

Sample	addition	T_c [K]	ΔT_c [K]	R (300 K) [Ω]	RRR	grain connectivity [%]	U_0/k_B (1 T) [K]
1	pure	8.35	2.35	0.03224	2.16	3.9	51.2
2	4 wt.-%	9.42	0.98	0.00930	4.49	6.52	171.2
3	5 wt.-%	9.28	1.18	0.00926	3.33	7.3	62.1
4	6 wt.-%	9.36	1.12	0.00718	2.41	11.3	99.3
5	7 wt.-%	9.18	0.94	0.00335	2.44	23.8	45.6

The FeSe material was produced using commercial powders of Fe (99.9% purity), and Se (99.5 % purity). Firstly, the powders were mixed in a nominal ratio Fe:Se =1:1 in a glove box and thoroughly ground using an agate mortar and loaded into a ceramic vial under high-purity argon atmosphere inside the glove box. The ball milling was performed in a glove box for 6 h. The as-milled powder mixture was then pressed into pellets of 5 mm in diameter and 2 mm thickness using an uniaxial pressure of 100 MPa, and sealed into an evacuated quartz tube and heated at 600 °C for 24 h. In a second step, after furnace cooling, the product was opened in the glove box. Then, the sintered pellets were once more thoroughly ground and mixed, and again ballmilled similar to the first step. The samples were sealed in quartz tubes and sintered at two different temperatures for 24 h, and finally, the temperature was lowered to room temperature at a cooling rate of 230 °C/h. Pure FeSe samples were prepared at 900 °C and 950 °C, while samples with Ag-addition recieved a treatment at a lower reaction temperature at 825 °C. The constituent phases of the samples were identified with a high-resolution automated X-ray powder

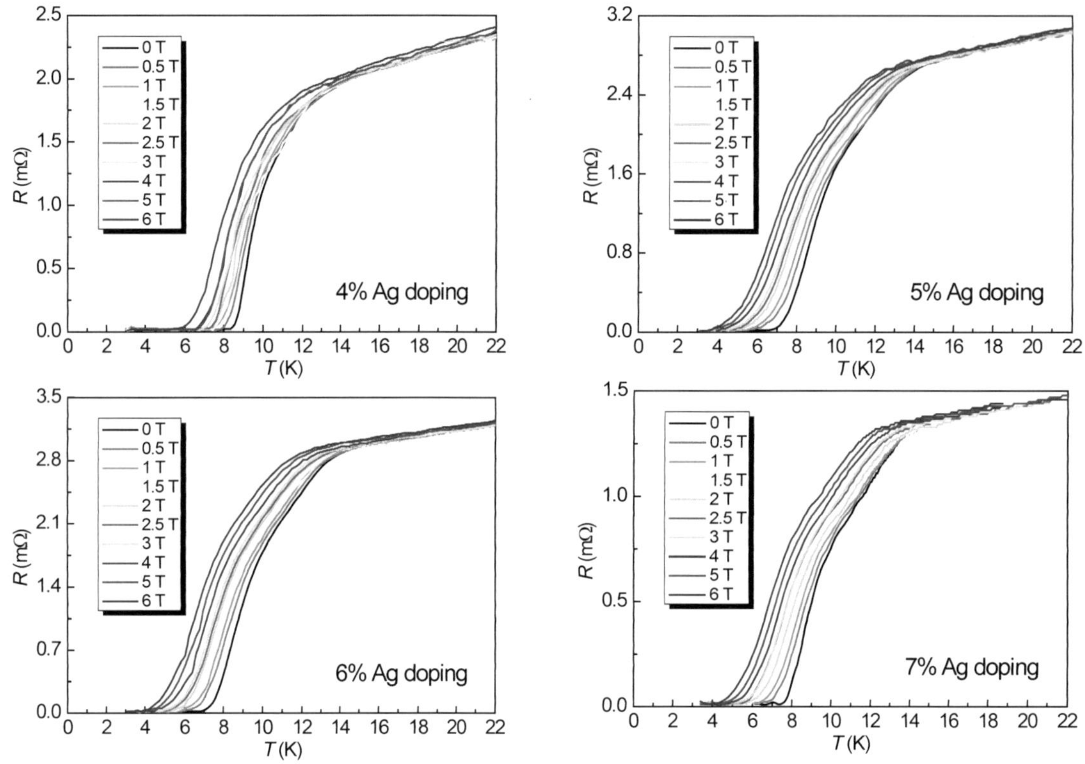

Figure 2. Resistance vs. temperature in applied magnetic fields (H ⊥ sample surface) for all Ag-containing FeSe-samples.

diffractometer (RINT2200), using Cu-K$^{\alpha}$ radiation generated at 50 kV and 300 mA. A detailed description of the microstructure of the samples is given elsewhere [6, 16]. We notice a slight shift of the 001 reflection in our samples with increasing Ag-content, similar to the obervation in Ref. [4], which may indicate that Ag is incorporated in the FeSe lattice.

3. Results and discussion

For the transport measurements, long bars with dimensions 1 × 2 × 5 mm^3, resulting in a distance of 3 mm between the voltage contacts, were cut from the bulk samples using a diamond saw. Thin copper wires of 100 μm in diameter were used as leads for low resistance contacts placed onto the samples by means of silver paste. The electric measurements were performed in an Oxford Instruments 10/12 T Teslatron cryostat in the temperature range 2 K – 300 K. The resistance was measured in four-probe configuration with a dc current of 2.5 mA applied perpendicular to the magnetic field direction. Field steps of 0.5 T (low fields) and 1 T (high fields) were used for the $R(T)$ measurements. The magnetic field sweep rate was 0.7 T/min, and temperature steps of 0.25 K were applied. The measurements were controlled by a MatLab program operating a stabilized current source (Keithley 2400) and a voltmeter (Keithley 2100 multimeter) [17].

Magnetic measurements were performed in a QD PPMS system with ±7 T applied field (perpendicular to the sample surface. For these measurements, cubic samples with dimensions 1 × 1 × 0.5 mm^3 were cut from the bulks. Table 1 summarizes all the properties of the samples

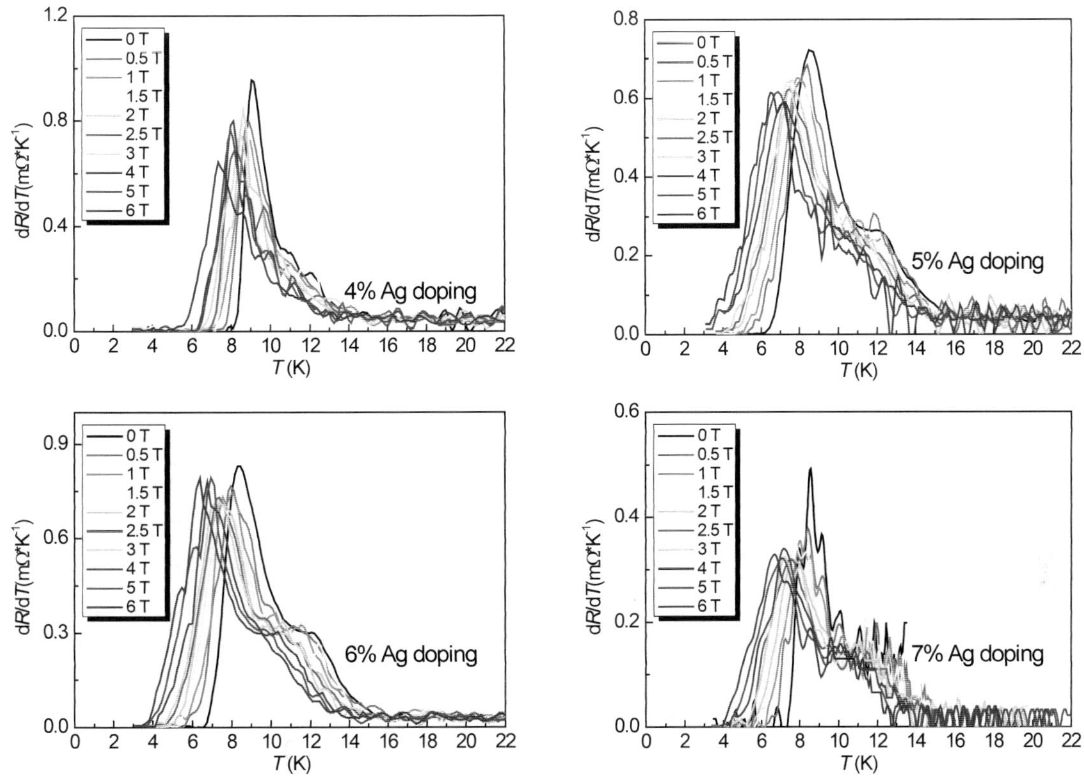

Figure 3. Resistance derivative, dR/dT, versus temperature for all Ag-containing samples.

investigated here.

Figure 1 (a) presents the resistance measurements on the Ag-containing samples in the temperature range 3 K $< T <$ 300 K. The change of resistance due to the Ag-addition is obvious from the graph. While the room temperature resistance for samples with 4 wt.-% and 5 wt.-% Ag-addition is similar, the further Ag-addition clearly reduced the resistance measured.

Figure 1 (b) gives a detailed view of the superconducting transitions of all Ag-containing samples. The main panel presents a comparison of the superconducting transitions normalized by the resistance value at 14 K (just above the superconducting transition). All data nearly fall on a common line except the data of the sample with 7 % Ag-addition at around 8 K, whereas the the sample with 4 wt.-% Ag-addition reveals the highest T_c of all samples measured. The inset to (b) shows the non-normalized superconducting transitions. The superconducting transition temperatures, T_c, were determined as indicated in the inset to (b). The transition widths, ΔT, were determined using a 10%-90% criterion. From these measurements, we determine the residual resistance ratio (RRR= $R(T = 300 \text{ K})/R(T = 14 \text{ K})$). The RRR is increased at the sample with 4 wt.-% Ag-addition, and then decrases again with increasing amount of Ag content. This behavior is similar to that seen in [4], except that in our case, the 4 wt.-% Ag sample was found to be the optimum in our sample series. Another important measure is the grain connectivity. The geometrical connectivity can be estimated using an approach describing different MgB_2 samples [18]. The connectivity κ is calculated using

$$\kappa = \Delta\rho_g/\Delta\rho \quad , \tag{1}$$

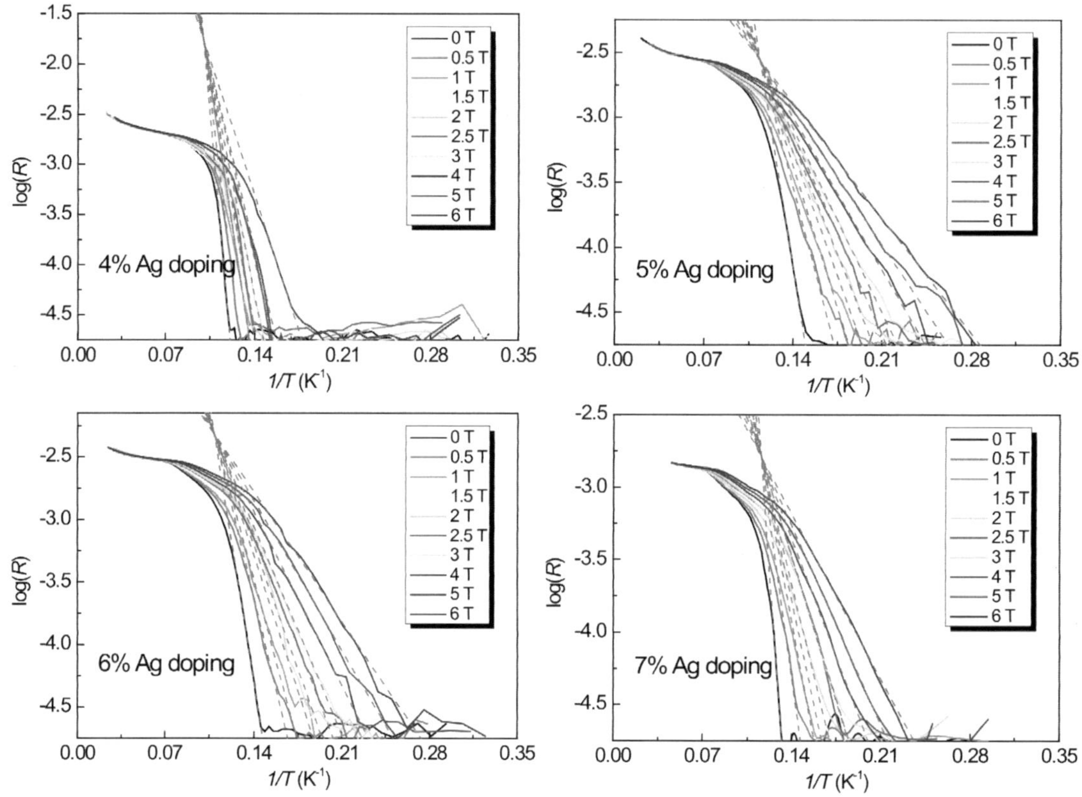

Figure 4. Arrhenius plots of all Ag-containing samples including the fitting to determine the pinning potential, U_0.

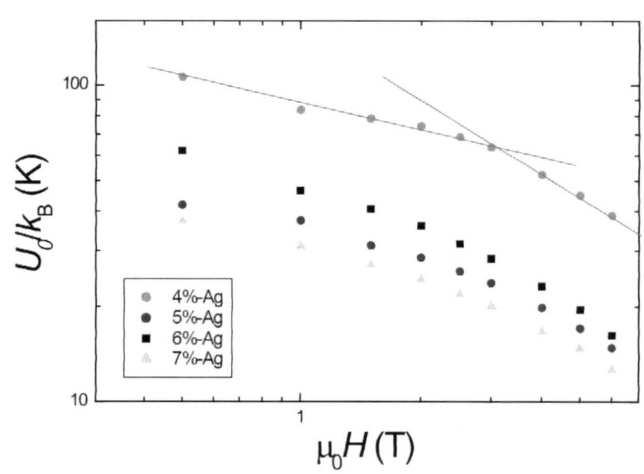

Figure 5. TAFF pinning potential, U_0, as function of the applied field in a double-log representation. The lines are indicating the change in field dependence above an applied field of 3 T.

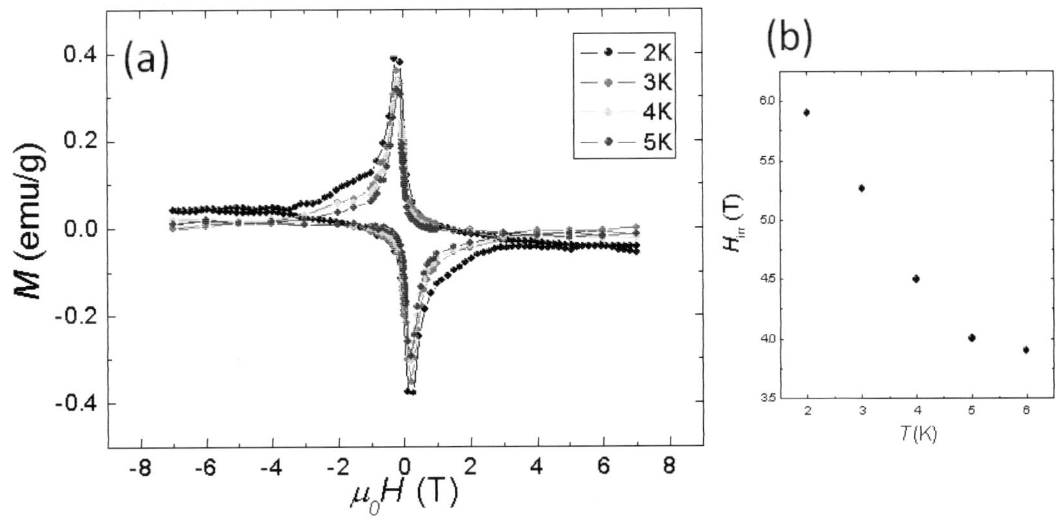

Figure 6. Magnetization loops for the sample with 4 wt.-% Ag-addition at various temperatures. The shape of the loops reveal the polycrystalline character of the sample. (b) shows the irreversibility line of the sample with 4 wt.-% Ag-addition.

while $\Delta\rho_g$ can be assumed to be 0.472 mΩcm for randomly oriented three dimensional FeSe samples [19,20] and $\Delta\rho = \rho(300\text{ K}) - \rho(14\text{ K})$. All these data are summarized in Table 1.

Figure 2 presents the resistance data in applied magnetic fields ($H \perp$ sample surface) for all Ag-containing samples. The superconducting transitions broaden on appliying magnetic fields. Also here, the 4 wt.-%-Ag sample shows the sharpest superconducting transitions. In Fig. 3, we present dR/dT-plots of the data. In all cases, we obtain a peak at the superconducting transition. No further steps in the derivative dR/dT at around 65 – 85 K were observed in the present samples, whereas the pure sample does show this step [19]. This indicates that the Ag-addition is suppressing the structural transition.

The resistance data can be converted into Arrhenius-plots by plotting the logarithm of the electric resistivity against the inverse temperature, which is shown in Fig. 4. From the slopes of the linear parts, the the thermally activated flux-flow (TAFF) pinning potential can be calculated [21]. The results of this procedure are summarized in Fig. 5, where the obtained pinning potentials, U_0, in the TAFF regime are presented as a function of the applied field, $\mu_0 H$. Here, it is remarkable that the sample with 4 wt.-% Ag-addition shows the highest pinning potential of all samples measured. All values of U_0 taken at 1 T applied field are listed in Table 1 to enable a comparison with the data of Ref. [4]. It should be noted here that our data show the same overall behavior as reported in [4], but are somewhat smaller indicating that the preparation temperature/time can still be further optimized. The values of U/k_B, where k_B is the Boltzmann constant, are obtained from the slopes of the $\ln\rho$ versus $1/T$ dependences at the different applied fields. The activation energy U_0 shows a weak field dependence at low fields, changing in a stronger one at $B \approx 3$ T. Both branches can be described by a power law behavior, $U \sim B^{-\beta}$, with $\beta \approx 0.5$ at low fields, and $\beta \approx 1.8$ at high fields. An explanation for the parabolic-like shape of the curves may be the presence of strong grain boundary pinning [22].

We now turn to the magnetic measurements. In Fig. 6 (a), $M(H)$-loops are given for the sample with 4 wt.-% Ag at temperatures between 2 K and 5 K. The field sweep rate was 0.35 T/min. All curves exhibit an asymmetric shape being1 a sign of granular behavior. There are no flux jumps observed, even at the lowest temperature. The critical current densities were

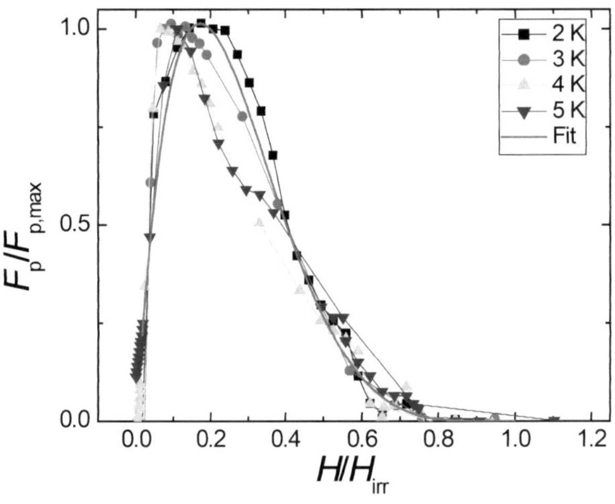

Figure 7. Pinning force scaling for the sample with 4 wt.-% Ag-addition, compared to the pure samples and other data of the literature. The red line is a fit to all data.

calculated using the extended Bean approach. The irreversibility fields, $H_{\rm irr}$, were determined using a criterion of 100 A/cm^{-2} from double-log plots [23]. The dependence of $H_{\rm irr}$ on temperature is illustrated in Fig. 6 (b). A remarkable result is further that the hysteresis loops do not change much when measuring with a field sweep rate of 0.7 T/min. This indicates that dynamic relaxation is only small for bulk FeSe. Figure 7 presents the results of the pinning force scaling ($F_p/F_{\rm p,max}$ vs. $h = H_a/H_{\rm irr}$ [24]) on the 4 wt.-% Ag-added FeSe sample. The pure FeSe samples exhibited a good scaling and a peak at $h_0 \sim 0.4$ [6, 25]. In contrast to this, the Ag-added samples are found to show a peak at $h_0 \sim 0.2$ considering all measured data (red line). This is similar to the pure FeSe sample prepared at 950 °C [6]. It should be noted that the data at 2 K and 3 K differ from the ones measured at 4 K and 5 K in the region just above h_0, which is a sign of different contributions to the flux pinning: At higher T, the grain boundary pinning is dominant, whereas at low T, the pinning at point defects inside the FeSe grains plays a more important role. In case the FeSe grains are well coupled, one has to consider a current flow through the entire sample perimeter. This current flow is then affected by percolation as well as the anisotropy due to the non-textured FeSe grains. In this case, the approach of Eisterer [26] for MgB_2 samples is to be considered for our samples. The consequence is then that all peak positions are appearing at lower h as in the standard approach. Other impurity phases which were seen in the X-ray data [6,16] are forming separate grains or are located at the grain boundaries, so no effect on the flux pinning is expected. Finally, from the comparison with data presented in Ref. [25], we can deduce that the Ag-added FeSe samples exhibit an improved flux pinning at the grain boundaries, which gets even more pronounced at temperatures close to T_c. Here, it is important to point out that the present polycrystalline FeSe samples were prepared using a classical sintering technique at relatively low reaction temperatures. This is in contrast to previous work, where the samples were subjected to much higher temperatures (1050 °C) to partially melt the material. In this case, it is much more likely that Ag enters the FeSe crystal structure as compared to the sintering process. Nevertheless, our data were found to coincide reasonably well with other data in the literature. The change of resistivity due to the Ag-addition in our samples is obvious, and the sample with the best superconducting properties was found to be the sample with 4 wt.-% Ag-addition.

4. Conclusions

We have performed resistivity measurements and magnetic measurements on Ag-containing polycrystalline, bulk FeSe samples. The resistance data clearly reveal the improvement of grain connectivity due to the Ag-addtion. Our data were found to coincide reasonably well with other data in the literature, even though the preparation route (sintering at lower reaction temperatures) was different. The magnetization loops exhibit an asymmetric shape which indicates granularity. The irreversibility fields determined do not vary much with temperature. The pinning force scaling demonstrates a peak position, h_0 ~0.2, indicating dominant pinning provided by point defects (low T) and grain boundaries (high T).

Acknowledgments

We would like to thank K. Berger and B. Douine (GREEN, Nancy) for valuable discussions. This work is supported by DFG-Project Ko2323/10, which is gratefully acknowledged.

References

[1] Hsu F-C Luo JY, Yeh, KW, Chen TK, Huang TW, Wu PM, Lee YC, Huang YL, Chu YY, Yan DC and Wu MK 2008 *Proc. Natl. Acad. Sci.* **105** 14262
[2] Ma Y W 2012 *Supercond. Sci. Technol.* **25** 113001
[3] Leo A, Grimaldi G, Guarino A, Avitabile F, Nigro A, Galluzzi A, Mancusi D, Polichetti M, Pace S, Buchkov K, Nazarova E, Kawale S, Bellingeri E and Ferdeghini C 2015 *Supercond. Sci. Technol.* **28** 125001
[4] Nazarova E, Balchev N, Nenkov K, Buchkov K, Kovacheva D, Zahariev A and Fuchs G 2015 *Supercond. Sci. Technol.* **28** 125013
[5] Galluzzi A, Polichetti M, Buchkov K, Nazarova E, Mancusi D and Pace S 2017 *Supercond. Sci. Technol.* **30** 025013
[6] Muralidhar M, Furutani K, Dinesh Kumar, Koblischka M R, Ramachandra Rao M S and Murakami M 2016 *Phys. Status Solidi* A **213** 3214
[7] Matsumoto G, Inoue K, Kikuchi A, Takeuchi T and Kiyoshi T 2006 *J. Phys.: Conf. Ser.* **43** 31
[8] Shimoyama J, Hanafusa K, Yamamoto A, Katsura Y, Horii S, Kishio K and Kumakura H 2008 *J. Phys.: Conf. Ser.* **97** 012255
[9] Muralidhar M, Inoue K, Koblischka M R and Murakami M 2016 *Adv. Eng. Mater.* **17** 831
[10] Dwir B, Affronte M and Pavuna D 1989 *Appl. Phys. Lett.* **55** 399
[11] Pavuna D, Berger H, Affronte M, Van der Maas J, Capponi J, Guillot M, Lejay P and Tholence J L 1988 *Solid State Commun.* **68** 535
[12] Koblischka M R, Schuster T and Kronmüller H 1993 *Physica* C 211 263
[13] Diko P, Antal V, Kavecansky V, Yang C and Chen I 2012 *Physica* C **476** 29
[14] E. Nazarova, K. Buchkov, S. Terzieva, K. Nenkov, A. Zahariev, D. Kovacheva, N. Balchev and G. Fuchs 2015 *J. Supercond. Nov. Magn.* **28** 1135
[15] Migita M, Takikawa Y, Sugai K, Takeda M, Uehara M, Kuramoto T, Takano Y, Mizuguchi Y and Kimishima Y 2013 *Physica* C **484** 66
[16] Fabitha K, Ramachandra Rao M S, Muralidhar M, Furutani K and Murakami M 2017 *J. Supercond. Nov. Magn.* **30** 3117
[17] Wiederhold A, Koblischka M R, Inoue K, Muralidhar M, Murakami M and Hartmann U 2016 *J. Phys.: Conf. Ser.* **695** 012004
[18] Yamamoto A, Shimoyama J, Kishio K and Matsushita T 2007 *Supercond. Sci. Technol.* **20** 658
[19] Böhmer A E, Taufour V, Straszheim W E, Wolf T and Canfield P C 2016 *Phys. Rev. B* **94** 024526
[20] Sedky A, Nazarova E, Nenkov K and Buchkov K 2017 *J. Supercond. Nov. Magn.* **30** 2751
[21] Lei H, Wang K, Hu R, Ryu H, Abexkoon M, Bozin E S and Petrovic C 2012 *Sci. Technol. Adv. Mater.* **13** 054305
[22] Pandya S, Sherif S, Sharath Chandra L S and Ganesan V 2011 *Supercond. Sci. Technol.* **24** 045011
[23] Koblischka M R and Murakami M 2000 *Supercond. Sci. Technol.* **13** 738
[24] Dew-Hughes D 1974 *Philos. Mag.* **30** 293
[25] Koblischka M R and Muralidhar M 2016 *Int. J. Mod. Phys.* B **30** 1630017
[26] Eisterer M 2008 *Phys. Rev. B* **77** 144524

IOP Conf. Series: Journal of Physics: Conf. Series **1054** (2018) 012019 doi:10.1088/1742-6596/1054/1/012019

Temperature insensitivity of Fermi surface in electron-doped high-temperature FeSe films on $SrTiO_3$

K. Shigekawa[1], K. Nakayama[1], G. N. Phan[1], M. Kuno[1], K. Sugawara[2,3], T. Takahashi[1,2,3], and T. Sato[1,3]

[1]Department of Physics, Tohoku University, Sendai 980-8578, Japan
[2]WPI Research Center, Advanced Institute for Materials Research, Tohoku University, Sendai 980-8577, Japan
[3]Center for Spintronics Research Network (CSRN), Tohoku University, Sendai 980-8577, Japan

k.shigekawa@arpes.phys.tohoku.ac.jp

Abstract. We have performed high-resolution angle-resolved photoemission spectroscopy study of electron-doped multilayer FeSe films on $SrTiO_3$. We determined the evolution of the electronic structure in a wide temperature range and found temperature-insensitive nature of the Fermi surface in contrast to the strong temperature dependence observed in non-doped FeSe. Such a striking doping dependence would provide a key to understanding the origin of unconventional fermiology and physical properties in FeSe.

1. Introduction

The discovery of iron-based superconductors [1] opened a new avenue for the study of high-temperature superconductivity. An intriguing character of iron-based superconductors is their multiband nature which is believed to play an important role for antiferromagnetic and nematic orders in the non-doped parent compound and for high-temperature superconductivity in the doped compound [2]. Theoretical calculations for iron-based superconductors generally predict the formation of hole-like and electron-like Fermi surfaces at the Brillouin-zone center and corner, respectively [3-5]. While such multiband nature has been confirmed by experiments such as angle-resolved photoemission spectroscopy (ARPES) studies [6-9], there are some disagreements between the theoretical and experimental results and their origin remains unclear. One of the major issues is the pronounced temperature dependence of the Fermi surface; namely, recent ARPES studies on FeSe revealed that the size of the Fermi surface strongly changes with temperature even in the normal state [10, 11]. This unusual temperature evolution lays the foundation for the emergence of nematic order which leads to the d_{xz}/d_{yz} orbital splitting [12, 13] and also for unconventional superconductivity in the Bardeen-Cooper-Schrieffer (BCS) and Bose-Einstein condensation (BEC) crossover regime [14], suggesting the importance of further systematic study on the electronic structure over a wide temperature range.

Recently, FeSe is also attracting considerable attention due to the discovery of high-temperature superconductivity in single-layer films on $SrTiO_3$ substrate [15]. The reported superconducting transition temperature (T_c) exceeds 65 K [16-19] which is significantly higher than T_c of ~8 K for bulk FeSe [20]. The importance of interfacial effects, such as the electron-phonon coupling across the interface between FeSe and $SrTiO_3$, has been proposed as the origin of the observed drastic T_c enhancement [15, 21, 22]. In addition to the interfacial effects, the electron doping from $SrTiO_3$ is essential for high-temperature superconductivity [16, 17]. In fact, electron doping by alkaline-metal deposition leads to high-temperature superconductivity above 40 K in multilayer FeSe films [23] and even in bulk FeSe [24], where the interfacial effects are absent. A question naturally arises as to how the electronic structure evolves with temperature in electron-doped high-temperature FeSe superconductor; is it similar to that of non-doped FeSe? Answering this question would provide a key for understanding the origin of unusual fermiology and physical properties of FeSe.

In this paper, we report high-resolution ARPES study on electron-doped multilayer FeSe films on $SrTiO_3$. We have succeeded in determining the electronic structure in a wide temperature range. We compare the obtained ARPES data with the previous results on non-doped FeSe, and found marked difference in the temperature dependence of the band shifts.

2. Experiment

The 20-monolayer (20-ML) FeSe films were fabricated on TiO_2-terminated Nb(0.05wt%)-doped $SrTiO_3$ substrate (SHINKOSHA) with the molecular-beam-epitaxy (MBE) technique in a vacuum better than 2×10^{-10} Torr [23]. After the fabrication, the film was transferred to the ARPES chamber without exposing to the air. Electron doping necessary for the realization of high-temperature superconductivity was achieved by the Cs deposition onto the film surface at room temperature using a Cs dispenser (SAES Getters) [25]. ARPES measurements were performed in an ultrahigh vacuum better than 5×10^{-11} Torr using a Scienta-Omicron SES2002 electron analyzer with a He discharge lamp ($h\nu$ = 21.218 eV) at Tohoku University. The energy and angular resolutions were set to 12-30 meV and 0.2°, respectively. The Fermi level (E_F) of the film was referenced to that of a gold film evaporated onto the sample holder.

3. Results and discussion

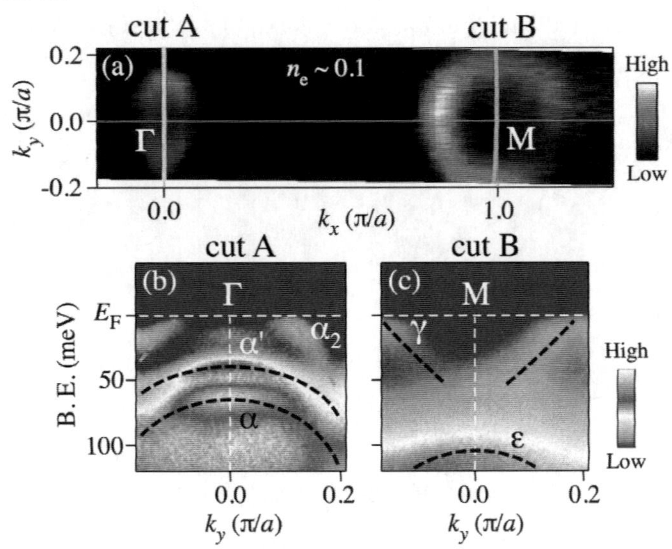

Figure 1. (a) ARPES-intensity mapping at T = 30 K for the Cs-deposited 20-ML FeSe film plotted as a function of a two-dimensional wave vector. The intensity is obtained by integrating the spectral intensity within ±5 meV of E_F. (b) and (c) ARPES-intensity plots near E_F as a function of binding energy (B.E.) and wave vector measured along the cuts A and B shown in (a), respectively. Black and orange dashed curves are a guide for the eyes to trace the band dispersions.

First, we show the electronic structure of Cs-deposited 20-ML FeSe films measured at T = 30 K (Fig. 1). Around the Brillouin-zone corner (M point), one can find a large electron-like Fermi surface originating from the γ band with the bottom of the dispersion around 50 meV below E_F [Figs. 1(a) and 1(c)]. The formation of the large electron pocket is a manifestation of heavy electron doping to the topmost FeSe layer from the Cs atoms. As a result, the hole-like bands (α and α') at the Brillouin-zone center (Γ point) do not cross E_F as seen in Fig. 1(b). The electron carrier concentration (n_e) estimated from the size of the electron pocket γ is ~0.1 electrons/Fe which is close to the optimal doping where the highest T_c of ~ 40 K is realized [25]. It is noted that, while there is another hole-like band (α_2) approaching E_F near the Γ point [see finite intensity near the Γ point in Fig. 1(a) and orange curve in Fig. 1(b)], it comes from interior FeSe layers beneath the topmost layer which remain nearly non-doped even after Cs deposition [25]. Therefore, in the following, we will focus only on the α, α', and γ bands originating from the electron-doped topmost FeSe layer to discuss the temperature-induced evolution of the band structure in the electron-doped region.

Figures 2(a) and 2(b) show the band dispersions around the Γ and M points, respectively, measured over a wide range of temperature from 13 to 250 K. At the Γ point, spectral intensities of the hole-like α and α' bands are gradually broadened with increasing temperature but are still visible even at the highest temperature of 250 K [Fig. 2(a)]. Similarly, the electron-like γ band is clearly observed around the M point irrespective of temperature [Fig. 2(b)]. The observed qualitative similarity of the overall band dispersions between 13 and 250 K indicates the absence of temperature-induced drastic electronic reconstruction such as the lifting of d_{xz}/d_{yz} orbital degeneracy by electronic nematicity found in the low-temperature phase of non-doped FeSe [12, 13].

To investigate the temperature dependence of the band structure in more detail, we have extracted the energy distribution curve (EDC) at the Γ point at each temperature as plotted in Fig. 3(a). One can identify two peaks at 13 K at binding energies of ~30 and ~60 meV, which correspond to the top of α' and α bands, respectively. This two-peaked structure is more clearly visible up to higher temperatures in the second-derivative intensity plot in Fig. 3(b), where one can recognize only a weak temperature dependence of the peak positions. The estimated temperature-dependent energy shifts of the α and α' bands are 10 meV or less [see red circles in Fig. 3(b)]. Next, to evaluate an energy shift in the γ band

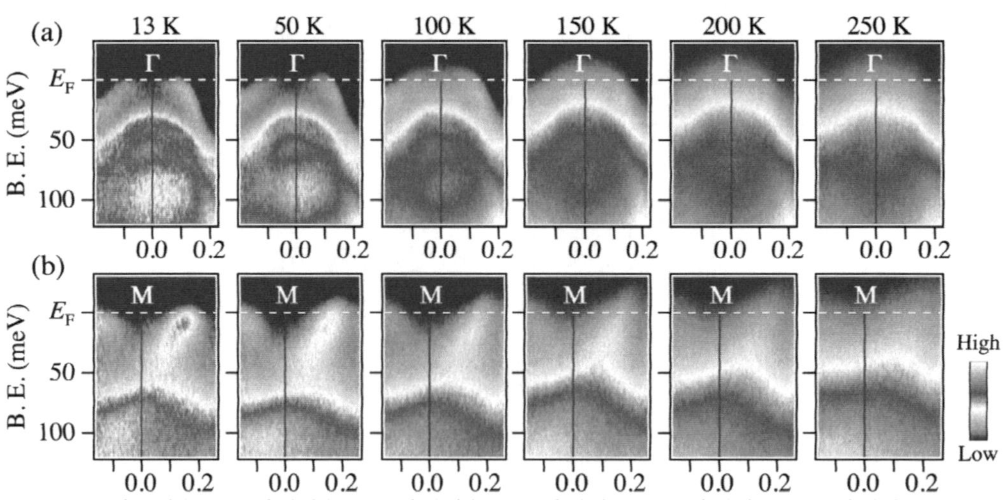

Figure 2. (a) and (b) Temperature dependence of the near-E_F ARPES intensity plotted as a function of binding energy and wave vector measured around the Γ and M points, respectively. These cuts nearly trace cuts A and B in Figs. 1(b) and 1(c), respectively.

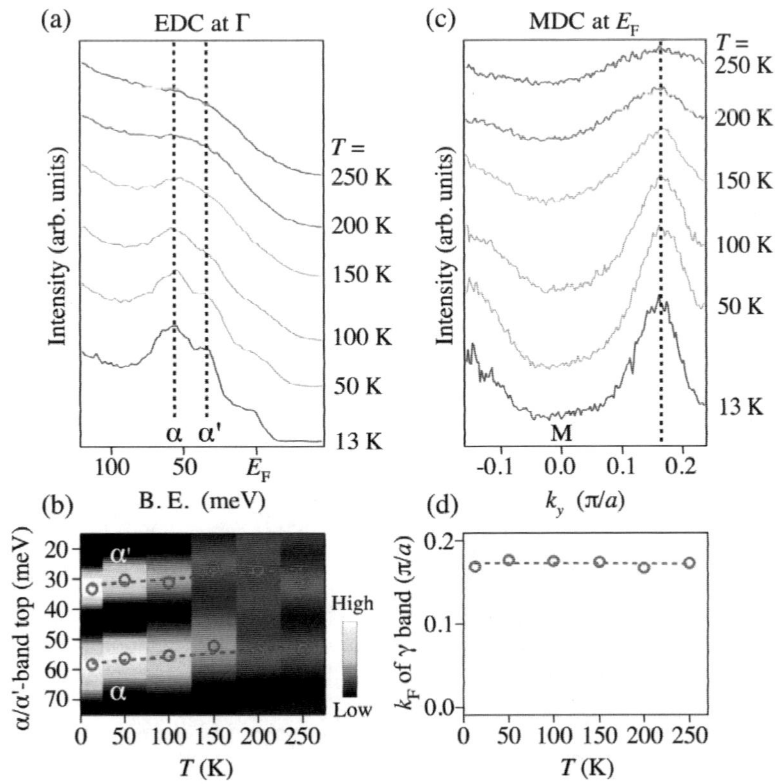

Figure 3. (a) Temperature dependence of the energy distribution curve (EDC) at the Γ point extracted from Fig. 2(a). (b) Corresponding second-derivative intensity plotted as a function of binding energy and temperature. In (b), the two-peaked structure corresponding to the α and α'-band tops is magnified. Also, the α- and α'-band tops estimated from the peak positions in the second-derivative intensity are shown in red circles. (c) The momentum distribution curve (MDC) at E_F at various temperatures extracted from Fig. 2(b). (d) Temperature dependence of the Fermi wave vector (k_F) for the γ band estimated from the numerical fittings to the MDCs in (c).

at the M point, we show the temperature dependence of the momentum distribution curve (MDC) at E_F in Fig. 3(c) (note that the bottom of the γ band is ill-defined in the EDC and hence we utilize MDC here). As clearly seen from Fig. 3(c), the peak position in the MDC which corresponds to the Fermi wave vector of the γ band is stationary to temperature (~0.18 π/a), as confirmed by the result of numerical fittings plotted in Fig. 3(d). This result demonstrates the temperature insensitivity of the Fermi surface in electron-doped FeSe.

The observed temperature insensitivity is in sharp contrast to the pronounced temperature dependence of the Fermi-surface volume reported in non-doped FeSe [10, 11]. In the previous reports, several scenarios such as the band renormalization effect by spin fluctuations, the Pomeranchuk effect, and the chemical potential effect have been intensively discussed as the origin of the pronounced temperature dependence in non-doped FeSe [10, 11]. Since these effects should be doping dependent (e.g. the coupling between electrons and spin fluctuations may be significantly weakened by the disappearance of the hole-like Fermi surface via electron doping and hence the renormalization effect would be reduced), the observed contrasting behavior of the electron-doped versus non-doped regions would provide a key to elucidating the origin of the unusual temperature evolution in non-doped FeSe. Future quantitative analyses on the doping dependence of these effects and the comparison with the present ARPES results are of crucial importance.

4. Summary

We reported high-resolution ARPES results on Cs-deposited 20-ML FeSe films grown on $SrTiO_3$ substrate. The systematic temperature-dependence study revealed the temperature insensitivity of the Fermi surface and the absence of the electronic reconstruction, both of which are in contrast to the recent observations in non-doped FeSe. The contrasting behavior between the electron-doped and non-doped regions would be a key to elucidate the underlying interactions which control the electronic structure and the physical properties of FeSe.

Acknowledgements

This work was supported by grants from JSPS KAKENHI Grant Numbers JSPS KAKENHI Grant Numbers JP17H04847, JP17H01139, JP15H02105, JP25287079, JP15H05853, JP25107003, the Program for Key Interdisciplinary Research, and the Murata Science Foundation.

References

[1] Kamihara Y *et al.* 2008 *J. Am. Chem. Soc.* **130** 3296
[2] Stewart G R 2011 *Rev. Mod. Phys.* **83** 1589-1652
[3] Singh D J and Du M H 2008 *Phys. Rev. Lett.* **100** 237003
[4] Kuroki K *et al.* 2008 *Phys. Rev. Lett.* **101** 087004
[5] Graser S *et al.* 2009 *New J. Phys.* **11** 025016
[6] Lu D H *et al.* 2008 *Nature* **455** 81-84
[7] Ding H *et al.* 2008 *Europhys. Lett.* **83** 47001
[8] Nakayama K *et al.* 2010 *Phys. Rev. Lett.* **105** 197001
[9] Umezawa K *et al.* 2012 *Phys. Rev. Lett.* **108** 037002
[10] Rhodes L C *et al.* 2017 *Phys. Rev. B* **95** 195111
[11] Kushnirenko Y S *et al.* 2017 *Phys. Rev. B* **96** 100504(R)
[12] Nakayama K *et al.* 2014 *Phys. Rev. Lett.* **113** 237001
[13] Shimojima T *et al.* 2014 *Phys. Rev. B* **90** 121111(R)
[14] Kasahara S *et al.* 2014 *Proc. Natl. Acad. Sci. USA* **111** 16309-16313
[15] Wang Q-Y *et al.* 2012 *Chin. Phys. Lett.* **29** 037402
[16] He S *et al.* 2013 *Nat. Mater.* **12** 605-610
[17] Tan S *et al.* 2013 *Nat. Mater.* **12** 634-640
[18] Ge J-F *et al.* 2015 *Nat. Mater.* **14**, 285-289
[19] Zhang Z *et al.* 2015 *Sci. Bull.* **60** 1301-1304
[20] Hsu F-C *et al.* 2008 *Proc. Natl. Acad. Sci.* **105** 14262-14164
[21] Xiang Y-Y *et al.* 2012 *Phys. Rev. B* **86** 134508
[22] Lee J J *et al.* 2014 *Nature* **515** 245-248
[23] Miyata Y *et al.* 2015 *Nat. Mater.* **14** 775-779
[24] Wen C H *et al.* 2016 *Nat. Commun.* **7** 10840
[25] Phan G N *et al.* 2017 *J. Phys.: Conf. Ser.* **871** 012017

IOP Conf. Series: Journal of Physics: Conf. Series **1054** (2018) 012020 doi:10.1088/1742-6596/1054/1/012020

Anisotropy of critical current densities in $Ba_{1-x}K_xFe_2As_2$ and $Ba(Fe_{1-x}Co_x)_2As_2$ with splayed columnar defects

N Ito[1], S Pyon[1], T Kambara[2], A Yoshida[2], S Okayasu[3], A Ichinose[4], T Tamegai[1]

[1]Department of Applied Physics, The University of Tokyo, 7-3-1 Hongo, Bunkyo-ku, Tokyo 113-8656, Japan

[2]Nishina Center, RIKEN, 2-1 Hirosawa, Wako, Saitama 351-0198, Japan

[3]Japan Atomic Energy Agency, Advanced Science Research Center, Tokai, Ibaraki 319-1195, Japan

[4]Central Research Institute of Electric Power Industry, Electric Power Engineering Research Laboratory, 2-6-1 Nagasaka, Yokosuka, Kanagawa 240-0196, Japan

E-mail: ito-nozomu815@g.ecc.u-tokyo.ac.jp

Abstract. Effects of bimodal splayed columnar defects in $Ba_{1-x}K_xFe_2As_2$ and $Ba(Fe_{1-x}Co_x)_2As_2$ single crystals through 2.6 GeV U irradiation and 220 MeV Au irradiation are investigated. Magnetic field profiles measured by magneto-optical imaging clarified the presence of in-plane anisotropy of the critical current density (J_c) in 2.6 GeV U irradiated crystals, where J_c parallel to the splay plane is larger than that perpendicular to the splay plane. However, the in-plane anisotropy of J_c is absent in 220MeV Au irradiated crystals. Possible origin of this discrepancy is discussed based on the structure of introduced defects.

1. Introduction

Columnar defects in type-II superconductors serve as artificial pinning centers, which lead to an enhancement of critical current density J_c [1, 2, 3]. It has been proposed that a further enhancement of J_c is possible by dispersing the direction of the columnar defects [4]. In such a system with splayed columnar defects, enhancement of J_c has been confirmed in 1.08 GeV Au irradiated $YBa_2Cu_3O_{7-\delta}$ single crystals [5] and 2.6 GeV U irradiated $Ba_{1-x}K_xFe_2As_2$ single crystals [6]. Due to the competing effects of suppression of vortex motion by unequal lengths of vortex segments between splayed columnar defects and detrimental effect of misalignment of vortices, the optimal splay angles were found to be $\approx \pm 5°$ in both of the above systems.

However, since these results were obtained through magnetization measurements, the estimated J_c is a weighted average of J_c's along two directions in the ab plane. Since directions parallel and perpendicular to the splay plane of columnar defects can be distinguished in a system with bimodal splayed columnar defects, J_c parallel to the splay direction ($J_{c\parallel}$) and J_c perpendicular to the splay direction ($J_{c\perp}$) can be different. Such an in-plane anisotropy of J_c in systems with splayed columnar defects has been confirmed through resistivity measurements in 1 GeV Au irradiated $YBa_2Cu_3O_{7-\delta}$ [7] and 3.9 GeV Au irradiated $YBa_2Cu_3O_{7-\delta}$ [8], where Ohmic dissipation was higher when the current is passed in the direction perpendicular to the splay plane, suggesting $J_{c||} > J_{c\perp}$. By contrast, magneto-optical (MO) images of 0.9 GeV Pb irradiated $DyBa_2Cu_3O_{7-\delta}$ with splayed columnar defects indicated $J_{c||} > J_{c\perp}$ at high fields with anisotropy increasing at higher fields, while $J_{c||} < J_{c\perp}$ at low fields [9]. For iron-based superconductors, MO images of 2.6 GeV U irradiated $Ba_{0.6}K_{0.4}Fe_2As_2$ with splayed columnar defects indicated $J_{c||} > J_{c\perp}$ at high temperatures close to T_c under the self-field [10]. However, for iron-based superconductors, no observation has been made on other materials or other ions.

Here, we report in-plane anisotropy of J_c in $Ba_{0.6}K_{0.4}Fe_2As_2$ single crystals and $Ba(Fe_{0.93}Co_{0.07})_2As_2$ single crystals with splayed columnar defects observed through MO imaging, and discuss the possible mechanism for the behavior.

2. Experimental Methods

$Ba_{0.6}K_{0.4}Fe_2As_2$ single crystals ($T_c \sim 38$ K) were synthesized by FeAs flux method [11, 12, 13], and $Ba(Fe_{0.93}Co_{0.07})_2As_2$ single crystals ($T_c \sim 24$ K) were synthesized by FeAs/CoAs flux method [14]. The crystals were shaped into rectangular parallelepiped with length and width of approximately 500 μm and with thickness below half of the projected range of ions. Bimodal splayed columnar defects were installed into the crystals through 2.6 GeV U irradiation or 220 MeV Au irradiation. 2.6 GeV U irradiation was performed using the RIKEN Ring Cyclotron, and 220 MeV Au irradiation was performed using the tandem accelerator in JAEA. The bimodal splay angle (θ_{CD}) is denoted by the angle between their c-axis and the ion beam. For 2.6 GeV U irradiation, all the samples were irradiated at a total dose of $B_\Phi = 8$ T. For 220 MeV Au irradiation, all the samples were irradiated at a total dose of $B_\Phi = 8 \text{ T} \times (\cos\theta_{CD})^{-1}$.

In order to estimate J_c, magnetization measurements were performed with a superconducting quantum interference device (SQUID) magnetometer. The single crystal was placed in a quartz sample holder and fixed with Apiezon N grease. J_c was calculated from the results of the magnetization measurements using extended Bean model.

In order to observe the in-plane anisotropy of J_c, MO imaging was performed. The single crystal was mounted on a copper sample holder and held together with Apiezon N grease to fix and to increase thermal conductivity to the sample holder. On top of the sample holder, a ferrimagnetic garnet indicator film with a reflective mirror was placed in direct contact with the crystal. The whole assembly was placed on a cold finger and shrouded by a radiation shield, then cooled by a He-flow cryostat. The light intensity versus flux density reference was obtained after locating the sample above the T_c between -500 Oe and 500 Oe using a copper solenoid. The reference were fitted with a second order polynomial, in which the coefficients were used to calculate the flux density. Remanent state images were obtained after applying a field of 1 kOe for 1 second.

In the usual MO imaging method using a ferrimagnetic garnet indicator, an in-plane field is applied to diminish the effect of in-plane magnetic domain formed in the indicator. However, an in-plane field can induce an in-plane anisotropy of J_c. In order to observe the pure in-plane anisotropy of the samples, MO imaging was performed without applying in-plane field, which leads to MO images with noises due to in-plane magnetic domains.

3. Results and Discussion

For detailed discussions, it is crucial to be aware of the structure of introduced defects through irraidation. Figures 1(a)-(c) show scanning transmission microscope (STEM) images of 2.6 GeV U irradiated $Ba_{0.6}K_{0.4}Fe_2As_2$ single crystal, 220 MeV Au irradiated $Ba_{0.6}K_{0.4}Fe_2As_2$ single crystal, and 2.6 GeV U irradiated $Ba(Fe_{0.93}Co_{0.07})_2As_2$ single crystal, respectively. The white arrows indicate the c-axis. It should be noted that individual defect structures are expected to be independent of splay angle as long as they are created by heavy ions injected from not too far from the c-axis. As shown in figure 1(b), in $Ba_{0.6}K_{0.4}Fe_2As_2$ single crystal, 220 MeV Au irradiation introduces discontinuous columnar defects with the length of ~10 nm. As shown in figure

IOP Conf. Series: Journal of Physics: Conf. Series **1054** (2018) 012020 doi:10.1088/1742-6596/1054/1/012020

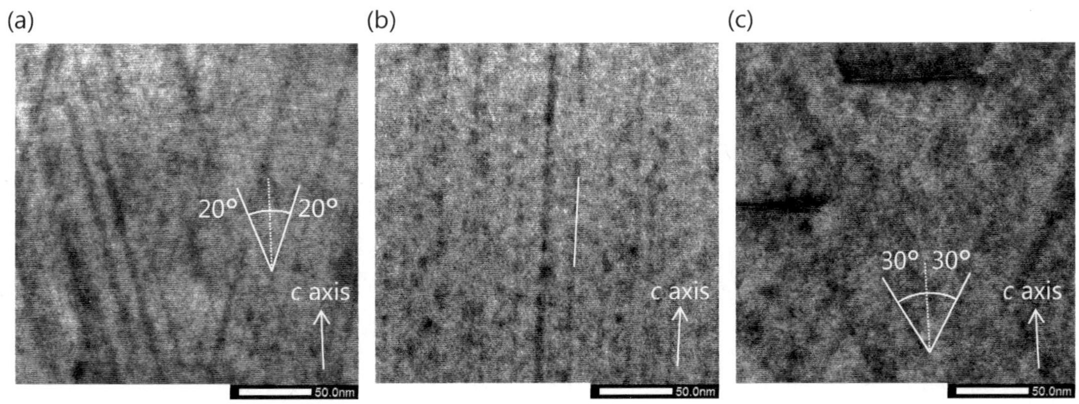

Figure 1. STEM images of cross-sections of (a) 2.6 GeV U irradiated $Ba_{0.6}K_{0.4}Fe_2As_2$ single crystal with splayed columnar defects of $\theta_{CD} = \pm 20°$, (b) 220 MeV Au irradiated $Ba_{0.6}K_{0.4}Fe_2As_2$ single crystal with columnar defects of CD$||c$, and (c) 2.6 GeV U irradiated $Ba(Fe_{0.93}Co_{0.07})_2As_2$ single crystal with splayed columnar defects with $\theta_{CD} = \pm 30°$. The white arrows indicate the c-axis.

1(c), in $Ba(Fe_{0.93}Co_{0.07})_2As_2$ single crystal, 2.6 GeV U irradiation also introduces discontinuous columnar defects with the length of ~100 nm. As shown in figure 1(a), in $Ba_{0.6}K_{0.4}Fe_2As_2$ single crystal, 2.6 GeV U irradiation introduces more continuous columnar defects. Structures perpendicular to the c-axis observed in 2.6 GeV U irradiated $Ba(Fe_{0.93}Co_{0.07})_2As_2$ single crystal (figure 1(c)) are considered to be stacking faults.

Figures 2(a)-(b) show magnetic field dependence of J_c in the pristine and irradiated $Ba_{0.6}K_{0.4}Fe_2As_2$ single crystals. As shown in figure 2(a), the self-field J_c at 2 K in the pristine $Ba_{0.6}K_{0.4}Fe_2As_2$ single crystal exhibits a value of 3.6 MA/cm^2. As shown in figure 2(b), splayed columnar defects introduced through 220 MeV Au irradiation enhance the self-field J_c at 2 K over 20 MA/cm^2.

Figures 2(c)-(e) show magnetic field dependence of J_c in the pristine and irradiated $Ba(Fe_{0.93}Co_{0.07})_2As_2$ single crystals. As shown in figure 2(c), the self-field J_c at 2 K in the pristine $Ba(Fe_{0.93}Co_{0.07})_2As_2$ single crystal exhibits a value of 1.2 MA/cm^2. As shown in figures 2(d)-(e), splayed columnar defects introduced through 2.6 GeV U irradiation and 220 MeV Au irradiation enhance the self-field J_c at 2 K over 2.5 MA/cm^2.

Figure 3(a) shows the remanent state MO image of 2.6 GeV U irradiated $Ba(Fe_{0.93}Co_{0.07})_2As_2$ single crystal with splayed columnar defects with $\theta_{CD} = \pm 5°$ after sweeping the field from 1 kOe back to zero field at 21 K. A critical state with a double Y-shaped current discontinuity line is formed. Unlike the isotropic case with a discontinuity line of an angle of 45° with respect to the sample edge, in-plane anisotropy of J_c appears as a skewed double Y-shaped discontinuity line. It is confirmed $J_{c||} > J_{c\perp}$ for 2.6 GeV U irradiated $Ba(Fe_{0.93}Co_{0.07})_2As_2$ single crystal with splayed columnar defects, which is consistent with the result for 2.6 GeV U irradiated $Ba_{0.6}K_{0.4}Fe_2As_2$ [10].

For quantitative analysis of the in-plane anisotropy of J_c, line profiles (figure 3(b), (c)) were taken along two lines parallel (white) and perpendicular (red) to the splay direction in figure 3(a). The distances of the flux density peak from the sample edge are not equal. Here, the in-plane anisotropy of J_c can be estimated from the ratio of the two distances. For 2.6 GeV U irradiated $Ba(Fe_{0.93}Co_{0.07})_2As_2$ single crystal with splayed columnar defects with $\theta_{CD} = \pm 5°$, the in-plane anisotropy is estimated as $J_{c||}/J_{c\perp} \sim 1.23$ at 21 K.

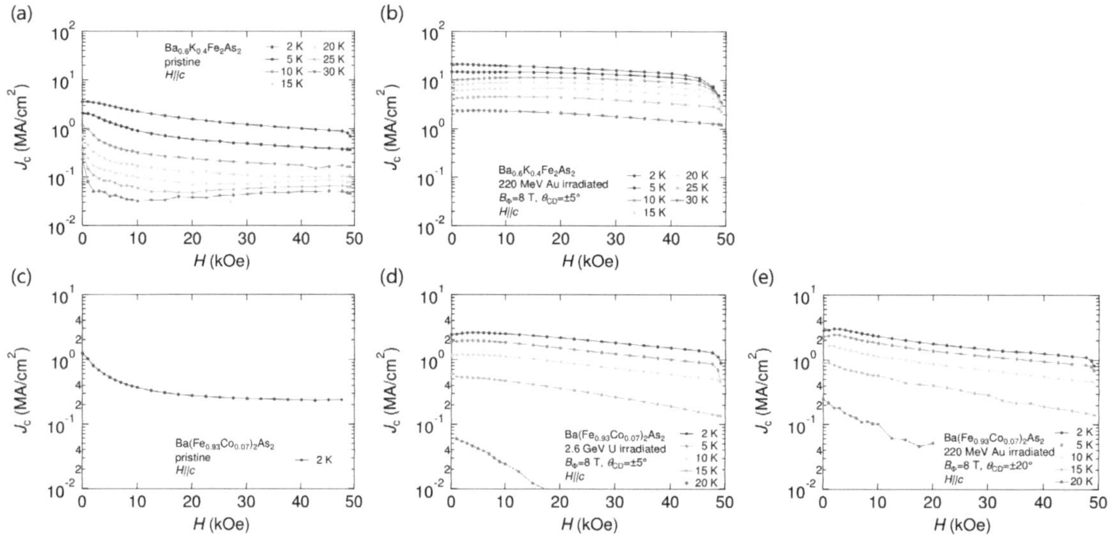

Figure 2. Magnetic field dependence of J_c in (a) pristine $Ba_{0.6}K_{0.4}Fe_2As_2$ single crystal, (b) 220 MeV Au irradiated $Ba_{0.6}K_{0.4}Fe_2As_2$ single crystal with splayed columnar defects with $\theta_{CD} = \pm 5°$, (c) pristine $Ba(Fe_{0.93}Co_{0.07})_2As_2$ single crystal, (d) 2.6 GeV U irradiated $Ba(Fe_{0.93}Co_{0.07})_2As_2$ single crystal with splayed columnar defects with $\theta_{CD} = \pm 5°$ and (e) 220 MeV Au irradiated $Ba(Fe_{0.93}Co_{0.07})_2As_2$ single crystal with splayed columnar defects with $\theta_{CD} = \pm 20°$.

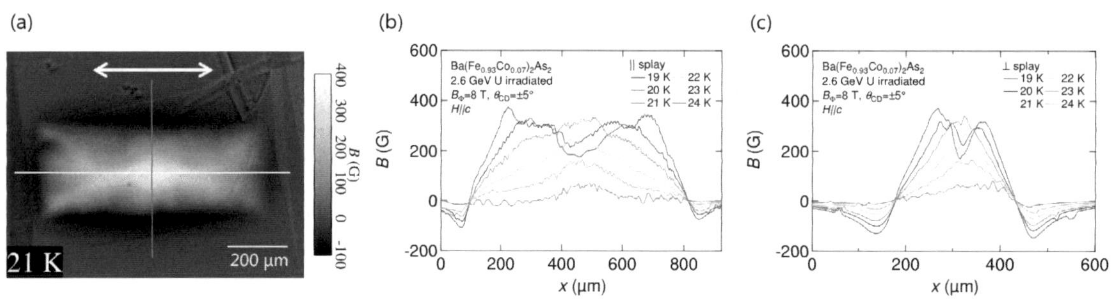

Figure 3. (a) The MO image of 2.6 GeV U irradiated $Ba(Fe_{0.93}Co_{0.07})_2As_2$ single crystal with splayed columnar defects with $\theta_{CD} = \pm 5°$ in the remanent state after applying a field of 1 kOe along the c-axis at $T = 21$ K. The white arrow indicates the splay direction. (b) Line profiles along the white line in (a). (c) Line profiles along the red line in (a).

Figure 4(a) shows the remanent state MO image of 220 MeV Au irradiated $Ba_{0.6}K_{0.4}Fe_2As_2$ single crystal with splayed columnar defects with $\theta_{CD} = \pm 5°$ after sweeping the field from 1 kOe back to zero field at 37 K. Since the crystal has a macroscopic defect in the left half, the in-plane anisotropy should be discussed focusing on the right half of the sample. Unlike 2.6 GeV U irradiated $Ba_{0.6}K_{0.4}Fe_2As_2$ and $Ba(Fe_{0.93}Co_{0.07})_2As_2$, the angle between the discontinuity line and the sample edge is nearly 45°, which indicates isotropic J_c. Figures 4(b)-(c) show line profiles taken along two axes as shown by the white (parallel to the splay direction) and red (perpendicular to the splay direction) lines in figure 4(a). For 220 MeV Au irradiated

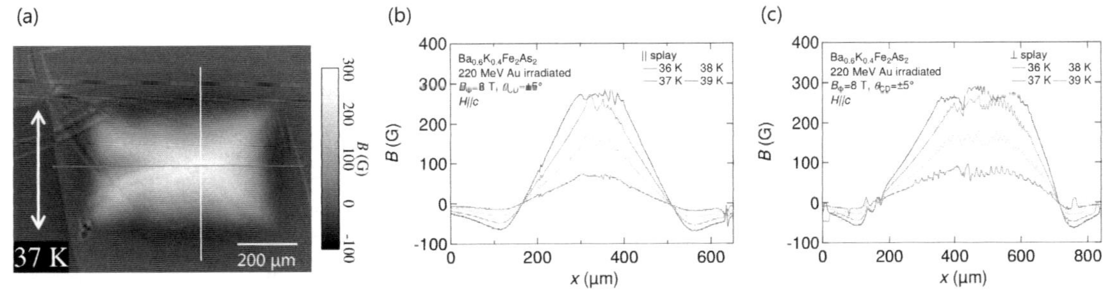

Figure 4. (a) The MO image of 220 MeV Au irradiated $Ba_{0.6}K_{0.4}Fe_2As_2$ single crystal with splayed columnar defects with $\theta_{CD} = \pm 5°$ in the remanent state after applying a field of 1 kOe along the c-axis at $T = 37$ K. The white arrow indicates the splay direction. (b) Line profiles along the white line in (a). (c) Line profiles along the red line in (a).

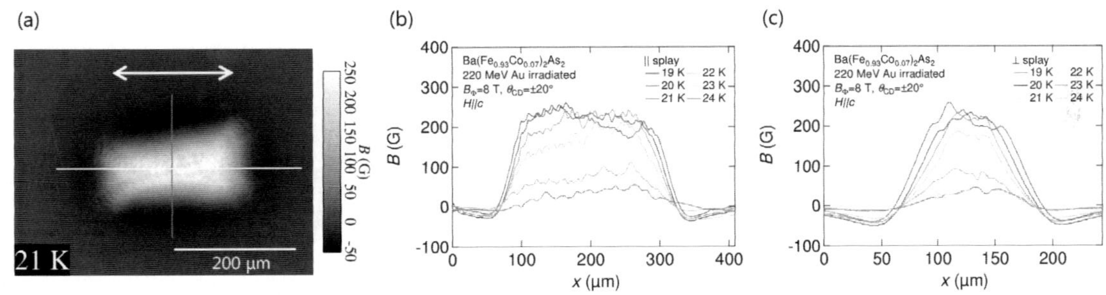

Figure 5. (a) The MO image of 220 MeV Au irradiated $Ba(Fe_{0.93}Co_{0.07})_2As_2$ single crystal with splayed columnar defects with $\theta_{CD} = \pm 20°$ in the remanent state after applying a field of 1 kOe along the c-axis at $T = 21$ K. The white arrow indicates the splay direction. (b) Line profiles along the white line in (a). (c) Line profiles along the red line in (a).

$Ba_{0.6}K_{0.4}Fe_2As_2$ single crystal with splayed columnar defects with $\theta_{CD} = \pm 5°$, the in-plane anisotropy is estimated as $J_{c\|}/J_{c\perp} \sim 1.08 \sim 1$ at 37 K.

Figure 5(a) shows the remanent state MO image of 220 MeV Au irradiated $Ba(Fe_{0.93}Co_{0.07})_2As_2$ single crystal with splayed columnar defects with $\theta_{CD} = \pm 20°$ after sweeping the field from 1 kOe back to zero field at 21 K. Unlike 2.6 GeV U irradiated $Ba_{0.6}K_{0.4}Fe_2As_2$ and $Ba(Fe_{0.93}Co_{0.07})_2As_2$, the angle between the discontinuity line and the sample edge is nearly 45°, which indicates isotropic J_c. Figure 5(b)-(c) show line profiles taken along two axes as shown by the white (parallel to the splay direction) and red (perpendicular to the splay direction) lines in figures 5(a). For 220 MeV Au irradiated $Ba(Fe_{0.93}Co_{0.07})_2As_2$ single crystal with splayed columnar defects with $\theta_{CD} = \pm 20°$, the in-plane anisotropy is estimated as $J_{c\|}/J_{c\perp} \sim 0.96 \sim 1$ at 21 K.

All the obtained results can be summarized as follows. 2.6 GeV U irradiated samples with splayed columnar defects show clear in-plane anisotropy with $J_{c\|} > J_{c\perp}$, while 220 MeV Au irradiated samples with splayed columnar defects have little in-plane anisotropy. These results are interpreted as follows. In systems with bimodal splayed columnar defects, the vortex motion is controlled by zig-zag type kinks for $J_{c\|}$, whereas the vortex motion is manifested by double kinks for $J_{c\perp}$, as suggested by López *et al.* [7], which leads to $J_{c\|} > J_{c\perp}$. One of the possible origins of the difference between 2.6 GeV U irradiation and 220 MeV Au irradiation is continuity of the introduced columnar defects. Columnar defects in 220 MeV Au irradiated $Ba_{0.6}K_{0.4}Fe_2As_2$ are discontinuous, while those in 2.6 GeV U irradiated $Ba_{0.6}K_{0.4}Fe_2As_2$ are more continuous. Splayed continuous columnar defects such as those introduced in $Ba_{0.6}K_{0.4}Fe_2As_2$ through 2.6 GeV U irradiation can strongly suppress the vortex motion for $J_{c\parallel}$, which leads to $J_{c\parallel} > J_{c\perp}$, while splayed discontinuous columnar defects such as those introduced in $Ba_{0.6}K_{0.4}Fe_2As_2$ through 220 MeV Au irradiation serve as relatively isotropic pinning centers, which leads to little in-plane anisotropy of J_c. Though columnar defects introduced in $Ba(Fe_{0.93}Co_{0.07})_2As_2$ through 2.6 GeV U irradiation are also discontinuous, each segment is much longer than those introduced in $Ba_{0.6}K_{0.4}Fe_2As_2$ through 220 MeV Au irradiation and can induce a larger in-plane anisotropy. Since 220 MeV Au irradiated $Ba(Fe_{0.93}Co_{0.07})_2As_2$ single crystal has little in-plane anisotropy, columnar defects introduced in $Ba(Fe_{0.93}Co_{0.07})_2As_2$ through 220 MeV Au irradiation are expected to be similarly discontinuous those introduced in $Ba_{0.6}K_{0.4}Fe_2As_2$ through 220 MeV Au irradiation.

4. Summary

We performed MO imaging to observe in-plane anisotropy of J_c in 2.6 GeV U irradiated $Ba(Fe_{0.93}Co_{0.07})_2As_2$, 220 MeV Au irradiated $Ba_{0.6}K_{0.4}Fe_2As_2$, and 220 MeV Au irradiated $Ba(Fe_{0.93}Co_{0.07})_2As_2$ single crystals with bimodal splayed columnar defects. We found obvious in-plane anisotropy as $J_{c\parallel} > J_{c\perp}$ in 2.6 GeV U irradiated crystals and no obvious in-plane anisotropy in 220 MeV Au irradiated crystals. The difference between 2.6 GeV U irradiated crystals and 220 MeV Au irradiated crystals can be understood as the difference of pinning efficiency for $J_{c\parallel}$ due to the difference of continuity of introduced columnar defects.

References

[1] Civale L, Marwick A D, Worthington T K, Kirk M A, Thompson J R, Krusin-Elbaum L, Sun Y, Clem J R and Holtzberg F 1991 *Phys. Rev. Lett.* **67** 648
[2] Nakajima Y, Tsuchiya Y, Taen T, Tamegai T, Okayasu S and Sasase M 2009 *Phys. Rev.* B **80** 012510
[3] Tamegai T, Taen T, Yagyuda H, Tsuchiya Y, Mohan S, Taniguchi T, Nakajima Y, Okayasu S, Sasase M, Kitamura H, Murakami T, Kambara T and Kanai Y 2012 *Supercond. Sci. Technol.* **25** 084008
[4] Hwa T, Doussal P L, Nelson D R and Vinokur V M 1993 *Phys. Rev. Lett.* **71** 3545
[5] Krusin-Elbaum L, Marwick A D, Wheeler R, Feild C, Vinokur V M, Leaf G K and Palumbo M 1996 *Phys. Rev. Lett.* **76** 2563
[6] Park A, Pyon S, Tamegai T and Kambara T 2016 *Physica* C **530** 58
[7] López D, Krusin-Elbaum L, Safar H, Vinokur V M, Marwick A D, Sun J Z and Feild C 1997 *Phys. Rev. Lett.* **79** 4258
[8] Kwok W K, Paulius L M, Vinokur V M, Petrean A M, Ronningen R M and Crabtree G W 1998 *Phys. Rev. B* **58** 14594
[9] Schuster T, Kuhn H, Indenbom M, Leghissa M, Kraus M and Konczykowski M 1995 *Phys. Rev. B* **51** 16358
[10] Park A, Pyon S, Ohara K, Ito N, Tamegai T, Kambara T, Yoshida A and Ichinose A 2018 *Phys. Rev. B* **97** 064516
[11] Kihou K, Saito T, Ishida S, Nakajima M, Tomioka Y, Fukazawa H, Kohori Y, Ito T, Uchida S I, Iyo A, Lee C H and Eisaki H 2010 *J. Phys. Soc. Jpn.* **79** 124713
[12] Taen T, Ohtake F, Akiyama H, Inoue H, Sun Y, Pyon S and Tamegai T 2013 *Phys. Rev.* B **88** 224514
[13] Taen T, Ohtake F, Pyon S, Tamegai T and Kitamura H 2015 *Supercond. Sci. Technol.* **28** 085003
[14] Nakajima Y, Taen T and Tamegai T 2009 *J. Phys. Soc. Jpn.* **78** 023702

IOP Conf. Series: Journal of Physics: Conf. Series **1054** (2018) 012021 doi:10.1088/1742-6596/1054/1/012021

Direct Current Measurement of Hall Effect in the Mixed State for the Iron-chalcogenide Superconductors

R. Ogawa, T. Ishikawa, M. Kawai, F. Nabeshima, A. Maeda

Dept. of Basic Science, the Univ. of Tokyo

Abstract. In order to investigate how the composition and the pinning affect the behavior of the Hall effect in the mixed state, we measure the Hall resistivity for iron-chalcogenide superconductor $FeSe_{1-x}Te_x$ films near the superconducting transition temperature. We observe a sign change of the Hall resistivity in $FeSe_{0.5}Te_{0.5}$ ($x = 0.5$) films, but $FeSe_{0.8}Te_{0.2}$ ($x = 0.2$) films do not show the sign reversal. Increasing applying current density, the sign change of the Hall resistivity in the film with $x = 0.5$ disappeared. We evaluated the strength of the pinning in samples from an activation energy and a scaling law, and found that the film with $x = 0.5$ has the stronger pinning than that of the film with $x = 0.2$. All of those results suggest that the pinning has related to the Hall resistivity anomaly in Fe chalcogenides.

1. Introduction

Properties of the type-II superconductors have attracted significant interest for many years. In particular, one of the interesting and puzzling phenomenon is the Hall effect of superconductors in the mixed state. It is known that the sign of the Hall resistivity in the mixed state is different from that in the normal state, for some high T_c superconductors and conventional superconductors, e.g. V and Nb [1-3], and it was recently reported that the Hall resistivity of an iron-based superconductor, $Ba(Fe_{1-x}Co_x)_2As_2$, also showed the sign reversal below T_c [4]. Moreover, a double sign reversal was observed in some cuprates, such as $Tl_2Ba_2CaCu_2O_8$ [5].

In general, the behavior of the Hall effect in the normal state reflects the electronic structure at Fermi surface. On the other hand, in the mixed state, a vortex motion contributes to the behavior of the Hall effect. However, such anomalous sign changes of the Hall resistivity cannot be explained by basic models of vortex motion. For example, in the Bardeen-Stephen model [6], where the vortex cores are considered to be in the normal state, the sign of Hall resistivity in the mixed state is the same as in the normal state. Likewise, the Nozières-Vinen model, in which vortex cores experience Magnus force, also leads to the same sign of Hall resistivity as in the normal state [7,8].

Since the discovery of the Hall anomalies, several theoretical approaches have been taken to explain this phenomenon, and as the origin of sign reversal, many possibilities were proposed. One is the influence of the superconducting fluctuations. Fukuyama, Ebisawa and Tsuzuki have shown that Hall conductivity appears as a result of the electron-hole asymmetry, in the time dependent Ginzburg-Landau (TDGL) equation [9]. Similarly, Aronov, Hikami and Larkin have shown that the sign of Hall conductivity is determined by a general gauge invariance requirement of the TDGL equation [10]. However, the application of those theories to cuprates gives the opposite result to the experiments [11]. Wang *et al.*, have shown that a backflow current due to the pinning has an influence on the Hall effect [12]. Kopinin and Vinokur have also shown that a sufficiently strong pinning can result in a sign reversal [13]. On the other hand, experiments in samples with columnar defects showed the contradictory results about whether a pinning modify the Hall conductivity or not [14,15]. In addition, other possibilities were pointed out, such as the vortex core charging [16], the vortex many-body correlation [17]. However, the origin of the Hall anomaly still remains controversial, and consensus regarding this subject is not reached yet.

Another interesting issue on the Hall effect in the mixed state is a scaling behavior, $|\rho_{xy}| \sim \rho_{xx}^{\beta}$. The scaling law was first discovered for YBCO films [18], and then it has been observed in other superconductors in the mixed state [19]. A phenomenological model considering the effect of pinning on the Hall resistivity gives a scaling exponent β of 2 [20]. Wang *et al.*, proposed a theory for the Hall effect in the mixed state including both the flux pinning effect and thermal fluctuations, and they demonstrated that the scaling exponent β decrease with increasing pinning strength [12].

In order to clarify the behavior of the Hall effect in the mixed state, we measured the Hall resistivity for the iron-chalcogenide superconductor $FeSe_{1-x}Te_x$ films near the transition temperature, and investigate how the composition and the pinning strength affect the Hall effect in the mixed state. We observed the sign reversal of Hall resistivity for $FeSe_{0.5}Te_{0.5}$ ($x = 0.5$) films, while $FeSe_{0.8}Te_{0.2}$ ($x = 0.2$) films do not show the sign anomaly. We also observed the relation of $|\rho_{xy}| \sim \rho_{xx}^{\beta}$, which indicate the validity of the scaling in those samples. Evaluation of the activation energy and the scaling exponent suggest that the Hall anomaly has deeply related to the vortex motion and pinning in the iron-chalcogenide superconductor.

2. Experiment

In this study, we measured $FeSe_{1-x}Te_x$ films with $x = 0.2$ and 0.5, which were grown by the PLD method with a KrF laser[21-23]. $FeSe_{1-x}Te_x$ polycrystalline pellets were used as targets and commercially available CaF_2 (100) substrates were used. Measurements were carried out with a Physical Property Measurement System of Quantum Design. The longitudinal resistance and Hall resistance were measured by the standard six-probe method using silver paste for contact with the magnetic field up to 9 T applied perpendicular to Fe-chalcogenide planes. Unless otherwise noted, the measurements were carried by applying a dc current density of 100 A/cm^2.

3. Results and discussions

Figure 1 shows the temperature dependence of the longitudinal resistivity, ρ_{xx}, and Hall resistivity, ρ_{xy}, for a film with $x = 0.2$ and a flim with $x = 0.5$. For the film with $x = 0.2$ near the superconducting transition temperature, the Hall resistivity approaches to zero from positive as temperature decreases and does not show a sign change . On the other hand, for the film with $x = 0.5$, we observed that the sign of the Hall resistivity changes from positive to negative as temperature decreases at $B = 0.5$ T. Figure 2 represents a current-voltage characteristics for the film with $x = 0.5$. The gradient of *I-V* curve is approximately unity in the region where the sign changes are observed, which means that the longitudinal resistivity is ohmic and Hall resistivity anomaly is observed in the thermally assisted flux-flow (TAFF) region.

The magnetic field dependences of the Hall conductivity are expressed as $\sigma_{xy} = \sigma_{xy}^{\mathrm{n}} + \sigma_{xy}^{\mathrm{SC}}$, where σ_{xy}^{n} and $\sigma_{xy}^{\mathrm{SC}}$ are the Hall conductivity due to the normal quasiparticles inside and around the vortex and the Hall conductivity due to the motion of vortices. σ_{xy}^{n} is proportional to B and $\sigma_{xy}^{\mathrm{SC}}$, according to TDGL theory, is proportional to $1/B$ [24]. Therefore, in low magnetic fields, the superconducting term becomes dominant and the Hall conductivity is approximately proportional to $1/B$. However, the Hall conductivity drops faster than $1/B$ for the film with $x = 0.5$ at lower temperatures as the magnetic field increases, which are shown in Figure 3.

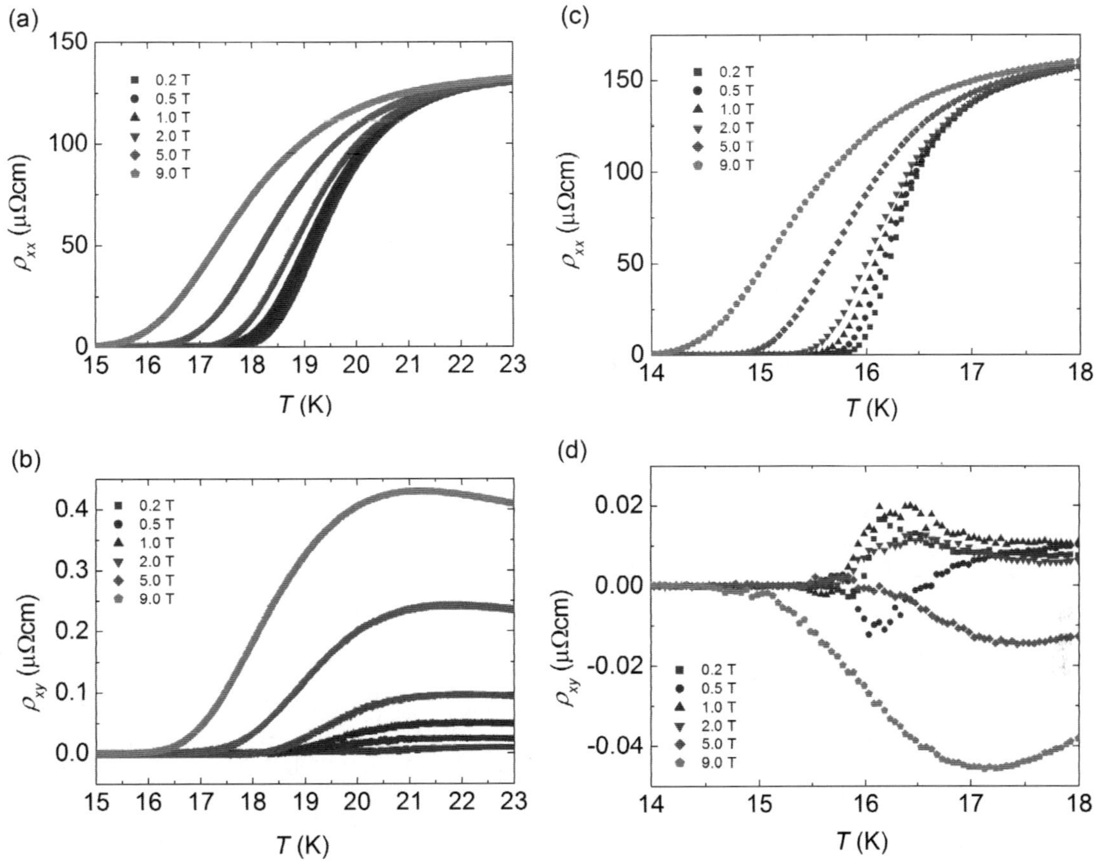

Figure 1. (a),(b) Temperature dependences of the longitudinal resistivity and the Hall resistivity for film with $x = 0.2$. (c),(d) Temperature dependences of the longitudinal resistivity and the Hall resistivity for film with $x = 0.5$.

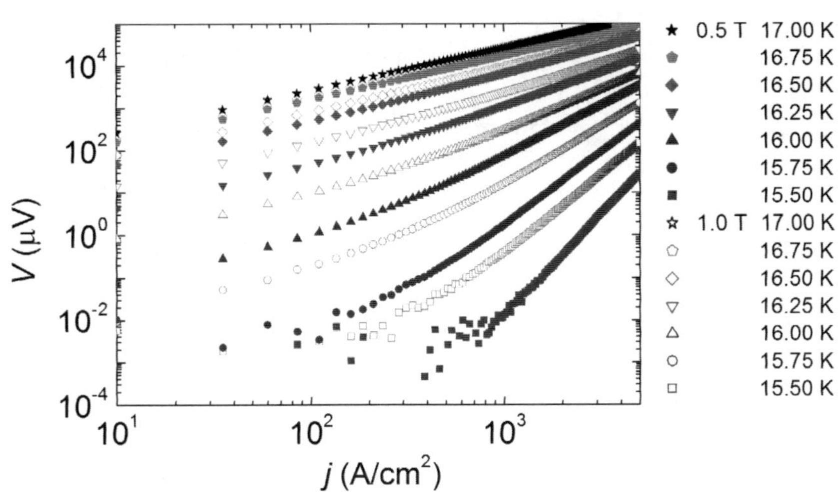

Figure 2. DC current-voltage characteristics of film with $x = 0.5$.

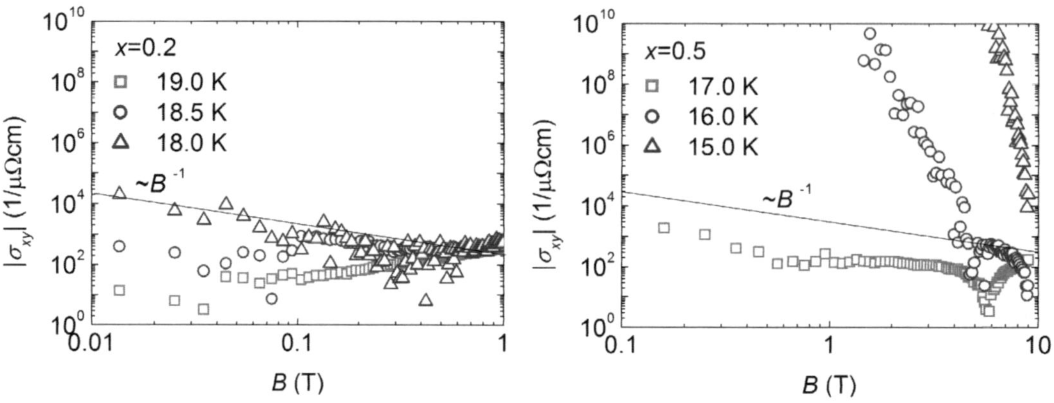

Figure 3. Magnetic field dependences of the absolute values of Hall conductivity for films with $x = 0.2$ and $x = 0.5$ respectively. Solid lines are proportional to $1/B$.

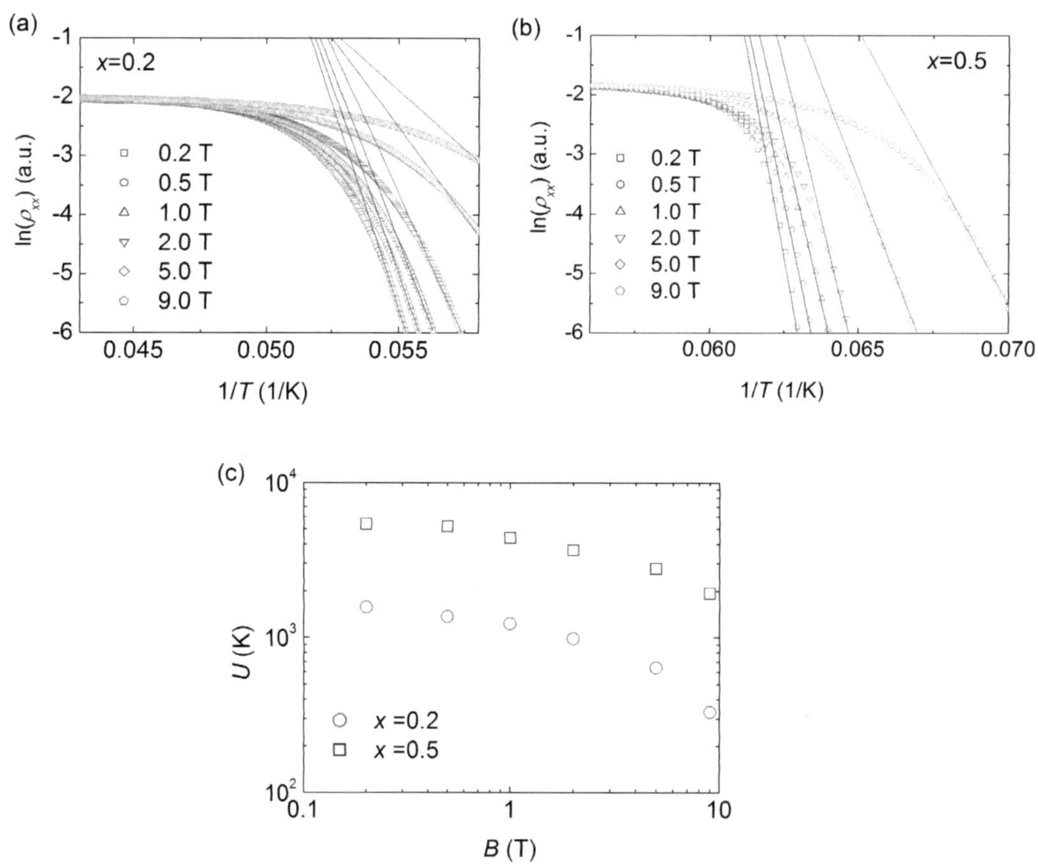

Figure 4. (a),(b) Arrhenius plots for films with $x = 0.2$ and $x = 0.5$. Solid lines are the fitted lines to obtain activation energies. (c) Calculated magnetic field dependence of the activation energy for $x = 0.2$ and $x = 0.5$, respectively.

To investigate the influence of the pinning on the Hall effect in the mixed state, we show Arrhenius plots for the films with $x = 0.2$ and $x = 0.5$ (Figure 4). The longitudinal resistivity in the TAFF region is expressed as $\rho_{xx} \sim \exp(-U/k_B T)$, where U is the activation energy [20]. From the gradients of Arrhenius plots in low temperature region, we find that the values of U are 1200 K for $x = 0.2$ at $B = 1.0$ T, and 4500 K for $x = 0.5$ at $B = 1.0$ T, respectively. Together with the already-published Hall data in other iron-based superconductors [25-28], we summarized the estimated U and the absence/ presence of the sign reversal of the Hall effect in Table 1. This suggests that the sign changes occur when the activation energy is relatively high. The result of our measurements for $FeSe_{1-x}Te_x$ films shows the same tendency and consistent with the previous studies in other iron-based superconductors.

To further investigate the influence of the pinning, we measured the Hall resistivity for film with $x = 0.5$ at different dc current densities from 10 A/cm^2 to 5000 A/cm^2. Figure 5 shows the temperature dependence of the longitudinal resistivity and the Hall resistivity at $B = 0.5$ T and $B = 1.0$ T. At $B = 0.5$ T, the sign reversal disappears as the current density become stronger. This result suggests that the strong current density decrease the pinning potential, leading to the disappearance of the sign change.

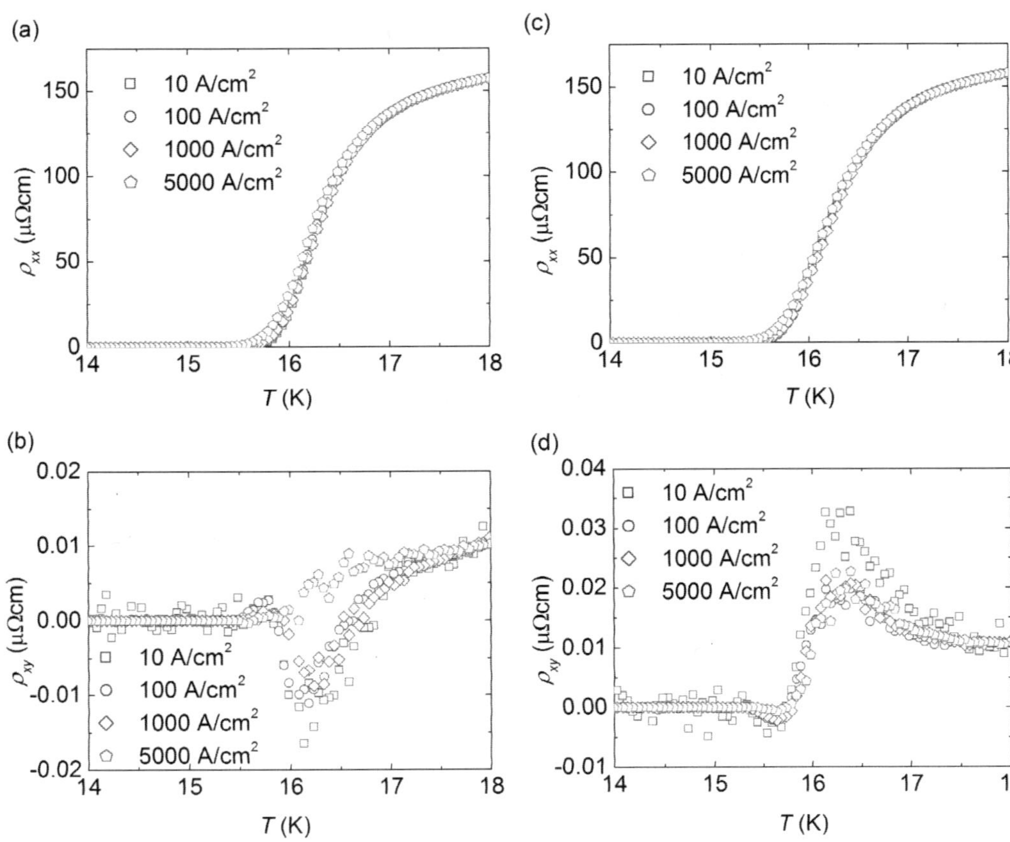

Figure 5. (a),(b) Temperature dependences of the longitudinal resistivity and the Hall resistivity at $B = 0.5$ T for a film with $x = 0.5$ by applying different current densities. (c),(d) Temperature dependences of the logitudinal resistivity and the Hall resistivity at $B = 1.0$ T for a film with $x = 0.5$ by applying different current densities.

Table 1. The relation between the activation energy and the sign reversal for various iron-based superconductors.

		$U(B = 1\ \mathrm{T})$ (K)	Sign reversal	Ref.
$Fe_{1+y}(Te_{1+x}S_x)_z$	single crystal	200	×	[25]
$Ba(Fe_{1-x}Co_x)_2As_2$ ($x = 0.08$)	single crystal	4700	×	[26]
$Ba(Fe_{1-x}Co_x)_2As_2$ ($x = 0.10$)	single crystal	6000	○	[26]
$BaFe_2As_2$	film	2300	×	[27]
$NaFe_{1-x}Co_{x2}As_2$ ($x = 0.0220$)	single crystal	1000	×	[28]
$NaFe_{1-x}Co_{x2}As_2$ ($x = 0.0205$)	single crystal	2000	×	[28]
$FeSe_{1-x}Te_x$ ($x = 0.2$)	film	1200	×	
$FeSe_{1-x}Te_x$ ($x = 0.5$)	film	4500	○	

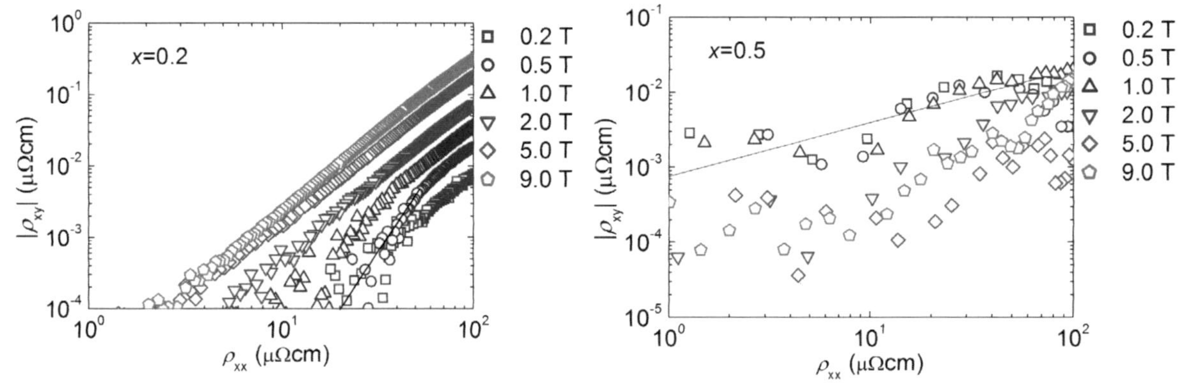

Figure 6. The relation between the longitudinal resistivity and the Hall resistivity for $x = 0.2$ and 0.5. Solid lines are the fitting lines to obtain scaling exponents β.

To evaluate the strength of pinning in other indicators, we plot the relation between the longitudinal resistivity and the Hall resistivity for $FeSe_{1-x}Te_x$ films, which are shown in Figure 6. We measured those values at fixed magnetic fields and sweeping temperatures. We observed $|\rho_{xy}| \sim \rho_{xx}^{\beta}$, which indicate the validity of the scaling in those samples, and we obtained, at $B = 0.5$ T, the scaling exponents β of 2.2 for film with $x = 0.2$, and of 0.7 for film with $x = 0.5$, respectively. According to Wang's theory [12], a larger β value implies weaker pinning. Therefore, those values are consistent with the activation energy analysis shown in Table I.

4. Conclusion

We measured the Hall resistivity of $FeSe_{1-x}Te_x$ films near T_c and observed that the sign of the Hall resistivity changes for films with $x = 0.5$, while the sign of the Hall resistivity of films $x = 0.2$ remains the same as that in the normal state. Increasing applying current density, the sign change of the Hall resistivity vanished. Evaluation of the activation energy and the scaling exponent clarify that the film with $x = 0.5$ has the relatively stronger pinning than that of the film with $x = 0.2$. All of those results suggest that the Hall anomaly has deeply related to the vortex motion and the pinning in the iron-chalcogenide superconductor.

References

[1] Van Beelen H, *et al.* 1967 *Physica* **36** 241
[2] Galffy M and Zirngiebl E 1988 *Solid state communications* **68** 929
[3] Blatter G, *et al.* 1994 *Rev. Mod. Phys.* **66** 1125
[4] Wang L M, *et al.*2011 *Phys. Rev.* B **83** 134506
[5] Hagen S J, *et al.*1991 *Phys. Rev.* B **43** 6246
[6] Bardeen J and Stephen M J 1965 *Phys. Rev.* **140** A1197
[7] Noziéres P and Vinen W F 1966 *Phys. Mag.* **14** 667
[8] Vinen W F and Warren A C 1967 *Proc. Phys. Soc.* **91**410
[9] Fukuyama H, Ebisawa H, and Tsuzuki T 1971 *Prog. Theor. Phys.* **46** 1028
[10] Aronov A G, Hikami S, and Larkin A I 1965 *Phys. Rev.* B **51** 3880
[11] Nagaoka T, *et al.* 1998 *Phys. Rev. Lett.* **80** 3594
[12] Wang Z D, Dong J, and Ting C S 1994 *Phys. Rev. Lett.* **72** 3875
[13] Kopnin N B, and Vinokur V M 1999 *Phys. Rev. Lett.* **83** 4864
[14] Kang W N., *et al.* 1996 *Phys. Rev. Lett.* **76** 2993.
[15] Samoilov A V, *et al.* 1995 *Phys. Rev. Lett.* **74** 2351
[16] Khomskii D I and Freimuth A 1995 *Phys. Rev. Lett.* **75** 1384
[17] Ao P 1998 *J. Phys. Cond. Matt.* **10** L677
[18] Luo J, *et al.* 1992 *Phys. Rev. lett.* **68** 690
[19] Samoilov A V 1993 *Phys. Rev. Lett.* **71** 617
[20] Vinokur V M, *et al.* 1993 *Phys. Rev. Lett.* **71** 1242
[21] Imai Y, *et al.* 2010 *Japanese Journal of Applied Physics* **49** 023101
[22] Imai Y, *et al.* 2010 *Applied Physics Express* **3** 043102
[23] Imai Y, *et al.* 2015 *Proceedings of the National Academy of Sciences* **112** 1937
[24] Dorsey A T 1992 *Phys. Rev.* B **46** 8376
[25] Lei H, *et al.* 2010 *Phys. Rev.* B **82** 134525
[26] Wang L M, *et al.* 2011 *Phys. Rev.* B **83** 134506
[27] Sato H, *et al.* 2013 *Phys. Rev.* B **87** 064504
[28] Wang L M, *et al.* 2013 *J. Phys. Cond. Matt.* **25** 395702

Anomalous Metal Interface Effect of Iron-based Superconductors

R Tajima[1], Y Miyamoto[1], T Fujii[2], and A Matsuda[1]

1. Department of Physics, School of Advanced Science and Engineering, Waseda University, Okubo 3-4-1, Shinjuku, Tokyo 160-8555, Japan
2. Cryogenic Research Center, the University of Tokyo

Abstract: We report the results of the Andreev spectroscopy of the (Co-doped Ba-122 single crystals)/metal/Pb junctions with a metal layer of Al, Pb, Ag, and Sn. We observed Andreev-like spectra for each metal layer and the junctions were identified to be (superconductor)/(normal metal) type. The spectra showed different behaviors for each metal layer. Especially in the case of Sn, the junction showed complex structure where different S/N type features were folded. We consider this behavior is due to a chemical reaction, in the interface between an iron-based superconductor and a layering metal.

Keywords : Iron-based superconductor, Tunnel junction, Conductance spectroscopy, Andreev reflection

1. Introduction

In iron-based superconductors (IBS), the pairing symmetry of the superconducting order parameter is particularly important to investigate its pairing mechanism. The Josephson effect measurements, which have actually been used to determine the symmetry of cuprate superconductors as d-wave [1], is one of the most effective methods for determining the pairing symmetry.

We have studied the junctions, in which one electrode was a Fe[Te,Se] single crystal and the counter electrode was metal/(metal oxide)/Pb double (triple) layers, because the proximity effect between IBS and a conventional metal is expected to be another sensitive tool to investigate the pairing symmetry of IBS. However, an abnormally large gap was observed in some combination of the electrodes [2]. Then, we estimated this structure as the interface effect between IBS and a specific metal (in this case, Pb). Since the proximity effect is strongly affected by the condition of the interfaces between electrodes, it is very important to clarify what is happing in the interfaces. It is also important for the Josephson effect criteria to be effective. Moreover, the existence of a large gap in Fe[Te,Se] case may suggest the stabilization or the formation of the higher T_c phase/material due to the interface, especially in the FeSe system.

In this paper, we report the results of the Andreev spectroscopy of the Co-doped $Ba(FeAs)_2$ (Ba-122) single crystals, using a junction structure same as described in our previous report [2]. Here, a Ba-122 crystal surface has a direct contact with a proximity layer metal. Ba-122 has been known to give good contact with the metal electrode and show the Josephson effect in the combination with Pb [3]. Therefore, it is suitable to investigate the material dependence of the proximity layer. We report here the results of Al, Pb, Sn, and Ag metal layers. In addition to conventional Andreev spectra, we will show the very anomalous spectra both in the structure and in the temperature dependence in a junction with a metal contact.

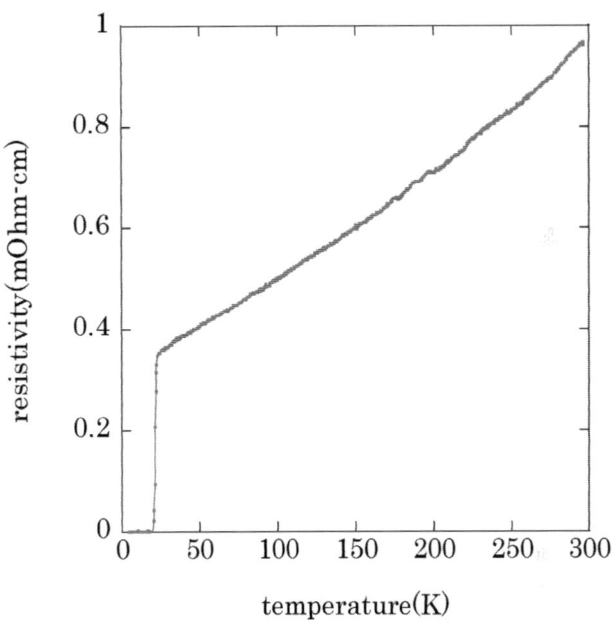

Figure 1. Temperature dependence of resistivity for $Ba(Fe_{0.88}Co_{0.12})_2As_2$ single crystal.

2. Experiment

In this work, we used the single crystals of $Ba(Fe_{0.88}Co_{0.12})_2As_2$ grown by a flux method. 'Figure 1' shows a typical temperature dependence of the resistivity of a crystal. The T_c onset occurred at ~ 22 K, and the transition width was ~ 3 K.

A $Ba(Fe_{0.88}Co_{0.12})_2As_2$ crystal was first set on the sapphire substrate and cleaved in a vacuum. Next, an Al, Ag or Sn proximity layer with the thickness of ~2 nm was deposited on the clean surface immediately after the cleavage. Then after taking the substrate out of the vacuum chamber, the junction window was formed by a resin in the air. Finally, the Pb counter electrode with a thickness of ~800 nm was deposited to form a cross-shape via a metal mask. The typical size of the junction area was ~0.3×0.3 mm^2. During depositions, a pressure was kept under ~10^{-6} torr. In this process, we did not intentionally oxidize the metal surface before depositing Pb electrode, so that oxide layer was not expected on the proximity layer. 'Figure 2' shows schematic view of our junction.

We measured *I-V* characteristics and their temperature dependences using a four-probe geometry. The differential conductance was obtained by a numerical differentiation of *I-V* characteristics.

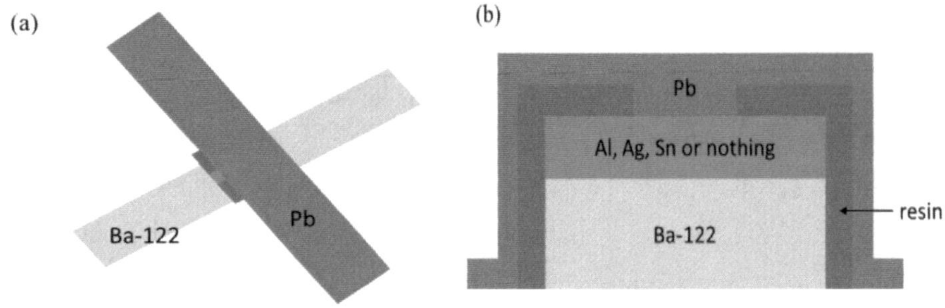

Figure 2. (a) Top of view, (b) cross sectional view of the junction

3. Result and discussion

'Figure 3' shows the results of *dI/dV-V* characteristics for junctions with various proximity layer materials. Each curve was normalized by the data above T_c, where the all the *dI/dV-V* curves were folded. In 'figure 3(d)', the origins of the data were shifted in a vertical direction according to their temperature. Our junctions were considered to be a (superconductor)/(normal metal) type, because the effect of the superconducting transition of the Pb counter electrode was not clearly observed. One reason can be that the conductance between the Ba-122 and the proximity layer is expected to be much lower than that between the proximity layer and the counter electrode. Another reason is that the gap value of Pb is very small compared to that of Ba-122 and can be neglected if major gap structures are concerned.

We found the features, which may be accounted by the Andreev reflection, in each junction. However, the detailed shapes were evidently dependent on the material of the proximity layer. In the case of the Al layer in 'figure 3(a)', a typical Andreev spectrum was obtained. The spectra showed a small gap structure at ~3 mV and a large gap structure at ~8 mV. These values are approximately consistent with the previous reports [5]. However, above 10 mV no structure was observed. Since they might be well interpreted by the two gap BTK model [5], we assure that our device is well described as an S/N junction.

In the case of Pb and Ag, in 'figure 3(b)' and 'figure 3(c)', we have observed Andreev spectra again. In the Pb junction, we could observe a small gap at ~ 5 mV and a large gap-like feature at ~10-12 mV, but not a sign of a peak splitting around 0 mV. In the Ag junction, we also saw a two gap feature with gap energies of ~3 mV and ~5 mV, so that large gap value seemed to be reduced compare to the Al and Pb case. In addition, a sharp zero bias peak was observed. Although the spectra are "typical" compared to our Sn layer case in 'figure 3(d)' or the case of FeSeTe/Pb junction in our previous report [2], there are clear material dependence in the spectra. At this stage, we could not well describe the reason of these variations. Since they originate from the difference in the layering materials, it is natural to attribute them to the chemical effect to the IBS at the interface. The interfacing metal may modify the complex band structure of IBS different way and modify a set of gap values associated to each band

and/or the weights of electronic density of states. In some case, the interface metal may work as a carrier dopant at the interface.

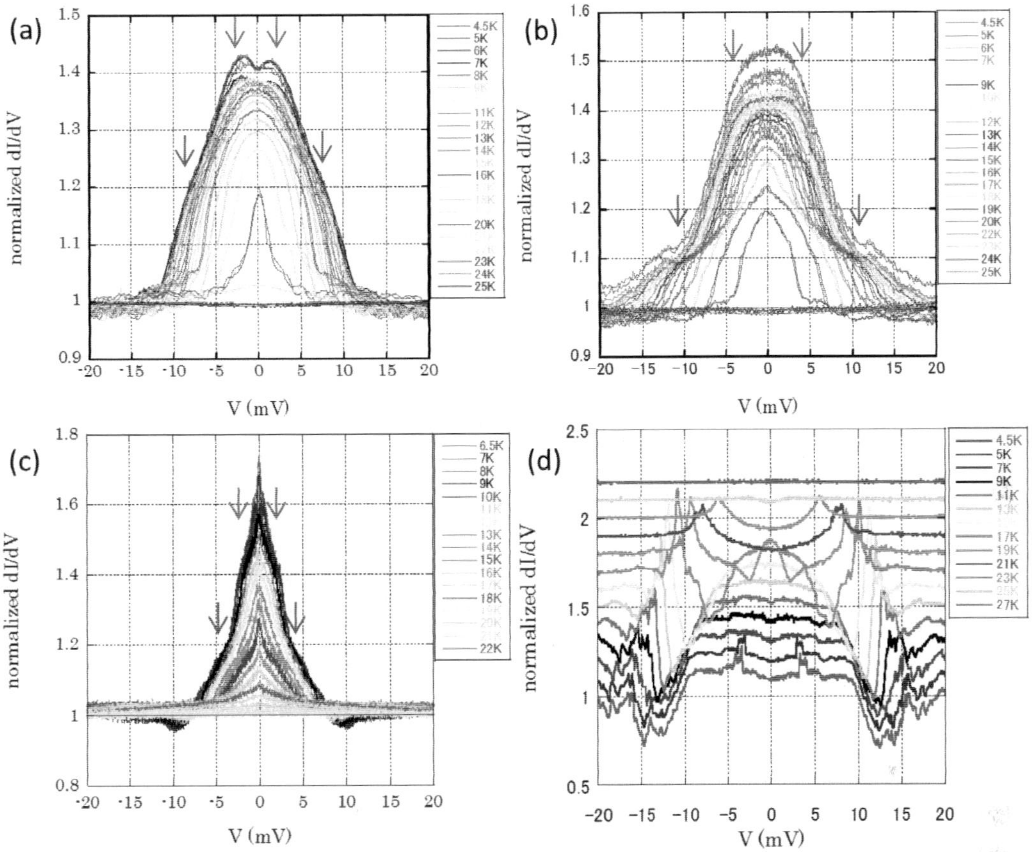

Figure 3. Temperature dependence of the junction conductance normalized by the normal value using an (a) Al, (b) Pb, (c) Ag and (d) Sn electrode

One thing which characterizes interface effect and can be evaluated is the magnitude of contact resistance. The Al junction showed highest resistivity of ~ 20 Ω/cm^2, the Pb and the Sn medium ~10 Ω/cm^2, the Ag lowest. The IBS/Ag interface provided the best metallic contact. Thus, the observed zero bias peak in the Ag junction can be understood as a trace of a Josephson current [3]. On the other hand, pure Pb did not yield a sufficient contact with an IBS. This explains why we could not observe Josephson current.

Compared to the cases of Al, Pb, and Ag junctions, that of Sn in 'figure 3(d)', were very anomalous both in the structure and in the temperature dependence. There are three characteristic features; (1) the small peak structures observed below 8 K, (2) the conventional conductance enhancement due to the Andreev reflection, (3) the large peak with dip structure in the highest energy range. The feature 3 can be identified to that of the simple S/I/N tunnel junction, where the peak energy corresponds to the gap value. In 'figure 4', the temperature dependence of these three features was plotted. Since the feature 1 disappears at around a T_c of Pb, it might be related to Pb superconductivity. The feature 2 disappeared at 19 K. Since T_c of a bulk Ba-122 is around this temperature, we can identify it as an Andreev reflection spectrum due to bulk Ba-122 superconductivity, with a gap value about ~8 mV. On the other hand, the feature 3 disappeared at ~25 K, indicating the existence of a higher T_c entity in this system. Actually, above 20 K, only feature 3 can be observed.

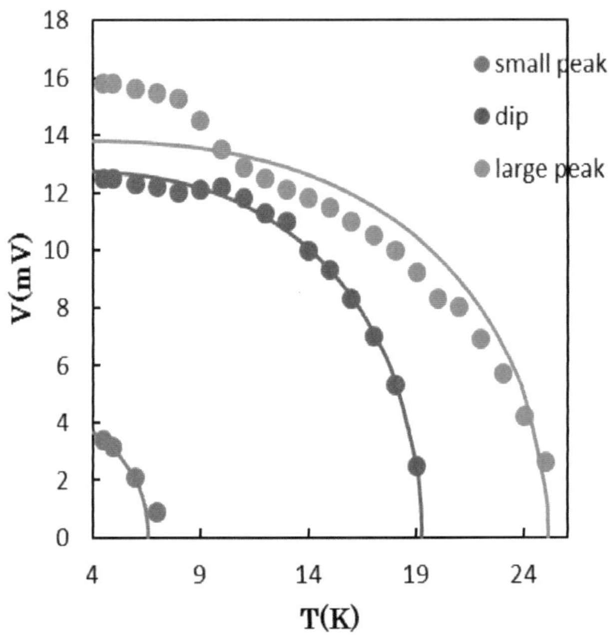

Figure 4. The temperature dependence of characteristic features for the Ba-122/Sn/Pb junction.

The Sn layer may glow on the Ba-122 crystal as an island. Then 2 nm of Sn layer would not fully cover the crystal surface. Then our junction can be interpreted as a parallel connection of a Ba-122/Sn/Pb junction and a Ba-122/Pb junction. The *I-V* characteristics of latter Pb junction can be same as that in Fig.3 (b), therefore, it would give the feature 2 part of the Sn junction. Then the feature 3 can come from the *I-V* characteristics of the "true" Ba-122/Sn/Pb junction. Then one hypothesis is that the deposited Sn reacts with a Ba-122 at the surface and makes some higher T_c phase and an insulating layer on it. Our results of Fe[Te,Se]/Pb junctions suggested that the deposited Pb reacted with Fe[SeTe] and gave a lage gap-like feature [2]. It is natural to consider that the similar reaction take place in the Ba-122 case. However, here, Pb is replaced by Sn. Higher T_c may be due to an extra doping caused by a metal contact.

In this scenario, we assumed the existence of the higher T_c phase. At present, only supporting fact is that temperature dependence of the associated gap follows BCS-like curve. To confirm superconductivity directly, it is important to measure electrode's resistivity in the present configuration that has not yet done. The heating effect may account for the nonlinear *I-V* curve of this type. However, because the junction resistance is as high as that of Pb junction and our Pb junction did not show any such effect, it may not be the case.

4. Summary

We have studied the temperature dependence of the conductance spectra of S/N/S' junctions with a Co-doped Ba-122 single crystals as a base electrode. The spectra were N material dependent and indicated the existence of an interface effect between an iron based superconductor and a layering metal. Here, the interface effect means some chemical reaction, the carrier doping effect, and so on, between adjacent materials. Especially, for Sn, the effect is very large and the formation of a phase with T_c of 25 K has been suggested.

5. References

[1] Sigrist M and Rice T M 1992 *J. Phys. Soc. Jpn.***61** 4283
[2] Fujioka N 2005 *Physica C* **518** 28
[3] Zhang X 2009 *Phys. Rev. Lett.* **102**, 147002
[4] Tortello M 2010 *Phys. Rev. Lett.* **105**, 237002
[5] Terashima K 2009 *Proc. Natl. Acad. Sci. U S A* **106**, 7330

Transport properties of FeSe epitaxial thin films under in-plane strain

M Kawai[1], F Nabeshima[1] and A Maeda[1]

[1] Department of Basic Science, The University of Tokyo, 3-8-1 Komaba, Meguro-ku, Tokyo 153-8902, Japan

E-mail: kawai@maeda1.c.u-tokyo.ac.jp

Abstract. We demonstrate successful fabrication of FeSe films with various magnitude of lattice strain, from tensile to compressive. The superconducting transition temperature, T_c, shows the systematic change by in-plane strain. It reaches 12 K for the films with compressive strain whereas the superconductivity disappears for the films with large tensile strain. The structural transition temperature, T_s, decreases slightly as the in-plane strain is more compressive. In other words, there is a negative correlation between T_c and T_s. It is also found that both the hole and the electron carrier densities evaluated by transport properties increase as the in-plane strain is more compressive. Therefore, there is a clear correlation between T_c and the carrier densities, which suggests that it is essential for the T_c enhancement of iron-based superconductors to increase carrier densities.

1. Introduction

Discovery of iron-based superconductors(FeSCs) has attracted much attention to both fundamental studies and applications[1]. An iron-chalcogenide superconductor, FeSe[2], is one of the most suitable material for understanding the superconductivity of the FeSCs because it has the simplest crystal structure among the FeSCs, composed of conducting planes alone. The superconducting transition temperature, T_c, of the bulk crystal FeSe at ambient pressure is 9 K. However, it is increased up to 38.3 K under high pressure[3, 4, 5]. In single-layer FeSe films on $SrTiO_3$ substrates, T_c increased dramatically up to about 65 K[6, 7] or even higher[8]. From these point of view, FeSe has the high-T_c potential. Therefore, it is important to reveal the superconductivity mechanism of FeSe.

As was demonstrated by the pressure effect[3, 4, 5], T_c of FeSe is sensitive to the lattice strain. Previously, we succeeded in obtaining $FeSe_{1-x}Te_x$ thin films for all x values, which is impossible in bulk crystal, and T_c of these films reaches 23 K at $x = 0.2$[9]. Since Te-substitution introduces a negative chemical pressure to FeSe, this result shows that T_c of FeSe was strongly affected by the isotropic pressure. On the other hand, even in the samples of $FeSe_{1-x}Te_x$ with the same x value, there is a negative correlation between the a-axis and c-axis parameters[9, 10, 11], suggesting that the anisotropic strain may play some role. Since the essential feature of the T_c enhancement in FeSe films is that the films are compressed along the a-axis[12, 13], the separation of the effect of the anisotropic pressure from the isotropic pressure is important for the correct understanding of the pressure effect of iron-chalcogenides. Thus, we focused on the role of the anisotropic strain, and prepared FeSe films with various magnitude of in-plane strain, including both compressive and tensile, where the transport properties were investigated.

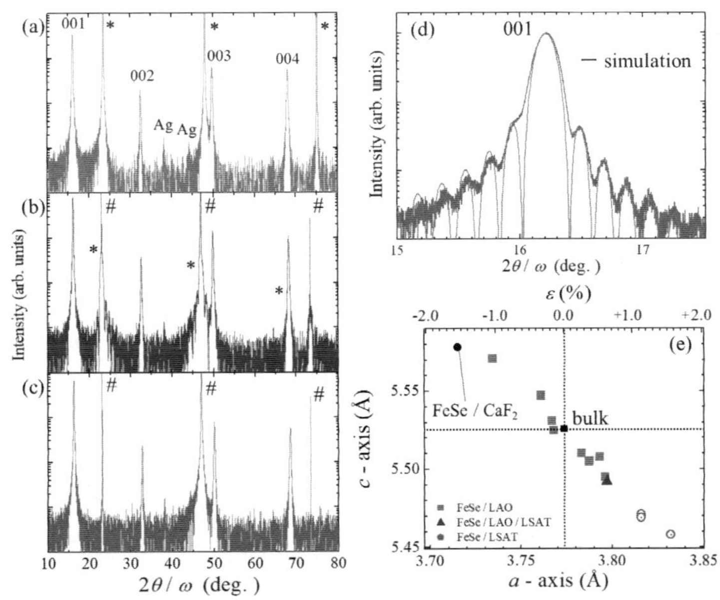

Figure 1. XRD patterns for FeSe thin films on (a) LAO, (b) LSAT with LAO buffer (c) LSAT. Asterisks and number signs represent the peaks resulting from the LAO and LSAT respectively. (d) the enlarged plots of (c) around 001 peak. (e) Relation between the a-axis and c-axis lengths in the FeSe films. Those of a bulk FeSe[14] and FeSe film on CaF_2 [12] are also plotted. The $\varepsilon = (a - a_{\text{bulk}})/a_{\text{bulk}}$ is in-plane strain parameter.

2. Method

FeSe films were grown under vacuum ($\sim 10^{-8}$ Torr) by the pulse laser deposition (PLD) method using a KrF laser[15, 16]. An $Fe_{1.1}Se$ polycrystalline pellet was used as a target. The single-crystalline substrates used in this study were $LaAlO_3$(LAO, a=3.789 Å) and $(LaAl)_{0.7}$-$(SrAl_{0.5}Ta_{0.5})_{0.3}O_3$(LSAT, $(1/2)a$=3.868 Å). We also prepared FeSe on LSAT with LAO buffer, to control the magnitude of strain. The a-axis and the c-axis lengths of the films were determined from the (204) and (00l) reflections in the X-ray diffraction (XRD) measurements. The thicknesses of the grown films were 30$\sim$75 nm, measured by a Dektak 6 M stylus profiler. For transport measurements, these films were deposited in a six-terminal shape using a metal mask. The measured area was 1.2 mm long and 1.0 mm wide. The electrical resistivity and the Hall effect were measured using Physical Property Measurement System (PPMS, Quantum Design, Inc.) from 2 to 300 K.

3. Result and Discussion

Figure 1(a)-(c) show the XRD patterns of the grown FeSe thin films on the three different substrates. These films showed only the (00l) reflections of the tetragonal PbO structure except for peaks from Ag paste, which indicates that the films were well-oriented along the c-axis. A four-fold symmetry of the phi-scan of the (101) reflection was also confirmed. Figure 1(d) shows the enlarged plots of Fig. 1(c) around the (001) peak. The Laue fringe is clearly observed and the intensity of each peak also well agrees with theoretically expected ones, which demonstrate that the grown films have flat surface. Figure 1(e) shows the relation between the a-axis and the c-axis lengths of the films, which shows both a and c parameters change continuously for a wide

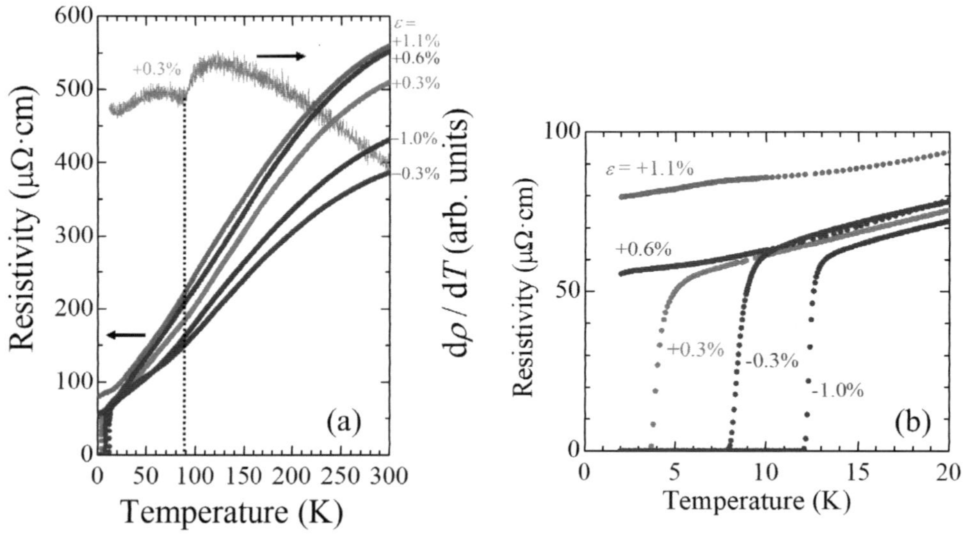

Figure 2. (a) Temperature dependence of the electrical resistivity, ρ, of the FeSe thin films and (b) the enlarged plots of (a) around T_c.

range with the negative correlation. The fact that the films with longer a has shorter c can be understood simply by the Poisson effect. Therefore, all of those demonstrate that we succeeded in obtaining a series of single-crystalline films with various magnitude of anisotropic strain. The in-plane parameter, $\varepsilon = (a - a_{\rm bulk})/a_{\rm bulk}$, is varied from -1.0% to +1.5% in this study.

Figures 2(a) and (b) show the temperature dependence of the electrical resistivity, ρ. The "residual" resistivity ratio, $RRR(= \rho(300\,{\rm K})/\rho(0\,{\rm K}))$, reaches $\sim$18, which shows that our films have good crystalline quality. Note that a kink anomaly which corresponds to the structure transition is observed at about 90 K in the ρ-T curve as bulk FeSe[17]. The structural transition temperature, T_s, can be determined as a position of the kink anomaly in the dρ/dT-T curve, and the in-plane strain dependence of T_s is discussed soon later. It is noteworthy that T_c changes systematically along with the change of the in-plane strain parameter, ε. T_c reaches 12 K for the films with compressive strain whereas the superconductivity disappears for the films with large tensile strain.

Figure 3 shows the temperature dependence of the Hall coefficient, R_H, of the films. In each film, the behavior above 100 K are similar. R_H is rather small, and with decreasing temperature, R_H changes from negative to positive below 120 K. On the other hand, with further decreasing temperature, R_H starts to depend on ε. As ε becomes larger, R_H become larger at low temperatures.

Figures 4(a)-(c) show T_s, T_c and R_H at 20 K as a function of ε, respectively. As was already described, when ε increases, T_c decreases systematically and disappears rapidly in the films with the large tensile strain. On the other hand, T_s does not strongly depend on ε, but slightly decreases as ε becomes smaller. In other words, T_c and T_s has a negative correlation. This negative correlation is consistent with the sudden increase of T_c at the disappearance of the structural transition in Te-substituted films[11]. Since the structural transition is thought to be the so-called nematic transition, this negative correlation between T_c and T_s suggests that the

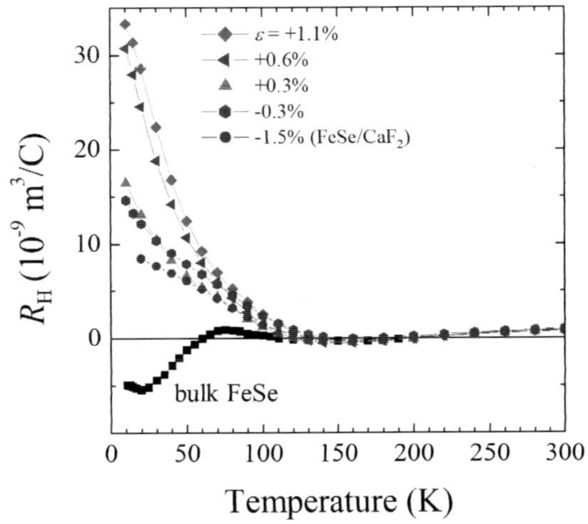

Figure 3. Temperature dependences of Hall coefficients, $R_{\rm H}$, of the FeSe films. A result of bulk FeSe[18] and FeSe film on CaF_2 [12] are also plotted.

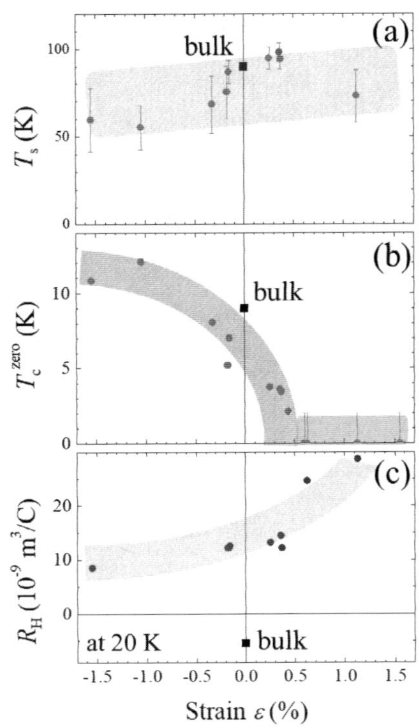

Figure 4. The in-plane strain parameter (ε) dependence of (a) $T_{\rm s}$ (b) $T_{\rm c}$ and (c) $R_{\rm H}$ at 20 K. A result of bulk FeSe[18] are also plotted. The hatched curves are guides for the eye.

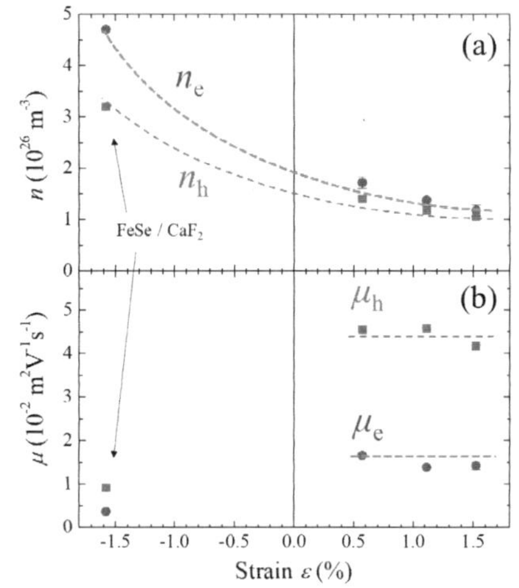

Figure 5. The in-plane strain parameter, ε, dependence of (a) carrier density and (b) mobility. The dashed curves are guides for the eye.

electronic nematicity is unfavorable to raise T_c in iron-chalcogenide.

The systematic change of $R_H(20\text{ K})$ as a function of ε in our films was also observed. Note that $R_H(20\text{ K})$ of the bulk FeSe[18] does not fit to our result in films. This difference may be caused by the difference of the chemical composition or the difference in the local structural parameters such as bond angle and anion height between films and bulk crystal. In any case, we believe that this difference in the low-T R_H behavior between films and bulk singles crystal is not important for superconductivity because T_c of the bulk crystal is fitted to our data in films.

In multiple-band systems like iron-chalcogenides, R_H is not related to the carrier density in a simple form. To discuss the R_H data in more detail, we apply a text-book approach for multiple-band materials. In order to evaluate the carrier density and the mobility of the films, we also measured the magneto-resistance. In a classical two-band model, the resistivity tensor is expressed as

$$\rho_{xx}(0) = \frac{1}{e(n_e\mu_e + n_h\mu_h)}, \quad (1)$$

$$\rho_{yx}(B) = \frac{-n_e\mu_e^2 + n_h\mu_h^2}{e(n_e\mu_e + n_h\mu_h)^2}B, \quad (2)$$

$$\frac{\rho_{xx}(B)}{\rho_{xx}(0)} = \frac{n_e n_h \mu_e \mu_h (\mu_e + \mu_h)^2}{(n_e\mu_e + n_h\mu_h)^2}B^2, \quad (3)$$

where μ_h, μ_e, n_h, and n_e are the hole mobility, the electron mobility, the hole density and the electron density, respectively. Because there are four parameters for three equations, we assumed the carrier density ratio (n_h/n_e) based on an angle-resolved photoemission spectroscopy (ARPES) data[19].

The obtained μ_h, μ_e, n_h and n_e are summarized in Fig. 5. Both the hole and the electron carrier densities increase systematically as ε becomes smaller. Recently, the ARPES study using our films also showed the increase of the hole and the electron densities in an in-plane compressed sample[19]. This agreement between our Hall data and the ARPES study demonstrate the validity of the conclusion we obtained. Namely, the increase of the carrier densities is essential for the increase of T_c. On the other hand, both the hole and the electron mobility are almost unchanged in the films on oxide substrates. In addition, those of FeSe on CaF_2 is rather small. In the FeSe films on CaF_2, TEM observation showed that F^- ions penetrated into the films from the CaF_2 substrates[20, 21, 22], which may be the reason for small mobilities of FeSe on CaF_2. The fact that T_c is the highest in films with the small mobilities suggests that the mobility is not important for superconductivity. Rather, our result shows that the most essential factor to realize high T_c is the increase of the carrier densities in iron-chalcogenides.

4. Conclusion

We succeeded in fabricating a series of FeSe films with various magnitude of the in-plane strain, including both tensile and compressive. T_c and R_H show the systematic change by in-plane strain, ε, whereas T_s does not strongly depend on ε and decreases slightly as ε becomes smaller. This negative correlation between T_c and T_s suggest that the electronic nematicity is unfavorable to raise T_c. The carrier densities evaluated by transport properties also increase systematically as ε becomes smaller. We demonstrate a clear correlation between T_c and the carrier densities, which suggests that it is essential for the T_c enhancement of iron-chalcogenides to increase the carrier densities.

References

[1] Kamihara Y, Watanabe T, Hirano M and Hosono H 2008 *J. Am. Chem. Soc.* **130** 3296
[2] Hsu F C, Luo J Y, Yeh K W, Chen T K, Huang T W, Wu P M, Lee Y C, Huang Y L, Chu Y Y, Yan D C and Wu M K 2008 *Proc. Natl. Acad. Sci. U.S.A.* **105** 14262
[3] Masaki S, Kotegawa H, Hara Y, Tou H, Murata K, Mizuguchi Y and Takano Y 2009 *J. Phys. Soc. Jpn.* **78** 063704 (*Preprint* `http://journals.jps.jp/doi/pdf/10.1143/JPSJ.78.063704`) URL `http://journals.jps.jp/doi/abs/10.1143/JPSJ.78.063704`
[4] Medvedev S, McQueen T M, Troyan I A, Palasyuk T, Eremets M I, Cava R J, Naghavi S, Casper F, Ksenofontov V, Wortmann G and Felser C 2009 *Nature Materials* **8** 630 EP – URL `http://dx.doi.org/10.1038/nmat2491`
[5] Sun J P, Matsuura K, Ye G Z, Mizukami Y, Shimozawa M, Matsubayashi K, Yamashita M, Watashige T, Kasahara S, Matsuda Y, Yan J Q, Sales B C, Uwatoko Y, Cheng J G and Shibauchi T 2016 *Nat. Commun.* **7** 12146
[6] Qing-Yan W, Zhi L, Wen-Hao Z, Zuo-Cheng Z, Jin-Song Z, Wei L, Hao D, Yun-Bo O, Peng D, Kai C, Jing W, Can-Li S, Ke H, Jin-Feng J, Shuai-Hua J, Ya-Yu W, Li-Li W, Xi C, Xu-Cun M and Qi-Kun X 2012 *Chinese Physics Letters* **29** 037402 URL `http://stacks.iop.org/0256-307X/29/i=3/a=037402`
[7] Liu D, Zhang W, Mou D, He J, Ou Y B, Wang Q Y, Li Z, Wang L, Zhao L, He S, Peng Y, Liu X, Chen C, Yu L, Liu G, Dong X, Zhang J, Chen C, Xu Z, Hu J, Chen X, Ma X, Xue Q and Zhou X J 2012 *Nature Communications* **3** 931 EP – URL `http://dx.doi.org/10.1038/ncomms1946`
[8] Ge J F, Liu Z L, Liu C, Gao C L, Qian D, Xue Q K, Liu Y and Jia J F 2014 *Nature Materials* **14** 285 EP – URL `http://dx.doi.org/10.1038/nmat4153`
[9] Imai Y, Sawada Y, Nabeshima F and Maeda A 2015 *Proc. Natl. Acad. Sci. U.S.A.* **112** 1937–1940 URL `http://www.pnas.org/content/112/7/1937.abstract`
[10] Imai Y, Sawada Y, Asami D, Nabeshima F and Maeda A 2016 *Physica C* **530** 24 ISSN 0921-4534 URL `http://www.sciencedirect.com/science/article/pii/S0921453416300375`
[11] Imai Y, Sawada Y, Nabeshima F, Asami D, Kawai M and Maeda A 2017 *Sci. Rep.* **7** 46653 URL `http://dx.doi.org/10.1038/srep46653`
[12] Nabeshima F, Imai Y, Hanawa M, Tsukada I and Maeda A 2013 *Appl. Phys. Lett.* **103** 172602 URL `http://dx.doi.org/10.1063/1.4826945`
[13] Maeda A, Nabeshima F, Takahashi H, Okada T, Imai Y, Tsukada I, Hanawa M, Komiya S and Ichinose A 2014 *Appl. Sur. Sci.* **312** 43 URL `http://dx.doi.org/10.1016/j.apsusc.2014.02.124`
[14] McQueen T M, Huang Q, Ksenofontov V, Felser C, Xu Q, Zandbergen H, Hor Y S, Allred J, Williams A J, Qu D, Checkelsky J, Ong N P and Cava R J 2009 *Phys. Rev. B* **79**(1) 014522 URL `https://link.aps.org/doi/10.1103/PhysRevB.79.014522`
[15] Imai Y, Tanaka R, Akiike T, Hanawa M, Tsukada I and Maeda A 2010 *Jpn. J. Appl. Phys.* **49** 023101 URL `http://stacks.iop.org/1347-4065/49/i=2R/a=023101`
[16] Imai Y, Akiike T, Hanawa M, Tsukada I, Ichinose A, Maeda A, Hikage T, Kawaguchi T and Ikuta H 2010 *Appl. Phys. Express* **3** 043102 URL `http://stacks.iop.org/1882-0786/3/i=4/a=043102`

[17] Böhmer A E, Hardy F, Eilers F, Ernst D, Adelmann P, Schweiss P, Wolf T and Meingast C 2013 *Phys. Rev. B* **87**(18) 180505 URL http://link.aps.org/doi/10.1103/PhysRevB.87.180505

[18] Kasahara S, Watashige T, Hanaguri T, Kohsaka Y, Yamashita T, Shimoyama Y, Mizukami Y, Endo R, Ikeda H, Aoyama K, Terashima T, Uji S, Wolf T, von Löhneysen H, Shibauchi T and Matsuda Y 2014 *Proc. Natl. Acad. Sci. U.S.A.* **111** 16309–16313 URL http://www.pnas.org/content/111/46/16309.abstract

[19] Phan G N, Nakayama K, Sugawara K, Sato T, Urata T, Tanabe Y, Tanigaki K, Nabeshima F, Imai Y, Maeda A and Takahashi T 2017 *Phys. Rev. B* **95**(22) 224507 URL https://link.aps.org/doi/10.1103/PhysRevB.95.224507

[20] Ichinose A, Nabeshima F, Tsukada I, Hanawa M, Komiya S, Akiike T, Imai Y and Maeda A 2013 *Supercond. Sci. Technol.* **26** 075002 URL http://stacks.iop.org/0953-2048/26/i=7/a=075002

[21] Ichinose A, Tsukada I, Nabeshima F, Imai Y, Maeda A, Kurth F, Holzapfel B, Iida K, Ueda S and Naito M 2014 *Appl. Phys. Lett.* **104** 122603 (*Preprint* http://dx.doi.org/10.1063/1.4869961) URL http://dx.doi.org/10.1063/1.4869961

[22] Tsukada I, Ichinose A, Nabeshima F, Imai Y and Maeda A 2016 *AIP Advances* **6** 095314 URL http://dx.doi.org/10.1063/1.4963646

FABRICATION OF GRAIN BOUNDARY JUNCTIONS USING NdFeAs(O,F) SUPERCONDUCTING THIN FILMS

T Omura[1], T Matsumoto[2], T Hatano[1,2], K Iida[1,2], H Ikuta[1,2]

[1] Department of Crystalline Materials Science, Nagoya University, Furo-cho, Chikusa-ku, Nagoya 464-8603, Japan
[2] Department of Materials Physics, Nagoya University, Furo-cho, Chikusa-ku, Nagoya 464-8603, Japan

E-mail: oomura.taito@j.mbox.nagoya-u.ac.jp

Abstract. We report on the growth of NdFeAs(O,F) thin films on [001]-tilt MgO bicrystal substrates with misorientation angle θ_{GB}=6°, 12°, 24° and 45°, and their inter- and intra-grain transport properties. X-ray diffraction study confirmed that all our NdFeAs(O,F) films are epitaxially grown on the MgO bicrystals. The θ_{GB} dependence of the inter-grain critical current density J_c shows that, unlike Co-doped $BaFe_2As_2$ and Fe(Se,Te), its decay with θ_{GB} is rather significant. As a possible reason of this result, fluorine may have diffused preferentially to the grain boundary region and eroded the crystal structure.

1. Introduction

It is very important to examine the current-limiting effect of grain boundaries (GBs) for fabricating superconducting wires and tapes based on high critical temperature (T_c) superconductors. For cuprate superconductor $YBa_2Cu_3O_{7-\delta}$ (YBCO), the J_c characteristic across the grain boundary (J_c^{inter}) has been intensively investigated using a bicrystal substrate in which two single crystals with different orientations are thermaly jointed [1]. Although J_c^{inter} of YBCO does not decrease with increasing the misorientation angle θ_{GB} until a critical angle θ_c (3°~5°), J_c^{inter} attenuates exponentially with θ_{GB} above θ_c. Therefore, for practical use as superconducting tapes, biaxially textured YBCO conductors whose in-plane orientation distribution is smaller than θ_c are required. However, this demands a costly fabrication process and delayed significantly conductor applications.

Iron-based superconductors (FBS) form a material class that has a high T_c next to the cuprates and a high upper critical field (H_{c2}) exceeding 100 T with a moderate H_{c2} anisotropy [2]. Regarding the grain boundary characteristics, Katase *et al.* reported that θ_c is as large as 9° for Co-doped $BaFe_2As_2$, which is about twice larger than that of YBCO [3]. In addition, the attenuation of J_c^{inter} above θ_c was more moderate than YBCO. As a result, J_c^{inter} of Co-doped $BaFe_2As_2$ was higher than that of YBCO at 4 K and $\theta_{GB} \geq 20°$. Our group reported that P-doped $BaFe_2As_2$ showed also a slow decay of J_c^{inter} with θ_{GB} and a high J_c^{inter} of 10^6 A/cm^2 for θ_{GB}=24° at 4 K, which is the highest value ever reported for YBCO and FBS for this misorientation angle [4]. Similar to the $BaFe_2As_2$ system, the transition from strong-link to weak-link occurs at θ_c ~ 9° for Fe(Se,Te) grown on $SrTiO_3$ bicrystals [5]. All these results are very promising for high field conductor applications.

Among the various FBSs discovered to date, *Ln*Fe*Pn*(O,F) (*Ln*: lanthanoid, *Pn*: pnictogen, and 1111) has the highest T_c up to 58 K [6,7], indicative of potentially high J_c. However the characteristics of grain boundary has not yet been examined in detail since the crystal growth of this system is very difficult. We have succeeded in growing high quality epitaxial NdFeAs(O,F) thin films by molecular beam epitaxy (MBE) [8,9]. This opened the opportunity to fabricate 1111 films on bicrystal substrates. Here, we report on the transport properties of NdFeAs(O,F) films grown on bicrystal substrates with various misorientation angles.

2. Experimental details

NdFeAs(O,F) thin films were grown by MBE. Solid sources of Fe, Fe_2O_3, As, NdF_3, and Ga were charged in Knudsen cells. Here, Ga was used as a F-getter to control the amount of fluorine [10]. The films were grown on 6°, 12°, 24°, and 45° [001]-tilt MgO bicrystal substrates and the substrate temperature was fixed at 800°C. We have first grown a NdFeAsO parent phase layer, and then NdOF subsequently. This resulted in a superconducting 1111 film with a high T_c as fluorine diffused into the 1111 phase from the top fluoride layer [10,11]. The film thickness of NdFeAsO was about 160 nm and that of NdOF about 50 nm. Reflection high energy electron diffraction (RHEED) was used for in-situ monitoring the surface structure during the thin film growth. Phase purity and crystalline quality of the obtained thin films were examined by X-ray diffraction (XRD) using Cu Kα radiation. Microstructural characterization was performed by transmission electron microscope (TEM).

After the NdOF over-layer was removed by Ar-ion milling, the films were photolithographically patterned and etched by Ar-ion milling to form micro-bridges for transport measurements. As shown in Figure 1, the films were patterned so as to measure the transport properties of both grain boundary and intra-grain on the same sample. The bridges were 20-40 μm-wide and 1 mm-long across the grain boundary and 0.25 mm-long within the grain. The current–voltage characteristics and the temperature dependence of resistivity were measured by a four-probe method. A criterion of 1 μV/cm was employed for determining intra-grain J_c. On the other hand, inter-grain J_c was defined as the intersection between $E = 0$ and a linear fit to the non-ohmic linear differential region (i.e., the region where the electric field E depended linearly on the current density J).

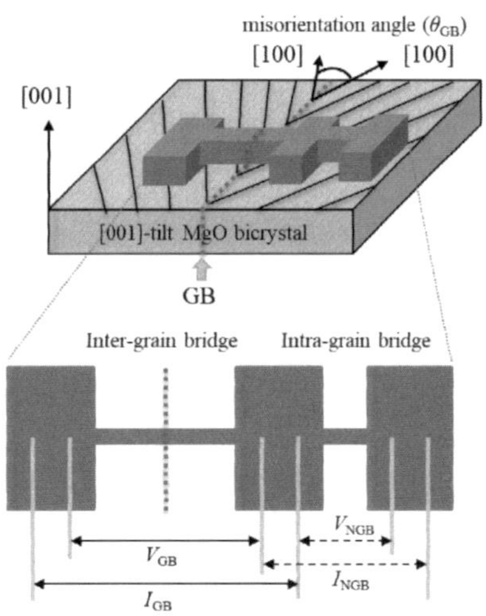

Figure 1. A schematic illustration of the micro-bridge for transport measurements. V_{GB} and I_{GB} are the voltage and current contacts for the GB junction, while V_{NGB} and I_{NGB} are those for the intra-grain bridge.

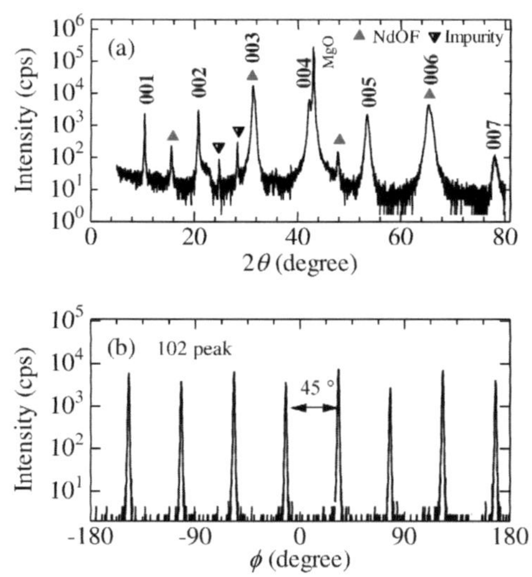

Figure 2. (a) θ-2θ scan of the NdFeAs(O,F) film on [001]-tilt MgO bicrystal with θ_{GB} = 45° measured with the Bragg-Brentano geometry using Cu Kα radiation. (b) ϕ-scan of the 102 NdFeAs(O,F) peak.

3. Results and discussion

Figure 2 shows the result of structural characterization by XRD of the NdFeAs(O,F) thin film on [001]-tilt MgO bicrystal with θ_{GB} = 45°. The NdFeAs(O,F) film has grown *c*-axis oriented as revealed by the θ-2θ scan, Figure 2(a). Impurity phase was observed as the by-product of chemical reaction between NdOF and NdFeAsO. The ϕ-scan of the 102 NdFeAs(O,F) peak showed an eightfold pattern with a 45° separation as shown in Figure 2(b). Hence, the NdFeAs(O,F) film has grown epitaxially on the [001]-tilt MgO bicrystal. We confirmed that all films on the bicrystal substrates with various θ_{GB} were biaxially textured.

Figure 3 compares the resistivity curves for the inter-grain bridges with various θ_{GB}. Here θ_{GB} = 0° is represented by the intra-grain bridge that was fabricated from the NdFeAs(O,F) film on the θ_{GB} = 6° bicrystal substrate (see Figure 1). The intra-grain bridge had an onset T_c (T_c^{onset}) of 47 K and zero T_c (T_c^0) of 46 K. On the other hand, for inter-grain bridges, both T_c^{onset} and T_c^0 decreased with increasing the misorientation angle. Additionally, a finite resistivity even below T_c was observed for the inter-grain bridges with $\theta_{GB} \geq$ 24° as shown in Figure 3(b), whereas the resistivity of the other bridges dropped well below the instrumental limitation.

To clarify the reason why the bridges with higher misorientation angles had a finite resistivity, we performed microstructural characterization by TEM. Shown in figure 4(a) is the cross-sectional view of the NdFeAs(O,F) thin film with θ_{GB} = 24° in the vicinity of GB region. Clearly, both the NdFeAs(O,F) layer and the MgO substrate around GB are damaged. Elemental mappings acquired from the same area showed the absence of Fe and As, and the segregation of Mg (figure 4(b)). Other than the GB region, we did not see such damage to the NdFeAs(O,F) film and the substrate. These results show that F preferentially diffused through GB and thereby eroded NdFeAs(O,F) and MgO. It was reported for a YBCO/Ca-doped YBCO bilayer film that Ca diffused from the over-layer into the grain boundary region of the bottom YBCO layer, causing hole overdoping and consequently increased the inter-grain J_c [12]. Similarly, fluorine had probably diffused more easily along the grain boundary in our films, but unlike YBCO, the excess fluorine gave a negative impact on the structural and transport properties because of its strong reactivity.

Shown in Figure 5(a) is the *E*-*J* characteristics of the inter-grain bridges for various θ_{GB} measured at 4.2 K. For comparison, the data of the intra-grain bridge is also plotted. As can be seen, the intra-grain curve showed a typical power-law behavior (i.e., $E \sim J^n$), suggesting that J_c is limited by the intra-grain itself. On the other hand, the inter-grain bridges for θ_{GB} = 6° and 12° showed a linear behavior at low *E* due to flux flow along the GB, indicative of J_c limitation by GB (Figure 5(b))[13,14]. With further increasing θ_{GB}, a finite resistivity was observed as shown in Figure 3(b), and accordingly, the inter-grain J_c could not be evaluated.

Figure 6 shows the inter-grain J_c for θ_{GB} = 6° measured at various temperatures without applying a magnetic field. The intra-grain J_c of the same film is also plotted for comparison. At low temperatures,

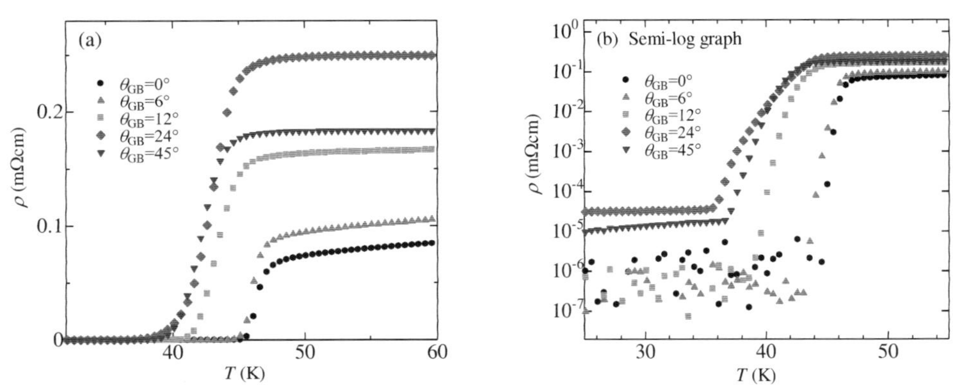

Figure 3. (a) ρ-T curves of the inter-grain bridges with various misorientation angles and the intra-grain bridge. (b) Semi-logarithmic plots of the same ρ-T curves.

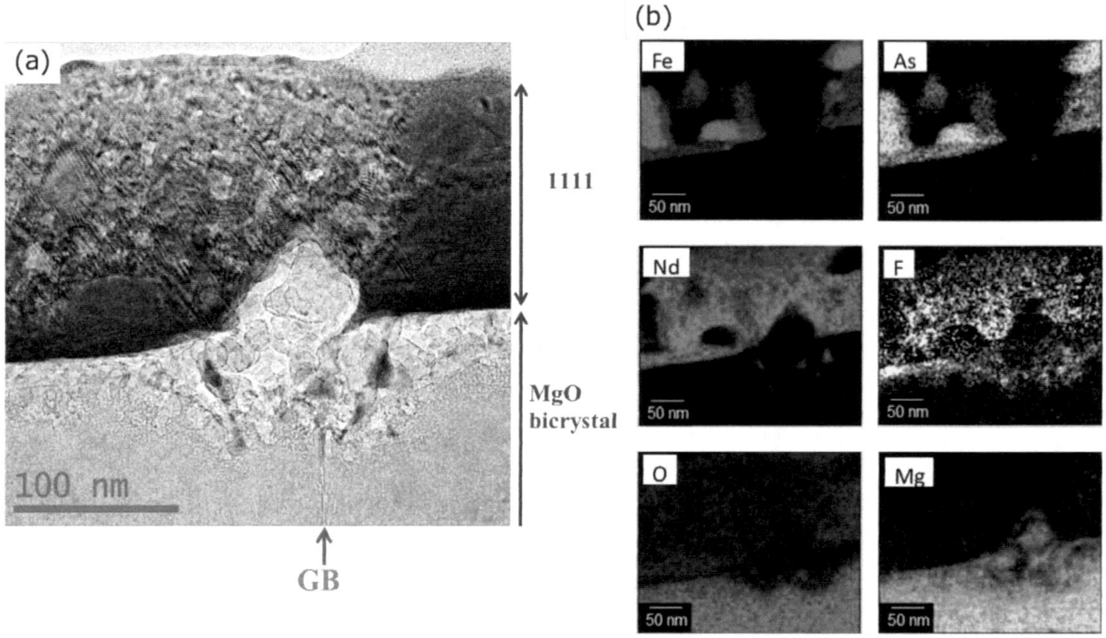

Figure 4. (a) Cross-sectional view of the NdFeAs(O,F) thin film with $\theta_{GB} = 24°$ in the vicinity of the grain boundary region. (b) Elemental mappings acquired from the same area.

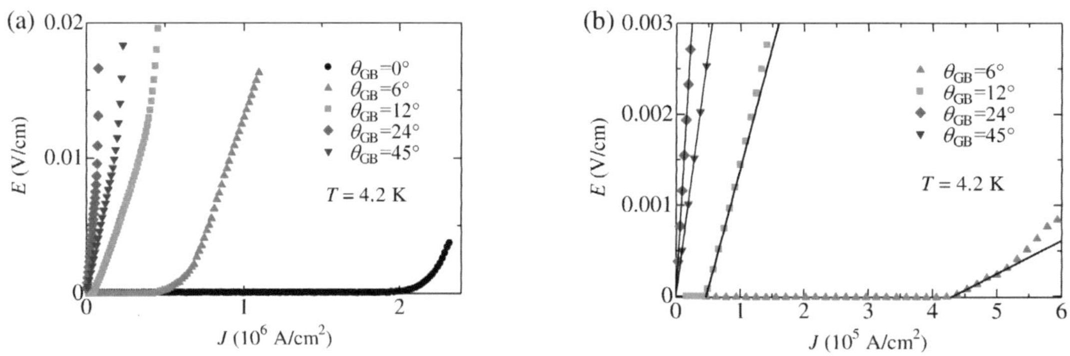

Figure 5. (a) *E-J* characteristics of the inter-grain bridges for various misorientation angles and the intra-grain bridge. (b) Enlarged view of the *E-J* curves at low *E*.

the self-field J_c of the intra-grain exceeded 1 MA/cm^2. However, J_c^{inter} reduced by almost 30% compared to J_c^{intra}. Hence, the critical angle is less than 6°, which is different from Co-doped $BaFe_2As_2$ and Fe(Se,Te).

Figure 7 summarizes the ratio of inter- and intra-grain J_c as a function of misorientation angle. For comparison, the data for Co-doped $BaFe_2As_2$ [3] and Fe(Se,Te) [5] are also plotted. As can be seen, the ratio is nearly unity until $\theta_{GB} \sim 9°$ for both Co-doped $BaFe_2As_2$ and Fe(Se,Te), indicative of strong-coupling between the grains. On the other hand, our NdFeAs(O,F) film showed a weak link behavior even at $\theta_{GB} = 6°$. However, the present result may not reflect the intrinsic GB properties of NdFeAs(O,F). As stated above, the inter-grain bridges for $\theta_{GB} \geq 24°$ had a finite resistivity due to the erosion of GB region by F. Although we observed a non-zero J_c for $\theta_{GB} = 6°$ and 12°, it is well likely

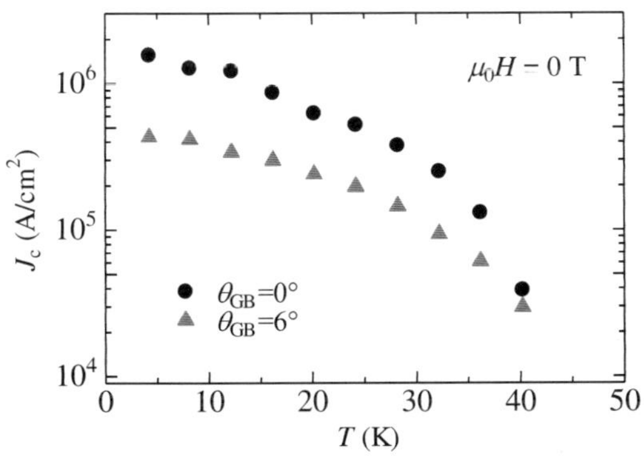

Figure 6. Self-field intra- and inter-grain ($\theta_{GB} = 6°$) J_c as a function of temperature.

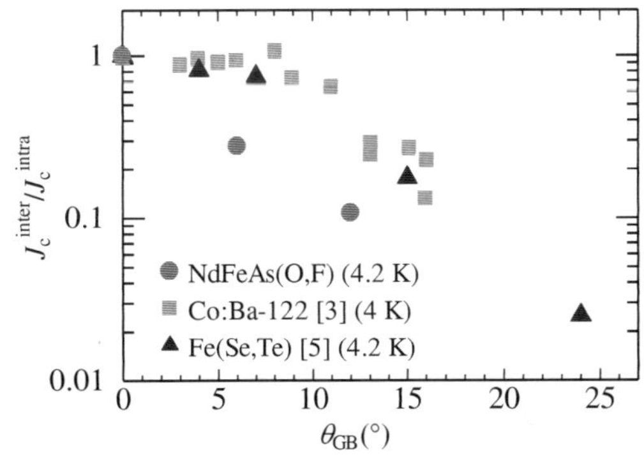

Figure 7. The ratio of inter- and intra-grain J_c (J_c^{inter}/J_c^{intra}) as a function of θ_{GB}. For comparison, the data of Co-doped $BaFe_2As_2$ [3] and Fe(Se,Te) [5] are also shown.

that fluorine had diffused in excess along the GB region in these films as well and caused some damages. Hence the present result of inter-grain J_c is probably an underestimation.

To characterize the intrinsic nature of grain boundary for NdFeAs(O,F), the excess F-diffusion to GB should be prevented. In our earlier studies, the growth temperature was 670°C for both NdFeAsO and NdOF [10,11]. Later we found that increasing the growth temperature improved the crystallinity of the 1111 phase [15], and we have grown both the 1111 and NdOF phases at 800°C in the present study. However, the growth temperature of NdOF over-layer would not influence the crystallinity of the 1111 layer. Hence, suppressing the excess diffusion of fluorine without compromising the crystallinity of the 1111 phase might be achieved by lowering the growth temperature of NdOF while keeping that of NdFeAsO at 800°C, which would be the direction of a future study. That method would also be useful to improve the performance of 1111-based coated conductors. Previously, we reported the field (H) dependence of J_c for NdFeAs(O,F) coated conductors [16]. Although T_c was around 43 K, almost twice as high as that of Co-doped $BaFe_2As_2$, the J_c - H performance was inferior to that of Co-doped $BaFe_2As_2$. However, this NdFeAs(O,F) coated conductor was fabricated at 800°C. Based on the results of the present study, we think that the grain boundaries of the coated conductor had suffered damages from excess fluorine, and the suppression of excess F-diffusion would improve the J_c - H properties.

4. Summary

NdFeAs(O,F) thin films were grown on 6°, 12°, 24°, and 45° [001]-tilt MgO bicrystal substrates and their electrical transport properties were investigated. Structural characterization by X-ray diffraction revealed that all films have grown epitaxially. Finite resistivity was observed for the inter-grain bridges with higher misorientation angles ($\theta_{GB} \geq 24°$). Additionally, by comparing with Co-doped $BaFe_2As_2$ and Fe(Se,Te) grain boundary junctions, the inter-grain J_c showed a larger decay with θ_{GB}. As a possible reason of this result, fluorine may have diffused preferentially to the grain boundary region and degraded the structural and transport properties. The suppression of excess F-diffusion to GB would therefore be the key for studying the intrinsic grain boundary properties.

Acknowledgements

This work was partially supported by the JSPS Grant-in-Aid for Scientific Research (B) Grant Number 16H04646. We acknowledge Kimitaka Higuchi of High Voltage Electron Microscope Laboratory, Institute of Materials and Systems for Sustainability, Advanced Characterization Nanotechnology Platform, for TEM observation.

References

[1] Hilgenkamp H and Mannhart J 2002 *Rev. Mod. Phys.* **74** 485

[2] Putti M, Pallecchi I, Bellingeri E, Cimberle M R, Tropeano M, Ferdeghini C, Palenzona A, Tarantini C, Yamamoto A, Jiang J, Jaroszynski J, Kametani F, Abraimov D, Polyanskii A, Weiss J D, Hellstrom E E, Gurevich A, Larbalestier D C, Jin R, Sales B C, Sefat A S, McGuire M A, Mandrus D, Cheng P, Jia Y, Wen H H, Lee S and Eom C B 2010 *Supercond. Sci. Technol.* **23** 034003

[3] Katase T, Ishimaru Y, Tsukamoto A, Hiramatsu H, Kamiya T, Tanabe K and Hosono H 2011 *Nat. Commun.* **2** 409

[4] Sakagami A, Kawaguchi T, Tabuchi M, Ujihara T, Takeda Y and Ikuta H 2013 *Physica C* **494** 181

[5] Si W, Zhang C, Shi X, Ozaki T, Jaroszynski J and Li Q 2015 *Appl. Phys. Lett.* **106** 032602

[6] Miyazawa K, Kihou K, Shirage P M, Lee C H, Kito H, Eisaki H and Iyo A 2009 *J. Phys. Soc. Jpn.* **78** 034712

[7] Fujioka M, Denholme S J, Ozaki T, Okazaki H, Deguchi K, Demura S, Hara H, Watanabe T, Takeya H, Yamaguchi T, Kumakura H and Takano Y 2013 *Supercond. Sci. Technol.* **26** 085023

[8] Kawaguchi T, Uemura H, Ohno T, Watanabe R, Tabuchi M, Ujihara T, Takenaka K, Takeda Y and Ikuta H 2009 *Appl. Phys. Express* **2** 093002

[9] Kawaguchi T, Uemura H, Ohno T, Tabuchi M, Ujihara T, Takenaka K, Takeda Y and Ikuta H 2010 *Appl. Phys. Lett.* **97** 042509

[10] Kawaguchi T, Uemura H, Ohno T, Tabuchi M, Ujihara T, Takeda Y and Ikuta H 2011 *Appl. Phys. Express* **4** 083102

[11] Uemura H, Kawaguchi T, Ohno T, Tabuchi M, Ujihara T, Takeda Y and Ikuta H 2012 *Solid State Commun.* **152** 735

[12] Hammerl G, Schmehl A, Schulz R R, Goetz B, Bielefeldt H, Schneider C, Hilgenkamp H and Mannhart J 2000 *Nature* **407** 162

[13] Díaz A, Mechin L, Berghuis P and Evetts J E 1998 *Phys. Rev. B* **58** R2960

[14] Verebelyi D T, Christen D K, Feenstra R, Cantoni C, Goyal A, Lee D F and Paranthaman M 2000 *Appl. Phys. Lett.* **76** 1755

[15] Chihara M, Sumiya N, Arai K, Ichinose A, Tsukada I, Hatano T, Iida K and Ikuta H 2015 *Physica C* **518** 69

[16] Iida K, Kurth F, Chihara M, Sumiya N, Grinenko V, Ichinose A, Tsukada I, Hänisch J, Matias V, Hatano T, Holzapfel B and Ikuta H 2014 *Appl. Phys. Lett.* **105** 172602

AC Resistance of Driven Vortices of Superconductors Measured by Microwave Technique

H Kurokawa, F Nabeshima and A Maeda

Department of Basic Science, The University of Tokyo, 3-8-1, Komaba, Meguro-ku, Tokyo 153-8902, Japan

kurokawa@maeda1.c.u-tokyo.ac.jp

Abstract. To investigate the dynamics of driven vortices in superconductors, microwave resonance characteristics were measured under dc driving force in Nb. DC driving current was supplied to the resonator, and the current dependence of the resonance spectra was investigated in the vortex state. When dc driving current was increased across the critical current, a parabolic dependence of the center frequency on dc current as well as an increase of the inverse of the quality factor, Q^{-1}, were observed. In addition, the ac resistance resulting from the motion of vortices showed an increase around the dc critical current, and the increase depended on frequency. This result suggests that the pinning force and the resultant depinning of the vortices is frequency dependent even near the crossover frequency from the low-frequency reactive response to the high-frequency dissipative response.

1. Introduction

Dynamic pinning force, F_p, acting on driven vortices of superconductors, namely, the dissipation due to the pinning in the moving state, has attracted a lot of attention not only as a source of the dissipation in the moving vortices, but also as a model system for kinetic friction at the solid-solid interface[1]. However, the microscopic process resulting in F_p has not been understood yet. The understanding of F_p is challenging, since it treats the nonequilibrium systems with nonlinearity and randomness. Thus, the elucidation of the origin of F_p is important not only for vortex physics, but also for fundamental understandings of deformable driven systems with nonlinearity and randomness.

The concept of F_p was first introduced by Kim in the equation of force balance for flux flow[2], which is expressed as

$$F_v + F_p = F_d, \tag{1}$$

where F_v is the viscous drag force, and F_d is the driving force. Several theoretical attempts were made to explain the origin of F_p. For instance, F_p was calculated by treating pinning force as a perturbation to the equation of motion[3]. Later, applying field theoretical method[4], Grundberg and Rammer calculated F_p both analytically and numerically[5]. However, in these treatments, F_p was not derived from the microscopic motion of each vortices. On the other hand, Yamafuji and Irie attributed F_p to the relaxation of the elastic deformation of the vortex lattice[6,7]. On the basis of that theory, Lowell tried to explain the velocity dependence of F_p, but the theory was limited to the large velocity limit, and could not explain experimental results in a unified manner[8]. Despite those theoretical challenges, to our knowledge, no experimental studies have examined the microscopic process of F_p.

Thus, we planned to reveal the microscopic dissipation mechanism resulting in F_p experimentally.

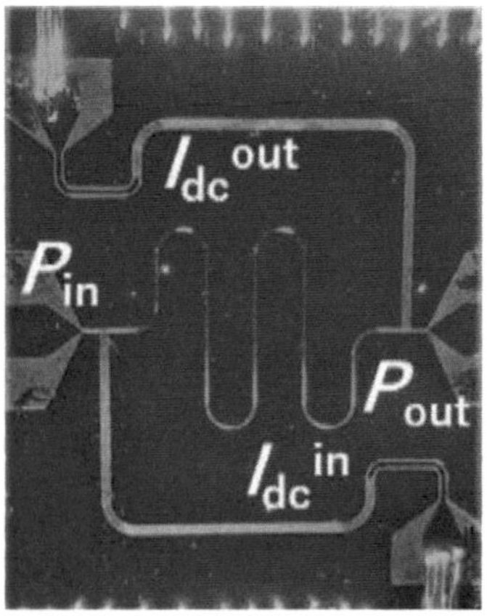

Figure 1. The optical microscope image of the 7.8 GHz resonator.

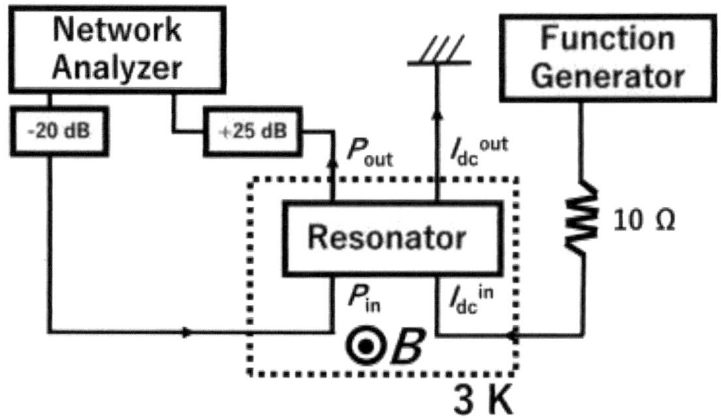

Figure 2. The block diagram of the measurement.

To investigate the dynamic properties of the moving state, we measured the ac resistance of driven vortices in the Nb films around the crossover frequency, f_{cr}, from the reactive response to the dissipative response[9]. As a result, we observed the increase of the ac resistance when the vortices are dc driven, which was found to be frequency dependent.

2. Methods

Microwave transmission line resonance method was used to measure the driving force dependence of the ac resistance of the vortex lattice. The resonators were made of Nb films, which were deposited on 5 mm × 5 mm × 0.5 mm MgO substrates by rf sputtering. Deposition rate was monitored by a quartz oscillator. The typical film thickness was approximately 100 nm, and T_c of the film was 8.7 K, and the resistivity at 10 K was 7 μΩ cm. The upper critical field, B_{c2}, at 3 K was 15000 G and the lower critical field, B_{c1}, was estimated to be much less than 10 G, considering the demagnetization factor of the thin strip[10–12]. The dc critical current, I_c, at 3 K, 30 G was around 72 mA, and f_{cr} of the Nb films was estimated to be 10~20 GHz[13].

To measure the center frequency, f_0, and the quality factor, Q, under dc driving current, coplanar waveguide λ/2 resonators were patterned on the Nb films by the photolithography technique (Figure 1). The patterned films were etched by reactive ion etching with CF_4. The width of the center strip was 10 μm and the gap between the center strip and the ground was 5 μm. The resonator was capacitively coupled to the external circuit through a 4 μm gap. The characteristic impedance of the resonators was about 50 Ω. DC bias current was applied to the resonator through leads forming λ/2 filters[14]. These filters suppress the loss due to the coupling to the external circuit. Two resonators were fabricated, f_0 of which are 3.7 GHz and 7.8 GHz, and the quality factors were 5000 and 9000 at 3 K, 0 G, respectively.

The block diagram of the measurements is shown in figure 2. Magnetic field was applied perpendicular to the Nb film. To avoid hysteresis, current was increased to 75 mA after the field cooling. In order to suppress Joule heating, rectangular pulsed current was supplied to the resonator to drive vortices. By synchronizing the network analyzer with the function generator, the transmission power, S_{21}, was measured at the flat part of the pulsed current around f_0. The pulse width and the duty cycle were 25 ms and 0.1, respectively, which was limited by the measurement time of the network analyzer (~15 ms). The amplitude of the pulsed current was estimated by measuring the voltage drop at a 10 Ω resistance in series with the sample. The maximum current on the resonance, I_{max}, were roughly estimated by $I_{max} \approx (P_{circ} /2\pi f_0 L)^{1/2}$, following Bothner *et al.*[15]. Here, $P_{circ} = 4P_{in}r(1 - r)Q/\pi$ is the rf power circulating in the resonator, P_{in} (P_{out}) denotes the input (output) power at the resonator, and $r = (P_{out}/P_{in})^{1/2}$. Assuming the loss at the cables to be 5~10 dB, P_{in} and P_{out} was estimated to be 1×10^{-7} W,

1×10^{-9} W, respectively. With P_{circ} of 1.1×10^{-3} W and the total line inductance, $L = 3.8 \times 10^{-8}$ H, I_{max} was calculated to be 7.7×10^{-5} A, which was two orders of magnitude smaller than I_c.

The measured resonance spectra were analyzed as follows. In the mixed state, several factors contribute to Q^{-1} as

$$\frac{1}{Q} = \frac{1}{Q_{\mathrm{super}}} + \frac{1}{Q_{\mathrm{vor}}} + \frac{1}{Q_{\mathrm{die}}} + \frac{1}{Q_{\mathrm{rad}}}, \quad (2)$$

where Q_{super}^{-1}, Q_{vor}^{-1}, $Q_{die}^{-1}$and Q_{rad}^{-1} are the superconductor loss, the vortices loss, the dielectric loss and the radiation loss, respectively. In a superconducting state, Q^{-1} is known to be less than 10^{-4} at 4.2 K[15], and Q_{vor}^{-1} is absent for $B = 0$. Q_{die}^{-1} of MgO is considered to be less than 10^{-4}[16]. Additionally Q_{rad}^{-1} is typically much less than 10^{-4} in the coplanar waveguide resonator undercoupled to the external circuit[17]. However, in this experiment, the extra leads suppling the dc bias current were attached to the resonators, leading to a finite coupling to the external circuit. Hence, Q_{rad}^{-1} was considered to become larger than 10^{-4}, and it mainly contributed to Q^{-1} without magnetic field, which was 1.2×10^{-4} at 7.8 GHz. When magnetic field was applied, the changes of Q_{die}^{-1} and Q_{rad}^{-1} are assumed to be small compared with that of Q_{vor}^{-1}. Since applied magnetic field was much less than B_{c2}, the change in Q_{super}^{-1} was also neglected. Then, the changes in the ac resistance due to the vortex motion can be calculated as

$$\Delta R(B,I) = \omega L\left(\frac{1}{Q}(I,B) - \frac{1}{Q}(0\,\mathrm{mA}, 0\,\mathrm{G})\right). \quad (3)$$

In addition to $\Delta Q^{-1}(B, I)$, the total inductance L was necessary to obtain ΔR by equation (3). $L = L_m + L_k$ is calculated as

$$L_{\mathrm{m}} = \frac{\mu_0}{4}\frac{K(k')}{K(k)}, \quad L_{\mathrm{k}} = \mu_0 \frac{\lambda^2}{dw} g(s,w,d), \quad (4)$$

$$g(s,w,d) = \frac{1}{2k'^2 K(k)^2}\left[-\ln\left(\frac{d}{4w}\right) - \frac{w}{w+2s}\ln\left(\frac{d}{4(w+2s)}\right) + \frac{2(w+s)}{w+2s}\ln\left(\frac{s}{w+s}\right)\right], \quad (5)$$

where L_m is the magnetic inductance of a coplanar waveguide, L_k is the kinetic inductance of a superconductor, $K(k)$ is the complete elliptic integral of the first kind with the modulus $k = w/(w+2s)$, $k' = (1 - k^2)^{1/2}$, μ_0 is the vacuum permeability, d is the film thickness, w is the width of the center strip, and s is the gap between the ground and the center strip[18]. Penetration depth was estimated to be 96 nm by the following equation[19],

$$\lambda = 1.05\, (\rho(T_c)/T_c)^{1/2} \times 10^{-3}\ \mathrm{m}. \quad (6)$$

Consequently, L was obtained, thus the ac resistance due to vortices was calculated by equation (3).

3. Results and Discussion

Figure 3 shows the dc I-V characteristics of the 7.8 GHz resonator measured at 3 K, 30 G by the four-probe method. From this data, I_c was estimated to be 72 mA ± 2 mA. The uncertainty of I_c comes from the voltage fluctuation due to the strong pinning at the low magnetic fields compared with B_{c2}. Indeed, the I-V characteristics at low magnetic fields is known to show hysteresis as well as the abrupt voltage jump because of the pinning[20]. Figure 4 shows the resonance spectra of the 7.8 GHz resonator with increasing dc bias current. The shape and the position of the resonance spectra changed as the current increased from 2 mA to 75 mA. As the current amplitude became larger, Q^{-1} increased, while f_0 continuously decreased. First, we focus on the decrease of f_0. Parabolic current dependence in f_0, as shown in the inset of figure 5, was already observed by Gittleman *et al.*[21], which was explained considering that the order parameter of the superconductor is suppressed by current. The solution of Ginzburg-Landau equation for sufficiently thin films is

$$|\varphi|^2 = \left(\frac{\lambda_0}{\lambda}\right)^2\left[1 - \frac{8\pi^2}{c^2}\frac{\lambda_0 d}{H_0^2} f_\infty(t) j^2\right] = \left(\frac{\lambda_0}{\lambda}\right)^2\left[1 - k^2 j^2\right], \quad (7)$$

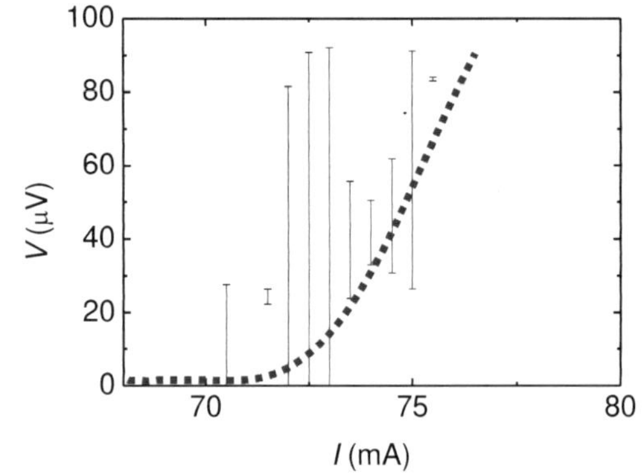

Figure 3. The dc *I-V* characteristic of the 7.8 GHz resonator measured at 3 K, 30 G. The dotted line is given as a guide to the eye.

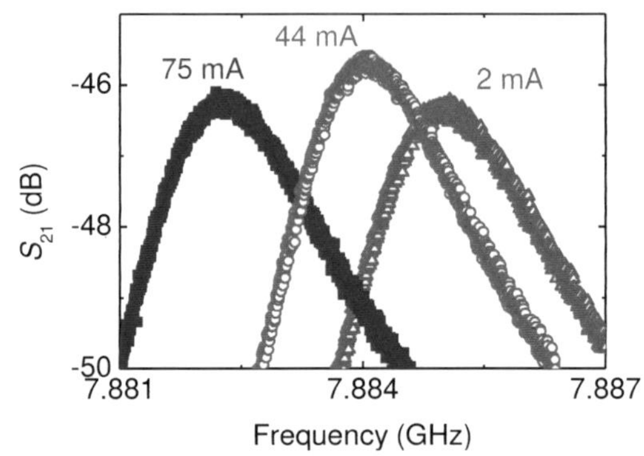

Figure 4. The resonance spectra of the 7.8 GHz resonator measured at 3 K, 30 G.

in the limit $((2e^2/\hbar^2c^2)H_c^2\lambda^4)^{1/2} \sim \infty$, where φ is the order parameter, λ is the penetration depth, λ_0 is the penetration depth at absolute zero, d is the thickness of the film, $H_0 = (4\pi/c)jd$ is the magnetic field at the film surface, $f_\infty(t) = (1 - t^2)^{-2}(1 - t^4)^{-1}$, and t is reduced temperature[21]. Thus, the current density $\mathbf{j}$ can be expressed as

$$\mathbf{j} = -\left(\frac{c}{4\pi\lambda_0^2}\right)|\varphi^2|\mathbf{A} = -\left(\frac{c}{4\pi\lambda^2}\right)\left[1-k^2j^2\right]\mathbf{A}\,. \tag{8}$$

Substituting $j = j_{dc} + j_{ac}$ and using $E_{ac} = -(1/c)(\partial A/\partial t)$, we obtain the imaginary part of the complex conductivity,

$$\sigma_2(j,t) = \left(\frac{c^2}{4\pi\omega\lambda^2}\right)\left[1-k^2j^2\right]. \tag{9}$$

Assuming the uniform current density and $\sigma_2 >> \sigma_1$, the inductance change can be written as

$$\Delta L \propto \sigma_2(j) - \sigma_2(0) \propto j^2\,. \tag{10}$$

Thus, from equations (7) and (10), the parabolic dependence of f_0 on j can be understood in terms of the pair breaking effect by the dc driving current.

Next, we discuss the change in Q^{-1}. Using equation (4), the change in the ac resistance per unit length was estimated (figure 6), and two remarkable features were observed. First, the ac resistance decreased slightly in the low current regime. Although we do not understand the origin of this decrease, this may be due to the anharmonicity in the pinning potential. As the current density increases, vortices move from initial positions to different equilibrium positions where the pinning force and the driving force balance with each other[22]. The oscillation of vortices in the pinning potential can be described by the equation of motion, $\eta v + kx = j_{ac}\Phi_0$, where η is the viscous drag coefficient, Φ_0 is the flux quantum. The pinning constant, k, is $k = \partial^2U_p/\partial r^2$, where U_p is the pinning potential[23]. If U_p was harmonic everywhere, k would be a constant and the resistance would be the same even when vortices change their position. The second remarkable feature is that the ac resistance increases gradually in the high current regime, which is in sharp contrast to the dc behavior shown in figure 3. The difference between the dc response and the ac response is understood by remembering that vortices oscillate even for the smaller current than I_c, leading to the finite resistance. As for the increase of the ac resistance with increasing the dc driving force, a qualitative explanation is as follows. Vortices suffer from the pinning even around f_{cr}, which means that, the ac responses of the vortices are still reactive to some extent at these frequencies[9]. On the other hand, moving vortices are known to feel weaker pinning force than pinned ones[24]. Hence, if the effect of the pinning is reduced by the translational motion of the vortices, the ac response is expected to be more resistive (less reactive), leading to the increase of the ac resistance.

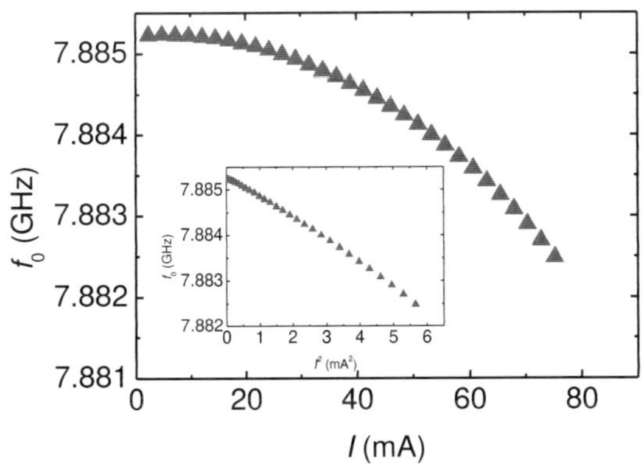

Figure 5. The current dependence of f_0. The inset shows I^2vs f_0 plot.

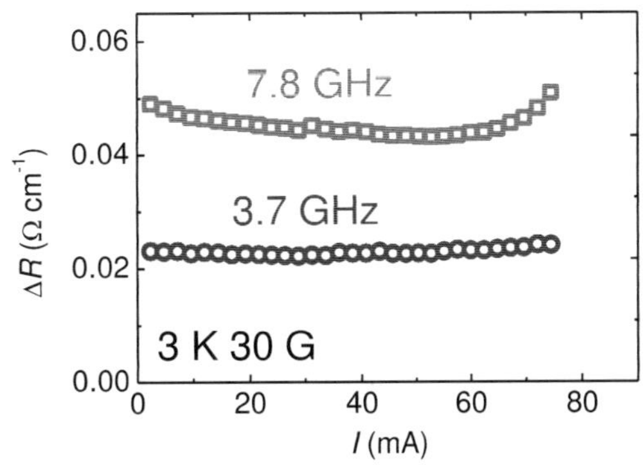

Figure 6. The current dependence of ΔR measured at 3.7 GHz and 7.8 GHz.

Finally, we discuss the frequency dependence. Figure 6 shows ΔR measured at 3.7 GHz and 7.8 GHz. ΔR of 7.8 GHz per unit length increases from 0.043 Ω cm^{-1} at 53 mA to 0.051 Ω cm^{-1} at 75 mA, while that of 3.7 GHz increases 0.002 Ω cm^{-1} at most. We estimated the ratio of the change in ΔR, (ΔR_{max} - ΔR_{min})/ΔR(2 mA), where ΔR_{max} (ΔR_{min}) is the maximum (minimum) value of ΔR. It was 8 % for 3.7 GHz and 16 % for 7.8 GHz, respectively. These differences suggest that the depinning process depends on frequency[9]. To obtain the further detailed picture, experiments at other frequencies are indispensable, which are now in progress.

Shklovskij and Dobrovolskiy numerically solved the equation of motion for a single vortex in the tilted washboard potential[25]. They calculated the current density dependence of the resistivity, ρ, at the normalized frequency $\Omega = 0.1$, which is one-tenth of f_{cr}. Their calculation showed a sharp rise of the real part of ρ at dc critical current density which was very different from a continuous increase observed in our experiment. Although there are several differences in the situations between their calculation and our experiment, such as the frequency, the ac current amplitude and the definition of the critical current density, the most significant difference seems to be the presence/the absence of the randomness in the pinning potential. Indeed, the periodic sinusoidal potential was assumed in their calculation, whereas the realistic pinning potential is with an extensive amount of disorder. More elaborate treatment is needed urgently even theoretically.

4. Conclusion

In conclusion, we investigated the microwave responses of the driven vortices with superconducting coplanar waveguide resonators. From the driving force dependence of the resonance spectra, first, the parabolic current dependence of the center frequency was understood considering the current dependence of the order parameter. Second, the ac resistance decreased in the low current regime, while it increased around the dc critical current. The increase was understood by considering that driven vortices feel weaker pinning force than the pinned vortices. Finally, from the frequency dependence, the depinning process was found to depend on frequency even near the crossover frequency. To understand the microscopic origin of the dynamic pinning force, further studies including measurements of the ac response at other frequencies, voltage noise measurements and numerical simulations are needed, which are now in progress.

References

[1] Maeda A, Inoue Y, Kitano H, Savel'ev S, Okayasu S, Tsukada I, Nori F 2005 *Phys. Rev. Lett.* **94** 077001
[2] Kim Y B, Hempstead C F, Strand A R 1965 *Phys. Rev.* **139** 1163
[3] Schmid A, Hauger W 1973 *J. Low Temp. Phys.* **11** 667
[4] Müllers J, Schmid A 1995 *Phys. Rev. Lett.* **75** 136
[5] Grundberg S, Rammer J 2000 *Phys. Rev. B* **61** 699
[6] Yamafuji K, Irie F 1967 *Phys. Lett. A* **25** 387
[7] Matsushita T, Kusayanagi E, Yamafuji K 1979 *J. Phys. Soc. Japan* **46** 1101
[8] Lowell J 1970 *J. Physics C* **3** 712
[9] Gittleman J I, Rosenblum B 1968 *J. Appl. Phys.* **39** 2617
[10] Finnemore D K, Stromberg T F, Swenson C A 1966 *Phys. Rev.* **149** 231
[11] Osborn J A 1945 *Phys. Rev.* **67** 351
[12] Stan G, Field S B, Martinis J M 2004 *Phys. Rev. Lett.* **92** 097003
[13] Požek M, Grbić M S, Janjušević D, Dulčić A, Paar D, Wagner T 2007 *Phys. C Supercond.* **460** 1291
[14] Li S X, Kycia J B 2013 *Appl. Phys. Lett.* **102** 242601
[15] Bothner D, Gaber T, Kemmler M, Koelle D, Kleiner R, Wünsch S, Siegel M 2012 *Phys. Rev. B* **86** 014517
[16] Krupka J, Geyer R G, Hinken J H 1994 *IEEE Trans. Microw. Theory Tech.* **42** 1886
[17] Göppl M, Fragner A, Baur M, Bianchetti R, Filipp S, Fink J M, Leek P J, Puebla G, Steffen L, Wallraff A 2008 *J. Appl. Phys.* **104** 113904
[18] Yoshida K, Watanabe K, Kisu T, Enpuku K 1995 *IEEE Trans. Appiled Supercond.* **5** 1979
[19] Orlando T P, McNiff E J, Foner S, Beasley M R 1979 *Phys. Rev. B* **19** 4545
[20] Danckwerts M, Goñi A, Thomsen C 1999 *Phys. Rev. B* **59** R6624
[21] Gittleman J, Rosenblum B, Seidei T E, Wicklund A W 1965 *Phys. Rev.* **137** A527
[22] Campbell A M 1971 *J. Phys. C* **4** 3186
[23] Golosovsky M, Tsindlekht M, Davidov D 1996 *Supercond. Sci. Technol.* **9** 1
[24] Pierre L D, Giamarchi T 1998 *Phys. Rev. B* **57** 11356
[25] Shklovskij V A, Dobrovolskiy O V 2008 *Phys. Rev. B* **78** 104526

Paramagnetic and Glass States of a YBCO Film Containing Nanorods at Low Magnetic Fields

H Deguchi[1], A Harada[1], T Yamada[1], M Mito[1], T Horide[1] and K Matsumoto[1]

[1]Faculty of Engineering, Kyushu Institute of Technology, Kitakyushu 804-8550, Japan

E-mail: deguchi@mns.kyutech.ac.jp

Abstract. The magnetic properties of $YBa_2Cu_3O_{7-x}$ (YBCO) multilayered film containing $BaHfO_3$ (BHO) nanorods at zero and small fields have been investigated. The dc magnetization and the ac susceptibility were measured with a SQUID magnetometer. The paramagnetic magnetization is shown after the field-cool-warming below the transition temperature $T_c = 87$ K. The nonlinear susceptibility had a large peak around T_c, which reflects a glass-transition. An aging effect occured in the relaxation of zero-field-cooled magnetization below the T_c. Paramagnetic behavior and magnetic-glass properties were observed in the YBCO multilayered film containing BHO nanorods. The results suggest that the YBCO multilayered film is a novel magnetic-glass system at low fields.

1. Introduction

Recently, multilayered films comprising $YBa_2Cu_3O_{7-x}$ (YBCO) with nanorods as artificial pinning centers have been extensively studied for their flux pining properties in strong magnetic fields. The transport properties of YBCO films are strongly enhanced by columnar pins, which are produced by the introduction of compound defects, e.g., $BaHfO_3$(BHO), $BaZrO_3$, and $BaSnO_3$ nanorods.[1,2] YBCO films with these nanorods effectively improve the critical current density (J_c) in strong magnetic fields. High-temperature superconductors exhibit rich phase diagrams with vortex glass, and Bragg and Bose glass phases. They are actively studied to further understand the behavior of the glass-liquid transition.[3] However, there are few experimental studies concerning the magnetic properties of YBCO films that have investigated the low field, which corresponds to the Meissner state of the bulk YBCO. Even at such low field, the magnetic flux can be trapped in the nanorods of the YBCO films, and the vortex state of a small number of vortices confined in the nanorods would be different from that of the bulk samples. In addition, the geometrical disorder due to the random location of the nanorods in the films corresponds to random flux in the glass system such as the chiral-glass one in the ceramic cuprate superconductors.[4-7] In this work, we have investigated the magnetic properties of a YBCO multilayered film containing BHO nanorods at zero and small fields to elucidate the magnetic states of random vortices in the system. Paramagnetic and glass states were observed at low fields below the T_c.

2. Experimental

Multilayered films comprising three YBCO layers containing BHO and four pure YBCO layers were prepared on a $SrTiO_3$ substrate via alternating the ablation of the YBCO stoichiometric target and YBCO-BHO mixed target.[1] The concentration of BHO in the premixed targets is 6 wt%. The thickness

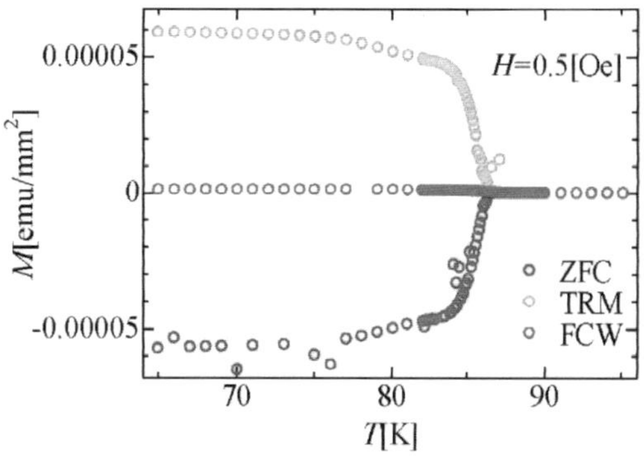

Figure 1. Temperature dependence of the ZFC, FCW, and TR magnetization at $H = 0.5$ Oe.

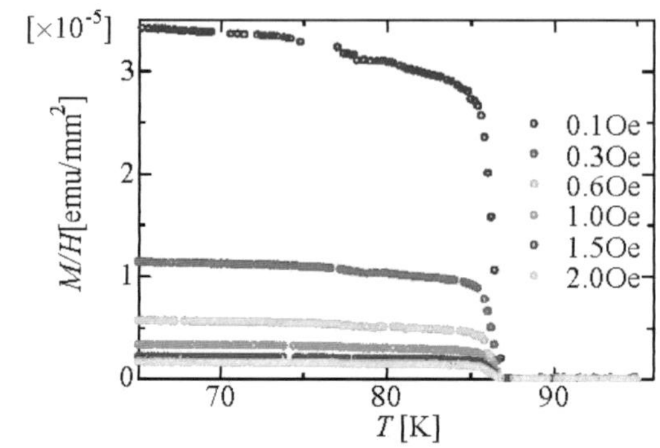

Figure 2. Temperature dependence of the FCW susceptibility at different dc applied fields.

of each layer was 30 nm, and the total film thickness was 220 nm. Transmission electron microscopy was used to characterize the microstructure of the film. The BHO nanorods with 7 nm diameter grew almost parallel to the *c*-axis direction of the film.

The dc magnetization and the ac magnetic response were measured with a SQUID magnetometer (Quantum Design MPMS-7) using the ultra-low-field option at the temperature sweep mode. The sample space of the magnetometer was shielded with μ-metal. As a result, the residual field was reduced to less than 10 mOe. The nonlinear susceptibility was derived from the harmonics in-phase Fourier component for the ac-field response. An external field was applied perpendicular to the film plane. The magnetizations and the susceptibilities were normalized with the in-plane area of the film.

3. Results and Discussion

The temperature dependence of the dc magnetization and ac susceptibility was measured to investigate the glass behavior at low fields. Figure 1 shows the temperature dependences of the zero-field-cooled (ZFC), field-cool-warming (FCW), and thermoremanent (TR) magnetizations at $H = 0.5$ Oe. The transition at $T_c = 87$ K was identified as the superconducting ordering, in which diamagnetism due to the Meissner effect appears in ZFC magnetizations. Moreover, the FCW magnetization was much lower than the ZFC one and was positive below the T_c. The paramagnetic signal of the FCW magnetization is known as the paramagnetic Meissner effect (PME) or Wohlleben effect.[8] Magnetic-glass behaviors such as the presence of TR magnetization and the discrepancy between the FCW and ZFC magnetization at the T_C were observed at low field. Below the T_c, $M_{ZFC} - M_{FCW} = M_{TRM}$ is adequately satisfied for all temperatures.

In Figure 2, the FCW susceptibilities (M_{FCW}/H) are plotted as a function of temperature at dc magnetic fields between 0.1 and 2.0 Oe. Below the T_c, the FCW susceptibility increased with decreasing field and decreasing temperature, and was nearly constant at low temperatures. The field dependence of the FCW susceptibilities at 66 K is shown in Figure 3. The FCW susceptibilities at 66 K could be fitted using the equation $M_{FCW}/H = \chi_0 + m/(H+H_0)$, where $\chi_0 = -1.17\times10^{-6}$ emu, $m = 3.72\times10^{-5}$ emu and $H_0 = 7.31\times10^{-4}$ Oe. Such field dependence is observed in ceramic samples of $Bi_2Sr_2CaCu_2O_8$ (Bi2212) superconductors, and suggests a paramagnetic Meissner behavior.[9]

Next, the temperature dependence of the in-phase (m') and of out-of-phase (m") ac magnetic response was measured at a frequency of 1 Hz and an ac-field amplitude of 0.1, 0.3, and 0.5 Oe under zero external field. The in-phase response m' abruptly changed at the T_c and showed diamagnetism below it. This behavior was similar to that of the ZFC dc-magnetization shown in Figure 1. The diamagnetism of m' saturated below *ca.* 70 K, and the saturated value linearly increased with the ac-

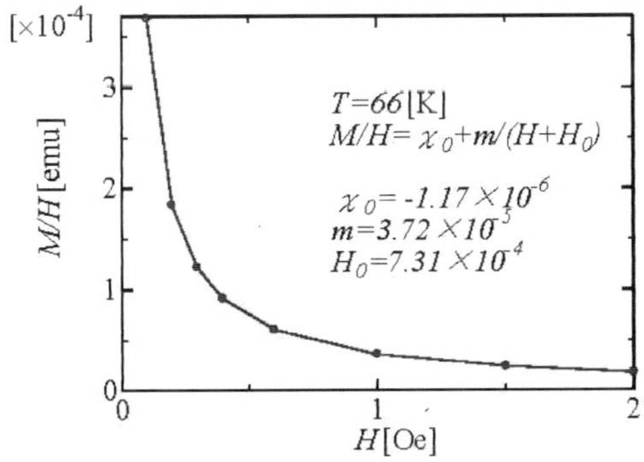

Figure 3. Field dependence of the FCW susceptibility M_{FCW}/H at 66 K. The line represents the fit using the equation given in the text.

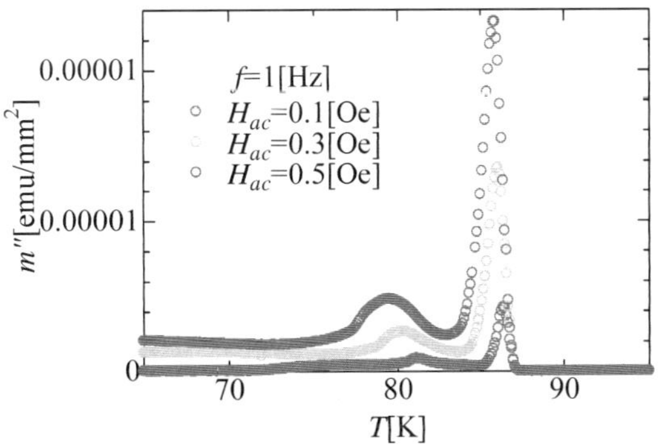

Figure 4. Temperature dependence of the out-of-phase ac magnetic responses m" at a frequency of 1 Hz with an ac-field amplitude of 0.1, 0.3, and 0.5 Oe at $H = 0$ Oe.

field amplitude, which is due to the Meissner effect resulting from the superconductivity of the film. Figure 4 shows the temperature dependence of the out-of-phase response m". The large peak corresponding to m" was observed at the T_c. m" decreased at about 85 K. Then, m" formed a double maximum. The magnitude of the peak depended on the ac-field amplitude, and the results exhibit dissipation even in the low-frequency region. The nonlinear susceptibilities estimated from the first term of the series of in-phase odd-harmonic responses at 1 Hz with ac-field amplitudes of 0.1, 0.3, and 0.5 Oe are shown in Figure 5. A positive and a negative peak of nonlinear susceptibility were respectively observed around the T_c and 85 K; the intensity did not depend on the ac-field amplitude. The former peak corresponded to the superconducting transition and the latter to the glass-transition.[4,7]

The relaxation of the ZFC magnetization was observed via the following procedures. The sample was cooled down from 100K above the T_c to the measuring temperature ($T_s = 82$ K) under zero field at a cooling rate of 4 K/min; then, it was maintained at the T_s for a certain time t_w. The ZFC magnetization was then measured as a function of time under a 0.3 Oe field. The relaxation was logarithmically slow in time and the ZFC magnetization towards the value of the FC one. In Figure 6, the time dependence of the ZFC magnetization measured at $T_s = 82$ K is shown. Three different waiting times, $t_w = 1000$, 3000, and 10000 s, were used. The relaxation curves depended on the waiting time t_w. The three curves were different, and the relaxation rate decreased with increasing t_w. Such influence of the waiting time t_w on the dynamics is called the aging effect, which has been observed for frustrated systems such as spin-glasses [10] and the granular Bi2212 superconductor [11] with the PME. The aging effect of the granular Bi2212 was understood by modeling the polydomain microstructure of the superconducting grains as Josephson-junction networks.

Luzhbin *et al.*[12] investigated the paramagnetic magnetization in field-cooled YBCO thin films. The experimental results of the PME are different from the results of this work. They observed a PME over a wide magnetic-field range ($5 < H < 10^4$ Oe), and the paramagnetic value of magnetization steadily increased with decreasing temperature, indicating the absence of saturation. However, the magnetic-glass behavior was not observed. They concluded that the origin of the PME was in good agreement with the model of vortex compression proposed by Koshelev and Larkin.[13] Koblischka *et al.*[14] observed the PME in an artificially granular YBCO thin film which was patterned into a hexagonal close-packed lattice of disks with a diameter of 50 μm. The field and temperature dependence of the PME is similar to our results. The authors concluded that the PME response was due to flux trapping effects in the spaces between the superconducting disks.

Li [15] reviewed the experimental results, and the simulations and theories related to the PME and

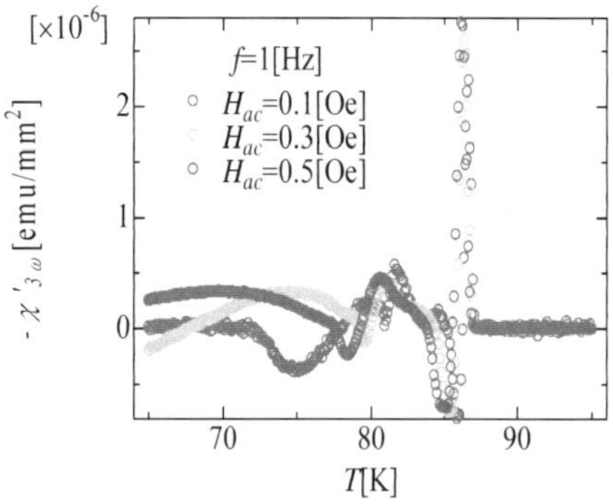

Figure 5. Temperature dependence of the nonlinear susceptibility at a frequency of 1.0 Hz with an ac-field amplitude of 0.1, 0.3, and 0.5 Oe at H = 0 Oe.

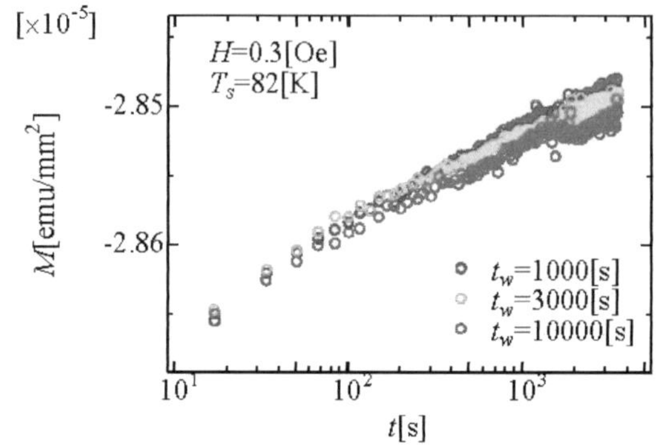

Figure 6. Observation time dependence of the ZFC magnetization for H = 0.3 Oe at T= 82 K with waiting times of 1000, 3000, and 10000 s.

dynamical phenomena such as the glass behavior and aging effect of ceramic cuprate superconductors at zero and low fields. Our results were consistent with those of said superconductors. Recently, Granato [16] studied the magnetic flux disorder in nanohole thin films via numerical simulation of a Josephson-junction array model. He considered the geometrical disorder owing to the random location of nanoholes in the film as the random flux in the model. Therefore, the vortex due to the random locations of nanorods in the YBCO film may correspond to the random flux in the cuprate ceramic superconductors.

4. Summary

The PME was observed in the FCW magnetization of a YBCO film containing BHO nanorods below the T_c. The presence of TR magnetization and the discrepancy between the FCW and ZFC magnetization at T_c are shown at low field. The large peaks of the out-of-phase response m" and nonlinear susceptibility were observed around the T_c at zero-field. The slow relaxation and aging phenomena of the ZFC magnetization were observed at 82 K. All these observations suggest that the YBCO multilayered film has a novel paramagnetic and glass state at low field, which is different from the vortex phases existing above the Meissner state.

Acknowledgements

The SQUID measurement was conducted at the Center for Instrumental Analysis, Kyushu Institute of Technology.

References

[1] Matsumoto K, Tanaka I, Horide T, Mele P, Yoshida Y and Awaji S 2014 *J. Appl. Phys.* **116** 163903
[2] Horide T, Matsumoto K, Mele P, Yoshida Y, Ichinose A, Kita R, Horii S and Mukaida M 2009 *Phys. Rev. B* **79** 092504
[3] Blatter G, Feigel'man M V, Geshkenbein V B, Larkin A I and Vinokur V M 1994 *Rev. Mod. Phys.* **66** 1125
[4] Deguchi H, Warabino R, Ka S, Mito M, Hagiwara M and Koyama K 2017 *J. Phys. Conf. Series* **871** 012011
[5] Deguchi H, Ashida T, Shoho T, Kato Y, Mito M, Takagi S, Hagiwara M and Koyama K 2011 *J.*

Phys. Conf. Series **320** 012076
[6] Deguchi H, Syudo M, Ashida T, Sasaki Y, Mito M, Takagi S, Hagiwara M and Koyama K 2013 *Phys. Procedia* **45** 129
[7] Kawamura H 2010 *J. Phys. Soc. Jpn.* **79** 011007
[8] Braunish W, Knataev N, Neuhausen S, Gruz A, Kock A, Roden B, Khomskii D and Wohlleben D 1992 *Phys. Rev. Lett.* **68** 1908
[9] Braunish W, Knauf N, Bauer G, Kock A, Becker A, Freitag B, Gruz A, Kataev N, Neuhausen S, Roden B, Khomoskii D, Wohlleben D, Bock J and Preisler E 1993 *Phys. Rev. B* **48** 4030
[10] Lundgren L, Svedlindh P, Nordblad P and Beckman O 1983 *Phys. Rev. Lett.* **51** 911
[11] Papadopoulou E L, Nordbald P, Svedlindh P, Schoneberger R and Gross R 1999 *Phys. Rev. Lett.* **82** 173
[12] Luzhbin D A, Pan A V, Komashko V A, Flis V S, Pan V M, Dou S X and Esquinzai P 2004 *Phys. Rev. B* **69** 024506
[13] Koshelev A E and Larkin A I 1995 *Phys. Rev. B* **52** 13559
[14] Koblischka M R, Pust L, Chikumoto N, Murakami M, Nilsson B and Claeson T 2000 *Physica B* **284-288** 599
[15] Li M S 2003 *Physics reports* **376** 133
[16] Granato E 2016 *Phys. Rev. B* **94**, 060504(R)

IOP Conf. Series: Journal of Physics: Conf. Series **1054** (2018) 012027 doi:10.1088/1742-6596/1054/1/012027

Effects of chirality of a helical magnetic field on a superconductor

Saoto Fukui[1], **Masaru Kato**[1] **and Yoshihiko Togawa**[2], **Osamu Sato**[3]

[1] Department of Mathematical Sciences, Osaka Prefecture University, 1-1, Gakuencho, Nakaku, Sakai, Osaka 599-8531, Japan

[2] Department of Physics and Electronics, Osaka Prefecture University, 1-1, Gakuencho, Nakaku, Sakai, Osaka 599-8531, Japan

[3] Osaka Prefecture University College of Technology, 26-12, Saiwaicho, Neyagawa, Osaka 572-8572, Japan

E-mail: st110035@edu.osakafu-u.ac.jp

Abstract. We have investigated vortex states in a superconductor under a helical magnetic field from a chiral helimagnet. We solve the Ginzburg-Landau equations with three-dimensional finite element method. In this model, two antiparallel vortices appear under the helical magnetic field. There directions are normal to the largest surface of the superconductor.

1. Introduction

A chiral helimagnet has a long periodic helical magnetic structure [1]. The magnetic structure is shown in figure 1 (a). Under an applied magnetic field, the chiral helimagnetic structure transforms into a periodic soliton structure, which is called a chiral soliton lattice shown in figure 1 (b). An origin of this chiral helimagnetic structure is a competition of two interactions between nearest neighbor spins; the Dyzaloshinsky-Moriya (DM) interaction [2, 3] and a ferromagnetic exchange interaction.

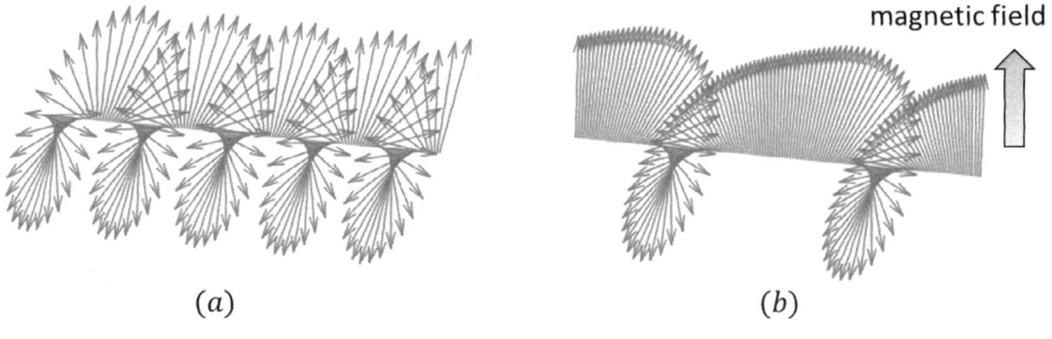

Figure 1. Spin configurations of (a) a chiral helimagnetic structure, and (b) a chiral soliton lattice under an applied magnetic field H_{appl}.

It is known that the chiral helimagnet gives various physical properties, for example, a giant magnetoresistance and discrete changes of the resistance [4]. We focus on effects of the chiral helimagnet on a superconductor. There are many investigations of effects of a ferromagnet on the superconductor [5, 6, 7, 8]. In the ferromagnet / superconductor bilayer system, vortices appear in the superconductor spontaneously under an uniform magnetic field from a magnetic domain of the ferromagnet. Also, the pinning effect occurs due to ferromagnetic dots on the superconductor. On the other hand, the chiral helimagnet creates the helical magnetic field. We consider the possibility of novel influences of the chiral helimagnet on vortex structures in the superconductor. In our previous study [9, 10, 11, 12, 13], we investigated vortex structures in a two-dimensional superconductor under a helical magnetic field instead of the chiral helimagnet / superconductor bilayer system (figure 2(a)). Then, the effect of the helical magnetic structure in the chiral helimagnet was replaced by the oscillating magnetic field. Solving the Ginzburg-Landau equations with the two-dimensional finite element method (FEM), we obtained various vortex structures.

So far, we have restricted ourselves to the two-dimensional superconductor. However, in order to consider effects of chirality of the helical magnetic field on the vortex structure, we must consider a three-dimensional superconductor. In this report, we show the three-dimensional vortex structures in a three-dimensional superconductor under the three-dimensional helical magnetic field from the chiral helimagnet.

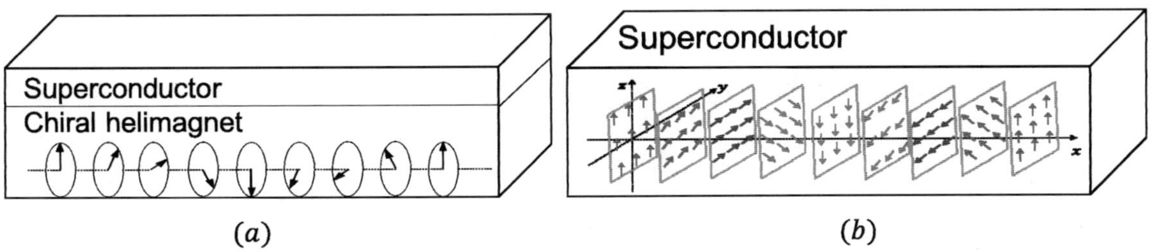

Figure 2. (a) A chiral helimagnet / superconductor bilayer system. (b) A three-dimensional superconductor under a helical magnetic field.

2. Methods

We consider a three-dimensional superconductor under the helical magnetic field from the chiral helimagnet, which is shown in figure 2(b). In order to obtain stable vortex structures in the superconductor, we solve the Ginzburg-Landau equations with the three-dimensional FEM. We start from the Ginzburg-Landau free energy for the three-dimensional system,

$$\mathcal{F}(\psi, \boldsymbol{A}) = \int_V \left(f_n + \alpha|\psi|^2 + \frac{\beta}{2}|\psi|^4 \right) dV + \int_V \left\{ \frac{1}{2m_s} \left| \left(-i\hbar\nabla - \frac{e_s \boldsymbol{A}}{c} \right) \psi \right|^2 + \frac{|\boldsymbol{h}|^2}{8\pi} - \frac{\boldsymbol{h}\cdot\boldsymbol{H}}{4\pi} \right\} dV, \tag{1}$$

where ψ is a superconducting order parameter and f_n is a free energy density of the normal state. $\alpha(T)$ is a function of temperature, $\alpha(T) = \alpha'(T - T_c)$. α' and β is a positive constant, T_c is a critical temperature of the superconductor. m_s and e_s is an effective mass and an effective charge of electrons in the superconductor, respectively. $\boldsymbol{H}$ is an external magnetic field, $\boldsymbol{h} = \nabla \times \boldsymbol{A}$ is a local magnetic field, and $\boldsymbol{A}$ is a magnetic vector potential. ψ and $\boldsymbol{A}$ are normalized as,

$$\tilde{\psi} = \frac{\psi}{\sqrt{\alpha/\beta}}, \quad \tilde{\boldsymbol{A}} = \frac{2\pi}{\Phi_0}\boldsymbol{A}, \tag{2}$$

IOP Conf. Series: Journal of Physics: Conf. Series **1054** (2018) 012027 doi:10.1088/1742-6596/1054/1/012027

where $\Phi_0 = ch/2e$ is the quantum flux. In the three-dimensional FEM, we use a tetrahedron as a finite element, which is shown in figure 3. There are four volume coordinates for $e-$th element,

$$N_i^e = (a_i + b_i x + c_i y + d_i z)/6V_e \quad (i = 1,\ 2,\ 3,\ 4), \tag{3}$$

where V_e is a volume of the tetrahedron for $e-$th element and a_i, b_i, c_i, and d_i are given in the Appendix. Using these volume coordinates, the order parameter $\tilde{\psi}^e$ and the vector potential $\tilde{\boldsymbol{A}}^e$ in $e-$th element are expanded as,

$$\tilde{\psi}^e(x,y,z) = \sum_i \tilde{\psi}_i^e N_i^e, \quad \tilde{\boldsymbol{A}}^e(x,y,z) = \sum_i \tilde{\boldsymbol{A}}_i^e N_i^e \quad (i = 1,\ 2,\ 3,\ 4), \tag{4}$$

where $\tilde{\psi}_i^e$ and $\tilde{\boldsymbol{A}}_i^e$ are values of the order parameter and the magnetic vector potential at $i-$th vertex in $e-$th element. After Frechét derivatives of the free energy (equation (1)) about $\tilde{\psi}$ and $\tilde{\boldsymbol{A}}$, then substitute equation (4). Therefore, we obtain these equations,

$$\sum_j \left[P_{ij}(\{\tilde{A}\}) + P_{ij}^{2R}(\{\tilde{\psi}\})\right] \mathrm{Re}\tilde{\psi}_j^e + \sum_j \left[Q_{ij}(\{\tilde{\boldsymbol{A}}\}) + Q_{ij}^2(\{\tilde{\psi}\})\right] \mathrm{Im}\tilde{\psi}_j^e = V_i^R(\{\tilde{\psi}\}), \tag{5}$$

$$\sum_j \left[-Q_{ij}(\{\tilde{\boldsymbol{A}}\}) + Q_{ij}^2(\{\tilde{\psi}\})\right] \mathrm{Re}\tilde{\psi}_j^e + \sum_j \left[P_{ij}(\{\tilde{\boldsymbol{A}}\}) + P_{ij}^{2I}(\{\tilde{\psi}\})\right] \mathrm{Im}\tilde{\psi}_j^e = V_i^I(\{\tilde{\psi}\}), \tag{6}$$

$$\sum_j R_{ij}(\{\tilde{\psi}\})\tilde{A}_{jx} + \sum_j S_{ij}^{xy}\tilde{A}_{jy} + \sum_j S_{ij}^{xz}\tilde{A}_{jz} = T_i^x - U_i^x, \tag{7}$$

$$\sum_j R_{ij}(\{\tilde{\psi}\})\tilde{A}_{jy} + \sum_j S_{ij}^{yx}\tilde{A}_{jx} + \sum_j S_{ij}^{yz}\tilde{A}_{jz} = T_i^y - U_i^y, \tag{8}$$

$$\sum_j R_{ij}(\{\tilde{\psi}\})\tilde{A}_{jz} + \sum_j S_{ij}^{zx}\tilde{A}_{jx} + \sum_j S_{ij}^{zy}\tilde{A}_{jy} = T_j^z - U_i^z. \tag{9}$$

Coefficients are also given in Appendix. We solve these equations self-consistently including nonlinear terms.

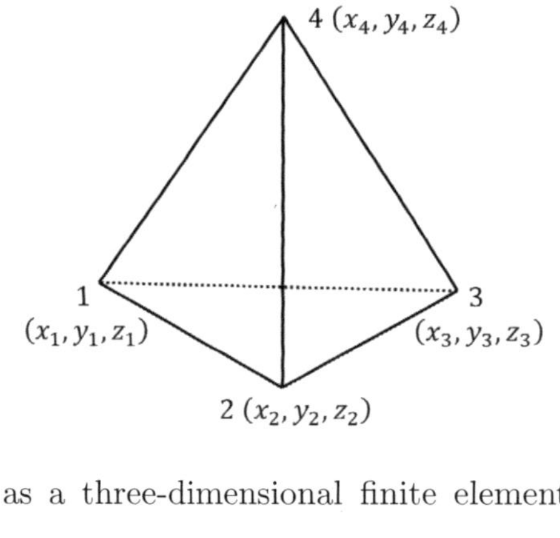

Figure 3. A tetrahedron as a three-dimensional finite element. 1, 2, 3, and 4 are volume coordinates.

We assume that the effect of the chiral helimagnet on the superconductor is taken as a helical magnetic field. The magnetic field from the chiral helimagnet is obtained from the Hamiltonian for the chiral helimagnet,

$$\mathcal{H} = -J\sum_n \boldsymbol{S}_n \cdot \boldsymbol{S}_{n+1} + \boldsymbol{D} \cdot \sum_n \boldsymbol{S}_n \times \boldsymbol{S}_{n+1} + 2\mu_B \boldsymbol{H}_{\mathrm{appl}} \cdot \sum_n \boldsymbol{S}_n, \tag{10}$$

where $J(>0)$ is a coefficient of the ferromagnetic exchange interaction and $\boldsymbol{D} = (D, 0, 0)$ is a DM vector [14]. $\boldsymbol{S}_n$ is the spin of the n-th site. μ_B is the Bohr magneton and $H_{\text{appl}} = (0, 0, H_{\text{appl}})$ is the homogeneous applied magnetic field. $\boldsymbol{S}_n$ is expressed by the polar coordinate as

$$\boldsymbol{S}_n = S(\sin\theta_n \cos\phi,\ \sin\theta_n \sin\phi,\ \cos\theta_n). \tag{11}$$

S is a magnitude of the spin. In the monoaxial chiral helimagnet, $\phi = \pi/2$. From the Hamiltonian [equation (10)] in the continuum limit, we obtain a distribution of angles of spins $\theta(x)$ [14];

$$\theta(x) = 2\sin^{-1}\left[\text{sn}\left(\frac{\sqrt{H^*}}{k}x \mid k\right)\right] - \frac{1}{2}\pi, \tag{12}$$

where $H^* = 2\mu_B H_{\text{appl}}/(a^2S^2\sqrt{J^2 + D^2})$ is a normalized applied magnetic field and k is the modulus of the Jacobi's elliptic function $\text{sn}(u|k)$. k is determined by,

$$\frac{\pi \tan^{-1}(D/J)}{4\sqrt{H^*}} = \frac{E(k)}{k}, \tag{13}$$

where $E(k)$ is the complete elliptic integral of the second kind. We assume a lattice constant $a \sim \xi_0$ (ξ_0 is a coherence length at zero temperature). Finally, from equations (11) and (12), the external magnetic field $\boldsymbol{H} = ((\boldsymbol{H}_{\text{ext}})_x, (\boldsymbol{H}_{\text{ext}})_y, (\boldsymbol{H}_{\text{ext}})_z)$ in equation (1) is given by,

$$(\boldsymbol{H}_{\text{ext}})_x(x) = 0, \tag{14}$$

$$(\boldsymbol{H}_{\text{ext}})_y(x) = H_0 \sin\left(2\sin^{-1}\left[\text{sn}\left(\frac{\sqrt{H^*}}{k}x \mid k\right)\right] - \frac{1}{2}\pi\right), \tag{15}$$

$$(\boldsymbol{H}_{\text{ext}})_z(x) = H_0 \cos\left(2\sin^{-1}\left[\text{sn}\left(\frac{\sqrt{H^*}}{k}x \mid k\right)\right] - \frac{1}{2}\pi\right) + H_{\text{appl}}. \tag{16}$$

H_0 is a magnitude of the magnetic field from the chiral helimagnet. In our previous study, we use only $(\boldsymbol{H}_{\text{ext}})_z(x)$ (equation (16)) in the two-dimensional superconductor[5, 6, 7, 8]. In the present study, we consider three components of the external magnetic field for the three-dimensional superconductor. We calculate the Ginzburg-Landau equations (5)-(9) using the magnetic field in equations (14)-(16).

3. Results and Discussions

We show vortex configurations in the superconductor under the helical magnetic field in equations (14)-(16) with $H_{\text{appl}} = 0$. We take the temperature $T = 0.3T_c$ and the Ginzburg-Landau parameter $\kappa = \lambda_0/\xi_0 = 5$, where λ_0 is a penetration length at zero temperature. The ratio between the ferromagnetic exchange interaction and the DM interaction is taken as $D/J = 0.16$, which is adequate for $CrNb_3S_6$[15]. The system size are $1.0L'\xi_0 \times 15\xi_0 \times 13\xi_0$. Here, $L'\xi_0 = (2\pi/\tan^{-1}(D/J))\xi_0$ is a helical period of the helical magnetic field for $H_{\text{appl}} = 0$. When $D/J = 0.16$, L' is approximately 39.2699. In the system, the superconducting region is surrounded by the vacuum region shown in figure 4. The size of the superconductor is $(1.0L' - 3.0)\xi_0 \times 12\xi_0 \times 10\xi_0$. At first, we show the distribution of the magnetic field in figure 5. In equations (14)-(16), we consider the case of $|H_0/(\Phi_0/\xi_0^2)| = 0.15$. The vortex configuration under this helical magnetic field is shown in figure 6. These figures represent configurations of vortices in $xy-$planes at $z = 1.5\xi_0$, $6.5\xi_0$, and $11.5\xi_0$. Vortices appear in the region $|(\boldsymbol{H}_{\text{ext}})_y/(\Phi_0/\xi_0^2)| \sim 0.00$ and $|(\boldsymbol{H}_{\text{ext}})_z/(\Phi_0/\xi_0^2)| \sim 0.15$, where the external magnetic field are parallel to the $z-$direction (see figures 5 (c) and (d)). There are antiparallel with each other. In the finite system, a vortex appear at the center in the system under the homogeneous

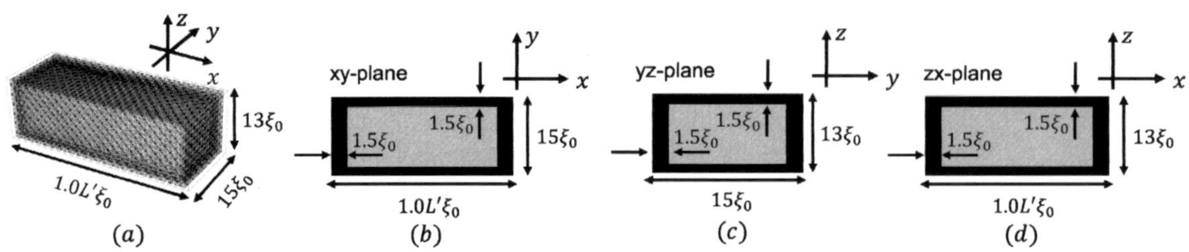

Figure 4. (a) Three-dimensional superconductor model. The system size is $1.0L'\xi_0 \times 13\xi_0 \times 15\xi_0$. The distance between the normal region and the superconducting region is $1.5\xi_0$. $xy-$plane, $yz-$plane, and $zx-$plane are (b), (c), and (d).

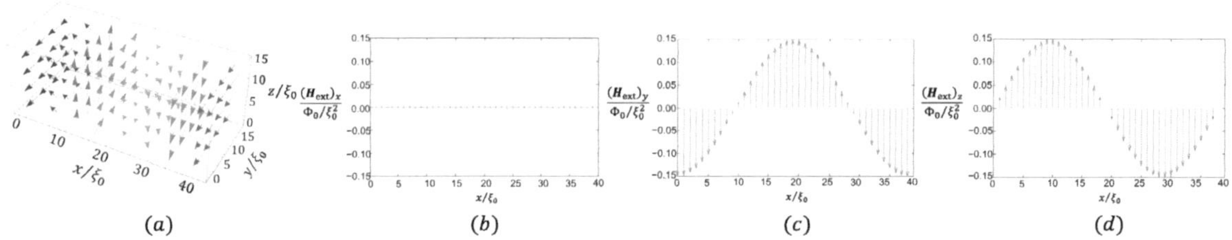

Figure 5. (a) The distribution of the helical magnetic field, (b) $x-$component of the magnetic field, (c) $y-$component of the magnetic field, (d) $z-$component of the magnetic field. The magnitude of the magnetic field $|H_0/(\Phi_0/\xi_0^2)| = 0.15$.

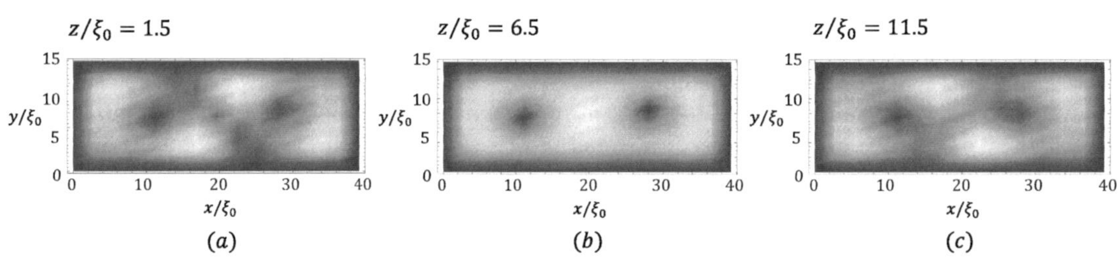

Figure 6. Configurations of the vortices in the $xy-$plane at (a) $z = 1.5\xi_0$, (b) $z = 6.5\xi_0$, and (c) $z = 11.5\xi_0$.

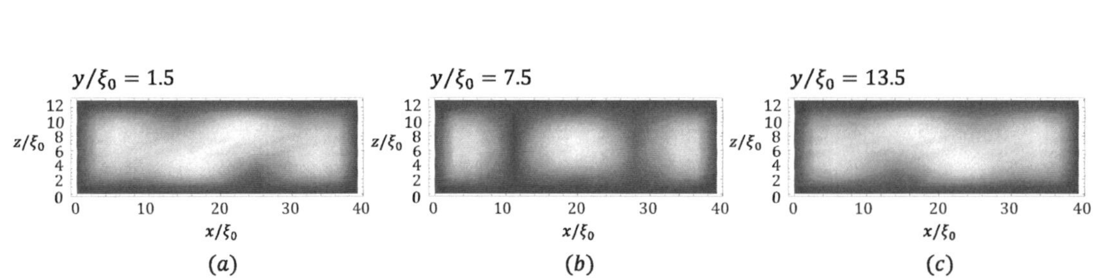

Figure 7. Configurations of the vortices in the $zx-$plane at (a) $y = 1.5\xi_0$, (b) $y = 7.5\xi_0$, and (c) $y = 13.5\xi_0$.

magnetic field [16]. In this system, the vortex doesn't appear at the center ($x/\xi_0 \sim 20.0$ and $y/\xi_0 \sim 7.5$), where $|(\boldsymbol{H}_{\rm ext})_z/(\Phi_0/\xi_0^2)| = 0.00$. Under the helical magnetic field, vortices appear at the largest magnetic field that in normal to the surface of the superconductor. Therefore, the configuration of vortices depends on the shape of the superconductor. This dependence of the configuration doesn't occur under the homogeneous magnetic field.

Moreover, two vortices appear antiparallel with each other. In our previous study, we investigated vortex structure under the oscillating magnetic field normal to a two-dimensional superconductor [12]. Under the oscillating magnetic field, vortices prefer to appear in all largest magnetic field regions except for ends of the superconductor. Present results agree with previous results. Two vortices normal to $z-$direction appear in the three-dimensional system.

Finally, we focus on effects of the $y-$component magnetic field. We show distributions of the order parameter in $zx-$plane at $y = 1.5\xi_0$, $7.5\xi_0$, and $13.5\xi_0$ in figure 7. At $y = 7.5\xi_0$ (see figure7(b)), we found that two vortices parallel to $z-$axis appear. Although $y-$component magnetic field also oscillates with the same amplitude to $z-$component magnetic field, vortices don't appear. When the system size is small, the magnetic field avoids to penetrate into the superconductor due to the Meissner effect. The $zx-$plane is smaller than $xy-$plane, so vortices affected by $y-$component of the magnetic field don't appear due to the Meissner effect or demagnetizing effect. Therefore, the appearance of vortices affects on the system size and when the system size becomes larger, vortices will appear in both $xy-$plane and $zx-$plane simultaneously. Moreover, vortices will appear parallel to the helical magnetic field, which doesn't appear in the case of the homogeneous magnetic field and the two-dimensional system.

4. Conclusions

We have obtained vortex structures in the three-dimensional superconductor under the helical magnetic field from the chiral helimagnet. Solving the Ginzburg-Landau equations in the model shown in figure 4, we found that two antiparallel vortices along the $z-$direction appear. The magnetic field along $z-$direction, $(\boldsymbol{H}_{\rm ext})_z$ changes spatially and these vortices appear at the largest $(\boldsymbol{H}_{\rm ext})_z$. However, when the system size is small, vortices don't appear due to the Meissner effect. In the future, we will investigate vortex structure under the helical magnetic field in the larger system and clarify effects of the chirality of the helical magnetic field on the superconductor.

Appendix

Coefficients in Eqs.(5)-(9) are given as,

$$P_{ij}(\{\boldsymbol{A}\}) = \sum_{\alpha} K_{ij}^{\alpha\alpha} + \sum_{i_1,i_2} I_{i_1 i_2 ij} \sum_{\alpha} A_{i_1\alpha}^e A_{i_2\alpha}^e - \frac{1}{\xi(T)^2} I_{ij},$$

$$P_{ij}^{2R}(\{\psi\}) = \frac{1}{\xi(T)^2} \sum_{i_1 i_2} I_{i_1 i_2 ij} (3\mathrm{Re}\psi_{i_1}^e \mathrm{Re}\psi_{i_2}^e + \mathrm{Im}\psi_{i_1}^e \mathrm{Im}\psi_{i_2}^e),$$

$$P_{ij}^{2I}(\{\psi\}) = \frac{1}{\xi(T)^2} \sum_{i_1 i_2} I_{i_1 i_2 ij} (\mathrm{Re}\psi_{i_1}^e \mathrm{Re}\psi_{i_2}^e + 3\mathrm{Im}\psi_{i_1}^e \mathrm{Im}\psi_{i_2}^e), \quad Q_{ij}(\{\boldsymbol{A}\}) = \sum_{i_1} \sum_{\alpha} (J_{ji_1 i}^{\alpha} - J_{ii_1 j}^{\alpha}) A_{i_1}^{\alpha},$$

$$Q_{ij}^2(\{\psi\}) = \frac{2}{\xi(T)^2} \sum_{i_1 i_2} I_{i_1 i_2 ij} \mathrm{Re}\psi_{i_1}^e \mathrm{Im}\psi_{i_2}^e, \quad R_{ij}(\{\psi\}) = \kappa^2 \xi(T)^2 \sum_{\alpha} K_{ij}^{\alpha\alpha} + \sum_{i_1 i_2} I_{i_1 i_2 ij} \psi_{i_1}^{e*} \psi_{i_2}^e,$$

$$S_{ij}^{\alpha\beta} = \kappa^2 \xi(T)^2 (K_{ij}^{\alpha\beta} - K_{ij}^{\beta\alpha}), \quad T_i^{\alpha} = \sum_{i_1 i_2} J_{i_1 i_2 i}^{\alpha} \mathrm{Im}(\psi_{i_1}^{e*} \psi_{i_2}^e), \quad U_i^{\alpha} = \kappa^2 \xi(T)^2 \frac{2\pi}{\Phi_0} \left(J_i^{\beta} H_{\gamma} - J_i^{\gamma} H_{\beta} \right),$$

$$V_i^R(\{\psi\}) = \frac{2}{\xi(T)^2} \sum_{i_1 i_2 i_3} I_{i_1 i_2 i_3 i} \mathrm{Re}(\psi_{i_1}^e \psi_{i_2}^{e*}) \mathrm{Re}\psi_{i_3}^e, \quad V_i^I(\{\psi\}) = \frac{2}{\xi(T)^2} \sum_{i_1 i_2 i_3} I_{i_1 i_2 i_3 i} \mathrm{Re}(\psi_{i_1}^e \psi_{i_2}^{e*}) \mathrm{Im}\psi_{i_3}^e,$$

where α and $\beta = x,\ y,\ z$ and $I_{ij},\ I_{i_1i_2i_3i_4},\ J_i^{x_i},\ J_{i_1i_2i_3}^{x_i},\ K_{ij}^{x_ix_j}$ are integrals. (α, β, γ) is a cyclic permutation of (x, y, z). These integrals are given by,

$$I_{ij} = \int_{V_e} N_i^e N_j dV,\quad I_{i_1i_2i_3} = \int_{V_e} N_{i_1}^e N_{i_2}^e N_{i_3}^e dV,\quad I_{i_1i_2i_3i_4} = \int_{V_e} N_{i_1}^e N_{i_2}^e N_{i_3}^e N_{i_4}^e dV,$$
$$J_i^{x_i} = \int_{V_e} \frac{\partial N_i}{\partial x_i} dV,\quad J_{i_1i_2i_3}^{x_i} = \int_{V_e} \frac{\partial N_{i_1}^e}{\partial x_i} N_{i_2}^e N_{i_3}^e dV,\quad K_{i_1i_2}^{x_ix_j} = \int_{V_e} \frac{\partial N_{i_1}^e}{\partial x_i}\frac{\partial N_{i_2}^e}{\partial x_j} dV \qquad (x_i,\ x_j = x,\ y,\ z).$$

Coefficients $a_i,\ b_i,\ c_i,\ d_i$ in Eq.(3) are given by,

$$\begin{aligned}
a_i &= \epsilon_i \left\{x_l(y_m z_n - y_n z_m) + x_m(y_n z_l - y_l z_n) + x_n(y_l z_m - y_m z_l)\right\},\\
b_i &= \epsilon_i \left\{y_l(z_n - z_m) + y_m(z_l - z_n) + y_n(z_m - z_l)\right\},\\
c_i &= \epsilon_i \left\{z_l(x_n - x_m) + y_m(z_l - z_n) + y_n(z_m - z_l)\right\},\\
d_i &= \epsilon_i \left\{x_l(y_n - y_m) + x_m(y_l - y_n) + x_n(y_m - y_l)\right\}.
\end{aligned}$$

Here $\epsilon_i = 1\ (i = 1, 3)$ or $\epsilon_i = -1\ (i = 2, 4)$, and $(i,\ l,\ m,\ n)$ is a cyclic permutation of (1, 2, 3, 4) and V is a volume of a tetrahedron.

References

[1] Togawa Y, Koyama T, Takayanagi K, Mori S, Kousaka Y, Akimitsu J, Nishihara S, Inoue K, Ovschinnikov A S and Kishine J 2012 *Phys. Rev. Lett.* **108** 107202
[2] Dzyaloshinskii I E 1958 *J. Phys. Chem. Solids* **4** 241
[3] Moriya T 1960 *Phys. Rev.* **120** 91
[4] Togawa Y, Kousaka Y and Inoue K, 2016 *J. Phys. Soc. Jpn* **85** 112001
[5] Milorad V M and Francois M P 2004 *Phys. Rev. Lett.* **93** 267006
[6] Milorad V M and Francois M P 2005 *Europhys. Lett.* **70** 670
[7] Tamegai T, Nakao Y, Mohan S and Nakajima Y 2011 *Supercond. Sci. Technol.* **24** 024015
[8] Lyuksyutov I F and Pokrovsky V L 2005 *Adv. Phys.* **54** 67
[9] Fukui S, Kato M and Togawa Y 2016 *Supercond. Sci. Technol.* **29** 125008
[10] Fukui S, Kato M and Togawa Y 2017 *J. Phys. Conf. Ser.* **807** 052010
[11] Fukui S, Kato M and Togawa Y 2016 *Physica C* **530** 51
[12] Fukui S, Kato M and Togawa Y 2017 *J. Phys. Conf. Ser.* **871** 012026
[13] Kato M, Fukui S, Sato O and Togawa Y 2017 *Physica C* **533** 137
[14] Kishine J, Inoue K and Yoshida Y 2005 *Prog. Theor. Phys. Sppl.* **159** 82
[15] Benjamin C, Alexander B, Nirmal G, David M and Minbhyea L 2014 *Appl. Phys. Lett.* **105** 072405
[16] Kokubo N, Okayasu S, Nojima T, Tamochi H and Shinozaki B 2014 *J. Phys. Soc. Jpn.* **83** 083704

Size effects of critical temperatures and order parameter distributions under four-fold symmetry of nano-structured superconductors

Masaki Umeda[1], **Masaru Kato**[1] **and Osamu Sato**[2]

[1] Department of Mathematical Sciences, Osaka Prefecture University, 1-1 Gakuen-cho, Naka-ku, Sakai, Osaka, 599-8531, Japan
[2] Osaka Prefecture University College of technology, 26-12 Saiwai-cho, Neyagawa-city, Osaka, 572-8572, Japan

Abstract. We investigate size and shape dependences of critical temperatures T_c's and order parameter distributions for nano-structured superconductors. We use the finite element method to solve the Gor'kov equations. We find that order parameter distributions hold four-fold symmetry for square superconductors. It is different from order parameter distributions of rectangular superconductors, which have two-fold symmetry. We show that the difference comes from symmetry of the superconductor and degeneracy of states of Cooper pairs, by investigating size dependences of order parameter distributions and those of eigen-energies of quasi-particles.

1. Introduction

Bulk superconducting material has own critical temperature T_c. However, T_c for a nano-structured superconductor depends on size and shape of the superconductor. Parmenter showed that T_c increases monotonically with decreasing size of a cubic superconductor, using the BCS theory [1]. Experimentally Nishizaki showed that the high pressure torsion [2-3] makes many fine grains in bulks of Nb, and enhances T_c, because these fine grains become nano-structured superconductors [4]. T_c for the nano-structured superconductor depends on surface modes of phonon spectrum [5] and quantum size effects on electrons [6-9]. We study the quantum size effects for nano-structured superconductors, by solving Gor'kov equations. We use the finite element method (FEM) [10] to incorporate effects of shapes of superconductors and real space order parameter distributions into numerical calculations. In previous study, we obtained size dependences of T_c, eigen-energies and order parameter distributions [11] for rectangular and square superconductors. Then we found that T_c shows saw-tooth behaviour as a function of size. Also we found T_c's for rectangular superconductors ($L \times 2L$; L is a length of a shorter side of the system) are higher than those of square superconductors. In this paper we discuss size effects on T_c, eigen-energies, and order parameter distributions for square plates. We show differences between square plates and rectangular plates.

2. Methods

We use the Gor'kov equations and a self-consistent equation, and the FEM to determine T_c, eigen-energies and order parameter distributions.

$$\left\{i\hbar\omega_n - \frac{1}{2m}\left(\frac{\hbar}{i}\nabla\right)^2 + \mu\right\}\mathcal{G}\left(\boldsymbol{x},\boldsymbol{x}',\omega_n\right) + \Delta\left(\boldsymbol{x}\right)\mathcal{F}^\dagger\left(\boldsymbol{x},\boldsymbol{x}',\omega_n\right) = \hbar\delta\left(\boldsymbol{x}-\boldsymbol{x}'\right) \tag{1}$$

$$\left\{-i\hbar\omega_n - \frac{1}{2m}\left(-\frac{\hbar}{i}\nabla\right)^2 + \mu\right\}\mathcal{F}^\dagger\left(\boldsymbol{x},\boldsymbol{x}',\omega_n\right) - \Delta^*\left(\boldsymbol{x}\right)\mathcal{G}\left(\boldsymbol{x},\boldsymbol{x}',\omega_n\right) = 0 \tag{2}$$

$$\Delta^*\left(\boldsymbol{x}\right) = -\frac{g}{\hbar\beta}\sum_{\omega_n}^{|\hbar\omega_n|<E_c} e^{-i\omega_n\eta}\mathcal{F}^\dagger\left(\boldsymbol{x},\boldsymbol{x},\omega_n\right) \tag{3}$$

Where, $\mathcal{G}$ is the Green's function, $\mathcal{F}$ is the anomalous Green's function, Δ is a order parameter, E_c is the cut-off energy of the BCS theory and $\omega_n = \left(2n+1\right)k_BT/\hbar$ is the Matsubara frequency. We expand Green's functions and the order parameter using area coordinates $\{N_j^e\}$

$$\mathcal{G}\left(\boldsymbol{x},\boldsymbol{x}',\omega_n\right) = \sum_{e,i}\sum_{e',j}\mathcal{G}_{ij}^{ee'}\left(\omega_n\right)N_i^e\left(\boldsymbol{x}\right)N_j^{e'}\left(\boldsymbol{x}'\right) \tag{4}$$

$$\mathcal{F}\left(\boldsymbol{x},\boldsymbol{x}',\omega_n\right) = \sum_{e,i}\sum_{e',j}\mathcal{F}_{ij}^{ee'}\left(\omega_n\right)N_i^e\left(\boldsymbol{x}\right)N_j^{e'}\left(\boldsymbol{x}'\right) \tag{5}$$

$$\Delta\left(\boldsymbol{x}\right) = \sum_{e,i}\Delta_i^e N_i^e\left(\boldsymbol{x}\right) \tag{6}$$

where subscripts and superscripts are node numbers and element numbers, respectively.

$$\begin{aligned}
&\sum_{i_2 i_3}\left[\left\{\left(i\hbar\omega_n + \mu\right)I_{ii_2}^e - \frac{\hbar^2}{2m}\sum_\alpha K_{ii_2}^{e\alpha\alpha}\right\}\mathcal{G}_{i_2 i_3}^{ee'}\left(\omega_n\right) + \sum_{i_1} I_{ii_1i_2}^e\Delta_{i_1}\mathcal{F}_{i_2i_3}^{\dagger ee'}\left(\omega_n\right)\right]I_{i_3j}^{e'} = \hbar\delta_{ee'}I_{ij}^e,\\
&\sum_{i_2 i_3}\left[\left\{\left(-i\hbar\omega_n + \mu\right)I_{ii_2}^{e'} - \frac{\hbar^2}{2m}\sum_\alpha K_{ii_2}^{e\alpha\alpha}\right\}\mathcal{F}_{i_2 i_3}^{\dagger ee'}\left(\omega_n\right) - \sum_{i_1} I_{ii_1i_2}^e\Delta_{i_1}^{*e}\mathcal{G}_{i_2i_3}^{ee'}\left(\omega_n\right)\right]I_{i_3j}^{e'} = 0,\\
&\sum_{i_1}\Delta_{i_1}^{*e}I_{ii_1}^e = \frac{g}{\beta\hbar}\sum_{\omega_n}e^{-i\omega_n\eta}\sum_{i_1i_2}\mathcal{F}_{i_1i_2}^{\dagger ee'}\left(\omega_n\right)I_{ii_1i_2}^e,
\end{aligned} \tag{7}$$

where

$$I_{ij}^e = \int N_i^e\left(\boldsymbol{x}\right)N_j^e\left(\boldsymbol{x}\right)d\boldsymbol{x} \tag{8}$$

$$K_{ij}^{e\alpha\beta} = \int\frac{\partial N_i^e\left(\boldsymbol{x}\right)}{\partial x_\alpha}\frac{\partial N_j^e\left(\boldsymbol{x}\right)}{\partial x_\beta}d\boldsymbol{x}. \tag{9}$$

In order to determine T_c, we set temperature T tends to T_c. Then we replace the Green's function by the normal metal Green's function. We expand the anomalous Green's function in power series of the order parameter. We approximate it up to the first order. Then the self-consistent equation becomes

$$\sum_i\left[I_{ij}^e - \frac{g}{\beta}\sum_{\omega_n}e^{-\omega_n\eta}\sum_{j_1j_2}\sum_{i_1i_2}\left(D^{-1}\right)_{j_1i_1}^e I_{ii_1i_2}^e\left(C^{-1}\right)_{i_2j_2}^e I_{jj_1j_2}^e\right]\Delta_i^{e*} = 0. \tag{10}$$

The secular equation for equation (10) becomes

$$\det\left[I^{e}_{ij}-\frac{g}{\beta}\sum_{\omega_n}e^{\omega_n\eta}\sum_{j_1j_2}\sum_{i_1i_2}\left(D^{-1}\right)^{e}_{j_1i_1}I^{e}_{ii_1i_2}\left(C^{-1}\right)^{e}_{i_2j_2}I^{e}_{jj_1j_2}\right]=0. \quad (11)$$

T_c is the maximum T that satisfies equation (11). Here

$$C_{ii_2}=\left(i\hbar\omega_n+\mu+\frac{i\omega_n\hbar}{2\left|\omega_n\right|\tau}\right)I^{e}_{ii_2}-\frac{\hbar^2}{2m}\sum_{\alpha}K^{e\alpha\alpha}_{ii_2} \quad (12)$$

$$D_{ii_2}=\left(-i\hbar\omega_n+\mu-\frac{i\omega_n\hbar}{2\left|\omega_n\right|\tau}\right)I^{e'}_{ii_2}-\frac{\hbar^2}{2m}\sum_{\alpha}K^{e\alpha\alpha}_{ii_2}. \quad (13)$$

In addition, the chemical potential μ is determined by the equation as follows,

$$N(T,V,\mu)=\sum_{\omega_n}\int d^3x\mathrm{tr}\mathcal{G}\left(\omega_n,x\right). \quad (14)$$

3. Results and Discussions

We consider superconducting square ($L \times L$) and rectangular ($L \times 3L$) plates. We impose boundary conditions that wave functions of electrons are zero at edges of plates. Therefore order parameters become zero at edges of these plates. We set that $k_F\xi = 3.0$, and $\Delta_0 / E_c = 0.2$, and densities of electrons and interaction constants g are fixed. Figure 1 shows models of the square and rectangular nano-structured superconductors for the FEM. Figure 2 shows size dependences of T_c's for superconducting square and rectangular plates. Here, we compare size dependences of T_c's between superconducting plates with same areas. From this result, we can see that smaller superconductors show higher T_c, T_c has saw-tooth behavior as a function of size, and the square superconductor shows smaller T_c comparing with the rectangular superconductor. These results are similar to our previous study [11].

Figure 3 shows size dependences of T_c and eigen-energies from the chemical potential $\xi_k = \varepsilon_k - \mu$. Figure 4 shows order parameter distributions at the sizes of the superconductors (a)-(f) in figure 3. From these results, we can see that drops of T_c and changes of order parameter distributions occur when one of eigen-energies becomes larger than E_c, as we discussed in previous study [11]. For example, from (a) to (b) in figure 3 (A), drop of T_c occurs, eigen-energy becomes larger than the cut off energy in figure 3 (B) and order parameter distribution changes from (a) to (b) in figure 4. Next we see changes of order parameter distributions. Figure 5 (a-b) shows a difference between figure 4 (a) and figure 4 (b), figure 5 (b-c) shows a difference between figure 4 (b) and figure 4 (c), and so on. Differences in figures 5 (a-b), 5 (c-d) and 5 (e-f) show distributions of Cooper pairs, which cease to contribute to superconductivity. Differences in figures 5 (b-c) and 5 (d-e) are negligible comparing with other differences in figure 5. This is because these two distribution differences represent concentration of Cooper pairs around center of the superconductor.

In figure 6 and figure 7, we show order parameter distributions and changes of order parameter distributions for the rectangular superconductor, as in figures 4 and 5 for the square superconducting plate. We can see that all of changes of order parameter distributions for the rectangular superconductor in figure 7 has well-ordered peaks. However, changes of order parameter distributions for the square plate in figures 5 (a-b) and 5 (c-d) don't have such simple structures. These differences come from symmetries of superconductors. Since the rectangular superconductor has two-fold symmetry, all of order parameter distributions and changes of order parameter distributions have two-fold symmetry. Also superconducting electron states, with same number of peaks in rectangular superconductor, have different eigen-energies when numbers of the peaks along shorter side and longer side are opposite. However, for the square superconductor, such electron states degenerate. Figure 8 (a) shows sum of differences of order parameter distributions in figures 7 (a-b) and 7 (e-f). Figure 8 (b) is a deformation of figure 8 (a) with aspect ratio 1:1. From comparing figure 8 (b) and figure 5 (c-d), we can see that they have same structure. This result shows that two Cooper pair states, with same number of peaks of witch number of peaks along shorter side and longer side are opposite, degenerate for the square superconductor because of its four fold-symmetry. Therefore, we can see that the number of cusps of T_c as a function of size of superconductor depends on the symmetry of superconductor. In addition, we expect more symmetric superconductor shows less cusps of T_c, because of degeneracy of Cooper pair states.

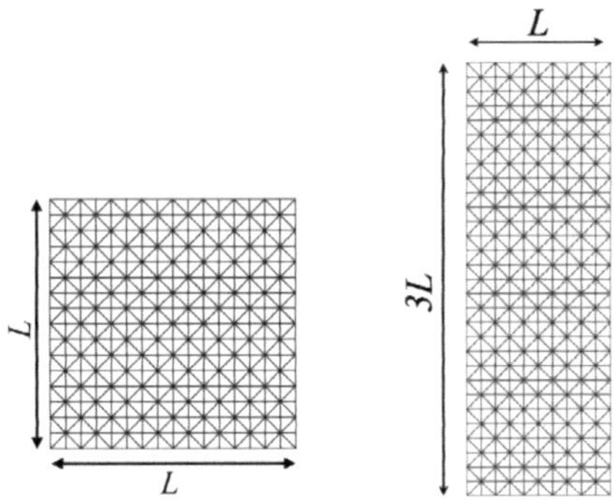

Figure 1. The systems of square and rectangular ($L \times 3L$) nano-structured superconducting plates for the FEM

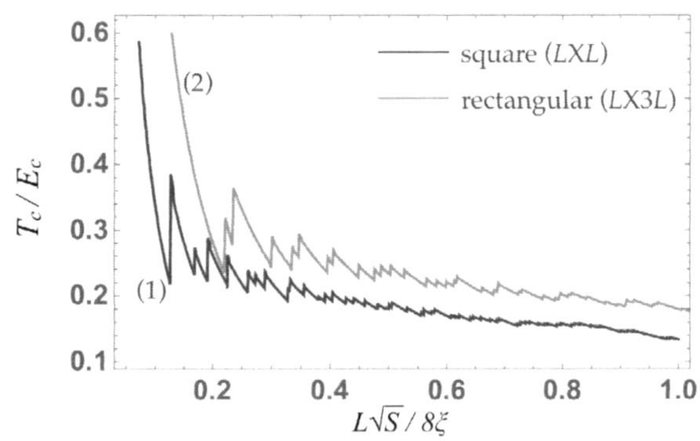

Figure 2. Size dependence of T_c for square (1) (S = 1) and rectangular ($L \times 3L$) (2) (S = 3) nano-structured superconductor with same area. Here ξ is the coherence length at $T = 0$.

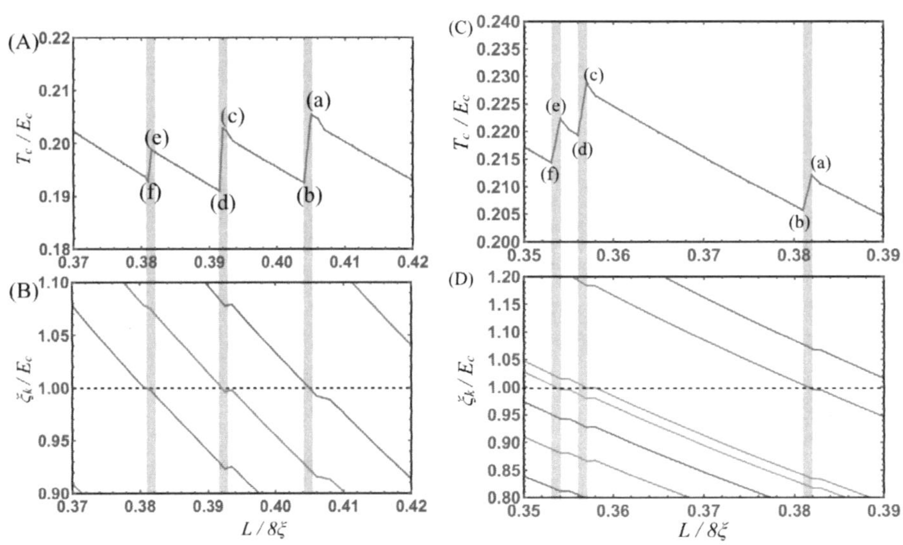

Figure 3. Size dependence of T_c (A) and eigen-energies (B) for a square superconductor and size dependence of T_c (C) and eigen-energies (D) for a rectangular ($L \times 3L$) superconductor where ξ is the coherence length at $T = 0$.

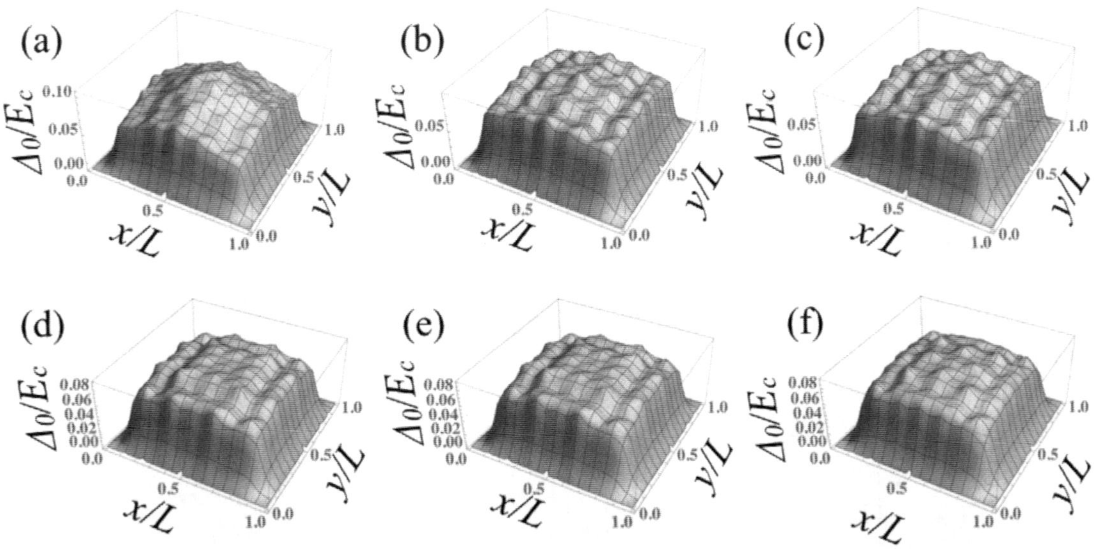

Figure 4. Distributions of amplitudes of order parameter Δ_0, for a nano-square superconductor. Where (a)-(f) correspond to the (a)-(f) in figure 3 (A). (L/ξ = 0.4050 (a), 0.4040 (b), 0.3920 (c), 0.3915 (d), 0.3815 (e) and 0.3810(f)).

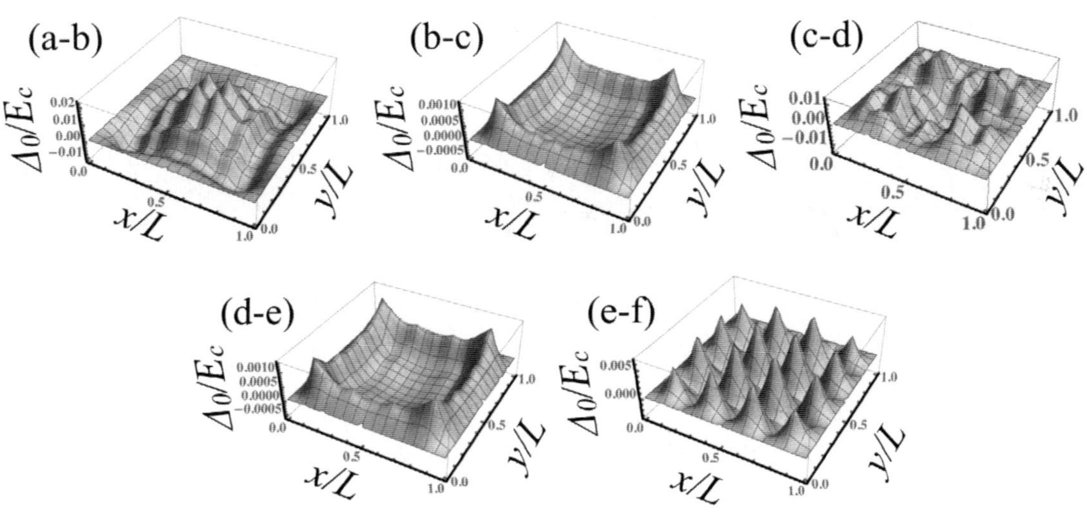

Figure 5. Differences of order parameter distributions, for nano-square superconductor. Figure (a-b), (b-c), (c-d), (d-e) and (e-f) show differences of order parameter distributions between figure 4 (a) and (b), figure 4 (b) and (c) and so on.

IOP Conf. Series: Journal of Physics: Conf. Series **1054** (2018) 012028 doi:10.1088/1742-6596/1054/1/012028

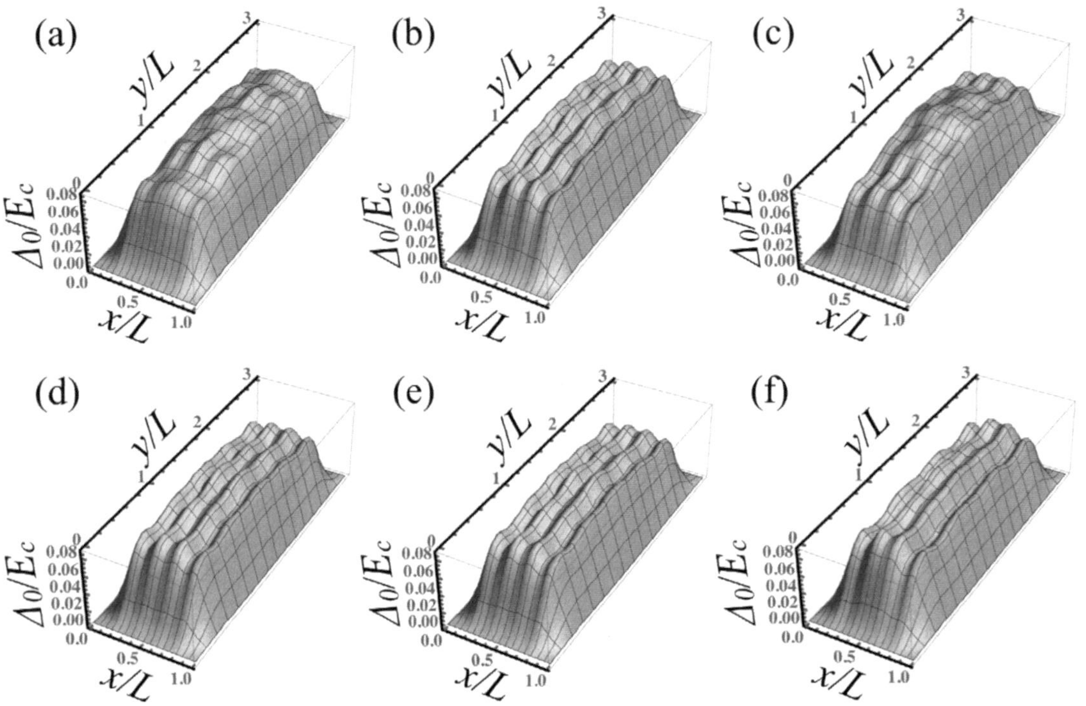

Figure 6. Distributions of amplitudes of order parameter Δ_0, for a nano-rectangular ($L \times 3L$) superconductor. Where (a)-(f) correspond to the (a)-(f) in figure 3 (C). ($L / \xi = 0.3820$ (a), 0.3810 (b), 0.3570 (c), 0.3560 (d), 0.3540 (e), and 0.3530 (f)).

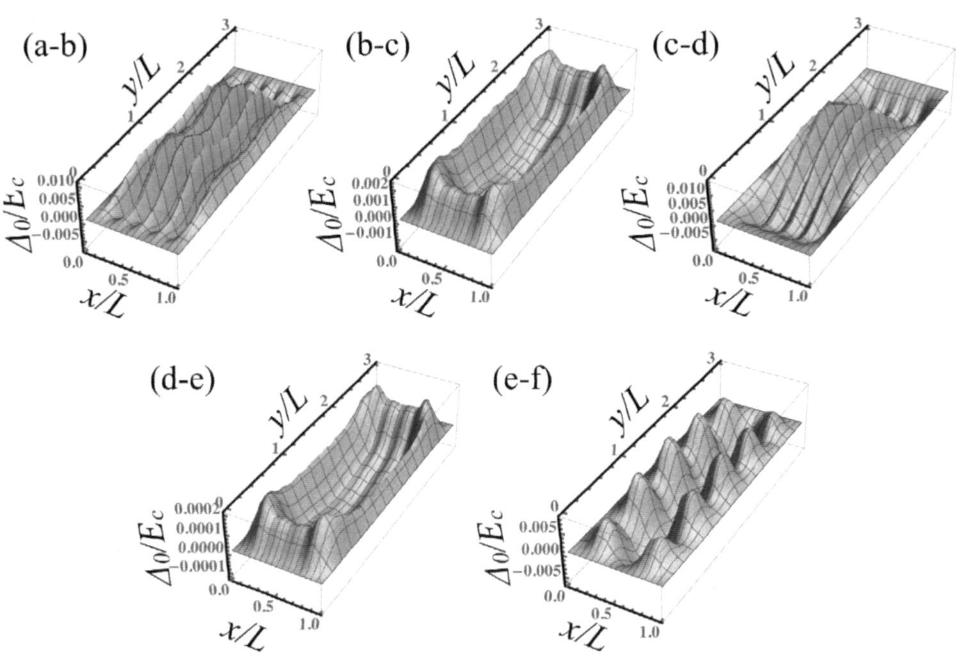

Figure 7. Difference of order parameter distributions for the nano-rectangular ($L \times 3L$) superconductor. Figure (a-b), (b-c), (c-d), (d-e) and (e-f) show differences of order parameter distributions between figure 6 (a) and (b), figure 6 (b) and (c) and so on.

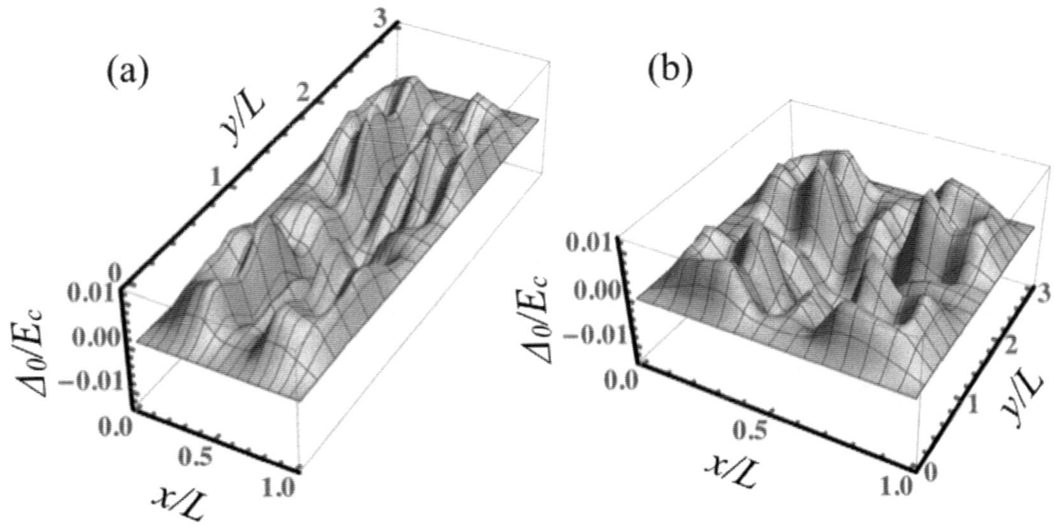

Figure 8. Sum of differences of order parameter distributions figure 6 (a-b) and figure 6 (e-f), for the nano-rectangular ($L \times 3L$) superconductor (a). (b) is a deformation of (a) with aspect ratio 1:1.

4. Summary

We have obtained T_c, eigen-energies, and order parameter distributions for superconducting square plates, by solving the Gor'kov equations with the FEM numerically. We have found that characteristic structure of order parameter distributions and changes of order parameter distributions for the superconducting square plate, which are different from those for the rectangular superconductor. These differences comes from difference of symmetries of systems, because all of order parameter distributions and changes of order parameter distributions for square and rectangular superconductors have four and two-fold symmetry, respectively. In addition, we have shown that Cooper pair states in square superconductor are degenerate because of the four-fold symmetry.

Acknowledgments
We thank for, T. Ishida, T. Nishizaki, Y. Nagai, T. Tamegai, S. Ooi, Y. Higashi, S. Fukui, N. Fujita and M. Kashiwagi for useful discussions. This work was supported in part of KAKENHI Grant Number JP 26400367, and program for leading graduate schools of ministry of education, culture, sports, science and technology-Japan.

References

[1] Parmenter R H 1968 *Phys. Rev.* **166** 392-6
[2] Bridgman 1943 *J. Appl. Phys.* **14** 273-83
[3] Lee S, Edalati K and Horita Z 2010 *Mater. Trans.* **51** 1072-9
[4] Nishizaki T, Lee S, Horita Z, Sasaki T and Kobayashi N 2013 *Physica C* **493** 132-5
[5] Dickey J M and Paskin A 1968 *Phys. Rev. Lett* **21** 1441-3
[6] Anderson P W 1959 *J. Phys. Chem. Solids* **11** 26-30
[7] von Delft J and Ralph D C 2001 *Phys. Rep.* **345** 61-173
[8] Shanenko A A, Croitoru M D and Peeters F M 2007 *Phys. Rev. B* **75** 014519
[9] Suematsu H, Kato M and Ishida T 2009 *J. Phys. Conf. Ser.* **150** 052250
[10] Kato M, Ishida T, Koyama T and Machida M 2012 *Superconductors— Materials, Properties and Applications* (Rijeka, Croatia: InTech) ed A Gabovich pp 319-42
[11] Umeda M, Kato M and Sato O 2016 *IEEE Trans. Appl. Supercond.* **26** 8600104

IOP Conf. Series: Journal of Physics: Conf. Series **1054** (2018) 012029 doi:10.1088/1742-6596/1054/1/012029

Critical states in a superconducting plate: Effects of heat generation from flux motion

Shinsuke Ooi and Masaru Kato

Department of Mathematical Sciences, Osaka Prefecture University, 1-1, Gakuencho, Nakaku, Sakai, Osaka 599-8531, Japan
E-mail: sxb01036@edu.osakafu-u.ac.jp

Abstract. We have studied vortex penetration into a superconducting plate under a perpendicular magnetic field. Solving the heat transport equation and the Maxwell equations with the current-voltage relation for the superconductor simultaneously, current density and temperature distribution due to vortices motion are obtained. Taking the effect of heat generation into account, there is a size dependence of temperature rise.

1. Introduction

When a perpendicular external magnetic field, which is larger than the lower critical field, is applied to a type II superconducting plate, the superconductor comes to be in a vortex state. Vortices penetrate to the superconducting plate from its edge. There are two kinds of vortex penetrations: a critical state and a vortex avalanche [1]. In the critical state, the Meissner shielding currents flow along an edge of the superconducting plate, and the Lorentz force due to the current drives vortices into the inside of the superconducting plate. When microscopic defects are present in the superconductor, vortices may be pinned by these defects. The pinning force from defects acts against the driving Lorentz force. The local balance between two forces creates a metastable equilibrium state, which is the critical state. In the critical state, magnitude of the shielding current density is adjusted to the critical current density.

On the other hand, the vortex avalanche is caused by thermal fluctuation at finite temperature. The vortex, which is pinned by the pinning center, jumps to another pinning center because of the driving force and the thermal fluctuation. This movement generates heat, and the temperature locally rises. Then superconductivity is weakened. Therefore, vortices further penetrate into the superconductor. Repeating this process, there appears the vortex avalanche.

In this study, we focus on the critical state. The critical states have been studied widely [1, 2, 3]. Although the influence of heat transport has not been taken into consideration, we investigate influence of heat transport on the vortex motion as in the vortex avalanche study [4]. We solved the heat transport equation and the Maxwell equation with a current-voltage relation for the superconductor with three-dimensional finite element method. We obtain distribution of temperature and current density when vortices penetrate into the superconducting plate.

2. Method

In this study, we solve following three equations. The first is the heat transport equation,

$$C\frac{\partial T}{\partial t}=\frac{\partial X}{\partial x}+\frac{\partial Y}{\partial y}+\frac{\partial Z}{\partial z}-\alpha(T-T_0)+\mathcal{J}\cdot\boldsymbol{E}\,. \tag{1}$$

Here, C is a specific heat, $T(x,y,z,t)$ is temperature of the superconductor, and $X(T)$, $Y(T)$, and $Z(T)$ are the heat flow in x, y, and z directions, respectively. The forth term on the right hand side is the heat transfer from the superconductor to a substrate, where α is a heat transfer rate, and T_0 is temperature of the substrate. The fifth term on the right hand side is the heat generation due to the creep motion of vortices, where $\mathcal{J}$ is a current density, and $\boldsymbol{E}$ is an electric field. These $X(T)$, $Y(T)$, and $Z(T)$ are given as follows,

$$\begin{cases} X(T)=\kappa_{11}\frac{\partial T}{\partial x}+\kappa_{12}\frac{\partial T}{\partial y}+\kappa_{13}\frac{\partial T}{\partial z} \\ Y(T)=\kappa_{21}\frac{\partial T}{\partial x}+\kappa_{22}\frac{\partial T}{\partial y}+\kappa_{23}\frac{\partial T}{\partial z} \\ Z(T)=\kappa_{31}\frac{\partial T}{\partial x}+\kappa_{32}\frac{\partial T}{\partial y}+\kappa_{33}\frac{\partial T}{\partial z} \end{cases}, \tag{2}$$

where κ is a heat conductivity tensor.

The second is the Maxwell equations,

$$\begin{cases} \mathrm{rot}(\mathrm{rot}\boldsymbol{A}-\boldsymbol{H}_a)=\frac{4\pi}{c}\mathcal{J} \\ \boldsymbol{E}=-\frac{1}{c}\frac{\partial \boldsymbol{A}}{\partial t} \end{cases}, \tag{3}$$

where $\boldsymbol{A}$ is a magnetic vector potential, $\boldsymbol{H}_a$ is an external magnetic field, and c is the speed of light.

The third is the current-voltage relation for the superconductor [4],

$$\boldsymbol{E}=\rho(\mathcal{J})\mathcal{J}\,. \tag{4}$$

Here $\rho(\mathcal{J})$ is a resistivity given by,

$$\rho(\mathcal{J})=\begin{cases} \rho_0(\mathcal{J}/\mathcal{J}_c)^{n-1} & \mathcal{J}\le\mathcal{J}_c, T\le T_c \\ \rho_0 & \mathcal{J}>\mathcal{J}_c, T\le T_c \\ \rho_n & \mathcal{J}>\mathcal{J}_c, T>T_c \end{cases}, \tag{5}$$

where n is a creep exponent, ρ_0 is a resistivity constant of superconducting state, ρ_n is a resistivity of normal state, $\mathcal{J}_c$ is a critical current density, and T_c is the critical temperature. The flux creep exponent is given by $n-1=n_0(1-T/T_c)$, where n_0 is a constant. This current-voltage relation represents the fact that due to the current flow, vortices creeps and the resistance rises. The current and temperature dependences of the resistivity in the superconductor are shown in figure 1.

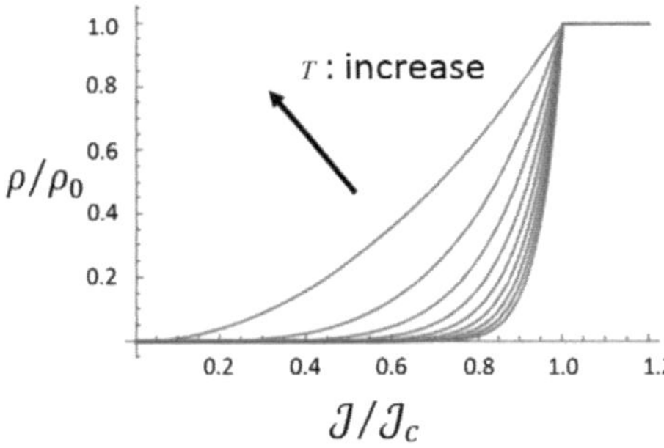

Figure 1. $\mathcal{J}$ and T dependences of resistivity.

3. Results and Discussion

In this study, we consider two sizes of superconducting plates. The sizes of superconductors were 8×16×0.2 μm^3 and 80×160×2 μm^3, for smaller and larger superconducting plates, respectively. The superconductors were surrounded by the substrate shown in figure 2. System sizes were 10×18×0.4 μm^3 and 100×180×4 μm^3 for smaller and larger superconducting plates, respectively.

We use material parameters corresponding to MgB_2, $C = 35\ \mathrm{kJ/Km^3} \times (T/T_c)^3$, $\alpha = 220\ \mathrm{kW/Km^2} \times (T/T_c)^3$, $\kappa = 0.17\ \mathrm{kW/Km} \times (T/T_c)^3$, $\rho_n = \rho_0 = 7\ \mu\Omega\mathrm{cm}$, $T_c = 39\ \mathrm{K}$, $\mathcal{J}_{c0} = 10\ \mathrm{mA/m^2}$, $n_0 = 19$. The external applied magnetic field is ramped up from $H_a = 0$ kG at a constant rate, $\dot{H}_a = 5\times 10^{15}$ kG/s. The initial temperature of the total system is 10 K. Although the temperature of the superconductor rises, the temperature of the substrate is kept constant 10 K.

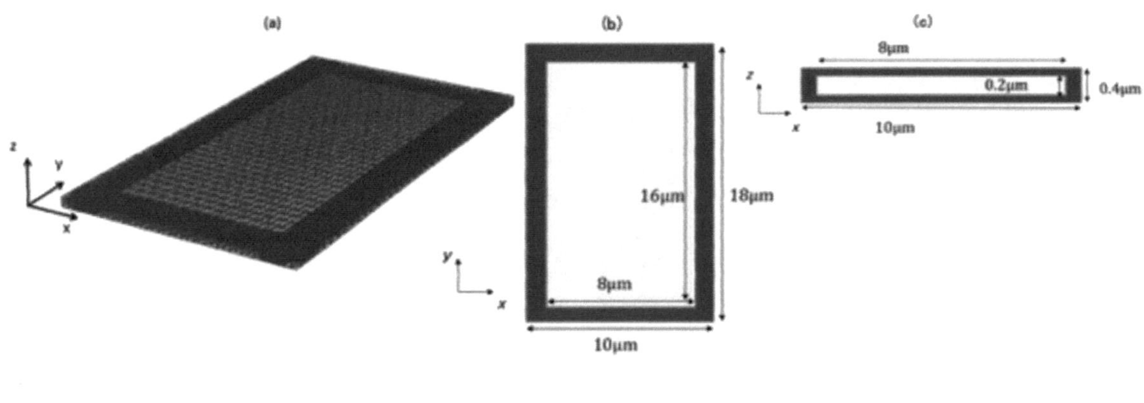

Figure 2. Three-dimensional model for smaller superconducting plate (a), a cross-sectional view in the x-y plane (b), and a cross-sectional view in the x-z plane (c).

Figure 3 shows a three-dimensional distribution of the current density in the smaller superconducting plate, when the external magnetic field is $H_a = 5\times 10^7$ kG. The current density is large around corners of the superconducting plate. Figures 4(a) and (b), and figures 4(c) and (d) show the development of distributions of x component of the current density $\mathcal{J}_x$ for smaller and larger superconducting plates, respectively. The external magnetic fields are raised from $H_a = 5\times 10^7$kG, $H_a = 15\times 10^7$kG, $H_a = 20\times 10^7$kG and $H_a = 25\times 10^7$kG in figures 4(a) and (c), and from $H_a =$

23×10^7kG, $H_a = 24\times10^7$kG to $H_a = 25\times10^7$kG in figures 4(b) and (d). Comparing figures 4(a) and (c), we find that the current density in the smaller superconducting plate is larger than that in the larger superconducting plate. In figure 4(b), superconductivity is destroyed for $H_a = 25\times10^7$kG where the current density is zero.

Next, we show temperature distribution in *x*-*y* plane in figure 5. Figures 5(a)-(d), and figures 5(e)-(h) show the temperature distributions for smaller and larger superconducting plates, respectively. The external magnetic field is $H_a = 5\times10^7$kG for (a) and (e), $H_a = 15\times10^7$kG for (b) and (f), $H_a = 20\times10^7$kG for (c) and (g), $H_a = 25\times10^7$kG for (d) and (h). These temperature distributions are dynamically depicted in each magnetic field, which is gradually increased.

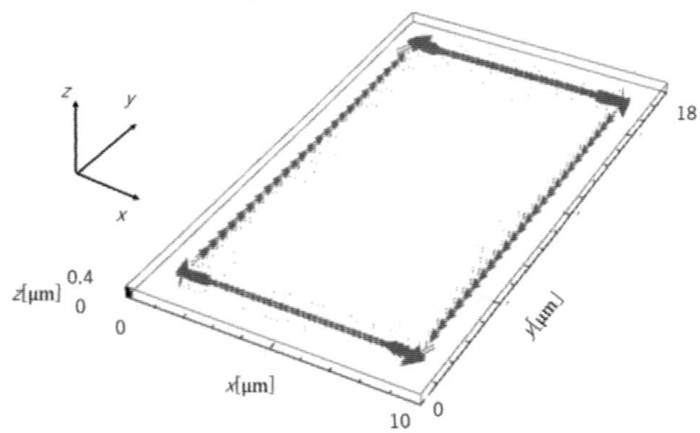

Figure 3. Three-dimensional current density for smaller superconducting plate

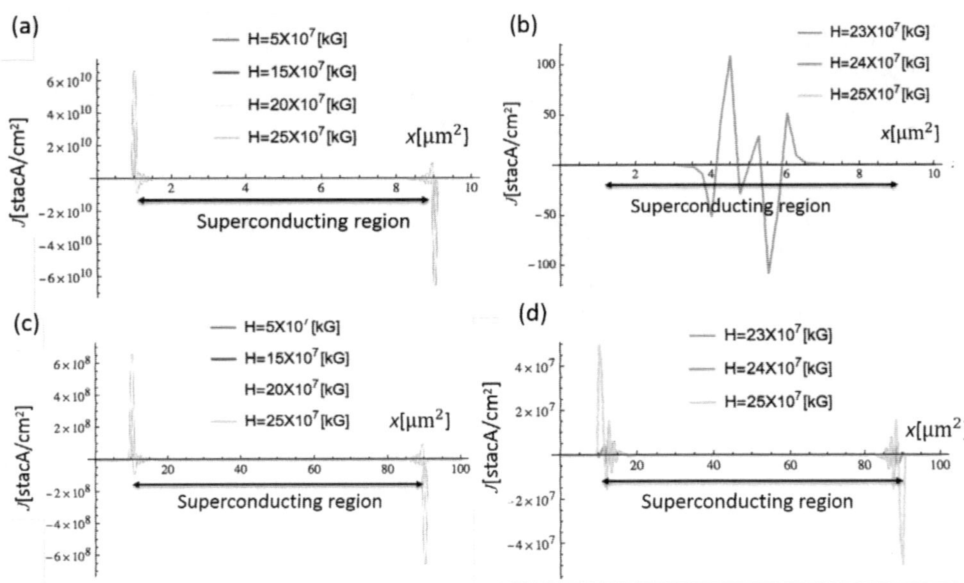

Figure 4. Current density in *x* direction; (a) and (b) the superconductor size is $8\times16\times0.2$ μm^3, (c) and (d) the superconductor size is $80\times160\times2$ μm^3

IOP Conf. Series: Journal of Physics: Conf. Series **1054** (2018) 012029 doi:10.1088/1742-6596/1054/1/012029

We find temperature rise at corners in superconducting plate in figures 5(b) and (f). In figure 5(d), where the external magnetic field is $H_a = 25\times10^7$ kG, the superconductivity is completely destroyed in the smaller superconducting plate since temperature exceeds critical temperature 39 K. On the other hand, in figure 5(h), superconductivity is destroyed only along edges in the larger superconducting plate at same magnetic field.

We can explain the change of current density and temperature rise on critical state in these results, as follows. Current density is large at corners of superconducting plate where temperature begins to rise. After that, heat propagates isotropically, the temperature rises around short sides because of short propagation distance. In the smaller superconducting plate, when the temperature exceeds the critical temperature 39 K, the superconductivity is destroyed and the current density becomes zero. While in the larger superconducting plate, the current density has finite value and superconductivity is only partially destroyed. But these results are a little strange. Considering the effect of demagnetization, the magnetic field locally becomes large at the edge of the superconducting plate and vortices are likely to penetrate. So larger superconducting plate generally lose superconductivity than smaller one. In order to solve this contradiction, we should consider following point: Total amounts of current flowing inside of both of the superconductors are same, but for smaller superconductor, the current density becomes larger. Therefore, the heat generation also increases, and the smaller the superconducting plate, superconductivity is easily destroyed. So, in this model, as we take the heat generation by current into account, these results are reasonable.

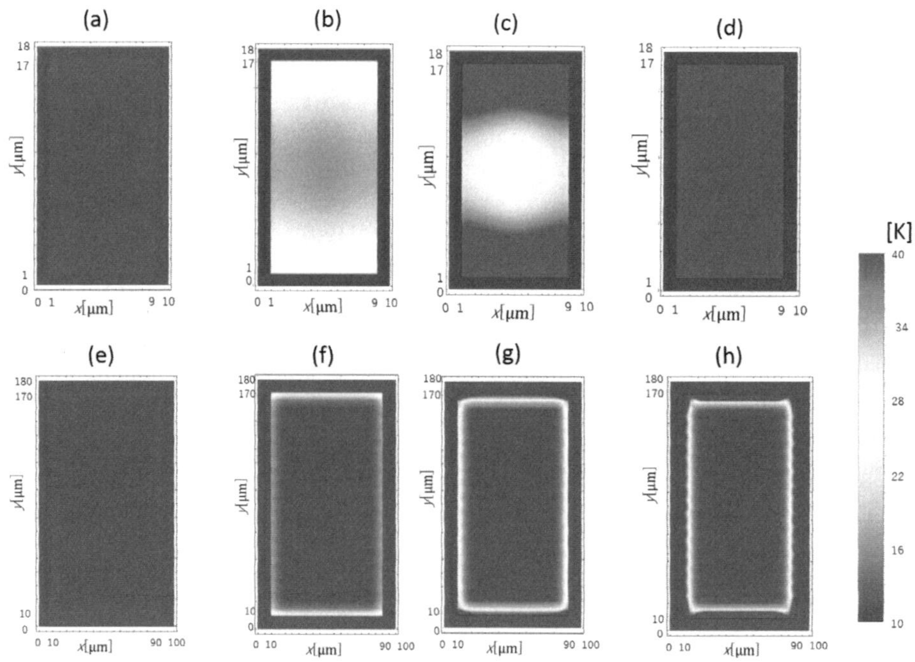

Figure 5. Temperature distribution in *x-y* plane. Sizes of superconductors are 8×16×0.2 μm^3 for (a), (b), (c) and (d) and 80×160×2 μm^3 for (e), (f), (g) and (h). The external field is $H_a = 5\times10^7$kG for (a) and (e), $H_a = 15\times10^7$kG for (b) and (f), $H_a = 20\times10^7$kG for (c) and (g), $H_a = 25\times10^7$kG for (d) and (h).

4. Summary and Future

We have studied critical state for superconducting plates. Solving the heat transport equation and the Maxwell equation with current-voltage relation for the superconductor by using three-dimensional finite element method, we obtained current density and temperature distribution. We found the effect of heat generation of superconductivity in small system for critical state and destruction of superconductivity. There is a size dependence of temperature rise, and superconducting state in smaller system is destroyed easier than that larger one.

In this study in order to obtain maximum thermal effects on the critical state, we consider a situation with the magnetic field up to $H_a = 25\times10^7$kG, which is much higher than today's status experimental techniques. But even for lower magnetic field, similar but weak effects will appear, we think. Such effects on the critical state as well as vortex avalanches are future problems.

References

[1] Altshler E and Johansen T H 2004 *Rev. Mod. Phys.* **76** 471
[2] Brandt E H 1992 *Phys. Rev.* B. **46** 8628
[3] Brandt E H 1995 *Phys. Rev. Lett.* **74** 3025
[4] Vestgården J I, Shantsev D V, Galperin Y M and Johansen T H 2011 *Phys. Rev.* B. **84** 054537

Fabrications of Small and High-quality Intrinsic Josephson Junctions by Combinatorial Method of Ar-ion and Focused Ga-ion Etchings

Shumpei Umegai, Ayami Yamaguchi, Yoshihiro Kakizaki, Daiki Kakehi and *Haruhisa Kitano

Department of Physics and Mathematics, Aoyama Gakuin University, Fuchinobe 5-10-1, Chuo-ku, Sagamihara, Kanagawa 252-5258, Japan

hkitano@phys.aoyama.ac.jp

Abstract. We present a study on the fabrication of the intrinsic Josephson junctions (IJJs), which are made of $Bi_2Sr_2CaCu_2O_y$ (Bi2212) single crystals, by using the focused Ga-ion etching and the Ar-ion etching. We made sure that the current-voltage (*I*-*V*) characteristics of small IJJs was never influenced by the direction of a Ga-ion beam in the focused ion beam (FIB) etchings. In addition, it was found that Ar-ion etching after FIB process is quite useful to improve the performance of the *I*-*V* curves of IJJs. By using a substrate with a small hole, we succeeded in employing the double-side Ar-ion etching after the FIB microfabrication. These results strongly support that the combinatorial method of FIB and Ar-ion etchings is very important to fabricate the small and high-quality IJJs.

1. Introduction

The discovery of the intrinsic Josephson junctions (IJJs) of high-T_c cuprate superconductors have stimulated many studies on the macroscopic quantum tunneling (MQT) observed for the phase switches from the zero voltage state [1,2] and finite voltage state [3-5], and those on the interlayer tunnel spectroscopy probing the superconducting gap and the pseudogap [6-8]. For such studies, the fabrication of small and high-quality IJJs is crucially important. The previous study on ultrasmall IJJs (a lateral area was 0.3 μm^2), which were fabricated from $Bi_2Sr_2CaCu_2O_{8+\delta}$ (Bi2212) by the focused ion beam (FIB) etching, suggested that the current-voltage (*I*-*V*) characteristics of small IJJs was degraded by the damage generated in the FIB etching processes [9]. Another study on the submicron IJJ stacks made of $La_{2-x}Sr_xCuO_4$ (La214) reported that an argon ion etching after FIB etching was useful to control the thickness of IJJs [10]. Recently, the transmission electron microscope (TEM) study of Bi2212-IJJs suggested that the amorphous damage produced by FIB etchings was successfully removed by an additional argon ion etching [11]. Thus, the appropriate combination of FIB and Ar ion etching techniques is expected to improve the performance of small stacked IJJs.

Here, we present a study on the fabrication of Bi2212-IJJs by using FIB and Ar-ion etchings. First of all, we confirmed that the *I*-*V* characteristics of the bridge-type IJJs, where IJJs were sandwiched by two slits fabricated on the side wall of a microbridge (See also Fig. 1), was never influenced by the direction of a Ga-ion beam in the FIB etching processes. This shows a good agreement with the recent report that the thickness of FIB damage is independent of the direction of the Ga ion beam [11]. We also demonstrate that Ar ion etching after FIB process is quite useful to improve the *I*-*V* characteristics of IJJs. Furthermore, we succeeded in employing the double-side etching by the irradiation of Ar ions after the FIB fabrication, by using a substrate with a micro-hole. These results strongly support the importance of the combinatorial method of FIB and Ar-ion etchings for the fabrication of small and high-quality IJJs.

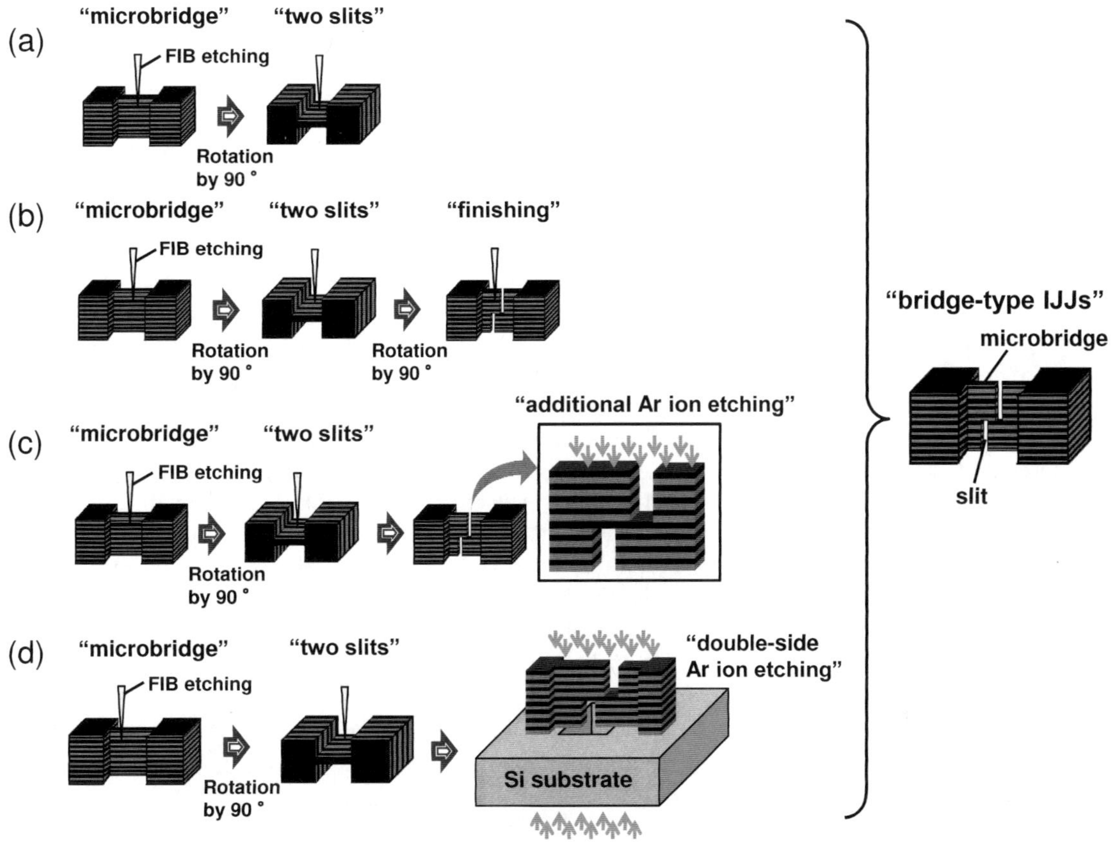

Figure 1. (Color online) Schematic of the FIB or Ar-ion etching processes to fabricate the bridge-type Bi2212-IJJs in (a) Method 1, (b) Method 2, (c) Method 3, and (d) Method 4.

2. Experiments

Bi2212 single crystals were grown by the floating zone method and annealed to be optimal doped. We fabricated the bridge-type Bi2212-IJJs by using the following four methods. Method 1 is the conventional FIB method where two slits nearby IJJs are fabricated after the fabrication of the microbridge, as illustrated in Fig. 1(a). Details of this method are described in elsewhere [11]. Method 2 is similar to a method investigated by Matsumoto *et al* [9]. In this method, as shown in Fig. 1(b), the finishing fabrication of the microbridge is done after the fabrication of two slits, in contrast to Method 1. Method 3 is a combinatorial method of FIB and Ar ion etchings, which was proposed by Kubo *et al* [10]. In this method, one of slits is finished by Ar ion etching after FIB etching, as shown in Fig. 1(c). Method 4 was newly developed in this work, based on the achievement of Method 3. In this method, both slits are finished by the additional Ar ion etchings, as shown in Fig. 1(d). Instead of a usual sapphire substrate, we use a silicon substrate with a square hole (50×50 μm^2), which has often been prepared as a nanopore to detect DNA molecules [12].

The FIB and Ar ion etchings were performed by using a JEOL JEM-9310FIB at 30 kV and an ELIONIX EIS-200ER at 1100 V, respectively.

3. Results and discussion

In this work, we measured the *I-V* curves of 22 devices fabricated by 4 methods shown in Fig. 1. Figure 2 shows the typical *I-V* characteristics of the IJJs (the lateral area is 1.2×0.98 μm^2), fabricated by using Method 1. We estimated the magnitude of the switching current I_{SW} and of the voltage jump V_{jump}, for the first and the maximum-ordered switching events, as shown in Fig. 2. We compared the results obtained for Method 1 with those for Method 2. Figures 3(a) and 3(b) show the plots of the switching current density j_{SW}, which is the switching current for a unit junction area, for the 1st and the maximum–ordered switches as a function of the lateral area *S*, respectively. Figures 3(c) and 3(d) show the plots of V_{jump} versus *S* for the 1st and the maximum–ordered switches, respectively. These data were obtained in a low temperature region ($T/T_c \sim 0.05$ to 0.1), where the temperature dependence of j_{SW} and V_{jump} was found to be negligibly small. We also found that there was a distribution in the measured values of T_c for the IJJ devices, determined by the onset temperature of the superconducting transition in the dc resistance, as shown in Fig. 3(e) [13]. This indicates that the FIB etching processes give the damage degrading the superconducting properties to the IJJ devices. However, in contrast to the previous report by Matsumoto *et al.* [9], we did not observe any advantage of Method 2.

Independent of the difference between Method 1 and Method 2, our results show a common feature that the magnitude of j_{SW} and V_{jump} is decreased with decreasing *S*, except for j_{SW} for the maximum-ordered switches ($j_{SW}{}^{max}$). We found that the magnitude of $j_{SW}{}^{max}$ was not decreased very well even for IJJs with small *S*. These results strongly suggest that the *I-V* characteristics of the bridge-type IJJs is never influenced by the direction of a Ga-ion beam in the FIB etching processes. It shows a good agreement with our resent TEM observations that the thickness of FIB damage (~30 nm for a beam emitted at 50 pA and 30 kV) is independent of the incident direction of the Ga-ion beam [11].

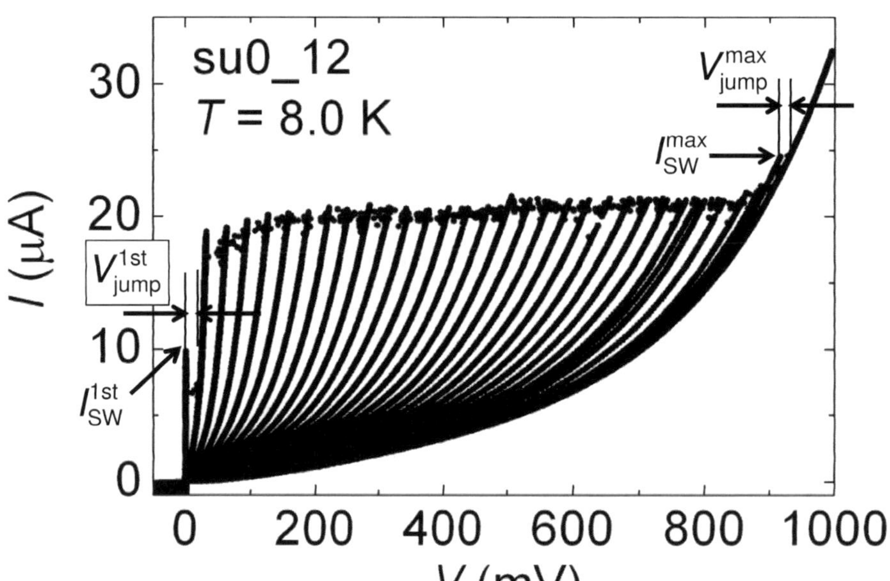

Figure 2. *I-V* characteristics of the IJJs (the lateral area is 1.2×0.98 μm^2), which were fabricated by Method 1.

Next, we performed similar estimations of the *I-V* characteristics for 5 IJJ devices, which were fabricated by using Method 3. Although this method has originally been developed to control the junction number involved in the La214-IJJs [10], we applied it to the removal of the damage region produced in the FIB processes. Figures 4(a) and 4(b) show the plots of j_{SW} and V_{jump} as a function of the total irradiation time of Ar-ion etching, respectively. They were obtained in the low temperature region ($T/T_c \sim 0.05$ to 0.1). We confirmed that both of j_{SW} and V_{jump} roughly showed the increasing behavior with increasing the Ar-ion irradiation time, although there was a wide range in the data obtained at the same irradiation time. The difference between data obtained at the same irradiation time is considered to be due to the difference in the influences of FIB damages before Ar-ion etching, as suggested by the plots of T_c shown in Fig. 4(c). The recent TEM study also indicates that the additional Ar-ion etching after the FIB process successfully trims the FIB-induced damage [11]. Thus, it is concluded that the additional Ar-ion etching after the FIB etching plays an important role to improve the performance of the *I-V* curves of the bridge-type IJJs.

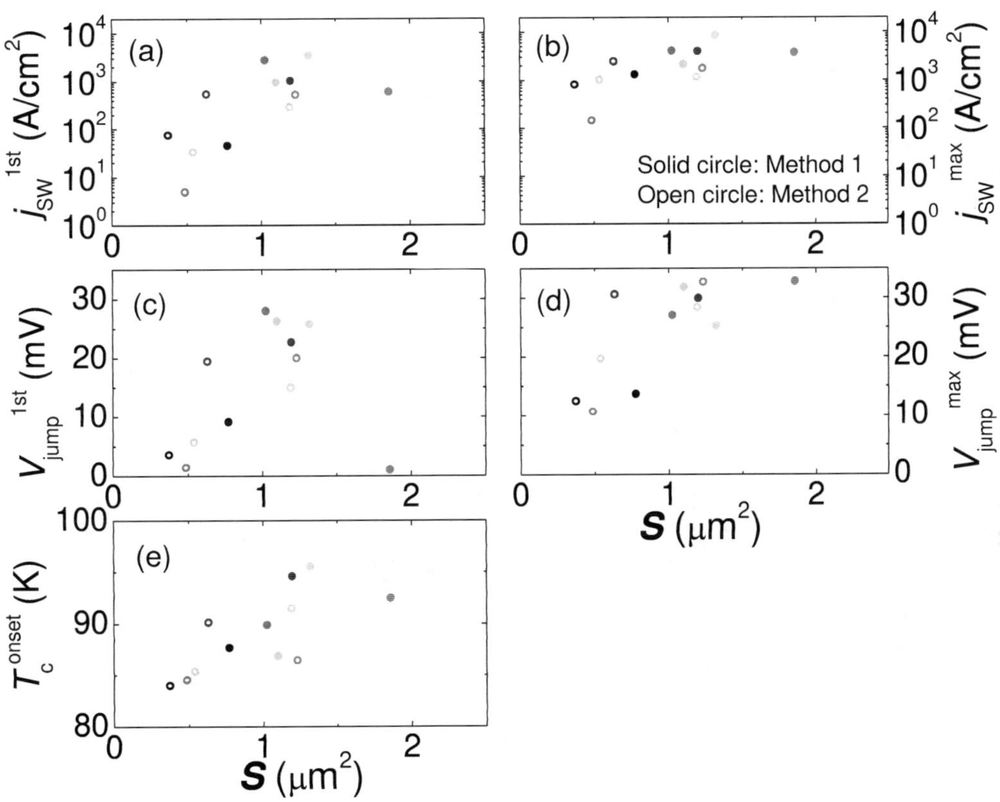

Figure 3. (Color online) Comparison between Method 1 (solid circles) and Method 2 (open circles). Symbols with the same color represent the data for the same device. (a) Plots of j_{SW}^{1st} versus S. (b) Plots of j_{SW}^{max} versus S. (c) Plots of V_{jump}^{1st} versus S. (d) Plots of V_{jump}^{max} versus S. (e) Plots of T_c^{onset} versus S.

Finally, based on the experimental results obtained by Method 3, we tried to develop the new fabrication method (Method 4), which pursued the additional Ar-ion etching after the FIB processes in order to remove the FIB damage more completely. In Method 4, both slits are additionally etched by Ar-ion irradiation, while only one of slits is additionally etched in Method 3.

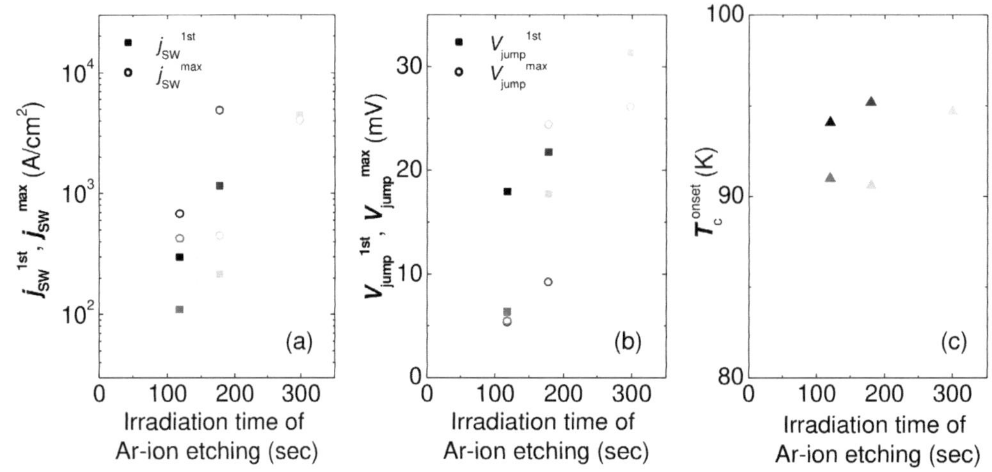

Figure 4. (Color online) (a) $j_{SW}{}^{1st}$ and $j_{SW}{}^{max}$ for 5 IJJs fabricated by Method 3 as a function of the irradiation time of Ar-ion etching. (b) Similar plots of $V_{jump}{}^{1st}$ and $V_{jump}{}^{max}$. (c) Similar plots of $T_c{}^{onset}$. Note that symbols with the same color represent the data for the same device.

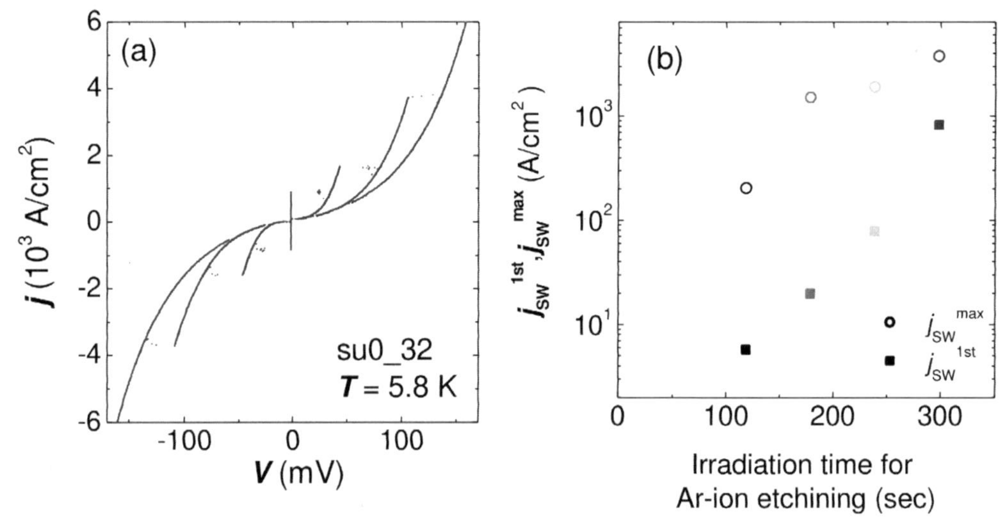

Figure 5. (Color online) (a) *I-V* characteristics of the IJJs (the lateral area is 1.3×1.1 μm^2), which were fabricated by Method 4. The total irradiation time of the Ar-ion etching was 5 min for each slit. (b) Plots of $j_{SW}{}^{1st}$ and $j_{SW}{}^{max}$ for 4 IJJs fabricated by Method 4 as a function of the irradiation time of Ar-ion etching for each slit.

Figure 5(a) shows the *I-V* characteristics of the IJJs (the lateral area is 1.3×1.1 μm^2), fabricated by Method 4. The total irradiation time of the Ar-ion etching was 5 min for each slit, which was the same as the maximum value in Method 3. As shown in Fig. 5(b), we confirmed that $j_{SW}{}^{1st}$ was monotonously increased with increasing the irradiation time of the Ar-ion beam, although the magnitude of $j_{SW}{}^{1st}$ was not as largely enhanced as that shown in Fig. 4(a). More systematic investigation is needed to demonstrate the advantage of Method 4.

4. Conclusion

We present the detailed comparison between the characteristics of Bi2212-IJJs fabricated by four different methods. From the results obtained by Methods 1and 2, we concluded that the incident Ga-ion beam direction in the FIB processes did not influence to the *I-V* characteristics. We also demonstrated that the additional Ar-ion etching after the FIB processes, which was applied in Methods 3 and 4, was quite useful to improve the performance of the *I-V* characteristics measured for the bridge-type IJJs. This effect is probably because the damage region due to the FIB etching processes is successfully removed by the Ar-ion etching. Thus, the combinatorial method using both the Ar-ion and the focused Ga-ion beams has a beneficial effect on the fabrication of the high quality IJJs.

Acknowledgments

We thank T. Mitsui for preparing Si substrates with micro-holes. This work was partly supported by Grants-in-Aid for Scientific Research from JSPS and MEXT of Japan (Grant Nos. 21560020, 24560017) and MEXT-Supported Program for the Strategic Research Foundation at Private Universities (2013-2017). FIB microfabrication performed in this work was supported by Center for Instrumental Analysis, College of Science and Engineering, Aoyama Gakuin University.

References

[1] Inomata K, Sato S, Nakajima K, Tanaka A, Takano Y, Wang H B, Nagao M, Hatano H and Kawabata S 2005 *Phys. Rev. Lett.* **95** 107005
[2] Jin X Y, Koval Y, Lukashenko A, Ustinov A V, and Müller P 2006 *Phys. Rev. Lett.* **96** 177003
[3] Ota K, Hamada K, Takemura R, Ohmaki M, Machi T, Tanabe K, Suzuki M, Maeda A and Kitano H 2009 *Phys. Rev.* B **79** 134505
[4] Nomura Y, Mizuno T, Kambara H, Nakagawa Y and Kakeya I 2015 *J. Phys. Soc. Jpn.* **84** 013704
[5] Takahashi Y, Kakehi D, Takekoshi S, Ishikawa K, Ayukawa S and Kitano H 2016 *J. Phys. Soc. Jpn.* **85** 073702
[6] Suzuki M, Watanabe T and Matsuda A 1999 *Phys. Rev. Lett.* **82** 5361
[7] Krasnov V M 2009 *Phys. Rev. B* **79** 214510
[8] Suzuki M, Hamatani T, Anagawa K and Watanabe T 2012 *Phys. Rev. B* **85** 214529
[9] Matsumoto T, Kashiwaya H, Shibata H, Eisaki H, Yishida Y and Kashiwaya S 2008 *Appl. Phys. Express* **1** 101701
[10] Kubo Y, Yamaguchi T, Tanaka T, Ueda S, Ishii S, Tsuda S, Islam ATM N, Tanaka I and Takano Y 2011 *J. Appl. Phys.* **109** 033912
[11] Kakizaki Y, Koyama J, Yamaguchi A, Umegai S, Ayulawa S and Kitano H 2017 *Jpn. J. Appl. Phys.* **56** 043101
[12] Kato Y, Sakashita N, Ishida K and Mitsui T 2017 *J. Phys. Chem. B* (*in press*) DOI: 10.1021/acs.jpcb.7b06932
[13] We note that the slightly higher values of T_c (exceeding 92 K) for the optimal doped Bi2212 are due to the non-negligible contribution of the large superconducting fluctuation above T_c.

IOP Conf. Series: Journal of Physics: Conf. Series **1054** (2018) 012031 doi:10.1088/1742-6596/1054/1/012031

Dynamics of Phase Switch in the Intrinsic Josephson Junctions Made of Bi2212 with Perfectly-stoichiometric Cation Compositions

Yuji Watabe, Shumpei Umegai, Haruka Ohnuma, Ayami Yamaguchi, Jun-ichi Shimoyama, and *Haruhisa Kitano

Department of Physics and Mathematics, Aoyama Gakuin University, Fuchinobe 5-10-1, Chuo-ku, Sagamihara, Kanagawa 252-5258, Japan

hkitano@phys.aoyama.ac.jp

Abstract. We study the phase switching dynamics in the intrinsic Josephson junctions (IJJs) of $Bi_2Sr_2CaCu_2O_y$ (Bi2212) cuprate superconductors, focusing on the effect of stoichiometry of cation composition. The current-voltage characteristics for the IJJs with the cation composition approaching to be perfectly stoichiometric (Bi:Sr:Ca:Cu=2:2:1:2) show the highly-underdamped nature with a very small retrapping current. We found that the switching current distribution $P(I)$ for the IJJs with perfectly stoichiometric composition and less damage had double peaks or broadened. This suggests that the phase switched junction is not always specified in the case of IJJs with stoichiometric composition, since more than one junction with almost the same switching current is associated with the observed switching events. We obtained the magnitude of a critical current density J_c and an effective temperature T_{esc} for the phase escape composed of a single component, by analysing the bias current dependence of the switching rate $\Gamma(I)$. The results indicate that the perfectly-stoichiometric cation composition can greatly enhance the magnitude of J_c, although a crossover temperature to the macroscopic quantum tunnelling (MQT) for the higher-ordered phase switches is not influenced by the enhancement of J_c, in contrast to the conventional MQT theory.

1. Introduction

The intrinsic Josephson junctions (IJJs) of $Bi_2Sr_2CaCu_2O_y$ (Bi2212) cuprate superconductors have great advantage such as a perfectly smooth tunnel junction on an atomic scale and a high critical current density J_c, which remarkably increase a crossover temperature T_{cr} to the macroscopic quantum tunneling (MQT) state in the current-biased IJJs [1]. However, the underlying properties of IJJs have not been investigated in terms of nonstoichiometry of cation compositions, which is empirically known to induce disorder and constrict the magnitude of J_c and the superconducting transition temperature T_c [2]. For instance, the previous magnetization measurements for Bi2212 crystals with approaching the cation compositions to be stoichiometric suggested that the value of J_c in the vortex state rather than T_c was unexpectedly enhanced, implying the importance of the stoichiometric cation compositions for the enhancement of J_c and flux-pinning [3]. According to the conventional MQT theory [4], it is expected that T_{cr} is proportional to a square root of J_c in a Josephson junction. Thus, the control of cation stoichiometry in Bi2212 crystals is quite important to extract the excellent properties underlying the IJJs.

In this paper, we present a study on the phase switching properties in a small stack of IJJs with the cation compositions approaching to be perfectly stoichiometric (Bi:Sr:Ca:Cu=2:2:1:2). We found that the switching current distribution *P(I)* often had double peaks or broadened. This suggests that more than one junction with almost the same switching current is associated with the observed switching events. We also tried to obtain the magnitude of J_c and an effective temperature T_{esc} for the data showing a sharp-single peak in *P(I)*. The results indicate that the perfectly-stoichiometric cation composition favors the great enhancement of J_c. This effect is expected to increase T_{cr}, as was predicted in the conventional MQT theory. However, the comparison between the previous results using the Bi2212-IJJs with slightly nonstoichiometric compositions [5,6] indicated that the value of T_{cr} for the higher-ordered phase switching events was not influenced by the increase of J_c, suggesting that the MQT rate for the higher-ordered switches was dominated by another contribution which was not included in the conventional theory.

2. Experiments and analyses

Bi2212 single crystals with the cation compositions approaching to be perfectly stoichiometric were grown by the floating zone method in 5% O_2 / Ar and annealed at 600 °C in air to be optimal doped. The cation composition was determined as Bi:Sr:Ca:Cu=2.0:2.0:1.0:2.0 by using the inductively-coupled-plasma (ICP) analyses. Other details of the crystal growth are described elsewhere [7]. The bridge-type Bi2212-IJJs were fabricated by using the focused ion beam (FIB) milling techniques [5]. The superconducting transition temperature T_c of the IJJs after the FIB fabrication was estimated as ~90 K by the zero resistance. This value is slightly larger than that for the previous IJJs [5] with slightly nonstoichiometric compositions (Bi:Sr:Ca:Cu=2.1:1.9:1.0:2.0).

The switching current distribution *P(I)* was measured at several temperatures below 35 K, by applying a bias current with a constant rate dI/dt=10 or 30 mA/s under 10,000 switching events. The switching rate $\Gamma(I)$ was obtained from the measured *P(I)* [8] and analyzed by considering both processes of the thermally-activated (TA) escape and the multiple phase-retrapping (PR) in a tilted washboard potential. Other details of the analyses of $\Gamma(I)$ is described elsewhere [6].

3. Results and discussion

Figures 1 and 2 show the *I-V* characteristics, which were measured at 3.15 K by using triangular ac voltage waveform with a repetition frequency of 23 Hz and a load resistor of 10 kΩ, for two devices of small IJJs with the perfectly-stoichiometric composition (ayf42 and ypf13, respectively). The lateral sizes of IJJs were 1.55×0.94 μm^2 and 1.22×1.15 μm^2, respectively. We observed one superconducting (zero-resistive) and four resistive branches in the *I-V* characteristics for both devices, as shown in Figs. 1 and 2. For the former device (ayf42), the switching current density for the first to the fourth switch (1st SW to 4th SW) was 3.5-7×10^3 A/cm^2 at 3.15 K, moderately larger than the previous results for the IJJs with slightly nonstoichiometric compositions [5]. On the other hand, the switching current density for the third switch (3rd SW) of the latter device (ypf13) was found to exceed 1.6×10^4 A/cm^2 at the same temperature. These results strongly suggest that the stoichiometric cation composition in Bi2212 crystals is quite important to extract the underlying properties of IJJs as well as the pinning properties in the vortex state. We also observed that the retrapping current density to the zero voltage state was less than 100 A/cm^2 for both devices, as small as the previous results for the IJJs with nonstoichiometric compositions. This suggests that the stoichiometric cation composition enhances a quality factor in IJJs, which keeps them in the highly underdamped region.

In addition, we observed that each resistive branch in the former device (ayf42) was almost equally spaced, showing a slightly larger voltage jump than that for the previous Bi2212-IJJs, while there were two branches closely apposed in the latter device (ypf13). The previous numerical study on the *I-V* characteristics in IJJs suggested that the voltage spacing between branches was strongly influenced by the strength of capacitive coupling between Josephson junctions or the existence of non-equivalent configuration in the voltage state of IJJs [9]. Although the observed features in the voltage spacing might be attributed to the stronger capacitive coupling in the IJJs with stoichiometric cation composition, we do not discuss it in this paper. In order to discuss such effects more quantitatively, the more precise control of the junction number included in IJJs (for instance, the same 4 junctions as the previous numerical simulations [9]) is required for a quantitative comparison with the numerical simulations.

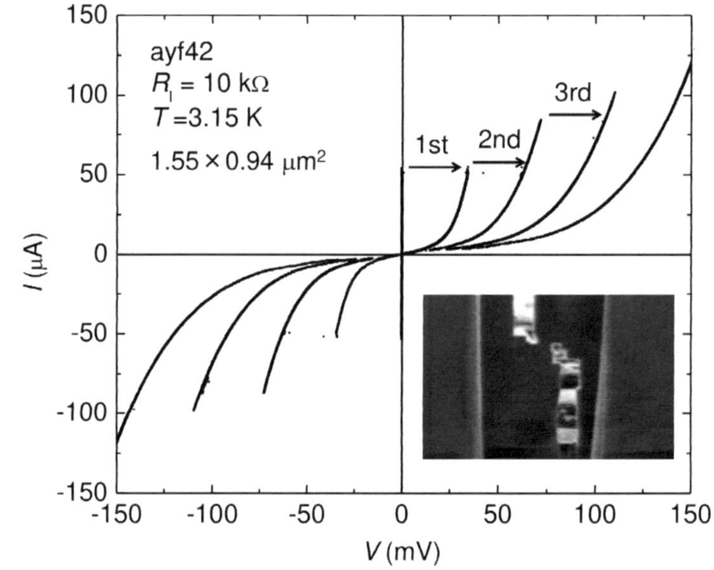

Figure 1. *I-V* characteristics of the bridge-type Bi2212-IJJs (ayf42) with the perfectly-stoichiometric composition, measured at 3.15 K. The inset shows an image of scanning ion microscope of the IJJ device.

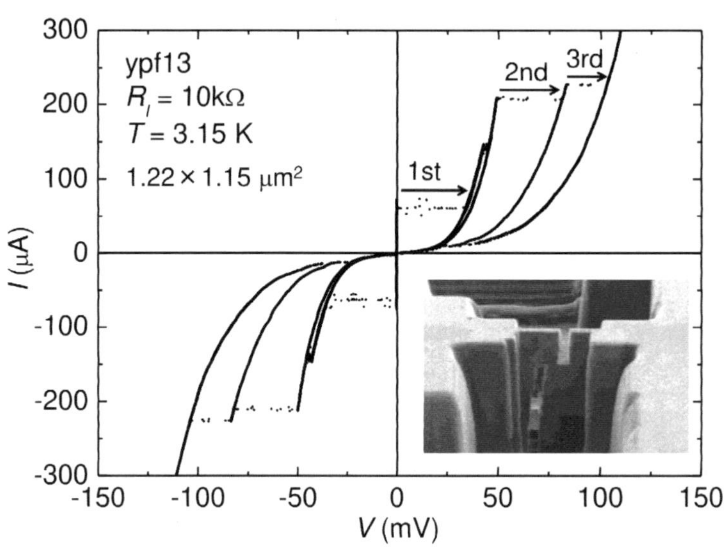

Figure 2. *I-V* characteristics of the bridge-type Bi2212-IJJs (ypf13) with the perfectly-stoichiometric composition, measured at 3.15 K. The inset shows an image of scanning ion microscope of the IJJ device.

Next, we performed the measurements of $P(I)$ at several temperatures. Figures 3 and 4 show the bias current dependence of $P(I)$, measured at 3 K, for the 1st SW to 3rd SW of the two devices (ayf42 and ypf13, respectively). Surprisingly, we found that $P(I)$ had double peaks or broadened, suggesting that more than one junction with similar switching currents was associated with the observed phase switching events. In the previous measurements using IJJs with nonstoichiometric composition [6], we have observed that the existence of more than one junction with slightly different J_c influenced the bias current dependence of $P(I)$ and $\Gamma(I)$, although the contribution of such effects was much smaller than that observed in this work. We also found that there was no double-peak in $P(I)$ for other IJJs with a lower T_c, which were considered to be more damaged by the FIB fabrication, even if the cation composition was perfectly stoichiometric. These results imply that the microscopic disorders in the insulating layers of IJJs, which are produced by the nonstoichiometric composition or the slight damage in the FIB fabrication, strongly affect the sequence of phase switches in IJJs, since a junction with the weaker Josephson coupling in IJJs is switched to the voltage state at a smaller bias current. Thus, we consider that the phase switches measured in the IJJs with perfectly-stoichiometric composition and less damage tend to include the contributions from more than one junction, leading to the double-peaked or broadened structure in $P(I)$.

As for the broadened $P(I)$ observed in the 2nd SW of the latter device (ypf13), the contribution of large junction effects is also considered as another candidate [10,11]. However, as will be discussed below, we concluded that a large junction effect due to the decrease of the Josephson penetration depth followed by the enhancement of J_c was not so strong to influence the phase switching dynamics.

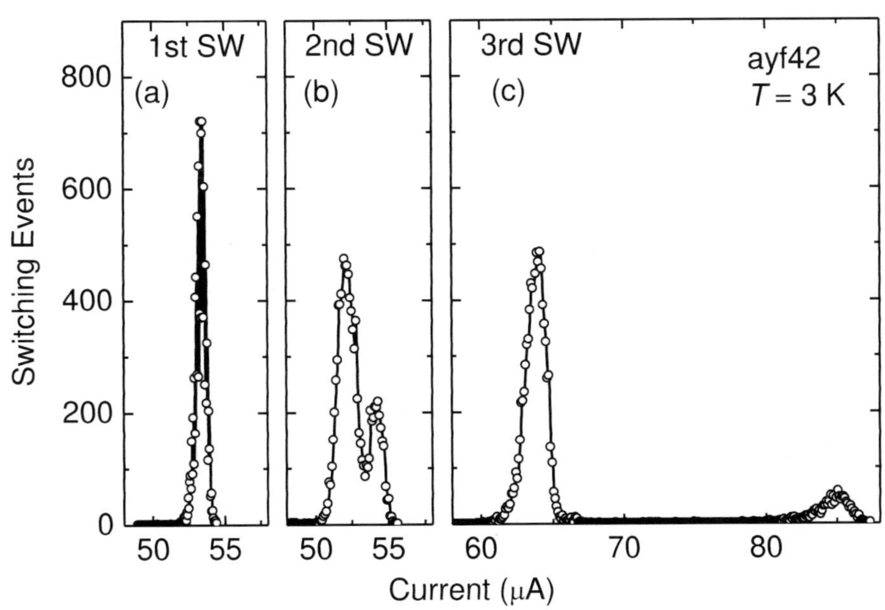

Figure 3. The switching current distribution for (a) for the first switches, (b) for the second switches, (c) for the third switches, which was measured at 3 K in the IJJs (ayf42).

Since our fitting model of $P(I)$ and $\Gamma(I)$ is based on the phase switches occurring in a *single* Josephson junction [6], we cannot analyse the measured $P(I)$ where more than one junction is associated with the phase switches unless the contribution from each junction is specified. Although the double peak structure observed in the 1st SW for the latter device (ypf13) seemed to be safely separated, this feature is found to be dependent on the value of dI/dt, suggesting a possibility that there are multiple components in each peak of the structure. Thus, in this work, we only analysed the data for the 1st SW of the former device (ayf42) and the 3rd SW of the latter device (ypf13), where only a single component seems to be included in $P(I)$.

Figures 5(a) and 5(b) show the fitting results of $\Gamma(I)$ for the 1st SW (ayf42) and the 3rd SW (ypf13), respectively. The deviation from the fitting curve, which is more prominent in $\Gamma(I)$ at higher bias current region and at lower temperatures in Fig. 5(a), is considered to be due to the small contribution from different junctions. The magnitude of J_c is 3.8×10^3 A/cm^2 for the 1st SW (ayf42) and 1.8×10^4 A/cm^2 for the 3rd SW (ypf13). The former value for the 1st SW (ayf42) is similar to that for the previous IJJs with nonstoichiometric composition, suggesting that the former device (ayf42) was slightly damaged by the FIB fabrication. On the other hand, the latter value of for the 3rd SW (ypf13) is much larger than the previous results [6], demonstrating that the stoichiometric composition with less damage can greatly enhance the magnitude of J_c in Bi2212-IJJs. We note that such an enhancement of J_c is not so seriously influenced to the estimation of the Josephson penetration depth, since it is inversely proportional to a square root of J_c. This leads to a conclusion that the Josephson penetration depth corresponding to J_c=1.8×10^4 A/cm^2 is roughly one-third or half of that for the Bi2212-IJJs measured previously. Thus, we can rule out a possibility that the phase dynamics in the IJJs with perfectly stoichiometric composition is dominated by the dynamics of fluxons in a large junction where the Josephson penetration depth is much smaller than a lateral size of the Josephson junction.

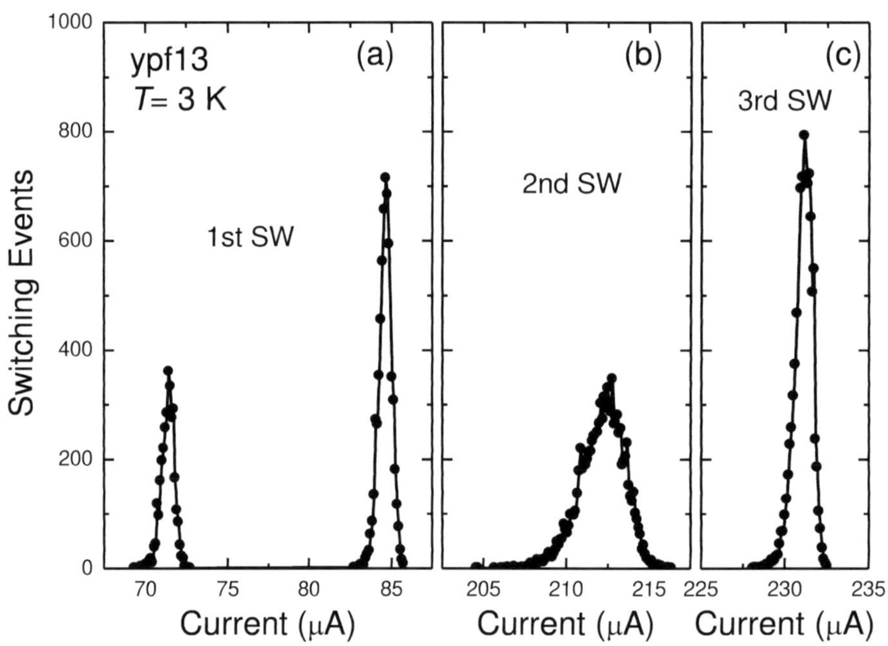

Figure 4. The switching current distribution for (a) for the first switches, (b) for the second switches, (c) for the third switches, which was measured at 3 K in the IJJs (ypf13).

Figure 5(c) shows the plots of the effective escape temperature T_{esc} in the phase switches versus the bath temperature T_{bath}. We found that T_{esc} for the 1st SW (ayf42) deviated from T_{bath} at 2 K while T_{esc} for the 3rd SW (ypf13) deviated from T_{bath} below 8 K, suggesting a sign of the crossover to the MQT state. According to the conventional theory of MQT phenomena [4], T_{cr} is expected to be proportional to a square root of J_c. Although the measurements at lower temperatures than 2 K are required for the confirmation of T_{cr} for the 1st SW (ayf42), the T-independent behavior of T_{esc} for the 3rd SW (ypf13), surviving up to 8 K, is clearly beyond the conventional prediction, since it only gives ~2.3 K as an estimate of T_{cr}. Such an unusual result has often been discussed in the higher-ordered phase switches in Bi2212-IJJs [5,6,12,13]. We found that the temperature (~8 K) below which T_{esc} for the 3rd SW was independent of T_{bath} was almost the same as the previously-reported value in the higher ordered SWs of the IJJs, in spite of a large difference in the estimates of J_c. This strongly suggests that the value of T_{cr} for the higher-ordered SWs is determined by a novel mechanism, which has not been included in the conventional theory but proposed in our previous paper [6]. Recently, we also obtained a similar result in a study of the dependence of T_{cr} for the higher-ordered SWs on the carrier concentration in the bridge-type IJJs made of Y-doped Bi2212 [14]. Thus, both results show a good agreement with each other.

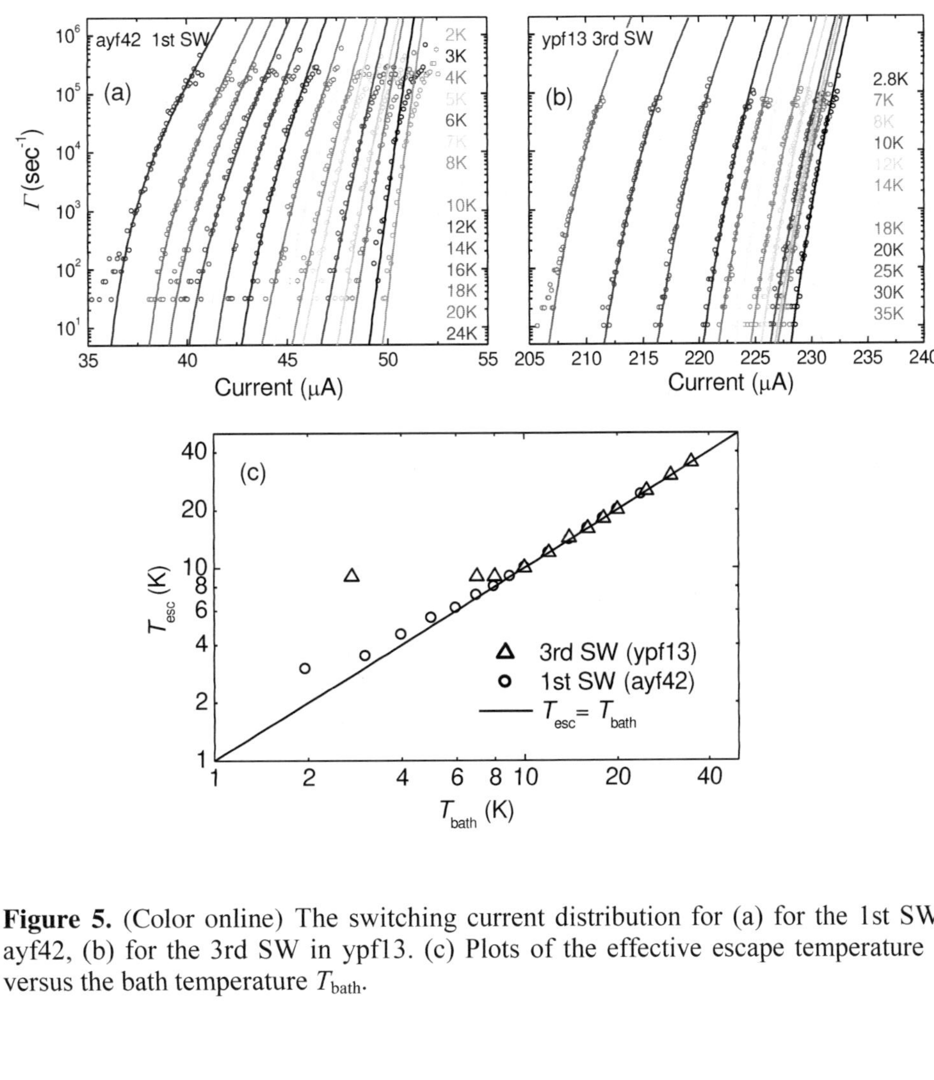

Figure 5. (Color online) The switching current distribution for (a) for the 1st SW in ayf42, (b) for the 3rd SW in ypf13. (c) Plots of the effective escape temperature T_{esc} versus the bath temperature T_{bath}.

4. Conclusion

We present the study on the measurements of *I-V* curves and *P(I)* in the Bi2212-IJJs, focusing on the effect of stoichiometry of cation compositions. Our results clearly indicate that the switching current density in the IJJs with perfectly stoichiometric composition is greatly enhanced, keeping the retrapping current density very low. We also observed that *P(I)* had double peaks and broadened, probably due to the multiple contributions from several junctions for the observed phase switches. Since such a feature made the fitting analyses of *P(I)* and *Γ(I)* more difficult, we obtained the magnitude of J_c and T_{esc} for the data where only a single component was contributed. The results show that the magnitude of J_c, which determines the strength of Josephson coupling in IJJs, can be greatly enhanced as the cation composition becomes perfectly stoichiometric, expecting an increase of T_{cr} to the MQT state. However, the temperature dependence of T_{esc} for the higher-ordered SWs was found to show almost the same behavior as that reported previously in the IJJs with nonstoichiometric compositions, in spite of a large difference in J_c. These results suggest that the excellent properties underlying IJJs such as J_c can be extracted by the precise control of the cation stoichiometry, while the crossover to the MQT state for the higher-ordered SWs is not influenced by the enhancement of J_c.

Acknowledgments

We thank Y. Kubo for fruitful discussions. This work was partly supported by Grants-in-Aid for Scientific Research from JSPS and MEXT of Japan (Grant Nos. 21560020, 24560017) and MEXT-Supported Program for the Strategic Research Foundation at Private Universities (2013-2017). FIB microfabrication performed in this work was supported by Center for Instrumental Analysis, College of Science and Engineering, Aoyama Gakuin University.

References

[1] Inomata K, Sato S, Nakajima K, Tanaka A, Takano Y, Wang H B, Nagao M, Hatano H and Kawabata S 2005 *Phys. Rev. Lett.* **95** 107005
[2] Hobou H, Ishida S, Fujita K, Ishikado M, Kojima K M, Eisaki H and Uchida S 2009 *Phys. Rev.* B **79** 064507
[3] Uchida S, Shimoyama J, Makise T, Horii S and Kishio K 2006 *J. Phys: Conference Series* **43** 231
[4] Caldeira A O and Leggett A J 1981 *Phys. Rev. Lett.* **46** 211
[5] Kakehi D, Takahashi Y, Yamaguchi H, Koizumi S, Ayukawa S and Kitano H 2016 *IEEE Trans. Appl. Supercond.* **26** (3) 1800204
[6] Kitano H, Takahashi Y, Kakehi D, Yamaguchi H, Koizumi S and Ayukawa S 2016 *J. Phys. Soc. Jpn.* **85** 054703
[7] Makise T, Uchida S, Horii S, Shimoyama J and Kiishi K 2007 *Physica* C **460-462** 772
[8] Fulton T A and Dunkleberger L N 1974 *Phys. Rev.* B **9** 4760
[9] Machida M and Koyama T 2004 *Phys. Rev.* B **70** 024523
[10] Kitano H, Ota K and Maeda A 2006 *Supercond. Sci. Technol.* **20** S68
[11] Kubo Y, Sboychakov A O, Nori F, Takahide Y, Ueda S, Tanaka I, Islam A T M N and Takano Y 2012 *Phys. Rev.* B **86** 144532
[12] Takahashi Y, Kakehi D, Takekoshi S, Ishikawa K, Ayukawa S and Kitano H 2016 *J. Phys. Soc. Jpn.* **85** 073702
[13] Nomura Y, Okamoto R and Kakeya I 2017 *Supercond. Sci Technol.* **30** 105001
[14] Kitano H, Yamaguchi A, Takahashi Y, Umegai S, Watabe Y, Ohnuma H, Hosaka K and Kakehi D 2018 *J. Phys: Conference Series* **969** 012065

IOP Conf. Series: Journal of Physics: Conf. Series **1054** (2018) 012032 doi:10.1088/1742-6596/1054/1/012032

Improvement of uniformity of I_c distributions in long REBCO with BMO coated conductors by in-plume PLD method

A Ibi, T Machi, K Nakaoka, T Izumi

National Institute of Advanced Industrial Science and Technology (AIST)
1-2-1 Namiki, Tsukuba, Ibaraki 305-8564 Japan

akira-ibi@aist.go.jp

Abstract. Long $REBa_2Cu_3O_X$ (REBCO, RE: rare earth element) with $BaMO_3$ (BMO, M: metal) coated conductors have been expected for the industrial and commercial applications at high temperatures in magnetic fields. Especially, the multi-filamentary structure long REBCO with BHO coated conductors are effective for control of the shielding current of coated conductor excited by applied magnetic fields and alternating-current loss. We fabricated long $EuBa_2Cu_3O_X$ (EuBCO) with $BaHfO_3$ (BHO) coated conductors by the IBAD/PLD (conventional PLD) method, which showed the high in-field I_c values. However, in order to realize REBCO with BMO coated conductors for industrial and commercial applications, the much higher uniformity of not only longitudinal and but transversal I_c distributions of long coated conductors with high in-field performance is required. We have tried to develop the long EuBCO with BHO coated conductors by the in-plume (plane-plume) PLD method with high deposition rate of about 24 nm/s to obtain high in-filed performance and low production cost. More recently, we fabricated the 105 m long EuBCO with BHO coated conductors with high uniformity of longitudinal I_c distributions of 2.54 % (STDEV) by the in-plume (plane-plume) PLD method comparing with coated conductors by the conventional PLD method. We measured each I_c value of 8 filaments of 581 μm in width and slot of 20 μm in width for 0.6 m long EuBCO with BHO coated conductors by in-plume (plane-plume) PLD method. I_c distribution in a width direction of the EuBCO with BHO coated conductors by in-plume (plane-plume) PLD method is more uniform than that of the conventional PLD method.

1. Introduction

$REBa_2Cu_3O_{7-x}$ (REBCO, RE: rare earth element) with $BaMO_3$ (BMO, M: metal) coated conductors [1, 2] have been expected for the industrial and commercial applications at high temperatures in magnetic fields, such as magnetic resonance imaging (MRI) and heavy ion medical accelerator. Recently, we have found that $EuBa_2Cu_3O_{7-x}$ (EuBCO) with $BaHfO_3$ (BHO) coated conductors by IBAD (ion-beam assisted deposition) [3] and PLD (pulsed laser deposition) methods show higher critical current densities (J_c) and critical currents (I_c) at high temperatures in self and magnetic fields than those of other REBCO with BMO coated conductors. The obtained I_c(min) values in the all angles of applied magnetic field of 3 T at 77 K is 133.9 to 141.2 A/cm-w [4]. EuBCO with BHO layer were about 3.6 μm in thickness. As a result of 3D-SEM (Scanning Electron Microscope) images, the volume percentage of random oriented grains of EuBCO, CuO and voids etc. which prevent a superconducting current flow in the EuBCO layers were much lower than those in the $GdBa_2Cu_3O_{7-x}$ (GdBCO) layers [4]. Moreover, BHO nanorods in REBCO layers were formed short and thin with uniform distribution in the whole REBCO layers thickness in comparison with $BaZrO_3$ (BZO) and $BaSnO_3$ (BSO) nanorods in REBCO layers [5]. We successfully faricated a 93 m long EuBCO with BHO coated conductor which showed equivalent performance to that of short samples. However, the characteristics of uniformity of I_c distributions and the I_c-B-θ profile of long length EuBCO with BHO coated conductors have not been yet achieved for the industrial and commercial applications.

The in-plume PLD method is performed shortening the target-substrate distance to increase the deposition rate [6, 7]. Moreover, we developed the plane-plume method, which could make the plane area for deposition by scanning the irradiation points of pulsed laser. This method is called as plane-plume PLD and can realize high uniformity of I_c distributions in long REBCO coated conductors by in-plume PLD method can be expected, since it minimize the influence of change of plume conditions such as tilt and swing of plume.

We have tried to develop the long EuBCO with BHO coated conductors by the in-plume (plane-plume) PLD method. As a result, we fabricated the long EuBCO with BHO coated conductors with high uniformity of I_c distributions comparing with those by the conventional PLD method.

2. Experimental procedure

The EuBCO with BHO layer were deposited by in-plume (plane-plume) PLD method with vapor-liquid-solid (VLS) [8] mode on 10 mm wide CeO_2 / $LaMnO_3$ / IBAD-MgO / Y_2O_3 / $Gd_2Zr_2O_7$ / Hatelloy C-276 substrates. The value of in-plane and out-of-plane texturing degrees ($\Delta\phi$ and $\Delta\omega$) of CeO_2 cap layer were about 2 to 3 and 1 to 2 degrees, respectively. A 200 W industrial XeCl excimer laser with a wavelength of 308 nm was used at a pulse repetition rate of 300 Hz and a pulse energy of 600 to 700 mJ. The optical system is synchronized to the laser pulse and scanning of the laser beam to on a bulk EuBCO target containing BHO was controlled as well. The laser repetition rate at 300 Hz is divided into 18 plumes which are almost the same in size and shape. Typically, the pulse repetition rate of 300 Hz and a pulse energy of 600 to 700 mJ lead the deposition rate of about 24 nm/s for a EuBCO with BHO layer. Commercially available targets of sintered off-stoichiometric $EuBa_{1.8}Cu_{3.0\ or\ 3.2}O_x$ containing 3.5 mol% doped BHO with a diameter of 6 inch were used for deposition of the EuBCO with BHO layer. The deposition temperature was about 930 ° C for a substrate tape with a transferring speed of 60 m/h. The oxygen pressure was maintain to be at about 800 mTorr with a flow of 10 sccm oxygen. The T-S distance (distance between the target and the substrate) was set about 70 mm. The thickness of EuBCO layer with BHO nanorods was about 800 nm for one time deposition with the substrate tape transferring speed of 60 m/h as a typical value. I_c values was measured by using the conventional four-terminal method with the criterion of 1 μV/cm. The multi-filamentary structure long EuBCO with BHO coated conductors was fabricated by using the KrF excimer laser with the wavelength of 248 nm irradiation with Reel-to-Reel system [9]. To examine the angular dependence of I_c values on applied magnetic field angle, which was changed from 0 to 180 degrees to *c*-axis (180 degree corresponds to a magnetic field paralleled to the *c*-axis direction).

3. Results and discussions

Figure 1 shows (a) the schematic diagrams of our conventional (out-of-plume) and (b) in-plume (plane-plume) PLD system. The in-plume PLD is performed shortening the target-substrate distance to increase the deposition rate. However, the in-plume PLD is difficult to control the deposited REBCO layer composition and the increased supersaturation. To solve these problems, we used the Ba-deficient off-stoichiometric REBCO target and increase the number of multi-plume with scan of X-Y axes directions (plane-plume PLD).

Figure 2 indicates the longitudinal I_c distributions at 77 K and self-field of (a) a 93 m long EuBCO with BHO coated conductor by conventional PLD method with deposition rate of 4.2 nm/s and (b) a 105 m long EuBCO with BHO coated conductor by the in-plume (plane-plume) PLD method with deposition rate of 24 nm/s. The EuBCO with BHO layer were about (a) 3.6 μm and (b) 0.8 μm in thickness, respectively. The I_c was measured every 0.6 m. The STDEV (standard deviation) of longitudinal I_c distributions is improved to 2.54 from 4.02 % by using the in-plume PLD method. The transferring substrate through the inside of multi-plume with scan of X-Y axes directions (plane-plume) is less affected by change of plume conditions such as tilt and swing of plume. Therefore, we thought that the longitudinal I_c distributions shows uniform since few changes in compositional shift of EuBCO layer from target, deposition rate of EuBCO layer (with change optimized deposition temperature), and EuBCO layer thickness etc. Figure 3 shows the longitudinal n-values at 77 K and self-field of (a) a 93 m long EuBCO with BHO coated conductor by the conventional PLD method and (b) a 105 m long EuBCO with BHO coated conductor by the in-plume (plane-plume) PLD method. The n-value was measured every 0.6 m. The STDEV of longitudinal n-values distributions is also improved to 2.49 from 8.42 % by using the in-plume PLD method.

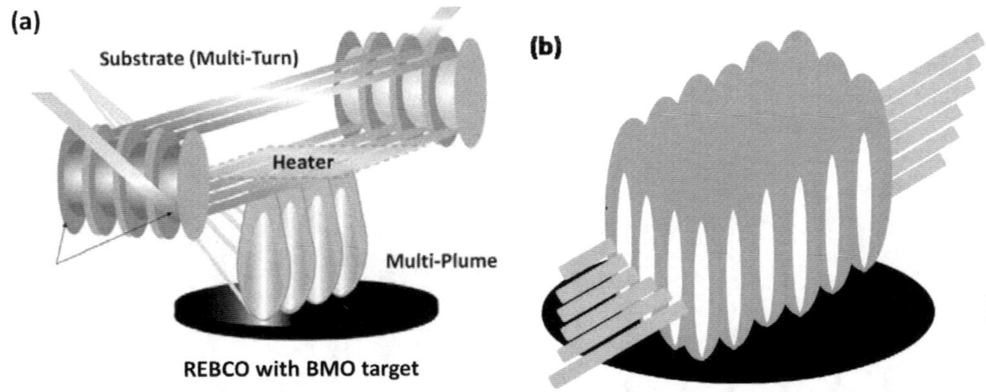

Fig. 1 (a) schematic diagrams of our conventional (out-of-plume) and (b) in-plume (plane-plume) PLD system.

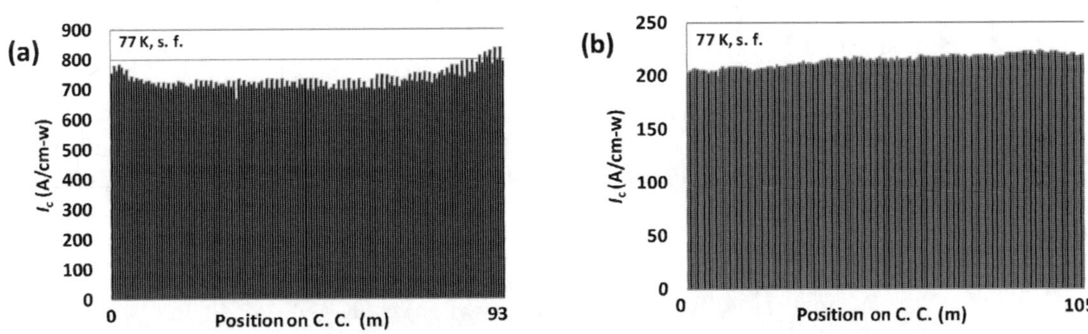

Fig. 2 Longitudinal I_c distributions at 77 K and self-field of (a) a 93 m long EuBCO with BHO coated conductor by conventional PLD method and (b) a 105 m long EuBCO with BHO coated conductor by in-plume (plane-plume) PLD method.

Figure 4 indicates the I_c distributions of (a) 0.6 m long EuBCO with BHO coated conductor of 8 filaments of 440 μm in width by the conventional (out-of-plume) PLD method and (b) 0.6 m long EuBCO with BHO coated conductor of 8 filaments of 581 μm in width by the in-plume (plane-plume) PLD method. We measured each I_c value of 8 filaments of 440 or 581 μm in width and slot of 20 μm in width for 0.6 m long EuBCO with BHO coated conductors. The variation of each I_c value were (a) 10.4 and (b) 4.3 %, respectively ($\{(I_{c\ ave.}) / (I_{c\ min.})\} / I_{c\ ave.} \times 100$ (%)). I_c distribution in a width direction of the EuBCO with BHO coated conductors by in-plume (plane-plume) PLD method is more uniform than that of the conventional PLD method.

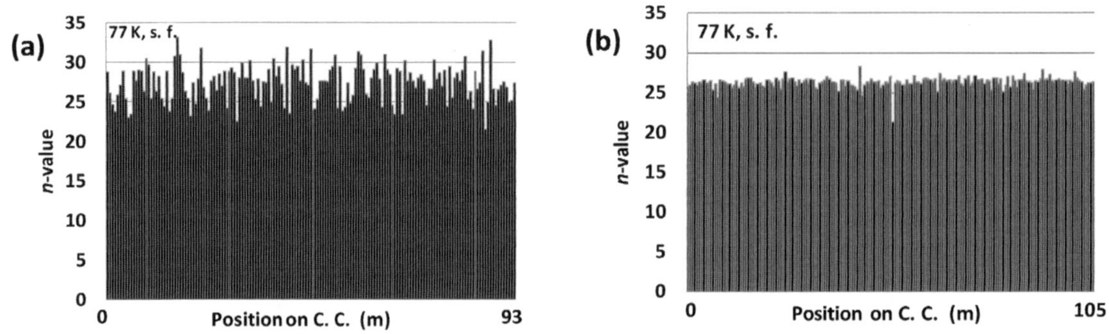

Fig. 3 Longitudinal n-values at 77 K and self-field of (a) a 93 m long EuBCO with BHO coated conductor by conventional PLD method and (b) a 105 m long EuBCO with BHO coated conductor by in-plume (plane-plume) PLD method.

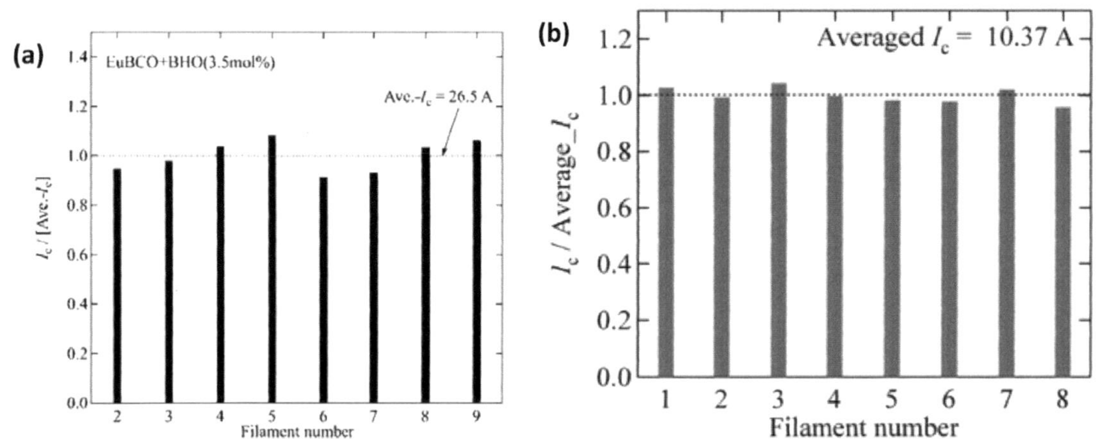

Fig. 4 Normalized I_c distributions of (a) 0.6 m long EuBCO with BHO coated conductor of 8 filaments of 440 μm in width by conventional PLD method and (b) 0.6 m long EuBCO with BHO coated conductor of 8 filaments of 581 μm in width by in-plume (plane-plume) PLD method.

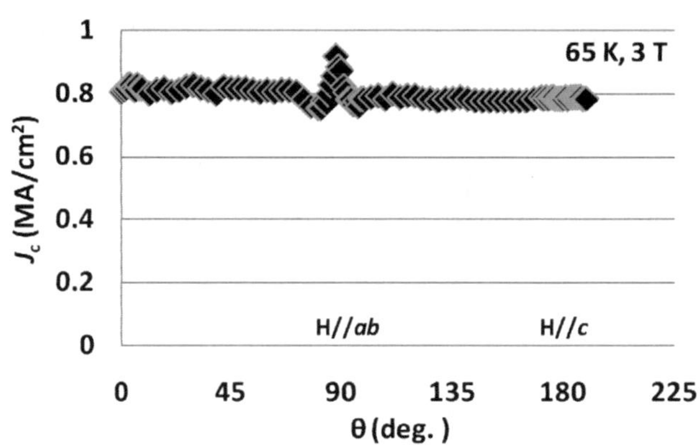

Fig. 5 Angular dependence of applied magnetic field of 3 T on I_c values for EuBCO with BHO coated conductor at 65 K by in-plume (plane-plume) PLD method.

Figure 5 shows the angular dependence of the applied magnetic field of 3 T on the J_c values at 65 K for EuBCO with BHO coated conductor by in-plume (plane-plume) PLD. The EuBCO with BHO layer was about 1.6 μm in thickness. The in-plume (plane-plume) PLD is effective to improve J_c anisotropies in magnetic field at high temperature. It is supposed that the BHO becomes much shorter and random direction in EuBCO layer by the in-plume (plane-plume) PLD method comparing with the conventional PLD method because of supersaturation should increase with increasing deposition rate. In addition, the in-plume PLD method also is considered naturally to introduce lots of stacking faults by the result of TEM images.

4. Conclusion

We have been investigating the long REBCO with BMO coated conductors by IBAD / conventional (out-of-plume) PLD and in-plume PLD method. The in-plume PLD is effective to increase the deposition rate with high superconducting characteristics. Moreover, the in-plume (plane-plume) PLD method is also effective to obtain high uniformity of longitudinal and transversal I_c and longitudinal *n*-values distributions in long EuBCO with BHO coated conductors because of the transferring substrate through the inside of multi-plume with scan of X-Y axes direction (plane-plume) which less affected by change of plume condition such as tilt and swing of plume.

We evaluated the J_c-B-θ profile at 65 K and 3 T of EuBCO with BHO coated conductor by in-plume PLD method. The in-plume PLD method is also effective to improve J_c anisotropies in magnetic field at high temperature.

Acknowledgements
This work is supported by the New Energy and Industrial Technology Development Organization (NEDO).

References

[1] Goyal A, Kang S, Leonard K J, Martin P M, Gapud A A, Varela M, Paranthaman M, Ijaduola A D, Specht E D, Thompson J R, Christen D K, Pennycook S J, List F A 2005 Supercond. Sci. Technol. **18** 1533
[2] Yamada Y, Takahashi K, Kobayashi H, Konishi M, Watanabe T, Ibi A, Muroga T, Miyata S 2005 Appl. Phys. Lett. **87** 132502
[3] Iijima Y, Tanabe N, Kohno O, Ikeno Y 1992 Appl. Phys. Lett. **60** 769
[4] Yoshida T, Ibi A, Takahashi T, Yoshizumi M, Izumi T, Shiohara Y 2014 Physica C: Superconductivity **504** 42-46
[5] Tobita H, Notoh K, Higashikawa K, Inoue M, Kiss T, Kato T, Hirayama T, Yoshizumi M, Izumi T, Shiohara Y 2012 Supercond. Sci. Technol. **25** 062002
[6] Lee S, Chikumoto N, Yokoyama T, Machi T, Nakao K, Tanabe K 2009 IEEE Trans. Appl. Supercond. **19** 3192
[7] Chikumoto N, Lee S, Nakao K, Tanabe K 2009 Physica C **469** 1303
[8] Ibi A, Yoshida T, Izumi T, Shiohara Y, Yokoe D, Kato T, Hirayama T 2016 Physics Procedia **81** 97-100
[9] Katayama K, Machi T, Nakamura T, Takagi Y, Nakaoka K, Yoshizumi M, Izumi T, Shiohara Y 2014 Physics Procedia **58** 142

IOP Conf. Series: Journal of Physics: Conf. Series **1054** (2018) 012033 doi:10.1088/1742-6596/1054/1/012033

Development of Scribing Technique by using Multiple-laser Beams for Multi-filamentary Coated Conductors

Takato Machi, Akira Ibi and Teruo Izumi

Department of Energy and Environment, National Institute of Advanced Industrial Science and Technology (AIST), 1-2-1 Namiki, Tsukuba, Ibaraki 305-8564 Japan

t.machi@aist.go.jp

Abstract. The $REBa_2Cu_3O_{7-d}$ (REBCO) coated conductors (CCs) with the multi-filamentary structure is very useful for controlling the AC-loss and the shielding current. We have developed laser scribing technique to fabricate multi-filamentary structure by irradiating a rectangular beam using excimer laser source (KrF, λ~248nm). Since the reflectance of stabilized silver and copper at the wavelength of Nd: YAG or fiber laser (~ 1 μm) is 98% or more, destructive thermal processing must be required for processing in the wavelength range, which was the reason why we chose the excimer laser as light source. Actually, we successfully fabricated scribing structure with high precision. On the other hand, the improvement of the processing speed has been an important issue to be solved. Although the multi-laser beam system is effective for this purpose, the uniform intensity distribution in the width direction has to be realized. There is no strong Gaussian component in the excimer laser compared with a solid laser. The uniformity is useful for a single beam, however, it is not enough for the multi-beam system. In order to solve this issue, we have modified the beam intensity distribution by using homogenizer. Concretely, the beam intensity in the minor axis direction was flattened by a homogenizer using a plurality of horizontally elongated cylindrical lenses. This modification made the uniform range of laser intensity wider and it was confirmed that multi-laser beams can be applicable as a high-speed scribing technique.

1. Introduction

Reduction of the AC loss and control of the influence of shielding magnetic field are indispensable for applying CCs to superconducting magnet. For that purpose, we have been developing the thinning process of coated conductors by the laser scribing method which digs stabilizing and superconducting layers into a slot shape[1]. We have developed a long scribing equipment using excimer laser (KrF), which has an excellent controllability of slot width almost without deterioration of performance[2]. Additionally, the process can suppress degradation of peel strength. And the applicability to the long tapes was already confirmed. Many other research institutes have been also trying scribing machining [3,4], but only we have made it possible to perform scribing processing for a long CC of 100 m class[5]. The next technical problem on this excimer laser scribing method is to improve the processing speed.

In our previously developed equipment, it was the biggest factor limiting the processing speed that only one beam could be irradiated at a time. Although the coherent excimer laser used as a laser source is a gas laser, it does not have a stronger Gaussian component than a solid laser, but the laser beam shaped into a rectangle by the optical mask is not uniform, then only one beam could be used in the one scribing path.

Therefore, we attempted to make the beam intensity uniform in wider area on the mask surface by introducing a beam homogenizer in the direction corresponding to the width direction of the CCs. Then, the dual beam irradiation was taken place in the obtained uniform beam, and the effectiveness of this modification was evaluated.

2. Experimental

As shown on the left of fig. 1, the intensity distribution of the laser beam is not uniform, and in general, it has a Gaussian distribution with high intensity at the center of the beam. In order to produce a plurality of rectangular beams from one beam, it is necessary to create a top-hat type distribution having uniform intensity on the mask surface as shown in the right side of fig. 1. After designing and modification of the equipment, we carried out real scribing process of REBCO and evaluated it.

Figure 1 Schematic view of beam profiles before modification(left) and expected shape by the modification(right).

2.1. *New design of optics of laser scribing equipment*

For dual beam formation, it is necessary to equalize the laser beam intensity in the width direction (minor axis direction) of the CCs. Since the CCs are transported at a constant speed under the irradiation of rectangular shape pulsed beams, there is no need on uniformity in the longitudinal direction of the CCs. A conceptual diagram of the optical system for the modification is shown in fig.2.

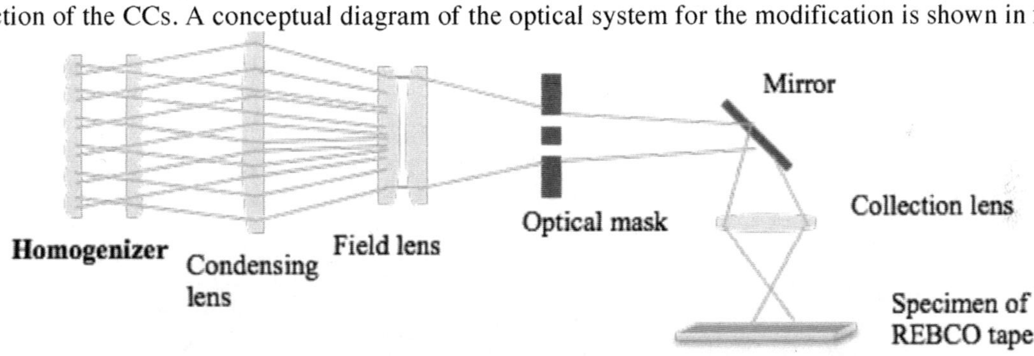

Figure 2 A concept of optical design for widely uniform laser beam

As a method of homogenizing the beam in the minor axis direction, a system (homogenizer) in which two sets of several cylindrical lenses were combined in the minor axis direction was adopted. Also, a condensing lens was provided to make the size of the field lens surface constant before the mask. A new field lens was also provided to irradiate the beam that passed through the mask. The uniform range on the tape surface was designed to be 0.6 mm or more in the width direction.

2.2. *Observation of beam uniformity and processing for CCs*

The beam intensity distribution on the surface of CCs was observed by a beam profiler (LaserCam-HR, COHERENT). We used EuBCO CCs for processing, the thickness of Ag was 8 μm on the top of EuBCO layer and the thickness of EuBCO layer was 0.8 μm. The stainless steel optical mask (thickness: 0.2 mm) was designed to form two rectangular beams of 20 μm wide and 1 mm long on the surface of the CCs. Since the magnification of the condensing lens is 15 times, the rectangle opened of the mask was 300 μm wide and 15 mm long. The distance between the two rectangles was designed to be 450 μm on the tape surface. Using this mask, a scribing process was carried out to divide the 10 mm width tape into 21 filaments. Because the dual beams scribe two slots at one-path, the irradiation position was shifted by 940 μm steps, and the tape was divided into 21 by 10 times of processing. The processed length was 60 cm. The depths of the slots formed in the EuBCO CCs were observed with a scanning laser microscope(VK8700, KEYENCE).

3. Results and discussion

Figure 3 shows the observed results of the laser intensity distribution on the CCs surface. Figure 3 (a) shows the beam-intensity distribution before the modification, showing that the center of the beam was strong and a Gaussian type. In the case of (b) which is the beam intensity distribution after modification, it was found that a uniform beam profile of 600 μm in the width direction was obtained which is the same width in the design. In the case of (c) in which the mask (100 μm width) was inserted for producing a dual beam, it succeeded in obtaining dual beams with similar intensity.

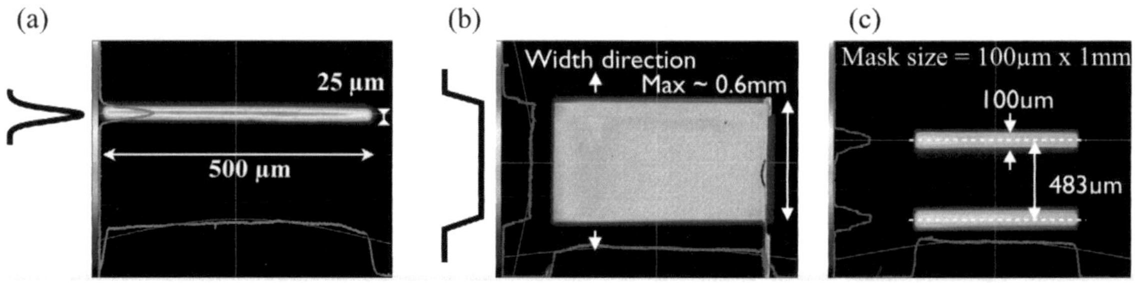

Figure 3 Beam profiles of laser beams on the position of specimens. (a) 2D-profile of laser beam intensity before modification, (b) 2D-profile without mask after modification, (c) 2D-profile of dual beams after modification.

Figure 4 shows the observation results of the two slots scribed by the dual beams by a scanning laser microscope. From this depth distribution, line profiles could be calculated and data on the depth and width of the slots could be obtained. It was confirmed that the average of the widths of the 19 filaments excluding the two at both edges was 450 μm and which corresponded to the filament width was as designed. Also, the groove width was 20 μm on average, and it was also found that slots were formed as intended.

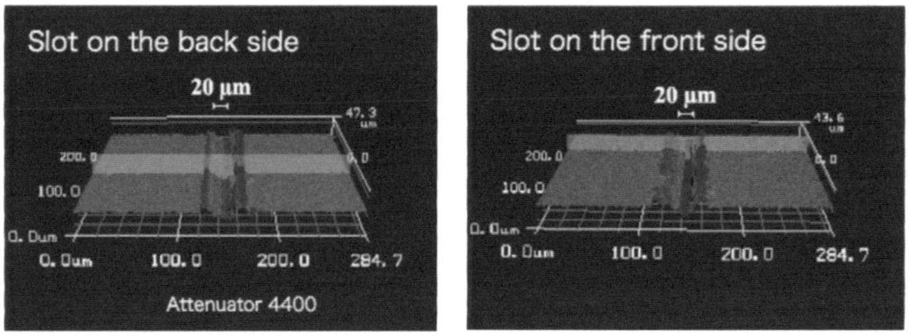

Figure 4 Observation result by scanning laser microscope for dual slots at back side (left) and front side (right).

Fig. 5 shows the energy density dependence of the depth of slots scribed by dual beams. The energy density is controlled by attenuating the beam intensity before reaching the mask surface by the attenuator. Therefore, the horizontal axis of fig. 4 shows the count number of the stepping motor that rotates the attenuator, the energy density increases as the count number increases, and the 5500 count corresponds to about 100 mJ / mm^2. The depth distributions of the dual beams in the front and back sides were almost the same, and it was shown that dual scribing processes can be simultaneously performed as designed.

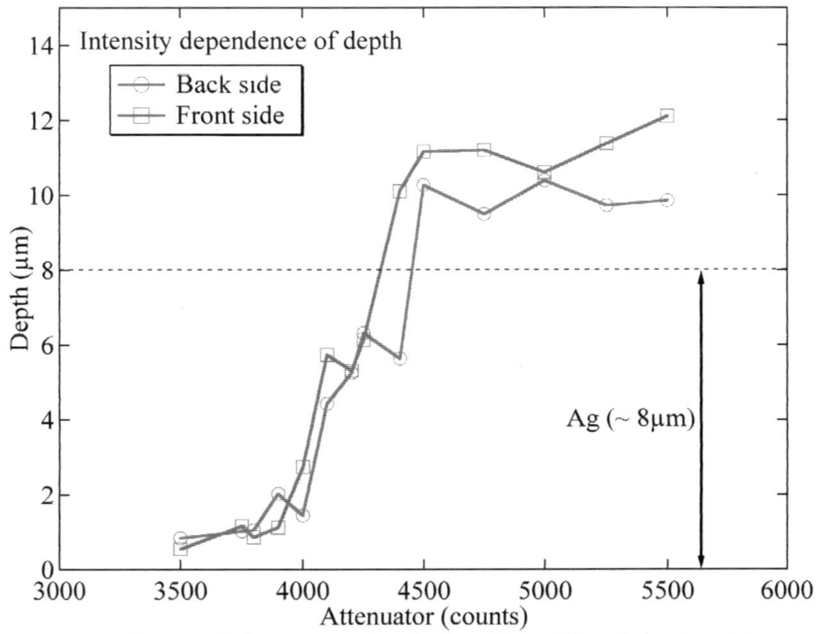

Figure 5 The energy density profiles of the slope depth in simultaneous laser scribing processing.

It was confirmed that the scribing time was shortened to about half the time than before. The sheet resistance between the filaments was as low as 0.63 Ω / sq. It is assumed that the poor electrical isolation was arisen from the re-adhering debris which increased by spouting from the two slots. Therefore, it is necessary to establish an optimum assist gas spraying for blowing debris. The I_c of each of the 21 filaments was measured and summed up to be 173.7 A. As I_c before processing was 195 A, the reduction rate of I_c was about 11%. The width of the slot was 20 μm, and since there were 20 slots, the area reduction rate was 8%. Therefore, the reduction rate of I_c due to scribing was about 3%. This value is equivalent to the rate of reduction when the slots were scribed one by one irradiation, and deterioration did not increase by scribing with the dual beam.

4. Summary

In order to make the scribing speed higher, we have improved the uniformity of beam intensity distribution for dual beam irradiation at one-path processing. By introducing a homogenizer in the optics of equipment, the beam intensity in the minor axis direction was flattened, and we have succeeded in dual beam scribing. In the future, we will develop techniques to further increase the number of beams.

Acknowledgement
This work was supported by METI, AMED and NEDO.

References

[1] Machi T *et al.* 2013 *Supercond. Sci. Technol.* **26** 105016
[2] Machi T *et al.* unpublished
[3] Majkic G *et al.* 2011 *IEEE Trans. Appl. Superconduct.* vol. **21** No.**3** 3297
[4] Nast R *et al.* 2014 *Journal of Physics: Conference Series* **507** 022023.
[5] Katayama K *et al.* 2014 *Physics Procedia* **58** 142 – 145

Development of surface planarization process using MOD-Gd-Zr-O layer

K Nakaoka[1], Y Takahashi[2], Y Usui[1], A Ibi[1], T Machi[1] and T Izumi[1]

[1] National Institute of Advanced Industrial Science and Technology, 1-2-1 Namiki, Tsukuba, Ibaraki, Japan
[2] SWCC Showa Cable System Co. Ltd., 4-1-1 Minami-Hashimoto, Sagamihara, Kanagawa, Japan

koichi.nakaoka@aist.go.jp

Abstract. $REBa_2Cu_3O_y$ (RE: rare-earth element, REBCO) coated conductors (CCs) have a high critical current density (J_c) under a magnetic field at the liquid nitrogen temperature and a potential performance for high-magnetic field applications. In order to prepare a biaxially highly oriented REBCO, a highly-textured buffer layer is required. The ion-beam assisted deposition (IBAD) technique is one of the promising techniques for preparation of the highly-textured layer even on the non-textured metallic substrate such as Hastelloy™. However, excellent surface smoothness and uniformity are indispensable to the non-textured metallic substrate using the IBAD technique. Therefore, an inexpensive surface planarization process has been required for fabrication of REBCO CCs with low-cost. In this study, we have investigated the surface planarization using the metal-organic deposition (MOD) method for the IBAD-MgO texturing. The Gd-Zr-O (GZO) planarization layer was prepared by MOD method on roughly-polished Hastelloy™ C-276 substrate. The surface roughness values (R_a) of MOD-GZO planarization layer became smaller than that of the Hastelloy™ C-276 with increase in coating times. Minimum R_a of the MOD-GZO planarization layer achieved less than 2 nm, which is smooth enough for obtaining the highly-textured IBAD-MgO film. The in-plane grain alignment of CeO_2 film deposited on sputter-$LaMnO_3$ / IBAD-MgO / MOD-GZO / Hastelloy™ C-276 substrate by the pulsed laser deposition was 4°, and high J_c value of 2.5 MA/cm^2 (1.8 μm-*t*) was obtained for MOD-YBCO film fabricated on it. The CCs using MOD-GZO planarization layer provided the delamination strength over 70 MPa, which was compatible with conventional structure of CCs.

1. Introduction

Ion-beam assisted deposition (IBAD) technique gives us a highly-textured layer for deposition of a biaxially oriented $REBa_2Cu_3O_y$ (RE: rare-earth element, REBCO) film on the non-textured metallic substrate. IBAD-MgO textured layer is favorable for low-cost production of REBCO coated conductors (CCs) because of its high-production rate [1-3]. The thickness of IBAD-MgO layer necessary to obtain highly-textured is only ~10 nm [1,4]. The thinning of IBAD textured layer leads to shortening of processing time drastically. However, excellent surface smoothness and uniformity are required to the non-textured metallic substrate in using the IBAD technique. We have used the Hastelloy™ C-276 as the non-textured metallic substrate, and mechanical polished it to surface roughness value (R_a) of 2 nm. Since mechanical polishing is quite expensive, an inexpensive surface planarization technique has been required. A planarization process using metal-organic deposition

(MOD) derived Y_2O_3 was developed by J. Hanisch *et al.* [5] and L. Civale *et al.* [6], which could help reducing the mechanical polishing cost. In this study, we investigated the surface planarization by the MOD planarization layer on the Hastelloy^{TM} C-276 substrate tape for the IBAD-MgO texturing. We selected Gd-Zr-O (GZO) as the material of the MOD planarization layer, because of the conventional architecture of the buffer layer is pulsed laser deposition (PLD)-CeO_2 / sputter-$LaMnO_3$ / IBAD-MgO / ion beam sputter deposition (IBS)-GZO / Hastelloy^{TM} C-276. In the conventional architecture, IBS-GZO acts as a diffusion barrier layer. Therefore, MOD-GZO layer is expected to play a role as the planarization and diffusion barrier layer.

2. Experimental

MOD-GZO starting solution was prepared by dissolving Gd and Zr salts with a 1: 1 (mole ratio) in the organic solvent and controlled to have a concentration of 0.15 M. This starting solution was dip-coated at withdrawal rate of 10 m/h on the roughly (electrically) polished Hastelloy^{TM} C-276 substrates with R_a = 3.5 nm. The coated films were heat-treated at 500, 550, 600, and 650℃. The once-coated MOD-GZO film thickness was 20 nm after heat-treatment. To obtain desired thickness of MOD-GZO layers, we repeated dip-coating and heat-treatment steps several times. For comparison, 110 nm thick IBS-GZO layer was prepared on the fine-polished Hastelloy^{TM} C-276 substrate with R_a = 2 nm. The IBAD-MgO layer was deposited on the MOD-GZO / roughly-polished Hastelloy^{TM} C-276 and IBS-GZO / fine polished Hastelloy^{TM} C-276. Then, $LaMnO_3$ and CeO_2 layers were deposited on the IBAD-MgO layers by RF sputtering and PLD, respectively. Superconducting YBCO layer was fabricated by the trifluoroacetate (TFA)-MOD method [7].

The surface roughness was evaluated by atomic force microscopy. Critical current (I_c) was measured at 77 K under self-field by the DC 4-probe transport technique with a criterion of 1 μV/cm. Delamination strength of CCs was measured by stud pull method using a 2.7 mm diameter stud pin.

3. Results and Discussion

Figure 1 shows the R_a of the MOD-GZO layers with multiple coating and heat-treatment at 500℃ on the roughly-polished Hastelloy^{TM} C-276 substrates with R_a = 3.5 nm. R_a of the MOD-GZO layers decreased with increasing the number of coatings. The decrease of R_a values was in good agreement with the calculated line (indicated with a dotted line in figure 1) obtained from the following equation [6];

$$R_a = R_0 \times (1 - \frac{t}{t_0})^n \qquad (1)$$

where R_0 is starting surface roughness (3.5 nm), t/t_0 is shrinkage of the coated film (0.1), and n is the number of coatings. The value of shrinkage obtained in this work (t/t_0 = 0.1) is significantly smaller than the value of 0.85 reported by L. Civale *et al.* [6]. This might be caused by using the low-concentration MOD-GZO starting solution for obtaining high coating rate and thin film thickness by the dip-coating method. R_a values of MOD-GZO layer on roughly-polished Hastelloy^{TM} C-276 substrate approach 2 nm by 4–5 time coatings, which value is almost same as that of fine-polished Hastelloy^{TM} C-276 substrate. PLD-CeO_2 texturing using the MOD-GZO planarization layer was characterize by x-ray diffraction. A PLD-CeO_2 / sputter-$LaMnO_3$ / IBAD-MgO layer was deposited on the MOD-GZO planarization layer. The film thickness of

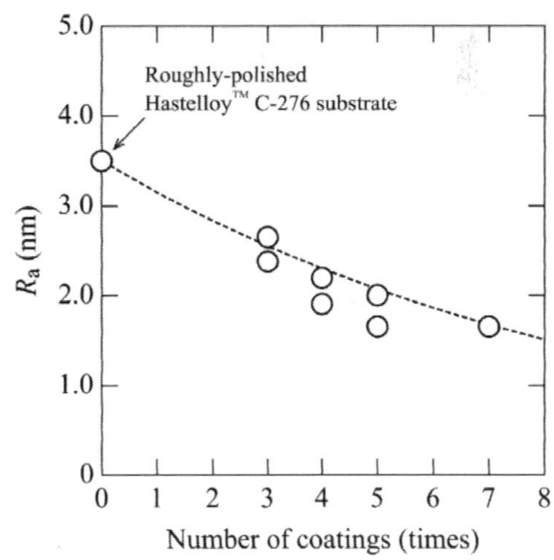

Figure 1. Relationship between R_a values and the number of coatings for the MOD-GZO planarization layer deposited on the roughly-polished Hastelloy^{TM} C-276 substrate tape with R_a = 3.5 nm.

MgO, $LaMnO_3$, and CeO_2 layers were 5, 10, and 400 nm, respectively.

Figure 2 shows dependence of the number of coatings for the MOD-GZO planarization layers on $\Delta\varphi$ value of PLD-CeO_2 films. The $\Delta\varphi$ value of PLD-CeO_2 films on MOD-GZO planarization layers was independent of the numbers of coatings over 3 times. The $\Delta\varphi$ value of PLD-CeO_2 films deposited on sputter-$LaMnO_3$ / IBAD-MgO / MOD-GZO / roughly-polished Hastelloy™ C-276 substrate reached 4° , which was as same as that on the conventional PLD-CeO_2 / sputter-$LaMnO_3$ / IBAD-MgO / IBS-GZO / fine-polished Hastelloy™ C-276 substrate. These results indicate that the requirement of R_a of the surface on which the IBAD-MgO layer is deposited for obtaining good texturing of the CeO_2 layer is about 2.5 nm and the in-plane alignment of the CeO_2 become less sensitive when R_a is smaller than 2.5 nm. According these results, it was confirmed that MOD-GZO layer effectively works of the planarization layer for the IBAD-MgO texturing.

YBCO film of 1.8 μm thickness was deposited on the PLD-CeO_2 / sputter-$LaMnO_3$ / IBAD-MgO / MOD-GZO / roughly-polished Hastelloy™ C-276 and PLD-CeO_2 / sputter-$LaMnO_3$ / IBAD-MgO / IBS-GZO / fine-polished Hastelloy™ C-276 substrates by the TFA-MOD method. Highly in-plane textured YBCO film was obtained on the PLD-CeO_2 / sputter-$LaMnO_3$ / IBAD-MgO / MOD-GZO / roughly-polished Hastelloy™ C-276 substrate as well as that on the PLD-CeO_2 / sputter-$LaMnO_3$ / IBAD-MgO / IBS-GZO / fine-polished Hastelloy™ C-276 substrate. The I_c and critical current density (J_c) values of the YBCO film on the PLD-CeO_2 / sputter-$LaMnO_3$ / IBAD-MgO / MOD-GZO / roughly-polished Hastelloy™ C-276 substrate were 446 A/cm-w and 2.5 MA/cm^2, which were almost the same as those on the conventional PLD-CeO_2 / sputter-$LaMnO_3$ / IBAD-MgO / IBS-GZO / fine-polished Hastelloy™ C-276 substrate (I_c = 470 A/cm-w, J_c = 2.6 MA/cm^2). These results show, the MOD-GZO planarization layer also works as a diffusion barrier layer effectively. It was confirmed that MOD-GZO layer could play roles both of planarization and diffusion barrier.

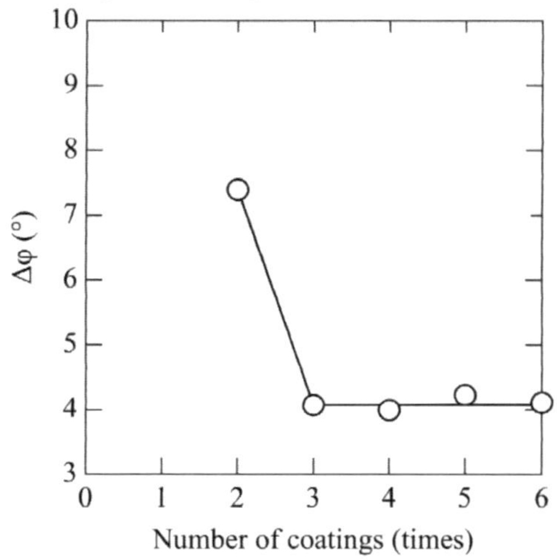

Figure 2. Relationship between $\Delta\varphi$ value of CeO_2 and the number of coatings for MOD-GZO planarization layer.

Figure 3 shows the relationship between delamination strength of CCs deposited on PLD-CeO_2 / sputter-$LaMnO_3$ / IBAD-MgO / MOD-GZO / roughly-polished Hastelloy™ C-276 substrate and heat-treatment temperature for MOD-GZO planarization layers. The CCs with heat-treatment temperature of 500℃ had low delamination strength below 43 MPa, where the delamination might start in MOD-GZO planarization layer due to the SEM-EDX observation. The delamination strength was increased with increasing heat-treatment temperature. The CCs with the heat-treatment temperature of over 550℃ had high delamination strength over 70 MPa, which values were no difference in the CCs deposited on conventional PLD-CeO_2 / sputter-$LaMnO_3$ / IBAD-MgO / IBS-GZO / Hastelloy™ C-276 substrate. This delamination was started in the superconducting layer. In addition, there

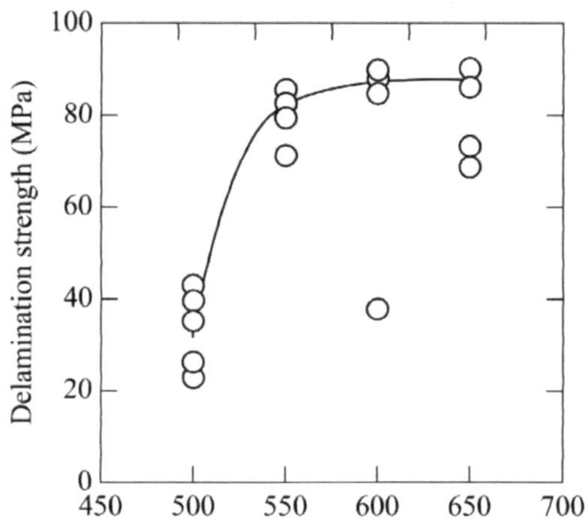

Figure 3. Relationship between delamination strength of CCs and heat-treatment temperature for MOD-GZO planarization layer.

was no difference in the J_c values of these CCs using MOD-GZO planarization layers. This result is believed that there is no problem in practical application of CCs deposited on PLD-CeO_2 / sputter-$LaMnO_3$ / IBAD-MgO / MOD-GZO / roughly-polished Hastelloy^{TM} C-276 substrate.

4. Conclusion

We investigated the surface planarization process using the MOD-GZO. Multiple dip-coated and heat treated MOD-GZO films were planarized with increasing coating times on the roughly-polished Hastelloy^{TM} substrates. R_a values were lower than 2 nm over 4-5 times coating, which approached that of the fine-polished Hastelloy^{TM} C-276 substrate. Highly in-plane textured PLD-CeO_2 / sputter-$LaMnO_3$ / IBAD-MgO layer was obtained on the MOD-GZO planarized Hastelloy^{TM} C-276 substrate. High J_c YBCO film of 2.5 MA/cm^2 was obtained by the TFA-MOD process on the PLD-CeO_2 / sputter-$LaMnO_3$ / IBAD-MgO / MOD-GZO / Hastelloy^{TM} C-276 substrate. MOD-GZO planarization layer was proved to work as a planarization layer and diffusion barrier layer on metallic substrate. High delamination strength was exhibited by CCs using MOD-GZO planarization layer with high-temperature heat-treatment (>550℃). MOD-GZO planarization layer has advantages of can reducing polishing cost of the substrate and lower apparatus cost. We will apply the MOD-GZO planarization layer for the long tape.

Acknowledgments

This study was supported by the Ministry of Economy, Trade and Industry (METI), and the New Energy and Industrial Technology Development Organization (NEDO).

References

[1] Yamada Y, Miyata S, Yoshizumi M, Fukushima H, Ibi A, Izumi T and Shiohara Y 2009 *Physica C* **469** 1298
[2] Xiong X, Lenseth K P, Reeves J L, Rar A, Qiao Y, Schmidt R M, Chen Y. Li Y, Xie Y and Selvamancikam V 2007 *IEEE Trans. Appl. Supercond.* **17** 3375
[3] Yoshizumi M, Miyata S, Ibi A, Fukushima H, Yamada Y, Izumi T and Shiohara Y 2009 *Physica C* **469** 1361
[4] Arendt P N and Foltyn S R 2004 *MRS. Bull.* **29** 543
[5] Hanich J, Matias V, Rowley J, Steenwelle R, Gibbons B J, Crilli F and Ashworth S, 2007 *IEEE Trans. Appl. Supercond.* **17** 3577
[6] Civale L and Xie Y Y 2009 *Superconductivity for Electric Power Systems Annual Peer Review* Alexandria August 4-6
[7] Nakaoka K, Matsuda J, Kitoh Y, Goto T, Yamada Y, Izumi T and Shiohara Y 2007 *Physica C* **469** 519

Angular dependence of J_c in YBCO films with *c*-axis correlated nano-rods and in-plane-distributed nano-particles

T Sueyoshi, M Suenaga, T Furusawa, S Matsuyuki and T Fujiyoshi

Department of Computer Science and Electrical Engineering, Kumamoto University, 2-39-1 Kurokami, Chuo-ku, Kumamoto 860-8555, Japan

E-mail: tetsu@cs.kumamoto-u.ac.jp

Abstract. The quasi-multilayered films consisting of $YBa_2Cu_3O_y$ layers with $BaSnO_3$ nano-rods and the pseudo layers of in-plane-distributed $BaSnO_3$ nano-particles were fabricated using a PLD method, in order to clarify the pinning landscape simultaneously improving the critical current densities, J_cs, both at $B \parallel c$ and at $B \parallel ab$. The insertion of the pseudo layers into $YBa_2Cu_3O_y$ films contributes to the enhancement of J_c at $B \parallel ab$ but it is a tendency to reduce the J_c at $B \parallel c$. When the density of the nano-particles per layer is decreased, by contrast, the J_c at $B \parallel ab$ can be enlarged without the reduction of the *c*-axis peak. This is attributed to the fragmentation of the channel for flux creep motion through the pseudo layers. Furthermore, the J_c in tilted magnetic fields off the *c*-axis enhances as well as *c*-axis peak in a high magnetic field, when $BaSnO_3$ nano-particles are increased.

1. Introduction

To improve a critical current density J_c in magnetic field for $REBa_2Cu_3O_y$ (REBCO, RE: rare earth elements) based coated conductors, an introduction of additional crystalline defects and impurities into REBCO films is a promising way, where nano-sized defects work as pinning centers (PCs) and effectively arrest flux lines penetrating in superconducting materials [1]. Nano-rods, which are formed through self-assembled stacks of $BaMO_3$ (M = Zr, Sn, Hf, etc.) along the *c*-axis into REBCO films, belongs to one-dimensional (1D) PCs and effectively interrupt the motion of flux lines when the magnetic field is applied parallel to the *c*-axis [2, 3]. Planar defects consisting of continuous impurity layers along the *ab*-plane are two-dimensional (2D) PCs and act as *ab*-correlated PCs, leading to a weak field dependence of J_c at $B \parallel ab$ [4]. Nano-particles classified as three-dimensional (3D) PCs, on the other hand, have the morphology with no correlated orientation for flux pinning, leading to the isotropic pinning force against any orientations of magnetic field [5].

In recent years, further improvement of flux pinning including a reduction of the anisotropy of J_c has been attempted by using a combination of different types of PCs, which is referred to as hybrid flux pinning [6, 7]. The hybrid flux pinning by using nano-rods and nano-particles enhances not only the J_c peak centered at $B \parallel c$ but also the J_c in a wide angular range of magnetic field [7], where nano-particles play an important role in the prevention of the sliding motion of flux lines for magnetic fields tilted off the direction of nano-rods and in the suppression of the motion of double kinks of flux lines due to thermal fluctuation [8]. The combination of 1D and 3D PCs, however, does not sufficiently reduce the anisotropy of J_c, where the *c*- axis J_c-peak is larger than the J_c peak at $B \parallel ab$ in most cases. The combination of 1D and 2D PCs consisting of *c*-axis-aligned nano-rods and *ab*-plane-aligned planar

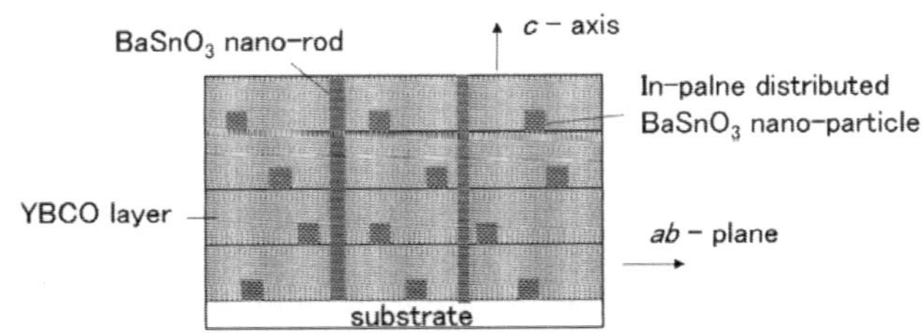

Figure 1. Schematic illustration of quasi-multilayered films consisting of YBCO layers with $BaSnO_3$ nano-rods and the pseudo layers of in-plane-distributed $BaSnO_3$ nano-particles.

nanostructures, on the other hand, can contribute to the improvement of J_c both at $B \parallel c$ and at $B \parallel ab$ [6, 9]. The insertion of planar defects along the *ab*-plane, however, tends to reduce the *c*-axis peak in the J_c anisotropy caused by nano-rods, since flux lines along the *c*-axis easily move along the *ab*-aligned nanostructure layers for $B \parallel c$ [4] or the structure of flux lines changes from conventional Abrikosov vortices to pan-cake vortices [10].

In this work, we investigated the hybrid flux pinning structure which improves the J_c in all magnetic field orientations, i.e. at $B \parallel c$, at $B \parallel ab$, and in the intermediate-angle region, by preparing the quasi-multilayered films consisting of YBCO layers with $BaSnO_3$ nano-rods and the pseudo layers of in-plane-distributed $BaSnO_3$ nano-particles using a pulsed laser deposition (PLD) method. Figure 1 schematically illustrates distribution of the nano-rods and nano-particles in the multilayered films. The nano-rods consisting of $BaSnO_3$ contribute to the enhancement of J_c at $B \parallel c$, while the in-plane-distributed $BaSnO_3$ nano-particles act as PCs with features of both 2D and 3D PCs: the nano-particles aligned along the *ab*-plane can be expected to enhance the J_c at $B \parallel ab$ and in the intermediate-angle region, but without reducing the *c*-axis peak.

2. Experimental details

The quasi-multilayered films consisting of YBCO layers with $BaSnO_3$ nano-rods and the pseudo layers of in-plane-distributed $BaSnO_3$ nano-particles were fabricated on (100) $SrTiO_3$ substrates by a PLD technique alternating a 2.0 vol.% $BaSnO_3$ doped YBCO target and a $BaSnO_3$ target. The repetition rate of the laser was 2 Hz for the deposition of YBCO layers with $BaSnO_3$, while the pseudo layers of $BaSnO_3$ were deposited at 1 Hz. The growth of $BaSnO_3$ nano-rods in the YBCO layer would be continuous along the film-thickness direction even in the multilayering process [11]. The deposition of $BaSnO_3$ by several laser pulses, on the other hand, provides not a $BaSnO_3$ layer but in-plane-aligned $BaSnO_3$ nano-particles [12]. During the deposition, the substrate was heated at 760 °C and the oxygen partial pressure was 300 mTorr. After the deposition, the films were cooled down naturally from 760 °C to 500 °C and from 200 °C to room temperature in 600 Torr oxygen, whereas the cooling rate was controlled at 10 °C/min from 500 °C to 200 °C. A series of the quasi-multilayered films in this work is listed in Table 1. We refer to a sample as B(*m*, *n*), where *m* and *n* denote the number of laser pulses on the $BaSnO_3$ target and the total number of $BaSnO_3$ / YBCO+$BaSnO_3$ bilayers, respectively. A layer of YBCO+$BaSnO_3$ was always deposited on the top of the multilayered films and the total number of laser pulses on the $BaSnO_3$ doped YBCO target was kept constant at 3600 pulses, resulting in the film thickness of about 300 nm.

Table 1 Samples in this work.

Sample	Number of laser pulses on $BaSnO_3$ target, *m*	Number of bilayers, *n*	T_c [K]
B(3, 5)	3	5	89.2
B(6, 5)	6	5	89.8
B(3, 10)	3	10	89.7
B(6, 10)	6	10	90.0
BSO2	-	-	89.4

Simultaneously, a 2.0 Vol.% $BaSnO_3$ doped YBCO film was prepared as a reference sample, which are referred to as BSO2.

Transport properties were measured using a four-probe method, where the films were patterned into bridge geometry with about 40 μm width and 1 mm length using a standard photolithography. The value of J_c was defined by electric field criterion of 1 μV/cm. The angular dependence of J_c was evaluated as a function of the angle θ between the magnetic field and the *c*-axis, where the magnetic field was always applied perpendicular to the direction of the current (the maximum Lorentz force configuration).

3. Results and discussion

Firstly, we compare the angular behaviour of J_c between the multi-layered films with different *m*, in order to clarify the thinning-out effect of the in-plane-distributed nano-particles on the *c*-axis peak formed by the nano-rods. Figure 2(a) shows the J_c as a function of magnetic field orientation at 77 K and 1 T for the 5-bilayer films with different *m*, compared with BSO2. Two peaks can be observed in the angular dependences of J_c for all the samples. One of the peaks emerges at $\theta = 0°$, which arises from the *c*-axis correlated pinning due to the nano-rods. Another is the steep peak at $\theta = 90°$, which is generally attributed to the stacking faults or to the intrinsic pinning due to the CuO_2 planes. The J_c peak at $\theta = 90°$ is also affected by the structure of multilayered films [5] and is more evolved with increasing *m*, as shown in Figure 2(a). This is because the $BaSnO_3$ nano-particles aligned along the in-plane direction of the film effectively work as correlated PCs along the *ab*-plane. It is noteworthy that the multilayared films with larger *m* more enlarge the J_c in the intermediate angle region between $B \parallel c$ and $B \parallel ab$. This suggests that the in-plane-distributed $BaSnO_3$ do not grow into layers but into particles: each nano-particles is isolated enough to work as 3D-PCs.

The insertion of the impurity layers aligned along the *ab*-plane, on the other hand, tends to weaken the *c*-axis peak [10, 13]. For B(6, 5), the J_c peak at $\theta = 0°$ is lower than that for BSO2. This is because the in-plane-distributed $BaSnO_3$ nano-particles for B(6, 5) easily induce the flux creep motion of the interstitial flux lines between nano-particles [4], resulting in the negative effect for the flux pinning at $B \parallel c$ (see figure 3). Note that the dimensionality of the YBCO system, which significantly affects the flux pinning by linear defects [10], would not be modified by the insertion of the impurity layers in this work, since the thickness of the bilayer is about 30-60 nm and the impurity deposition did not form complete non-superconducting-layers.

When *m* is decreased, by contrast, no deterioration in the *c*-axis peak is identified for B(3,5), resulting in the upward shift in the J_c over the entire magnetic field directions compared to BSO2. The influence of *m* on the J_c becomes more evident when the number of bilayers is increased to $n = 10$, as shown in Figure 4. These results suggest that the thinning of in-plane-distributed nano-particles effectively suppresses the flux creep motion between the nano-particles, resulting in the improvement of the J_c at

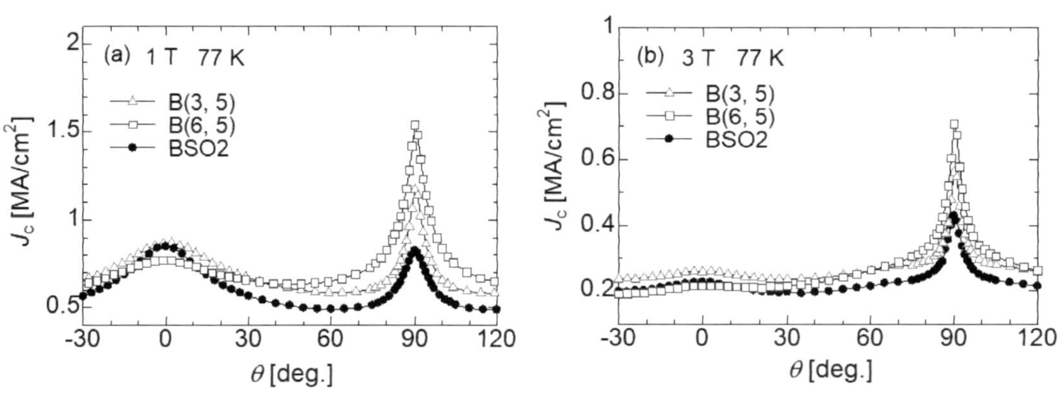

Figure 2. Angular dependence of J_c for multilayered films with 5 bilayers and different laser pulses on $BaSnO_3$ target at 77 K for (a) 1 T and (b) 3 T.

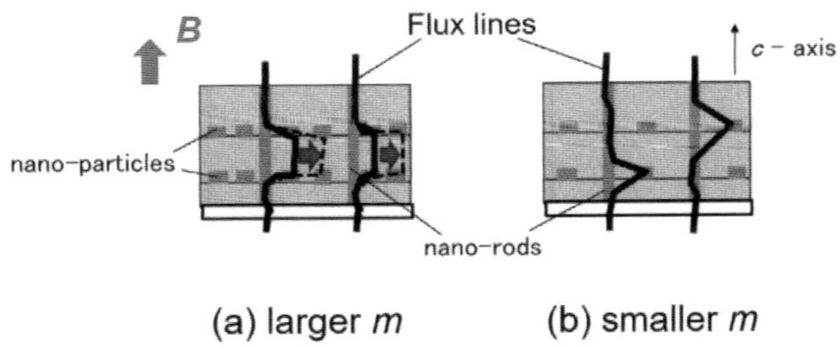

Figure 3. Schematic illustrations of possible flux pinning states in the presence of nano-rods and in-plane-distributed nano-particles. As the number of m is larger, the density of the nano-particles per layer increases, which easily induces the flux creep motion between the nano-particles as shown by blue arrows.

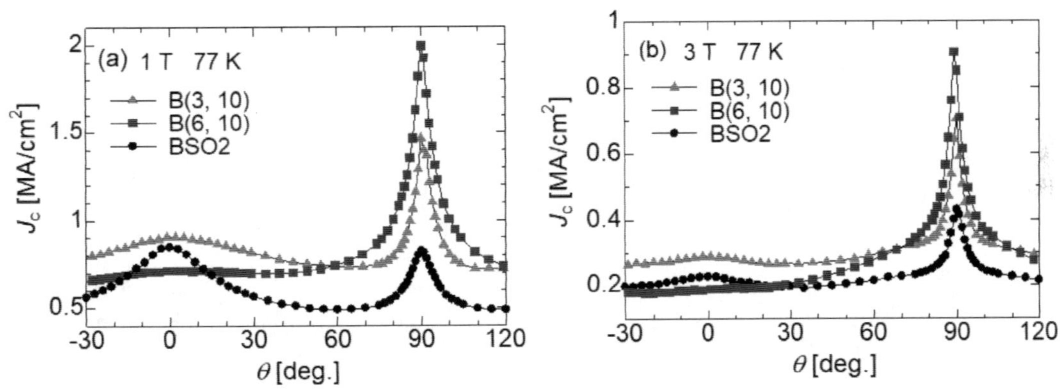

Figure 4. Angular dependence of J_c for multilayered films with 10 bilayers and different laser pulses on $BaSnO_3$ target at 77 K for (a) 1 T and (b) 3 T.

$B \parallel ab$ without a reduction of the c-axis peak. In increasing the magnetic field, the influence of the nano-particle pinning becomes apparent even on the J_c at $B \parallel c$: the J_c around $\theta = 0°$ for the multilayered film with $m = 3$ exceeds that for BSO2. When flux lines outnumber the nano-rods in high magnetic field, interstitial flux lines existing between the nano-rods are effectively trapped by the in-plane-distributed $BaSnO_3$ nano-rods [13].

To explore further optimization for the flux pinning effect of the multilayered films, we investigated the influence of the bilayer number n on the angular behaviour of J_c for the multilayered films with $m = 3$, as shown in figure 5. The multilayered films with $m = 3$ show upward shift in J_c compared to BSO2 in overall orientations of magnetic field, although the c-axis peak for the multilayered films is equal to or slightly higher than that of BSO2 for 1 T. When the number of bilayer n increases, the upward shift in J_c becomes more remarkable at $B \parallel ab$ and in the intermediate angle region. The increase in J_c at $B \parallel ab$ is due to a large number of the bilayers which work as pseudo planar PCs. In the intermediate angle region, on the other hand, the nano-particles play a role in 3D PCs [7, 13], which can pin the segments of flux lines released from the nano-rods along the c-axis. Therefore, a large number of bilayers is effective for the improvement of J_c both in high magnetic field and in the intermediate angle region.

In Figure 6, we compare the angular dependence of J_c between the multilayered films with $m \times n = 30$, which means the same amount of nano-particles regardless of the different multilayer structures. The J_c in a wide range of magnetic field orientations can be improved by thinning out the nano-particles on each layer and by increasing the number of bilayers, even if the amount of nano-particles is the same. In other words, tuning of spatial distribution of nano-particles would be one of the keys to further enhance the overall J_c for the hybrid flux pinning. Note that the J_c peaks at $B \parallel ab$ for B(3, 10) and

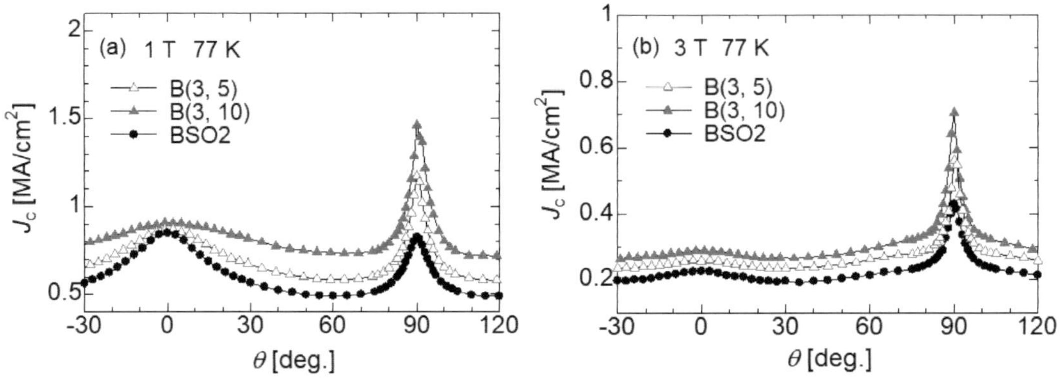

Figure 5. Angular dependence of J_c for multilayered films with m = 3 and different numbers of bilayers n at 77 K for (a) 1 T and (b) 3 T.

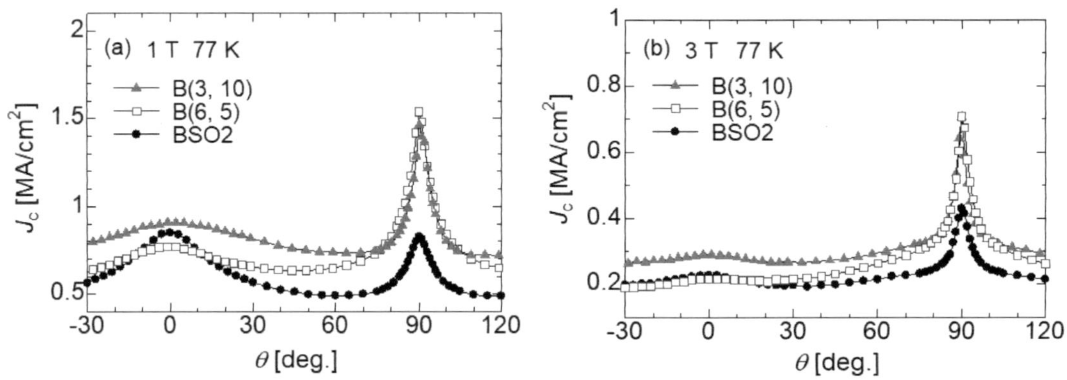

Figure 6. Angular dependence of J_c for multilayered films with $m \times n = 30$ at 77 K for (a) 1 T and (b) 3 T.

B(6, 5) nearly overlap each other for both 1 T and 3 T. The in-plane-distributed nano-particles contribute to effective flux pinning at $B \parallel ab$. The same peak behaviour of J_c, however, is not always provided for $B \parallel ab$ by the different multi-layered structures even if the amount of nano-particles is the same [13].

4. Conclusions

In order to clarify the pinning structure simultaneously attaining high J_c both at $B \parallel c$ and at $B \parallel ab$, we investigated the influence of the structure of the multilayered YBCO films with $BaSnO_3$ nano-rods and in-plane-distributed $BaSnO_3$ nano-particles on the angular behaviour of J_c, where the number of nano-particles per layer and the number of bilayers were controlled. The in-plane-distributed nano-particles can contribute not only to improving the J_c at $B \parallel ab$ but also to keeping the c-axis peak due to the nano-rods by thinning out the nano-particles per layer. In addition, the J_c can be further evolved in high magnetic field and in a wide range of magnetic field orientation by increasing the number of bilayers consisting of YBCO layer and the nano-particles. The formation of the in-plane-distributed nano-particles for the hybrid flux pinning structure is one of the important factors for the improvement of J_c in all magnetic field orientations

5. Acknowledgement

This work was supported by KAKENHI (16K06269) from the Japan Society for the Promotion of Science.

References

[1] Matsumoto K and Mele P, 2010 *Supercond. Sci. Technol.* **23** 014001
[2] Mele P, Matsumoto K, Horide T, Ichinose A, Mukaida M, Yoshida Y, Horii S and Kita R, 2008 *Supercond. Sci. Technol.* **21** 032002
[3] Tobita H, Notoh K, Higashikawa K, Inoue M, Kiss T, Kato T, Hirayama T, Yoshizumi M, Izumi T and Shiohara Y, 2012 *Supercond. Sci. Technol.* **25** 062002
[4] Yang H, Wang H, Maiorov B, Lee J, Talbayev D, Hinton M J, Feldmann D M, MacManus-Driscoll J L, Taylor A J, Civale L, Lemberger T R and Jia Q X, 2009 *J. Appl. Phys.* **106** 093914
[5] Gapud A A, Kumar D, Viswanathan S K, Cantoni C, Varela M, Abiade J, Pennycook S J and Christen D K, 2005 *Supercond. Sci. Technol.* **18** 1502
[6] Wu J Z, Shi J J, Baca1 F J, Emergo R, Wilt J and Haugan T J, 2015 *Supercond. Sci. Technol.* **28** 125009
[7] Horide T, Kawamura T, Matsumoto K, Ichinose A, Yoshizumi M, Izumi T and Shiohara Y, 2013 Supercond. Sci. Technol. **26** 075019
[8] Maiorov B, Baily S A, Zhou H, Ugurlu O, Kennison J A, Dowden P C, Holesinger T G, Foltyn S R and Civale L, 2009 *Nat. Mater.* **8** 398
[9] Ishikawa K, Ichino Y, Yoshida Y, Yoshizumi M, Izumi T and Kato T, 2015 *J. Cryo. Super. Soc. Jpn.* **50** 218 (in Japanese)
[10] Holzapfel B, Kreiselmeyer G, Kraus M, Bouffard S, Klaumünzer S, Schultz L and Saemann-Ischenko G, 1993 *Phys. Rev. B* **48** 600
[11] Tsuruta A, Yoshida Y, Ichino Y, Ichinose A, Matsumoto K and Awaji S, 2014 *Supercond. Sci. Technol.* **27** 065001
[12] Huhtinen H, Schlesier K and Paturi P, 2009 *Supercond. Sci. Technol.* **22** 075019
[13] Sueyoshi T, Kotaki T, Uraguchi Y, Suenaga M, Makihara T, Fujiyoshi T and Ishikawa N, 2016 *Physica C* **530** 72

Electrical Characteristics of Twisted Soldered-Stacked-Square (3S) Wires Consisting of HTS Narrow Tapes with 1mm Width

Y S Luo[1], M Y Yin[2], Y K Zhou[3], M Song[1], N N Hu[1], Z Y Li[3], Z Hong[3], Z Jin[3]

[1] Institute of superconducting technology, Electric Power Research Institute of Guangdong Power Grid Co.,Ltd., Shui Jun Gang No. 8 Dongfengdong Road, Yuexiu District, Guangzhou, 510080, China.
[2] Sino-Korean School of Multimedia Design, Shanghai University of Engineering Science, 333 Long Teng Road, Songjiang District, Shanghai 201620, China.
[3] Department of Electrical Engineering, Shanghai Jiao Tong University, No.800 Dongchuan Road, Minhang District, Shanghai, 200240, China.

E-mail: lys1017@126.com

Abstract**—**To reduce the effect of external magnetic field, we proposed a novel wire with a square cross section of 1 mm × 1 mm which is named Soldered-Stacked-Square (3S) HTS wire. 3S HTS wire has great advantages in HTS applications due to its small square dimension, but it is inevitable to endure twisting. To study the twist structure of 3S HTS wire, a series of twisted 3S HTS wire samples are prepared, and we evaluate the typical electrical characteristics: twist pitch, critical current, self-field AC loss and so on. It is shown that the critical current of twisted 3S wire is constant up to a twist pitch of 80 mm, and the minimum twist pitch is approximately 50 mm. Measured AC losses are independent of frequency and are in good agreement with the thin strip theoretical values. Specially, measured AC losses are a little smaller than the thin strip theoretical value when peak transport current is reaching critical current. Besides, the measured AC losses of twisted 3S wire and 3S wire are almost identical.

1. Introduction

External magnetic field would influence the performance of HTS wires including critical current, AC loss and so on, especially when the magnetic field is perpendicular to the HTS wire's surface [1]. In order to reduce the effect of external magnetic field, many studies have been done, which mainly focus on stack or twist multifilament structures: tapes stacked and transposed according to the Roebel Assembled Coated Conductors technique [2], tapes stacked and twisted along the longitudinal direction [3], round strand composed of coated conductor cable [4] and so on. However, these studies are usually based on HTS tapes with more than 4 mm width.

Different from above studies, we proposed a novel wire with a square cross section of 1 mm × 1 mm which is named Soldered-Stacked-Square (3S) HTS wire [5]. Due to the low width-to-thickness ratio and small square dimension, 3S HTS wire has great advantages in cabling, but it is inevitable to endure twisting in actual applications. Therefore, it is necessary to have further study on the twist structure of 3S HTS wire.

In this paper, we investigate the basic electrical characteristics for the twisted 3S HTS wires: twist pitch, critical current, self-field AC loss and so on. We firstly fabricate several 3S HTS wire samples, (2+7b)-wire samples through narrowing, soldering and stacking process, and then twist the samples with different twist pitches. The entire fabrication process will be introduced in detail. Then we study the critical current characteristics versus twist pitch of the (2+7b)-T-wire. Furthermore, we investigate the self-field AC loss characteristics of the (2+7b)-T-wire, including dependence on frequency and twist pitch.

2. Experiment

2.1 Fabrication process of twisted 3S HTS wire

Fig. 1 shows the schematic view of the fabrication process of the twisted Soldered-Stacked-Square (3S) HTS wire. As depicted in Fig. 1, an original second generation (2G) HTS tape with 4 mm width is incised into four narrow tapes with 1 mm width. This process is accomplished by a modified commercial cutting device which is widely used to flat copper plate cutting. Then several 2G HTS narrow tapes and brass narrow tapes are immersed together into a solder bath and undergone stacking and soldering process simultaneously ensuring that the cross section is square. Finally, the 3S HTS wire is twisted by a torsion device. In this work, we used the original 2G wires with 4 mm width provided from SuNAM company from Korea.

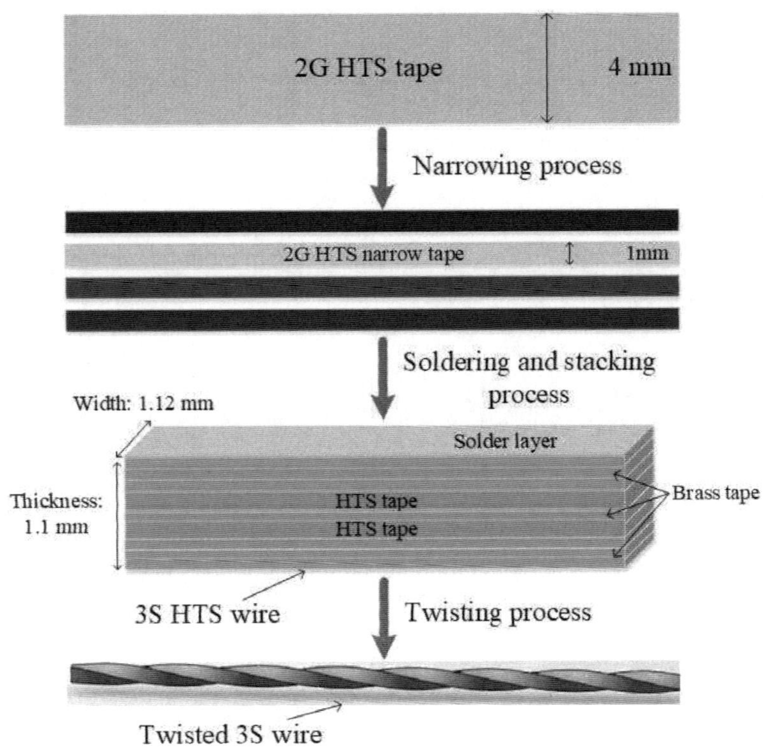

Fig. 1. Fabrication process of twisted 3S HTS wire.

In this study the prepared 3S HTS wire samples consist of two layers of HTS narrow tapes with 1 mm width and seven layers of brass narrow tapes (7b). This kind of 3S HTS wire is named (2+7b)-wire. Here the thicknesses of HTS and brass narrows tapes are 0.15 mm and 0.10 mm, respectively. In the fabrication of (2+7b)-wire samples, two pieces (piece 1 and 2) of HTS narrow wires are used, whose uniformity of critical current along length direction is shown good, as shown in Fig. 2. The main specifications of the HTS narrow tape and (2+7b)-wire are listed in Table I.

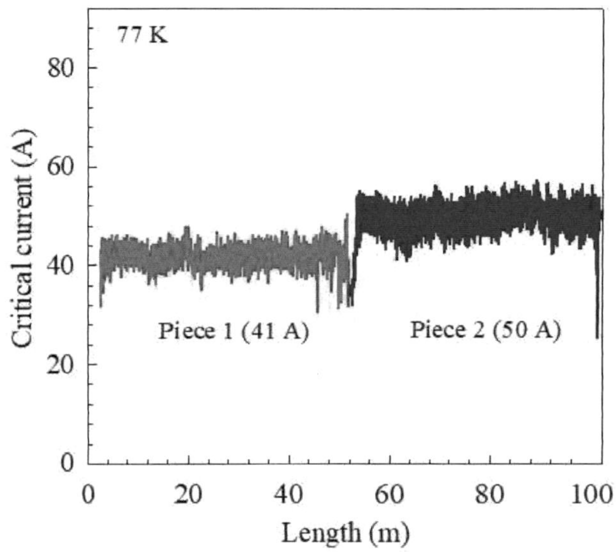

Fig. 2. Uniformity along length direction of critical current distribution for HTS narrow tapes used in fabrication of (2+7b)-wire samples.

TABLE I
SPECIFICATIONS OF 3S HTS WIRES

HTS narrow tape	
Width and thickness	1.0 mm × 0.15 mm
Thickness of superconducting layer	~1.3 μm
Thickness of silver layers	~1 μm
Thickness of substrate layer (including buffer)	~108 μm
Thickness of copper plating layer, each side	~20 μm
3S (2+7b)-wire	
Width and thickness of 3S wire	1.12 mm × 1.10 mm
Width and thickness of brass tape	1.0 mm × 0.1 mm
Number of HTS narrow tapes	2
Number of brass narrow tapes	7

2.2 Measurement of twisted 3S HTS wire

In this study, the critical current of the HTS narrow tapes were firstly measured by magnetic measurement method through hall sensors and calibrated by the value of critical current for the short length sample of piece 1 and piece 2 measured from a standard electrical four-point method, and the result is shown in Fig. 2.

The AC loss is measured by pickup coil method. The major equipment of the experiment setup includes waveform generator, AC current transducer, oscilloscope, amplifier and pickup coil. AC loss could be worked out by integrating product of transport current and the voltage picked up after compensating the inductive component using pickup coil.

Fig. 3 shows the photos of the torsion device and twisted 3S wire samples. Ends of the 3S HTS wire sample are fixed on the plate connected to current leads. The rotation shaft can be rotated at any angle to apply a twist to the 3S HTS wire with specified twist pitch. At the bottom of the device, there are three springs connected to a moveable plate, so that we can test the twisted 3S wire samples in changeable lengths. The allowed length of 3S wire samples is 300 ~ 400 mm. The whole torsion device will be put into the LN_2 vessel after fixing the 3S HTS wire. The (2+7b)-wire through twisting process is named (2+7b)-T-wire.

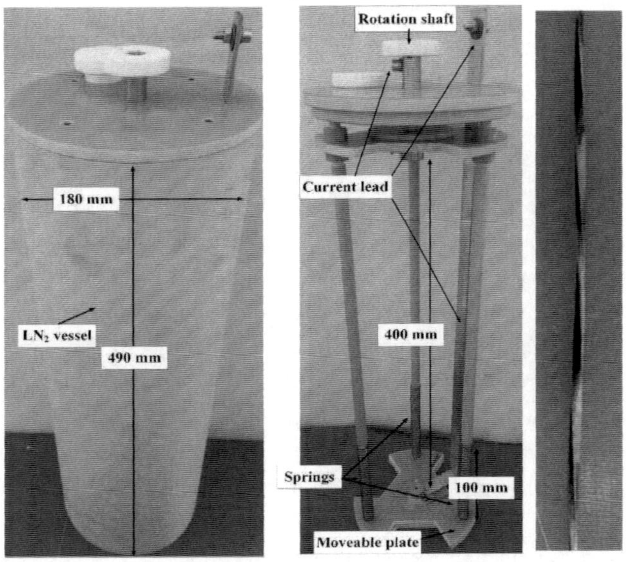

Fig. 3. Photos of the torsion device for short 3S samples and twisted 3S sample, (2+7b)-T-wire.

3. Test results of twisted 3S HTS wire

We investigate the basic characteristics for the twisted 3S HTS wire samples: twist pitch, critical current, self-field AC loss and so on. Here, all 3S wires of (2+7b)-wire are fabricated using the narrow tapes of piece 1 and 2 shown in Fig. 2. All performance tests are carried out at liquid nitrogen temperature of 77 K.

3.1 Twist pitch vs. critical current

To check the allowable twist pitch of the (2+7b)-wire in terms of critical current, we first measure the critical currents of three virgin wires without twisting process, and record their curves of voltage and current. Fig. 4 shows the measured critical currents of three (2+7b)-wire samples 1, 2, and 3. And the length between voltage taps is 220 mm. As shown in Fig. 4, the measured critical currents of samples 1, 2, and 3 are 86, 90, and 94 A, respectively. From the test results we can see the curves of voltage and current for three samples are typical ones of HTS wires.

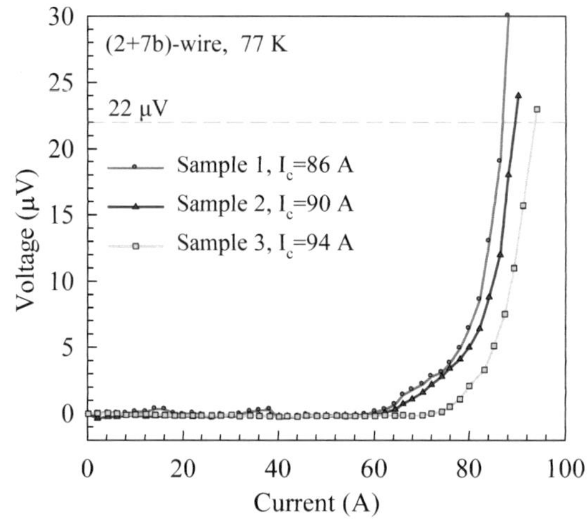

Fig. 4. Measured critical currents of three 3S HTS wire samples.

In Fig. 5, critical current measurement under various twist pitches is performed on the (2+7b)-T-wire sample 1 in Fig. 4, as a typical case. Here, we choose the reprehensive curves for four different twist pitches from whole measured data to show them more clearly. It can be seen from Fig. 5 that the curves of (2+7b)-T-wire with different twist pitches show typical superconducting characteristics. Besides, the critical current of sample 1 degrades rapidly when the twist pitch approaches 55 mm.

Fig. 6 shows the critical current versus twist pitch of (2+7b)-T-wire samples. Data shown in Fig. 6 are obtained from the three samples, whose length of voltage taps is 220 mm length. The measured critical currents after twisting process are normalized by the initial critical currents (I_{c0}), and the criterion of minimum twist pitch is defined as 95% retention of the initial critical current. As seen in Fig. 6 the critical current is constant up to a twist pitch of approximately 80 mm, and the minimum twist pitch of (2+7b)-T-wire is approximately 50 mm. In order to keep a certain margin for the following self-field AC loss tests, we set the twist pitch as 100 mm for fabricating the (2+7b)-T-wire samples.

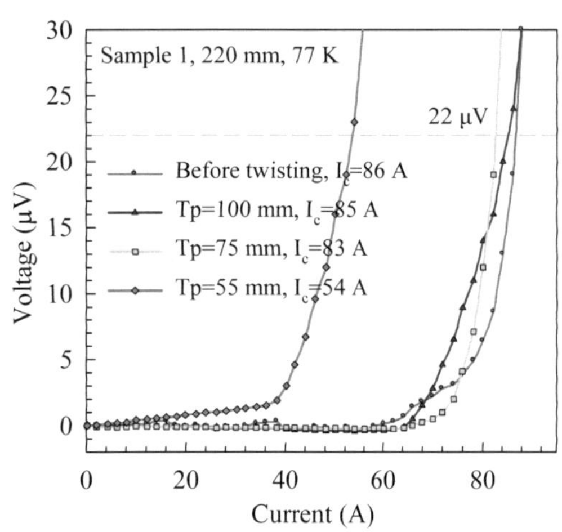

Fig. 5. Measured critical currents of (2+7b)-T-wire, sample 1, under various twist conditions.

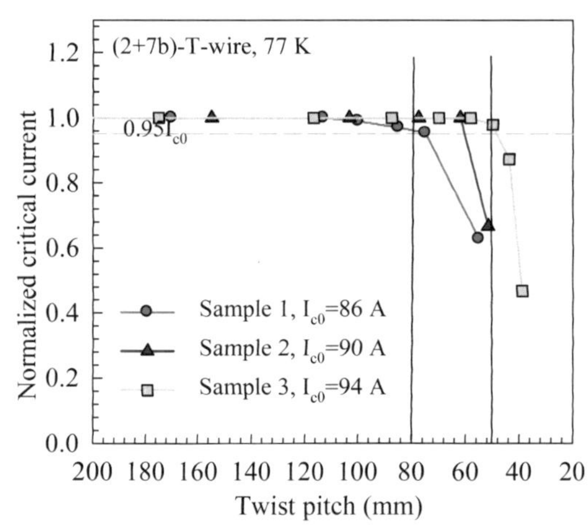

Fig. 6. Critical current vs. twist pitch of (2+7b)-T-wire samples.

3.2 Self-field AC loss characteristics

In this section, we test the basic characteristics of self-field AC loss for twisted 3S HTS wire, (2+7b)-T-wire, which is another important reference data for electrical characteristics of HTS tapes in superconducting applications. The AC loss is evaluated by integrating the product of transport current and voltage after compensating the inductive component during one period [6]. In addition, the self-field AC loss are compared with two theoretical values from the ellipse and thin strip equations of Norris model.

Fig. 7 shows the results of frequency dependence on AC loss of (2+7b)-T-wire sample 1. The (2+7b)-T-wire sample is tested with the frequency, 50, 100, and 200 Hz, respectively. As depicted in Fig. 7, the measured self-field AC loss of (2+7b)-T-wire is independent of the frequencies. This means that hysteresis loss is a major component in self-field loss and the eddy current loss generated from brass layers is so small that can be ignored. Besides, the self-field AC loss of (2+7b)-T-wire matches very well with the thin strip theoretical values of Norris model. When the transport current is low, the measured result is slightly above the thin strip theoretical values. With the increasing of transport current, measured result becomes a little smaller than the thin strip theoretical value.

Fig. 8 indicates the comparison of the self-field AC loss between (2+7b)-wire and (2+7b)-T-wire sample 1. The 3S wire sample is tested with the frequency 100 Hz. It can be seen that the measured AC losses of (2+7b)-T-wire and (2+7b)-wire are almost identical, from which we can see that the twist structure cannot affect the self-field loss of the 3S HTS wires.

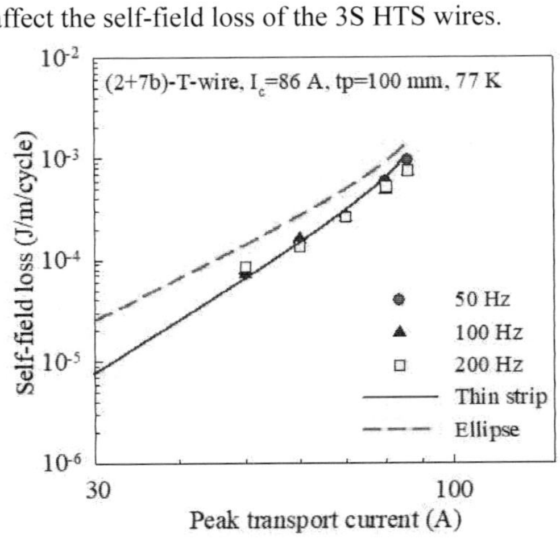

Fig. 7. Frequency dependence on AC loss of twisted 3S HTS wire, (2+7b)-T-wire.

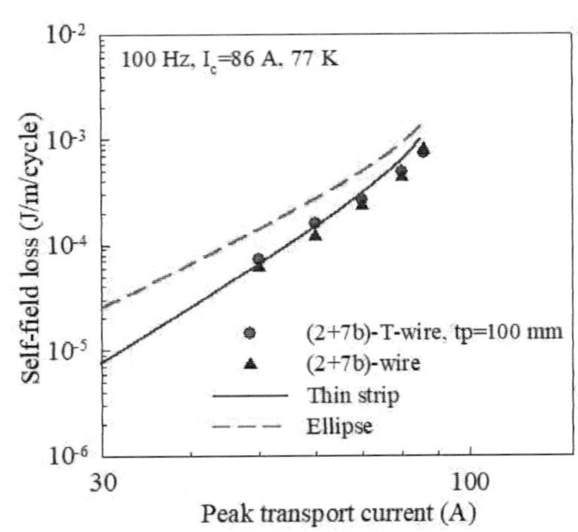

Fig. 8. Comparison of AC loss between (2+7b)-wire and (2+7b)-T-wire.

Fig. 9 shows the comparison of the self-field AC loss among three (2+7b)-T-wire samples. The (2+7b)-T-wire samples are tested with the twist pitch of 100 mm and the frequency 100 Hz. As seen in Fig. 9, the measured AC loss of three twisted 3S wire samples are almost identical.

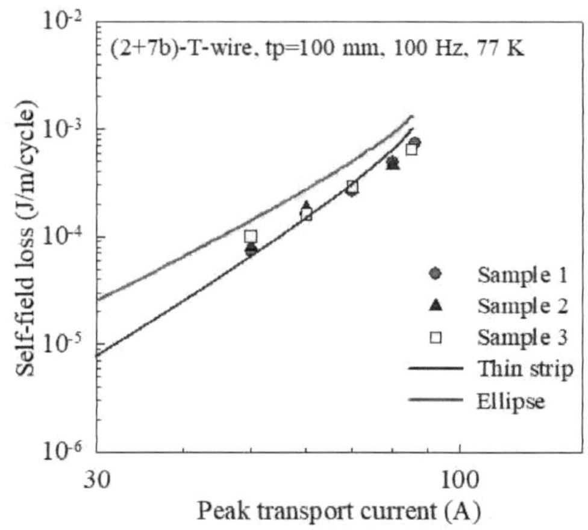

Fig. 9. Comparison of AC loss among three (2+7b)-T-wire samples.

4. Conclusion

In this paper, we firstly fabricate several 3S HTS wire samples through narrowing, soldering and stacking process, and then twist the samples with different twist pitches. Furthermore, we investigate the basic electrical characteristics for the twisted 3S HTS wire samples: twist pitch, critical current, self-field AC loss and so on.

From the test results, we can see that the critical current of the (2+7b)-T-wire is constant up to a twist pitch of approximately 80 mm, and the minimum twist pitch is approximately 50 mm. Besides, measured AC losses are independent of frequency and are in good agreement with the thin strip theoretical values. This means that hysteresis loss is a major component in self-field loss and the eddy current loss generated from brass layers is so small that can be ignored, and the measured AC losses are a little smaller than the thin strip theoretical value when peak transport current is reaching critical current. Furthermore, the measured AC losses of (2+7b)-T-wire and (2+7b)-wire are almost identical. Therefore, we can see that twist structure cannot affect the self-field loss of the 3S HTS wires.

Acknowledgement

This work supported by Science and Technology Project of China Southern Power Grid Company Limited with Number GDKJXM20172607 (036100KK52170017).

References

[1] Iwakuma M, Nigo M, Inoue D, et al. AC loss properties of YBCO superconducting tapes exposed to external AC magnetic field[J]. IEEE Transactions on Applied Superconductivity, 2005, 15(2):1562-1565.

[2] J. Fleiter, A. Ballarino, L. Bottura, and P. Tixador, "Electrical characterization of REBCO Roebel cables," Supercond. Sci. Technol., vol. 26, no. 6, pp. 065014–065015, Jun. 2013.

[3] Takayasu M, Chiesa L, Bromberg L, et al. HTS twisted stacked-tape cable conductor[J]. Superconductor Science Technology, 2011, 25(1):014011.

[4] D. Uglietti, R. Wesche, and P. Bruzzone, "Fabrication trials of round strands composed of coated conductor tapes," IEEE Trans. Appl. Supercond., vol. 23, no. 3, Jun. 2013, Art. ID 4802104.

[5] Li Z Y, Hu D, Zhang L, et al. Development of a Novel Soldered-Stacked-Square (3S) HTS Wire Using 2G narrow tapes with 1 mm Width[J]. IEEE Transactions on Applied Superconductivity, 2017, PP(99):1-1.

[6] Li Z Y, Li J, Wang Y, et al. A Study on Critical Current and AC Loss Characteristics of Novel 2G HTS Narrow Wires[J]. IEEE Transactions on Applied Superconductivity, 2016, 26(4):1-4.

AC Loss Property of Stacked REBCO Superconducting Multifilamentary Tapes under Perpendicular Magnetic Field

H Sasa[1], T Ito[1], M Iwakuma[1], S Miura[1], T Izumi[2], T Machi[2] and A Ibi[2]

[1]Research Institute of Superconductors Science and Systems, Kyushu University, Fukuoka 819-0395, Japan
[2]Research Institute of Advanced Industrial Science and Technology, Ibaraki 305-8564, Japan

E-mail: sasa@sc.kyushu-u.ac.jp

Abstract. For AC applications of high-temperature superconductors, such as $REBa_2Cu_3O_y$ (REBCO, RE = rare earth) superconducting tapes, an AC loss is a great part of total heat load of cooling systems. In the case of conduction cooling, it may cause a serious temperature rise of the superconducting systems. Therefore, it is very important to reduce the AC loss in superconducting windings and grasp its properties under various conditions. For AC loss reduction, REBCO tapes should be scribed into a multifilamentary structure by the laser-scribing technique. In this study, AC loss properties in the non-scribed and scribed tapes were investigated, especially for $B_m < B_p$, where B_p is the breaking point of the AC loss curve and corresponds to the penetration field where the magnetic flux penetrates to the center of the tape or filaments. Further, the dependences of B_p on the number of stacking tapes and filaments were investigated from the viewpoint of demagnetization effect. It was cleared that the AC loss for $B_m < B_p$ decrease and B_p increase with increasing number of stacked tapes.

1. Introduction

Many projects for AC applications of $REBa_2Cu_3O_y$ (REBCO, RE = rare earth) superconducting tapes are in progress globally. Various experimental and theoretical studies on AC loss of REBCO tapes have been conducted [1]-[4]. REBCO superconducting tapes can be used even at high temperature and high magnetic field because of its excellent critical current properties and high critical temperature. For AC use, the reduction of AC loss is the most important problem. The AC loss of REBCO superconducting tape is proportional to the width of the tape as predicted by conventional theory. Our research group has developed the laser-scribing technique to make a REBCO tape into a multifilamentary structure with little degradation of the critical current [5]. Denoting the applied magnetic field amplitude as B_m and the penetration field where the magnetic flux penetrates to the center of the tape or filaments as B_p, the AC loss of a REBCO tape with N filaments decrease to $1/N$ of those of the non-scribed one for $B_m > B_p$. B_p corresponds to the breaking point of AC loss vs B_m curve. On the other hand, the details of the AC loss properties for $B_m < B_p$ have not been investigated yet. In our previous papers, it was revealed that the AC losses for $B_m > B_p$ coincide regardless of the number of stacked tapes when the number of filaments is one or four [6]. On the other hand, the AC losses for $B_m < B_p$ decrease as the number of stacked tapes increases [6]. In this paper, focusing on the difference in AC loss of scribed and non-scribed tapes especially for $B_m < B_p$, we investigated the AC loss properties of 8-filament tapes besides non-scribed and 4-filament tapes. Moreover, the AC loss properties of the REBCO tapes stacked into 1, 3 and 6 layers were measured to investigate the dependences of B_p on the number of stacked tapes and the number of filament of a tape from the viewpoint of the demagnetizing effect.

2. Experimental setup

2.1. Parameters REBCO superconducting tapes

The parameters of sample REBCO superconducting tapes are listed in table 1. The REBCO tapes were fabricated by IBAD and the PLD technique. By doping $BaHfO_3$ to the $EuBa_2Cu_3O_y$ superconducting matrix as pinning centers, the critical current properties have been improved [7]. For AC reduction, the superconducting layers were divided into four or eight filaments by laser-scribing technique. The details of laser-scribing technique were reported in [5]. The tapes were stacked into 1, 3 and 6 layers perpendicularly. Every tape was covered with kapton tape to insulate with each other. In the case of stacked tapes, the distance between the neighboring superconducting layers was 208.4 μm. The I_C of the 4- and 8-filament tapes calculated from measured *M-B* curves were 305.7 A and 201.6 A respectively. There was no degradation by scribing.

Table 1. Parameters of the EuBCO tapes.

	non-scribed and 4-filament tapes	8-filament tapes
Dimensions of the tape	5 mm in width, 112 μm in thickness	5 mm in width, 106 μm in thickness
Substrate	Hastelloy (100 μm)	Hastelloy (100 μm)
Buffer layer	$CeO_2 + LaMnO_3 + MgO + Y_2O_3 + Gd_2Zr_2O_7$ (1 μm)	$CeO_2 + LaMnO_3 + MgO + Y_2O_3 + Gd_2Zr_2O_7$ (0.7 μm)
Superconducting layer	$EuBa_2Cu_3O_y + BaHfO_3$ [3.5 mol%] (3.6 μm)	$EuBa_2Cu_3O_y + BaHfO_3$ [3.5 mol%] (0.7 μm)
I_C of the tape at 77 K, self-field	300 A	200 A

2.2. Experimental Equipment

In this study, a saddle-shaped pick-up coil was used to measure the magnetization and AC loss of the superconducting tapes [6]. The external magnetic field was applied in perpendicular to the tape face by using a NbTi superconducting magnet. The range of the amplitude and the frequency were from 0.004 to 4.3 T and from 0.01 to 0.2 Hz, respectively. Temperature ranged from 25 to 77 K. The AC loss, *W*, per cycle and unit volume was evaluated by integrating the magnetization, *M*, by the applied magnetic field, *B*, for a single period. It corresponds to the inner area of the *M-B* curves.

$$W = \int_0^t B dM = -\oint M dB \tag{1}$$

We confirmed in the previous study that AC loss of REBCO tapes has no frequency dependence [8]. So the combination of frequency and amplitude was not cared.

3. Results and discussion

3.1. Behaviour of AC loss for $B_m < B_p$

The AC loss of REBCO tapes increases monotonically with increasing temperature [4]. Figure 1 shows the observed field amplitude, B_m dependences of AC losses of the non-scribed and scribed tapes which were stacked into 1, 3 and 6. The arrows in figure 1 indicate each B_p.

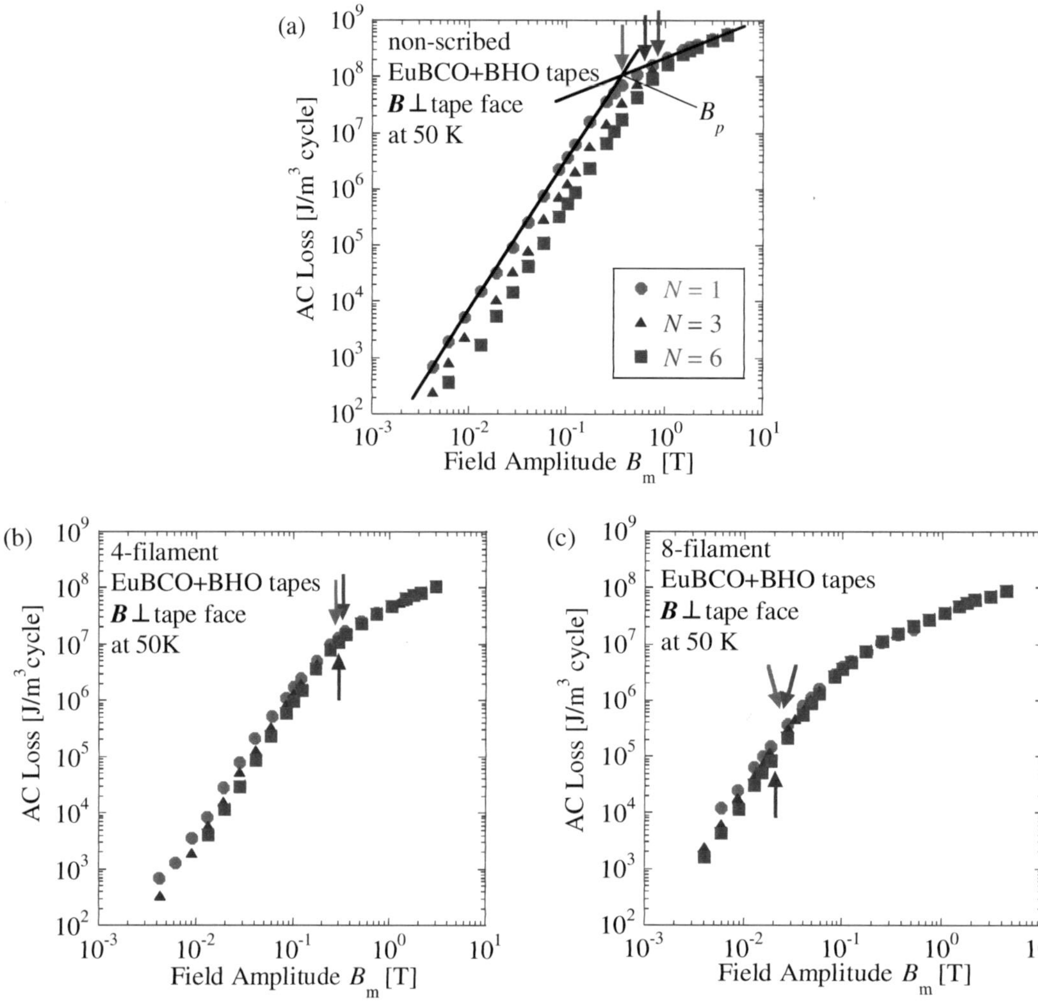

Figure 1. Observed B_m dependences of the AC losses of (a) non-scribed (b) 4-filament (c) 8-filament tapes at 50 K. N indicates the number of the stacked tapes.

B_p was determined from the intersection point of two asymptote lines for simplicity as shown in figure 1(a). As for the non-scribed tapes, the AC losses decrease with increasing number of the stacked tapes for $B_m < B_p$. This property can be explained from the viewpoint of demagnetizing effect as follows. Generally, the effectively applied magnetic field to superconducting tape, B_{eff}, is different from the external applied magnetic field, B_{ext}, because of the diamagnetic field, B_d, which is caused by the shielding current in the tape. The relation between them are expressed as

$$B_{eff} = B_{ext} + B_d \tag{2}$$

In the Meissner state, using the demagnetizing coefficient, ν, the diamagnetic field is expressed as

$$B_d = \nu B_{eff} \tag{3}$$

Substituting equation (3) to equation (2), we can obtain

$$B_{eff} = B_{ext}/(1-\nu) \tag{4}$$

On the other hand, using the magnetization, M, the inner magnetic field of the superconductor, B_{in}, is expressed as

$$B_{\text{in}} = B_{\text{eff}} + M \tag{5}$$

Considering the case of $B_{in} = 0$, i.e. $B_m < B_p$, following equation is obtained from above equations.

$$-M = B_{\text{eff}} = B_{\text{ext}}/(1 - \nu) \tag{6}$$

ν is determined by the shape of the superconducting material and the direction of the applied field. In the case that the field is applied in parallel to a superconducting infinite slab, ν = 0. On the other hand, in the case that the field is applied in perpendicular to the slab, ν is close to one. When ν is close to one, B_{eff} is much larger than B_{ext}. So, the AC loss is larger than in the case smaller ν.
When the superconducting tapes were stacked into multilayer, ν becomes smaller and B_{eff} becomes smaller. So, the AC loss decreases with increasing number of the stacked tapes. On the other hand, for $B_m > B_p$, the AC losses roughly agree with each other regardless of the stacking number of tapes. ν for $B_m > B_p$ is almost zero regardless of the shape of the tapes because the field penetrates throughout the tape. Therefore, the AC losses also have no dependence on the number of stacked tapes.

Next, the AC losses shown in figure 1 are replotted in figure 2 for comparison between the filament number dependences. In figure 2, for simplicity, the smaller graphs in each graph shows the enlarged view for large field amplitude and the arrows indicate each B_p.

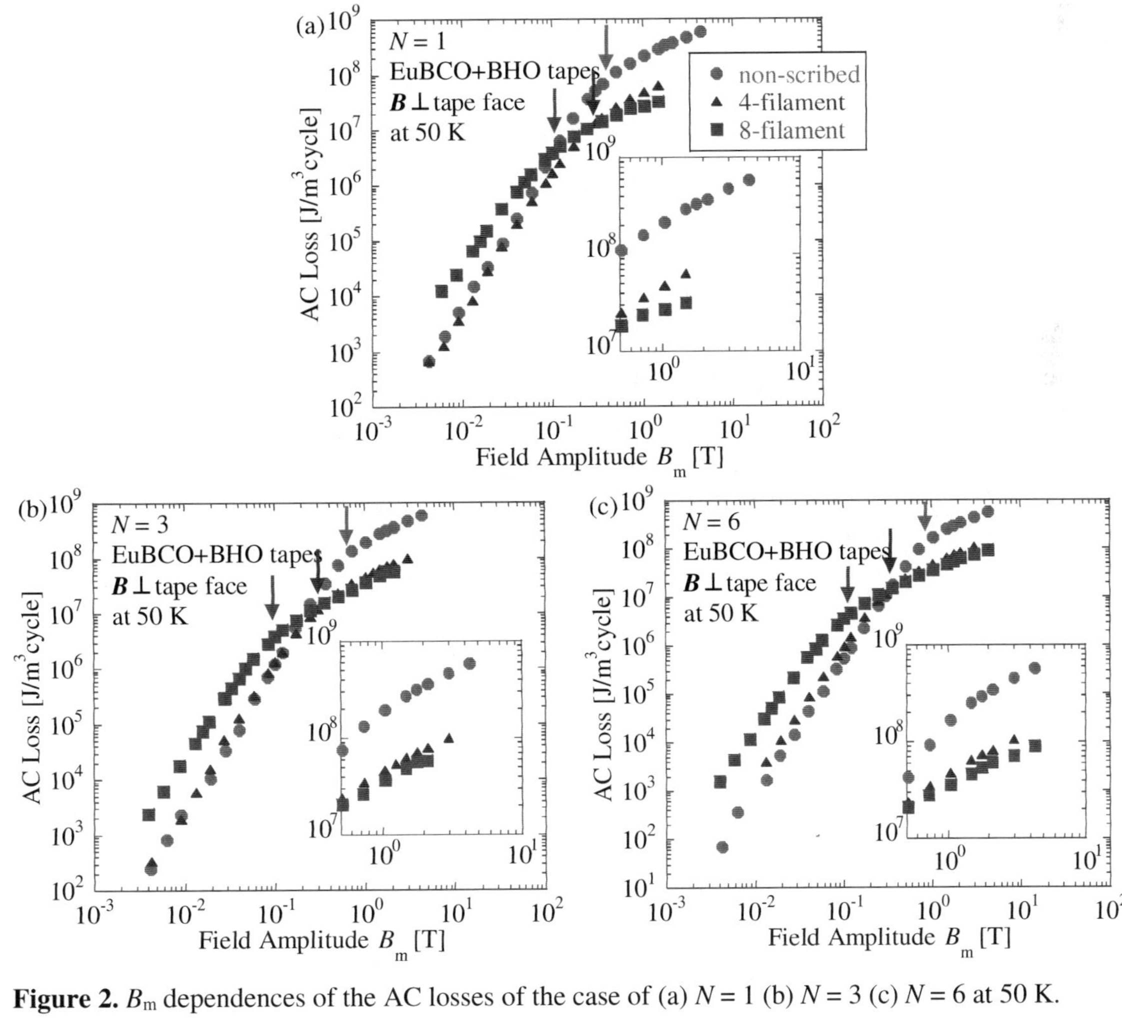

Figure 2. B_m dependences of the AC losses of the case of (a) $N = 1$ (b) $N = 3$ (c) $N = 6$ at 50 K.

For $B_m > B_p$, the AC losses decrease with increasing number of filaments. That can be expected from the conventional theory [5]. On the other hand, for $B_m < B_p$, the AC losses increase with increasing number of filament. For $B_m < B_p$, in the case of scribed tapes, the applied field penetrates the tape more easily than the case of non-scribed tapes. Therefore, the AC losses of the scribed tapes are larger than those of the non-scribed tapes. That can be also predicted from the conventional theory.

Figure 3 shows the field amplitude dependences of the ratios of AC losses in the non-scribed, 4-filament and 8-filament tapes to those in the one-layer tape at 50 K for $B_m < B_p$.

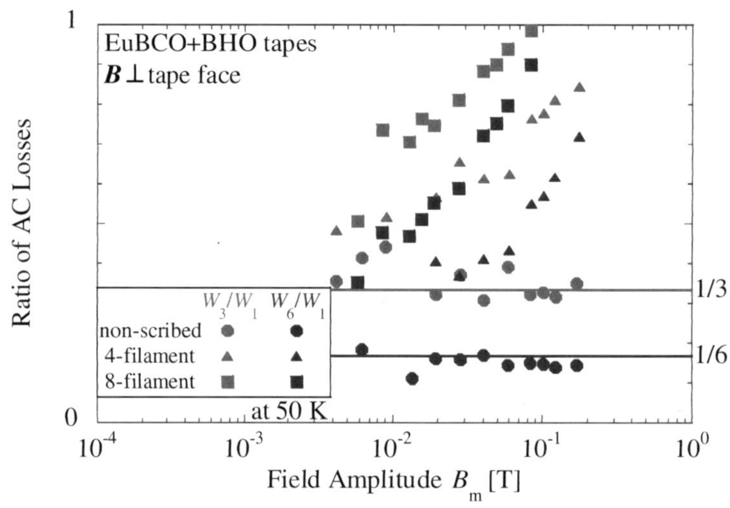

Figure 3. Comparison between field amplitude dependences of ratio of AC losses in non-scribed, 4-filament and 8filament tapes at 50 K for $B_m < B_p$. W_N indicate the AC loss of the N-layer stacked tapes.

In the case of the non-scribed tapes, the ratios roughly correspond to approximately $1/N$. Increasing N, the whole shape of the tapes becomes longer perpendicularly, and then the demagnetizing coefficient decreases. Then, from equation (6), the magnetization decreases and AC loss decreases. From equation (1), it means that the AC loss of the N-layer stacked tapes decreases to $1/N$ compared with that of one-layer tape. In the case of the multifilamentary tapes, the ratios are higher than $1/N$ and increase as increasing the filaments and B_m. For $B_m < B_p$, in the case of multifilamentary tapes, the applied field penetrates each filament more easily than the case of non-scribed tapes because the shielding current is directly proportional to the width of the filament [4]. Thus, the AC losses increase with increasing filaments.

3.2. *Behaviour of B_p*

Figure 4 shows that the temperature dependences of B_p for each number of the filaments and the stacked tapes.

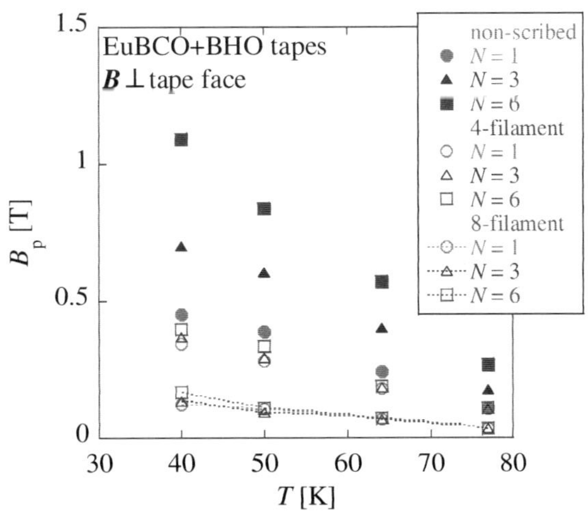

Figure 4. Temperature dependences of B_p for non-scribed, 4-filament, 8-filament tapes, N = 1, 3 and 6.

B_p decreases with increasing T in every case. Increasing T, the critical current decrease and the magnetic field can penetrate the tapes more easily. Consequently, B_p decreases. B_p increases as increasing the stacking number. Increasing N, the whole shape of the tapes becomes longer perpendicularly, and then the demagnetizing coefficient decreases. Then, from equation (6), the magnetic field applied effectively to the tapes is small. Therefore, the applied field requires larger amplitude to penetrate to the center of the tapes, i.e. B_p increases as increasing the stacking number. The number of filaments dependences of B_p can be explained from the viewpoint of the shielding current. B_p decreases as increasing the number of filaments. It can be explained by the shielding current dependence of the AC loss mentioned above. For multifilamentary tapes, especially for 8-filament tapes, B_p barely depends on N. In the case of multifilamentary tapes, the width of the filaments is much smaller than that of the non-scribed tapes. Therefore, the magnetic flux can penetrate more easily than the case of the non-scribed tapes. It means that the demagnetizing coefficient of the multifilamentary tape is much smaller than that of the non-scribed tapes. Consequently, B_p barely depends on N.

4. Conclusion

The AC loss properties of EuBCO superconducting tapes were investigated under various conditions with temperature, the number of stacked tapes and the number of filaments as a parameter. The AC losses for $B_m < B_p$ decreased with increasing number of the stacked tapes. The ratios of AC losses in the stacked tapes to those in one-layer tape depended on the number of filaments. That can be explained by the difference in demagnetizing coefficient. B_p was influenced by temperature, the number of stacked tapes and the number of the filaments. These are also caused by the difference in demagnetizing coefficients of the tapes for each condition. The laser-scribing technique is useful to decrease B_p and reduce the AC loss of EuBCO superconducting tapes for $B_m > B_p$.

Acknowledgement

This research was partially supported by the Japan Science and Technology Agency (JST): Advanced Low Carbon Technology Research and Development Program (ALCA) and the Japan Society for the Promotion of Science (JSPS): Grant-in-Aid-for Scientific Research (JP26249042 and JP17H06931).

References

[1] Majoros M, Glowacki B, Campbell A, Levin G, Barnes P and Polak M 2005 *IEEE Trans. Appl. Supercond.* **15** 2819-22

[2] Ogawa J, Shiokawa M, Ciszek M and Tsukumoto T 2003 *IEEE Trans. Appl. Supercond.* **13** 1735-8
[3] Kasai S and Amemiya N 2005 *IEEE Trans. Appl. Supercond.* **15** 2855-8
[4] Iwakuma M, Nigo M, Inoue D, Miyamoto N, Funaki K, Iijima Y, Saitoh T, Yamada Y, Izumi T and Shirohara Y 2006 *Supercond. Sci. Technol.* **19** 350-8
[5] Katayama K, Machi T, Nakamura T, Takagi Y, Nakaoka N, Yoshizumi M, Izumi T and Shiohara Y 2014 *Physics Procedia* **58** 142-5
[6] Ueno T, Iwakuma M, Ito T, Yun K, Adachi K, Tomioka A, Hase Y, Konno M, Izumi T, Machi T and Ibi A 2017 *IEEE Trans. Appl. Supercond.* **27** 4701806
[7] Tobita H, Notoh K, Higashikawa K, Inoue M, Kiss T, Kato T, Hirayama T, Yoshizumi M, Izumi T and Shiohara Y 2012 *Supercond. Sci. Technol.* **25** 062002
[8] Iwakuma M, Hayashi H, Okamoto H, Tomioka A, Konno M, Saito T, Iijima Y, Suzuki Y, Yoshida S, Yamada Y, Izumi T and Shiohara Y 2009 *Phys.C.* **469** 1726-32

IOP Conf. Series: Journal of Physics: Conf. Series **1054** (2018) 012038 doi:10.1088/1742-6596/1054/1/012038

Superconducting joint of REBCO wires for MRI magnet

Shinichi Mukoyama[1], **Akinobu Nakai**[1], **Hisaki Sakamoto**[1], **Shinji Matsumoto**[2], **Gen Nishijima**[2], **Mamoru Hamada**[3], **Kazuyoshi Saito**[3], **Yasuyuki Miyoshi**[3]

[1] Furukawa Electric Co, Ltd., 6, Yawatakaidan-dore, Ichihara, Chiba 290-8555, Japan
[2] National Institute for Materials Science, Tsukuba 305-0033, Japan
[3] Japan Superconductor Technology, Inc., Kobe 651-221, Japan

<shinichi.mukoyama@furukawaelectric.com>

Abstract. A high temperature superconducting wire (HTS wire) is promising for various superconducting magnet applications because it operates at a higher temperature than a low temperature superconducting wire (LTS wire) is. Particularly, a MRI magnet using the HTS wires (HTS-MRI magnet) is expected to obtain light-weight, compact and low operating cost compared to a LTS-MRI magnet. The MRI magnets are generally operated by a persistent current mode. A magnet of the persistent current mode consists of multiple superconducting magnets and persistent current switch (PCS) that connects with superconducting joints.. However, the superconducting joint of the HTS wires has not been realized stably at this time. We have developed a superconducting joint by using a commercial REBCO wire. The HTS-MRI magnet requires that resistivity of less than 10^{-12} ohm and current capacity of more than 100 A is achieved by a direct-joint between superconducting layers of the two REBCO wires. Moreover, measurement equipment for low joint resistivity was prepared as measuring the decay of the magnetic field in the one-turn loop. In our R&D, the joint resistance and the critical current were achieved with 155 A and $5.3\times10^{-13}\Omega$.

1. Introduction

A MRI magnet using high temperature superconducting (HTS) wires will be operated at higher temperature than a conventional MRI magnet fabricated of low-temperature superconducting (LTS) wires. The HTS-MRI magnet is cooled by a conduction cooling with a cryo-cooler, and is expected to obtain light-weight, compact and low operation cost. The LTS-MRI magnet is operated in a persistent current mode, which doesn't need an electrical current source to keep a constant magnetic field. NbTi wires used in the LTS-MRI magnet are joined with less than $10^{-13}\Omega$-digits resistivity. Recently, several groups have tried and succeeded the superconducting joint for $REBa_2Cu_3O_{7-\delta}$(REBCO) coated conductors [1]-[4], but the joint has not been realized stably and good reproducibly yet. The interface in superconducting joint of REBCO wires is required to crystallize into a superconducting structure. Optimization and stabilization of the process condition are important for the reproducibility. In our ongoing project, we aim to contribute the development of a HTS-MRI magnet system, which is operated at 50 K by a cryo-cooler cooling. The joint resistivity will be characterized by temporal magnetic field decay of the magnet. If the static magnetic field fluctuates in time in a MRI system, or if the intensity of the magnetic field is not uniform, the phase of the signal measured in response to the variation changes. This change, distortion and deformation of the image in the image to be re-formed, appear to become uneven tint change. In particular, the speed of recent images, for high performance, susceptible to these fluctuations, the time stability of 0.1 ppm/hour is required.

2. Superconducting joint of HTS wire

2.1. REBCO wire

The REBCO wire has better current loading performance under high magnetic field than other HTS wires do, such as BSCCO and MgB2. Moreover, in the case of the persistent current mode operation, a flux flow resistivity in the wire is needed to be extremely low[5]. Since the REBCO wire has the high n-value as almost same as the LTS wires, the REBCO wire is suitable for MRI magnets. In figure 1, a structure of the commercial REBCO wire that produced by SuperPower Inc. The REBCO wire constructs of a Hastelloy® substrate, intermediate layers such as MgO by the IBAD plating, and a superconducting layer of GdYBCO by the MOCVD process. In addition, the silver layer is deposited on the superconducting layer for a chemical stabilization, and the entire wire is covered with a copper layer by electrical plating for the current loading stabilization.

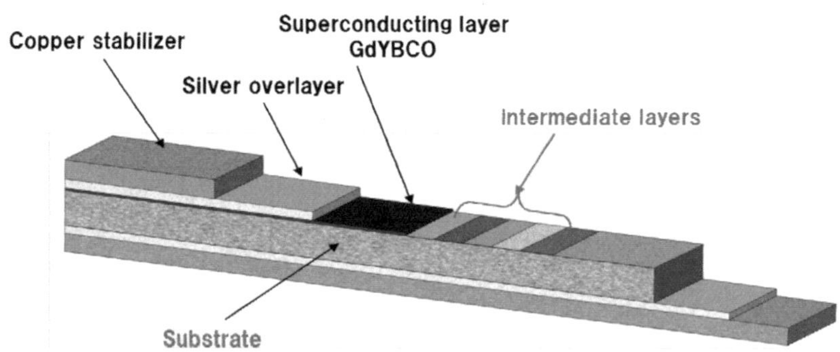

Figure 1. Structure of REBCO wire （SuperPower）

2.2. Experiment of superconducting joint

In order to join the superconducting wires with almost zero resistivity, it is necessary to contact the bared surfaces of the superconducting layers and to change the boundary interface into a superconductor. The REBCO wires without a Cu stabilizer were prepared to establish the joint process and the process condition,. Table 1 shows the specifications of the REBCO wire used for the joint test.

Table 1. Specification of REBCO sample tape for superconducting joint

Sample	SuperPower Inc. (SF6050 AP)
Width of tape	6 mm
Thickness of Substrate	50 μm
Thickness of SC layer	1.6 μm
Thickness of Ag layer	2 μm
Minimum critical current	190 A

The process of the superconducting joint consists of 4 steps. In the first step, the silver overlayer of about 1 cm length is dissolved by a chemical etching to expose the superconducting layer. In the second step, the liquid solution containing the superconducting materials of the REBCO is applied on the surface of the superconducting layers. The two tapes are pre-calcined at about 500 degree-C. In the third step, the two tapes are contacted together as shown in figure 2. The two tapes are compressed by the mechanical pressure, and bake it in a furnace at about 800 degrees-C to crystallize the interface. In the fourth step, a process of oxygen annealing is conducted under heating in high oxygen atmosphere.

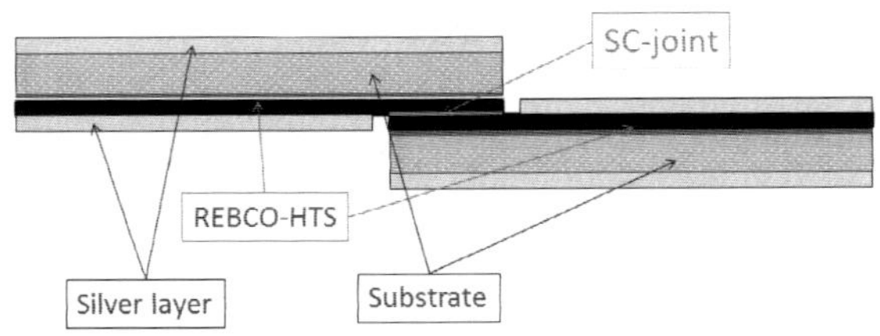

Figure 2. Schematic structure of superconducting joint

By the processes mentioned above, the joint samples were fabricated, and its properties were examined. The joint sample was examined by two kinds of tests. First, a property of current and voltage (I-V) of the joint sample was measured by a four-terminal method in liquid nitrogen. The result of the I-V test is shown in figure 6. Since critical current of HTS wires is generally defined at 1μV/cm criterion, Ic of the joint was evaluated as 155 A in the criterion of the voltage tap length with 5 cm. Also, the straight line of 10^{-8} Ω is shown in the figure, it is understood from this figure that the resistivity value of the joint is less than 10^{-8} Ω at 150 A.

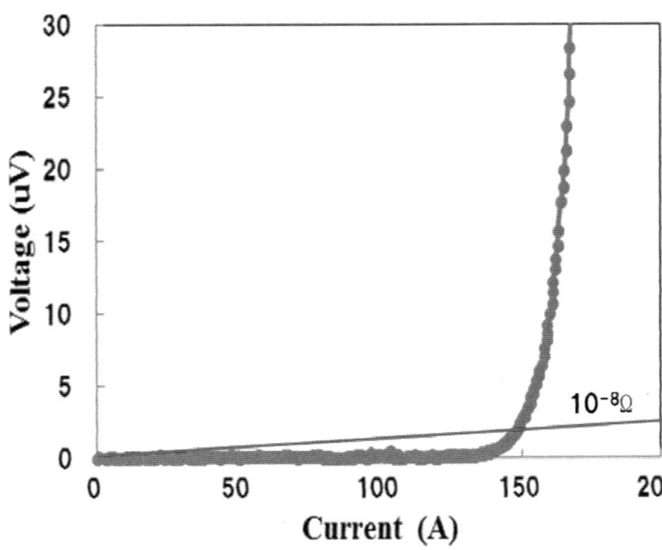

Figure 3. I-V characteristic of superconducting joint at 77K

3. Persistent Current Decay Measurement

The extremely lower resistivity than 10^{-12}Ω is not measured by the four-terminal method, but by the current decay method [6]. A schematic diagram of the measuring device is shown in figure 4. A one-turn loop is fabricated with the REBCO wire by joining both end of the wire. Persistence current flows in the loop, and the magnetic field generates by the current in the loop. The magnetic field is measured with a hall sensor. The joint resistivity is derived from equation (1).

$$R = L/t \times ln\frac{B}{B_0} \qquad (1),$$

where R is resistivity, L is inductance of the loop, t is measurement time, and B/B_0 is decay ratio of the magnetic field. A photograph of test equipment is shown in figure 5. The test equipment consists of the one-turn loop with the joint, the exciting coil made of Cu wire, and the heater that raise temperature at a part of the loop to the normal state temperature. The equipment is set on a cold head of a GM refrigerator to measure a resistivity at various temperates from 77 K to 30 K.

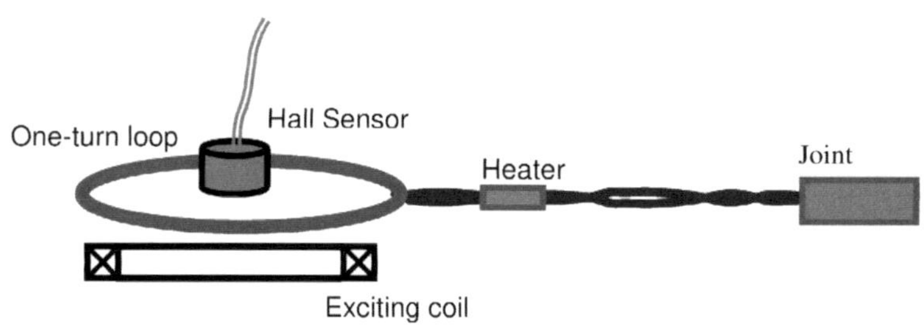

Figure 4. Structure of low resistivity measurement system using one-turn loop

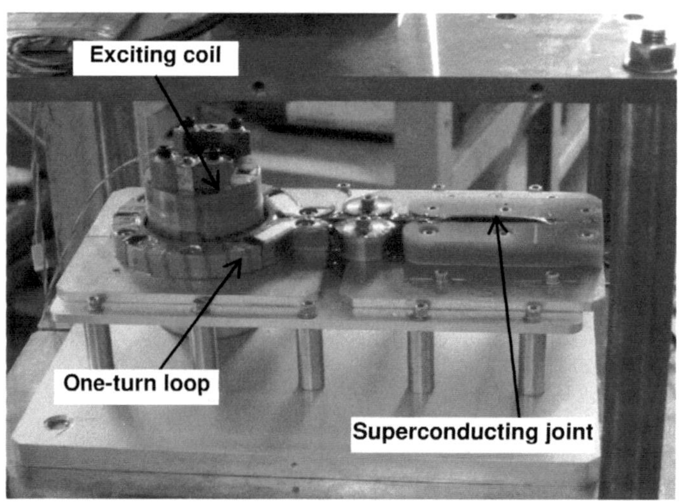

Figure 5. Low-resistivity measuring equipment using one-turn loop

One example of the measured results is shown in figure 6. The magnetic field of persistent current in the loop was measured at 30 K. The magnetic field of the loop was decreased immediately after demagnetization of the exciting coil, but the subsequent attenuation became stable and low after 3 hours later. The inductance of the one-turn loop was 0.3 μH by the analysis of a circle current. The resistivity of the joint was calculated by the equation (1), and the resistivity was confirmed to be 9.7 ×

10^{-13} Ω. The current of the loop was calculated by Ampere low, and the magnetic field of 1.7 mT was corresponded to about 70 A.

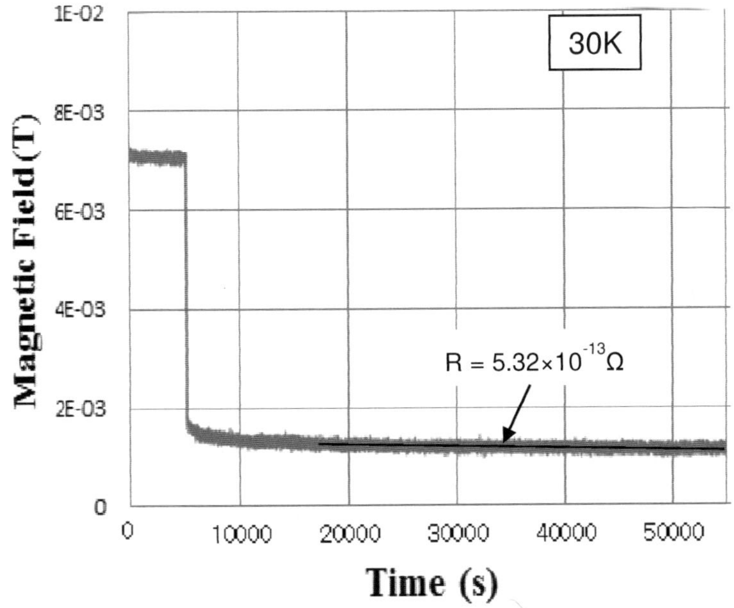

Figure 6. Result of magnetic field measurement with one-turn loop at 30 K

4. Conclusion

The superconducting joint of REBCO wires was developed, and 150A of Ic characteristic was achieved in short length of the joint samples. The measuring system of the persistent current with a one-turn loop was prepared, and we measured the one-turn loop sample with a superconducting joint. The joint resistivity was measured from the field decay rate and was achieved to 5.3×10^{-13}Ω.

In the NEDO project, we are planning to investigate the superconducting joints by using the HTS solenoid magnet systems. The HTS magnet will consist of multiple coils and a persistent current switch (PCS) that connected with the superconducting joints. The HTS magnet will be fabricated by Japan Superconductor Technology Co., Ltd, (JASTEC) and the evaluation will be conducted by the National Institute for Materials Science (NIMS). Since it is necessary to make many superconducting joints in high quality without fail in the solenoid magnet system, we will obtain the superconducting joint skill with stability and productivity. Now we have continued the R&D to realize stable superconducting joints until the fabrication of the HTS magnet.

Acknowledgment

This work is conducted as a part of "Technology Development to Promote Commercialization of High-Temperature Superconductivity" sponsored by New Energy and industrial technology Development Organization (NEDO). Authors would like to thank Dr. Kitaguchi of NIMS for providing access to his experimental set up.

References

[1] Park Y J, Lee M W, Oh Y K and Lee H G 2014 'Laser drilling: enhancing superconducting joint of $GdBa_2Cu_3O_{7-\delta}$ coated conductors', *Superconductor Science and Technology* **27**, 085008

[2] Furukawa Electric Co., Ltd. news release 2016 Development of Persistent Current Technology Using Rare Earth Superconducting Materials *https://www.furukawa.co.jp/en/release/2016/kenkai_160427.html*

[3] Jin X, Yanagisawa Y, Maeda H and Takano Y 2014 Development of a superconducting joint between a $GdBa_2Cu_3O_{7-\delta}$ coated conductor and $YBa_2Cu_3O_{7-\delta}$ bulk: towards a superconducting joint between RE (Rare Earth) $Ba_2Cu_3O_{7-\delta}$ coated conductors *Physics Procedia* **58 pp** 268-271.

[4] Ohki K, Nagaishi T, Kato T, Yokoe D, Hirayama T, Ikuhara Y, Ueno T, Yamagishi K, Takao T, Piao R, Maeda H, and Yanagisawa Y 2017 Fabrication, microstructure and persistent current measurement of an intermediate grown superconducting (iGS) joint between REBCO-coated conductors *Supercond. Sci. Technol.* **30** 115017

[5] Hase T, Awaji S, Watanabe K, Oguro H and Yamano S 2016 The effect of flux creep in the closed loop coil using REBCO tape *Conference proceeding of CSSJ 2016 spring*. 1A-p4, (in Japanese)

[6] Leupold M J and Iwasa Y 1976 Superconducting joint between multifilamentary wires 1. Joint-making and joint results, *Cryogenics* **16** Issue 4 pp 215-216

Numerical Simulation on Coupling Current for Multifilamentary HTS Wire

T Koizumi[1], E Morikawa[1], S B Kim[1] and H Ueda[1]
[1] Okayama University, Okayama, Japan

E-mail address: hiroshi.ueda@okayama-u.ac.jp

Abstract. Since a high temperature superconducting (HTS) wire such as Bi-2223 ($Bi_2Sr_2Ca_2Cu_3O$) and REBCO((RE)Ba_2Cu_3O) tapes indicates good superconducting characteristics under high magnetic field, ultra-high field magnets wound HTS wire are applicable to a nuclear magnetic resonance (NMR) spectrometer and magnetic resonance imaging (MRI). The large and long-time-constant screening current is induced in the HTS wire, which is the tape shape and not twisted, and the magnetic field generated by screening current deteriorated the field quality such as temporal stability and spatial homogeneity. Because NMR and MRI requires highly accurate field on temporal stability and spatial homogeneity, it is necessary to investigate the influence of the screening current-induced field. In REBCO tape, the screening current can be reduced by dividing the superconductor layer. However, filaments are electrically connected because they are covered with copper due to strength and thermal stability. On the other hand, a Bi-2223 is wire which multiple superconducting filaments are covered with silver or a silver alloy, therefore, the screening current is smaller than that in a REBCO tape. However, in a Bi-2223 tape, a coupling current flows because of electrical bridge between the filaments. In this study, we discuss coupling current distribution from numerical simulation on the multifilamentary HTS coil which is given the local electrical contact between filaments. The screening-current field decreases with increasing the interval distance. In addition, the current distribution is different depending on the interval distance.

1. Introduction

High temperature superconducting (HTS) wires indicates such as Bi-2223 ($Bi_2Sr_2Ca_2Cu_3O$) and REBCO((RE)Ba_2Cu_3O) tapes good superconducting characteristics under high magnetic field. Therefore, the research and development on the application of HTS coil to the high field magnets for NMR, MRI, and accelerator and so on are progress [1-12]. In these application, the magnets need to create highly homogeneous and temporally stable field. However, the screening currents lead to the serious problem in HTS magnets for the applications required very high field quality. In coils wound with HTS tapes, large screening currents are induced by the radial component of the magnetic field because of tape shape and no twist. Then, the total current distributions in the tape consists of the transport and screening currents, the total current flow inhomogeneously in tape. As a result, an irregular magnetic field is generated by the screening currents. The screening-current field affects the magnetic field distribution in the following issues: (1) field reduction, (2) no-repeatability of field, (3) temporal stability, and (4) deterioration of field quality. These effects are very critical in NMR, MRI, accelerator applications. Therefore, the prediction and reduction of screening-current field are needed.

The filamentization of REBCO tape shows enables us to reduce the possibility of reducing the screening current induced magnetic field in REBCO coils [13][14]. The multi-filamentary REBCO tape is usually plated with copper for mechanical strength and thermal stability. The behaviour and diffusion process of screening current depend on the geometry treatment of the coil winding and the effective transverse resistivity between filaments in in REBCO tape. On the other hand, a Bi-2223 is wire which multiple superconducting filaments are covered with silver or a silver alloy, therefore, the screening current is smaller than that in a REBCO tape [15]. However, in a Bi-2223 tape, a coupling current flows because of electrical bridge between the filaments. Since these are current distributions different from the transport current, they are considered to affect the temporal stability and spatial homogeneity. In this study, we focused on the bridging between filaments in HTS tape and investigated the current distribution in the multi-filamentary HTS tape by using our developed simulation code. We discuss on the effects of bridging of multi-filamentary HTS tape on the diffusion process of screening current and the behaviour of the magnetic field.

2. Numerical simulation

The screening current, which passes from end to end in HTS coil, is affected by the coil winding configuration, such as pancake winding and layer winding, Therefore, The developed three-dimensional electromagnetic-field numerical simulation code for the screening-current field have been developed [16]. This simulation is based on a finite element method (FEM) and a fast multipole method (FMM). The governing equation formulated based on thin-film approximation, Biot-Savart law, Faraday's law of induction, and Ohm's law is expressed as

$$\{\boldsymbol{\nabla} \times \rho(\boldsymbol{\nabla} T \times \boldsymbol{n})\} \cdot \boldsymbol{n} + \frac{\mu_0 d}{4\pi} \frac{\partial}{\partial t} \int_S \frac{(\nabla T' \times \boldsymbol{n}') \times \boldsymbol{R}}{R^3} \cdot \boldsymbol{n} dS' = -\frac{\partial \boldsymbol{B}_a}{\partial t} \cdot \boldsymbol{n} \quad (1)$$

where T is the normal component to the current vector potential of the tape's wide face, n is the unit normal vector to the tape's wide face, ρ is the resistivity of the superconductor, μ_0 is the space magnetic permeability, $\boldsymbol{B}_a$ is the external magnetic field, $\boldsymbol{R}$ is the vector toward the field point from the source point, and d is the thickness of the superconducting layer. The resistivity of the superconductor, ρ is defined by the power law and Ohm's law

$$\rho = \frac{E_c}{J_c(B,\theta)} \left(\frac{|\boldsymbol{J}|}{J_c(B,\theta)} \right)^{n(B,\theta)-1} \quad (2)$$

where $\boldsymbol{J}$ is the current density in the superconducting layer. The validity of the numerical simulation for the screening-current field was confirmed [17][18].

In this simulation, the dependence of the I_c-B-θ characteristic at 4.2 K as shown in Figure 1. And the n-value was set to be constant at 16.

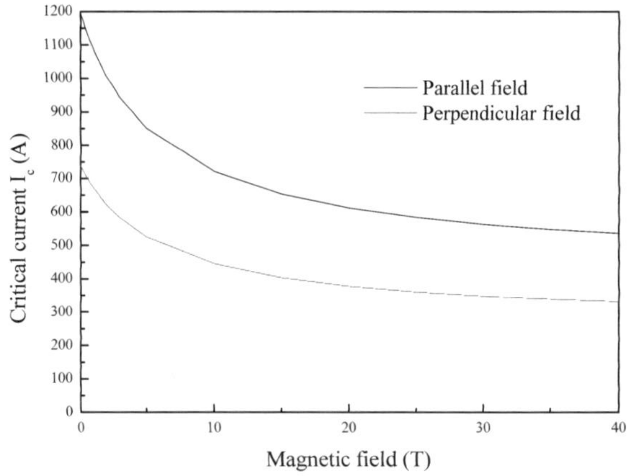

Figure 1. I_c-B-θ characteristic.

Figure 2 shows the schematic drawing of the model of multi-filamentary HTS wire with bridging. In this study, the interval of bridging between filaments is assumed to be non-random and equal distance in order to investigate the fundamental behavior of coupling current in the multi-filamentary HTS wire with bridging. Table 1 and 2 list the specifications of the model HTS wire and HTS coil, respectively, in this simulation. The resistivity values is determined from characteristics of silver and copper at 4.2 K [19][20]. Table 3 lists three cases of the interval distance and the model of the multifilamentary wire shows as Figure 2.

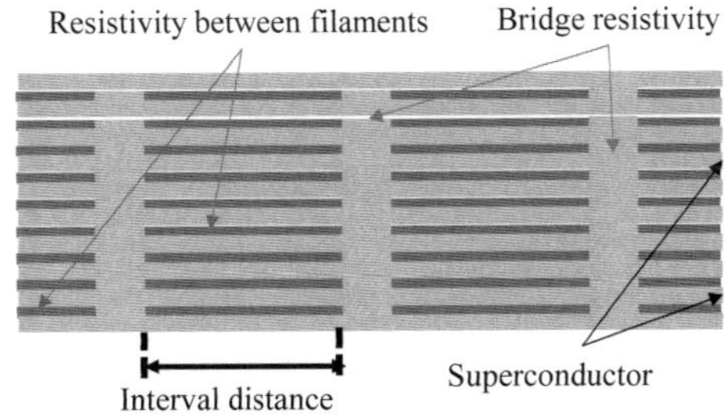

Figure 2. Model of the multifilamentary HTS wire.

Table 1. HTS wire specifications

Superconductor layer thickness (μm)	1.0
Tape width (mm)	4.5
Number of filaments	10
Gap between filaments (μm)	100
Resistivity between filaments (Ωm)	1.0×10^{-7}
Bridge resistivity (Ωm)	1.0×10^{-10}

Table 2. Coil specifications

Inner diameter (mm)	80
Outer diameter (mm)	108.08
Height (mm)	101
Turns / Single pancake coil	39
Number of Single pancake coils	20
Gap between coils (mm)	1.0
Transport current (A)	295.72
Magnetic field @ center (T)	2.0

Table 3. Interval distance between bridges

	Case I	Case II	Case III
Interval distance (mm)	16.75~22.64	205.2~277.3	2089.8~2823.8

3. Simulation results and discussion

In this study, the transport current was 259.72 A at sweep rate of 1.0 A/s and 5.0 A/s. The calculation results of screening-current field at the coil center at 1.0 A/s and 5.0 A/s are shown as Figure 3. The screening current field is defined as the result of subtracting the ideal magnetic field which does not consider the screening current from the result used in this simulation. As shown in Figure 3, the screening-current field increases during charge the coil because the area in which screening current flows is expanded. And the screening-current field decreases with increasing the interval distance of the bridge resistivity. When we compared with 1.0 A/s and 5.0 A/s just after the excitation to 295.72 A, the difference between the Case I and III about the screening-current field is 3.7 mT and 1.8 mT.

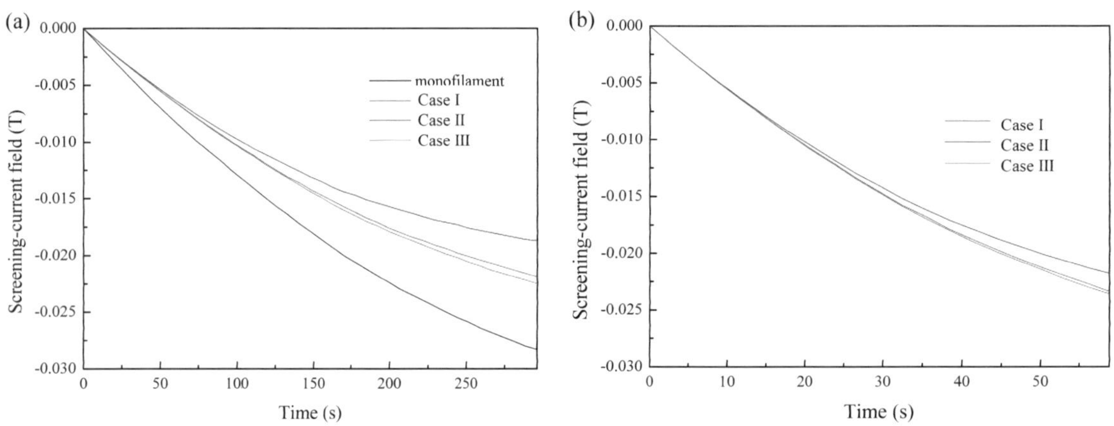

Figure 3. Behavior of screening-current field time dependency at coil center in the (a) 1.0 A/s and (b) 5.0 A/s.

The current distributions of the topmost single-pancake coil in 20-stacked coil just after the excitation to 295.72 A are shown in Figure 4. This figure shows the expanding view from the start of winding to the end of winding. The screening current in pancake-wound coils mostly flows at the edge of the coil because the component of the magnetic field perpendicular to the wide face of the tape is large. In the pancake-wound coil, because the tapes at the edges of the coil are exposed completely to the perpendicular component of the magnetic field, the path of the screening current circulates from end to end of the tape. As shown in Figure 4, the coupling currents between filaments decrease with increasing the interval distance between bridges. And at the sweep rate of 5.0 A/s, the coupling currents between filaments increase because the magnetic field is hard to penetrate in multimilitary HTS wire. Figure 5 shows schematic drawing of the local-loop current in the multifilamentary HTS wire with bridging. As seen in Figure 5(a), when the interval distance is shorter, the screening current flows in the local short loop and the magnetic field penetrates from the side of wire. The current distribution as shown in Figure 4(a) and (d) are explain in the case of Figure 5(a). On the other hand, when the interval distance is longer, the magnetic field penetrates from the end of wire as shown in Figure 5(b). The current distribution as shown in Figure 4(c) and (f) are explain in the case of Figure 5(b). Therefore, the deviation of the current across the tape is relaxed, the screening-current field decrease as shown in Figure 3. Also, the change in the magnetic field at a fixed time increases as the sweep rate becomes faster, so that the influence of the interval distance becomes smaller.

4. Conclusion

In this paper, we numerically investigated the influence on the screening-current field and current distribution for multifilamentary HTS wire with bridging. The screening-current field decreases with increasing the interval distance. And the influence of the interval distance is decrease when the sweep rate becomes faster. In addition, the current distribution is different because of local current loop depending on the interval distance.

In the future, we will establish the model of Bi2223 and investigate the effect of screening-current for ultra-high-field NMR and accelerator. And we will evaluate the field homogeneity and temporal stability for shimming.

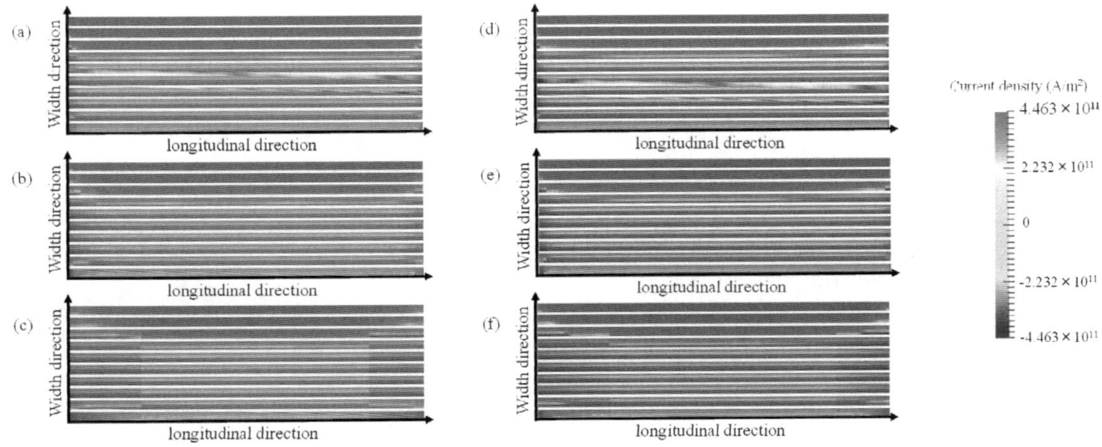

Figure 4. Current distribution at charging complete on the tape winding of top single pancake coil in the (a) Case I, (b) Case II, (c) Case III (1.0 A/s), (d) Case I, (e) Case II, (f) Case III (5.0 A/s).

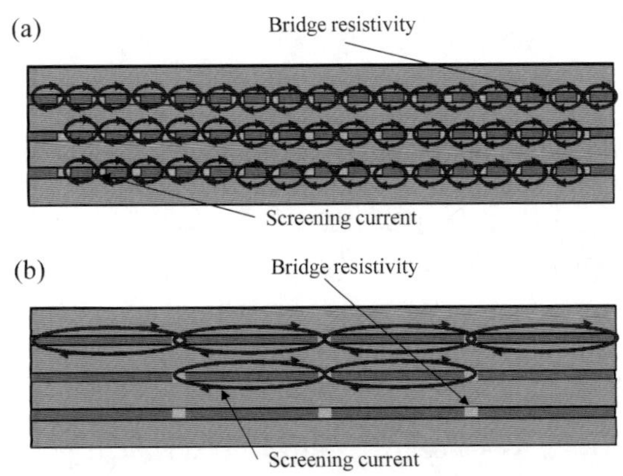

Figure 5. Schematic drawing of the local-loop current in the multifilamentary HTS wire with bridging of (a) short interval distance and (b) long interval distance.

Acknowledgments
The part of this work was supported by Grant-in-Aid for Scientific Research (C), the Ministry of Education, Science, Sports and Culture (No.16K06222).

References

[1] Parkinson B J , Slade R and Mallett M J D , Chamritski V 2013 Development of a cryogen free 1.5 T YBCO HTS magnet for MRI *IEEE Trans. Appl. Supercond.* **23** 4400405
[2] Terao Y, Ozaki O, Ichihara C, Kawashima S, Hase T, Kitaguchi H, Kobayashi S, Sato K, Nakajima I, Oonishi N, Poole M, Takeda K, Urayama S and Fukuyama H 2013 Newly designed 3 T MRI magnet wound with Bi-2223 tape conductors *IEEE Trans. Appl. Supercond.* **23** 4400904
[3] Maeda H and Yanagisawa Y 2014 Recent developments in high-temperature superconducting magnet technology (review) *IEEE Trans. Appl. Supercond.* **24** 1-12
[4] Slade R A, Parkinson B J and Walsh R M 2014 Test results for a 1.5 T MRI system utilizing a cryogen-free YBCO magnet *IEEE Trans. Appl. Supercond.* **24** 4400705

[5] Hashi K, Ohki S, Matsumoto S, Nishijima G, Goto A, Deguchi K, Yamada K, Noguchi T, Sakai S, Takahashi M, Yanagisawa Y, Iguchi S, Yamazaki T, Maeda H, Tanaka R, Nemoto T, Suematsu H, Miki T, Saito K and Shimizu T 2015 Achievement of 1020 MHz NMR *J. Magn. Reson.*, **256** 30-33

[6] Iwasa Y, Bascunan J, Hahn S, Voccio J, Kim Y, Lecrevisse T, Song J and Kajikawa K 2015 A High-Resolution 1.3-GHz/54-mm LTS/HTS NMR Magnet *IEEE Trans. Appl. Supercond.* **25** 4301205

[7] Bascunan J, Hahn S, Lecrevisse T, Song J, Miyagi D and Iwasa Y 2016 An 800-MHz all-REBCO Insert for the 1.3-GHz LTS/HTS NMR Magnet Program-A Progress Report *IEEE Trans. Appl. Supercond.* **26** 4300205

[8] Park D, Bascunan J, Michael P C, Lee J, Hahn S and Iwasa Y 2017 Construction and Test Results of Coils 2 and 3 of a 3-Nested-Coil 800-MHz REBCO Insert for the MIT 1.3-GHz LTS/HTS NMR Magnet *IEEE Trans. Appl. Supercond.*

[9] Parkinson B J, Bouloukakis K and Slade R A 2017 A compact 3 T all HTS cryogen-free MRI system *Supercond. Sci. Technol.* **30** 125009

[10] Tosaka T, Miyazaki H, Iwai S, Otani Y, Takahashi M, Tasaki K, Nomura S, Kurusu T, Ueda H, Noguchi S, Urayama S and Fukuyama H 2016 R&D Project on HTS Magnets for Ultrahigh-Field MRI Systems *IEEE Trans. Appl. Supercond.* **26** 402505

[11] Yokoyama S, Lee J, Imura T, Matsuda T, Eguchi R, Inoue T, Nagahiro T, Tanabe H, Sato S, Daikoku A, Nakamura T, Shirai Y, Miyagi D and Tsuda M 2017 Research and Development of the High Stable Magnetic Field ReBCO Coil System Fundamental Technology for MRI *IEEE Trans. Appl. Supercond.* **27** 4400604

[12] Yanagisawa Y, Piao R, Iguchi S, Nakagome H, Takao T, Kominato K, Hamada M, Matsumoto S, Suematsu H, Jin X, Takahashi M, Yamazaki T and Maeda H 2014 Operation of a 400MHz NMR magnet using a (RE:Rare Earth)Ba2Cu3O7-x high-temperature superconducting coil: Towards an ultra-compact super-high field NMR spectrometer operated beyond 1 GHz *J Magn Reson.* **249** 38-48

[13] Yanagisawa Y, Xu Y, Jin X, Nakagome H and H Maeda 2015 Reduction of Screening Current-Induced Magnetic Field of REBCO Coils by the Use of Multi-Filamentary Tapes *IEEE Trans. Appl. Supercound.* **25** 6603705

[14] Ueda H, Saito H, Ariya Y, Mochida A, Wang T, Wang X, Agatsuma K and Ishiyama A 2015 Reduction of Irregular Magnetic Field Generated by Screening Current in REBCO Coil *IEEE Trans. Appl. Supercound.* **25** 6603205

[15] Yanagisawa Y, Xu Y, Iguchi S, Hamada M, Matsumoto S, Nishijima G, Nakagome H, Takao T, Suematsu H, Oshima Y, Jin X, Takahashi M and Maeda H 2015 Combination of high hoop stress tolerance and a small screening current-induced field for an advanced Bi-2223 conductor coil at 4.2K in an external field *Supercond. Sci. Technol.* **28** 125005

[16] Ueda H, Fukuda M, Hatanaka K, Michitsuji K, Karino H, Wang T, Wang X, Ishiyama A, Noguchi S, Yanagisawa Y and Maeda H 2014 Measurement and Simulation of Magnetic Field Generated by Screening Currents in HTS Coil *IEEE Trans. Appl. Supercound.* **24** 4701505

[17] Ueda H, Ishiyama A, Ariya Y, Wang T, Wang X, Agatsuma K, Miyazaki H, Tosaka T, Nomura S, Kurusu T, Urayama S and Fukuyama H 2015 Evaluation of Magnetic-Field Distribution by Screening Current in Multiple REBCO Coils *IEEE Trans. Appl. Supercound.* **25** 4700705

[18] Ueda H, Imaichi Y, Wang T, Ishiyama A, Noguchi S, Iwai S, Miyazaki H, Tosaka T, Nomura S, Kurusu T, Urayama S and Fukuyama H 2016 Numerical Simulation on Magnetic Field generated by Screening Current in 10-T class REBCO Coil *IEEE Trans. Appl. Supercond.* **26** 4701205

[19] Smith D R and Fickett F R 1994 *Cryogenic Properties of Silver* National Institute of Standards and Technology Boulder

[20] Cryogenic Society of Japan 1993 *Handbook on Supercondcutivity and Cryogenics (Chodendo Teionkohgaku Handbook in Japaneses)* (Tokyo: Ohmsha, Ltd.) pp 1088

The microstructure characterization and phase composition analysis of (Bi,Pb)-2223 Ag/tapes with SnO, MgO and Ag_2O mix-doping

X Y Lu[1], D Yi[2], T Fujino[1] and A Nagata[1]

[1]Faculty of Engineering and Resource Science, Akita University, Akita, Japan
[2] School of Material Science and Engineering, Central South University, Changsha,China

E-mail: xiaoyelu@gipc.akita-u.ac.jp

Abstract. The Ag/tapes with a nominal composition of $Bi_{1.8}Pb_{0.4}Sr_{1.9}Ca_{2.1}Cu_{3.5}O_y$ + x wt% SnO + y wt% MgO + z wt%Ag_2O (x = 0, 0.2, 0.4; y = 0, 0.2; z = 0, 0.2) were prepared by sintering at 835℃ for 120 h after partial-melting at 845℃ for 1 h. The tape with 0.2wt%SnO, 0.2wt%MgO and 0.2wt%Ag_2O mix-doping shows the highest proportion of Bi-2223 phase and the highest critical current density. A major Bi-2223 phase coexisted with a few Bi-2212 phase, and large $(Sr,Ca)_2CuO_3$, $(Sr,Ca)CuO_2$ (~10 μm), and fine MgO particles (~1 μm) in the tape with 0.2 wt% SnO, 0.2 wt% MgO and 0.2 wt% Ag_2O mix-doping. The Sn and Pb amount in the Bi-2223 crystals is larger than that in the Bi-2212 crystals. It suggests that the SnO in the mix-doping with Ag_2O and MgO is favorable for the formation of Bi-2223 phase, and the MgO doping can increase J_c due to improving the flux pinning.

1. Introduction

Among all the high-temperature superconductors discovered to date, Bi-2223 oxide superconductor is the most promising material for a tape used in high-current applications. However, critical current density (J_c) is strongly influenced by the presence of pinning defects and their possible matching with the cores of the vortex lines [1]. The defect structure is essential for flux pinning enhancement, which is in particular, the most important problem for the Bi-based superconductors [2,3]. In order to increase flux pinning, the Bi-2212 phase was doped by Al, Pb or transition metal elements to produce point defects [3,4], while in the Bi-2223 phase, Sr was partially substituted by Ba [5] or Sn [6] for the same purpose. In previous work [7, 8], we reported the effect of MgO, Ag_2O, PtO_2 particles on the microstructure and pinning properties of Ag sheathed Bi-2223 tapes. The flux pinning was not clearly improved by the individual doping of MgO, Ag_2O or PtO_2 particles, respectively. However, MgO doping did not affect the formation rate of Bi-2223 phase, and could suppress the growth of secondary phases. Ag_2O or PtO_2 additions resulted in coarse secondary phases, and slowed the formation rate of Bi-2223 phase. Recently, we have reported the effect of SnO, MgO and Ag_2O mix-doping on the formation and the superconducting properties of Bi-2223 phase of the Ag-sheathed tapes in the partial-melting and sintering process. The tape with 0.2wt% SnO, 0.2wt% MgO and 0.2wt%Ag_2O mix-doping has shown the highest proportion of Bi-2223 phase and the highest critical current density [9]. In the present work, we studied the microstructure characterization and phase composition distribution of (Bi,Pb)-2223 Ag/tapes with SnO, MgO and Ag_2O mix-doping.

2. Experimental

The precursor with a nominal cation ratio of Bi:Pb:Sr:Ca:Cu=1.8:0.4:1.9:2.1:3.5 was prepared by high purity (>99.9%) Bi_2O_3, PbO, $SrCO_3$, $CaCO_3$ and CuO powders. After calcination at 800℃ for 12 h, the powder was divided into six batches in which one was taken as the pure sample. 99.99% purity SnO, 99.9% MgO and 99.8% Ag_2O powders with an average size of 0.5 μm were added to the other five batches in five different mixture of x wt% SnO + y wt% MgO + z wt% Ag_2O (x = 0, 0.2, 0.4; y = 0, 0.2; z = 0, 0.2). Each powder was deformed to a pellet with a size of W 2 mm× H 3 mm ×L 150mm by cold pressing at a pressure of 233 MPa. These samples placed on a silver sheet were partially melted in air at 850℃ for 1 h, then furnace-cooled to room temperature. In this work, a specific method for the preparation of the Ag-sheathed tapes was used. First, the bulk samples were cut into 1×1×15mm^3 sizes, and then wrapped with three turns of 0.1 mm thick Ag sheets. The wrapped samples were rolled to tapes with an overall thickness of 0.2 mm and a width of 3 mm. Finally, the resultant tapes were sintered at 835℃ for 120 h after partial-melting at 845℃ for 1 h. The microstructure characterization and phase composition analysis were performed by means of a JXA-8230 Electron Probe Microanalyzer (EPMA). The critical current I_c was measured at 77 K in self-field using standard four-probe technique with the criterion of 1 μVcm^{-1}. The critical current density J_c was determined from I_c and a cross-section of the oxide core. The formation of the Bi-2223 phase was analyzed by XRD with Cu Kα radiation.

3. Results and discussion

Fig. 1 shows the volume fraction of Bi-2223 phase of the tapes. The individual SnO doping (b and c), SnO andAg_2O mix-doping (d), and SnO and MgO mix-doping (e) all decrease the conversion of Bi-2212 phase to Bi-2223 phase by comparison with the un-doping (a). The tape with individual 0.4wt% SnO doping shows the lowest conversion. However, the tape with 0.2wt%SnO, 0.2wt%MgO and 0.2wt%Ag_2O mix-doping (f) shows the highest proportion of Bi-2223 phase.

Fig. 2 shows the J_c of tapes un-doped (a), 0.2wt%SnO doped (b), 0.4wt%SnO doped (c), 0.2wt% SnO + 0.2wt%Ag_2O mix-doped (d), 0.2wt%SnO + 0.2wt%MgO mix-doped (e), and 0.2wt%SnO + 0.2wt%MgO + 0.2wt%Ag_2O mix-doped (f). The SnO and MgO mix-doped tape (e) shows higher J_c value than the un-doped tape (a) and the SnO and Ag_2O mix-doped tape (d) although the volume fraction of Bi-2223 phase in SnO and MgO mix-doped tape (e) is lower than that in the un-doped tape (a) and the SnO and Ag_2O mix-doped tape (d) as shown in Fig. 1. The tape with individual 0.4wt%SnO doping shows the lowest J_c value. However, the tape with 0.2wt%SnO, 0.2wt%MgO and 0.2wt%Ag_2O mix-doping shows the highest J_c value.

In order to correlate the critical current density to the SnO, MgO and Ag_2O mix-doping, the microstructure characterization and phase composition analysis were performed by means of EPMA. Fig. 3 shows a SEM micrograph of the tape with 0.2wt% SnO, 0.2wt% MgO and 0.2wt%Ag_2O mix-doping. EPMA measurements have shown that major dark-gray regions are the Bi-2223 crystals, and a few light-gray regions are the Bi-2212 crystals. The fine black regions (~1 μm) are MgO particles, and the large black regions (~10 μm) are $(Sr,Ca)_2CuO_3$ (21) and $(Sr,Ca)CuO_2$ (11), respectively. The fine MgO particles may be favorable for flux pining.

Fig. 4 shows an EPMA compositional map of Pb (red) and Bi (blue). The Pb amount in the Bi-2223 grains is larger than that in the Bi-2212 crystals. Fig. 5 shows an EPMA compositional map of Sn (red), Ag (blue) and Mg (green). The Sn amount in the Bi-2223 grains is larger than that in the Bi-2212 crystals, and the Ag amount in the Bi-2223 grains seems no difference with that in the Bi-2212 crystals. Fig. 6 shows an EPMA compositional map of Sr (red), Ca (blue) and Cu (green).

From Fig. 1, the individual doping of SnO (Fig. 1(a) and (b)) decreases the proportion of Bi-2223 by comparison with the no-doping tape (Fig. 1(a)). However, the mix-doping of SnO, Ag_2O and MgO (Fig. 1(f)) increases the proportion of Bi-2223. Above results suggest that the SnO in the mix-doping with Ag_2O and MgO is favorable for the formation of the Bi-2223 phase, and the MgO doping can influence on J_c due to improving the flux pinning.

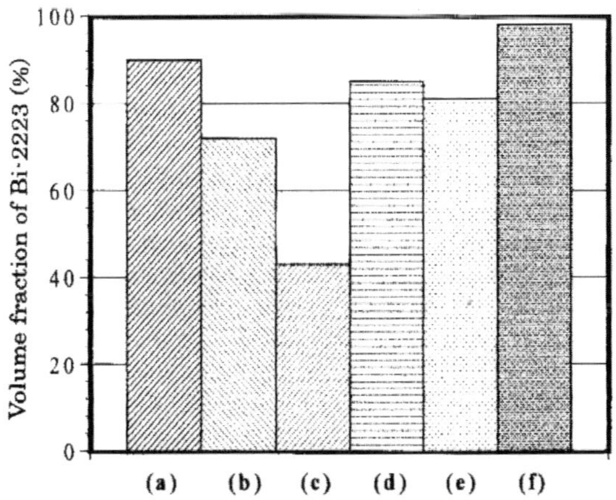

Fig. 1. The volume fraction of Bi-2223 phase of tape un-doped (a), 0.2wt% SnO (b), 0.4wt%SnO (c), 0.2wt%SnO + 0.2wt%Ag_2O (d) 0.2wt%SnO + 0.2wt%MgO (e), 0.2wt%SnO + 0.2wt%MgO + 0.2wt%Ag_2O (f).

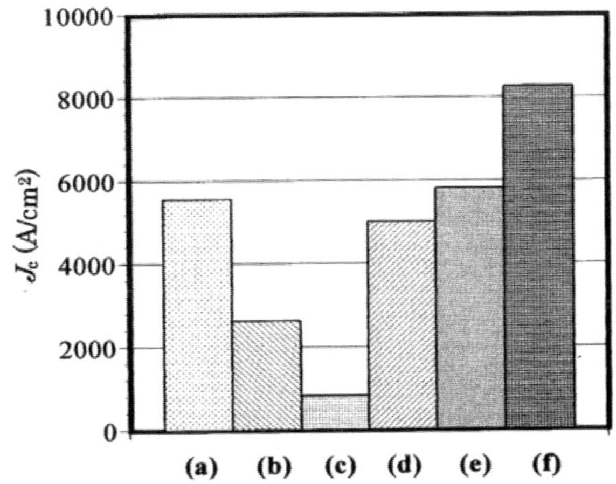

Fig. 2. The Critical current density (J_c) of tapes un-doped (a), 0.2wt% SnO (b), 0.4wt%SnO (c), 0.2wt%SnO + 0.2wt%Ag_2O (d) 0.2wt%SnO + 0.2wt%MgO (e), 0.2wt%SnO + 0.2wt%MgO + 0.2wt%Ag_2O (f).

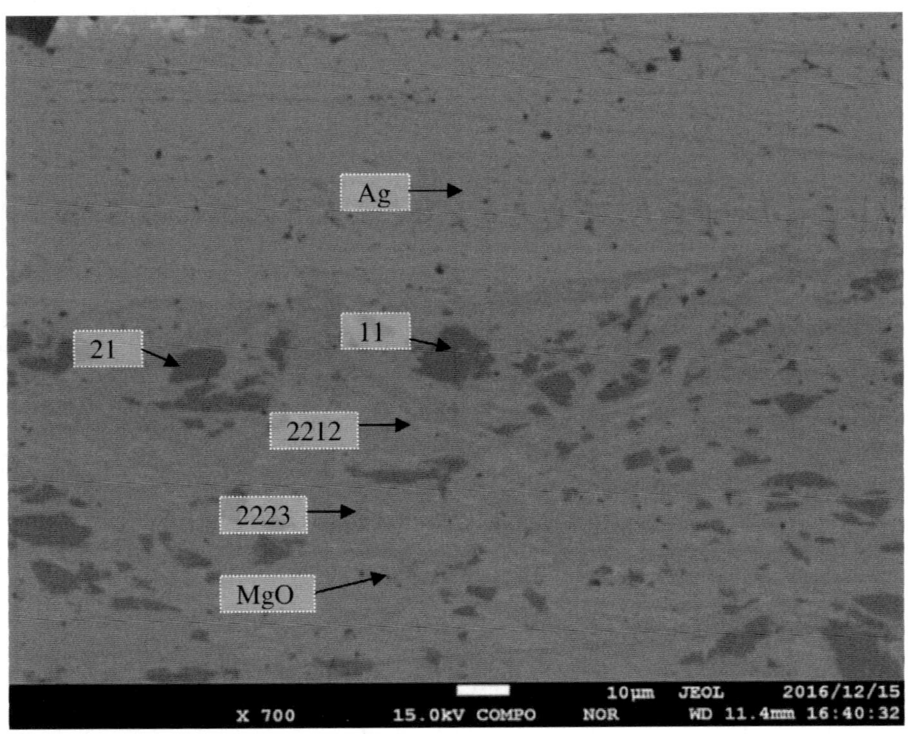

Fig. 3. SEM micrograph for the tape with 0.2wt% SnO,0.2wt% MgO and 0.2wt% Ag_2O mix-doping.

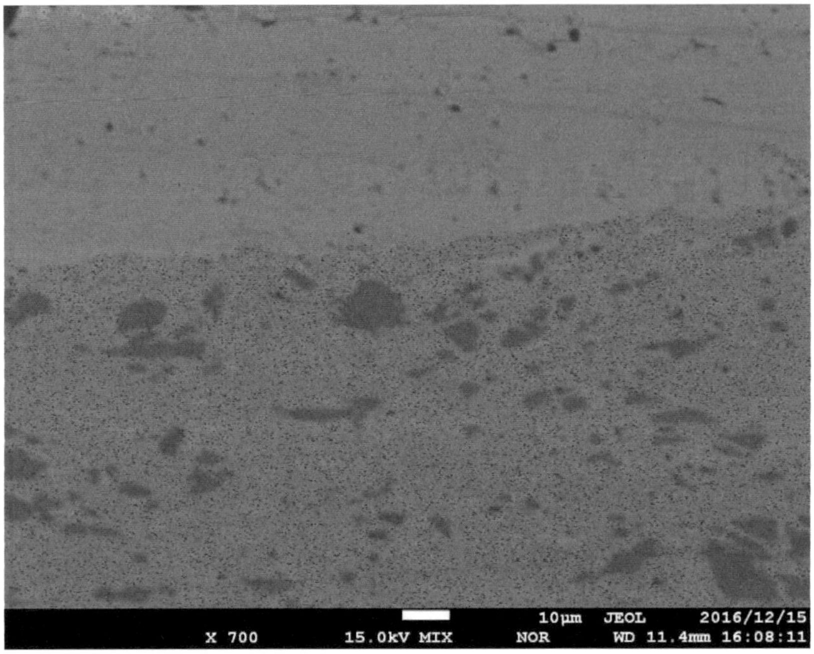

Fig. 4. EPMA compositional map of Pb (red) and Bi (blue) for the tape with 0.2wt% SnO, 0.2wt% MgO and 0.2wt% Ag_2O mix-doping.

Fig. 5. EPMA compositional map of Sn (red), Ag (blue) and Mg (green) for the tape with 0.2wt% SnO, 0.2wt% MgO and 0.2wt% Ag_2O mix-doping.

Fig. 6. EPMA compositional map of Sr (red), Ca (blue) and Cu (green) for the tape with 0.2wt% SnO, 0.2wt% MgO and 0.2wt% Ag_2O mix-doping.

4. Conclusion

The Ag/tapes with a composition of $Bi_{1.8}Pb_{0.4}Sr_{1.9}Ca_{2.1}Cu_{3.5}O_y$ + x wt%SnO + y wt%MgO + z wt% Ag_2O (x = 0, 0.2, 0.4; y = 0, 0.2; z = 0, 0.2) were prepared by sintering at 835℃ for 120 h after partial −melting at 845℃ for 1 h. The tape with 0.2wt%SnO, 0.2wt%MgO and 0.2wt%Ag_2O mix-doping shows the highest proportion of Bi-2223 phase and the highest critical current density. A major Bi-2223 phase coexisted with a few Bi-2212 phase, and large $(Sr,Ca)_2CuO_3$, $(Sr,Ca)CuO_2$ (~10 μm) and fine MgO particles (~1 μm) in the tape. The Sn and Pb amount in the Bi-2223 crystals is larger than that in the Bi-2212 crystals. It suggests that the SnO in the mix-doping with Ag_2O and MgO is favorable for the formation of Bi-2223 phase, and the MgO doping can increase J_c due to improve the flux pinning.

References

[1] Goupil C, Ruyter A, Hardy V and Simon C 1997 *Phyica C* **278** 23-30
[2] Majewski P, Elschner S, Hettich B, Lang C, Kaesche S and Aldinger F 1994 *Supercond. Sci. Technol.* **7** 514-517.
[3] Kazin P E, Poltavets V V, Tretyakov Y D, Jansen M, Freitag B and Mader W 1997 *Physica C* **280** 253-265
[4] Paulose P L, Patil S, Frank H and Guntherodt G 1993 *Physica C* **208** 11-17
[5] Ikeda H, Yoshizaki R and Yoshikawa K 1994 *Physica B* **194** 2205-2206
[6] Seyoum H M, Habib J M, Bennet L H, Wong W, Shapiro A J and Swartzendruber L J 1990 *Supercond. Sci. Technol.* **3** 616-621
[7] Lu X Y, Nagata A, Sugawara K and Kamada S 2000 *Physica C* **335** 51-55
[8] Lu X Y, Nagata A, Sugawara K and Kamada S 1996 *Advances in Cryogenic Engineering (Materials)* **42** 723-730
[9] Lu X Y, Yi D and Chen H 2015 *Advanced Materials Research* **1096** 251-255

Effects of intermediate annealing on the microstructure and transport properties of Bi-2223 wires and tapes

Xiaobo Ma[1,2], Shengnan Zhang[2], Zeming Yu[2], Guoqing Liu[2], Huiling Zheng[2], Chengshan Li[2], Jinshan Li[1], Pingxiang Zhang[1,2]

[1] State Key Laboratory of Solidification Processing, Northwestern Polytechnical University, Xi'an 710072, China
[2] Northwest Institute for Nonferrous Metal Research (NIN), Xi'an 710016 China

E-mail:yequ198508@163.com

Abstract. Monofilament and 37-filamentary Bi-2223 tapes were fabricated by powder in tube (PIT) process. And the mono-filamentary wires were drawn from ϕ12mm to hexagon 1.51 mm with one and three times intermediate annealing, respectively. The influences of different intermediate annealing times on the core density, morphology and transport properties of Bi-2223 wires and tapes were systematically investigated. It was noticed that the density and the Vickers micro-hardness values of superconducting core on the cross-section of one time annealing wires were higher than those of three times annealing wires. Meanwhile, it was observed that the Ag/oxide core interfaces of the wire with one time annealing were more smooth than those of the three times annealing wire. With the intermediate annealing time decreasing from three to one, the critical current density J_c increased from 17.9 kA/cm^2 to 20.5 kA/cm^2. And the enhancement of Jc-B properties suggests the improvement of intergrain connections due to the increasing core density. Meanwhile one time annealing procedure was used to Ag-Au sheathed tape, J_c reached 16.3 kA/cm^2, this value was also higher than those traditional wires in our previous experiments, which suggests that the decreasing of annealing times during drawing process is also effective in AgAu sheath tapes for the enhancement of current capacity.

1. Introduction

Since the first discovery of Bi-based superconductors in 1986, $Bi_2Sr_2Ca_2Cu_3O_{10+\delta}$(Bi-2223) has become one of the most promising high temperature superconductors (HTS) for practical applications under liquid nitrogen temperature (77 K), due to the high superconducting transition temperature (T_c~110 K) and high critical current density (J_c) [1–3]. During the powder in tube (PIT) process, deformation of Bi-2223/Ag composites is very important from the viewpoint of Bi-2223/Ag superconducting tape performance considering the density and the final microstructure of the superconducting cores are two of the dominating factors for the current capacity [4-7]. Core density influences the kinetics of 2223 phase creation and finally the quality of inter-grain connectivity: a higher core density results in a higher J_c [8,9]. The interface between the Ag sheath and the oxide core is very important for the current flow and texture formation, considering that the interface areas carry higher current than the core area. The wavy interface will severely interfere the texture formation of Bi-2223 grains along the metal sheath, and reduce the high current capacity areas [10-12].

During drawing, the Ag sheath experiences work hardening, therefore intermediate annealing is necessary in order to prevent breaking of the wire. Thus, it is quite necessary to optimize the intermediate annealing of drawing process in the Ag sheathed tapes. A significant number of studies related to the drawing process of the Bi-2223/Ag composites have been published previously. It has been demonstrated that softening of the Ag sheath by annealing will enhance the formation of sausages [13] and decrease the core density, The authors argue that this is because that increasing number of anneals or annealing temperature and time will enhance the difference of deformation resistance between Ag and the oxide core. Wang *et al* [14] deduce that the effect of annealing procedure on sausages is more significant than that of particle size of precursor powder.

In the present study, aiming at the enhancement of current capacity of Bi-2223 tapes, the influences of intermediate annealing times during drawing on the monofilament density, Ag/oxide core interface morphology and critical current properties of Ag sheathed Bi-2223 tapes were systematically discussed. Due to the formation of more smooth interfaces and high core density, an enhancement of J_c over 14% was obtained by decreasing the annealing times from three to one. Meanwhile, the one time annealing procedure was also applied to Ag-Au sheathed tapes and a higher critical current density was achieved.

2 Experimental

Bi-2223 precursor powders with nominal composition of $Bi_{1.76}Pb_{0.34}Sr_{1.93}Ca_{2.02}Cu_{3.06}O_x$ were prepared through co-precipitation process [15, 16] with the initial phase assemblage of (Bi, Pb)-2212, Ca_2CuO_3 and CuO. Mono filament wires were fabricated first by packing the precursor powders into Ag sheath. Then a series swaging and drawing process were performed with different intermediate annealing times to draw the wires from ϕ12 mm to hexagon 1.51 mm. 1# sample was annealed in long tube furnace at 400 °C for 0.5 h at the diameter of ϕ12, ϕ3.95, ϕ2.02 mm, and 2# sample was annealed only at the diameter of ϕ2.02mm, 3# sample was AgAu sheathed (Bi, Pb)-2223 (inner and outer sheaths are both AgAu), and was also annealed only at ϕ2.02mm. The intermediate annealing profile of different samples is shown in Table 1. 37-filamentary Ag sheathed (Bi, Pb)-2223 tapes were fabricated by reassemble the monofilament wires into AgMgNi sheath [17]. And after a series of swaging, drawing and rolling process, three different final tapes with the cross section dimensions of ~0.29×4.4 mm were obtained. The as-rolled tapes with the length of 20 cm were then thermo-mechanically processed in 7.5% O_2/Ar atmosphere by three heat treatments at 822, 826, and 780°C, respectively. An intermediate rolling step was adopted to develop a dense and aligned Bi-2223 phase structure.

The morphology of wires were observed by optical microscope after polishing. The core density of the wires were calculated as

$$d_{core} = \frac{m\text{-}d_{Ag}A_{Ag}l}{A_{core}l}$$

where m is the overall weight of a l long piece of wire, A_{Ag} and A_{core} are the cross section area of Ag and core component, respectively, and d_{Ag} and d_{core} are the density of Ag and core, respectively. Micro-Vickers hardness HV0.25 (100 g, 15 s) was measured on the cross-section of wires to monitor the core density. The critical currents (I_c) at 77 K, 0~0.6 T with the field parallel to the rolling plane of the tape, were measured using a standard four-probe DC method with the criterion of 1 μV/cm. Critical current density, J_c was calculated as $J_c = I_c/A$, where A is the cross section area of superconducting core.

Table 1. Intermediate annealing profile of different samples.

Sample number	Sheaths	Annealing times	Annealing size(mm)
1#	Ag+AgMgNi	3	ϕ12, ϕ3.95, ϕ2.02
2#	Ag+AgMgNi	1	ϕ2.02
3#	AgAu+AgAu	1	ϕ2.02

3 Results and discussion

3.1 Density and hardness changes during drawing

Density measurements of monofilament wire with different size (from ϕ3.95 mm to ϕ1,64 mm) of the samples are shown in figure 1, where D_0 refers to the starting size of monofilament wire (ϕ12mm), D refers to actual size. For 1#、 2# and 3# samples, the initial packing density is mostly the same, about 3.06g/cm^{-3}. It is clearly seen that the density of 1# sample is always the lowest among all the samples, especially at the beginning of drawing process. It reaches the maximum value of 4.93 g/cm^{-3} at the diameter of ~ϕ2.11 mm. The densities of 2# and 3# sample are almost the same along the drawing process, which both reach the maximum value of ~5.06 g/cm^{-3} simultaneously at the diameter of ϕ2.02 mm. However, the final density of 3# is obviously higher than that of 2# after the final annealing process at ϕ2.02 mm. The 1# wire is annealed for three times as marked in figure 1, the annealing process can obviously lead to the softening of the sheath. Therefore, the constrain stress applied on ceramics core is smaller after annealing and cause the decrease of core density. The maximum stress applied during drawing is limited by the strength of the Ag sheath. Thus the adoption of AgAu alloys could make it possible to apply larger drawing stress and therefore increase the final density without breaking the wire, so the density of 3# sample is highest at the final dimension.

There was a common fact that the density fluctuated for three samples, this appeared to be real, and not caused by uncertainty in the measurements. The reason was that the critical density was reached. If the powder-flow model was used, further deformation will not increase the density, but produced cracks due to the brittleness of the BiSrCaCuO materials [18]. If the density is evaluated from the change of volume this crack formation will result in an apparent decrease in density. These cracks can be healed by further drawing so the density increases again.

Micro-Vickers hardness values of monofilament wires with different sizes (from ϕ3.95 mm to ϕ1.64 mm) of different samples are shown in figure 2. Micro-Vickers hardness values are measured at the center of ceramic core. It is clearly seen that the hardness of 1# sample is the lowest, and reaches the maximum value of 80.6 also at the diameter of ~ϕ2.11 mm. The hardness values of 2# and 3# wires are almost the same before the annealing at ϕ2.02 mm, and both reached the maximum values of 85.0 and 94.3 at the diameter of ϕ2.02 mm. And after annealing the hardness values of 3# are all higher than those of 2#. The changing trend of hardness is consistent with that of core density. Figure 3 shows the Micro-Vickers hardness values of core as a function of core density. A linear relationship can be observed between the core density and hardness, which verifies the both results in figure 1 and figure 2.

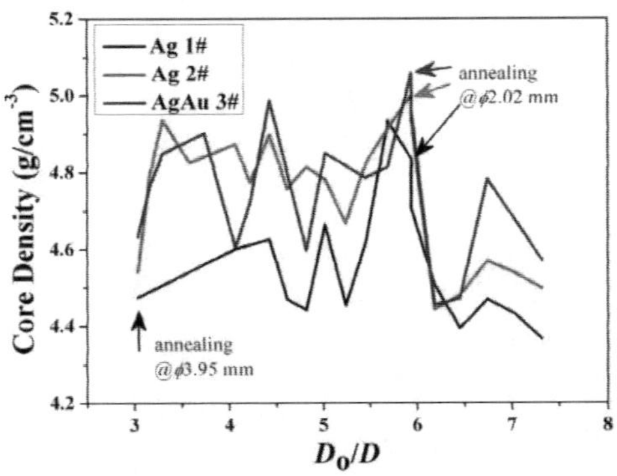

Figure 1. Density *vs* wire size for 1#, 2#, 3# samples

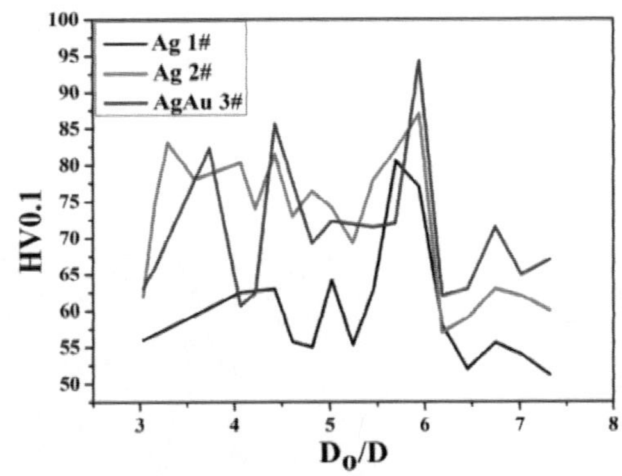

Figure 2. Micro-Vickers hardness *vs* wire size of different samples

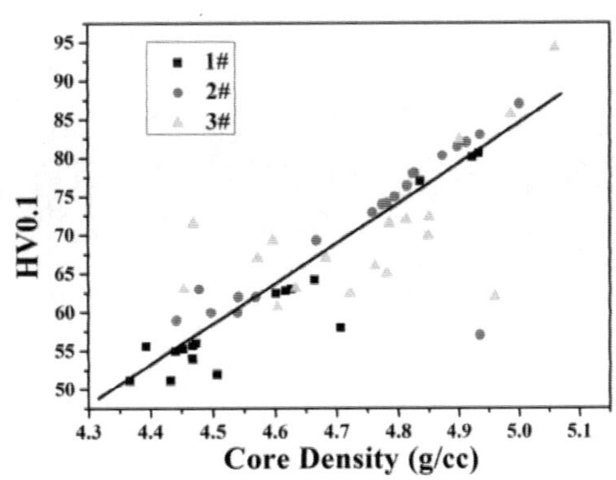

Figure 3. Micro-Vickers hardness as a function of core density

3.2 Cross-section morphology of monofilament

Cross-section images of monofilament wires with different sizes are shown in figure 4. It reveals that the interfaces between the Ag sheath and the oxide core is rough for 1# sample, while the interfaces for 2# and 3# samples are much more smooth. It is because that the strength of the sheath material has an obvious influence on the Ag/oxide core interfaces. Considering that the yield strength of Ag is lower than that of ceramic powders, the deformation behaviors of metal sheath and ceramic core are different during the drawing process. Sample 1# has been annealed for three times, and the strength of Ag sheath will decrease after every annealing, but the strength of powder essentially constant. Thus the deformation difference between sheath and ceramic powder is larger than that of 2# and 3# wires. Therefore, the deformation at Ag/core interface is more uniform in 2# and 3# sample and the formation of wavy interface is successfully aviod. For 3# sample, the sheath is AgAu alloy, the strength of which is higher than that of Ag, so the interface is smoother and the final core density is higher.

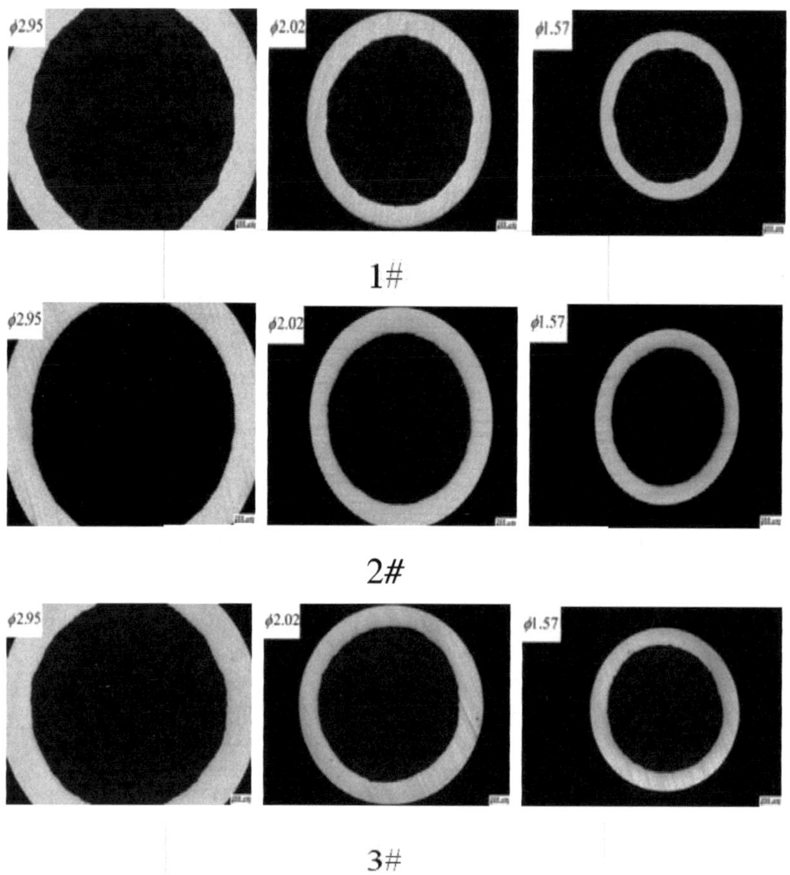

Figure 4. Optical micrographs of cross-section of monofilament wire with different size

3.3 Critical current density

Figure 5 shows the J_c values (77 K, 0 T) of Bi-2223 tapes of 1#, 2# and 3# samples after heat treatment. It can be noticed that the J_c values increase from 17.9 kA/cm^2 of 1# sample to 20.5 kA/cm^2 of 2# sample with the decrease of intermediate annealing from three to one time. The 15% enhancement of J_c suggests that the optimized intermediate annealing of drawing process for our Ag sheathed Bi-2223 wire is one time. Meanwhile, one time annealing procedure has also been adopted on Ag-Au sheathed tape. And a J_c value of 16.3 kA/cm^2 is achieved, which is higher than the results obtained in our previous experiments (the phase evolution process of AgAu sheathed Bi-2223 tapes is different with that of Ag sheathed tapes attributed to the slower oxygen penetration rate of AgAu alloy, Nowadays, the obtained current capacity of AgAu sheathed tapes is still ~30% lower than that of Ag sheathed tapes).

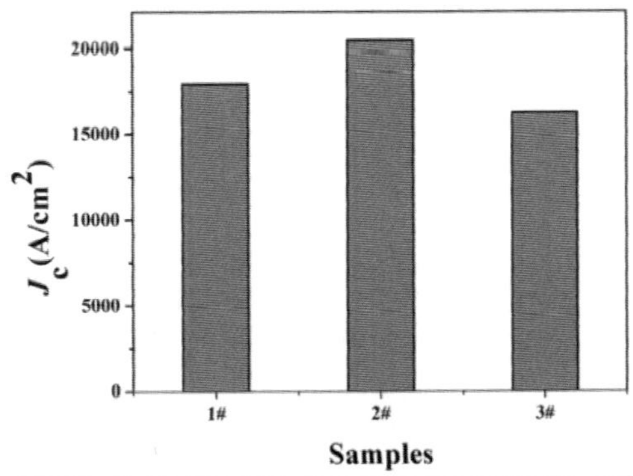

Figure 5. J_c values (77 K, 0 T) of 37-filament tapes of 1#, 2# and 3# samples

Fig 6 showed normalized $J_c/J_c(0)$ value in magnetic field applied parallel to the tape surface. It can be noticed that the $J_c/J_c(0)$ value of 2# sample is higher than that of 1# sample at low magnetic field. And they are mostly the same with the magnetic field increases up to 0.3 T. It implies that inter-grain connections of tapes are improved with the application of one time intermediate annealing, attributed to the increased density of ceramic core.

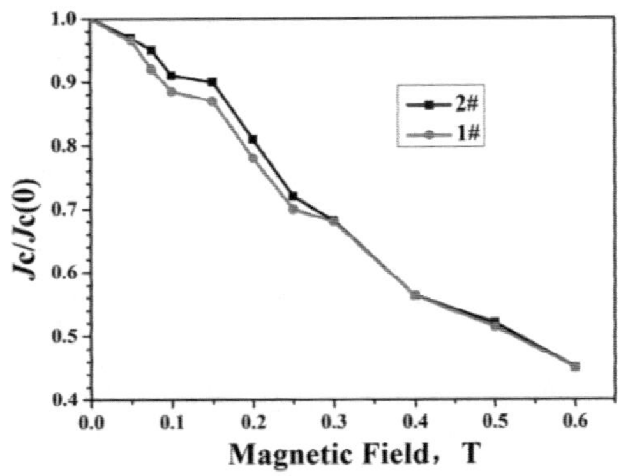

Figure 6. Normalized $J_c/J_c(0)$ value in magnetic field applied parallel to the tape surface

General speaking, there are two factors which mainly affect the critical current density of Bi-2223 tapes in our study, namely the superconducting core density and the Ag/superconducting core interface morphology. The density of the superconducting core should be sufficiently high. A higher density core directly correlated to more effective current carrying areas and a smaller crack density. Crack inside the superconducting core could limit J_c or even block the current completely. Such a core has more effective current carrying areas and good connectivity, so it always results in a higher J_c. And low density area should be smaller to increase the effective current carrying area. Therefore, the high density core of 2# sample lead to the higher current capacity.

On the other hand, the interface morphology is another important factor which can influence the formation of textures along the interface and effective current carrying area. The interface is crucial for the textural alignment of Bi-2223 grains near interface. Meanwhile the textural alignment is better near Ag/superconducting core interface than in the center of the superconducting core. Therefore, a rough interface may have an undesired influence on the alignment of Bi-2223 grains. Moreover wavy interface will reduce the actual cross-section area of the superconducting core and, therefore reduce the overall critical current. So the smooth interface in 2# sample with one time intermediate annealing is benefit to the high current capacity.

4. Conclusion

In this study, the influences of intermediate annealing times of drawing process on the core density, morphology and transport properties of Bi-2223 tapes were systematically investigated. Experimental results showed that with the decrease of annealing times from three to one, both the core density and the Vickers micro-hardness values of superconducting core increased consistently. Meanwhile, the Ag/oxide core interfaces became more smooth. Attributed to the higher superconducting core density and flat Ag/core interface, the maximum J_c value of 20.5 kA/cm^2 and higher $J_c/J_c(0)$ value was obtained on the Bi-2223 tapes with one times intermediate annealing, which is ~15% higher than that of the wires with three times intermediate annealing. Moreover, the one time annealing procedure was also effective to Ag-Au sheathed tape, and a J_c value of 16.3 kA/cm^2 has been achieved.

Acknowledgments

This study was financially supported by the National Key Project of Magneto-Constrained Fusion Energy Development Program under Grant 2015GB115001.

References

[1] Nassi M 2000 *Supercond. Sci.Technol.* **13** 460-463
[2] Sato K, Kato T, Ohkura K, Kobayashi S, Fjuino K, Ohmatsu K, Hayashi K 2000 *Supercond. Sci. Technol.* **13** 18-22
[3] Paul W, Lakner M, Rhyner J, Unternahrer P, Baumann Th, Chen M, Widenhorn L, Guerig A 1997 *Supercond. Sci. Technol.* **10** 914-918
[4] Grasso G, Jeremie A , Flukiger R 1997 *Supercond. Sci. Technol.* **8** 827-830
[5] Husek I, Kovac P, Pachla W 1995 *Supercond. Sci. Technol.* **8** 617-625
[6] Kovac P, Husek I, Pachla W 1997 *IEEE Trans. Appl. Supercond.* **7** 2098-2101
[7] Husek I, Kovac P, Kopera L 1996 *Supercond. Sci. Technol.* **9** 1066-1070
[8] Yamada Y, Sato M, Murase S, Kitamura T, Kamisada Y 1993 *Advances in Superconductivity.* (Springer Japan, Tokyo) p 152
[9] Parrell J A, Dorris S E, Larbalestier D C 1994 *Physica C.* **231** 137-146
[10] Li Q, Brodersen K, Hjuler H A, Freloft T 1993 *Physica C.* **217** 360-366
[11] Haldar P, Hoehn J G Jr, Rice J A, Molowidlo L R 1992 *Appl. Phys. Lett.* **60** 495-497
[12] Han Z, Freltoft T 1994 *Appl. Supercond.* **2** 201-215
[13] Zeimetz B, Pan A, Dou S X 1995 *Physica C.* **250** 170-174
[14] Wang W G, Horvat J, Liu H K, Dou S X 1996 *Supercond. Sci. Technol.* **9** 875-880
[15] Dorris S E, Prorok B C, Lanagan M T, Sinha S, Poeppel R B 1993 *Physica C.* **212** 66-74
[16] Dorris S E, Prorok B C, Lanagan M T, Browning N B, Hagen M R 1994 *Physica C.* **223** 163-172
[17] Yoo J M, Chung H S, Ko J W, Kim H D, Sha J 1996 *Physica C.* **267** 53-58
[18] Korzekwa D A, Bingert J F, Podtburg E J, Miles P 1994 *Appl. Supercond.* **2** 261-270

Effect of grinding method on the precursor powder of Bi2223 and properties of strip

G Q Liu[1], Q B Hao[1], H L Zheng[1], S N Zhang [1], X Y Xu[1], G F Jiao[1], L J Cui [2], P F Wang[1,2], C S Li[1]

1 Northwest Institute for Nonferrous Metal Research, Xi'an,710016,China
2 The College of Materials, Northwestern Polytechnical University, Xi'an,710072,China

E-mail:wpf_chenxi@163.com

Abstract. In this paper, the two kinds of precursor powders of BiPb2212 and $CaCuO_2$ were prepared by double powder method and an improved coprecipitation method, and then the BiPb2212 powders were fabricated by mechanical grinding and hand grinding. The BiPb2212 powders and $CaCuO_2$ powders were prepared by two kinds of grinding methods mixed with a certain proportion of grinding, the Bi2223 superconducting tape with 37 filaments were prepared by removing the carbon, filling tube and annealing. The XRD analysis, SEM observation, laser particle size analysis and the tap density test of the precursor BiPb2212 powders were carried out respectively by mechanical grinding and hand-grinding. The results showed that the phase composition, microstructure, particle size distribution and tube density detection were all significant differences by using two kinds of grinding methods. Finally, the superconductivity test of the treated strip results show that the critical current from the field for the Bi2223 multi-core strip at 77K is improved from 86A (hand-grinding method) to 127A(mechanical grinding method). Therefore, the Bi2223 precursor powder by using mechanical grinding method is more conducive to obtain high-performance.

1. Introduction

$Bi_{2-x}Pb_xSr_2Ca_2Cu_3O_y$ (Bi2223) superconducting tapes is one of the high-temperature superconducting materials. Powder-in-tube (PIT) is the mainstream technology for preparing Bi2223 superconducting tapes. High-quality precursor powder is the basis for the preparation of high-performance Bi-2223 tapes, the main characteristics of the parameters including the atomic ratio of the elements, phase composition and content, distribution of elements, particle size and distribution and carbon content and so on. When the atomic ratio is close, the phase composition, content, particle size distribution and elemental distribution of precursor powder have a great influence on the superconductivity of Bi2223 tape.

At present, the preparation methods of precursor powder include five kinds of solid phase mixing methods, such as sol-gel method [3], spray-drying method [4], chemical coprecipitation method [5] and spray pyrolysis method [6]. The main disadvantage of solid-phase mixing is that it is difficult to achieve uniform mixing, the sol-gel preparation method is complicated and difficult to control, and the spray drying method has the advantages of simple equipment, high preparation efficiency and easy introduction of impurities. For ordinary coprecipitation method, due to the difference in solubility of various substances, resulting in a larger deviation of the atomic ratio of the final powder, spray pyrolysis method requires more expensive equipment support, and the current use of spray pyrolysis equipment made in China have not yet prepared high quality superconducting precursor powder.

Ordinary coprecipitation method by using water as a solvent, due to the solubility of metal oxalate in water is relatively large, it is likely to loss, and the solubility of each element is vary greatly, resulting in the final powder with atomic order. The ratio of deviations is larger, reducing the ultimate current capacity of superconducting tape. Improved coprecipitation method by choosing ethanol as solvent, and by strictly controlling the ratio of ethanol and water in the precipitation system, so the solubility of metal oxalate can be greatly reduced, and the precursor powder with similar design ratio can be obtained, while the ethanol can increase the filtration rate of coprecipitated slurry and shorten the drying time.

Bi2223 precursor powder can be synthesized in one step, which is called single powder process and is currently used by most research institutes. BiPb2212 and $CaCuO_2$ powder (composed of Ca_2CuO_3 and CuO) can be synthesized first, and then the two powders can be mixed and grinded, known as the dual-powder process [7,8]. Based on the improved coprecipitation method, Bi2223 precursor powder was prepared by dual powder process route. BiPb2212 was pulverized by different grinding methods. The effects of different grinding methods on phase composition (XRD analysis), Microstructure (SEM observation), particle size distribution and density of tube, and the final Bi2223 phase velocity and current carrying capacity were tested.

2. Experiments

The following are BiPb2212 and $CaCuO_2$ powder preparation process.

First, weigh a certain amount of oxalic acid dissolved in ethanol to prepare a precipitant solution, and then follow the $Bi_{1.8}Pb_{0.34}Sr_{1.9}Ca_{1.0}Cu_2O_x$ ratio of Bi_2O_3, CuO, $CaCO_3$, $SrCO_3$ and $Pb(NO_3)_2$ dissolved in concentrated nitric acid, and configured as a nitrate solution.

The coprecipitation process shown in Fig.1, the nitrate solution was added to the precipitant solution at a certain speed by stirring, while we use the ammonia to adjust the pH of the precipitation system, and control the pH value in the range of 1.5 ± 0.05 [9], Standing after precipitation filtration and drying, then the preparation of the pre-decomposition of the powder and the phase heat treatment, after each heat treatment, grinding the powder by using mechanical grinding and manual grinding methods.

$CaCuO_2$ powder preparation process:

First, weigh a certain amount of oxalic acid dissolved in ethanol to prepare a precipitant solution, followed by $Ca_{1.0}Cu_{1.02}O_x$ ratio of Cu, $CaCO_3$ dissolved in concentrated nitric acid, configured as a nitrate solution, the same coprecipitation as shown in Fig.1, the nitrate solution was added to the precipitant solution during stirring, while we use the aqueous ammonia to adjust the pH of the precipitation system within a suitable range, standing precipitation and filter drying preparation of the powder pre-decomposition and heat treatment, after each heat treatment, grinding the powder by manual grinding.

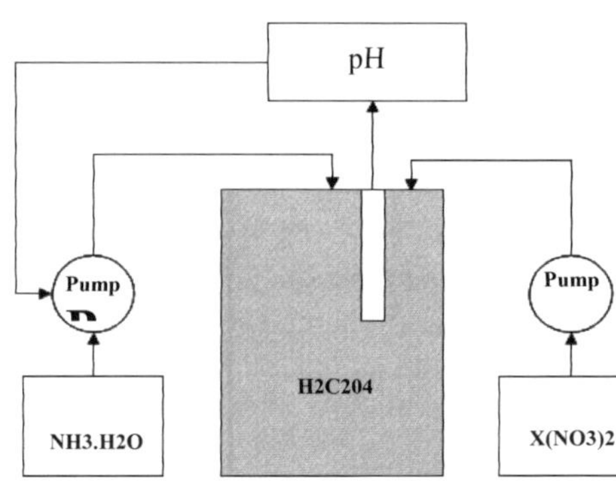

Fig.1. The draft of the improved co-precipitation

The BiPb2212 and $CaCuO_2$ powders were prepared by using two grinding methods with mixed and ground, according to the ratio of one to one substance, and the mixed grinding powder carbonized at high temperature, followed by manual packing, single-core wire processing, multi-core wire composite and processing, finally, prepared Bi2223 multi-core tape. The strip annealed twice in succession and annealed once for phase to obtain a Bi2223 multi-core tape with superconducting current-carrying capability.

The powders obtained by the two grinding methods were respectively subjected to the XRD analysis, the SEM observation, the laser particle size analysis and the tube tap density test. The final heat-treated strip was tested at 77 K using the standard four-lead method. The critical current is 1μv/cm.

3. Results and discussion

3.1 XRD analysis

BiPb2212 powder was prepared by the improved coprecipitation method after pre-decomposition and multiple sintering phase, after each heat treatment, we used manual grinding and mechanical grinding, it is found that the way can affect the final powder morphology, as shown in Fig.2, the main phase of the final powder is Bi2212 phase, but there is a difference in the second phase. In addition to the 1: 1AEC phase, the Bi2201phase exists in the hand-ground powder. We known the content and type of the second phase will affect the phase transformation process of Bi2223 in the subsequent strip and the superconducting current carrying capacity.

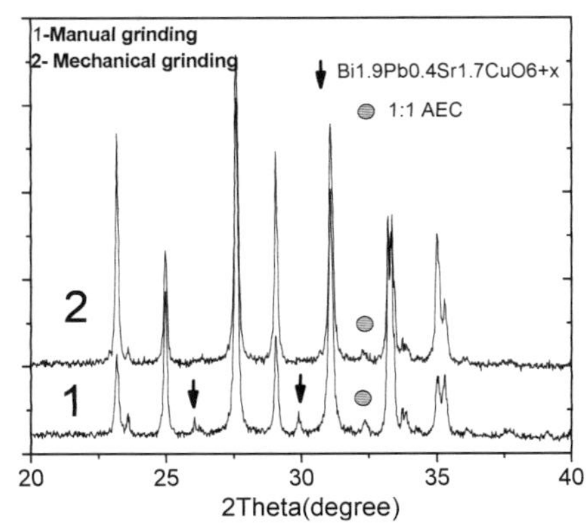

Fig.2. XRD patterns of BiPb2212 powders grinded by handwork and machine

3.2 SEM analysis

SEM photographs of powders prepared by different grinding methods are shown in Fig.3, the 1 and 2 are low magnification and high magnification of hand-milled powders, 3 and 4 are low and high-magnification pictures of mechanically milled powders. It is can be seen that the degree of dispersion of manual grinding powder is relatively low, the particles are larger, and there is obvious agglomeration, however, the mechanical grinding powder has a higher degree of dispersion, smaller particles and lower degree of agglomeration. The size of the powder particles by manual grinding in the high power electron microscope photograph is larger than 10 μm, and the particle boundaries are clear and clean, however, the powder particles by mechanical grinding are small, the size of the particle is less than 10 μm, the particle boundaries are fuzzy, not clean. The above comparison shows that the powder by using mechanical grinding played a better dispersion and broken due to high energy, stable output, and the output is unstable, and the grinding energy is low by manual grinding due to the existence of human factors, resulting in the degree of dispersion of the powder poor, the powder has not been well refined. Phase composition analysis and microscopic observation results show that mechanical grinding may be more conducive to obtain stable performance of multi-core superconducting tape

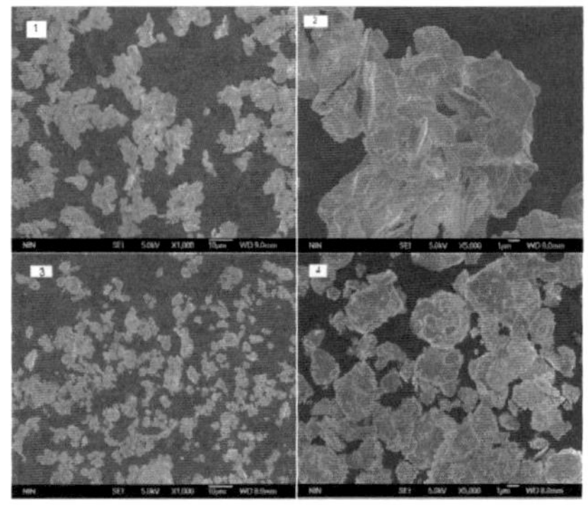

Fig.3. SEM photos of the BiPb2212 powders ground by handwork and machine

3.3 Performance analysis

The results of the laser particle size testing and analysis of powders prepared by different grinding methods are shown in Fig.4a and Fig.4b. It is found that the particle size of the hand-ground powder is larger than that of the mechanical grinding powder and the distribution of particle size is wide. The average particle size of the hand-ground powder is 16μm, the distribution range is from 5 μm to nearly 79 μm, compared with the mechanical grinding powder, average size of particle is 5 μm, the distribution range is from 2 μm to 14 μm, which is consistent with the SEM analysis. Using the mechanical grinding, the precursor powder with small particle size and narrow distribution can be obtained, it is important that the fine particle size can improve the uniformity of the powder. At the same time, the Bi2223 phase can be accelerated and shorten heat treatment time, reduce manufacturing costs due to the higher activity of the fine particles.

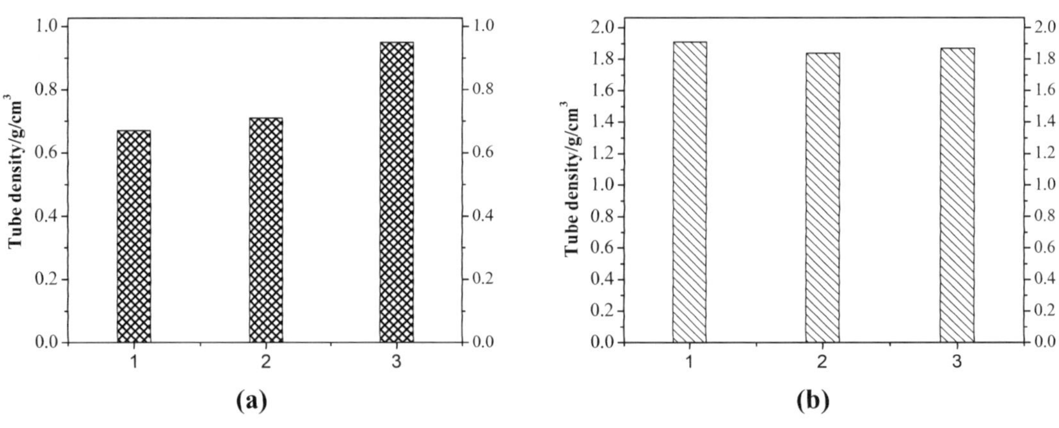

Fig.4. The granularity and distribution of the BiPb2212 powders (a) Grinded by handwork, (b) Grinded by machine

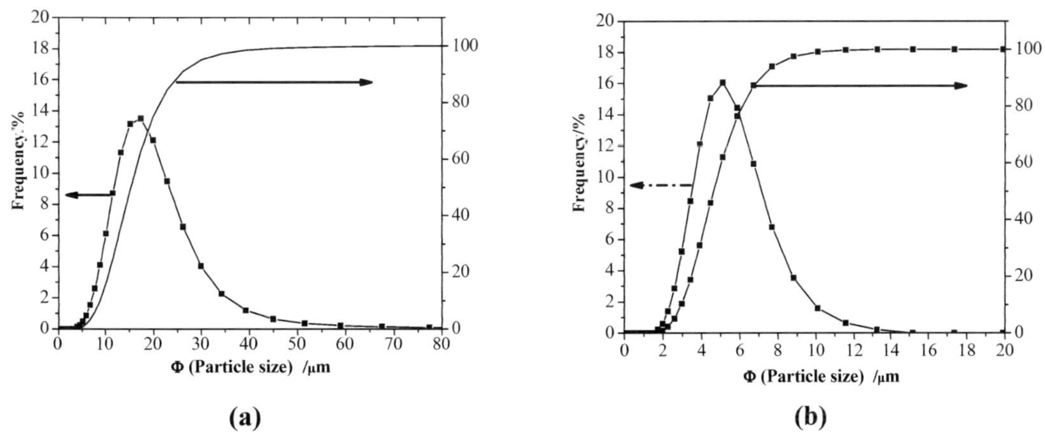

Fig.5. The powder density of the tube (a) by machine-grinding powder, (b) by hand-grinding powder

The BiPb2212 powder prepared by different grinding methods was mixed and ground in proportion with the $CaCuO_2$ powder to obtain the packed precursor powder. The two powders were vibrated and packed, and the tap density was calculated and carried out in three groups respectively. The results are shown in Fig.5a and Fig.5b. It has been found that the precursor powder with the hand-ground has a lower tap density and a poorer stability between batches, with a large fluctuation of about 20 to 30%. While the precursor powder with mechanically ground has a high tap density, high stability between batches, the fluctuations is less than 5%. The comparison results show that the mechanical grinding powder is more stable than manual grinding powder with little difference between sample batches, which is more favorable for the preparation of the strip.

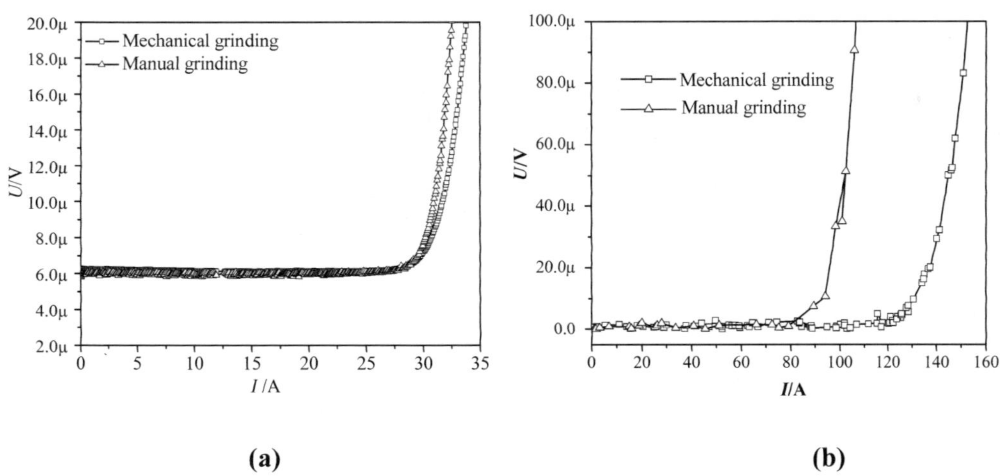

Fig.6. The *U-I* curves of Bi2223 tapes (a) after HT1 treatment,(b) final heat-treatment

Fig. 6a and Fig. 6b show the multi-core HT1 and the final UI curve after heat treatment with different grinding methods. Because of the heat treatment time of HT1 is longer, the difference between the powders is not shown. The test results of the current carrying capacity of the material showed that the I_c (critical current) of the strip prepared by manual grinding powder was 86A, and the I_c of the strip produced by the mechanical grinding powder was 127A. The I_c increased greatly compared with the hand-ground powder. The results also well validated the above stages to conclude that both the mechanical properties of the powders prepared in the mechanical grinding mode have the same phase composition, particle size distribution and reactivity, the stability between batches is more superior, and the prepared strip has higher current carrying capacity, it is more conducive to the practical development of Bi2223 superconducting tape.

4. Conclusion

In this paper, the dual-powder process was chosen to prepare the BiPb2212 and $CaCuO_2$ powders using the improved coprecipitation method respectively. The results showed that the two kinds of grinding methods had no significant effect on the final second phase type and content, however, the mechanical grinding method is more conducive to obtain a high degree of dispersion, small particles, and narrow particle size distribution of the powder. However, the tap density of the powder prepared by mechanical grinding is larger, and the variation of packing density is smaller between different batches. The precursor powder prepared by two kinds of grinding methods, the critical current (I_c) test of HT1 and the final heat treatment of the two strips results show that the mechanical grinding powder was used to prepare the strip. The I_c reached 127A at 77K, but the I_c of the strip was only 86A by hand grinding the powder. Therefore, the mechanical grinding is more conducive to high performance Bi2223 multi-core strip.

References:

[1] Zhou, L, *Research on the Development Strategy and Application of China HTS Materials* , Chemical Industry Press, Beijing,2008.
[2] Halim, S.A., Khawaldeh, S.A., Mohamed, S.B., Azhan, H., *Materials Chemistry and Physics* 61(1999), 251-259.
[3] Liu, R.S., Wang, W.N., Chang, C.T., Wu, P.T.,*Jpn. J. Appl. Phys*28(1989),L2155-L2155.
[4] Jiang, C. H., Yoo, J. M., Ko, J. W., Chung , H. S., Qiao, G.W., *Supercond. Sci. Technol*16(2003), 85–89.
[5] Shei, C.Y., Liu, R.S., Chang, C.T., Wu. P.T., *Inorg. Chem* 29(1990),3117-3120.
[6] Yoo, J., Kim, S.H. Ko, J.W., Kim, Y.K, *Supercond. Sci. Technol* 17(2004),S538–S542.
[7] Mao. C.B.,*Bi-based superconductor primary powder physical chemistry and high Tc phase and texture formation mechanism* , North eastern University, Shenyang,1997.
[8] Dorris, S.E., Prorok, B.C., Lanagan, M.T., Sinha, S., Poeppel. R B., *Physica C*212(1993), 66-74.
[9] Huang, Y.T., Shy, D.S., Chen. L.J.,*PhysicaC*294(1998),140–146.

Longitudinal magnetic field effect in critical current characteristics of Bi-2223 superconducting tape

Xuan Wu[1, 2] and Baorong Ni[1]

[1] Graduate School of Engineering, Fukuoka Institute of Technology, 3-30-1, Wajiro-higashi, Higashi-ku, Fukuoka 811-0295, Japan

[2] School of Electronic and Optical Engineering, Nanjing University of Science and Technology, Nanjing, Jiangsu Province 20094, China

E-mail : mam17102@bene.fit.ac.jp

Abstract. In this study, we focused on the longitudinal magnetic field effect in critical current densities and investigated the enhancement of the electrical conduction properties in a commercial Bi-2223 superconducting tape. Since the samples used in this study were polycrystalline and tape-shaped, the longitudinal magnetic field effect was expected to differ from that in single crystals or conventional superconducting alloys. Therefore, the critical current characteristics at magnetic fields of up to 5 T as well as at various angles between the magnetic field and current were measured in detail.

Keywords: Bi-2223 superconducting tape, critical current density, longitudinal magnetic field effect, anisotropy in upper critical magnetic field

1. Introduction

As one of the important high-T_c superconductors, Bi-2223 superconducting tape has a large potentiality of applications in electric power field. However, at the present stage, the critical current density (J_c) is still much lower than the practical level. In order to enhance the critical current properties, many researchers have done a lot of studies on improving J_c through enforcing and optimizing the flux pinning characteristic in superconductors [1], developing new and high-quality coated conductors [2], etc.

On the other hand, it is well known that the value of J_c in a longitudinal magnetic field is much larger than that in a transverse magnetic field (longitudinal magnetic field effect). For example, the difference of these two values could be over 100 times in a Ti-Nb alloy [3]. Recently, we focused on utilizing the longitudinal magnetic field effect to improve the critical current density in Bi-2223, and obtained some preliminary results insisting the validity of the effect. At the same time, some other researchers paid their attentions to the effect in YBCO coated conductors [4][5], and also obtained qualitatively similar results. In our previous study, unlike the results in conventional metallic superconductors, the value of J_c in the longitudinal magnetic field increased only by 30% compared with that in the transverse magnetic field. This is ascribed to the polycrystalline property in Bi-2223. In this study, more systematic investigation was carried out.

2. Experimental

2.1. Samples

Bi-2223 superconducting tape, as a commercial product fabricated by SUMIMOTO ELECTRIC company, was adopted in this study. The critical current of the tape is about 170 A at 77 K, and the cross-section of the superconducting phase is about 0.64 mm^2. The observed fracture surface of the tape was shown in figure 1. It can be found that Bi-2223 grains in tape are flake-like, *c*-axis orientated to a certain level and densely compacted in the silver-sheath. Because of the limitation on the value of the transport current in the sample in our measurement, which will be mentioned below, the cross-section of the sample was reduced to about 0.21 mm^2, i.e., 1.4 mm in width and 0.15 mm in thickness.

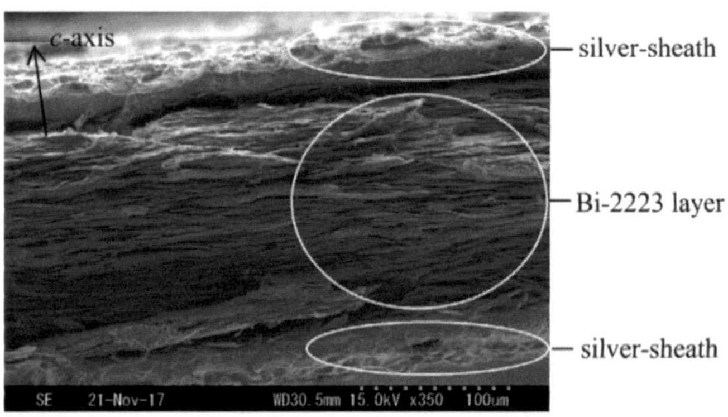

Figure 1. Fracture surface image of sample obtained by SEM.

2.2. Measurement

The cryogenic cooler was utilized in this study to control the temperature (T) of the sample in the range from 50 K to 77 K with a precision on the order of 1 K. The resistive method (four-probe method) was adopted for this study to measure the critical current (I_c) of the sample. Electric field criterion of 1 μV/cm was employed for the determination of I_c. The magnetic field (B) was applied up to 5 T.

2.3. Two kinds of variations in the direction of B

In this measurement, the magnetic field was applied in various directions to the current. The angle between the direction of the current and that of the applied magnetic field was defined as ϕ (i.e., $\phi = 0°$ denotes the case of the longitudinal magnetic field, and $\phi = 90°$ denotes the case of the transverse magnetic field). There are two kinds of variations of ϕ: one is the case that the direction of B is always parallel to the *ab* plane of grains while we change angle ϕ (mode A) as shown in figure 2 (a), the other

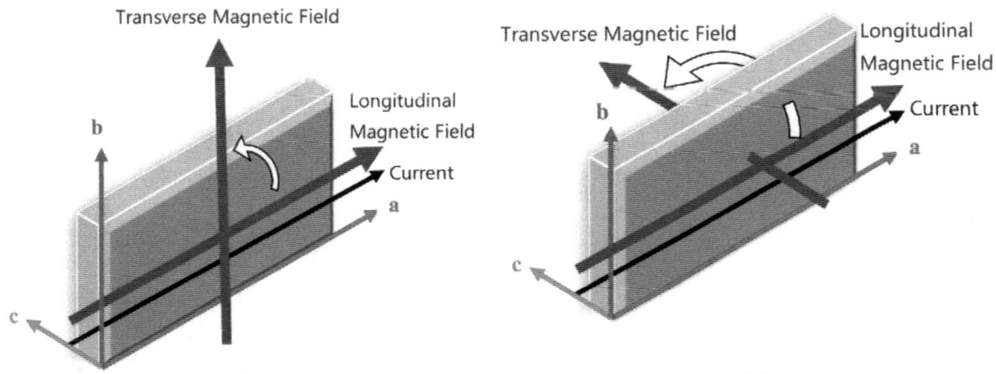

(a) (b)

Figure 2. Relation between current direction and applied magnetic field.

is that B is from the direction parallel to the ab plane to the direction parallel to the c axis (mode B), as shown in figure 2 (b). The essential difference of these two cases is that whether the upper critical field (B_{c2}) changes at various ϕ.

3. Results and Discussion

3.1. B dependence of J_c

Figure 3 illustrates the relationship between J_c and B at 50 K, while (a) and (b) represent that in mode A and in mode B respectively. It can be noted that the variation tendency of J_c in mode A is similar with that in mode B when ϕ is equal to 0°. However, in mode B, the variation of J_c is much more significant as ϕ is close to 90°, which is different from that in mode A. In addition, at 0.1 T (mode A), the enhancement of critical current density due to the longitudinal magnetic field effect is only about 1.4 times (i.e., from 4.29×10^8 A/m² in the transverse magnetic field to 5.93×10^8 A/m² in the longitudinal

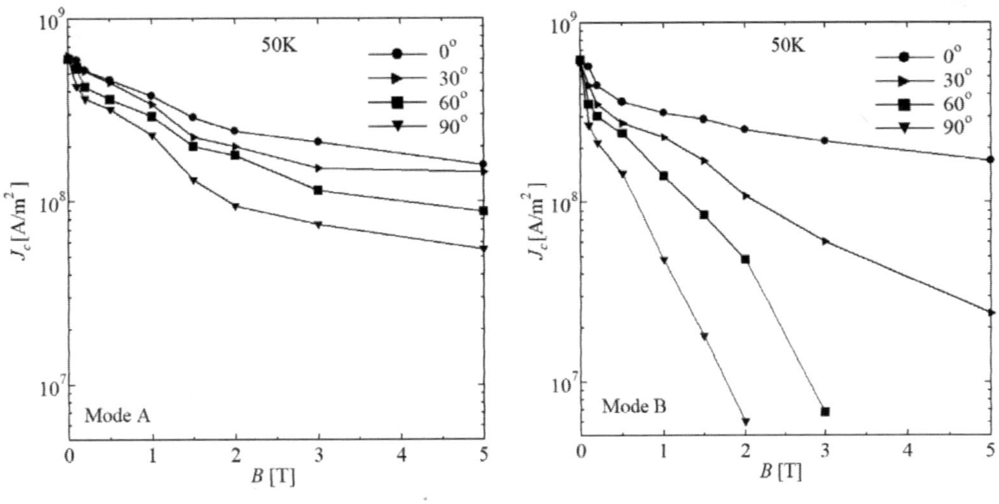

(a) (b)

Figure 3. B dependence of J_c. (a) and (b) are the behaviours in mode A and B, respectively.

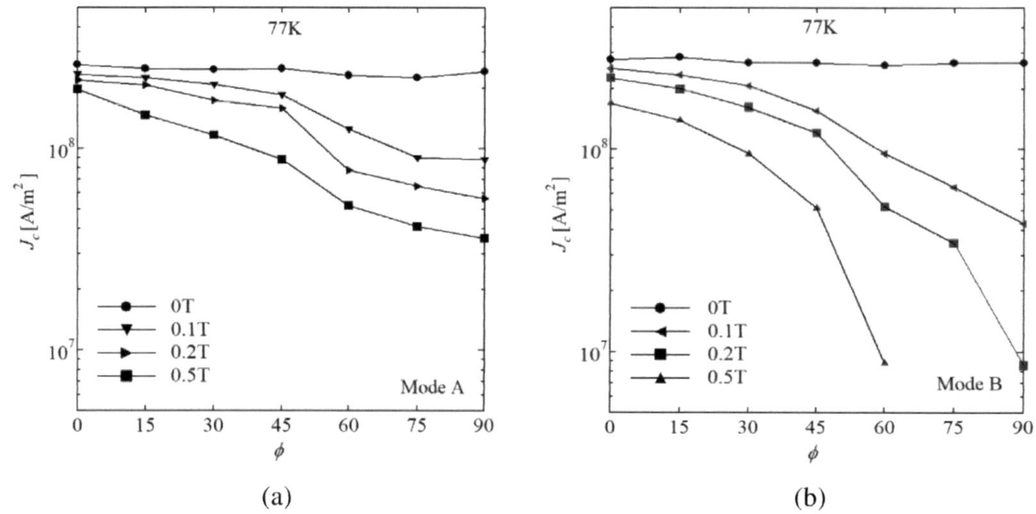

Figure 4. ϕ dependence of J_c. (a) and (b) are the behaviours in mode A and B, respectively.

magnetic field), while the ratio of enhancement can reach to 1.64 times at 1 T and the largest ratio is 2.9 times at 5 T. In mode B, different from that in mode A, the values of J_c are enlarged greatly by 1.15 times and 5.56 times at 0.1 T and 1 T, respectively, while the largest increase is about 39 times at 2 T.

The result insisted that J_c enhancement in mode B was larger than that in mode A. It is considered that J_c was affected only by the longitudinal magnetic field effect in mode A. For mode B, it was not only influenced by the longitudinal magnetic field effect, but also impaired by the reduction of the upper critical magnetic field while ϕ varies from 0° to 90°. It is well known that a significant anisotropy of B_{c2} exists in Bi-2223. For some temperatures and magnetic fields, the anisotropy could reach up to nearly a decade [6]. If B rotates around the b axis, as shown in figure 2 (b), the critical current density changed sharply with the increase of B. This can be explained by the fact that J_c was more significantly dominated by smaller B_{c2} in mode B.

3.2. *ϕ dependence of J_c*

Figure 4 demonstrates the relationship between J_c and ϕ at 77K, while (a) and (b) denote J_c vs. ϕ in mode A and mode B respectively. At 0.1 T (mode A), the value of J_c in the longitudinal magnetic field is 2.6 times compared with that in the transverse magnetic field, while the ratio is 3.86 times at 0.2 T. In mode B, the enhancement of J_c increase to 5.8 times at 0.1 T and 26 times at 0.2 T from the transverse magnetic field to longitudinal magnetic field, respectively. Comparing that in mode B, it can be observed that J_c changed slowly in mode A with the increase of ϕ except that at zero field. When magnetic field is zero, J_c keeps unchanged at all angles. This is because the longitudinal magnetic field effect does not exist at zero magnetic field.

3.3. *T dependence of J_c*

Figure 5 (a) and Figure 5 (b) demonstrates the relationship between J_c and T in mode A and B respectively. It shows that in the experimental temperature region, if the samples are applied by same B

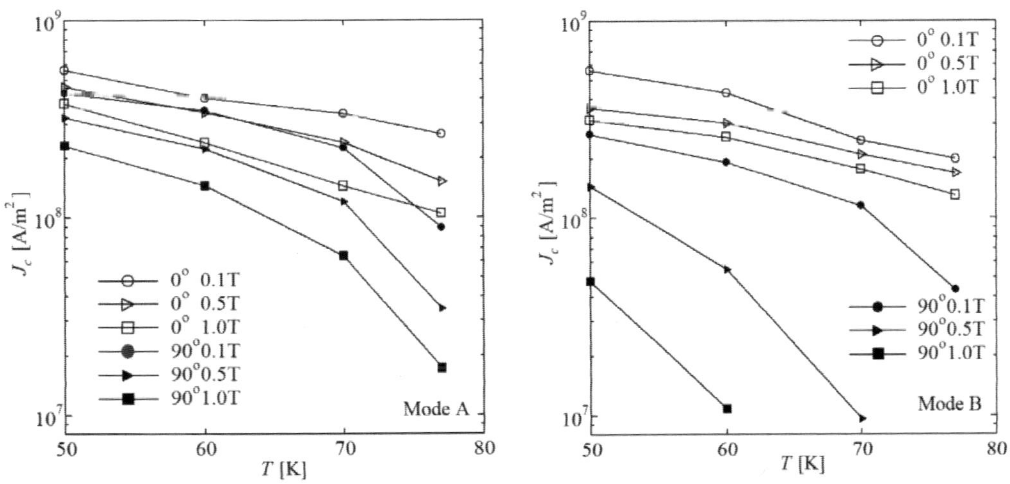

(a) (b)

Figure 5. T dependence of J_c. (a) and (b) are the behaviours in mode A and B, respectively.

and same T, J_c in the longitudinal magnetic field is always larger than that in the transverse magnetic field.

4. Conclusions

The critical current characteristics of Bi-2223 tape at various temperatures and magnetic fields were measured. The angle ϕ between the transport current and applied magnetic field was varied from 0° to 90°. The results show that, at 50 K, the longitudinal magnetic field effect can be estimated as about 1.4 (at 0.1 T) to 2.9 (at 5 T) times when the magnetic field is applied within the *ab* plane (mode A), while it increases to the level of over 40 times (at 2 T) when the influence of anisotropy of B_{c2} is taken into account (mode B). Furthermore, in the experimental temperature range, J_c under the longitudinal magnetic field at same B is always larger than that under the transverse magnetic field.

References

[1] Chen, W. M., Liu, H. K., Lin, F., Guo, Y. C., Jiang, S. S., & Dou, S. X. (2000). The flux pinning potential of Ag/Bi-2223 tapes for *H*//c-axis. *Physica C*, 341, 1353-1354.

[2] Shiohara, Y., Fujiwara, N., Hayashi, H., Nagaya, S., Izumi, T., & Yoshizumi, M. (2009). Japanese efforts on coated conductor processing and its power applications: new 5year project for materials and power applications of coated conductors (m-pacc). *Physica C*, 469(15), 863-867.

[3] Bychkov, Y. F., Vereshchagin, V. G., Karasik, V. R., Kurganov, G. B., & Mal'Tsev, V. A. (1969). Critical currents in a superconducting alloy with a rigid vortex lattice. *Practical Diabetes International*, 15(5), 141-142.

[4] Uglietti, D., Kitaguchi, H., Choi, S., & Kiyoshi, T. (2009). Angular dependence of critical current in coated conductors at 4.2 K and magnet design. *IEEE Transactions on Applied Superconductivity*, *19*(3), 2909-2912.

[5] Kiuchi, M., Koida, T., Minami, J., Takahashi, Y., Otabe, E. S., & Matsushita, T., et al. (2011). Field angle dependence of critical current density in YGdBCO coated conductor. *IEEE*

Transactions on Applied Superconductivity, 21(3), 3210-3213.

[6] Ginsberg, D. M. (Ed.). (1998). *Physical properties of high temperature superconductors I*. World Scientific.

Fabrication Process and Pressure Dependence of Critical Current Density in $Ba_{1-x}K_xFe_2As_2$ Superconducting HIP Wires

***Sunseng Pyon[1], Takahiro Suwa[1], Tsuyoshi Tamegai[1], Katsutoshi Takano[2], Hideki Kajitani[2], Norikiyo Koizumi[2] and Satoshi Awaji[3]**

1. Department of Applied Physics, The University of Tokyo, Hongo, Bunkyo-ku, Tokyo 113-8656, Japan

2 Naka Fusion Institute, National Institutes for Quantum and Radiological Science and Technology (QST), 801-1 Mukoyama, Naka-shi, Ibaraki 311-0193, Japan

3 High Field Laboratory for Superconducting Materials, Institute for Materials Research, Tohoku University, Sendai 980-8577, Japan

E-mail: pyon@ap.t.u-tokyo.ac.jp

Abstract. We demonstrate the fabrication of $(Ba,K)Fe_2As_2$ superconducting wires through a powder-in-tube (PIT) method and hot isostatic press (HIP) technique. Several fabrication processes with different sintering conditions of pressure and time were examined. The measurements of magnetization, *I-V* characteristics, filling factor, and Vickers hardness clarified that the improvement of fabrication process and sintering condition is effective for the enhancement of critical temperature T_c, which leads to the enhancement of critical current density J_c in HIP round wire. Wires which are fabricated in different processes show significantly different T_c and J_c. The core density, which can be enhanced by sintering at higher pressures, is the main key factor for enhancement of J_c in HIP wires, similar to pressed tapes. Changing the sintering time also slightly affect J_c.

1. Introduction

Iron-based superconductors (IBSs) are attractive candidates for future high-field applications of superconductors because they exhibit high critical temperature, T_c, large upper critical field, H_{c2}, and relatively low anisotropy compared with cuprate superconductors [1,2]. Among IBSs, $AE_{1-x}K_xFe_2As_2$ (AE = Ba, Sr) compounds have been extensively studied as promising candidate for application, because high critical current density, J_c, is realized in superconducting wires and tapes using $AE_{1-x}K_xFe_2As_2$ [3-14]. In these studies, it is demonstrated that uniaxial pressing for tapes [3-7] and hot isostatic pressing (HIP) for round wires [8-14] are effective methods to enhance J_c. Pressing superconducting wires and tapes using these methods is an effective method for enhancing the density and eliminating weak links between superconducting grains in the core of wires and tapes. It is expected that the values of J_c in these wires and tapes strongly depend on their fabrication and sintering processes. In the case of tapes, effects of fabrication process and pressing were revealed. For example, Gao *et al*. demonstrated that, final thickness of the pressed tapes and hardness of the core of the tapes strongly affect the J_c in the tapes [3]. They showed positive correlation between the J_c and hardness of the tape core [3]. The hardness, which is related to the core density, is controlled by the pressure of cold press. On the other hand, in the case of HIP round wire, how different fabrication processes and pressures affect the J_c have not been studied in detail. Reported HIP wires have been processed at high pressure of 100-200 MPa for 4-20 h [8,14]. However, sintering pressure and time dependence of J_c in the HIP wire have not been systematically studied yet. There is only one report where pressure dependence of J_c in the HIP wire was studied [12]. So the sintering condition for the HIP wire still has rooms for improvements. Furthermore, we have recently reported that drawing or rolling process for the wire has strong influence on J_c in the HIP wire [15]. Round wires drawn using dies with circular holes before groove-rolling ("drawn") have higher J_c than those fabricated using only a groove roller with square grooves ("rolled"). The effect of different fabrication process of the wire on J_c should also be studied.

In this work, we first investigate how different mechanical deformation processes affect J_c characteristics of HIP wires by comparing properties of wires that are fabricated by different processes. Second, we systematically investigate how the sintering conditions such as pressure and time affect the J_c performance of HIP wires, by comparing wires prepared at different pressures and sintered for various times.

2. Experiment

Superconducting wires of $(Ba,K)Fe_2As_2$ were fabricated by ex-situ powder-in-tube (PIT) method. Polycrystalline powders of $Ba_{0.6}K_{0.4}Fe_2As_2$ were prepared by the solid-state reaction. Used polycrystalline powder is the same as that used in the previous work [14], and details of powder preparation are described in ref. [14]. Ground powder was filled in a silver tube with an outer diameter of 4.5 mm and an inner diameter 3 mm. Here, three different fabrication processes were chosen. First, Ag tubes filled with powder were cold drawn into a round shape with a dimeter ~1.2 mm using dies with circular holes ("drawn"). Second, they were fabricated using only a groove roller with square grooves ("rolled"). Third, they were swaged using a rotary swaging machine ("swaged"). Obtained three kinds of wires were cut into short pieces. They are put into a 1/8 inch copper tube and redrawn into a square shape with a groove roller down to a diagonal dimension of 1.2 mm. After the drawing process, both ends of the wire were sealed by using an arc furnace. The sealed wires were sintered using the HIP technique. All wires were heated up to 700°C in argon atmosphere, and sintered under different pressures of 0.1-170 MPa and for 0.5-12 h. HIP process under the highest pressure 170 MPa were performed at National Institutes for Quantum and Radiological Science and Technology (QST), and others were performed at the University of Tokyo. In order to evaluate transport J_c, a DC electric current up to 150 A was delivered through the wire at a ramping rate of 50–100 A/min. Measurements were performed in liquid helium to minimize the effect of Joule heating at the current leads. The critical current measurements in high magnetic fields were carried out by using the 15T-SM at High Field Laboratory for Superconducting Materials, IMR, Tohoku University. Current–voltage (I–V) characteristics up to 140 kOe were measured by the four-probe method with solder for contacts. Bulk magnetization was measured to evaluate T_c and magnetic J_c by a superconducting quantum interference device (SQUID) magnetometer (MPMS-5XL, Quantum Design). Magnetic J_c is evaluated using the extended Bean model. The filling factor of the core was estimated from the volume and weight of cut and polished core of the wire. Vickers hardness, Hv, was measured on the polished surface of the wire core.

3. Results and discussion

Transport J_c and magnetic J_c of HIP wires sintered at various conditions are estimated. As shown in figure 1(a), "drawn" wire which was sintered at 170 MPa for 4 h shows the highest value of J_c. At the self-field, transport J_c exceeds 10^5 A/cm^2 which is the target value of J_c for application. At high field of 100 kOe, transport J_c reaches at 3.0 x 10^4 A/cm^2. Magnetic J_c also shows similar value and field dependence. This J_c value is larger than the previous world record of IBS round wire [14] and close to the value of the largest J_c, which was achieved recently [15]. The difference between this wire and the previous one [14] is just the fabrication process. This wire was "drawn" and previous one is "rolled". Such a fabrication method dependence of J_c is also observed in other wires. In figure 1(a) magnetic field dependence of magnetic J_c in "drawn", "swaged", and "rolled" wire are described. These three wires were sintered in the same condition at 9 MPa for 0.5 h. Magnetic J_c in "drawn" or "swaged" wire is larger than that in "rolled" wire. One of the possible reasons why fabrication process affects the J_c is the different degrees of degradation of superconducting core because of different wire fabrications. We have shown that superconducting properties of the wire are degraded during the wire fabrication process, and recovered during the sintering process [11]. It is possible that the degree of the degradation depends on the fabrication method. As shown in figure 1(b), T_c of "drawn", "swaged", and "rolled" HIP wire are different to each other. In "rolled" wire with the smallest J_c, T_c is also the lowest. The main factor for the suppression T_c is degradation of polycrystalline powders in the core of the wire during fabrication process [11]. Degraded superconducting property is partly recovered during sintering. The degree of recovery in the three kinds of wires must be the same because sintering temperature, pressure, and time is the same. So the degree of degradation should be different and they affect the J_c and T_c in the wire. It should also be noted that the highest T_c in these wires are realized in the "drawn" wire sintered at 170 MPa for 4 h. For recovery of superconducting properties, higher sintering pressure and longer sintering time are more effective. The other possible reason is the differences of grain orientations and texturing, which is reported in pressed tape [4,5]. Details of this possible reason are discussed in our former publication [15].

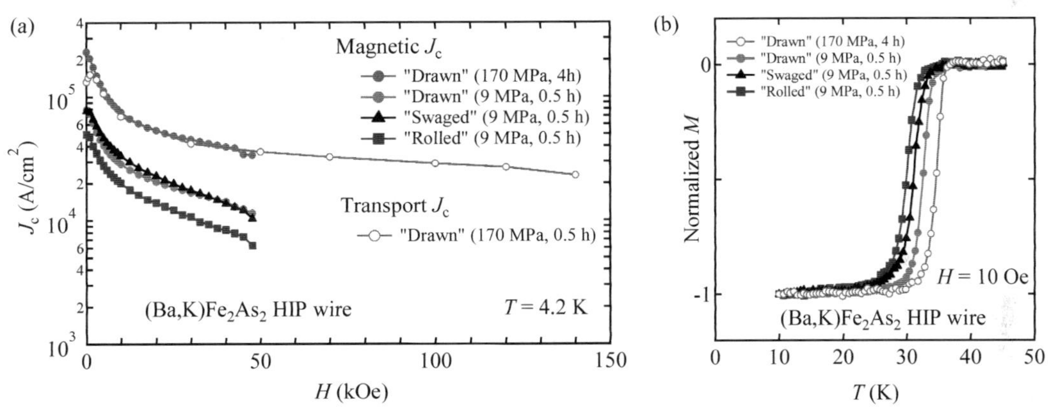

Figure 1. (a) Magnetic field dependence of transport J_c and magnetic J_c in the $(Ba,K)Fe_2As_2$ HIP wires processed in different conditions. (b) Temperature dependence of normalized magnetization at 10 Oe for the $(Ba,K)Fe_2As_2$ HIP wires.

Next, we focus on the effects of sintering pressure on physical properties in the HIP wire. To improve weak links between grains, higher density is advantageous. Sintering at higher pressure should be effective for that. As shown in figure 2(a), the filling factor shows clear positive correlation with the sintering pressure. At 170 MPa, the filling factor reaches 97%. Related to this correlation, Hv also shows positive correlation with the sintering pressure shown in figure 2(b). Increases in both the filling factor and Hv suggest the improvement of links in grains. Actually, J_c also shows a similar sintering pressure dependence as shown in figure 2(c). Hv dependence of J_c in HIP wires is summarized in figure 2(d). The positive correlation between J_c and Hv is observed. It is clearly confirmed for the first time systematically in IBS round wire that high core density plays a key role in achieving the practical level of J_c, as discussed in IBS pressed tape [4]. It is worth noting that the effect of fabrication process on J_c cannot be ignored, although the effect of sintering pressure is more significant. As shown in figure 2(c), when the sintering pressure is about 9 MPa, J_c values of the wires are not the same. As discussed above, J_c is enhanced not only by improvements of weak link between grains using high pressure but by controlling core states such as superconducting properties and orientation of grains with changing fabrication process. For further enhancement of J_c, both changing sintering pressure and fabrication process should be effective.

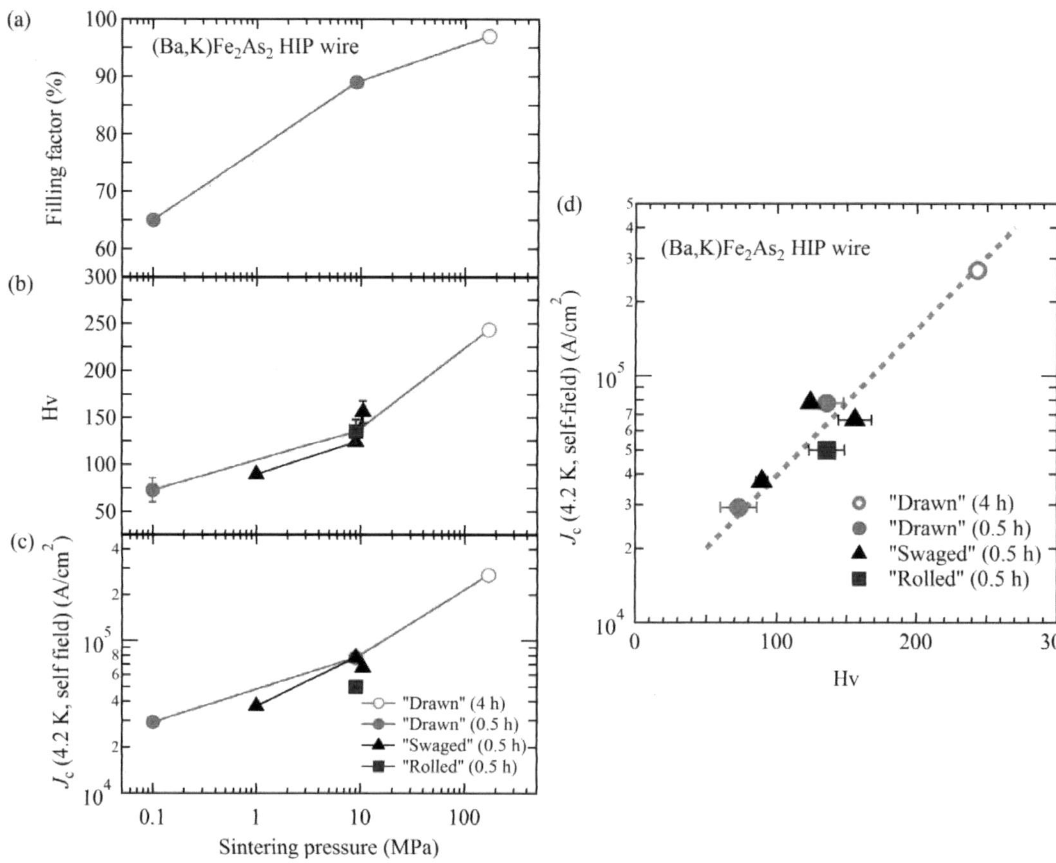

Figure 2. Sintering pressure dependence of (a) filling factor of the core, (b) Vickers hardness (Hv), and (c) magnetic J_c at 4.2 K under self-filed in the (Ba,K)Fe_2As_2 HIP wires processed in different conditions. (d) Hv dependence of magnetic J_c at 4.2 K under self-filed in the (Ba,K)Fe_2As_2 HIP wires.

We also investigated the effect of sintering time on the superconducting properties of the HIP wire. Sintering time dependences of J_c in "swaged" HIP wire is shown in figure 3. Fabrication processes and sintering temperatures and pressures are the same in these wires. Compared with the sintering pressure dependence on J_c as shown figure 2, the effect of sintering time on J_c is small. However, slight enhancement is found. For example, J_c of 7.8 x 10^4 A/cm^2 at sintering time of 0.5 h is enhanced to 10.1 x 10^4 A/cm^2 at sintering time of 8 h. Longer sintering time may be preferred for an increase of J_c in HIP wire [15]. Enhancement of J_c with increasing sintering time may be caused by higher sintering degree and improvement of superconducting properties and grain connections. This result suggests that further enhancement of J_c is expected by making sintering time longer. It should be noted that the optimal condition of sintering time should depend on sintering temperature. For example, Weiss *et al.* fabricated HIP wire at lower temperature of 600°C and longer time of 20 h [8]. Systematic analyses of sintering time dependence of J_c at different temperature has not been reported. Optimization of sintering time and temperature will realize certain degree of enhancement of J_c in the HIP round wire.

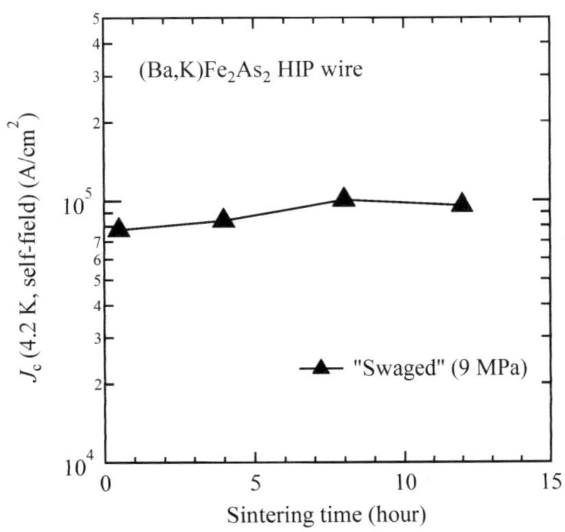

Figure 3. Sintering time dependence of magnetic J_c at 4.2 K under self-filed in the (Ba,K)Fe_2As_2 "swaged" HIP wires.

4. Summary

We have fabricated (Ba,K)Fe_2As_2 superconducting wires through a PIT method and HIP technique. Several fabrication processes such as "drawing", "swaging", and "rolling", and different sintering conditions of pressure and time were examined. The measurements of magnetization, *I-V* characteristics clarified that fabrication process significantly affects the enhancement of J_c in the HIP round wire. The estimation of filling factor of the core and Hv in several wires revealed that the core density plays a key role for the improvement of weak links between grains in the core, and it can be controlled by changing sintering pressure. This relation is systematically evaluated and clarified for the first time in IBS round wires, similar to the pressed tapes. Changing the sintering time also slightly affect J_c.

Acknowledgments

This work was partially supported by a Grant-in-Aid for Scientific Research (A) (17H01141), a Grant-in-Aid for Young Scientists (B) (No.16K17745), and the Japan-China Bilateral Joint Research Project by the Japan Society for the Promotion of Science (JSPS).

References

[1] Ma Y 2012 *Supercond. Sci. Technol.* **25** 113001
[2] Pallecchi I, Eisterer M. Malagoli A and Putti M 2015 *Supercond. Sci. Technol.* **28** 114005
[3] Gao Z, Togano K, Matsumoto A and Kumakura H 2014 *Sci. Rep.* **4** 4065
[4] Lin H, Yao C, Zhang X, Zhang H, Wang D, Zhang Q, Ma Y W, Awaji S and Watanabe K 2014 *Sci. Rep.* **4** 4465
[5] Lin H, Yao C, Zhang X, Dong C, Zhang H, Wang D, Zhang Q, Ma Y, Awaji S, Watanabe K *et al.* 2014 *Sci. Rep.* **4** 6944
[6] Zhang X, Yao C, Lin H, Cai Y, Chen Z, Li J, Dong C, Zhang Q, Wang D, Ma Y, Oguro H, Awaji S and Watanabe K 2014 *Appl. Phys. Lett.* **104** 202601
[7] Gao Z, Togano K, Matsumoto A and Kumakura H 2015 *Supercond. Sci. Technol.* **28** 012001
[8] Weiss J D, Tarantini C, Jiang J, Kametani F, Polyanskii A A, Larbalestier D C and Hellstrom E E 2012 *Nat. Mat.* **11** 682
[9] Pyon S, Taen T, Ohtake F, Tsuchiya Y, Inoue H, Akiyama H, Kajitani H, Koizumi N, Okayasu S and Tamegai T 2013 *Appl. Phys. Express* **6** 123101

[10] Pyon S, Tsuchiya Y, Inoue H, Kajitani H, Koizumi N, Awaji S, Watanabe K and Tamegai T 2014 *Supercond. Sci. Technol.* **27** 095002

[11] Pyon S, Yamasaki Y, Kajitani H, Koizumi N, Tsuchiya Y, Awaji S, Watanabe K and Tamegai T 2015 *Supercond. Sci. Technol.* **28** 125014

[12] Tamegai T, Pyon S, Tsuchiya Y, Inoue H, Koizumi N and Kajitani H 2015 *IEEE Trans. Appl. Supercond.* **25** 7300504

[13] Hecher J, Baumgartner T, Weiss J D, Tarantini C, Yamamoto A, Jiang J, Hellstrom E E, Larbalestier D C and Eisterer M 2016 *Supercond. Sci. Technol.* **29** 025004

[14] Pyon S, Suwa T, Park A, Kajitani H, Koizumi N, Tsuchiya Y, Awaji S, Watanabe K and Tamegai T 2016 *Supercond. Sci. Technol.* **29** 115002

[15] Tamegai T, Suwa T, Pyon S, Kajitani H, Takano K, Koizumi N, Awaji S, and Watanabe K IOP Conf. Ser. Mater. Sci. Eng. 279, 012028 (2017).

Enhancement of critical current density in AgSn-sheathed $(Sr,Na)Fe_2As_2$ superconducting tapes

T Suwa[1], S Pyon[1], T Tamegai[1] and S Awaji[2]

[1] Department of Applied Physics, The University of Tokyo, Hongo, Bunkyo-ku, Tokyo 113-8656, Japan
[2] High Field Laboratory for Superconducting Materials, Institute for Materials Research, Tohoku University, Sendai 980-8577, Japan

Email: tamegai@ap.t.u-tokyo.ac.jp

Abstract. We have fabricated $(Sr,Na)Fe_2As_2$ tapes by using AgSn alloy as a sheath material. In the tape sintered at 750°C, transport J_c reached 47 kA/cm² at 4.2 K and 50 kOe, which is larger than that of the Ag-sheathed $(Sr,Na)Fe_2As_2$ tape. This result indicates that we have succeeded in reducing the sintering temperature by more than 100°C keeping the high J_c value, which is very promising for practical applications. In the SEM images of polished longitudinal cross section of the tapes, the improvement of interface flatness by using AgSn alloy as a sheath material is confirmed.

1. Introduction

Superconductivity in layered iron-based superconductors was first discovered by Kamihara *et al.* in LaFeAs(O,F) in 2008 [1]. Soon after the discovery, several types of high temperature superconductors in iron pnictides were discovered. Among them, superconducting wires and tapes have been fabricated mainly by using $(AE,K)Fe_2As_2$ materials (AE = Ba or Sr), so called '122-type' [2-8], because of their high upper critical field $H_{c2}(0)$ [9, 10] and low anisotropy $\gamma(\sim 2)$ [10, 11], as well as their high transition temperature T_c (~36-38 K) [12, 13]. At present, the transport J_c for $(AE,K)Fe_2As_2$ tapes has reached 10^5 A/cm² [14, 15], which is the level for practical applications, at 4.2 K and 100 kOe.

Recently, a new material joined this kind of research, namely $(Sr,Na)Fe_2As_2$ [16, 17]. The reported transport J_c at 20 K up to 25 kOe is comparable to that in $(AE,K)Fe_2As_2$ tapes in the same condition. Following this study, we have fabricated $(Sr,Na)Fe_2As_2$ tapes by using Ag sheath and achieved transport J_c as high as 19 kA/cm² at 4.2 K and 140 kOe [18].

This year, excellent J_c characteristics in AgSn-sheathed $(Ba,K)Fe_2As_2$ tapes was reported [19]. In this report, J_c in AgSn-sheathed tapes sintered at 750 °C was as high as that in Ag-sheath tapes sintered at 850°C. Following this report, in the present work, we report the fabrication of $(Sr,Na)Fe_2As_2$/AgSn tapes and their J_c characteristics. In the tape sintered at 750°C, transport J_c reached 47 kA/cm² at 4.2 K and 50 kOe, which is larger than that of the Ag-sheathed $(Sr,Na)Fe_2As_2$ tape sintered at higher temperature of 875°C. We performed a scanning electron microscope observations and compositional analyses with an energy dispersive x-ray spectroscopy (SEM-EDX).

2. Experiments

First, we fabricated Ag-Sn alloy tube in the same way as reported in Ref. [19]. Sn concentration of the tube was 5%. In fabricating superconducting tapes of $(Sr_{1-x}Na_x)Fe_2As_2$, we basically followed the method reported in Ref. [17]. Since it is reported that a sample with Na concentration $x = 0.55$ exhibited the highest T_c (36.5 K), we selected x as 0.55 [16]. Polycrystalline samples were synthesized at 720°C for 18 h in a stainless steel pipe, where pellets of starting materials (SrAs, NaAs, Fe_2As) were placed. Then, the reacted materials were ground into powder, pelletized, and reacted at 720°C for 18 h again. The prepared polycrystalline materials were ground into powder with an agate mortar and pestle in a nitrogen filled glove box. The ground powder was tightly packed into a Ag-Sn alloy tube with outer and inner diameters of 4.0 and 2.6 mm, respectively. The tube was groove-rolled into a wire with 1.3 x 1.3 mm^2 cross section. Then, the wire was flat-rolled into a tape form with a thickness and width of 0.25 and 2.2 mm, respectively. The tape was cold-pressed under a pressure of 0.54 GPa. Finally, the tape was heated for sintering. Sintering temperature was 700°C, 750°C, 775°C, 800°C, 825°C, and 850 °C, and the sintering temperature dependence of J_c was studied.

Magnetic measurements were conducted using a commercial SQUID magnetometer (MPMS-XL5, Quantum Design), and magnetic J_c was estimated. Transport J_c was measured at 4.2 K by the standard four-probe method in magnetic fields up to 140 kOe. SEM images of the core were taken using SEM-EDX (S-4300, Hitachi High-Technologies equipped with EMAX x-act, HORIBA).

3. Results and discussion

3.1. Magnetic J_c estimated from magnetic hysteresis curves

We fabricated $(Sr,Na)Fe_2As_2$/AgSn tapes by sintering them at various temperatures, and estimated magnetic J_c through magnetic measurements with the field perpendicular to the flat surface of the tape. Figure 1(a) shows the magnetic field dependence of magnetic J_c in the tape sintered at 750°C, which recorded the highest J_c among tapes sintered at various temperatures. The obtained J_c at 4.2 K was as high as 110 kA/cm^2 and 65 kA/cm^2 at 50 kOe when the magnetic field was applied parallel and perpendicular to the tape surface, respectively. These values are higher than the record value for Ag-sheathed tapes sintered at 875°C. When we sintered a Ag-sheathed tape at 875°C, strong reaction between the core and Ag sheath was observed. However, when we fabricated a AgSn-sheathed tape by sintering at 750°C, no clear reaction was observed between the AgSn sheath and the core.

Figure 1(b) shows sintering temperature dependence of magnetic J_c at 4.2 K and 50 kOe for AgSn-sheathed and Ag-sheathed tapes. It is obvious that the best sintering temperature was lowered by more than 100°C in AgSn-sheathed tape. Similar behavior has been reported for $(Ba,K)Fe_2As_2$ tape in Ref. [19]. It was conjectured that using AgSn as a sheath material lead to the hardening of the core and enhancement of J_c. However, when the $(Sr,Na)Fe_2As_2$ tape was sintered at 850°C, the core reacted with the AgSn sheath severely. This was probably caused by the fact that the melting point of AgSn is about 900°C when Sn concentration is 5%, which is about 60°C lower than that of pure Ag.

3.2. Transport J_c measurements at 4.2 K in magnetic fields up to 140 kOe

Figure 2(a) shows *I-V* characteristics of $(Sr,Na)Fe_2As_2$ tape at 4.2 K up to 140 kOe, from which we evaluated transport J_c with the criterion of 1 μV/cm. Figure 2(b) shows magnetic field dependence of J_c in the tapes sintered at various temperatures. In the tape sintered at 750°C, transport J_c reached 47 kA/cm^2 at 50 kOe, which is larger than the largest J_c of the Ag-sheathed $(Sr,Na)Fe_2As_2$ tape. All AgSn tapes exhibited high and field-insensitive J_c. However, the transport J_c was lower than the magnetic J_c, and we can further improve J_c by clarifying the origin of this discrepancy.

First, these *I-V* measurements were conducted three months after fabricating these tapes. Then, deterioration of the $(Sr,Na)Fe_2As_2$ tapes possibly has something to do with residual Na in the core. When

IOP Conf. Series: Journal of Physics: Conf. Series **1054** (2018) 012045 doi:10.1088/1742-6596/1054/1/012045

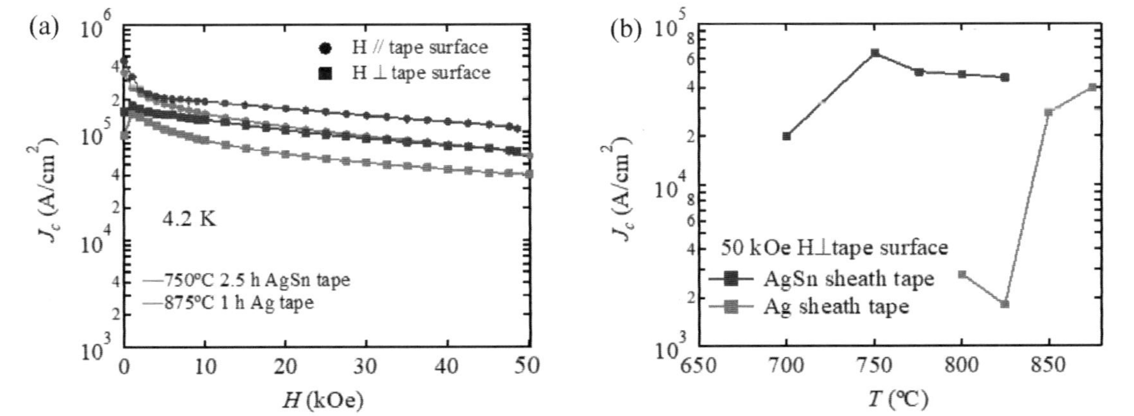

Figure 1. (a) Magnetic J_c versus magnetic field at 4.2 K for the $(Sr,Na)Fe_2As_2$/AgSn tape sintered at 750°C and Ag tape sintered at 875°C. (b) Magnetic J_c at 50 kOe when magnetic fields were applied perpendicular to the tape surface versus sintering temperature in AgSn tapes and Ag tapes.

this $(Sr,Na)Fe_2As_2$ tape is processed at high sintering temperatures, a part of Na segregates from the core. In fact, when we polish the surface of the tape in the air, we can observe bubbles are coming out of the core as shown in figure 3(b). These bubbles are most probably NaOH, which is caused by the reaction between the segregated Na in the core and H_2O in the air. This reaction can lead to deterioration of the core. The segregation of Na is possibly caused by the fact that the optimal doping level of Na in $(Sr,Na)Fe_2As_2$ is higher than that of K in $(Ba,K)Fe_2As_2$. So, there is a possibility that controlling the Na doping level leads to higher quality superconducting tapes which are more stable in the air.

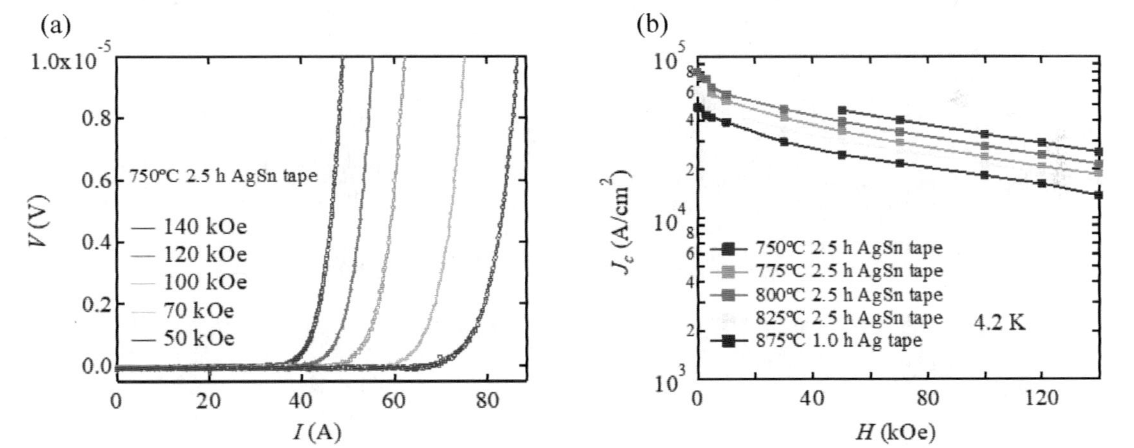

Figure 2. (a) *I-V* characteristics of $(Sr,Na)Fe_2As_2$ tape sintered at 750°C and 4.2 K. (b) Transport J_c as functions of magnetic field in $(Sr,Na)Fe_2As_2$ tapes sintered at various temperatures.

3.3. SEM images of the core

Figures 4(a)~(d) are SEM photographs of the interface between the core and the sheath on the polished longitudinal cross section of $(Sr,Na)Fe_2As_2$ tapes. Figures 4(a) and 4(c) are the Ag-sheathed tape, and figures 4(b) and 4(d) are the AgSn-sheathed tape. One observes that the shape of the core in the Ag-sheathed tape is distorted as shown in figure 4(a). Such distortion could be caused by the softness of the Ag. It is clearly observed in figure 4(c) that the interface between the core and the sheath is much less

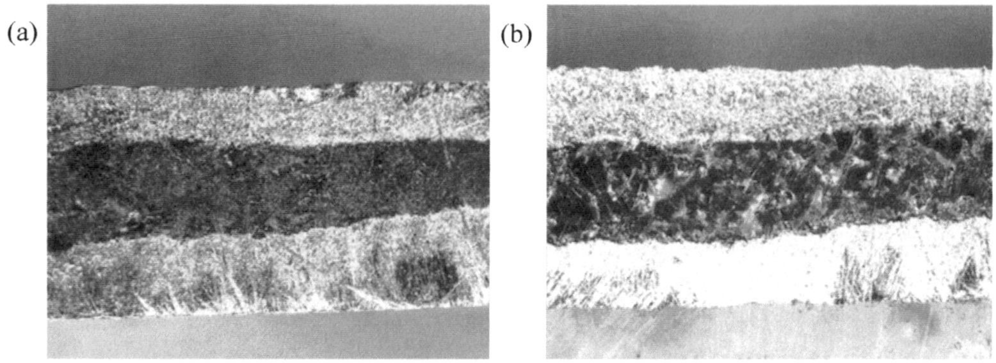

Figure 3. Optical images of the $(Sr,Na)Fe_2As_2/Ag$ tape sintered at 875°C on the polished longitudinal cross section of the tapes (a) before and (b) after NaOH bubbles coming out.

straight. On the other hand, one observes that the shape of the core in the AgSn-sheathed tape is more regular as shown in figure 4(b), and the interface between the core and the sheath is more straight as shown in figure 4(d). This is due to the hardness of the AgSn. These differences can lead to the enhancement and homogeneity of J_c. Such improvements of the interface morphology were also reported in the AgSn-sheathed $(Ba,K)Fe_2As_2$ tapes in Ref. [19].

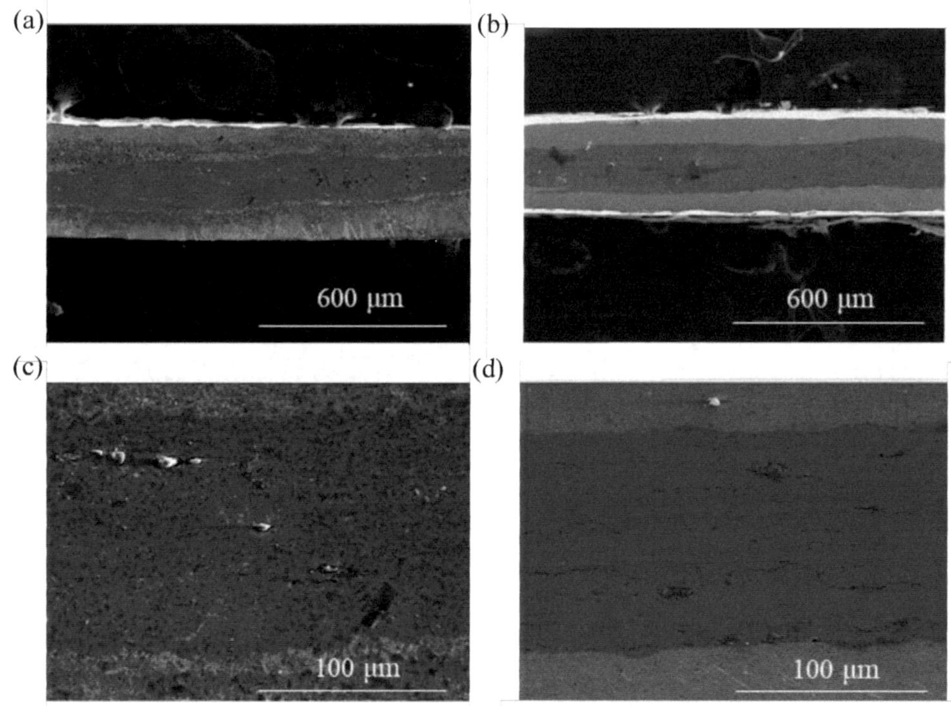

Figure 4. SEM images of the interface between the sheath and the $(Sr,Na)Fe_2As_2$ core of the Ag-sheathed tape ((a) and (c)) sintered at 875°C, and the AgSn-sheathed tape ((b) and (d)) sintered at 750°C. The observation was conducted on the polished longitudinal cross section of the tapes.

4. Summary

We have fabricated and characterized AgSn-sheathed (Sr,Na)Fe_2As_2 superconducting tapes prepared under different sintering temperatures including magnetic J_c estimated from magnetic hysteresis curves, transport J_c up to 140 kOe, and SEM images. In the tape sintered at 750°C, the value of magnetic J_c at 4.2 K was as high as 110 kA/cm^2 and 65 kA/cm^2 at 4.2 K and 50 kOe when the magnetic field was applied parallel and perpendicular to the tape surface, respectively. In the same tape, transport J_c reached 47 kA/cm^2 at 4.2 K and 50 kOe, which is larger than that of the Ag-sheathed (Sr,Na)Fe_2As_2 tape. These results indicate that we have succeeded in reducing the sintering temperature by more than 100 °C, keeping the high J_c value. SEM images of the core of the tapes clarified the improvement of smoothness of the interface between the core and the sheath by using AgSn alloy as a sheath material, which probably helped to improve the homogeneity of J_c in the core.

Acknowledgements

This work was supported by a Grant-in-Aid for Scientific Research (A) (17H01141), a Grant-in-Aid for Young Scientists (B) (16K17745), and the Japan-China Bilateral Joint Research Project by the Japan Society for the Promotion of Science (JSPS).

References

[1] Kamihara Y, Watanabe T, Hirano M and Hosono H 2008 *J. Am. Chem. Soc.* **130** 3296
[2] Ma Y W, Gao Z S, Qi Y P, Zhang X P, Wang L, Zhang Z Y and Wang D L 2009 *Physica* C **469** 651
[3] Togano K, Matsumoto A and Kumakura H 2011 *Appl. Phys. Express* **4** 043101
[4] Pyon S, Tsuchiya Y, Inoue H, Kajitani H, Koizumi N, Awaji S, Watanabe K and Tamegai T 2014 *Supercond. Sci. Technol.* **27** 095002
[5] Pyon S, Yamasaki Y, Kajitani H, Koizumi N, Tsuchiya Y, Awaji S, Watanabe K and Tamegai T 2015 *Supercond. Sci. Technol.* **28** 125014
[6] Weiss J D, Tarantini C, Jiang J, Kametani F, Polyanskii A A, Larbalestier D C and Hellstrom E 2012 *Nat. Mater.* **11** 682
[7] Pyon S, Suwa T, Park A, Kajitani H, Koizumi N, Tsuchiya Y, Awaji S, Watanabe K and Tamegai T 2016 *Supercond. Sci. Technol.* **29** 115002
[8] Tamegai T, Suwa T, Pyon S, Kajitani H, Takano K, Koizumi N, Awaji S and Watanabe K 2017 *IOP Conf. Ser. Mater. Sci. Eng.* **279**, 012028
[9] Putti M *et al.* 2010 *Supercond. Sci. Technol.* **23** 034003
[10] Wang Z S, Luo H Q, Ren C and Wen H H 2008 *Phys. Rev.* B **78** 140501R
[11] Yuan H Q, Singleton J, Balakirev F F, Baily S A, Chen G F, Luo J L and Wang N L 2009 *Nature* **457** 565
[12] Rotter M M, Tegel M, Johrendt D, Schellenberg I, Hermes W and Pöttgen R 2008 *Phys. Rev.* B **78** 020503
[13] Sasmal K, Lv B, Lorenz B, Guloy A M, Chen F, Xue Y Y and Chu C W 2008 *Phys. Rev. Lett.* **101** 107007
[14] Zhang X, Yao C, Lin H, Cai Y, Chen Z, Li J, Dong C, Zhang Q, Wang D, Ma Y, Oguro H, Awaji S and Watanabe K 2014 *Appl. Phys. Lett.* **104** 202601
[15] Gao Z, Togano K, Matsumoto A and Kumakura H 2014 *Sci. Rep.* **4** 4065
[16] Shinohara N, Tokiwa K, Fujihisa H, Gotoh Y, Ishida S, Kihou K, Lee C H, Eisaki H, Yoshida Y and Iyo A 2015 *Supercond. Sci. Technol.* **29** 115002
[17] Iyo A, Shinohara N, Tokiwa K, Ishida S, Tsuchiya Y, Ishii A, Asou T, Nishio T, Matsuzaki K, Takeshita N, Eisaki H and Yoshida Y 2015 *Supercond. Sci. Technol.* **28** 105007
[18] Suwa T, Pyon S, Park A, Tamegai T, Tsuchiya Y, Awaji S and Watanabe K 2017 *J. Phys.: Conf. Ser.* **871** 012062
[19] Togano K, Gao Z, Matsumoto A, Kikuchi A and Kumakura H 2017 *Supercond. Sci. Technol.* **30** 015012

IOP Conf. Series: Journal of Physics: Conf. Series **1054** (2018) 012046 doi:10.1088/1742-6596/1054/1/012046

History and recent progress of QMG ™ and QMG bulk magnets

M Morita[1]

[1]Nippon Steel & Sumitomo Metal Corporation
Advanced Technology Reseeach Laboratories, 20-1 Shintomi, Futtsu, Chiba Prefecture 293-8511 Japan

e-mail address: morita. d98. mitsuru@jp.nssmc.com

Abstract. The microstructure of QMG consists of a single crystal $REBa_2Cu_3O_x$ phase and fine RE_2BaCuO_5 particles. Y-based QMG was first produced in 1988 by the Quench and Melt Growth method. A bulk magnet made from QMG was proposed in 1989. QMG bulk magnets were realized by enlargement technology using RE substitute seed crystals. Based on these technological innovations, the development of NMR and MDDS, which are product-based applications of bulk magnets, has recently been reported from the research institutes of applied development. The history of the initial development of these QMG materials and QMG bulk magnets was reviewed. In addition, as progress of the recent QMG bulk magnet development, a new reinforcement method for large ring bulk magnets that can trap 10 T class strong magnetic field magnetization has been reported. For high field magnetization, the reinforcement of bulk magnets is essential to prevent cracking due to large hoop force. Compared to the conventional reinforcement method using only the outer metal ring, the new reinforcement method has been developed using thin QMG rings and metal ring sheets as the composite material with the inner and outer metal rings. In the case of the QMG ring reinforced by the conventional method, cracking occurred at 10T magnetization. Whereas the QMG ring created by the new reinforcement method successfully trapped approximately 10 T.

1. History of QMG and QMG bulk magnets

QMG is a bulk superconductor that has high J_c properties even in the high field. It consists of an RE (Rare Earth)-based oxide superconductor. Figure 1 shows the microstructure of QMG. Figure 2 shows the appearance of QMG. The main technologies characterizing QMG are indicated below.

1) Microstructure: RE_2BaCuO_5 (211) particles about 1 μm in size are finely dispersed in the single crystal $REBa_2Cu_3O_x$ (123) phase obtained by Pt or Ce addition for fine 211 particles.
2) Overall structure: QMG appears as a single crystalline structure obtained by the RE substituted seeding technique.
3) RE composition gradient method: This production method is mainly used to produce large QMG materials larger than φ 100 mm.

In this paper, only the origin of the microstructure was mainly reviewed as the history of QMG.

4) QMG bulk magnets: Figure 3 shows an example of the state of the generated magnetic field and superconducting current of the bulk magnet after magnetization.

QMG bulk magnets can generate a high magnetic field stronger than 10T by magnetization using field cooling.

Then the origin of the QMG bulk magnets was reviewed as the history of QMG bulk magnets. A material in which 211 particles 1μm in size are dispersed in a single crystalline 123 phase was manufactured for the first time using a Quench and Melt growth technique in 1988. First of all, the situation before 1988 will be described below.

In 1986 Bednorz & Müller reported the 30K class superconducting phenomenon with LaBaCuO-based material [1]. In 1987, M. K. Wu et al. reported a 90K class superconducting phenomenon with a YBaCuO-based material with T_c exceeding the liquid nitrogen temperature (77 K) [2]. Regarding J_c improvement of the YBaCuO-based material, initially, it was mainly studied using a sintered body. The sintered body is an assembly of crystal grains and contains many grain boundaries. Despite many kinds of investigations, J_c properties of the sintered body could not be improved especially in the presence of the magnetic field. Thus, the cause of the low J_c of the sintered body was explained in view of the weak link between grains at grain boundaries. Experimental evidence of weak link behavior has been demonstrated by Ekin et al. [3]. These weak link phenomena suggest that polycrystalline material such as the sintered body is unsuitable and that single crystals or sharply oriented materials are indispensable for high J_c

On the other hand, magnetization measurements of single crystals prepared by the flux method revealed that J_c in the single crystal is relatively high even in a high magnetic field as shown by Dinger et al. [4]. High J_c values of oriented polycrystalline bulk material were reported by Jin et al. [5]. They used a type of directional solidification called the MTG (Melt –Texture Growth) process. The microstructure is comprised of the alignment of plate-shaped 123 grains grown in the a-b direction. J_c as high as 17000 A/cm^2 in zero magnetic field and 4000 A/cm^2 in a field of 1T were reported. This paper asserts that the J_c property can be improved by eliminating the second phase such as coarse 211 particles and the CuO phase [6].

The motivation for the development of QMG was the improvement of J_c by single-crystallization in order to circumvent the weak-link. A single crystal was prepared by a self-flux method using a Pt crucible on the surface of a solidified sample. A single crystalline structure of the 123 phase having 211 grains about 10 μm in size was confirmed in the solidified sample also.

Furthermore, 123 melted in a Pt crucible was rapidly cooled by two Cu plates, the obtained melted and quenched body having a thickness of several mm was placed on a Pt net, and placed in a furnace heated to 1100°C. The crystal was grown by slowly cooling to around 1000°C. As a result, QMG (a material in which 211 particles about 1 μm in size are finely dispersed in a single crystal 123 phase) was obtained for the first time in January 1988 [7-9]. This was just before discovery of Bi-based superconducting material by Dr. Maeda et al. [10]. Later on, it was revealed that a small amount of Pt resulted in fine 211 particles about 1μm in size [11-13].

As described above, QMG was created from the single crystal production technology in order to circumvent the weak link problem. As the inventor of QMG, I assert that this invention was not created from the extension of the MTG method that is the process used to make the unidirectional solidified polycrystalline material. The technical concept of the production of single crystal was inherited by the invention of RE substituted seeding technology developed as the manufacturing method for large QMG [14]. By this seeding, large size QMG became manufacturable. As a result, various developments of QMG applications started.

QMG bulk magnets function as a permanent magnet was invented in 1989 [15]. The concept of the QMG bulk magnet was formed during the magnetization measurements for J_c evaluation of QMG of several mm^3 using a vibration sample magnetometer (VSM). The result of magnetic susceptibility measurement of the small QMG sample shows that if large QMG are obtained, high magnetic fields greater than 2T can be obtained [16]. At the same time as developing large QMG using seeding technology, the basic form of QMG and QMG bulk magnets was established.

In 1991, the generation of a magnetic field of 1.3 T in liquid nitrogen and the suppression of flux creep by additional cooling (to 63 K) after magnetization of 77 K were reported [17]. In 1998, by precision machining of the outer circumference of QMG and fitting a metal reinforcing ring, maximum flux density of 7.17 T at 40 K was obtained using a refrigerator for sample cooling and a liquid-He free 10T superconducting magnet [18-19]. Regarding the history of QMG and QMG bulk magnets, the progress that occurred in the first 10 years has been reviewed.

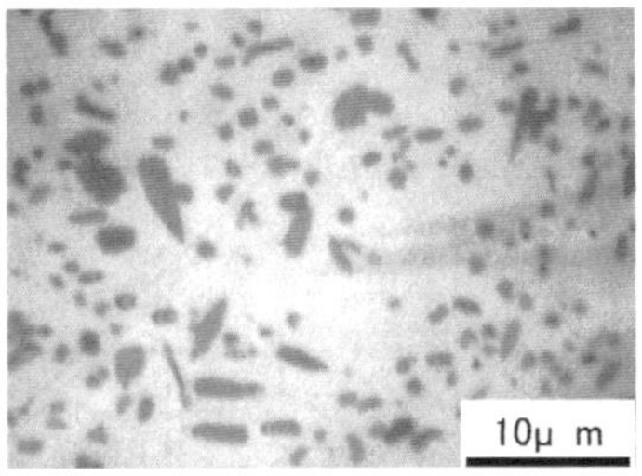

Figure 1. Microstructure of QMG.

Figure 2. Appearance of QMG.

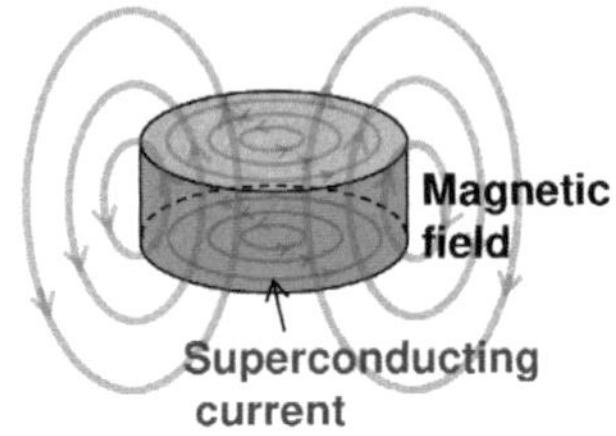

Figure 3. QMG bulk magnet

2. The recent progress of QMG bulk magnets for reinforcement

Recently, the development of NMR and MDDS applications using ring bulk magnets has progressed significantly [20, 21].In these applications, the generation of a strong magnetic field is necessary. In order to achieve higher performance, it is an important task to prevent cracks due to electromagnetic force. Meanwhile, there have been few reports on the mechanism responsible for material cracking, and the strain behavior of materials that causes fracturing is not well understood [22]. In this part of the paper, the progress in the reinforcement technology to prevent the QMG ring from fracturing is described.

2.1. Trapped flux density and strain behavior of ring-shaped QMG bulk magnets in the magnetization process. [23-24]

Figure 4 shows the temperature and field change profile in the magnetization process. Figure 5 shows the change of distribution on the induced superconducting current in the cross-section of the QMG ring in the magnetization process. The green area represents the induced current. When magnetization was performed at a relatively high temperature (in other words, low J_c) in a sufficient applied field, the superconducting current passage was observed on the entire cross-section of the ring, completely magnetizing the ring in phase ④. Conversely, when magnetization was performed at a relatively low temperature (in other words, high J_c), the ring was not completely magnetized. The magnetization was terminated leaving areas where no superconducting current flows as in ② and ③.

Stress (F) applied to the QMG ring is divided into the circumferential and radial directions. The stress causing cracking is the circumferential stress (hoop stress). Figure 6 shows the concept of the circumferential stress. The stress can be considered basically proportional to the product of the induced superconducting current (I) multiplied by magnetic field (B); and magnetic field (B) can be considered the sum of the external applied field (B_e) and trapped field (B_t) by the induced superconductivity current. As shown in Figure 6, a metal ring has been conventionally used for reinforcement. After fitting a QMG ring onto the metal ring with resin in the gap and then cooling, compressive force is obtained from the metal ring, thereby preventing the QMG ring from cracking.

Next, Figure 7 shows examples of the sample preparation for the measurement of the trapped flux density and the change of strain on the QMG ring in the magnetization process. A QMG ring sample with an outside diameter of 60 mm, inside diameter of 28 mm, and thickness of 20 mm was used for the test. It was fitted in a 10-mm thick stainless steel ring. Strain gauges were attached to locations near the inner and outer circumferences of the sample, so that change in the strain in the circumferential direction could be measured. Furthermore, a Hall effect sensor was placed in the center of the ring to measure changes in the trapped flux density. Figure 8 shows the change of the trapped flux density and strain in the magnetization process at 70 K and 4.5 T. Since the external field constituted the x-axis, in this figure, the start point of magnetization is 4.5 T. The magnetization process progressed toward the origin. For the change of strain, the increase of the elongation strain

IOP Conf. Series: Journal of Physics: Conf. Series **1054** (2018) 012046 doi:10.1088/1742-6596/1054/1/012046

from the amount of strain immediately before the magnetization was indicated as positive values. For the trapped flux density, values obtained by subtracting the externally applied magnetic field from the values indicated by the Hall effect sensor at the center of the sample were used.

In the early phases of the magnetization process (① to ②), since the superconducting current is induced in proportion to the amount of demagnetization in a high applied field, the change of strain and trapped flux density are increased almost proportionally. In the phase near the end of the magnetization process (④), the induced superconducting current is almost saturated. For this reason, the trapped magnetic flux density is almost constant as well. The change of strain is linearly decreased along with the reduction of the external field. However, the change of strain for the trapped field (B_t) remained in the end. The results described above indicate that depending on the circumstances, the change of strain would reach a peak in the magnetization process. In particular, when a sample is in the shape of a ring, it is easier to be magnetized completely, and the change in strain is considered to have a peak more frequently than in the case of a cylindrical sample. Furthermore, these results show that the change in strain near the inner circumference is larger than that near the outer circumference.

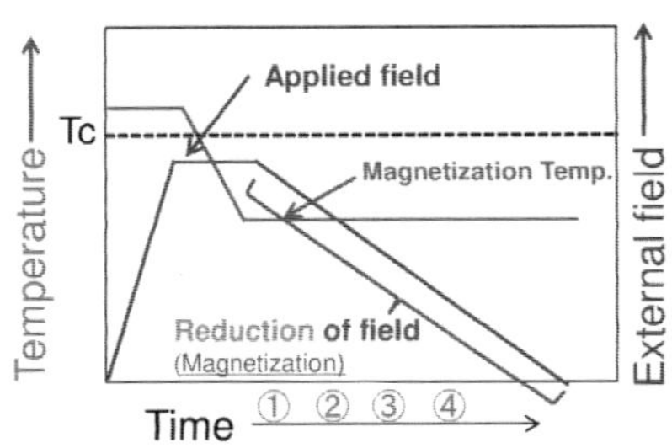

Figure 4. Temperature and field change in the magnetization.

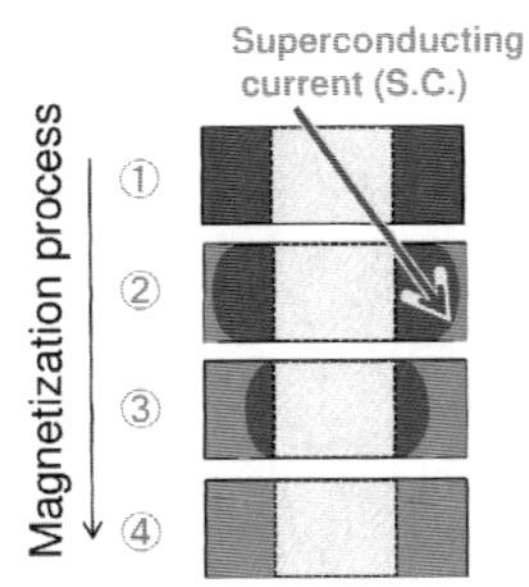

Figure 5. Distribution change of superconducting current in the cross section QMG ring.

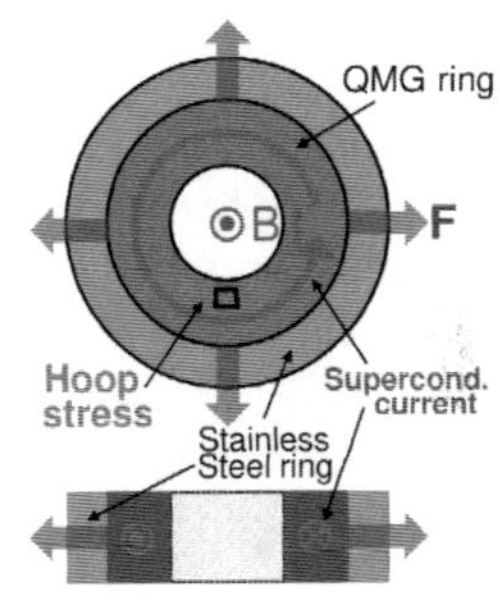

Figure 6. Hoop stress of magnetized bulk magnet with outer metal ring.

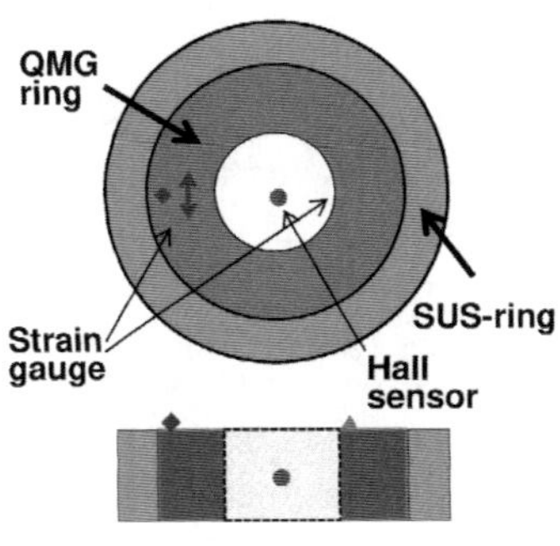

Figure 7. Reinforced sample with a metal ring and placement of sensors.

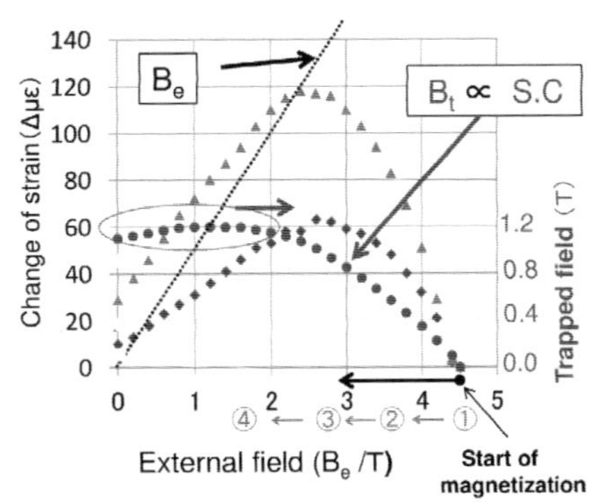

Figure 8. Change of stress and trapped field in the magnetization process.

Figures 9(a) and 9(b) compare the change in strain when a crack was caused in the magnetization process and that with no cracking during the magnetization. Figure 9(c) shows the sample used for the magnetization and the positions in which strain gauges were attached. As these figures show, while the change in strain near the inner circumference on the top face increased faster than that on the inner circumferential face, it reached a peak and took a downward turn quicker as well; and the value of the change in strain on the inner circumferential face rose higher at the end of magnetization.

In the case of magnetization at 55 K and 10 T, the increase of the strain amount was sharper than that at 60 K and 9 T due to the larger stress. When the strain near the inner circumferential face on the top face reached approx. 800 $\mu\varepsilon$, a crack occurred. This confirms that cracking started in the vicinity of the inner circumference in the magnetization process of a ring-shaped QMG bulk magnet.

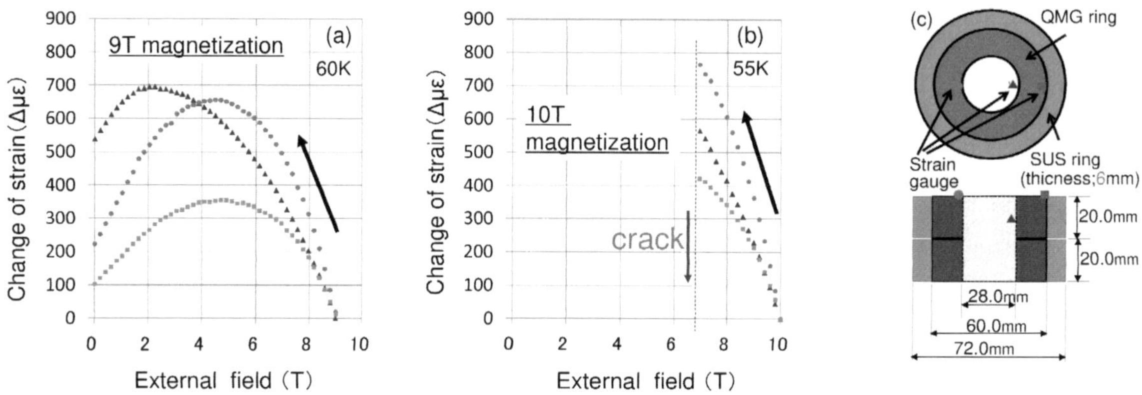

Figure 9 Change of strain in the magnetization on the piled reinforced rings (a) 9T at 60K magnetization, (b) 10T at 55K magnetization, (c) Sample and placement of sensors.

2.2. Concept of the new reinforcement method

These results indicate that the compression stress from the outer circumference alone would not be sufficient to reinforce the inner circumference to which a large amount of stress is applied. For a QMG bulk in the ring shape, as the largest strain works in the vicinity of the inner circumference, reinforcements of both the material itself and the inner periphery were considered to be effective. For this reason, a new reinforcement method of ring QMG bulk magnets was devised.

The method involved laminating QMG sheets and high-strength metal sheets into a high-strength composite material, as well as using an inner circumferential metal ring for reinforcing the inner circumference of the laminated QMG ring. The following describes an example of sample structures used for the new reinforcement.

Figure 10(a) shows a cross-sectional view of a ring-shaped QMG bulk magnet reinforced using the new reinforcement method. Figures 10(b) and 10(c) show an enlarged view of the center of the ring cross-section and that of the vicinity of the inner circumferential metal ring, respectively. First, I examined the method for making the composite material shown in Figure 10(b), and studied a method that involved making thinly-sliced QMG sheets and high-strength metal sheets into a composite material by laminating them. In order to improve the mechanical bonding between the metal rings and ring-shaped QMG sheets, Ag coating was applied to the QMG sheets before they were combined with the metal rings by soldering. Figure 10(c) shows the inner circumferential reinforcement. Due to the rate of thermal shrinkage of ordinary metals larger than that of QMG, an inner ring cannot be fixed by the compression force from cooling contrary to the outer circumferential metal ring. When the inner circumferential metal ring alone is attached as-is, it will shrink due to cooling and detach from the QMG rings. In view of this, the inner circumferential metal ring and metal sheets used to combine

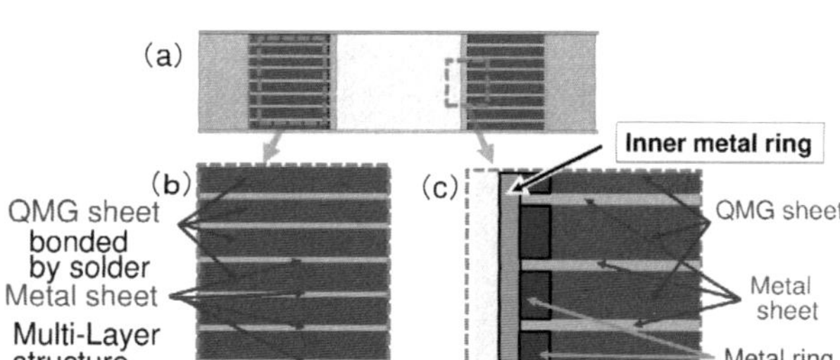

Figure 10 Schematic illustration of cross section of bulk magnet by new reinforcement method, (a) cross sectional view of whole the sample, (b) enlarged view of laminated structure, (c) enlarged view of the reinforcing structure near the inner ring

with the QMG rings were firmly soldered to maintain metallic bonding. This was performed to elicit the reinforcement effect of the inner circumferential metal ring work to prevent the QMG rings from deforming

2.3. Production of a QMG bulk magnet reinforced by the new method and evaluation of the reinforcement effect

A ring-shaped QMG bulk magnet with the new reinforcement method was actually created based on the concept described above. The strain and the trapped flux density were measured. Furthermore, the prototype was compared with a bulk magnet produced using the conventional reinforcement method. The following describes the details of the test production and evaluation.

2.3.1. Sample preparation

Figure 11 shows the parts used for the new reinforcement method. Figure 12 shows the assembling method. As shown in these figures, Ag coated QMG rings were arranged in the Cu ring. A stainless steel ring was arranged in the Ag coated QMG ring. Eight units thus prepared were alternatingly laminated with nine stainless steel sheets ring. The inner stainless steel ring and outer stainless steel ring were attached to the laminate. Each part was bonded by solder, and the QMG bulk magnet was created. Table 1 shows the dimensions of each reinforcement material. Figure 13 shows the appearance of the sample obtained through the assembly process. The sample was fixed in the center of the cold head of a refrigerator with grease. The outer circumferential ring was fixed with screws.

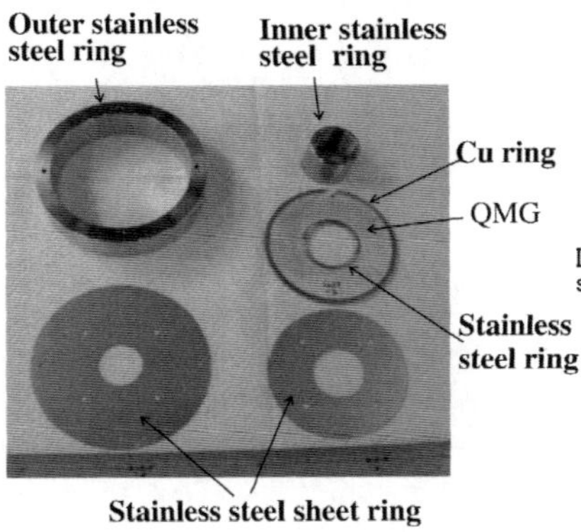

Figure11. The QMG ring and reinforcing members used .

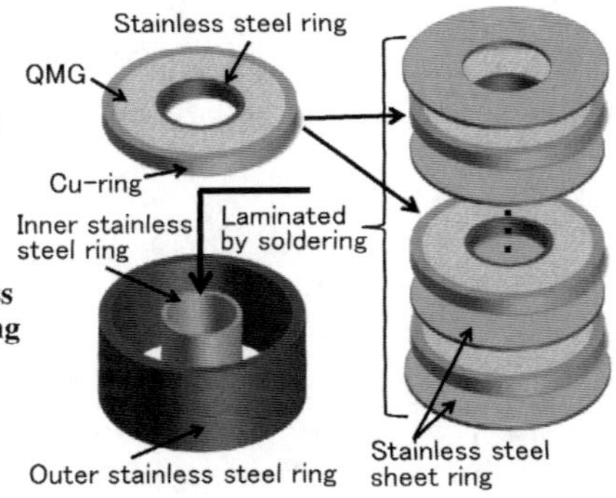

Figure 12. Assembly drawing

Figure 13. A sample prepared by the new reinforcing method.

Table 1 The dimensions of QMG ring and each reinforcement material

Parts of QMG ring magnet reinforced by new method	Outside diameter (mm)	Inside diameter (mm)	Thickness (mm)
QMG ring	60.0	28.0	2.0
Stainless steel sheet ring	64.0	24.0	0.5
Upper/lower stainless steel sheet ring	80.0	24.0	0.5
Outer stainless steel ring	80.0	64.0	20.0
Cu ring	64.0	60.0	2.0
Inner stainless steel ring	24.0	22.0	20.0
Stainless steel ring	28.0	24.0	2.0

2.3.2. Measurement method

The sample was cooled by He gas cooling using the refrigerator to maintain thermal uniformity of the sample. The sample was inserted into a superconducting magnet bore. After cooling the specimen

IOP Conf. Series: Journal of Physics: Conf. Series **1054** (2018) 012046 doi:10.1088/1742-6596/1054/1/012046

from room temperature to 95 K (non-superconducting state), a magnetic field with predetermined strength was applied to the specimen. Then, it was cooled to a predetermined temperature. When the temperature was sufficiently stable, the magnetic field was reduced at the rate of 0.03 T/min for demagnetization. The values of the strain gauges and flux density were recorded.

2.3.3. Results and discussion

Figure 14 shows the trapped flux density and change of strain when the magnetization was performed at 20 K and 10 T. The colors of the data plot lines in the figure correspond to the sensor colors indicated in sensors placement illustration in the figure. In this figure, the QMG bulk magnet is trapping approximately 10T. The slight decrease from 10T can be considered due to the aspect ratio of the sample. In the case of low temperature, that means high J_c, the induced superconducting current is concentrated on the outer circumferential face part of the sample. For the change of strain, in comparison with conventional reinforced bulk magnets, the same trend whereby change of strain at the inside is larger than that at the outside is observed.

Figure 15 shows a comparison of the change of strain in the magnetization at the same 60 K. Figure 15(b) shows the result of 9T magnetization for the bulk magnet with conventional reinforcement. Figure 15(a) shows the result of 10T magnetization for the sample with the new reinforcement described above. In the comparison of Figure 15(a) and Figure15 (b), firstly, in the case of the new method, The magnetic field of the peak of each strain curve is almost the same. This trend is thus considered due to the improved uniformity of material strength by the composite material structure.

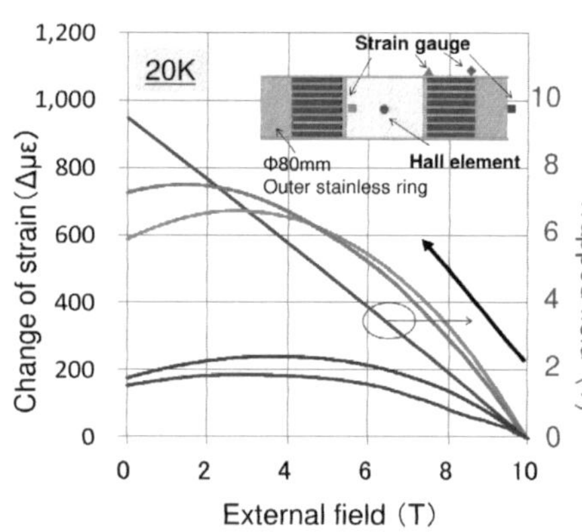

Figure14. change of trapped field and Strain during the magnetization process

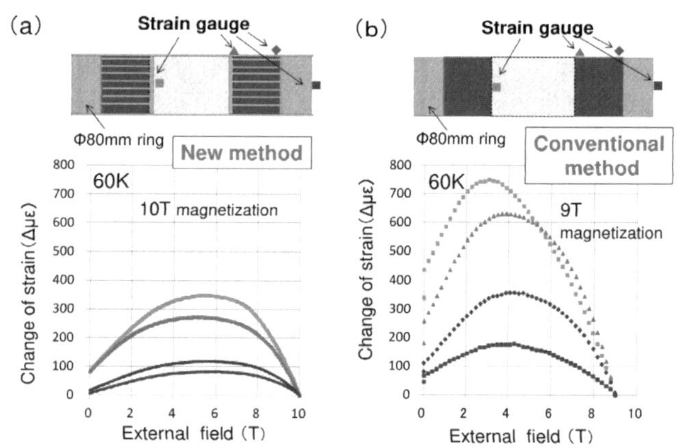

Figure 15 Comparison of strain changes of samples prepared by the new reinforcement method(a) and conventional method(b).

In a quantitative comparison, in the case of the conventional method, the strain reached a maximum of 700 $\mu\varepsilon$ on the inner circumferential face. In the case of the new method, the maximum strain value did not exceed 350 $\mu\varepsilon$. The maximum change of strain under the conventional method would reach 800 $\mu\varepsilon$, if the bulk magnet reinforced using the conventional method had been magnetized with 10 T not entailing a crack. From this result, the new reinforcement method is considered to have reduced the change of strain to about 40% of that working on the conventionally-reinforced bulk magnet. As seen from the comparison result, the new reinforcement method has an advantage over the conventional method. Moreover, in view of the strain maintained low when magnetizing with 10 T, the sample reinforced with the new method is very likely to endure magnetization using a magnetic field exceeding 10 T.

After these measurements, the distribution of trapped flux density was measured at 77 K. No change was observed in the distribution before and after the 10T magnetizations. For this reason, on the QMG sample reinforced using the new method, the sample was sufficiently prevented from being damaged from even the stress caused by repeated magnetization.

3. Conclusions

1) The first ten years of the history of QMG and QMG bulk magnets was reviewed.
2) As the recent progress of the ring QMG bulk magnets, a new reinforcement method was reported based on the following two concepts.

   One is to create a composite material of metal sheet and sliced QMG ring. The other is the inner circumference reinforcement by the metal rings to which metal sheets are bonded.

   For example, QMG rings and stainless steel rings were laminated into a composite material; the outer circumferential and inner circumferential metal rings were attached to the laminate by soldering. The obtained ring-shaped QMG bulk magnet trapped 10T without cracking. Compared with the conventional reinforcement method, the new reinforcement method enables significant reduction of the strain of the excited QMG bulk magnet.

References

[1] Bednorz J G, Müller K A 1986 *Z. Phys, B-Condensed Matter.* **64**, 189
[2] Wu, M. K. et al. 1987 *Phys. Rev. Letters.* **58** (9), 908
[3] Ekin J W, et al. 1987 *J. Appl. Pyhs.* **62** 4821
[4] Dinger T R, et al. 1987 *Phys. Rev. Lett.* **58** 2687
[5] Jin S, et al. 1987 *Appl. Phys. Lett.*, **51** 943
[6] Jin S, et al. 1988 *Appl. Phys. Lett.*, **52** 2074
[7] Morita M, et al. United States Patent, Patent Number 5,278,137, Priority Date: Jun.6, 1988
[8] Morita M and Matsuda S 1988 The Society of Non-Traditional Technology, Forum of Superconductivity Science and Technology, New Superconducting Materials Forum News. (10), 15 (in Japanise)
[9] Morita M, et al.: Physica C. 172, 383-387 (1990) (English Translation of Reference [8]
[10] Maeda H, et al.: U.S. Patent 7132388. 1988
[11] Ogawa N, et al. 1991 *Physica C* **177** 101
[12] Morita M, et al. 1991 *Jpn. J. Appl. Appl. Phys* **30** L813
[13] Kim J, 1992 *Mater. Res.* **2** 202
[14] Morita M, et al.1991 *Advances in Superconductivity*, p. 733
[15] Japanese Patent No. 2055511
[16] Morita M, et al. 1989 *Abstracts of CSJ Conference.* **41**, p. 14
[17] Morita M, et al.1992 *Advance in Superconductivity* **IV** (Springer-Verlag,Tokyo) 433
[18] Morita M, et al. 1998 Extended abstracts of The 1998 International Workshop on Superconductivity, July, 115
[19] Japanese Patent No. 4012311
[20] Nakamura T, et al.2011 *TEION KOGAKU,* **46** (3), 139-148
[21] Saho N, et al. 2012*TEION KOGAKU*, **47** (7), 431-435
[22] Miyamoto T 2001 Doctoral dissertation titled "Study on Mechanical Property Evaluation of Large Bulk Superconductors," Chapter 4, 2001
[23] Morita M, et al. 2015 *Abstracts of CSSJ Conference.* **91**, p. 170
[24] Morita M, et al. 2015 *Abstracts of CSSJ Conference.* **92**, p. 210

Collecting Ni-Sulfate Compound from Electroless Plating Waste by Magnetic Separation Technique with Use of HTS Bulk Magnets

T Oka[1], S Sasaki[1], H Sasaki[1], S Fukui[1], J Ogawa[1], T Sato[1], T Nakano[1], M Ooizumi[1], M Tsujimura[2] and K Yokoyama[3]

[1] Niigata University, 8050 Ikarashi-Ninocho, Nishi-Ward, Niigata 950-2181, Japan
[2] Aichi Giken Co., 50-1 Takeshita, Hitotugi-cho, Kariya, Aichi 448-0003, Japan
[3] Ashikaga Institute of Technology, 268-1 Ohmae-cho, Ashikaga, Tochigi 326-8558, Japan

E-mail: okat@eng.niigata-u.ac.jp

Abstract. The authors have developed a useful technique for extracting the Ni-sulfate compound from the "processed waste/fluid" of the electroless Ni-plating processes. After forming the fine $NiHPO_3$ precipitate from the waste, the coarse $NiSO_4$ crystals were synthesized through the reaction with the concentrated sulfuric acid. In the experiment, the open-gradient magnetic separation was employed to collect the $NiSO_4$ crystals from the muddy mixture composed of these compounds. The experiments were practically conducted with use of the Gd123-based HTS bulk magnets generating up to 3.99 T, which were activated by the field cooling magnetizing method operated at 35 K. Intense magnetic field emitted from the bulk magnet attracted the paramagnetic fluid and coarse $NiSO_4$ crystals on the bottom of the water channel, forming a giant drop of $NiSO_4$-saturated clear fluid and the slurry of nickel sulfate hexahydrate. This preferential collection suggests a feasible recycling process of Ni resource as a raw material in the plating processes.

1. Introduction

Since dense and hard Ni coatings uniformly cover even the complex material surfaces with its superior chemical tolerance against rusting or wearing by friction, the electroless nickel-plating process has a significant advantage, and the process has been widely used in numerous factories such as the electronic devices or car manufacturers. But, the waste water treatments after the plating processes cause the serious problems both on the environmental loads and the natural resource issues. Therefore, it is needed to establish the useful processes to collect and recycle the Ni resources as one of the rare earth metals.

We have proposed three feasible approaches in the past, such as the drain water purification which aimed to reduce the concentration of Ni and P from the drained water from factories [1]. The second trial targeted the deposition of Ni_3P which were generated by its unusual decomposition of plating fluid which contained Ni ions with high content in the waste liquid [2]. The third approach was a formation of $NiSO_4$ compound from the $NiHPO_3$ precipitate fabricated by adding the concentrated sulfuric acid to the slurry which contains Ni ions with high content [3].

Although the metallic Ni has, all we know, the ferromagnetic property, and it is easily attracted to magnets. But the fact is not well known that the chemical compound or the crystal of Ni-bearing compounds like a $NiSO_4$ crystal has weak magnetic moment and it is expected to be attracted or collected when we use extremely intense magnetic fields generated by superconducting devices. The high temperature superconducting (HTS) bulk magnets would be effectively capable of attracting these kinds of precipitates.

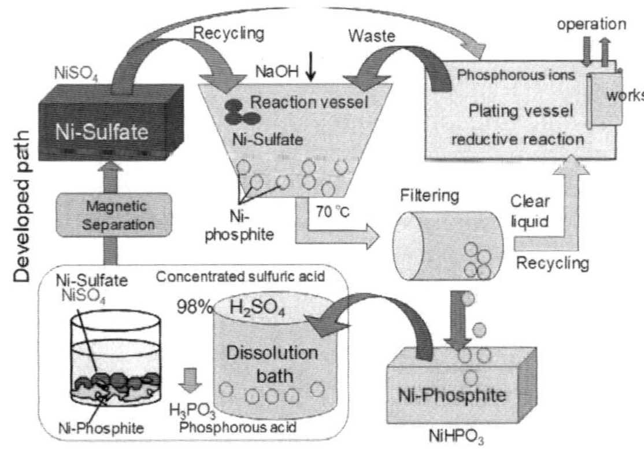

Figure 1. Recycling process of Ni-plating waste.

In the recycling process shown in Figure 1, after forming the fine $NiHPO_3$ (nickel-phosphite) precipitates by the heat treatment and pH controlling, we obtained the $NiSO_4$ crystals by adding concentrated sulfuric acid. As the mixed slurry composed of the $NiSO_4$ and $NiHPO_3$ are chemically unstable, and it is necessary to extract $NiSO_4$ crystals immediately after the deposition from the mixture.

In this study, we have tried to collect the $NiSO_4$ compounds by utilizing the slight difference between their magnetic properties from the processed waste mixture of these compounds. We aim to recycle the Ni resources through the magnetic separation processes with use of the HTS bulk magnet, and to elongate the lifetime of Ni-plating liquid, which must reduce the amount of the plating waste and resultant environmental loads.

2. Experimental procedure

2.1. Fabrication of Ni sulfate precipitates

Figure 1 shows the whole recycling process of electroless Ni-plating. The P ions formed by the reductive reaction changes Ni into the fine $NiHPO_3$ precipitates by adding NaOH at 70 °C. After the dissolution by adding the concentrated sulfuric acid, the $NiSO_4$ crystals deposit and form the mixed slurry of these compounds. $NiSO_4$ and $NiHPO_3$ crystals coexist in the sulfuric acid as a mixture.

The concentration-pH equilibrium diagram of these two compounds are shown in Figure 2 [4]. The different compounds crystalize with respect to the different regions of pH values, which precipitates on the bottom of the vessel, as shown in the figure. After the reaction, we call the mixture of $NiSO_4$ and $NiHPO_3$ crystals as "processed waste/fluid".

Figure 3 shows a photo of largely-grown crystals generated from this reaction. Strong attractive effect is expected on these crystals, because the magnetic force is proportional to the particle size. The magnetic force is shown in Eq (1), where V_p, χ_f and χ_p means the volume of the particles, susceptibility of the liquid and particles, respectively [5].

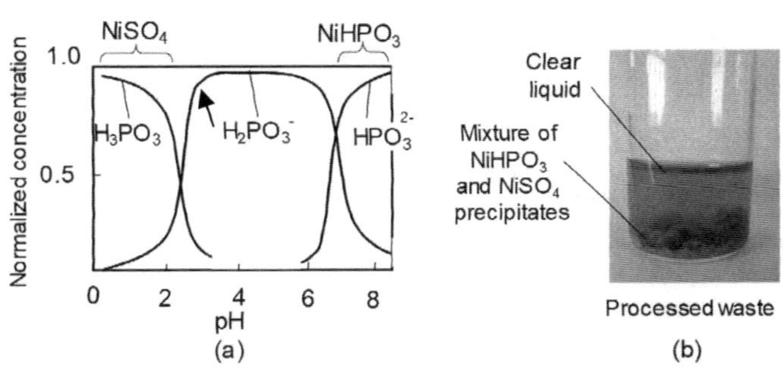

Figure 2. Formation of Ni-sulfate and Ni-phosphite (a) and processed fluid from the waste (b).

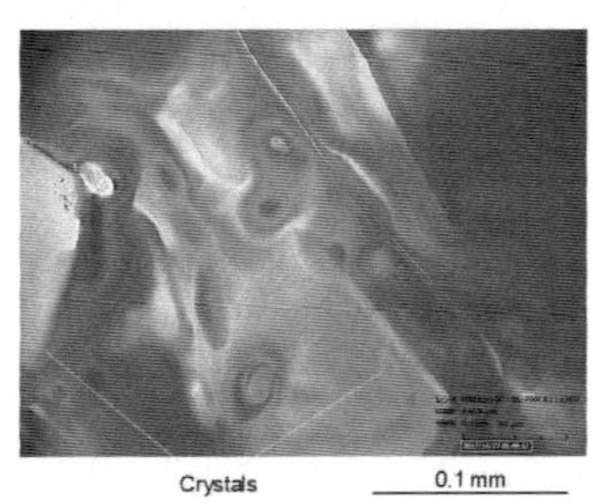

Figure 3. Crystals in the processed waste observed by the digital microscope.

$$F_m = \frac{4}{3}\mu_0 V_P \frac{9(\chi_p - \chi_f)}{(3+\chi_p)(3+\chi_f)} H \cdot \mathrm{grad}H \qquad (1)$$

The magnetic moment of the compounds and precipitates were estimated with use of SQUID magnetometer (Quantum Design, MPMS-XL).

2.2. HTS bulk magnet and its magnetic field distribution

Figure 4a shows a photo of HTS bulk magnet system used in the experiment, whereas Fig. 4b shows its magnetic field distribution measured at the surface of the magnetic pole, which indicates the most intense magnetic field of 3.99 T, emitting at the center surface of magnetic pole. The system loads a HTS Gd123-based bulk magnet with a size of 60 mm in diameter and 15 mm in thickness, which were activated at 35 K by field cooling (FC) method. The activation was operated from 6 T with use of JASTEC superconducting solenoid magnet [6, 7].

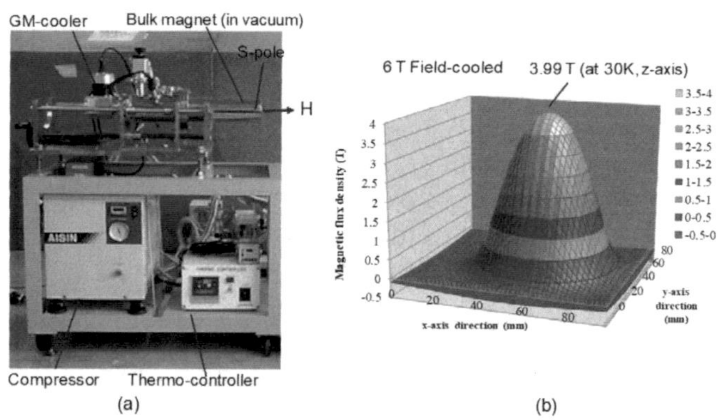

Figure 4. HTS bulk magnet system equipped by a strong magnetic pole on the table (a), and the magnetic field distribution measured on the pole surface, which generates 3.99 T.

2.3. Magnetic separation experiment

The magnetic separation experiments were conducted with respect to the $NiSO_4$ particles deposited from the processed fluid, where the fine $NiHPO_3$ slurry was dissolved by the concentrated sulfuric acid, and kept it at 58 °C under its supersaturated condition. Figure 5a and 5b show a view of the experiment

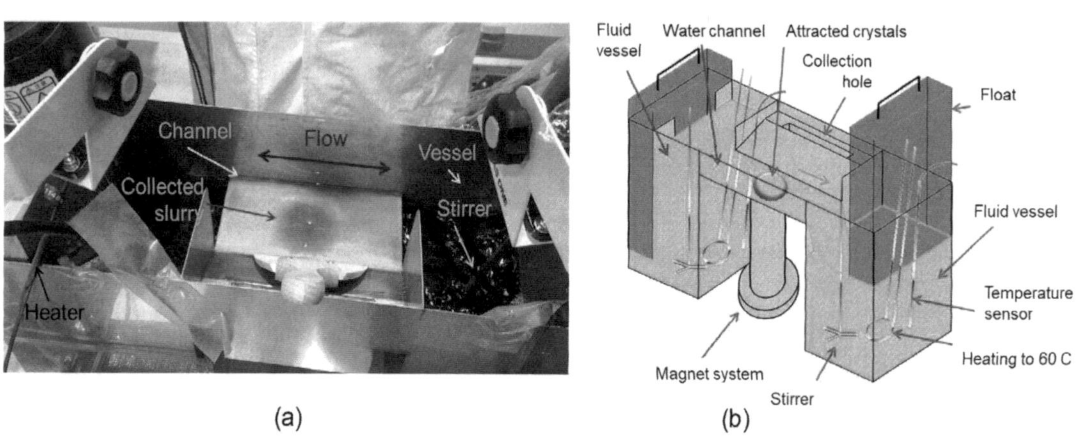

Figure 5. A view and construction of magnetic separation experiment, showing the green slurry attracted on the pole surface generating over 3 T (a), and the schematic illustration (b).

and its structure, respectively. The bulk magnet was located just under the water channel which is a path of the processed fluid. The fluid was fed from the vessel of 6 liter settled on both sides. A couple of floats are alternately sunk into the vessel to feed the fluid through the channel. The flowing processed fluid was exposed in the 3-T space the bulk magnet emits, and then, the slurry/crystals were attracted at the magnetic pole. The contents of Ni and P in the collected slurry were evaluated by ICP (SII, SPS1500V Plasma Spectrometer).

3. Results and Discussion

As shown in Figure 5a, we observed the apparent green slurry attracted on the bottom of the water channel fixed on the magnetic pole surface. A schematic illustration of the solution flow is shown in Figure 5b. The width of a part of the water channel was designed as 100 mm so that the fluid passes in the strong magnetic field. The processed fluid alternately flows between the vessels through the channel for about 1 minute. The maximum amount of collected crystal reached 29.4 g, and the amount of recovered amount corresponds to 8.4 kg a day, which inferred the feasible practical application to various actual industries.

Figure 6a shows the slurry collected on the bottom of channel. After the operation, the slurry was removed. One sees a drop of processed fluid on the bottom of the channel. Weakly magnetized paramagnetic liquid was attracted by the intense magnetic field to let the fluid form a dorm-like huge drop of solution with saturated concentration of Ni-sulfate. According to ICP analysis, we found that 85% of collected slurry were composed of $NiSO_4$ crystals.

Figure 6. A view of collected slurry in the channel (a) and the processed fluid attracted on the magnet after the flowing test (b).

Figure 7 shows the data of magnetic moment for the crystals of $NiSO_4$, $NiHPO_3$, and processed fluid. $NiSO_4$ crystal showed the highest moment. The data of processed fluid located between the rests. As magnetic moment for $NiSO_4$ is three times larger than that of $NiHPO_3$ might indicate the effective separation of $NiSO_4$ from the mixture.

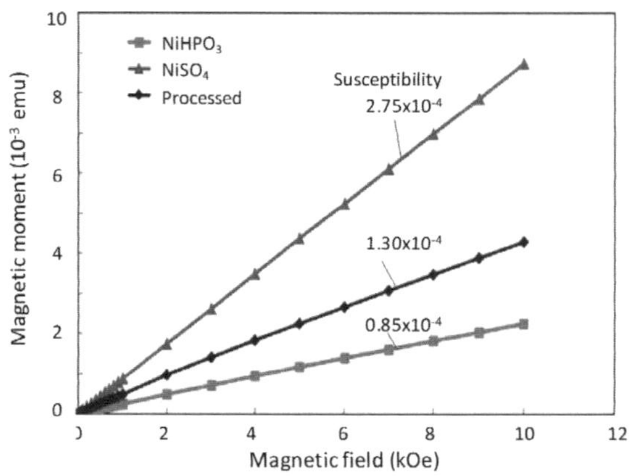

Figure 7. Magnetic property of the compounds and the processed fluid with the susceptibility data by each line. The susceptibility of $NiSO_4$ is as three times higher than that of $NiHPO_3$.

Figure 8 shows the result of XRD for the collected slurry after the experiment. The peaks of all were identified as major nickel sulfate hexahydrate crystals and the minor tetrahydrate, which means that the major content of collected crystal should be used as a recycled raw material of plating processes.

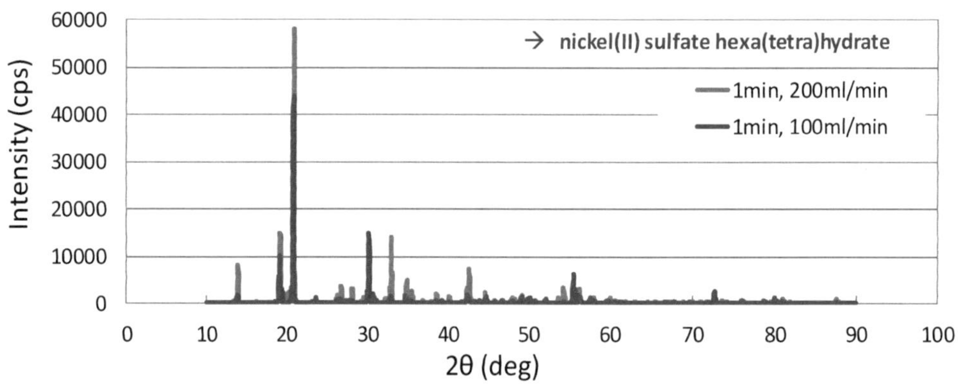

Figure 8. Characterization by XRD measurement with varying flow rates for the collected crystals.

4. Conclusion

We conducted the magnetic separation experiment on Ni-plating waste with use of the HTS bulk magnet generating 3.99 T, which was magnetically activated by field cooling method at 6 T. Intense magnetic field attracted the paramagnetic processed fluid on the bottom of the water channel, forming a giant drop of saturated fluid. The magnetic properties and the crystal structure of the collected slurry were estimated, which clarified the feasibility of separation technique. Since the crystal structure of collected $NiSO_4$ slurry was identified as the nickel sulfate hexahydrate, the slurry was found to be available to reuse as a raw material in the plating process after some refinement on its purity improvement.

Acknowledgment

The study has been partially supported by the grants-in-aid from MEXT of Japanese government, KAKENHI Grant (No. 15H03550) of Japan Society for the Promotion of Science, Japan Science and Technology Agency. The authors give them the greatest appreciation.

References

[1] Oka T, Kanayama H, Fukui S, Ogawa J, Sato T, Yamaguchi M, Ooizumi M, Terasawa T, Itoh Y and Yabuno R 2008 Application of HTS bulk magnet system to the magnetic separation techniques for water purification *Physica C* **468** 2128

[2] Oka T, Kimura T, Mimura D, Fukazawa H, Fukui S, Ogawa J, Sato T, Ooizumi M, Yokoyama K, Tsujimura M and Terasawa T 2013 Magnetic precipitate separation for Ni plating waste liquid using HTS bulk magnets *Physica C* **484** 325

[3] Oka T, Fukazawa H, Fukui S, Ogawa J, Sato T, Ooizumi M, Tsujimura M and Yokoyama K 2014 Collection of Ni-Bearing Material from Electroless Plating Waste by Magnetic Separation with HTS Bulk Magnet *Physica C* **496** 58

[4] Matsui F 1999 Recycling and Circulation of Electroless Nickel Plating Bath *Uemura Technical Report* **44** 21 (in Japanese)

[5] Watson J 1973 Magnetic filtration *J. Appl. Phys.* **44** 4209

[6] Oka T, Takayanagi Y, Machida S, Ichiju K, Fukui S, Ogawa J, Sato T, Ooizumi M, Tsujimura M and Yokoyama K 2016 Magnetic Separation for Recovering Ni-Compounds from Plating Waste with Use of HTS Bulk Magnets *IEEE Transactions on Applied Superconductivity* **26** 3700204

[7] Oka T, Hirayama E, Kanai T, Ogawa J, Fukui S, Sato T, Yokoyama K, Nakamura T 2014 Strong Magnetic Field Generators Containing HTS Bulk magnets and Compact Refrigerators and Their Field-Trapping Performances *IEEE Transactions on Applied Superconductivity* **24** 4900204

[8] Oka T, Kimura T, Mimura D, Fukazawa H, Fukui S, Ogawa J, Sato T, Ooizumi M, Tsujimura M and Yokoyama K 2011 Tokyo Magnetic Separation of Nickel from Electroless Plating Waste Fluid Using a HTS Bulk Magnet *TEION KOUGAKU* **46**, 655 (in Japanese)

IOP Conf. Series: Journal of Physics: Conf. Series **1054** (2018) 012048 doi:10.1088/1742-6596/1054/1/012048

Optimization of growth parameters for fabricating single grain (Gd, Dy)BCO bulk superconductors in top-seeded infiltration growth process

S Pavan Kumar Naik, M Muralidhar and M Murakami

Superconducting Materials Laboratory, Graduate School of Science and Engineering, Shibaura Institute of Technology, 3-7-5 Toyosu, Koto-ku, Tokyo 135-8548, Japan

Email: spavankumarnaik@yahoo.in

Abstract. Fabrication of single grain bulk $REBa_2Cu_3O_{7-\delta}$ (REBCO) superconductors with superior performance along with shape retaining is a recent topic. Mixed REBCO superconductors produced in melt growth (MG) technique had showed enhanced superconducting performance. However, this process offers many disadvantages such as macro-porosity, shrinkage in final products, inhomogeneous distribution of 211 secondary phase particles etc., which limit many practical applications. Infiltration growth (IG) process is advanced and superior to MG technique in several aspects. Recently, we fabricated (Gd, Dy)$Ba_2Cu_3O_{7-\delta}$ ((Gd, Dy)BCO) bulk superconductors through systematic addition of Dy_2BaCuO_5 (Dy-211) content. The addition of 20 wt.% of Dy-211 in GdBCO was found to be optimum which resulted in enhanced superconducting performance. In the present work, in order to determine a suitable temperature window for fabrication of large single grain (Gd, Dy)BCO bulk superconductors, isothermal experiments were carried out at several constant temperatures in top-seeded IG process in air atmosphere. Systematic microstructural and magnetic properties were assessed and analysed. The 211 secondary phase particles are enlarged to as high as ~ 25 μm when the sample assembly is dwelled at high temperatures and reduced to ~ 2 μm – 4 μm in the samples dwelled at lower temperatures. Main emphasis will be given on the growth rate progress and difficulties involved in IG processing of mixed REBCO superconductors.

1. Introduction

The high temperature $REBa_2Cu_3O_{7-x}$ (REBCO or RE-123, RE refers rare earth elements such as Nd, Sm, Gd, etc.) are potential for many practical applications due to their superior flux pinning abilities compared to YBCO superconductors. Among all REBCO composites, GdBCO superconductors are interested due to the ease of preparation. GdBCO materials can be synthesized in air atmosphere with very small fraction of formation of solid solutions of type $Gd_{1+x}Ba_{2-x}Cu_3O_y$, which are lower superconducting critical temperature (T_c) phases [1-4]. In addition, it is reported that the GdBCO materials produce very small sized Gd_2BaCuO_5 (Gd-211) phase particles, which are very essential for improving the critical current densities (J_c) especially at lower fields [4-6].

The mixed REBCO superconductors with binary and ternary RE elements are well-studied [1-5]. The results show that the mixed REBCO bulks are superior and promising candidates for many high field applications. The mixed REBCO materials produced in conventional melt growth (MG) technique are well established in literature. However, MG process offers the final products with many unwanted large shrinkage, macro-defects such as pores etc., [7]. Infiltration Growth (IG) process was developed to overcome the problems associated with MG process and is superior. The IG process offer to fabricate the bulk samples with negligible shrinkage and microstructure in which homogeneously distribute fine sized 211 phase particles without addition of any costly grain refiners such as Pt and CeO_2 [8-10]. Still, mixed REBCO materials processed in IG process are not studied well and very few reports are available in the literature [11-14]. Recently, we optimized the Dy- content in bulk GdBCO bulk superconductors and the addition of 20 wt.% of Dy-211 in GdBCO (GdDy-20) was found to be optimal composition which exhibited superior superconducting properties [15]. In this report, we study the growth, microstructural and superconducting properties of GdDy-20 samples, which were under-cooled at various constant temperatures.

2. Experimental details

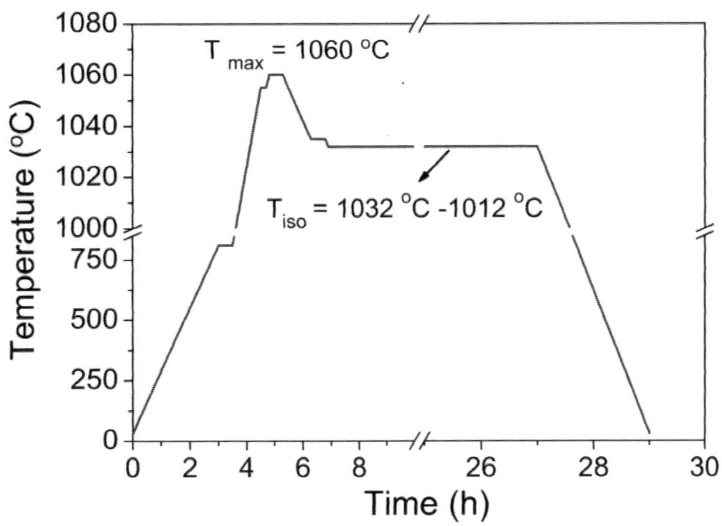

Fig. 1. The time-temperature schedule employed for under-cooling experiments.

The synthesis procedure followed for preparing the precursor powders of Gd-211, Dy-211 and liquid phases detailed in our previous works [15-17]. The preform pellets of GdDy-20-211 (mixture of 80 wt.% of Gd-211 and 20 wt.% of Dy-211) and liquid phase sources were pressed at an uni-axial pressure of 420 MPa. It is well-known fact that, the peritectic temperature (T_P) of a mixed REBCO system will be modified depending on the RE elements present in the mixture. To have the knowledge on the peritectic temperature of the GdDy)-20 composite, a simultaneous TG-DTA measurement was performed and the T_P was found to be 1027 °C ±3 °C [16]. The GdDy-20 sample assembly was heat treated to a maximum temperature (T_{max}) of 1060 °C and dwelled at various constant temperatures (T_{iso}) from 1032 °C to 1012 °C for 20 h for top-seeded IG (TS-IG) processing in air atmosphere. The heat treatment schedule followed for fabricating undercooling samples is shown in figure 1. A sub-specimen of size ~0.5x1.5x2 mm^3 was collected from the as processed bulk samples below ~ 1.5 mm of the seed and were oxygenated in flow of oxygen for 175 h at 450 °C to 400 °C. The final isothermal samples were referred as GdDy-1032, GdDy-1030 etc.

The microstructural features of the isothermal (Gd, Dy)BCO samples were investigated employing a scanning electron microscope (SEM). The temperature dependence of DC magnetic susceptibility $\chi(T)$ and MH loops at 77 K for estimating the field dependence of J_c were assessed employing a SQUID (Quantum Design, model MPMS-XL5minSITO). The field dependence of J_c was calculated utilizing the Extended Beans critical state model [18-19].

IOP Conf. Series: Journal of Physics: Conf. Series **1054** (2018) 012048 doi:10.1088/1742-6596/1054/1/012048

3. Results and discussion

Fig. 2. Top surface photographs of under-cooled (Gd, Dy)BCO samples. Growth has initiated from the Nd-123 seed at 1030 °C and the area of growth increased with lowering the under-cooling temperature.

The images of as processed under-cooled GdDy-1032 – GdDy-1012 samples are shown in figure 2. The sample under-cooled at 1032 °C show no growth indicating the temperature is higher than the peritectic temperature of (Gd, Dy)BCO system. The epitaxial growth has initiated in GdDy-1030 sample from the seed. As the under-cooling temperature decreased from 1030 °C-1022 °C, the area of growth increased. When the temperature is further reduced below 1022 °C, the random nucleation's (parasitic growth) has increased. Random nucleation's could occur in bulk REBCO materials due to various reasons such as doping, temperature differences especially at the edges of the sample, inhomogeneity in availability of liquid phases for peritectic reaction etc. From figure 2, it is evident that, as the parasitic growth increased, the epitaxial growth was inhibited largely and when the temperature reduced to 1012 °C, no growth was observed. From all these observations, we can conclude that the effective temperature window for fabricating large (Gd, Dy)BCO single grains is 1030 °C to 1022 °C.

The SEM micrographs of under-cooled GdDy-1030, GdDy-1028, GdDy-1026, GdDy-1024, GdDy-1023, GdDy-1022, GdDy-1021 and GdDy-1020 samples are shown in figure 3 (a)-(h) respectively. From a keen observation of microstructural features, it can be noticed that the 211 grains are refined as the under-cooling temperature reduced. This clearly indicates that, as the under-cooling temperature decreased, the peritectic reaction increased due to which the 211 phase particles consumed largely and becoming smaller sized. Larger sized 211 phase particles (~ 25 μm) were observed in GdDy-1030 – GdDy-1024 samples could be attributed to the Ostwald ripening phenomenon. As the temperature reduced, the 211 grain size are reduced to ~ 10 μm and 2 μm – 4 μm. Therefore, to produce fine sized 211 phase particles in the (Gd, Dy)BCO bulk samples, longer time slow cooling should be given between 1024 °C – 1015 °C.

IOP Conf. Series: Journal of Physics: Conf. Series **1054** (2018) 012048 doi:10.1088/1742-6596/1054/1/012048

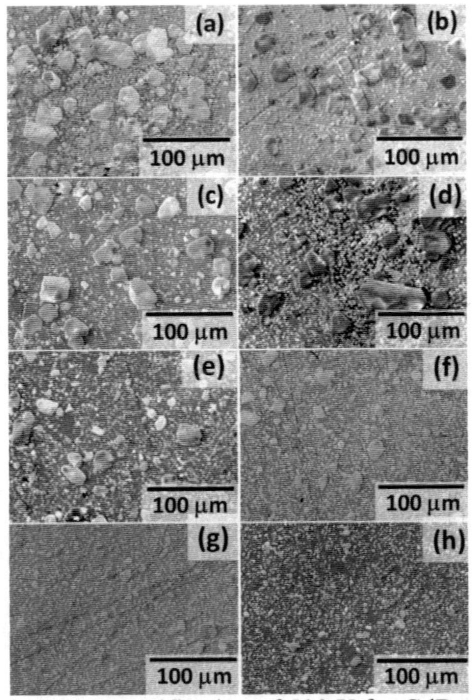

Fig. 3. The SEM images recorded at a magnification of 500 X for GdDy-1030, GdDy-1028, GdDy-1026, GdDy-1024, GdDy-1023, GdDy-1022, GdDy-1021 and GdDy-1020 are shown in panels (a)-(h) respectively.

The normalized DC magnetic susceptibility curves for under-cooled GdDy-1020, GdDy-1021, GdDy-1022, GdDy-1023, GdDy-1024 and GdDy-1026 samples are displayed in figure 4 panel (a). The onset of critical temperature is varied from 90 K to 92.5 K indicating all samples show superconducting nature. The superconducting transition width (ΔT_c) is varied from 3 K to 25 K. The broad width in $\chi(T)$ curves could be attributable to the compositional deviations occurred in different under-cooled composites. The field dependence of J_c for under-cooled GdDy-1020, GdDy-1021, GdDy-1022, GdDy-1023, GdDy-1024 and GdDy-1026 superconductors are determined at 77 K and the curves are compared in figure 4 (b). The self-field J_c values are of the order of 7.5 kA/cm^2 to 11.5 kA/cm^2. As the under-cooling temperature decreased, the growth rate was improved due to which the 211 phase was refined largely and supported higher J_c's compared to the samples dwelled at higher temperature.

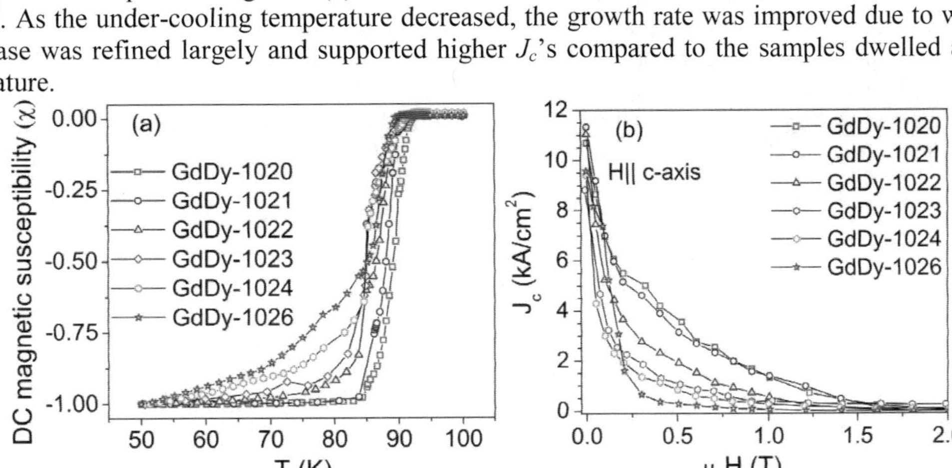

Fig. 4. (a) The normalized temperature dependence of DC magnetic susceptibility measurements for under-cooled GdDy-1026 to GdDy-1020 samples. Panel (b) shows the field dependence of J_c at 77 K.

4. Conclusions

In summary, isothermal experiments were conducted on optimized composition of 20 wt.% added Dy- content in GdBCO through top-seeded IG process. The temperature window for fabricating single grain (Gd, Dy)BCO materials was investigated. The non-superconducting 211 grains are grown to as large as ~ 25 μm sized in samples dwelled at elevated under-cooling temperatures (i.e., 1030 °C – 1024 °C) due to Ostwald ripening. The decrease in under-cooling temperature (below 1024 °C) resulted in fine sized 211 grains due to increased growth rate. The onset critical temperature for under-cooled samples was varied from 90 K to 92.5 K. The field dependence of J_c was superior in the samples dwelled at lower under-cooled temperature due to refined 211 phase particles. Utilizing the present results, one can fabricate larger sized single grain (Gd, Dy)BCO bulk superconductors with superior superconducting performance.

Acknowledgements
SPKN would like to acknowledge gratefully Shibaura Institute of Technology, Tokyo, Japan for supporting various facilities and financial support.

References

[1] Murakami M, Yoo S I, Higuchi T, Sakai N, Weltz J, Koshizuka N and Tanaka S 1994 *Jpn. J. Appl. Phys.*, **33** L715.
[2] Muralidhar M, Sakai N, Jirsa M, Murakami M and Hirabayashi I 2008 *App. Phys. Lett.*, **92** 162512.
[3] Murakami M 1992 *Supercond. Sci. Technol.*, **5** 185.
[4] Muralidhar M, Jirsa M, Sakai N, Hirabayashi I and Murakami M 2006 Studies of High Temperature Superconductors vol 50, ed. A V Narlikar (New York: Nova Science) p 229.
[5] Tomita M and Murakami M 2003 *Nature* **421** 517.
[6] Muralidhar M, Sakai N, Chikumoto N, Jirsa M, Machi T, Nishiyama M, Wu Y and Murakami M 2002 *Physical Review Letters* **89** 23.
[7] Cloots R, Kutzarova T, Mathieu J P and Ausloos M 2005 *Supercond. Sci. Technol.*, **18** R9.
[8] Chen Y L, Chan H M, Harmer M P, Todt V R, Sengupta S and Shi D 1994 *Physica C* **234** 232.
[9] Reddy E S and Rajasekharan T 1998 *J. Mater. Res.*, **130** 2472.
[10] Pavan Kumar Naik S, Devendra Kumar N, Swarup Raju P M, Rajasekharan T and Seshubai V 2013 *Physica C* **487** 72.
[11] Pavan Kumar Naik S, Swarup Raju P M, Rajasekharan T and Seshubai V 2016 *Supercond. Sci. Technol.*, **29** 115001.
[12] Pavan Kumar Naik S and Seshubai V 2017 *J. Phys. Chem. Sol.*, **101** 65.
[13] Pavan Kumar Naik S, Swarup Raju P M and Seshubai V 2016 *Mater. Chem. Phys.*, **182** 503.
[14] Pavan Kumar Naik S and Seshubai V 2017 *Cryogenics* **81** 47.
[15] Pavan Kumar Naik S, Muralidhar M and Murakami M 2018 *J. Supercond. Nov. Magn.,* **31** 981.
[16] Pavan Kumar Naik S, Muralidhar M, Jirsa M and Murakami M 2017 *J. Appl. Phys.*, **122** 193902.
[17] Pavan Kumar Naik S, Muralidhar M, Nakanishi Y and Murakami M 2018 *IEEE Trans. Appl. Superco.*, **28** 6801104.
[18] Chen D X and Goldfarb R B 1989 *J. Appl. Phys.,* **66** 2489.
[19] Bean C P 1962 *Phys. Rev. Lett.*, **8** 250.

IOP Conf. Series: Journal of Physics: Conf. Series **1054** (2018) 012049 doi:10.1088/1742-6596/1054/1/012049

The effect of cooling rate on critical current density and microstructure of single grain bulk $YBa_2Cu_3O_y$ superconductors grown by IG process

M. Santosh[1, 2], S Pavan Kumar Naik[1] and M. Murakami[1]

[1]Superconducting Materials Laboratory, Department of Materials Science and Engineering, Shibaura Institute of Technology, 3-7-5 Toyosu, Koto-ku, Tokyo 135-8548, Japan
[2]Faculty of Arts and Science, University of Toronto, E300 Huron Street, Toronto, M5S 3J6, Canada

santoshjuly6@gmail.com

Abstract. In the recent years, top-seeded infiltration-growth (TS-IG) process of $YBa_2Cu_3O_y$ (YBCO) had shown crucial and promising strengths as compared to melt growth process. The IG process lucidly clarified that a supply of liquid phase is more than essential for further growing large grains. Further, this experimentation aimed to enhance the performance of bulk YBCO superconducting materials, which were processed by IG process. The homemade Yb-123 and Y-211 were utilized in order to produce $YBa_2Cu_3O_y$ samples with the means of Yb-123+liquid (1:1) as a liquid source under a varied cooling rate. Moreover, single grain YBCO bulks were fabricated employing TS-IG process consisted of numerous varied cooling rates of 0.16 °C/h, 0.25 °C/h and 0.5 °C/h, respectively. In essence, all samples were oxygenated for 100 h with a constant pressure of 300 mL/min. Therefore, all samples' texture was determined by means of X-ray diffraction measurements. Microstructure analysis was performed by scanning electron microscopy for correlation of critical current density of TS-IG processed single grain YBCO bulks.

1. Introduction

The discovery of High Temperature $YBa_2Cu_3O_y$ (YBCO or 123) superconductors played a phenomenal role in modern science due to its improvement of superconducting properties, which transformed various industrial applications. YBCO superconductors are attractive for numerous technical applications mainly due to (i) critical superconducting temperature (T_c) above boiling temperature of liquid nitrogen 77 K and (ii) ease of synthesis [1]. High quality bulk YBCO superconductors are promising for many engineering and technological applications, such as levitation applications "Maglev", magnetic separation and superconducting motors and generators, etc., [2]. For the following practical applications, high critical current densities (J_c) and elevated trapped field were required [2-3]. The J_c and trapped field performance of a cuprate bulk superconductor was related to its final microstructure. Therefore, the control of superconducting microstructural properties is crucial for its final physical properties.

The Infiltration Growth (IG) method is superior among sintering and melt growth techniques. IG process allows fabrication near net bulk YBCO/REBCO superconductors along with fine sized 211 phase inclusions distributed homogeneously in the matrix [1, 4]. However, IG process involves complexity compared to former techniques. Many parameters need to be optimized for its high reproducibility. Among all, supplying sufficient amount of liquids and slow cooling time durations were its primary concerns. In the present work, we utilized Yb-123+$Ba_3Cu_5O_8$ as liquid phase (LP) source. The Yb-123 was chosen as LP source due to its lower melting point, which supplied abundant amount of liquid to porous 211 preform for peritectic reaction. We had fabricated single grain YBCO bulk superconductors by utilizing different slow cooling rates namely: 0.5 ^oC /h, 0.25 ^oC /h and 0.16 ^oC /h employing top-seeded IG process (TS-IG). In essence, structural, microstructural and superconducting properties were systematically investigated.

2. Experimental details

The precursor powders of Y_2BaCuO_5 (Y-211 or 211) and liquid phases of Yb-123 were prepared using high pure raw materials of Y_2O_3, Yb_2O_3, BaO and CuO in solid-state sintering method. The Yb-123+$Ba_3Cu_5O_8$ is used as (LP) source and preform pellets of Y-211 and LP source were pressed at an applied uni-axial pressure of 450 M.Pa. The ratio of 211:LP was 1:1 maintained for IG processing of all YBCO bulk samples. YBCO samples were fabricated employing TS-IG process with varying slow cooling rates of 0.5, ^oC /h, 0.25 ^oC /h and 0.16 ^oC /h and the corresponding final samples can be referred as Y-50, Y-100 and Y-150, respectively. The heat treatment schedule followed for synthesizing YBCO samples is shown in figure 1. The processed samples were oxygenated at 450 ^oC for 100 h for SQUID measurements with a constant pressure of 0.3 l/min.

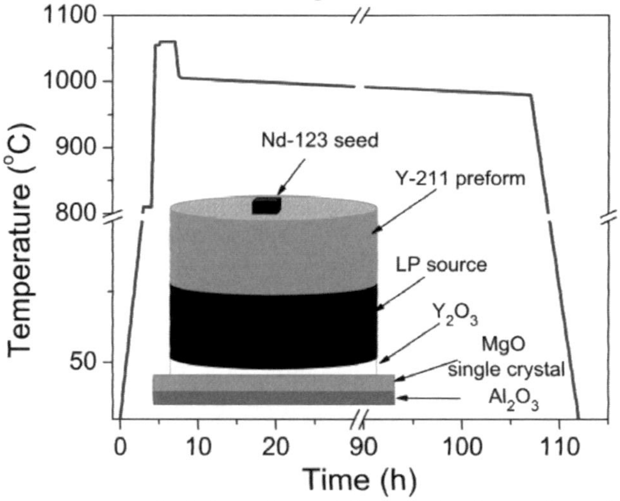

Figure 1. The time –temperature schedule followed for synthesizing the bulk YBCO samples. The samples assembly utilized for IG processing is also shown.

Structural analysis was carried out employing X-ray diffractometer (Rigaku made, λ= 1.54 $\mathring{A}$). Microstructural observations were conducted employing Scanning electron microscope (SEM, JEOL make, JSM-7100F model). The superconducting transition and field dependence properties were measured employing a SQUID (Quantum Design, model MPMS-XL5minSITO). The field dependence of J_c was calculated utilizing the Extended Beans critical state model [5-6]. J_c was estimated for rectangular specimens given by

$$J_c = 20\, \Delta M / [a^2 d\, (b - a/3)] \quad \ldots\ldots 1$$

Where d is the sample thickness, a , b are cross sectional dimensions, $b \geq a$, and ΔM is the difference of magnetic moments (M^+-M^-) during increasing and decreasing field in the M-H loop .

3. Results and Discussion

The top-view photograph of processed Y-100 bulk sample is displayed in figure 2 (a) as a citing example. The four-fold growth (in clear X shape) extension to its boundaries indicated that the sample was grown in a form of a single grain. Numerous reports suggested that REBCO bulk superconducting properties possess spatial dependency [7-8]. To overcome spatial dependence of properties, a bulk sub-specimen was cut 2 mm below NdBCO seed from each sample on which the measurements were done. The position of the sub-specimen where collected is illustrated in figure 2 (b).

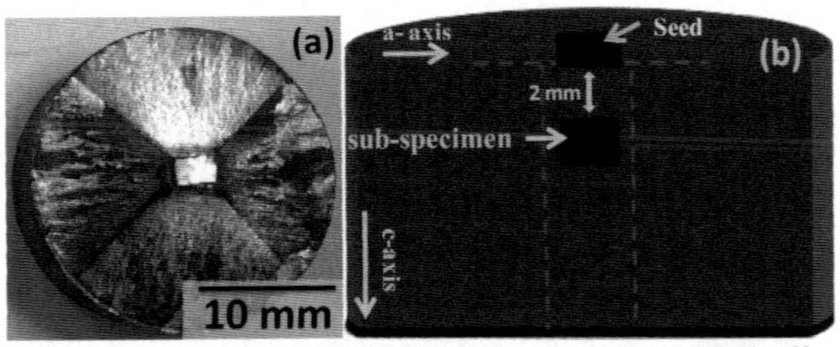

Figure 2. (a) The top-surface image of as TS-IG processed Y-100 bulk superconductor. Panel (b) shows the schematic illustration of position where a sub-specimen is collected.

The X-ray diffraction (XRD) patterns recorded for all bulk samples were displayed in figure 3. All samples were highly textured in c-axis orientation. The XRD peaks could be indexed to *(00l)* direction. It was also observed that the texture was improved with slow cooling duration. The intensity of the Bragg peaks, especially for (006) was increased with increase in slow cooling time.

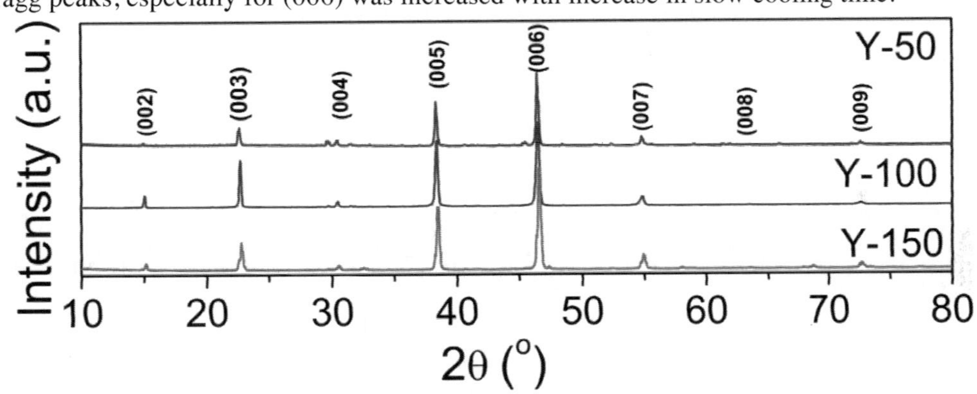

Figure 3. The XRD patterns of all bulk YBCO samples recorded at room temperature

The micrographs recorded for Y-50, Y-100 and Y-150 samples at 1000 X on ab-surface were displayed in figure 4 (a)-(c) respectively. The keen observation of microstructural features led to a conclusion that sample Y-100 was allowed to slow cool for 100 h that resulted in fine sized Y-211 phase particles among all. In the case of Y-50, the time duration is shorter for peritectic reaction due to which 211 phase particles were not completely utilized and resulted in larger sized particles in final sample. Whereas, in the case of Y-150 sample, by keeping 211 phase particles in the melt at elevated temperatures, Ostwald ripening had occurred and 211 phase particles was larger sized. The average size of the 211 phase particles in Y-50, Y-100 and Y-150 were estimated to be 1.8 μm, 1.3 μm and 2.3 μm respectively.

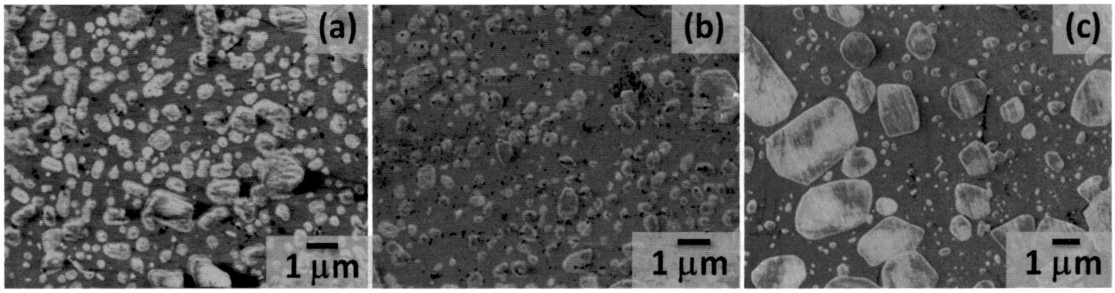

Figure 4. The micrographs of the Y-50, Y-100 and Y-150 are shown in panel (a)-(c) respectively. The Y-100 sample possesses fine sized 211 phase particles distributed homogeneously.

The superconducting transition curves for all bulk YBCO samples are shown in figure 5. A small magnetic field of 1 mT was applied normal to c-axis of each sample. The transition is very sharp indicting its high quality nature of the produced bulk samples. The onset of T_c for all samples was 91.5 K and its width of transition was 0.8 K – 1.3 K at tail. The broadness in the susceptibility curves at tail could be indicative of the presence of low T_c phases such as oxygen deficient phases.

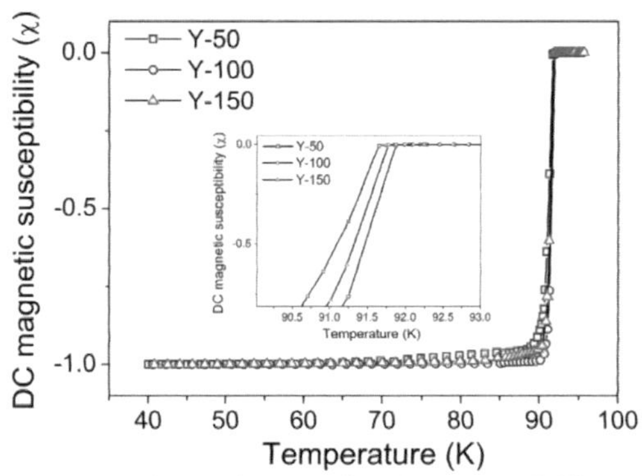

Figure 5. The normalized temperature dependence of DC susceptibility measurements of all YBCO samples. The curves are very sharp.

The field dependence of critical current density *($J_c(H)$)* was obtained at 77 K for all samples and $J_c(H)$ curves were indicated in figure 6. The zero field J_c of Y-50, Y-100 and Y-150 samples are 29 kA/cm^2, 45 kA/cm^2 and 36 kA/cm^2 respectively. The superior superconducting performance in Y-100 sample attributed to a homogeneous distribution of fine sized Y-211 particles evinced through SEM micrographs. High J_c observed, especially at lower fields, in the case of Y-100 superconductor can be justified by the enhanced flux pinning due to larger interfacial defect density of 211/123 boundaries. The secondary peak is observed in all of samples due to the presence of oxygen deficient phases, which were confirmed in DC susceptibility measurements. The secondary peak in $J_c(H)$ curves of YBCO bulk superconductors had begun at an applied magnetic field of ~0.75 T and extended up to ~ 3 T. If the secondary peak effect is compared among all bulks, the Y-50 sample is exhibiting inferior performance. The irreversibility field (H_{irr}) is defined, as the applied field required reaching the J_c value of 100 A/cm^2 [9-10]. In the present case, the H_{irr} value for all samples is observed to be ~ 3.8 T.

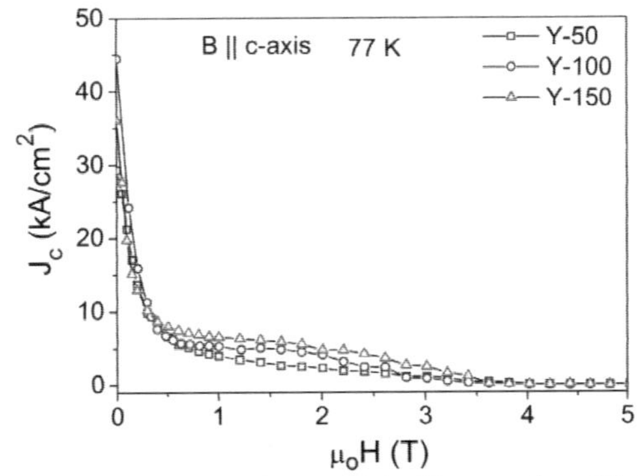

Figure 6. The $J_c(H)$ curves for Y-50, Y-100 and Y-150 bulk samples. Field was applied parallel to c-axis of all the samples. The Y-100 superconductor exhibiting superior superconducting performance.

4. Conclusions

Bulk YBCO superconductors were fabricated employing top Seeded Infiltration and Growth process under a time duration of slow cooling effect and its structural, microstructural and superconducting properties were investigated. The slow cooling time significantly modified the microstructure of final samples. The sample slow cooled for 100 h was optimum and resulted in a microstructure, which supported superior field dependence of critical current density. Longer duration of cooling resulted in coalescence of 211 phase particles due to its larger grown size. All samples were textured highly in *(00l)* direction and the texture of the samples was improved with increase in cooling time. The superconducting transitions were sharp that indicated its high quality nature of the produced YBCO samples. Field dependence of J_c is superior for Y-100 sample due to homogeneous distribution of fine sized non-superconducting 211 phase inclusions.

References

[1] Cloots R *et al* 2005 *Supercond. Sci. Technol.* **18** R9-R23
[2] Tomita M and Murakami M 2003 *Nature* **421** pp 517-520
[3] Muralidhar M *et al* 2008 *Appl. Phys. Lett.* **92** pp 162512-3
[4] Pavan Kumar Naik S *et al* 2013 *Physica C* **487** pp 72-76
[5] Bean C P 1962 *Phys. Rev. Lett.* **8** pp 250-253
[6] Gyorgy E M *et al* 1989 *Appl. Phys. Lett.* **55** pp 283-285
[7] Iida K *et al* 2005 *Supercond. Sci. Technol.* **18** pp 1421-7
[8] Pavan Kumar Naik S *et al* 2017 *Journal of Applied Physics* **122** pp 193902-10
[9] Kobayashi S *et al* 1996 *Phys. C* **258** pp 336-340
[10] Pavan Kumar Naik S *et al* 2017 *Journal of Physics and Chemistry of Solids* **101** pp 65–73

Trapping Large Magnetic Field by Suppression of Thermomagnetic Instability in Coated Conductor Stacks

T Hashimoto[1]**, S Pyon**[1] **and T Tamegai**[1]

[1] Department of Applied Physics, The University of Tokyo, 7-3-1 Hongo, Bunkyo-ku, Tokyo 113-8656, Japan

E-mail: th shin236@yahoo.co.jp

Abstract. A new type of bulk magnet can be fabricated by simply stacking short segments of coated conductors (CCs) of 123-type superconductors. To enhance the trapping field of the CC stack, we irradiated 800 MeV Xe ions to CCs. The fabricated CC stack consists of 80 pieces of irradiated CCs, and it was magnetized by field-cooling method in an applied field 9 T down to 5 K. When we tried to trap a large field by reducing the field at ramp rates faster than 0.5 mT/s, flux jumps occurred, which abruptly suppressed the trapped field. By reducing the magnetic field down to zero at a ramp rate slower than 0.5 mT/s, we have succeeded in trapping 9 T at the center of the CC stack. We also confirmed that the CC stack can sustain the high trapped field for more than 20 min, which demonstrates the practical usefulness of such a staked CC magnet.

1. Introduction

Coated conductors (CCs) of 123-type superconductors with a large critical current density (J_c) have been successfully developed for applications of power transmission and generation of high magnetic fields. However, such CCs can find applications in other purposes. One of them is an alternative method to generate a large magnetic field like bulk materials of high temperature superconductors [1, 2]. Similar techniques to obtain bulk magnets have been applied for MgB_2 [3] and iron-based superconductors [4]. In particular, in a recent study of trapping magnetic field in the bulk magnet of the iron-based superconductor, it is reported that the trapped field is only about 1 T [4]. When one attempts to magnetize such bulk materials, however, one faces problems of mechanical strength and thermomagnetic instability [5, 6]. In order to solve such problems of the bulk magnet, stacking short segments of CCs has been proposed and a modest field has been successfully trapped [7, 8, 9]. We have also fabricated a CC stack with two stacks of 130 pieces of GdBCO CCs and trapped 7.92 T at the center of the stacks [10]. In this study, we fabricated a model of a bulk magnet like square pillars by stacking square-cut CCs in the vertical direction. However, it was difficult to obtain constant magnetic fields over a wide range. To improve the homogeneity of the field distribution, stacking CCs obliquely has been attempted over a wide range [11].

In the present study, we aimed at enhancing the trapped magnetic field by fabricating the bulk magnet using GdBCO CCs with better J_c - H characteristics. For that purpose, GdBCO CCs produced by Fujikura were irradiated by 800 MeV Xe with $B_\Phi = 4$ T. Two stacks of each 40 pieces of GdBCO CCs were placed next to each other and miniature Hall probes for measuring the trapped field were placed at the center of the stacks. They were cooled down to 5 K in a magnetic field of 9 T, and the field was reduced to zero at different ramp rates between 1 mT/s and 0.2 mT/s. In this condition, we have succeeded in trapping 9 T at the center of the stacks. However, it was also found that as the ramp rate of the magnetic field was increased, flux jumps occurred and the trapped magnetic field was strongly suppressed. We also investigated the stability of the CC stacks by measuring the temporal evolution of the trapped field.

2. Experimental Methods

In the present study, we used GdBCO CCs manufactured by Fujikura Ltd., which were cut into 12 ×12 mm² square. The thickness of the superconducting layer was 2.14 μm and the silver overlayer was 3.11 μm. To enhance J_c of CCs, columnar defects were installed by irradiating 800 MeV Xe with $B_\Phi = 4$ T along the c-axis at National Institute of Radiological Sciences Heavy Ion Medical Accelerator in Chiba (NIRS, HIMAC) [12, 13]. 80 CCs were irradiated to form two piles of each 40 CC stack, and we measured the local magnetic field at the center of the two stacks with GaAs Hall probes made by Asahi Kasei Microdevices (HG-0711). The size of the Hall probe used as the magnetic sensor was 1.2 × 0.5 mm² and the thickness was 0.3 mm. It has been confirmed that ion-implanted GaAs Hall probes can be operated even at temperatures much lower than the specified temperature range and linearity with respect to the magnetic field is good due to low mobility [14]. Three Hall probes were soldered on a thin printed circuit board to be arranged at the intervals of 1.8 mm. Therefore, it was possible to measure the local field not only at the center but also at the positions 1.8 mm and 3.6 mm from the center. Since the thickness of one CC is 100 μm, the total height of the bulk magnet is only 8.5 mm including the part of Hall probes. For trapping magnetic field, we used Physical Properties Measurement System (PPMS). First, a magnetic field of 9 T was applied to the CC stacks at 100 K above T_c (~94 K), and the CC stacks was cooled to the target temperature. Finally, the external field was ramped down to zero slowly, leaving a trapped field in the stack. The CC was cooled via the helium gas.

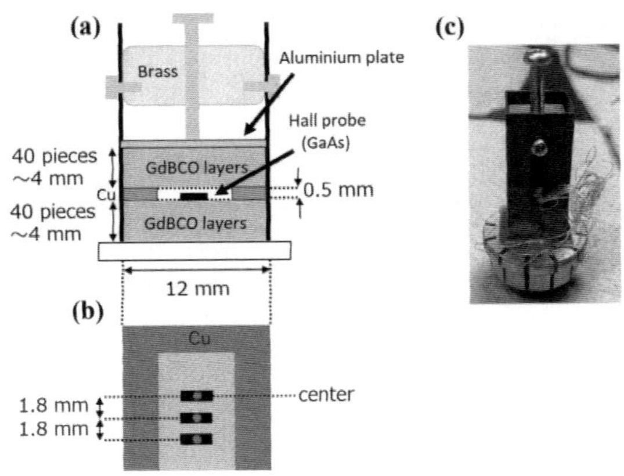

Figure 1. (a) Schematic vertical cross section of a bulk magnet consisting of double stack of GdBCO CCs. (b) Arrangements of Hall probes between the two CC stacks. (c) A photo of the bulk magnet on a puck of PPMS.

3. Results and Discussion

Figure 2 shows the external magnetic field dependence of J_c of the pristine and irradiated CCs used for the bulk magnet at 4.2 K. This graph shows that J_c of the GdBCO CC irradiated by 800 MeV Xe at B_Φ

= 4 T is about 1.5 times larger than that in the pristine sample. It suggests that columnar defects are successfully introduced into GdBCO CCs by Xe irradiation.

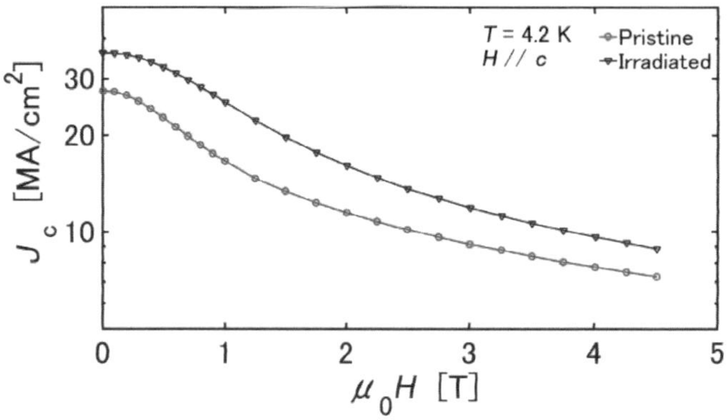

Figure 2. Magnetic field dependence of J_c at 4.2 K of the pristine and 800 MeV Xe irradiated GdBCO CCs.

The field trapped in the double stack of CC at different field-cooling temperatures are shown in Figure 3. It clearly demonstrates that as the temperature is decreased, the trapped field increases. In addition, it is found that the trapped field increased almost linearly. The ramp rate used for applied field were higher for higher temperatures and lower when approaching 5 K. In particular, 2.5 mT/s at 77 K, 50 K, 30 K, 0.5 mT/s at 10 K and 7.5 K. At 5 K, the applied field was ramped down at 0.5 mT/s from 9 T to 2 T, and 0.2 mT/s below 2 T.

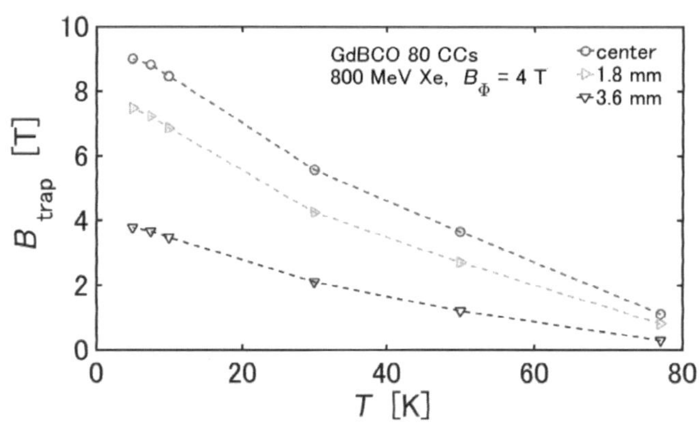

Figure 3. Temperature dependence of the trapped magnetic field in the stack.

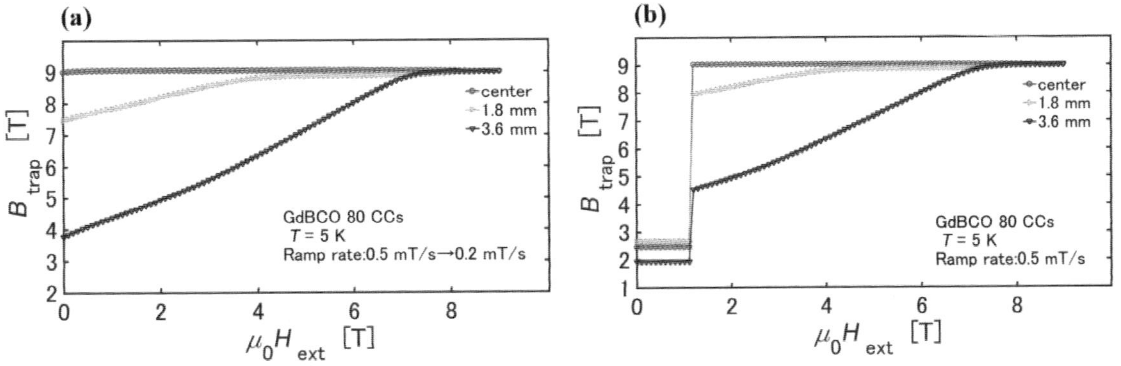

Figure 4. Local magnetic field at three locations of the bulk magnet after field cooling it at 9 T from above T_c down to 5 K (a) at 0.5 mT/s from 9 T to 2 T and 0.2 mT/s below 2 T, and (b) at 0.5 mT/s in the whole range.

The applied field dependence of the trapped field at 5 K at different ramp rates of the field are shown in Figure 4. From these figures, it is clear that we have succeeded in trapping 9 T in the center of CCs, which is the maximum magnetic field that can be trapped by 9 T PPMS system. This trapped field value is larger than 7.9 T trapped by 260 stacks of CCs reported in our previous study [10]. However, it took about 7 hours to trap 9 T to the bulk magnet. Also, from Figure 4(b), it is found that the trapped field is abruptly decreased around $\mu_0 H_{ext} = 1$ T when the ramp rate is higher. This is considered to be due to flux jump. Therefore, although it is confirmed that the thermomagnetic instability was suppressed by reducing the ramp rate, it is desirable to suppress the flux jump in order to obtain a high magnetic field quickly. In order to observe the stability of the trapped field, flux creep was measured for the trapped field up to 20 min after field cooling as shown in figure 5. From figure 5, even after 20 min, it can be seen that the trapped field at the center decreases only by about 0.3 %. It is necessary for the application to examine how much the trapped field is maintained after magnetizing the stack of CCs. So, confirmation of long-term stability of the trapped field at various temperature is necessary, when one attempts to use CC stacks as a permanent magnet.

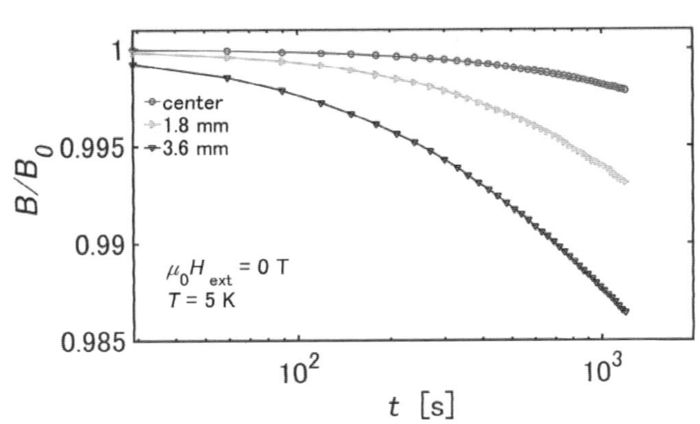

Figure 5. Flux creep measurements for trapped field for different positions in the first 20 min after magnetization. B_0 is the trapped field 3 s after magnetization.

4. Summary

We have fabricated a bulk magnet by stacking 80 pieces of GdBCO coated conductors which were irradiated by 800 MeV Xe with $B_{\Phi} = 4$ T. A large magnetic field of 9 T was successfully trapped in this bulk magnet. However, to avoid flux jumps, the magnetization process took more than 7 h. The trapped field was very stable and more than 99 % of the field was maintained at the center of the stack even 20 min after the magnetization.

Acknowledgements

We thank Fujikura Ltd. for supplying GdBCO coated conductors.

References

[1] Weijers H W, Markiewicz W D, Voran A J, Gundlach S R, Sheppard W R, Jarvis B, Johnson Z L, Noyes P D, Lu J, Kandel H, Bai H, Gavrilin A V, Viouchkov Y L, Larbalestier D C and Abraimov D V 2014 *IEEE Trans. Appl. Supercond.* **24** 4301805

[2] Durrell J, Dennis A, Jaroszynski J, Ainslie M, Palmer K, Shi Y, Campbell A, Hull J, Strasik M, Hellstrom E and Cardwell D 2014 *Supercond. Sci. Technol.* **27** 082001

[3] Fuchs G, Haler W, Nenkov K, Scheiter J, Perner O, Handstein A, Kanai T, Schultz L and Holzapfel B 2013 *Supercond. Sci. Technol.* **26** 122002

[4] Weiss J D, Yamamoto A, Polyanskii A A, Richardson R B, Larbalestier D C and Hellstrom E E 2015 *Supercond. Sci. Technol.* **28** 112001

[5] Tomita M and Murakami M 2000 *Supercond. Sci. Technol.* **13** 722

[6] Tomita M and Murakami M 2003 *Nature* **106** 104

[7] Hahn S, Beom Kim S, Cheol Ahn M, J. Voccio, Bascun J and Iwasa Y 2010 *IEEE Trans. Appl. Supercond.* **20** 1037

[8] Patel A, Filar K, Nizhankovskii V I, Hopkins S C and Glowacki B A 2013 *Appl. Phys. Lett.* **102** 102601

[9] Patel A, Hopkins S C and Growacki B A 2013 *Supercond. Sci. Technol.* **26** 032001

[10] Tamegai T, Hirai T, Sun Y and Pyon S 2016 *Physica C* **530** 20

[11] Mitchell-Williams T B, Baskys A, Hopkins S C, Kalitka V, Molodyk A, Glowacki B A and Patel A 2016 *Supercond. Sci. Technol.* **29** 085008

[12] Tamegai T, Taen T, Yagyuda H, Tsuchiya Y, Mohan S, Taniguchi T, Nakajima Y, Okayasu S, Sasase M, Kitamura H, Murakami T, Kambara T and Kanai Y 2012 *Supercond. Sci. Technol.* **25** 084008

[13] Ohtake F, Taen T, Pyon S, Tamegai T, Okayasu S, Kambara T and Kitamura H 2015 *Physica C* **518** 47

[14] Tamegai T, Iye Y, Oguro I and Kishio K 1993 *Physica C* **213** 33

Effects of SPS pressure on the mechanical properties of high packing ratio bulk MgB_2 superconductor

A Murakami[1], A Iwamoto[2] and J G Noudem[3]

[1] National Institute of Technology, Ichinoseki College, Takanashi, Hagisho, Ichinoseki, Iwate 021-8511, Japan
[2] National Institute for Fusion Science, 322-6 Oroshi-cho, Toki, Gifu 509-5292, Japan
[3] CRISMAT-CNRS UMR 6508/UNICAEN-ENSICAEN, 6 bd Marechal Juin, Caen Cedex 04, 14050, France

E-mail: amura@ichinoseki.ac.jp

Abstract. In order to investigate the effects of SPS pressure on the mechanical properties of MgB_2 bulk, MgB_2 bulk samples were processed by SPS under different pressures. Mechanical properties of these bulk samples were evaluated at room temperature through the bending tests for specimens cut from the bulk samples. Stress-strain behaviours of these bulk samples were almost linear until the fracture. No significant difference was observed for the Young's modulus values among the bulk samples. The average bending strength values of these bulk samples were also similar to each other. However, the bending strength data of the samples processed under higher pressure scattered widely in comparison with those of the sample processed under lower pressure. The bending strength values were related with the pore sizes observed on the tensile side fracture surfaces.

1. Introduction

Enlargement of bulk superconductor is effective in improving the performance of superconducting devices. Although the superconducting transition temperature of MgB_2 is lower than that of REBaCuO (RE: Y or rare-earth elements), both MgB_2 and REBaCuO bulks are promising for liquid-helium-free bulk superconductor. REBaCuO bulks have to be a single-grain to trap strong magnetic field. REBaCuO bulks are commonly fabricated by melt-processing using a seed crystal. It is difficult to obtain large single-grained REBaCuO bulk piece due to the undesirable nucleation apart from the seed crystal. On the other hand, MgB_2 bulk does not have to be a single-grain [1-12]. Thus, it is not so difficult to obtain large bulk piece for MgB_2 in comparison with REBaCuO. Sintering for the MgB_2 bulk is commonly carried out under the ambient pressure [1-4]. However, the fracture strength value of a low packing ratio MgB_2 bulk sintered under the ambient pressure [13] is lower than those of REBaCuO bulks [14,15]. Since bulk superconductors are subjected to electromagnetic force and thermal stress in the superconducting devices [16,17], improvements of the mechanical properties are important for the development of high performance superconducting devices. It has been reported that the superconducting properties of MgB_2 bulks processed by high-pressure sintering, such as hot-isostatic pressing (HIP) and spark plasma sintering (SPS), are excellent [5-12]. The present authors previously reported that the fracture strength of a high packing ratio MgB_2 bulk processed by SPS was significantly higher than those of MgB_2 bulks fabricated by other sintering processes [18]. Thus, spark plasma sintered MgB_2 bulk is promising for high-performance bulk superconductor. However, the effects of SPS conditions on the mechanical properties of MgB_2 bulk have not been understood extensively. The present authors previously investigated the effects of SPS temperature on the mechanical properties of MgB_2 bulk through the bending tests for specimens cut from MgB_2 bulk samples processed at 950, 1000 and 1100 ℃, respectively [19]. The mechanical properties of the bulk sample processed at 1100 ℃ were superior to those of other bulk samples [19]. In this study, the effects of SPS pressure on the mechanical properties were investigated through the bending tests for specimens cut from MgB_2 bulk samples processed at 1100℃.

2. Experimental procedure

Table 1 presents the SPS conditions for the fabrication of the MgB_2 bulk samples. In this study, two spark plasma sintered MgB_2 bulk samples were prepared under the pressure of 75 and 100 MPa. They are denoted as Samples 75 and 100, respectively, as shown in Table 1. Mechanical properties of these bulk samples are compared with those of a MgB_2 bulk sample processed in the previous study under the pressure of 50 MPa [19]. This bulk sample is denoted as Sample 50. Commercial MgB_2 powder from ABCR GmbH (Karlsruhe, Germany) was used for the fabrication of these bulk samples. The SPS process was carried out in the dynamic vacuum atmosphere (10^{-3} bar) under the pulsed electric current of 2000 A. A graphite die was used for the SPS process. The diameter and thickness of the bulk samples were about 30 and 8 mm, respectively. Details of the fabrication process are reported in [12].

Table 1. Spark plasma sintering conditions.

Sample	50[19]	75	100
Temperature [℃]	1100	1100	1100
Pressure [MPa]	50	75	100
Duration [min]	20	20	20

Bending test specimens were cut from the bulk samples as schematically shown in Figure 1. Wire-electrical discharge cutting machine was used for the cutting. Since spark plasma sintering process is carried out under the uniaxial pressure, two types of specimens were prepared to investigate the effects of the bending loading directions on the bending test results. These specimens are denoted as Specimens A and B, respectively. 4-point bending load was applied at room temperature in the 2.1 mm direction of the specimens as schematically shown in Figure 2. Outer supporting span and inner loading span of the 4-point bending test were 21 and 7 mm, respectively. While the loading direction for the bending tests of the Specimens A was parallel to the compression direction in the SPS process, the loading direction for the bending tests of the Specimens B was perpendicular to it. One strain gauge was glued to the specimens to measure the strain caused by the loading. The strain gauge length was 0.2 mm. The loading speed was 0.2 mm/min. 4-point bending stress σ was calculated by the following equation.

$$\sigma = \frac{3P(L-l)}{2wt^2} \tag{1}$$

Where P is applied load, L is outer supporting span (21 mm), l is inner loading span (7 mm), w and t are width and thickness of the specimens (2.8 and 2.1 mm), respectively. After the bending tests, fracture surfaces were observed by using a scanning electron microscope.

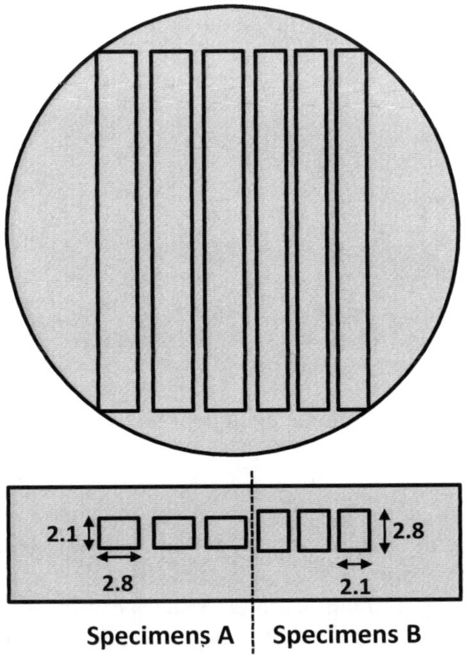

Figure 1. Schematic illustration of cutting out of bending test specimens.

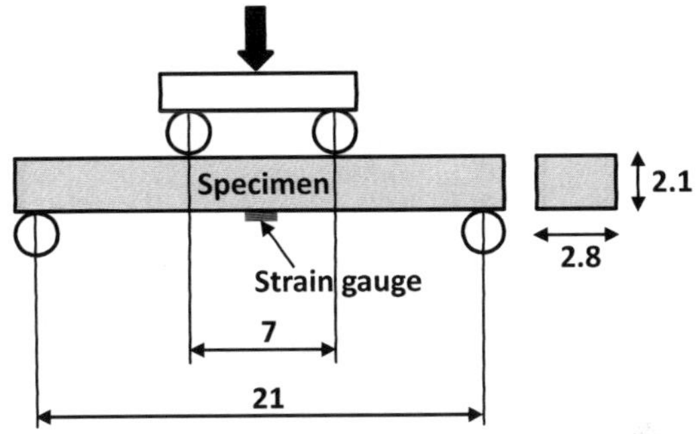

Figure 2. Schematic illustration of 4-point bending test.

3. Superconducting properties

To investigate the superconducting properties, the small specimens with dimensions of 2×2×2 mm^3 were cut from the bulk MgB_2 samples in order to measure their critical temperature (T_c) and record their magnetization hysteresis loops (*M-H* loops) at various temperature (5, 20 and 30 K) between –5 and +5 T with a SQUID magnetometer (Quantum Design, model MPMS5). The critical current density, J_c values of the sample processed at 75 MPa were estimated using the extended Bean critical state model for a rectangular sample [20].

The zero field cooled magnetization curve of Figure 3 performed on the sample processed at 75 MPa, shows a very narrow transition with an onset at 38.5 K. The critical current density, J_c, was estimated from magnetization, M-H cycle at different temperatures (Figure 4) on the basis of the extended Bean model [20] leading to an estimated critical current density $J_c \approx 2.2x10^5$ A/cm^2 in self field. While J_c decreases as the temperature and the external field increases respectively, its value with 1T applied field ($J_c = 1.1x10^5$ A/cm^2) is still interesting for practical applications. This high J_c value could be related to the excellent connectivity between grains and is similar to published results [11] in which the samples were prepared using the same processing conditions under high applied pressure (80 MPa).

4. Bending test results

Figure 5 presents stress-strain curves of the specimens cut from the Samples 75 and 100. Data points represent the fracture of the specimens. Stress-strain curves of the Specimens A and B are not distinguished in this figure. Since MgB_2 is commonly brittle and both the Samples 75 and 100 are quite dense as mentioned in the followings, stress-strain behaviours of these bulk samples are almost linear until the fracture. These stress-strain behaviours are similar to the behaviour observed for the

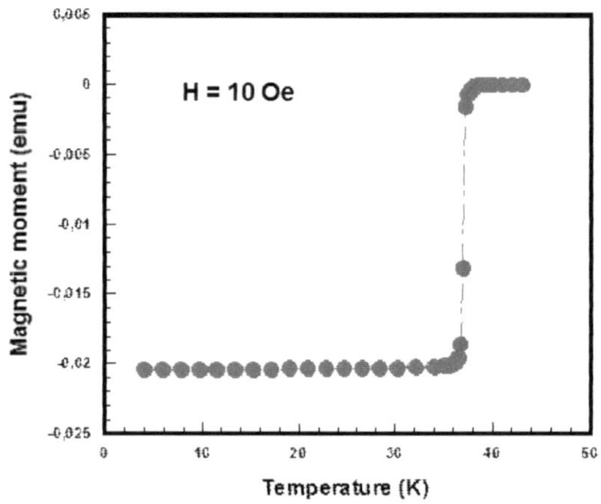

Figure 3. Magnetization curve vs. temperature.

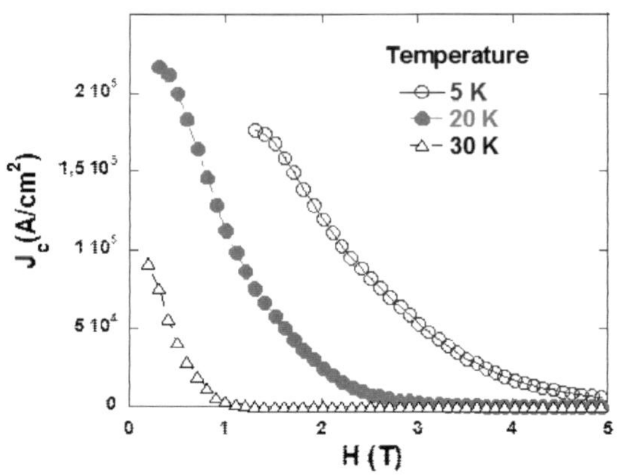

Figure 4. Magnetic field dependence of critical current density at different temperatures.

Sample 50 in the previous study [19]. On the other hand, it has been reported that stress-strain behaviours of the specimens cut from a low packing ratio MgB_2 bulk are not linear [13].

Figure 6 presents the relationship between the Young's modulus and SPS pressure for the Specimens A and B. The Young's modulus values were obtained from the slope of the stress-strain curves shown in Figure 5. Solid and open symbols represent the data of the Specimens A and B, respectively. While the loading direction for the bending tests of the Specimens A was parallel to the thickness of the bulk sample, the loading direction for the bending tests of the Specimens B was perpendicular to it (see Figures 1 and 2). The average Young's modulus values of the Sample 50 reported elsewhere [19] are also shown in Figure 6 for reference. There is no significant difference in the average Young's modulus values among the SPS pressures of 50, 75 and 100 MPa. It is mainly attributable to that the packing ratios of these bulk samples are similar to each other. Since the net cross-sectional area of the specimen increases with the increase of the packing ratio, the Young's modulus commonly increases with increase of the packing ratio. The increase of the net cross-sectional area commonly suppresses the deformation of the specimen and thus improves the Young's modulus. There is no significant difference in the Young's modulus values between the Specimens A and B of the high pressure Samples 75 and 100, which is similar to the bending test results of the low pressure Sample 50.

Figure 7 presents the relationship between the bending strength and SPS pressure. The bending strength means the stress at fracture. Figure 8 presents the Weibull plots of the bending strength data. Weibull plots are commonly carried out to evaluate the scatter of the data. The slope obtained through the linear fitting of the data points represents the Weibull coefficient m; larger Weibull coefficient value means smaller scatter of the data. Since the dependence of the bending strength data on the loading directions has not been observed distinctly in Figure 7, the data of the Specimens A and B are not distinguished in Figure 8. The average bending strength values of the Samples 50, 75 and 100 are similar to each other (see Figure 7). However, the bending strength data of the Samples 75 and 100 scatter widely in comparison with those of the Sample 50 (see Figure 8).

Figures 9 and 10 present scanning electron micrographs of the fracture surfaces of the Specimens A and B cut from the Sample 75. Bottom of the micrographs corresponds to the tensile side where the fatal crack initiates by bending loading. Although some extraordinary pores are observed on the fracture surfaces as marked by arrows, the specimens cut from the Sample 75 are quite dense. The sample 100 is also dense. There is no significant difference in the fracture surface morphologies between the Specimens A and B. Fracture surface morphologies of these specimens are typical of ceramic materials; the tensile side fracture surface is relatively flat. Based on the fracture surface morphology, it is deduced that unstable fracture occurred. Fracture surface morphologies of the

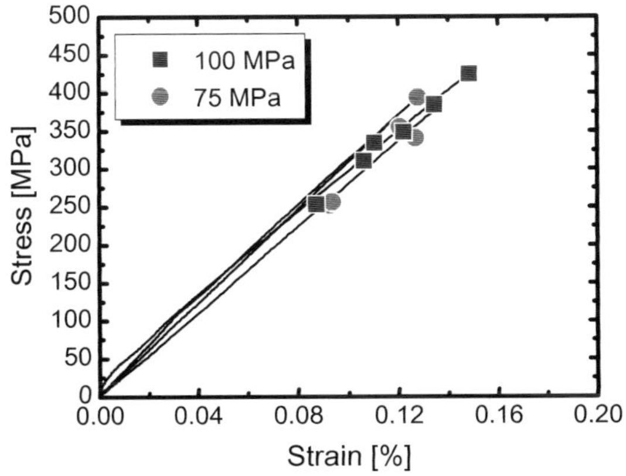

Figure 5. Stress-strain curves of specimens cut from bulk samples processed under 75 and 100 MPa, respectively.

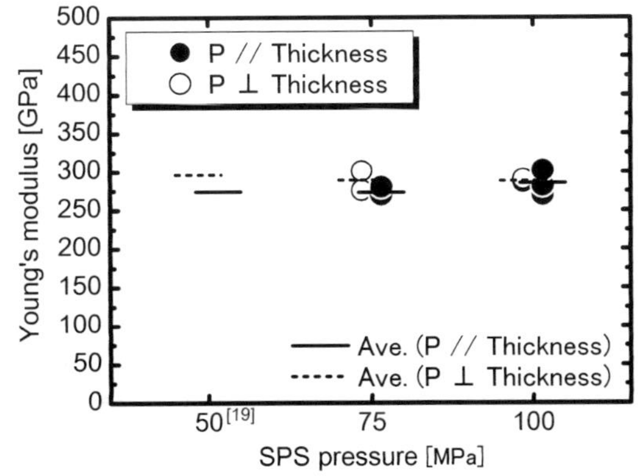

Figure 6. Relationship between Young's modulus and SPS pressure.

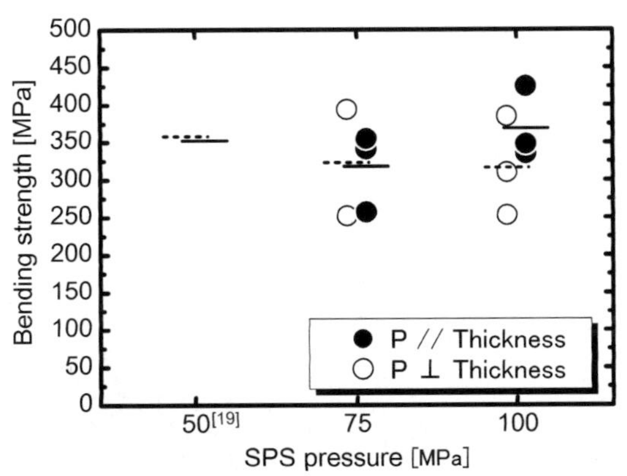

Figure 7. Relationship between bending strength and SPS pressure.

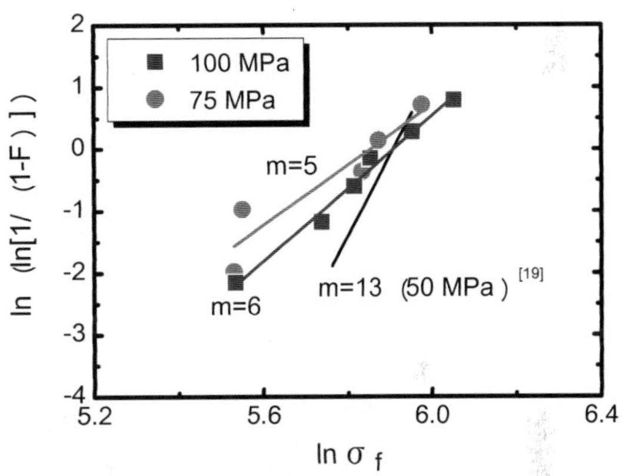

Figure 8. Weibull plots of bending strength data of specimens cut from samples processed under 75 and 100 MPa, respectively. Data of a sample processed under 50 MPa [19] are also shown for reference. F, σ_f and m represent cumulative fracture probability, bending strength and Weibull coefficient, respectively.

specimens cut from the Samples 75 and 100 are similar to each other. One extraordinary pore is clearly observed on the tensile side fracture surfaces of the specimens shown in Figures 9 (a) and 10 (a). The pore sizes of the specimens shown in Figures 9 (a) and 10 (a) are similar to each other. The bending strength values of these specimens are also similar to each other, 257 and 252 MPa, respectively. On the other hand, few pores are observed on the fracture surface of the specimen shown in Figure 10 (b). The bending strength value of this specimen is 394 MPa, which is the maximum value among the specimens of the Sample 75. One small pore is slightly observed on the fracture surface of the specimen shown in Figure 9 (b). The bending strength value of this specimen is slightly lower than that of the specimen shown in Figure 10 (b). Although it has not been clearly understood the reason why the bending strength data of the high pressure bulk samples (Samples 75 and 100) scatter widely in comparison with those of the low pressure bulk sample (Sample 50) at present, it is deduced that the bending strength values are related with the pore sizes observed on the tensile side fracture surfaces.

Figure 11 presents the relationship between the bending strength and porosity of MgB_2 bulk materials. Data of other MgB_2 bulk materials processed by SPS under different temperature [19] and by other sintering techniques such as capsule method CAP and HIP [13] are shown for reference. The packing ratio increases with increase of the SPS temperature. Quite dense bulk samples are obtained by SPS at 1100℃. The packing ratio largely affects the bending strength. The relationship between the average bending strength value and packing ratio can be approximated exponentially [13]. The average bending strength increases with increase of the packing ratio. In this study, in order to reduce the scatter of the bending strength data of the quite dense bulk, effects of the SPS pressure on the bending strength were investigated for the bulk samples processed at 1100℃. However, the bending strength data of the bulk samples processed under higher pressure, 75 and 100 MPa, scattered widely in comparison with the sample processed under 50 MPa as mentioned above. Future work will investigate the effect of dwell time during SPS process on the scatter of the strength data.

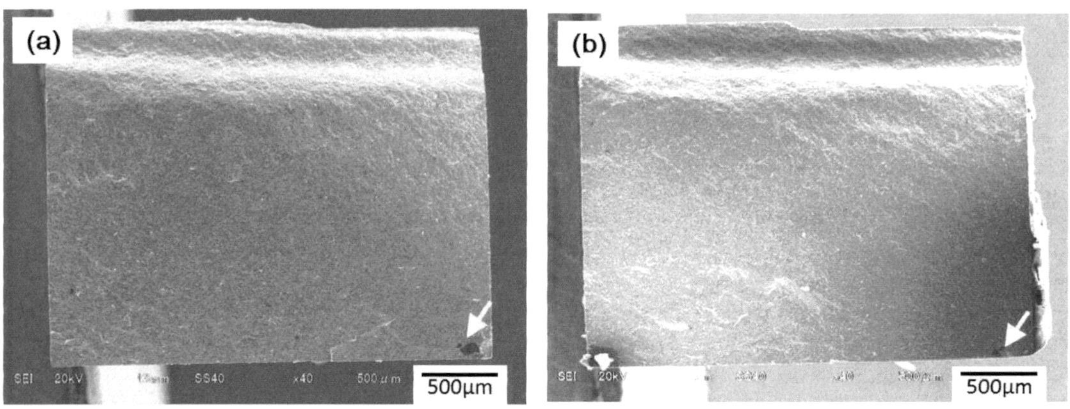

Figure 9. Fracture surfaces of Specimens A of Sample 75 fractured at (a) 257 MPa and (b) 355 MPa, respectively.

Figure 10. Fracture surfaces of Specimens B of Sample 75 fractured at (a) 252 MPa and (b) 394 MPa, respectively.

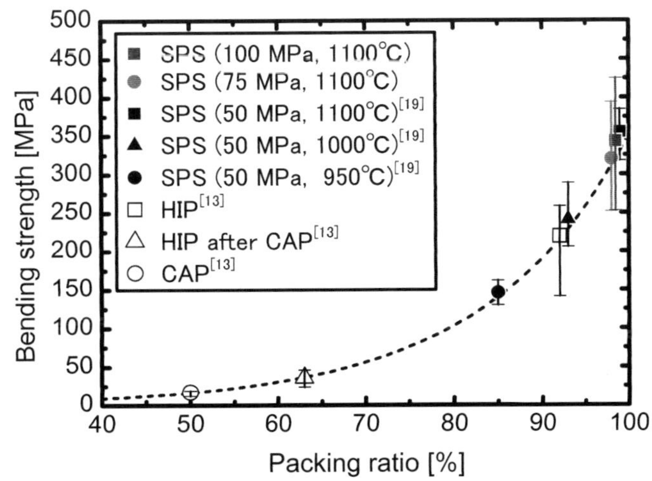

Figure 11. Relationship between bending strength and packing ratio. Data reported in [13] and [19] are also shown for reference. Broken line represents approximation curve obtained from data of bulk MgB_2 pieces processed by capsule method and HIP reported in [13].

5. Summary

In order to investigate the effects of SPS pressure on the mechanical properties of MgB_2 bulk, 4-point bending tests were carried out at room temperature for the specimens cut from the bulk samples. Dense bulk samples were obtained by the SPS process. Stress-strain behaviours of these bulk samples were almost linear until the fracture. The Young's modulus values were evaluated from the slope of the stress-strain behaviours of the specimens. No significant difference was observed for the Young's modulus values among the bulk samples. The average bending strength values of these bulk samples were also similar to each other. However, the bending strength data of the sample processed under higher pressure scattered widely in comparison with those of the sample processed under lower pressure. One extraordinary pore was observed on the tensile side fracture surfaces of the specimens with lower bending strength. The bending strength values were related with the pore sizes observed on the tensile side fracture surfaces.

Acknowledgements

The authors thank Mr. J. Lecourt of CRISMAT laboratory and students of National Institute of Technology, Ichinoseki College for their relevant technical support. This work was supported in part by KAKENHI (17K06825) and NIFS Collaboration Research Program (NIFS17KECA050).

References

[1] Zhao Y, Feng Y, Cheng C H, Zhou L, Wu Y, Machi T, Fudamoto Y, Koshizuka N and Murakami M 2001 *Appl. Phys. Lett.* **79** 1154
[2] Naito T, Sasaki T and Fujishiro H 2012 *Supercond. Sci. Technol.* **25** 095012
[3] Muralidhar M, Ishihara A, Suzuki K, Fukumoto Y, Yamamoto Y and Tomita M 2013 *Physica C* **494** 85
[4] Yamamoto A, Ishihara A, Tomita M and Kishio K 2014 *Appl. Phys. Lett.* **105** 032601
[5] Shields T C, Kawano K, Holdom D and Abell J S 2002 *Supercond. Sci. Technol.* **15** 202
[6] Tampieri A, Celotti G, Sprio S, Caciuffo R and Rinaldi D 2004 *Phys. C, Supercond.* **400** 97
[7] Durrell J H, Dancer C E J, Dennis A, Shi Y, Xu Z, Campbell A M, Hari Babu N, Todd R I, Grovenor C R M and Cardwell D A 2012 *Supercond. Sci. Technol.* **25** 112002

[8] Sasaki T, Naito T and Fujishiro H 2013 *Phys. Procedia* **45** 93
[9] Schmidt J, Schnelle W, Grin Y and Kniep R 2003 *Solid State Sci.* **5** 535
[10] Shim S H, Shim K B and Yoon J W 2005 *J. Amer. Ceram. Soc.* **88** 858
[11] Dancer C E J, Prabhakaran D, Basoglu M, Yanmaz E, Yan H, Reece M, Todd R I and Grovenor C R M 2009 *Supercond. Sci. Technol.* **22** 095003
[12] Noudem J, Aburras M, Bernstein P, Chaud X, Muralidhar M and Murakami M 2014 *J. Appl. Phy.* **116** 163916
[13] Murakami A, Teshima H, Naito T, Fujishiro H and Kudo T 2014 *Phys. Procedia* **58** 98
[14] Murakami A, Miyata H, Hashimoto R, Katgiri K and Iwamoto A 2008 *Physica C* **468** 1395
[15] Murakami A and Iwamoto A 2017 *J. Phys.: Conf. Ser.* **871** 012055
[16] Ren Y, Weinstein R, Liu J, Sawh R P and Foster C 1995 *Phys. C, Supercond.* **251** 15
[17] Miyamoto T, Nagashima K, Sakai N and Murakami M 2000 *Supercond. Sci. Technol.***13** 816
[18] Murakami A, Noudem J, Guesmi Z, Kudo T and Iwamoto A 2015 *Phys. Procedia* **65** 77
[19] Murakami A, Iwamoto A and Noudem J G *IEEE Trans. Appl. Supercond.* Submitted
[20] Chen D X and Goldfarb R B 1989 *J. Appl. Phys.* **66** 2489

Flux Pinning and Superconducting Properties of MgB_2-Diamond Nanocomposites

J Longji Dadiel, M Muralidhar*, M Murakami

Superconducting Material Laboratory, Graduate School of Science and Engineering, Shibaura Institute of Technology, 3-7-5 Toyosu, Koto-ku, Tokyo 135-8546, Japan

*miryala1@shibaura-it.ac.jp

Abstract. The present investigation focuses on a further performance improvement of disk-shaped bulk MgB_2 superconductors by means of a nanoscopic diamond powder, using a single-step solid-state reaction process. MgB_2 bulks were produced by in-situ solid state reaction in Ar gas using high purity commercial powders of Mg metal and amorphous B mixed in a fixed ratio of Mg:B = 1:2. Further, 0, 0.4, 0.8, and 1.2 wt% of nanoscopic diamond powder was added to improve flux pinning performance of the bulk MgB_2 material. All samples were sintered at 775 °C for 3 hours in Ar atmosphere. X-ray diffraction confirmed that the sample consisted mainly of MgB_2 phase and a small quantity of MgO phase. DC magnetization measurements showed a sharp superconducting transition with onset T_c at around 39.25 K and 37.42 K for the pure and 0.8 wt.% of nanoscopic diamond powder added samples, respectively. The highest self-field critical current density (J_c) around 300 kA/cm^2 and 105 kA/cm^2 were recorded at 20 K, in self-field and 1.5 T for the sample with 0.8 wt.% of nanoscopic diamond. The present results show that nanoscopic diamond powder is an effective pinning medium for bulk MgB_2 and might thus be an alternative to further improve the superconducting performance of the bulk MgB_2 material.

1. Introduction

Since the discovery of superconductivity in MgB_2 with the highest critical temperature, T_c of 39 K, a significant progress has been made concerning the development of processing techniques, flux pining, critical current density (J_c), large size MgB_2 bulk growth [1-5]. To utilize this material for industrial applications, including superconducting super-magnets, a further improvement of critical current density is crucial, especially in sintered bulks [6-7]. Several approaches were established for the processing and fabrication of bulk MgB_2 material. MgB_2 have been fabricated as a crystal, powder, thin film, macroscopic wire, nanowire or tape using various approaches such as solid state synthesis, mechanical alloying, sol-gel or vapor-transport process [8-9]. MgB_2 offers the possibility of wide engineering applications in the temperature range 20-30 K, where conventional superconductors such as Nb_3Sn and Ni-Ti alloy are not applicable due to their low T_c. Further, the field trapping capability for the practical applications of bulk MgB_2 magnets is similar to melt-textured $REBa_2Cu_3O_y$, which can be utilized as a trapped field magnet in nuclear magnetic resonance (NMR) and magnetic resonance imaging (MRI) [1-6]. In high magnetic fields MgB_2 exhibits a decrease in critical current density, J_c, due to its poor grain connectivity and lack of strong pinning centers [14-15]. Earlier studies indicated that an inclusion of nano-dopants enables control of MgB_2 microstructure by introducing nano-pinning centers into the superconducting matrix. Nanoscale diamond powder makes it possible to form a high density of nano-inclusions in MgB_2 matrix. To improve pinning performance of bulk MgB_2 materials, we tried to utilize a nanoscopic diamond powder in combination with optimum sintering temperature set in earlier studies [6]. In the present work, we study the effect a nanoscopic diamond powder on bulk MgB_2 material by means of X-ray diffraction (XRD), superconducting transition temperature and critical current density at 20 K. The best sample indicated a high critical current density of 298.9 $kAcm^{-2}$ at 20 K and the superconducting transition temperature around 37.5 K.

2. Experimental details

Bulk specimens of MgB_2 were prepared by an *in situ* solid state reaction. The precursor powder materials were Mg powder (99.9% purity, 325 meshes), sub-micron particle sized amorphous B powder (98.5% purity, ~250 nm), and diamond nanoparticles (<10 nm particle size) as nano-dopant. The mentioned powders were weighed 1.5 g in a glove box and thoroughly mixed under argon atmosphere in the same glove box. The Mg and B powders were mixed in the nominal ratio of 1:2 while varying the compositions of the nano-diamond for each sample produced at 0.0, 0.4, 0.8 and 1.2 wt.%. A uniaxial pressing machine was employed to press the precursors into pellets of 10 mm in diameter and 7 mm thickness. The resulting samples were then wrapped in titanium foils and subjected to heat treatment in a furnace, sintered at 775 °C for 3 h in flowing Ar atmosphere and finally cooled down to room temperature at a cooling rate of 100 °C h^{-1}.

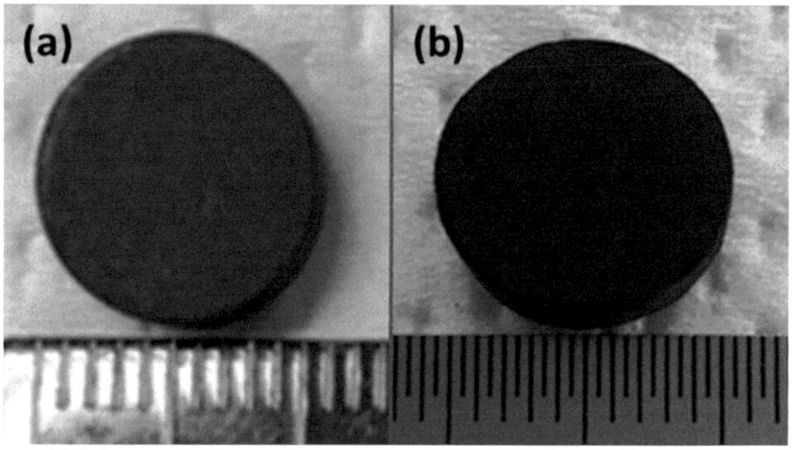

Figure 1. (a) Prepared MgB_2 pure bulk sample sintered at 775 °C (b) Prepared MgB_2-nD 0.8 wt.% sintered at 775°C. Both samples showed no cracks.

The samples were observed to have no defects when removed from the titanium foils (see fig. 1). We also observed that all samples were carefully fabricated with no indication of cracks. Microstructures of samples were also added and discussed later. The crystal structures and constituent phases were investigated using a high resolution X-ray powder diffractometer (Rigaku/Smart Lab), using Cu K_α radiation with wavelength of 1.5405 Å. The microstructure of samples were investigated using a scanning electron microscope (FESEM, JEOL make, JSM-7100 model). Smaller sized specimens of dimension 2.0×2.0×1.0 mm^3 were collected from the sintered pellets and subjected to measurements of the magnetization hysteresis loops (M-H loops) in field range from 0 to 5 T at 20 K and also the critical temperatures (T_c) using a SQUID magnetometer (MPMS5minSITO). J_c values were estimated based on the extended Bean's critical state model using the relation:

$$J_c = \frac{20\Delta M}{a^2 d\left(b - \frac{a}{3}\right)}, \quad (1)$$

where d is the sample thickness, a and b are the cross sectional dimensions with $b \geq a$, and ΔM is the difference of the magnetic moments during increasing and decreasing field in M-H loop [10].

3. Results and discussion

The X-ray diffraction patterns shown in Fig. 2 give information on the crystallographic phases at different doping levels. All three bulk MgB_2-nD samples consisted mainly of MgB_2 phase and a small quantity of MgO phase.

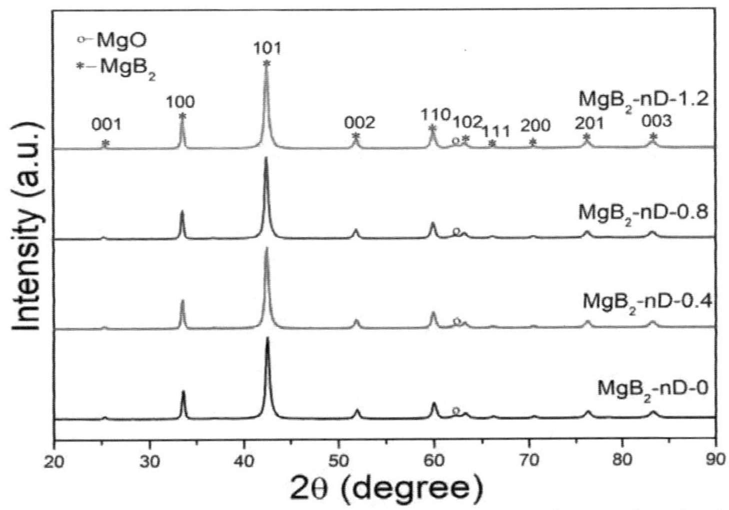

Figure 2. X-ray diffraction patterns of bulk MgB_2-diamond nanocomposites produced using the mixture of the amorphous boron powders in sintering process. All samples are sintered at 775 °C for 3h in argon atmosphere.

From previous research, we know that the nanoscale diamond should be reflected as an impurity phase in the MgB_2 material, with intensity depending on the initially added weight percentage of the nanoscale diamond. However, the main reflection of the nano-diamond (111) could not be seen in the XRD patterns due to an overlap with the MgB_2 (101) peak. Therefore, we were able to observe only MgO impurity by XRD, similar to our earlier reports [11-12].

Temperature dependence of magnetization curves was measured on the MgB_2 samples with a varying content of nano-diamond powder in a magnetic field of 1 mT. The critical temperature ($T_{c,onset}$) decreased with increasing nanoscale diamond content (Fig 3a). Note that a high T_c (onset) of 39.25 K was observed for a sample produced without nano-diamond. Here we obtained T_c (onset) around 37.3 K for 0.8 wt% of nano-diamond. This clearly indicates that we have to find an optimum between superconductivity degradation and pinning enhancement.

To see the effect of nano-diamond addition on critical current density, J_c was determined from magnetization hysteresis loops (*M-H* loops) detected by a SQUID magnetometer in applied magnetic fields from 0 to 5 T at 20 K. Fig 3(b) shows J_c values at 20 K as a function of applied magnetic field for the bulk MgB_2 material with varying content of nano-diamond. The MgB_2 sample with 0.8 wt% of nano-diamond powder exhibited higher critical current density than a pure MgB_2 material as well as the 0.4 wt% and 1.2 wt% samples. J_c increment at higher field has improved for the doped samples as shown on the graph (see inset Fig. 3c), which also supports the fact that the flux pinning mechanism has been enhanced through controlled nanodiamond doping. The self-field critical current density around 300 kA/cm^2 was observed at 20 K for the best sample as compared to the J_c of 278 kA/cm^2 for the pure sample. We can say that optimization was effective for the 0.8 wt.% nano-diamond added sample. The critical current density in high magnetic fields also improved (see inset Fig. 3b). In the MgB_2 sample with 1.2 wt% of nano-diamond powder, the critical current density decreased by a small value when compared with the pure sample. Thus, performance of the bulk MgB_2 material can be improved by optimizing the amount of nano-diamond powder in bulk MgB_2 material. The present study supports the suggestion that the nanoscale diamond can serve as a strong pinning medium to improve flux pinning of MgB_2 bulks at high magnetic fields.

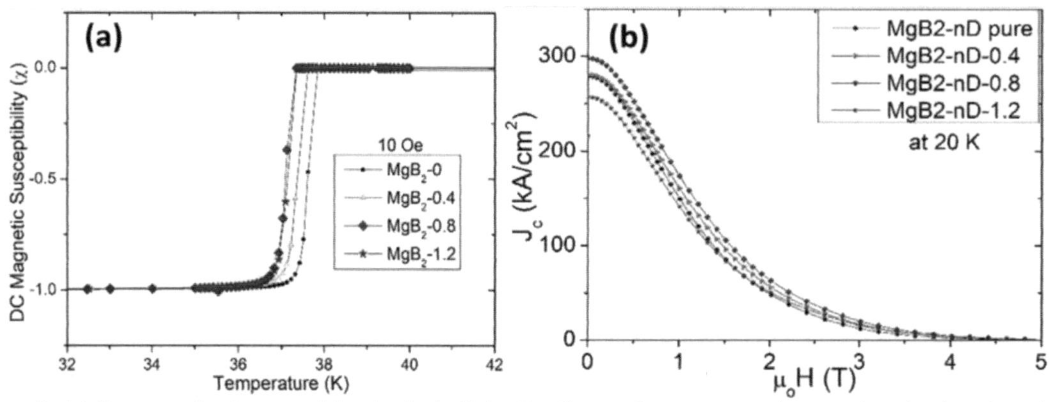

Figure 3. (a) Superconducting transition in the bulk MgB_2-diamond nanocomposites produced using the mixture of the amorphous boron powders in sintering process. (b) The field dependence of the critical current densities for the same material at $T = 20$ K. (c) The effect of pinning centers at higher fields. All samples are sintered at 775 °C for 3h in argon atmosphere.

The elementary pinning force of grain boundaries in MgB_2 is dependent on its nanostructural control. J_c at low field is simply determined by the connectivity, grain size, and elementary pinning force of the grain boundaries. The effects of grain connectivity and grain size on critical current density and flux pinning are analysed. There is qualitative agreement between grain-boundary pinning mechanism and the experimentally observed grain-size dependence of flux pinning in bulk MgB_2 [13].

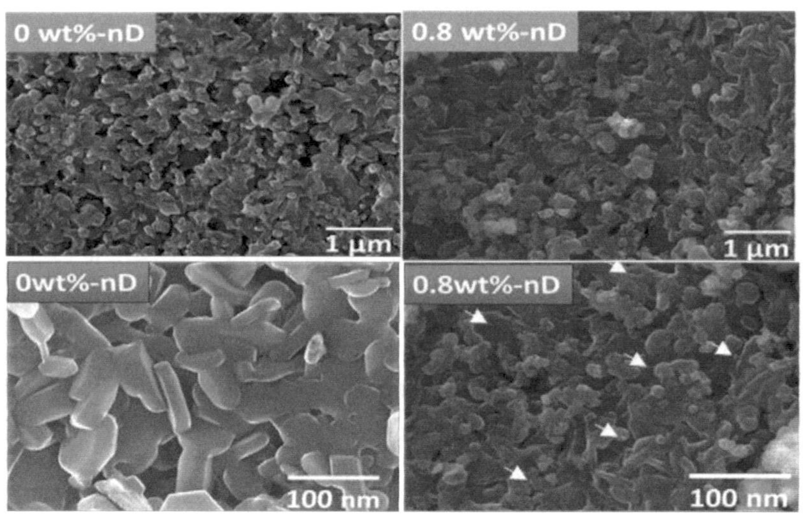

Figure 4. Micrograph of samples, both low and high magnification for diamond-added MgB_2 with different percentage contents.

The results of the microstructural analysis show the refined grains in the diamond added sample and we emphasize that these grains help in enhancing J_c. Although the percentage of the dopant added is low, this grain refinement behavior is in agreement with previous literature [15-16]. Therefore, the diamond nanoparticles may provide nucleation centers for MgB_2 and tightly bound to them [14]. These intergrain nano sized inclusions created by nanodiamond acts as strong pinning centres and are responsible for the improved flux pinning which are optimally distributed in the 0.8wt%-nD sample.

4. Conclusions

MgB_2-diamond nanocomposite was successfully synthesized by a solid state sintering process. X-ray diffraction indicated that all samples were single phase MgB_2 along with a small quantity of MgO. DC susceptibility vs. temperature measurements showed a sharp superconducting transition around 38 K for the doped samples. The magnetization measurements showed that 0.4 wt%, 0.8 wt% or 1.2 wt% of nanodiamond did not affect the critical temperature much as compared to pure MgB_2 materials. The grain morphology of the sample was improved. The critical current density and irreversibility field values were improved by nano-diamond addition to bulk MgB_2. The best J_c value at 20 K was around 300 kA/cm^2 in self-field and 175 kA/cm^2 at 1 T for the 0.8 wt% of the nano-diamond content. This work shows that doping by nano-diamond particles is a promising way to improve performance of MgB_2 bulk materials.

Acknowledgments

The paper was supported by Research Centre for Green Innovation and Grant-in-Aid FD research budget code: 112261, Shibaura Institute of Technology (SIT).

References

[1]. Vinod K, Abhilash Kumar R G, Syamaprasad U 2007 *Supercond. Sci. Technol.* **20** R1.
[2]. Perini E, Ginuchi G 2009 *Supercond. Sci. Technol.* **22** 045021.
[3]. Kajikawa K, Nakamura T 2009 *IEEE Trans. Appl. Supercond.* **19** 1669.
[4]. Gunchi G, Ripamonti G, Cavallin T, Bassani E 2014 *Cryogenics* **46** 23.
[5]. Kim J H, Oh S, Heo Y-U, Hata S, Kumakura H, Matsumoto A, Mitsuhara M, Choi S, Shimada, Maeda Y M, MacManus-Driscoll J L, Dou S X, 2012 *NPG Asia Materials*, **4** doi:10.1038/am.2012.3.
[6]. Muralidhar M, Inoue K, Koblischka MR, Tomita M, Murakami M 2014 *J. Alloys Compd.* **608** 102.
[7]. Muralidhar M, Nozaki K, Kobayashi H, Zeng X L, Koblischka-Veneva A, Koblischka MR, Inoue K, Murakami M 2015 *J. Alloys Compd.* **649** 833.
[8]. Lu J, Xiao Z, Lin Q, Claus H, Fang Z Z 2010 Journal of Nanometre 191058.
[9]. Kim J H, Oh S, Kumakura H, Matsumoto A, Heo Y.-U, Song K.-S, Kang Y.-M, Maeda M, Rindfleisch M, Tomsic M, Choi S, Dou S.X 2011 Adv. Mater. **23** 4942.
[10]. Chen D, Goldfarb R B 1989 *J. Appl. Phys.* **66** 2489.
[11]. Muralidhar M, Nozaki K, Kobayashi H, Zeng X L, Koblischka-Veneva A, Koblischka MR, Inoue K, Murakami M 2015 *J. Alloys Compd.* **649** 833.
[12]. Muralidhar M, Kenta N, Koblischka MR, Murakami M 2015 *Physica Status Solidi a*, **212** 2141.
[13] Hampshire and Jones 1987 *J. Phys. C* **21** 419.
[14] Zhao Y, Rui X F, Cheng C H, Zhang H, Munroe P, Zeng H M, Koshizuka N, Murakami M 2003 *Supercond. Sci. Technol.* **16** 10.
[15] Cheng C H, Yang Y, Munroe P and Zhao Y 2006 *Supercond. Sci. Technol.* **20** 296
[16] Gupta Anurag and Narlikar A V 2009 *Supercond. Sci. Technol.* **22** 125029

IOP Conf. Series: Journal of Physics: Conf. Series **1054** (2018) 012053 doi:10.1088/1742-6596/1054/1/012053

Evaluation of $YBa_2Cu_3O_{7-\delta}$-based microwave kinetic inductance detectors with rewound spiral resonators

K Sato[1], S Ariyoshi[1], S Negishi[1], S Hashimoto[1], H Mikami[1], K Nakajima[2] and S Tanaka[1]

[1] Toyohashi University of Technology, 1-1 Hibarigaoka, Tempaku-cho, Toyohashi, Aichi 441-8580, Japan
[2] Yamagata University, 4-3-16 Jonan, Yonezawa 992-8510, Japan

E-mail address: ariyoshi@ens.tut.ac.jp

Abstract. We propose a YBCO-based microwave kinetic inductance detector with rewound strip structure (spiral-MKID) suitable for imaging applications. The superconducting rewound strip acts as a microwave resonator and broadband antenna. A 25-pixel MKID array with a width of 10 μm and slightly varying length of each of the resonators was fabricated on a 10×10 mm^2-MgO substrate. Microwave resonance characteristics of the array were evaluated by measuring one of the scattering-matrix elements, S_{21}, using a vector network analyser. The 25 resonance dips were clearly observed around 5 GHz; the frequency spacing between adjacent pixels was 13 MHz, with a standard deviation of 5 MHz. Each resonance frequency was shifted by 170 MHz, owing to the temperature change from 11 K to 50 K. We showed that the YBCO-based spiral-MKIDs with 200 nm-thick films have loaded quality factors of 1,200 at 11 K and 700 at 50 K. We irradiated visible light to a single pixel on the array. The obtained noise equivalent power was on the order of 10^{-9} $W/Hz^{1/2}$, while the response time was smaller than 30 ms at 13 K. The responsivity of the array at 70 K was six times larger than that at 13 K.

1. Introduction

Terahertz waves in the range of 0.1 THz to 10 THz are promising for applications in various non-destructive and non-invasive imaging techniques [1]. Semiconductor bolometers are commonly used as broadband terahertz-wave detectors. However, high-speed terahertz imaging arrays that have sensitivities equal to those of bolometers are not yet commercially available. Microwave kinetic inductance detector (MKID), a superconducting photon detector, is a promising candidate to achieve high-speed and high-sensitivity terahertz imaging owing to the simple fabrication of large format arrays and its frequency-domain multiplexing capability [2]. Developed MKIDs, reported by other groups, provide high sensitivity for astronomical observations; however, they utilise low transition-temperature (T_c) materials such as metals (Al and Nb [3, 4]) and metallic nitride (TiN_x and NbN [5, 6]) superconductors, which operate at temperatures of several kelvins.

In this study, we propose an MKID array with rewound spiral resonators (spiral-MKIDs) [7, 8], using $YBa_2Cu_3O_{7-\delta}$ (YBCO) thin films. The YBCO-based MKIDs do not require large and expensive cooling systems, as required for conventional MKIDs at extremely low temperatures. Therefore, they can operate using an affordable He-refrigerator or liquid nitrogen (77 K). We fabricated a 25-pixel array, which consisted of 25 rewound spiral resonators. The superconducting rewound strip acts as a high quality-factor (Q) microwave resonator and as a broadband antenna. It is expected that the Q

IOP Conf. Series: Journal of Physics: Conf. Series **1054** (2018) 012053 doi:10.1088/1742-6596/1054/1/012053

values are on the order of thousands at temperatures larger than 70 K. Then, we characterise and discuss the electrical and optical performances of the developed YBCO-based MKIDs.

2. Detector design and fabrication

Electromagnetic (EM) numerical simulations are a powerful tool for effective optimization of the detector design. Prior to the fabrication of the detector, we calculated the microwave properties of the MKIDs using an EM simulator Sonnet. Figure 1 shows a schematic of the spiral-MKID. The prototype array consists of a single through-line and 25 spiral resonators. Coplanar waveguide (CPW) through-line is employed for the excitation and readout. The line and gap widths of the through-line were 40 μm and 10 μm respectively, which corresponded to a specific impedance of 50 Ω. Both line and gap widths of the rewound spiral resonators were 10 μm. The resonator lengths are approximately 12 mm, designed to be consistent with the effective guided half-wavelength; the resulting resonance frequencies are expected to be about 5 GHz using an MgO substrate. The resonator lengths differed from each other by 30 μm, which corresponded to a frequency shift of approximately 14 MHz. Therefore, the array can provide frequency-domain multiplexing. A trace was introduced in each resonator circuit, adjacent to the through-line, to provide capacitive coupling. The MKID array had a two-layer structure of the YBCO film and MgO substrate; photons were directly incident on the YBCO film. This resonator geometry allows to detect broadband wavelengths, although the detectable

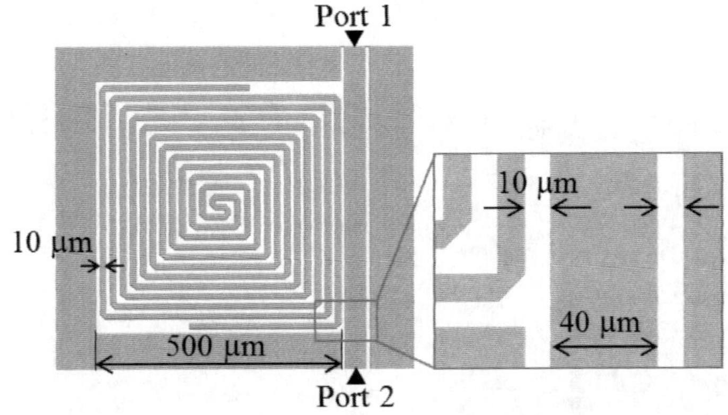

Figure 1. Schematic of the spiral MKID. The inset shows magnified view around the through-line. Dark and light regions are the YBCO film and bare MgO substrate, respectively.

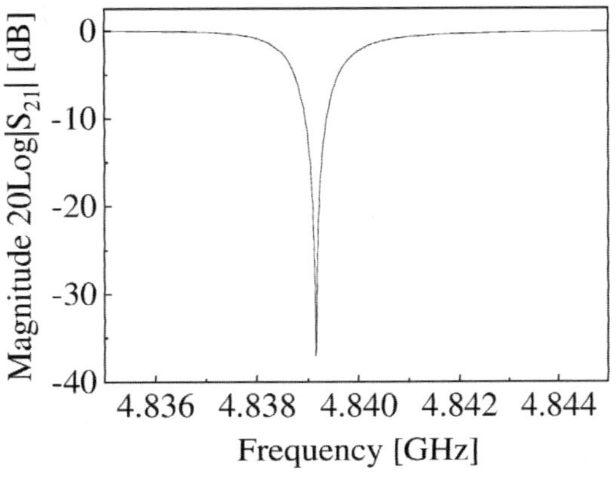

Figure 2. Simulated resonant curve of the YBCO-based spiral resonator.

frequency is limited to above 7 THz, owing to the superconducting energy gap of YBCO. Figure 2 shows a simulated resonant curve of the spiral resonator. The simulation parameters were assumed to be those of a YBCO film on an MgO substrate. The sheet inductance (L_s) and sheet resistance (R_{rf}) of the YBCO film were set to 1.0 pH/sq and 4.9×10^{-29} $\Omega\cdot Hz^{-1/2}$/sq at a temperature of 10 K, respectively [9]. MgO substrate that has a thickness of 0.5 mm, a relative permittivity of about 9.7 and a loss tangent of 3.5×10^{-6} was employed. For the analysis, the upper layer of the YBCO film was considered as an air with a thickness of 2.5 mm and a free space was applied as the upper boundary. The lower layer of the MgO substrate was considered as a lossless metal, which was consistent with the actual measurement environment. The calculated area was 5×5 mm^2, and the through-line and single resonator were arranged, while the length of the simulated resonator was set equal to the length of the resonator that had the lowest resonance frequency in the 25-pixel array. The simulations indicated that the resonance frequency is 4.839 GHz. Furthermore, a loaded quality factor (Q_L) of 3,800 at 10 K and a resonance dip of about 40 dB were estimated for the spiral resonator. Therefore, the YBCO-based MKID is expected to achieve high sensitivity, with Q_L smaller than that of the MKIDs based on low T_c superconductors.

Next, we deposited a 200 nm-thick YBCO film on a 10×10 mm^2-MgO substrate using radio frequency (RF) magnetron sputtering. The YBCO thin film had a T_c of approximately 84 K. The presence of an epitaxial film was confirmed by performing X-ray diffraction analysis. Then, the YBCO-based MKID was fabricated using standard photolithography and Ar-ion etching. Figure 3 shows the YBCO-based MKID array on an MgO substrate, and an enlarged image of a single pixel. Both sides of the through-line and edges of the resonators are grounded at the same plane.

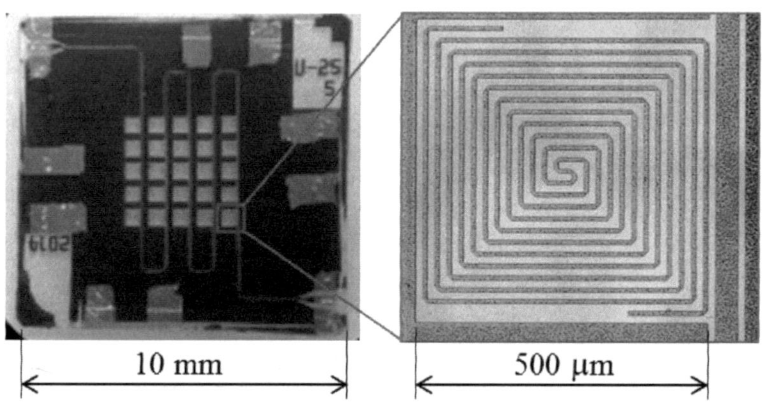

Figure 3. YBCO-based MKID array fabricated on an MgO substrate (left). Enlarged image of a single pixel (right).

3. Electrical evaluation

The fabricated chip of the YBCO-based MKID array was mounted in a cryostat (^{4}He refrigerator) and cooled down to 11 K. Microwave resonance characteristics of the array were evaluated by measuring one of the scattering-matrix elements, S_{21}, using a vector network analyser (VNA). The input power of the chip was set to −40 dBm. Amplifier, which had a gain of 28 dB, operated at room temperature, was connected to the output port of the MKID array to achieve low noise measurements. This array is unaffected by the bias power of less than −40 dBm at the through-line. At higher powers, the electric current distorted the resonance shape though non-linear effects [10]. Figure 4 shows the microwave resonance characteristics of the 25-pixel array, measured at 11 K, without calibration. A total of 25 resonance dips were clearly observed in the frequency range of 4.8 GHz to 5.2 GHz, which was in a good agreement with the simulated value (0.2 % error). The frequency interval between adjacent pixels was 13 MHz, with a standard deviation of 5.0 MHz. As shown in Figure 5 (a), the average resonance dip had an attenuation (A_{max}) of 10 dB and Q_L of approximately 1,200 at 11 K. The corresponding unloaded quality factor (Q_u) was approximately 3,000; Q_u was estimated using Q_L and A_{max} as $Q_u = Q_L / 10^{(-A_{max}/20)}$ [11]. The simulated Q_L and A_{max} of the resonator were 3,800 and 37 dB respectively; however the measured Q_L and A_{max} were smaller than simulated values. It is likely that this discrepancy emerged owing to the reduced quality and surface impedance of the YBCO film. At 50 K, 23 resonance dips in the 25 resonators were identified, and the Q_L was 700, while at 70 K, 13 resonance dips were identified. These resonance dips gradually disappeared at higher temperatures. The transmission of the through-line decreased to values smaller than 3 dB, at temperatures larger than 80 K. The resonance frequency was shifted by about 170 MHz by increasing the operating temperature from 11 K to 50 K.

Figure 5 (b) shows the temperature dependence of the resonance frequency. The frequency shift as a function of the temperature is:

$$\frac{\Delta f_r(T)}{f_r(T_{\min})} \propto \lambda_0 \left(1-\left(\frac{T}{T_C}\right)^2\right)^{-\frac{1}{2}} - \lambda(T_{\min}) \ , \qquad (1)$$

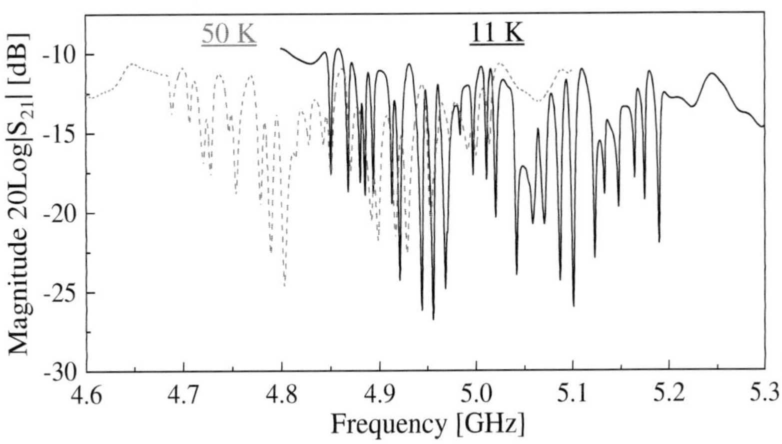

Figure 4. Microwave resonance characteristics of the array.

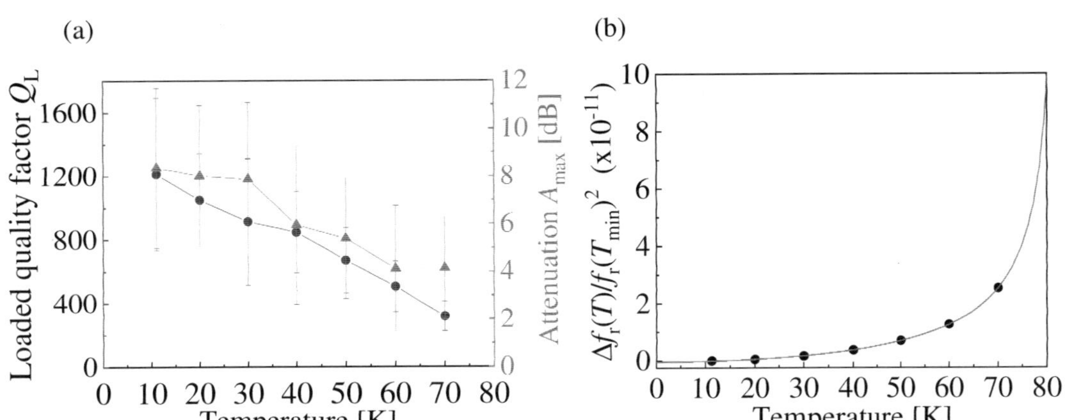

Figure 5. (a) Characteristics of the resonance dips. (b) Temperature dependence of the resonance frequency.

where $\Delta f_r(T) = f_r(T) - f_r(T_{\min})$ [12], and f_r, $T_{\min}$, and λ denote the resonance frequency, measured minimum temperature, and penetration depth, respectively. The penetration depth at zero temperature is $\lambda_0 \approx 200$ nm, obtained by extrapolation using the value at the minimum temperature. The kinetic inductance $L_k = \mu_0 \lambda(T)$ depends on the number density of Cooper-pairs. The temperature dependence of the resonance frequency is in good agreement with the dependence obtained using Equation (1). This result suggests that the resonance frequency shift is driven by the kinetic inductance.

4. Optical evaluation

As a first step of the optical evaluation, we investigated the response of the MKID array to pulsed visible light by measuring the microwave resonance shift. The MKID array was mounted in the cryostat and cooled down to 13 K. Then, it was irradiated by white visible light (power of several microwatts) using a light-emitting diode source, while the light source was turned on at 0.5 Hz with a 50 % duty cycle. The light source was placed outside the cryostat, transformed to parallel light through an optical lens, and then irradiated to the chip through the cryostat window. The irradiated power to one pixel was estimated from directional characteristics of the light source and the area of the pixel. Since the MgO substrate has high transmittance for visible light, most of the photon absorption occurs on the YBCO film. The shift of $|S_{21}|^2$ was measured using a network analyser. Figure 6 shows the response to visible light observed at 13 K. Responsivity of the MKID is defined as $d|S_{21}|^2/dP$, where

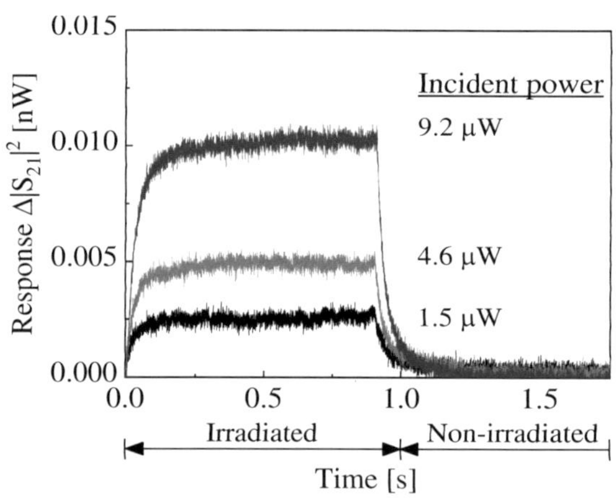

Figure 6. Optical response of the MKID at 13 K.

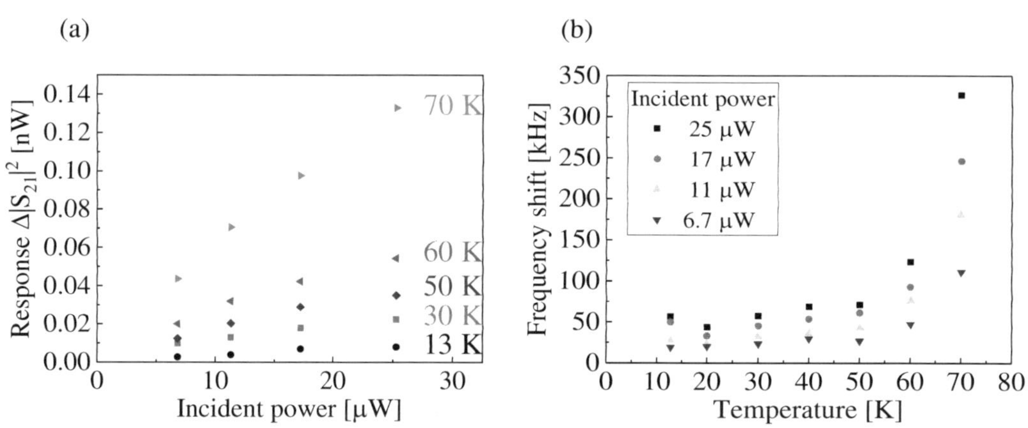

Figure 7. (a) Temperature dependence of the photoresponse. (b) Frequency shift as a function of the operating temperature.

$d|S_{21}|^2$ and dP denote the response power and incident light power, respectively. The responsivity was 1.1 μW/W, while the noise power (P_n) was about 2×10^{-4} nW. The noise equivalent power (NEP) was defined as follows:

$$\mathrm{NEP} = \frac{P_n}{\sqrt{\mathrm{d}f}}\left(\frac{\mathrm{d}|S_{21}|^2}{\mathrm{d}P}\right)^{-1} \quad [\mathrm{W} / \mathrm{Hz}^{1/2}], \tag{2}$$

where the frequency bandwidth (df) is defined as $\mathrm{d}f = 1/(2\Delta t_{\mathrm{int}})$ [Hz]. A time interval Δt_{int} was approximately 0.2 ms, depending on the specification of the VNA. The resulting NEP was on the order of 10^{-9} W/Hz$^{1/2}$ at 13 K. The obtained NEP value is limited by the fabrication process, not by the material properties. The measured response time was smaller than 30 ms. On the other hand, it was estimated that the resonator ring time $\tau_{\mathrm{res}} = Q_{\mathrm{L}}/\pi f_{\mathrm{r}}$ was on the order of microseconds [13], which is faster than the measured response time. Therefore, the measured value is thought to be due to not electric relaxation but thermal one. Optical adjustments as well as high quality YBCO film fabrication are expected to improve the performances of the detector.

We evaluated the photoresponse at higher temperatures in the range of 13-70 K, as shown in Figure 7 (a). The differential response is approximately a linear function of the incident power. The responsivity increased with the operating temperature; the responsivity at 70 K was six times larger than that at 13 K. Figure 7 (b) shows the frequency shift as a function of the operating temperature, for different values of the incident power. The frequency shift increased with the temperature. Furthermore, we observed the non-linear behaviour of the responsivity at around 70 K, which emerges owing to the large variations of L_{k} as a function of the incident power [14]. The sensitivity of the MKID depends on the Q value and frequency shift [15]. The value of Q becomes smaller at higher temperatures, while the responsivity becomes larger due to the large frequency shift. Therefore, higher sensitivity is expected by achieving good resonance properties at 77 K.

5. Conclusion

We proposed YBCO-based MKIDs with rewound strip structure suitable for imaging applications. A 25-pixel spiral MKID array with a width of 10 μm was fabricated. Its electrical and optical characteristics were evaluated. A total of 25 resonance dips were clearly observed at 11 K around 5 GHz. Each resonance frequency was shifted by about 170 MHz by increasing the temperature from 11 K to 50 K, owing to the kinetic inductance. We obtained an average Q_{L} of about 1,200 at 11 K and 700 at 50 K. Pulsed visible light was irradiated on the MKID array. The resulting NEP was approximately 10^{-9} W/Hz$^{1/2}$, while the response time was smaller than 30 ms at 13 K. Optical adjustments and high quality YBCO-film fabrication are expected to improve the performance of the detector at higher temperatures.

Acknowledgments

The authors would like to thank Professor Takashi Noguchi (National Astronomical Observatory of Japan) for his valuable suggestions for the development of the detector. This study was supported by the Tatematsu Foundation and by the Grant-in-Aid for Scientific Research (B) (No. 17H02809) from the Ministry of Education, Culture, Sports, Science and Technology (MEXT) of Japan.

References

[1] Kawase K, Ogawa Y, Watanabe Y and Inoue H 2003 *Opt. Express* **11** 2549-54
[2] Day P K, LeDuc H G, Mazin B A, Vayonakis A and Zmuidzinas J 2003 *Nature* **425** 817–21
[3] Naruse M, Sekimoto Y, Noguchi T, Miyachi A, Nitta T and Uzawa Y 2011 *J. Low Temp. Phys.* **167** 373–378
[4] Barends R, Baselmans J J A, Hovenier J N, Gao J R, Yates S J C, Klapwijk T M and Hoevers H F C 2007 *IEEE Trans. Appl. Supercond.* **17** 263–66
[5] Leduc H G, Bumble B, Day P K, Eom B H, Gao J, Golwala S, Mazin B A, McHugh S, Merrill

A, Moore D C, Noroozian O, Turner A D and Zmuidzinas J 2010 *Appl. Phys. Lett.* **97** 102509
[6] Ariyoshi S, Nakajima K, Saito A, Taino T, Tanoue H, Koga K, Hurukawa N, Yamada H, Ohshima S, Otani C and Bae J 2013 *Appl. Phys. Express* **6** 064103
[7] MA Z 2005 *IEICE Trans. Electron.* **E88-C** 1406–11
[8] Hayashi K, Saito A, Sawada T, Ogawa Y, Nakajima K, Yamada H, Ariyoshi S and Ohshima S 2013 *Phys. Procedia* **45** 213–16
[9] Delayen J R, Bohn C L and Roche C T 1990 *J. Supercond.* **3** 243–50
[10] Abdo B, Segev E, Shtempluck O and Eyal Buks 2006 *J. Phys.: Conf. Series* **43** 1346–49
[11] Kwok R S and Liang J F 1999 *IEEE Trans. Microw. Theory Techn.* **47** 111-114
[12] Lee J H, Yang W I, Kim M J, Booth J C, Leong K, Schima S, Rudman D and Lee S Y 2005 *IEEE Trans. Appl. Supercond.* **15** 3700–05
[13] Baselmans J, Yates S J C, Barends R, Lankwarden Y J Y, Gao J R, Hoevers H and Klapwijk T M 2008 *J. Low Temp. Phys.* **151** 524–29
[14] Atikian H A, Ghamsari B G, Anlage S M and Majedi A H 2011 *Appl. Phys. Lett.* **98** 081117
[15] Lindeman M A, Bonetti J A, Bumble B, Day P K, Eom B H, Holmes W A and Kleinsasser A W 2014 *J. Appl. Phys.* **23** 234509

IOP Conf. Series: Journal of Physics: Conf. Series **1054** (2018) 012054 doi:10.1088/1742-6596/1054/1/012054

Neutron signal features of Nb-based kinetic inductance detector with ^{10}B convertor

Yuya Miki[1], Hiroyuki Yamaguchi[1], Yuki Iizawa[1], Hiroaki Shishido[1,2], Kenji M Kojima[3,4], Kenichi Oikawa[5], Masahide Harada[5], Shigeyuki Miyajima[2,6], Mutsuo Hidaka[7], Takayuki Oku[5], Kazuhiko Soyama[5], and Takekazu Ishida[1,2]

[1]Department of Physics and Electronics, Osaka Prefecture University, Sakai, Osaka 599-8531, Japan
[2]NanoSquare Research Institute, Osaka Prefecture University, Sakai, Osaka 599-8531, Japan
[3]Muon Science Laboratory and Condensed Matter Research Center, Institute of Materials Structure Science, KEK, Tsukuba, Ibaraki 305-0801, Japan
[4]Department of Materials Structure Science, The Graduate University for Advanced Studies, Tsukuba, Ibaraki 305-0801, Japan
[5]Materials and Life Science Division, J-PARC Center, Japan Atomic Energy Agency, Tokai, Ibaraki 319-1195, Japan
[6]Advanced ICT Research Institute, NICT, Kobe, Hyogo 651-2492, Japan
[7]National Institute of Advanced Industrial Science and Technology (AIST), Tsukuba, Ibaraki 305-8568, Japan

Abstract. In our preceding works, we demonstrated successful neutron detection using a superconducting current-biased kinetic inductance detector (CB-KID), which is composed of two Nb-based superconducting meanderlines and the ^{10}B neutron absorption layer. The CB-KID with a ^{10}B absorption layer outputs the voltage pulses when it is irradiated by pulsed neutrons. We expected that the voltage V is proportional to a product of the bias current I_b and a time derivative of the local kinetic inductance $d\Delta L_k/dt$, and a pair of signals propagate along the Nb stripline as an electromagnetic wave at a certain fraction of the light velocity c toward end electrodes. It still remains to be revealed why the signal voltage shows such a continuum in the histogram of the signal height even if the incident energy of the light ion is apparently monochromatic. In the present work, we investigated the distribution of the height and width of the signal. We found a clear correlation between height and width, which might be a key of understanding the operating principle of our detector. We consider that the origin of the signal distribution is due to the positional dependence of the light ion bombardment with respect to the meandering Nb nanowire.

1. Introduction

Neutron beam has attracted a great deal of interests in view of fundamental researches and various useful applications in recent years. For example, neutron diffraction is a standard technique to determine the atomic or magnetic structure of new materials. It serves as a powerful tool in examining crystal structures in drug-development research and material sciences [1]. We can also observe the dynamics of carbons and water molecules in operating fuel cell and hydrocarbons in operating engine [2]. However,

spatial resolution is still not satisfactory enough to satisfy various needs. It typically remains on the sub-millimetre scale with the use of conventional neutron detectors such as a miniaturized gas counter [3] although it becomes on the scale of several tens μm in high-end neutron detectors such as a detector using a micro channel plate and optical readout [4]. Much faster response time is desired for counting high-intensity neutron beam for more effective use of neutrons. We developed a superconducting detector for sensing neutrons with high-spatial resolution and fast response time, and first succeeded in detecting the neutrons using an MgB_2 thin film in our preceding works [4-6]. We also developed a superconducting neutron detector using a Nb-based current-biased kinetic inductance detector (CB-KID) with a ^{10}B conversion layer [7-10]. Choosing Nb has an advantage in a fabrication process because it has been established well in the long history of superconductive electronics. Hence, Nb devices have merits to build them rather reproducibly using a service from a superconductive foundry. There is the reason why the response time of a CB-KID is very fast compared to other superconducting detectors. It depends on an excitation time from Cooper-pairs to quasi-particles in a superconductor, but is not sensitive to the recombination processes from quasiparticles to the Cooper pairs. The latter process includes the phonon-mediated electronic processes, and could be rather time consuming to complete the recombination. The electronic relaxation time of Nb is not so fast compared to other superconductors, but it would not be demerits of using Nb to fabricate the CB-KID since it does only see the earlier process of the Cooper-pair breaking.
Our detector is rather new and under development, and the details of the operating mechanism of CB-KID are not elucidated thoroughly to date. In this paper, we show the systematic investigation of the characteristic features of the signals from a neutron detector CB-KID to understand the operation.

2. Delay-Line Current-Biased Kinetic Inductance Detector

2.1. Operation Principle of CB-KID

A CB-KID consists of superconducting Nb-based meanderline. Figure 1 is a schematic diagram of CB-KID structure for providing imaging function. The CB-KID of an effective area 15 mm × 15 mm on a single Si chip (22 mm × 22 mm) is cooled to a temperature below a critical temperature T_c, and is usually operated at 4 K under a constant DC bias current I_b. When a mesoscopic portion of the meanderline exhibits an abrupt reduction in the density of the Cooper pairs, a transient change occurs in the local kinetic inductance $\Delta L_k = m_s \Delta \ell / n_s q^2$ of a hot spot candidate, where m_s is the effective mass of a Cooper pair, q_s is the effective electric charge of a Cooper pair, n_s is the Cooper pair density, $\Delta \ell$ is the length of the local hot spot candidate, and S is the cross-sectional area of the Nb wire. A temporary change in L_k generates a voltage V across the Nb nanowire expressed as follows:

$$V = RI_b + \frac{d(L_m + L_k + \Delta L_k)I_b}{dt} \approx I_b \frac{d(\Delta L_k)}{dt} \qquad (1)$$

Since the magnetic inductance L_m is described as $L_m = \mu l / 4\pi$, L_m does not alter with time. The factor $dI_b/dt = 0$ is kept as far as the detector is kept in the superconducting state. The kinetic inductance L_k of the entire nanowire of length ℓ is $L_k = m_s \ell / n_s q^2$, where we intend to exclude a tiny portion from the total length ℓ to estimate the kinetic inductance L_k. as a candidate of a possible hot spot. The response signal of CB-KID is supposed to be very fast because it depends on the time differentiation of Cooper pair density n_s. The output voltage of CB-KID is proportional to I_b, and the output voltage can be tuned at a desired level by adjusting I_b. We produced a neutron detector using Nb-based CB-KID coated with ^{10}B absorption layer.

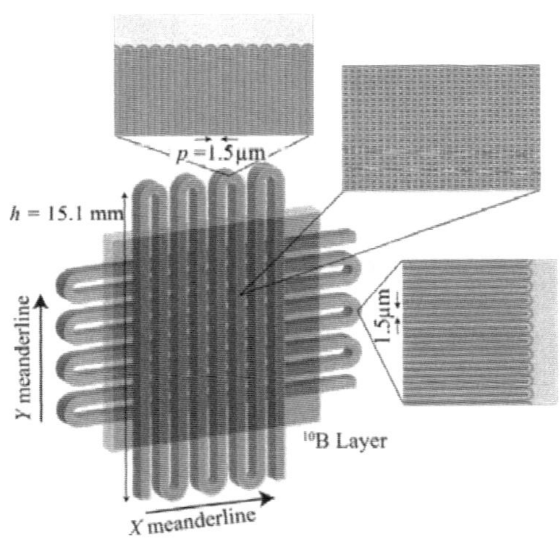

Figure 1. Schematic diagram of a CB-KID structure for imaging. The ^{10}B absorption layer is deposited on top of double Nb meanderlines.

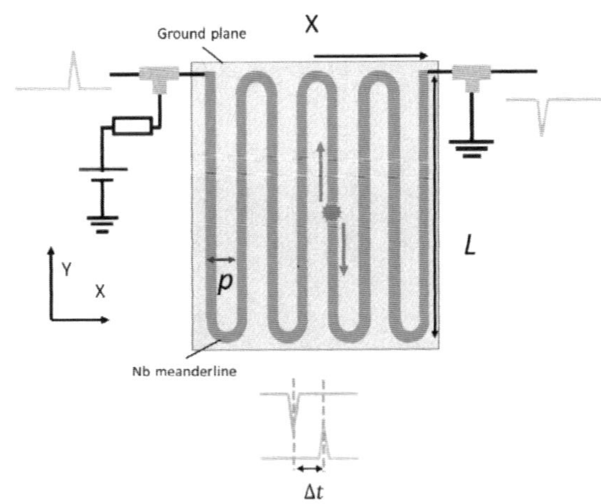

Figure 2. Diagram of signal generation and propagation along the stripline of CB-KID with a pitch p. A positive signal propagates toward the upstream of the DC bias current while it is reversed toward the downstream of the current.

2.2. Delay-line CB-KID

We describe the principle of delay-line CB-KID. In the delay-line CB-KID, the position of neutron nuclear reaction is detected from a time difference of the travelling signal caused by irradiating pulse neutrons on the ^{10}B layer. Figure 2 shows a hot spot on the Nb nanowire (stripline) and the signal transmission along the stripline. The signal is caused by the hot spot when the stripline is biased by the DC current while the hot spot is caused by the nuclear reaction between ^{10}B and neutron. The nuclear reaction $^{10}B(n, {}^{4}He){}^{7}Li$ mostly releases a ^{4}He particle of 1.47 MeV and a ^{7}Li particle of 0.88 MeV, and the local energy dissipation of each projectile is used to create a hot spot on the Nb nanowire stripline of the detector. When energy is added to the superconducting Nb meanderline, the Cooper pair density decreases locally. This leads to a generation of a pair of pulsed voltage signals according to equation (1), of which the sign (polarity) is dependent on the current direction. Each voltage signal propagates to both ends of the meander line as a pair of electromagnetic wave packets. Since the fractional lengths of the striplines, on which the signals move, are different from each other, the arrival timestamps of the signals at both ends are also different from each other. From the difference in the arrival timestamps of the signals at both ends of the meander line, we can calculate the position of the hot spot as

$$x = \frac{\Delta t}{2} \frac{p}{L} v_{\mathrm{X}} ,$$

on the basis of the ceiling function of x in units of the segment length L and a pitch p of line and space. Our system is able to identify a position of a local meso-excitation from a difference in arrival timestamps of the two signals of opposite polarities. This means that only four readout channels are enough to conduct a large area imaging, and is in marked contrast with the exiting other techniques.

3. Experimental principle and setup

Our neutron detector was cooled down to a temperature below T_c of Nb using a 4K Gifford–McMahon (GM) refrigerator. The signal was amplified by an ultra-low-noise amplifier (NF SA-430 F5) and was detected by a 2.5GHz sampling digital oscilloscope (Teledyne LeCroy HDO4104-MS). Neutron irradiation experiments were performed with the pulsed neutrons at the beam line BL10 in J-PARC. Figure 3 shows a schematic diagram of experimental setup. The cryostat has an anti-vibration mechanism to prevent the detector from mechanical displacements to ensure high resolution measurements. The whole system is controlled by the LabVIEW program on the windows OS. We store the whole wave form signals to perform data analyses after completing the data acquisition.

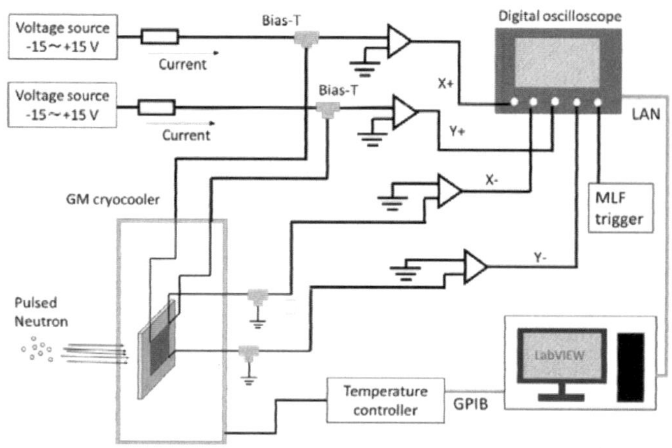

Figure 3. Schematic diagram of the experiment setup of neutron irradiation using CB-KID. The voltage across the sensor is amplified by a low-noise amplifier, and is observed by a digital oscilloscope.

4. Experimental Results

Figure 4 shows the typical output waveforms from detectors when the nuclear reaction event occurs by an incident neutron under the conditions of $T = 4$ K and $I_b = 150$ μA. Figure 4(a) is a typical quartet set of output waveforms from the *XY* detectors. The negative polarity signal is inverted by a differential amplifier. Fig. 4(b) is an enlargement of one of the signals of Fig. 4(a), where the peak voltage is 41.2 mV, and the full width at half maximum is 12.4 ns.

To clarify the signal characteristics of CB-KID, we focus on the peak voltage of the signal and full width at half maximum (HWHM). We investigated the distribution of the peak voltage and full width at half maximum of signals. Figure 5(a) shows the histogram of the peak voltage and Figure 5(b) shows the histogram of FWHM. These results are obtained by analysing 3,792,600 event signals. Despite of the energy for generating the mesoscopic excitation in the stripline is monochromatic determined by the *Q* value of the nuclear reaction, distribution of signal height and width is rather continuum. We consider that the origin of the signal distribution is due to the positional dependence of the meso-excitations with respect to the Nb stripline and the time-derivative dependent feature on the kinetic inductance. Especially, the logarithmic representation of the number of signals follows the linear relation, remarkably. This is a key to explain the origin of distribution in signal amplitude. However, the full width at half maximum seems to show the double components in the semi-logarithmic representation. The origin of the double components are not clear yet at this moment. We consider that it might be relevant to the fact whether a hot spot extends to the multiple Nb nanowires or not. If it were relevant to the multiple wires, the phonon processes must be involved in transferring the local excitations to the next. This may results in a prolonged elementary process of generating signal in addition to a short-time component. Detailed measurements of scanning a focused pulsed laser spot on the delay-line CB-KID would be useful to judge the validity of the present tentative explanations. The correlation between the height and width of the pulsed signals might be useful to discriminate the signals from noises while the detailed mechanism remains to be explored in our future work.

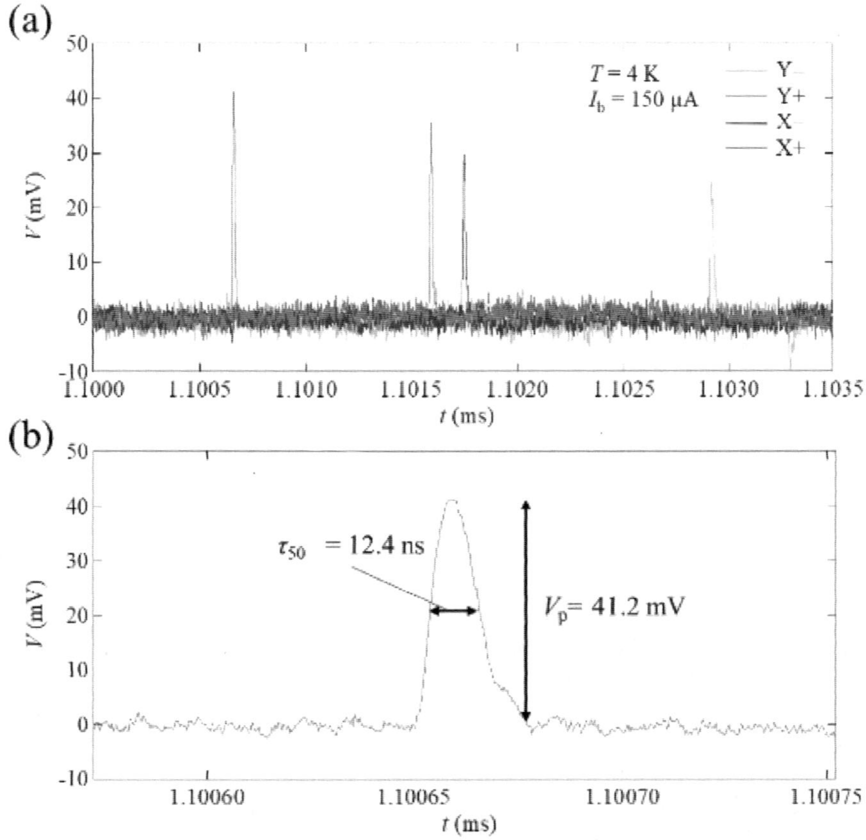

Figure 4. The typical output waveforms of detectors in detecting neutrons. These output signals are amplified by the amplifier with the gain of 46 dB. (a) A typical quartet set of output waveforms from the XY detectors. (b) One of the signals of Fig. 4(a) shows a pulse width of 12 ns and a pulse height of 41 mV.

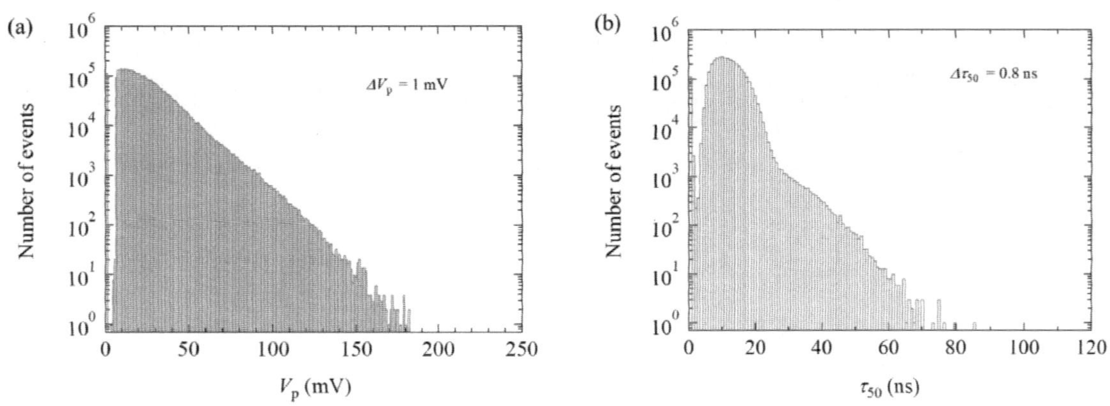

Figure 5. (a) The histogram of the peak voltage of signals. The bin width of voltage ΔV_p is 1 mV. (b) The histogram of the full width at half maximum of signals. The bin width of FWHM $\Delta\tau_p$ is 0.8 ns.

5. Conclusion

We investigated the characteristic features of a novel imaging system [6-10] proposed by our group on the basis of delay-line CB-KID method. Experimental observations of pulsed neutron signals from CB-KID with ^{10}B conversion layer were conducted systematically at the beam line (BL10) of Materials and Life Science Experimental Facility (MLF) of J-PARC. The CB-KID detects a neutron, and the response time of the signal is very fast on the order of tens of nanoseconds, and is much faster than other existing neutron detectors. We investigated the distribution of height and width of the output signals from neutron irradiation. The height and width of the signal formed continuum distribution in histograms and have a clear correlation between signal height and signal width. The correlation of signal might be a key of understanding the operating principle of our detector. Detailed origin of the observed profiles is not clear yet at this moment, but we plan to investigate a positional dependence of mesoscopic excitations with respect to nanowire stripline in our future work.

Acknowledgement

This work is supported by Grant-in-Aid for Scientific Research (S) No. 23226019, and Grant-in-Aid (A) No.16H02450 from JSPS. The devices were fabricated in the clean room for analog-digital superconductivity (CRAVITY). This work is partially benefitted by the use of VDEC, the University of Tokyo with the collaboration with Cadence Corporation. This work is supported of MLF program (Proposal No. 2015A0129, No. 2015P0301).

References

[1] Cheetham A K, Wilkinson A P 1992 *Angew. Chem. Int. Ed.* **31** 1557
[2] Manke I, Hartnig Ch, Grunerbel M, Kaczerowski J, Lehnert W, Kardjilov N, Hilger A, Banhart J, Treimer W and Strobl M 2007 *Appl. Phys. Lett.* **90**, 184101
[3] Ohshita H, Unoa S, Otomoa T, Koike T, Murakami T, Satoha S, Sekimoto M, Uchida T 2010 *Nucl. Instrum. Methods Phys. Res. Sect. A* **623** 126
[4] Tremsin A S, McPhate J B, Vallerga J V, Siegmund O H W, Feller W B, Lehmann E, Butler L G and Dawson M 2011 *Nucl. Instrum. Methods Phys. Res. Sect. A* **652** 400
[4] Ishida T, Nishikawa M, Fujita Y, Okayasu S, Katagiri M, Satoh K, Yotsuya T, Shimakage H, Miki S, Wang Z, Machida M, Kano T, and Kato M 2008 *J. Low Temp. Phys.* **151** 1074
[5] Yagi I, Yoshioka N, Shishido H, Yotsuya T, Miki S, Wang Z and Ishida T 2013 *IEEE Trans. Appl. Supercond.* **23** 2200904
[6] Yoshioka N, Yagi I, Shishido H, Yotsuya T, Miyajima S, Fujimaki A, Miki S, Wang Z and Ishida T 2013 *IEEE Trans. Appl. Supercond.* **23** 2400604
[7] Shishido H, Miyajima S, Narukami Y, Oikawa K, Harada M, Oku T, Arai M, Hidaka M, Fujimaki A and Ishida T 2015 *Appl. Phys. Lett.* **107** 232601
[8] Narukami Y, Miyajima S, Fujimaki A, Hidaka M and Ishida T 2015 *IEEE. Trans. Appl. Supercond.* **25** 2400904
[9] Miyajima S, Shishido H, Narukami Y, Yoshioka N, Fujimaki A, Hidaka M, Oikawa K, Harada M, Oku T, Arai M and Ishida T 2017 *Nucl. Instrum. Methods Phys. Res. Sect. A* **842** 71–5
[10] Shishido H, Miyajima S, Narukami Y, Oikawa K, Harada M, Oku T, Arai M, Hidaka M, Fujimaki A and Ishida T 2015 *Appl. Phys. Lett.* **107** 23260

IOP Conf. Series: Journal of Physics: Conf. Series **1054** (2018) 012055 doi:10.1088/1742-6596/1054/1/012055

Electrodynamic theory for the operation principle of a superconducting kinetic inductance stripline detector

T Koyama[1] and T Ishida[1,2]

[1]Department of Physics and Electronics, Graduate School of Engineering, Osaka Prefecture University, Sakai, Osaka 599-8531, Japan

[2]Nano-Science Research Institute, Osaka Prefecture University, Sakai, Osaka 599-8531, Japan

E-mail: tkoyama67@nifty.com

Abstract. Generation and transmission of voltage signals in a kinetic inductance detector made of a superconducting nanowire stripline is investigated on the basis of superconducting electromagnetism. We show that electric signals traveling along the stripline in this detector can be regarded as a Swihart mode. An equation which can describe the generation and transmission of voltage pulses in this detector is derived. A pair of voltage pulses with opposite polarities are created when a spatiotemporal variation in the density of the superconducting condensate occurs in a small region of this superconducting nanowire. The voltage pulses propagate along the stripline in the directions opposite to each other with the Swihart velocity. The characteristic of the radiation detector named *current-biased kinetic inductance detector* (CB-KID) is well described by our model.

1. Introduction

Superconductors are widely utilized in various high-performance devices with low energy consumption. It is also known that radiation detectors using superconducting circuits exhibit high sensitivity and resolution. Up to now several types of superconducting radiation detectors such as transition edge sensor (TES)[1, 2], superconducting nanowire single-photon detector (SNSPD)[3, 4], microwave kinetic inductance detector (MKID)[5] and also the one using S-I-S tunnel junctions [6, 7, 8] have been developed as the high-performance radiation detectors. Recently, a new type of detector using a superconducting stripline, which is named *current-biased kinetic inductance detector* (CB-KID), has been fabricated for application to a radiation detector that makes it possible to image the intensity distribution of incident radiations[9, 10, 11, 12, 13, 14]. The stripline detector is formed by a superconducting meander deposited on a substrate composed of a stack of insulating and superconducting layers, that is, the stripline has a S-I-S waveguide structure. Voltage pulses traveling on the current-biased stripline in CB-KID can be generated by a variation of the kinetic inductance of the stripline when the nanowire is locally heated. It has also been observed that a voltage pulse generated at a small hot spot on the current-biased stripline is split into two voltage pulses with opposite polarities and propagate toward the opposite directions with a constant velocity and reach the two ends of the stripline at different time depending on the place where the voltage pulse was created. Hence, by measuring the difference in the arrival times of the two voltage pulses at both ends of the stripline one can image the hot spot at which a variation of the superconducting current occurred. Since such a pair of voltage pulses can be generated by radiation incident on the detector as shown by Ishida et al.[9], CB-KID can be utilized as a radiation detector that can image incident radiation with high resolution. In fact CB-KID with an area of 1 cm^2 available now attains a resolution of 10^6 pixels[11].

In this paper the workings of CB-KID is theoretically investigated on the basis of superconducting electromagnetism. We show that the kinetic inductance variation caused by a transient hot spot on the current-biased superconducting nanowire can be described by an equation for the superconducting phase difference between the superconducting nanowire and the superconducting ground plane. The equation has a solution that two voltage pulses with opposite polarities excited by a local temporal variation of the density of the superconducting condensate propagate with the Swihart velocity[15] in the directions toward opposite electrodes.

2. Current-biased kinetic inductance detector

The radiation detector CB-KID, which is composed of a superconducting stripline, probes the time derivative of the kinetic inductance on the occasion that the superconducting nanowire in the stripline is locally heated by radiation. Since the kinetic inductance of a superconducting wire, L_k, is inversely proportional to the density of superconducting condensate n_s and the thickness s as $L_k \propto 1/n_s s$, a decrease in n_s by heating causes a large variation of the kinetic inductance in a thin superconducting wire. The workings of CB-KID have been analyzed by Shishido et al [11] on the basis of a superconducting circuit theory permitting the time variation of the kinetic inductance. In this circuit theory the voltage appearing across a hot spot in the current biased stripline is given as $V \simeq (dL_k/dt)I$, I being the bias current. In this paper we clarify the electrodynamic properties of a superconducting stripline on the basis of superconducting electromagnetism and provide a theoretical basis for the variation of the kinetic inductance by heating in CB-KID.

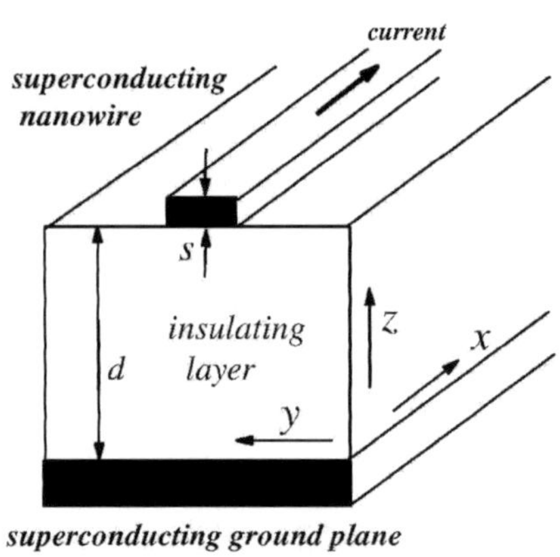

Figure 1. Schematical view of the stripline in CB-KID. The bias current I flows in the superconducting nanowire. The origin $z = 0$ is taken at the top surface of the superconducting ground layer.

2.1. S-I-S waveguide model

Let us construct a theory for the electrodynamic response of the superconducting stripline in CB-KID. The schematic view of a part of the stripline in this detector is depicted in Fig.1. It is seen that the present stripline has a S-I-S waveguide structure. Then, we first investigate the propagating mode in a S-I-S waveguide system. Suppose that the thickness of the superconducting nanowire s in this system is less than the London penetration depth λ_L, considering the fact that the thinner the thickness s is the better to obtain a large variation in the kinetic inductance in this system, while the insulating and superconducting layers forming the substrate are thick, for instance, $d > \lambda_L$ and $D \gg \lambda_L$, where d and D are, respectively, the thicknesses of the insulating and superconducting ground layers. Noticing this system has a S-I-S layer structure, we define the gauge-invariant phase differeence $\theta(x,t)$ between the two superconductors as follows,

$$\theta(x,t) = \varphi(x,d,t) - \varphi(x,z_s,t) - \frac{e^*}{\hbar c}\int_{z_s}^{d} dz A_z(x,z,t), \tag{1}$$

where $\varphi(x,z,t)$ is the superconducting phase at (x,z,t), $A_z(x,z,t)$ is the z component of the vector potential and $e^* = 2e$ is the charge of a Cooper pair. The coordinate axes in the present system are taken as shown in Fig.1. For simplicity the spatial dependence along the y-direction is neglected in this paper. Note that $\theta(x,t)$ is defined as the gauge-invariant phase difference between a point on the lower surface of the superconducting wire ($z = d$) and a point deeply inside the superconducting substrate($z = z_s < 0$ and $\lambda_L < |z_s| < D$). Following the procedure which is well-known in the theory of Josephson effect, one can derive the equation of motion for $\theta(x,t)$ as

$$\frac{\lambda_L + d}{d}\frac{\epsilon}{c^2}\partial_t^2\theta_y(x,t) - \partial_x^2\theta(x,t) = (\frac{e^*}{\hbar c})\partial_x[A_x(x,d,t) - \frac{\hbar c}{e^*}\partial_x\varphi(x,d,t)]. \tag{2}$$

Here, the term on the right-hand-side of Eq.(2) arises from the screening current flowing on the lower surface of the superconducting nanowire. To calculate this term one can utilize the London equation for the superconducting current flowing inside the superconducting nanowire, i.e.,

$$\mathbf{j}(x,z,t) = -\frac{c}{4\pi\lambda_L^2}(\mathbf{A}(x,z,t) - \frac{\hbar c}{e^*}\nabla\varphi(x,z,t)). \tag{3}$$

Substituting Eq.((3) into the Maxwell equations, one can derive the equation for the y-component of the magnetic field, $B_y(x,z,t)$, as

$$(\partial_x^2 + \partial_z^2)B_y(x,z,t) = \frac{1}{\lambda_L^2}B_y(x,z,t). \tag{4}$$

Consider the case where the scale of the spatial variation along the x-direction is much larger than λ_L. In this case Eq.(4) yields the solution in the presence of a bias current I as

$$B_y(x,z,t) = h(z)\tilde{B}_y(x,d,t) + \frac{e^{(2d+s-2z)/\lambda_L} - e^{-(2d+s-2z)/\lambda_L}}{e^{s/\lambda_L} - e^{-s/\lambda_L}}\frac{2\pi}{c}I, \tag{5}$$

where

$$h(z) = \frac{e^{(d+s-z)/\lambda_L} - e^{-(d+s-z)/\lambda_L}}{e^{s/\lambda_L} - e^{-s/\lambda_L}}. \tag{6}$$

In Eq.(5) $\tilde{B}_y(x,d,t)$ stands for the magnetic field component generated by the traveling signals at the boundary $z = d$. Using the above solution, one can derive the relation,

$$\partial_x(A_x(x,z,t) - \frac{\hbar c}{e^*}\partial\varphi(x,z,t)) = \lambda_L^2 h'(d)\partial_x B_y(x,d,t), \tag{7}$$

with $h'(d) = [\partial_z h(z)]_{z=d} = -\lambda_L^{-1}\coth(s/\lambda_L)$. Furthermore, since $B_y(x,z,t)$ is continuous at the interface $z = d$, the right-hand-side of Eq.(7) can be rewritten as

$$\lambda_L^2 h'(d)\partial_x B_y(x,d,t) = \lambda_L^2 h'(d)\frac{\epsilon}{c}\partial_t E_z(x,d,t) = -(\frac{\hbar c}{e^*})\frac{\lambda_L\coth(s/\lambda_L)}{d}\frac{\epsilon}{c^2}\partial_t^2\theta(x,t), \tag{8}$$

where ϵ is the dielectric constant of the insulating layer. Finally, from Eqs.(2), (7) and (8) it follows the wave equation,

$$\frac{1}{v^2}\partial_t^2\theta(x,t) - \partial_x^2\theta(x,t) = 0, \tag{9}$$

where the velocity v, which corresponds to the Swihart velocity in a Josephson junction [15], is given as

$$v = \frac{c}{\sqrt{\epsilon}}\sqrt{\frac{d}{d+\lambda_L(1+\coth(s/\lambda_L))}}. \tag{10}$$

Note that v is a function of the London penetration depth, which is dependent on temperature, that is, v is expected to vary with temperature. In Fig.2 we plot the velocity $v(T)/v(0)$ as a function of temperature T together with the experimental one, assuming $\lambda_L(T)$ obeys the temperature dependence in the two-fluid model, i.e., $\lambda_L(T) = \lambda_L(0)/\sqrt{1-(T/T_c)^4}$. It is seen that the temperature variation of the velocity of the voltage signal observed in the experiment is in good agreement with that predicted by Eq.(10). Hence, one may conclude that the electromagnetic propagation modes in CB-KID can be identified with those in the nanoscale S-I-S waveguide.

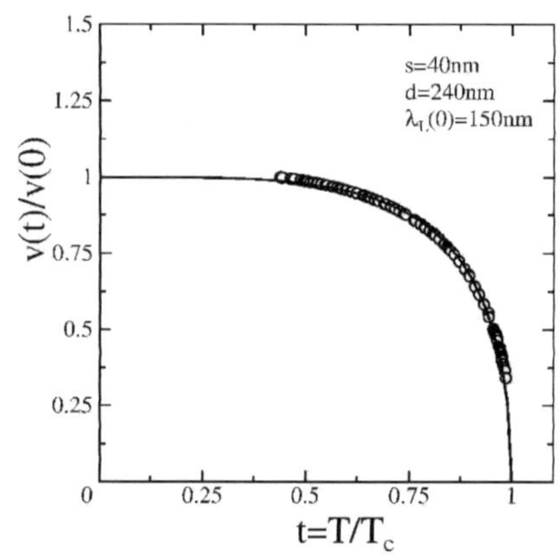

Figure 2. Temperature variation of the normalized velocity $v(t)/v(0)$. The circles indicate the experimental data obtained by Iizawa et al.[16] in the CB-KID using a Nb meander with a thickness s = 40nm which is deposited on an insulating layer with a thickness d = 240nm. The value $\lambda_L(0)$ = 150nm is chosen for fitting the theoretical result (solid line) to the experimental one.

2.2. Response to a local heating

Let us clarify the workings of CB-KID when a heat pulse is injected into this detector. Suppose that an incident heat pulse causes a local heating in CB-KID and then brings about a decrease in the density of superconducting condensate in a small heated region inside the current-biased superconducting nanowire. In the following the region in which the superconducting condensate is decreased is called a hot spot. Noting that λ_L^{-2} is proportional to the density of the condensate n_s, one may incorporate phenomenologically the effect of a hot spot into a theory by the replacement, $n_s \to n_s(x,t)$, i.e.,

$$\frac{1}{\lambda_L^2} = \frac{4\pi e^{*2}}{mc^2}n_s \to \frac{4\pi e^{*2}}{mc^2}n_s(x,t) = \frac{1}{\lambda_L^2}\cdot\frac{n_s(x,t)}{n_s} = \frac{\zeta(x,t)}{\lambda_L^2}, \tag{11}$$

where the function $\zeta(x,t) = n_s(x,t)/n_s$ is less than 1 only inside the hot spot. It is assumed that the hot spot is created at $t = 0$ and disappears after a short time interval τ. In this paper we investigate an ideal case in which the hot spot has a symmetric shape with a length x_h ($\gg \lambda_L$) along the x direction and its center is located at $x = 0$, i.e., $\zeta(-x,t) = \zeta(x,t)$, and thus $\zeta(x,t) < 1$ for $|x| < x_h/2$ and $0 < t < \tau$. On the basis of the above phenomenological consideration one can generalize the Maxwell-London equation as

$$\partial_z B_y(x,z,t) = \frac{\zeta(x,t)}{\lambda_L^2}[A_x(x,z,t) - \frac{\hbar c}{e^*}\partial_x\varphi(x,z,t)], \tag{12}$$

in the presence of a hot spot. Here, the quasi-particle current is neglected, which is justified in the low temperature region where the density of thermally excited quasi-particles is negligibly small. From Eq.(12) one can easily obtain the relation,

$$\partial_x[A_x(x,d,t) - \frac{\hbar c}{e^*}\partial_x\varphi(x,d,t)] = -\frac{\lambda_L}{\zeta(x,t)}\coth(\frac{s}{\lambda_L})\partial_x B_y(x,d,t)$$

$$+\frac{4\pi\lambda_L^2}{c}\frac{\partial_x\zeta(x,t)}{\zeta(x,t)^2}[\frac{c}{4\pi\lambda_L}\coth(\frac{s}{\lambda_L})\tilde{B}_y(x,d,t) + \frac{I}{s}]. \tag{13}$$

Here, use was made of the relation $B_y(x,z,t) = h(z)B_y(x,d,t)$ and Eq.(5), Since the spatial derivative, $\partial_x B_y(x,d,t)$, in Eq.(13) is expressed in terms of the phase difference as

$$\partial_x B_y(x,d,t) = \frac{\hbar c\epsilon}{e^* c^2}\partial_t^2\theta(x,t), \tag{14}$$

one can rewrite Eq.(2) as

$$[\frac{1}{v(x,t)^2}\partial_t^2 - \partial_x^2]\theta(x,t) = F(x,t). \tag{15}$$

in the presence of a hot spot, where $v(x,t)$ is the Swihart velocity,

$$v(x,t) = \frac{c}{\sqrt{\epsilon}}\sqrt{\frac{d}{d + \lambda_L(1 + \coth(s/\lambda_L)/\zeta(x,t))}}, \tag{16}$$

which is dependent on (x,t), and

$$F(x,t) = \frac{4\pi e^*\lambda_L^2}{\hbar c^2}\frac{\partial_x\zeta(x,t)}{\zeta(x,t)^2}[\frac{c}{4\pi\lambda_L}\coth(\frac{s}{\lambda_L})\tilde{B}_y(x,d,t) + \frac{I}{s}]. \tag{17}$$

Note that the function $F(x,t)$ has a finite value only inside the hot spot confined in a small spatiotemporal region, that is, $F(x,t)$ can be regarded as a source term which induces electromagnetic excitations propagating along the stripline, the Swihart mode. Hence, one may conclude that the emergence of a hot spot in the superconducting wire in CB-KID excites the Swihart mode inductively.

Let us now solve Eq.(15) in the presence of a hot spot, i.e., $F(x,t) \neq 0$ for $|x| < x_h/2$ and $0 < t < \tau$. Since we are interested in the asymptotic behavior of the solution for $|x| \to \infty$ and $t \to \infty$, we use the approximation, $v(x,t) \sim v$, Then, Eq.(15) is approximated as

$$[\frac{1}{v^2}\partial_t^2 - \partial_x^2]\theta(x,t) = F(x,t). \tag{18}$$

Suppose that a hot spot appears at $t = 0$. In this case the solution of Eq.(18) can be expressed as

$$\theta(x,t) = \theta_0(x,t) + \int_{-\infty}^{\infty} dx' \int_0^{\infty} dt' G(x - x'; t, t')F(x', t'), \tag{19}$$

where $\theta_0(x,t)$ is a solution of the homogenious equation, i.e., $[(1/v^2)\partial_t^2 - \partial_x^2]\theta_0(x,t) = 0$, and $G(x;t,t')$ is the Green function satisfying the equation,

$$[\frac{1}{v^2}\partial_t^2 - \partial_x^2]G(x;t,t') = \delta(x)\delta(t-t'). \tag{20}$$

In the present case we assume that $\theta_0(x,t)$ is the phase difference originating from the magnetic field induced by a constant bias current, that is, $\theta_0(x,t) = \theta_0(x) \propto x$, and $G(x;0,t') = 0$ because a hot spot is supposed to appear at $t = 0$. It is easy to solve Eq.(20) by using the Fourier transformation for the spatial variable x, i.e., $G(x;t,t') = \sum_k G_k(t,t')\exp(ikx)$. Here, the Fourier component, $G_k(t,t')$, is obtained as

$$G_k(t,t') = -\frac{v^2}{\omega_k}\Big\{\Theta(t'-t)\sin\omega_k t\cos\omega_k t' + \Theta(t-t')\cos\omega_k t\sin\omega_k t'\Big\}, \tag{21}$$

where $\omega_k = v|k|$ and $\Theta(t)$ is the Heaviside step function. Furthermore, using the Josephson relation $V(x,t) = -(\hbar/e^*)\partial_t\theta(x,t)$ (see Appendix), one can also derive the voltage difference from Eq.(19) as

$$V(x,t) = -\frac{\hbar}{e^*}\int_{-\infty}^{\infty} dx' \int_0^{\infty} dt' \partial_t G(x-x';t,t')F(x',t'). \tag{22}$$

The integral by x' in Eq.(22) can be performed analytically and then Eq.(22) yields the expression,

$$V(x,t) = V^{(-)}(x,t) + V^{(+)}(x,t), \tag{23}$$

where

$$V^{(-)}(x,t) = \frac{\hbar v^2}{4e^*}\Big\{\int_0^t dt'\Big[F(x+vt+vt',t') - F(x+vt-vt',t')\Big],$$
$$+\int_t^{\infty} dt'\Big[F(x+vt+vt',t') + F(x+vt-vt',t')\Big]\Big\} \tag{24}$$

and

$$V^{(+)}(x,t) = -\frac{\hbar v^2}{4e^*}\Big\{\int_0^t dt'\Big[F(x-vt+vt',t') - F(x-vt-vt',t')\Big]$$
$$+\int_t^{\infty} dt'\Big[F(x-vt+vt',t') + F(x-vt-vt',t')\Big]\Big\}. \tag{25}$$

Note that in the asymptotic region $t \to \infty$ the solution (23) is reduced to

$$V(x,t) = Y(x+vt) - Y(x-vt), \tag{26}$$

where

$$Y(z) = \frac{\hbar v^2}{4e^*}\int_0^{\infty} dt'\Big[F(z+vt',t') - F(z-vt',t')\Big]. \tag{27}$$

From this result one may conclude that a pair of voltage pulses with opposite polarities are created at a hot spot and they propagate in the directions opposite to each other with the Swihart velocity v. To clarify the polarity of the voltage pulses consider a case where the hot spot has a symmetric shape, i.e., $\zeta(-x,t) = \zeta(x,t)$, and the bias current I is large enough, i.e., $F(x,t) \simeq (4\pi e^*\lambda_L^2 I/\hbar c^2 s)[\partial_x\zeta(x,t)/\zeta(x,t)^2]$ (see Eq.(17)). In this case the spatial dependence of $F(x,t)$ satisfies the condition, $F(x,t) = -F(-x,t)$ and $F(x,t) \geq 0$ for $x > 0$. Thus, one finds

$$Y(0) = \frac{\hbar v^2}{4e^*}\int_0^{\infty} dt'\Big[F(vt',t') - F(-vt',t')\Big] = \frac{\hbar v^2}{2e^*}\int_0^{\infty} dt' F(vt',t') > 0, \tag{28}$$

which indicates that a positive voltage (negative) pulse travels in the negative (positive) direction with a constant velocity when the bias current flows in the positive direction. This result agrees with the polarities observed in CB-KID by Miyajima et al.[12], that is, our S-I-S waveguide model can correctly describe the operation of CB-KID. It is noted that the phase velocity derived from Eq.(15) shows no dispersion outside a hot spot, that is, a voltage pulse traveling in the stripline keeps its shape in the region outside the hot spot. This remarkable feature holds for a voltage pulse whose width is much greater than the inverse pair-breaking frequency, i.e., $\hbar/2\Delta$ with Δ and $\hbar$ being, respectively, the supercondcuting energy gap and the Planck constant, as discussed in [19]. Finaly, we also mention that the voltage given in Eq.(22) can be rewritten as

$$V \sim \frac{dL_k}{dt} \cdot I, \tag{29}$$

when the bias current is large. In Eq.(29) L_k is understood to be the kinetic inductance of the stripline and its explicit expression is given as

$$L_k(x,t) = \frac{4\pi\lambda_L^2}{c^2 s} \int_{-\infty}^{\infty} dx' \int_0^{\infty} dt' G(x-x', t-t') \frac{\partial_x \zeta(x',t')}{\zeta(x',t')}. \tag{30}$$

Thus, our theory also provides a microscopic basis of the circuit theory for CB-KID.

In conclusion we have constructed an electrodynamic theory for the generation and transmission of voltage pulses in the stripline in CB-KID, when the superconducting wire in the stripline is locally heated by irradiation. Electromagnetic signals traveling in the stripline have been identified with the Swihart mode in a S-I-S waveguide. The dynamical equation for the phase difference derived in this paper well describes the workings of CB-KID.

Acknowledgments
This work was partially supported by Grant-in Aid for Scientific Research (S) (No. 23226019) and (A) (No. 16H02450) from JSPS, Japan.

Appendix
Taking the time derivative of Eq.(1), we obtain the relation,

$$\begin{aligned} \partial_t\theta(x,t) &= \partial_t\varphi(x,d,t) - \partial_t\varphi(x,z_s,t) - \frac{e^*}{\hbar c}\int_{z_s}^{d} dz \partial_t A_z(x,z,t) \\ &= \frac{e^*}{\hbar}\int_{z_s}^{d} dz E_z(x,z,t) + \frac{e^*}{\hbar}[A_0^{\varphi}(x,d,t) - A_0^{\varphi}(x,z_s,t)], \end{aligned} \tag{31}$$

where

$$A_0^{\varphi}(x,z,t) = A_0(x,z,t) + \frac{\hbar}{e^*}\partial_t\varphi(x,z,t), \tag{32}$$

with $A_0(x,z,t)$ being the scalar potential. As discussed in [17], the gauge-invariant scalar potential A_0^{φ} can be neglected in conventional junction systems where the superconducting layers are much thicker than the Thomas-Fermi screening length. Then, defining the voltage difference as

$$V(x,t) = -\int_{z_s}^{d} dz E_z(x,z,t), \tag{33}$$

one finds the Josephson relation [18],

$$V(x,t) = -\frac{\hbar}{e^*}\partial_t\theta(x,t). \tag{34}$$

References

[1] Irwin K D, Hilton G C, Martinis J M, and Cabrera B 1995 *Rev. Sci. Instrum.* **66** 5311
[2] Chervenak A, Irvin K D, Grossman E N, Martinis J M, Reintsema C D, and Huber M E 1999 *Appl. Phys. Lett.* **74** 4043
[3] Goltsman J A, Okunev O, Chuikova G, Lipatov A, Semenov A, Smirnov K, Voronov B, Dzardanov C A, Williams C, and Sobolewski R 2001 *Appl. Phys. Lett.* **79** 705
[4] Miki S, Yamashita T, Fujiwara M, Sasaki M, and Wang Z 2010 *Optics Letters* **35** 2133
[5] Mazin B A, Day P K, Zmuidzinas J and Leduc H G 2002 *AIP Conf. Proc.* **1** 309
[6] Kurakado M and Mazaki H 1980 *Phys. Rev.* B **22** 168
[7] Kurakado M and Mazaki H 1981 *Nucl. Instr. Meth.* **185** 141
[8] Kurakado M and Mazaki H 1981 *Nucl. Instr. Meth.* **185** 149
[9] Ishida T, Yagi I, Yoshioka N, Huy H T, Yotsuya T, Satoh K, Uno M, Shimakage H, Miki S, Wang Z, Kato M, Machida M and Hojou H 2008 *J. Low Temp. Phys.* **151** 1074
[10] Shibata H 2014 *Appl. Phys. Express* **7** 103101
[11] Shishido H, Yoshida T and Ishida T 2015 *Appl. Phys. Express* **8** 113101
[12] Miyajima S, Shishido H, Narukami Y, Yoshioka N, Fujimaki A, Hidaka M, Oikawa K, Harada M, Oku T, Arai M and Ishida T 2017 *Nucl. Instr. Meth.* A **842** 71
[13] Narukami Y, Miyajima S, Fujimaki A, Hidaka M and Ishida T 2015 *IEEE. Trans. Appl. Supercond.* **25** 2400904
[14] Narukami Y, Oikawa K, Harada M, Oku T, Arai M, Hidaka M, Fujimaki A and Ishida T 2015 *Appl. Phys. Lett.* **107** 23260
[15] Swihart J 1961 *J. Appl. Phys.* **32** 461
[16] Iizawa T et al., this conference.
[17] Koyama T and Tachiki M 1996 *Phys. Rev.* B **54** 16183
[18] $e^* > 0$ is assumed in this paper.
[19] Kautz R L 1978 *J. Appl. Phys.* **42** 308

IOP Conf. Series: Journal of Physics: Conf. Series **1054** (2018) 012056 doi:10.1088/1742-6596/1054/1/012056

Physical characteristics of delay-line current-biased kinetic inductance detector

***Yuki Iizawa[1], Hiroyuki Yamaguchi[1], Yuya Miki[1], Kazuma Nishimura, Hiroaki Shishido[1,2], Kenji M. Kojima[3,4], Kenichi Oikawa[5], Masahide Harada[5], Shigeyuki Miyajima[2,6], Mutsuo Hidaka[7], Takayuki Oku[5], Kazuhiko Soyama[5], TomioKoyama[8], and Takekazu Ishida[1,2]**

[1]*Department of Physics and Electronics, Osaka Prefecture University, Sakai, Osaka 599-8531, Japan*

[2]*NanoSquare Research Institute, Osaka Prefecture University, Sakai, Osaka 599-8531, Japan*

[3]*Muon Science Laboratory and Condensed Matter Research Center, Institute of Materials Structure Science, KEK, Tsukuba, Ibaraki 305-0801, Japan*

[4]*Department of Materials Structure Science, The Graduate University for Advanced Studies, Tsukuba, Ibaraki 305-0801, Japan*

[5]*Materials and Life Science Division, J-PARC Center, Japan Atomic Energy Agency, Tokai, Ibaraki 319-1195, Japan*

[6]*Advanced ICT Research Institute, NICT, Kobe, Hyogo 651-2492, Japan*

[7]*National Institute of Advanced Industrial Science and Technology (AIST), Tsukuba*

[8]*Institute for Materials Research, Tohoku University, Sendai, Miyagi 980-8577, Japan*

Abstract Our preceding works reported the development of a high spatial and temporal resolution imaging system by using a current biased kinetic inductance detector (CB-KID) with the use of a delay-line technique. We called this system as a delay-line CB-KID, and succeeded in imaging of neutron events caused by the nuclear reaction and hot spots produced by using the delay-line CB-KID system. It was essentially important for our proposal to use a superconducting stripline to guide the pulsed signal where the signal propagates at a constant fast velocity along the stripline. In the present study, we intend to measure a propagation velocity of the signal along the stripline precisely to compare with the theoretical prediction of the signal propagation, which was recently developed with a superconducting waveguide S-I-S model by Koyama and Ishida [11]. Our present measurements showed a good agreement between the theoretical predictions and the experimental results on the propagation velocity as a function of temperature.

1. Introduction

Neutron imaging has been very useful for both scientific research and various applications. For example, neutron diffraction can determine the atomic structure or magnetic structure of various interesting materials [1]. We can observe the dynamics of carbons and water molecules in fuel cells and so forth [2]. There are still strong demands to improve the specifications of neutron detectors. On the other hand, it has been known to date that various sorts of superconducting detectors have been quite successful compared to conventional detectors. For example, a transition edge sensor (TES) [3], a superconducting nanowire single photon detector (SNSPD) [4], and a microwave kinetic conductance detector (MKID) [5] are recognized to work quite excellently in the communities. Generally, it is well known that superconducting detectors have features of high sensitivity, fast response and high-energy resolution. We may say that it was not so often to utilize a superconducting detector in measuring neutrons.

Recently, a TES with a B neutron absorption layer was proposed to use as a neutron detector [6]. Our group also proposed a superconducting neutron detector for intending a high spatial and temporal resolution imaging system of neutrons using a current biased kinetic inductance detector (CB-KID) with a ^{10}B neutron absorption layer [7,8,9] and the idea of a delay-line CB-KID technique [10]. A delay-line CB-KID is to use a signal transmission along the superconducting stripline as a delay line, where information on position can be derived from a difference in arrival timestamps. We need two conditions so as to obtain reliable positional data. One is to satisfy a constant propagation velocity for conducting imaging during the CB-KID measurements because the spatial resolution depends on the accuracy of the time and the propagation velocity. The other is to understand the physical basis of the delay-line CB-KID operation.

Recently, a theory of the signal propagation on a superconducting waveguide having an S-I-S structure has been proposed by Koyama and Ishida [11] within the framework of the London-Maxwell theory. The theory gave us a basis of operating mechanism of the CB-KID, and predicted a propagation velocity as a function of temperature. This model can be tested by the experiment. In the present work, for this purpose, we experimentally determined the signal propagation velocity along the CB-KID stripline as a function of temperature to test the validity of the Koyama-Ishida theory [11].

2. Current biased kinetic inductance detector

We briefly explain our CB-KID [7,8,9], which consists of a superconducting Nb ground plane, a superconducting Y meanderline of Nb wire, a superconducting X meanderline of Nb wire, and a ^{10}B conversion layer from neutrons to charged particles. The nuclear reaction ^{10}B(n, ^{4}He)^{7}Li mainly emits a ^{4}He particle of 1.47 MeV and a ^{7}Li particle of 0.88 MeV, and the local energy dissipation to the meanderline provided by each projectile is used to create a hot spot on the Nb nanowire stripline. The signal is caused by the hot spot when the stripline is biased by the DC current while the hot spot is caused by the nuclear reaction between a ^{10}B nucleus and a single neutron.

The primitive explanation of the operating principle of the CB-KID is as follows. A meanderline is cooled down to a temperature lower than a transition temperature T_c of the Nb wire. The inductance L in superconducting meanderline is expressed as

$$L = L_m + L_k + \Delta L_k \ , \tag{1}$$

where L_m is the magnetic inductance, L_k is the kinetic inductance but for a possible hot spot area, and ΔL_k is a tiny fraction of kinetic inductance from a region, where a hot spot will be created. The magnetic inductance L_m is given by

$$L_m = \frac{\mu \ell}{4\pi} \ , \tag{2}$$

where μ is the magnetic permeability and ℓ is the total length of the stripline. The kinetic inductance L_k is given by

$$L_k = \frac{m_s \ell}{n_s q_s^2 S} \ , \tag{3}$$

where m_s is the effective mass of the Cooper pair, q_s is the electric charge of the Cooper pair, n_s is the Cooper pair density, ℓ is the total length of stripline, and S is the cross-sectional area of stripline. When some amount of energy is given to the superconducting meanderline, it produces a hot spot in a tiny fraction ($\Delta l \ll \ell$) of the meanderline. In the hot spot, the Cooper pairs are destroyed in a restricted regime of the stripline, and this induces a local change in the kinetic inductance (ΔL_k) because the number of the Cooper pairs are instantaneously reduced in the limited region. If a DC-bias current flows through the entire meanderline, a transient change in the kinetic inductance causes a pair of voltage pulses, which propagate as a pair of electromagnetic waves propagating toward both ends of the Nb stripline with opposite polarities. A voltage pulsed signal $V(t)$ initially appears at the hot spot as expressed by

$$V(t) = RI_b + \frac{d(L_m + L_K + \Delta L_k)I_b}{dt} \approx I_b \frac{d(\Delta L_k)}{dt} \tag{4}$$

where the magnetic inductance L_m does not alter with time because L_m is described by the permeability as in equation (2), and the DC bias current does not change at all ($dI_b/dt \cong 0$) as far as the detector is kept in the superconducting zero-resistance state. However, the sign of the local voltage signal depends on the direction of the bias current I_b in equation (4).

In our previous work [7,8,9], we fabricated a neutron detector using Nb-based CB-KID coated with a ^{10}B conversion layer. The reaction of $^{10}B(n,\alpha)^7Li$ releases energy by 1.02 MeV or 0.84 MeV with a ^7Li particle and by 1.78 MeV or 1.47 MeV with an α particle. The superconducting meanderline generates a pair of voltage signal pulses when the superfluid density changes by the nuclear reaction in a limited tiny segment.

3. Imaging from a quartet set of travelling signals from delay-line CB-KID

The delay-line CB-KID is able to image the distribution of hot spots on the detector. Since the present studies are related with the imaging capability of the CB-KID, we will explain the principle of imaging. Figure 1 (a) shows the Nb nanowire in CB-KID. One can see a clear image of the crossing area of the *X* and *Y* meanderlines. Both the *X* and *Y* meanderlines consist of 10000 stripline segments of 0.9-μm width, 15.1-mm length, and 0.6-μm spacing. The repetition period is 1.5μm and total length of meanderline reaches 151 m.

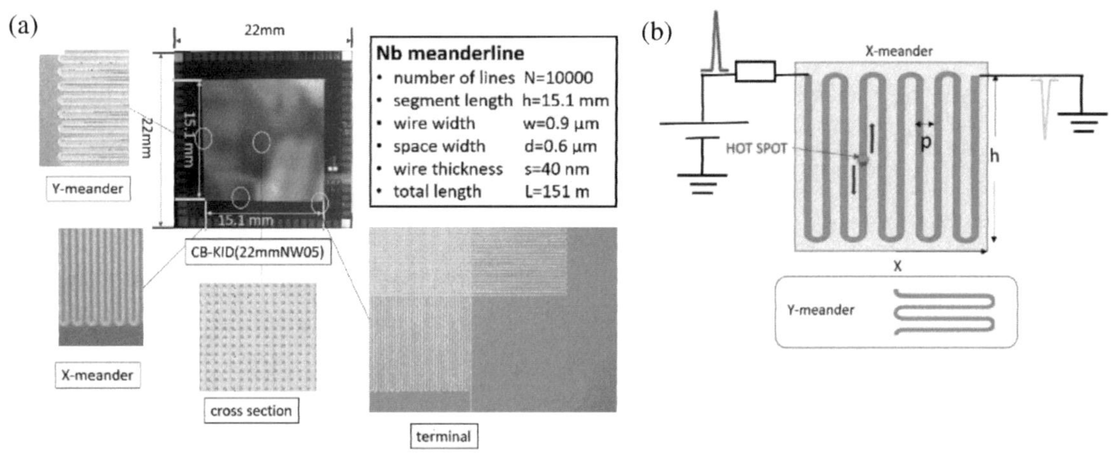

Figure 1. (a) The photograph of the entire CB-KID device. We also show the laser-microscope photographs of the edge of *Y* meanderline, the edge of *X* meanderline, the crossing region of the *X* and *Y* meanderlines, and the region of the end electrode of *X* meanderline. We also list the fundamental parameters of our CB-KID. (b) Schematic view of a delay-line CB-KID imaging system. The positive signal goes to the upstream side of the bias current while the negative signal goes toward the downstream side.

Figure 1 (b) shows schematic view of our delay-line CB-KID imaging system. As we explained in section 2, a pair of voltage pulses with opposite polarities appear when a hot spot occurs in the meanderline and propagate as electromagnetic wave toward both ends along the Nb stripline. A negative pulse propagates along the same direction to the DC-bias current flow while a positive pulse propagates to the opposite direction. Both pluses propagate through meanderline and arrive at one of the ends of the meanderline. Both arrival times at the ends of meander are different from each other depending on the travelling lengths of the two signals. We call this deferent time as a delay time Δt. In case of figure 1 (b), an arrival time of positive pulse is t_1, an arrival time of negative pulse is t_2, and a delay-time Δt

is $t_1 - t_2$. We can calculate a position from a delay time Δt and a propagation velocity v. A hot spot position in the meanderline can be determined as follows:

$$x = \mathrm{ceil}\left\{\frac{\Delta t/2 \times v}{h}\right\} p \,, \quad (5)$$

where h is the length of each segment of the meanderline, p is a repetition pitch for the meanderline. The "ceil" is a ceiling function, which works to convert the signal propagation length to the number of segments required for signal traveling. The Y position also can be determined in a similar manner, and hence we can image the positions of the mesoscopic excitations in the two-dimensional plane by using the very limited number (four) of leads for the readout circuits.

According to equation (5), we recognize that the propagation velocity v of the meanderline is directly relevant to the accuracy of the X positon obtained in our method. Therefore, it is very important to know the propagation velocity as precisely as possible. In addition, we have to understand the physical mechanism of the signal generation and transmission in CB-KID.

4. S-I-S structure of delay-line CB-KID and propagation velocity

Recently, Koyama and Ishida gave a theory of the signal generation and the propagation in CB-KID, and will be reported elsewhere [11]. We briefly summarize an essence of the theory to compare with the present experimental results.

A delay-line CB-KID consists of a superconducting Nb meanderline layer, a SiO_2 insulator layer, and a superconducting Nb ground plane layer. According to this configuration, the signal transmission is very efficient along the stripline due to the impedance matching in superconducting circuit. The schematic modelling view of a part of the stripline in our delay-line CB-KID is depicted in Figure 2. We regard this stripline as an S-I-S waveguide structure. On the basis of this modelling, Koyama and Ishida treated the system to obtain the signal generation and propagation on the basis of the London-Maxwell theory, and obtained an analytic formula of propagation velocity as a function of temperature [11]. The CB-KID is modeled as a structure of S-I-S while no tunneling currents are flowing due to a thicker layer of the insulator. Therefore, the system is very different from the Josephson junction (JJ) where the Cooper

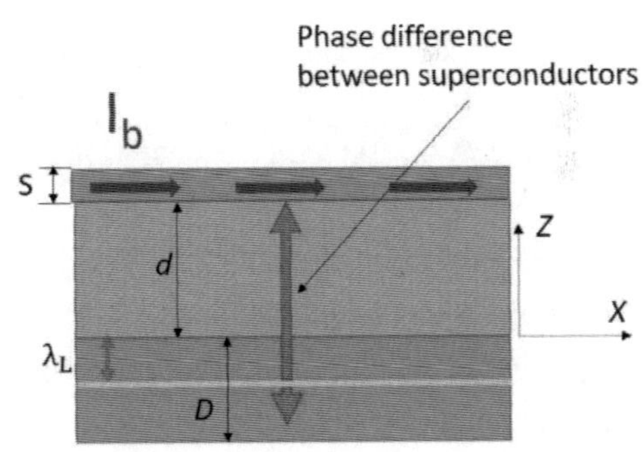

Figure 2. S-I-S structure schematic view of our delay-line CB-KID.

pairs tunneling is important, but it is useful to treat the dynamics of the system in terms of a superconducting phase difference as in JJ. The gauge-invariant phase difference $\theta(x,t)$ between the stripline and the ground plane is defined by

$$\theta(x,t) = \varphi(x,d,t) - \varphi(x,-\infty,t) - \frac{e^*}{\hbar c}\int_{-\infty}^{d} dz A_Z(x,z,t) \quad , \tag{6}$$

where $\varphi(x,z,t)$ is the superconducting phase at (x,z,t), $A_Z(x,z,t)$ is the z component of the vector potential, d is a thicknesses of insulating and superconducting layers and $e^* = 2e$ is the charge of a Cooper pair. $\theta(x,t)$ is the gauge-invariant phase difference between a point on the lower surface of the superconducting wire ($z = d$) and a point at a deep inside the superconducting ground plane. Following a procedure, which is often used in the theory of Josephson effect, one can drive the equation of motion for $\theta(x,t)$ as follows:

$$\frac{\lambda_L + d}{d}\frac{1}{c^2}\frac{\partial^2\theta(x,t)}{\partial t^2} - \frac{\partial^2\theta(x,t)}{\partial x^2} = \frac{e^*}{\hbar c}\frac{\partial}{\partial x}\left[A_x(x,d,t) - \frac{\hbar c}{e^*}\frac{\partial\varphi(x,d,t)}{\partial x}\right] \tag{7}$$

where the right-hand-side term of equation (7), $\frac{\partial}{\partial x}\left[A_x(x,d,t) - \frac{\hbar c}{e^*}\frac{\partial}{\partial x}\varphi(x,d,t)\right]$, arises from the current flowing in the lower surface of the superconducting nanowire. Using the London equation and the Maxwell equations, they obtained the wave equation from equation (7) as follows:

$$\frac{1}{v^2}\frac{\partial^2\theta(x,t)}{\partial t^2} - \frac{\partial^2\theta(x,t)}{\partial x^2} = 0 \tag{8}$$

where the velocity v is given as

$$v = \frac{c}{\sqrt{\varepsilon}}\sqrt{\frac{d}{\{d+\lambda_L(1+\coth(s/\lambda_L))\}}} \quad , \tag{9}$$

where c is the speed of light in a vacuum, ε is the dielectric constant of insulating layer, s is the thickness of superconducting nanowire and λ_L is the London penetration depth. Note that the velocity v is a function of the London penetration depth λ_L, which is dependent on temperature. Therefore, the propagating velocity v is expected to vary with temperature. Assuming a typical BCS superconductor, we consider that the London penetration depth λ_L approximately obeys the temperature dependence of the two-fluid model as follows:

$$\lambda_L(T) = \frac{\lambda_L(0)}{\sqrt{1-(T/T_C)^4}} \tag{10}$$

where T_c is a transition temperature. Note that equation (9) is different from the naïve prediction of the velocity in the stripline as

IOP Conf. Series: Journal of Physics: Conf. Series **1054** (2018) 012056 doi:10.1088/1742-6596/1054/1/012056

$$v = \sqrt{\frac{1}{L_k^0 \, C_s}} \quad , \tag{11}$$

where L_k^0 is the unit-length kinetic inductance and C_s is the unit-length capacity of the microstripline. Therefore, the precise experiment to determine the propagation velocity is necessary to provide the detailed data to compare.

5. Experimental apparatus to evaluate propagation velocity

5.1 Measurement system of propagation velocity

We plan to conduct systematic measurements of the propagation velocity to examine the validity of the London-Maxwell equation of the CB-KID [11]. Figure 3 shows the schematic view of measurement system of propagation velocity. Because our detector operates only at temperatures lower than T_c , CB-KID was cooled down by using a Gifford-McMahon (GM) cryocooler. A Cernox thermometer was placed in the neighborhood of the detector. The detector temperature was controlled by using a temperature controller, and was monitored the temperature and its stability through a GP-IB interface with the aid of a personal computer. We need a pulse signal of a narrow width to measure the propagation velocity precisely. For this purpose, we used a passive differentiation circuit to generate a sharp pulse from a rectangular wave of moderate width. The doublet pulse with a positive polarity and a negative polarity were generated by this method. We use a signal splitter to feed the pulses simultaneously to the *X*-meanderine as well as to the *Y*-meanderline. This signal is also fed into a Nb short wire on the same chip of CB-KID. The third signal is measured to compensate a lag time when the signal propagates along the semi-rigid cables wired inside of the cryostat, where all the semi-rigid cables were installed so as to have the same length. The output signals after transmitting the meanderline of CB-KID were amplified by the identical low-noise amplifiers located at room temperature. The gain of amplifier was nominally 46 dB with the bandwidth from 1 kHz to 100 MHz. The output signal amplified by the preamplifier goes to a digital oscilloscope with a 0.4-ns sampling period, and the whole signals were stored in the storage hard disk of the computer. A time difference between the signal passed through meanderline and the signal passed through the reference Nb short nanowire is determined by the post data analysis. Note that this method reasonably compensates a possible systematic error coming from a time in propagating the semi-rigid cable in the

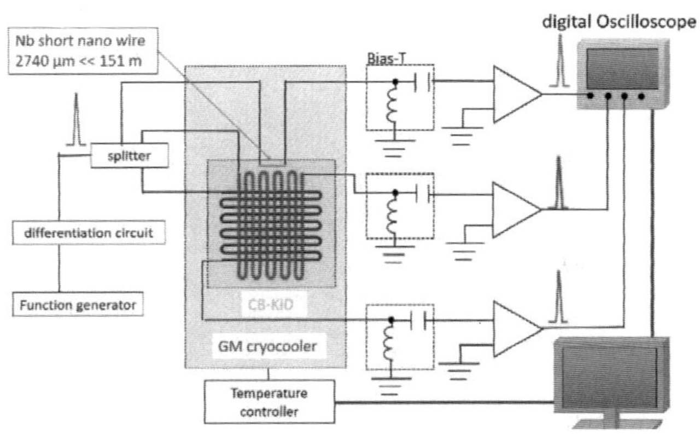

Figure 3. Schematic view of measurement system of propagation velocity SQUID microscope. (b)

cryostat. The propagation velocity was thus obtained from a time difference and the total length of the meanderline.

5.2 Experimental results of propagation velocity

Figure 4 shows results of propagation velocity measurements and data points are fitted by the theoretical equations (9) by assuming the two-fluid model of equation (10).

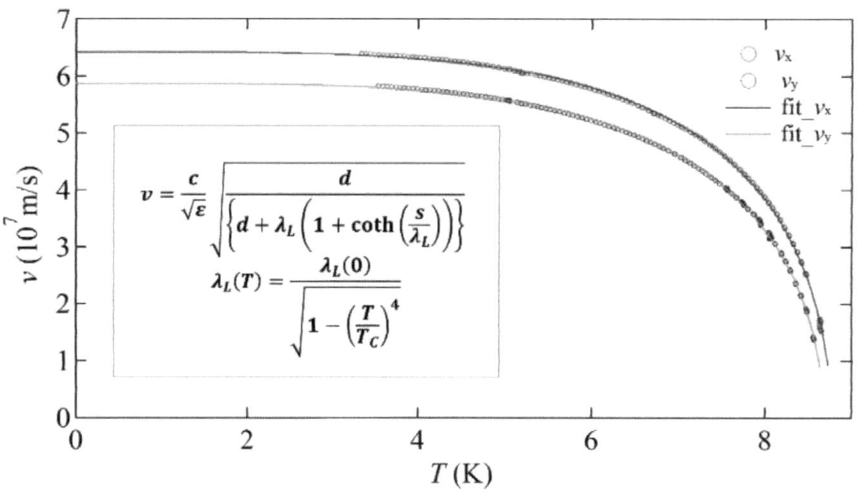

Figure 4. Propagation velocity as a function of temperature. The data points are fitted well by the least-squares method to the formula (9) of the Koyama-Ishida theory [11]. We list the parameters used in the least-squares fitting as the fixed parameters (c=2.99792458×10^8 m/s, d_x=350 nm, d_y=300 nm, s =40 nm) and the fitting parameters (T_c = 8.7643±0.0007 K, $\lambda(0) = 2.6366 \pm 0.00147 \times 10^{-7}$ m, $\epsilon = 2.8271 \pm 0.0261$ for the X detector, T_c=8.6799±0.0008 K, $\lambda(0) = 3.00241 \pm 0.00241 \times 10^{-7}$ m, $\epsilon = 2.31159 \pm 0.0323$ for the Y detector.)

In the least squares method, the speed of light in vacuum c, the thickness of the insulating layer d, and the thickness of the superconducting nanowire s were taken from the design parameters, the transition temperature T_c, the London penetration depth $\lambda(0)$ at 0 K, and the dielectric constant ϵ of insulating layer were taken as fitting parameters. We find that the propagation velocity depends on temperature. The experimental data obtained by the propagation velocity measurements were fitted quite nicely as predicted by Koyama and Ishida [11]. The agreement is much better than the prediction of equation (11). Parameters to fit the data are listed in Figure 4. We conclude that the London-Maxwell theory developed by Koyama and Ishida provides sound explanations of the operation principle of our CB-KID.

We notice that Zhao *et al.* [12] reported on delay-line imaging for single photon detector. Their system has a length of 19.7 mm and the signal propagates at a 2% speed of light velocity, which are in marked contrast with our work. Our system extend to 151 m more than 7500 times longer than their stripline, and the signal propagates at a 20-% of light velocity. Our system needs a fabrication technique enough to obtain a reliable device. Fast propagation velocity is also related with achieving a high-speed electronics in the future. Besides, our system works at the superconducting state while their system was supposed to be relevant to the resistive transient change in producing the systems. We have the double layers of the *X* detector and the *Y* detector employed in the present study. This is an essential difference from the method of Zhao *et al.* [12], and is inevitable to achieve a higher spatial resolution.

Further studies are in progress to investigate the bias current dependence of the propagation velocity to understand the oprating mechanism. We consider that the CB-KID operation would be closely related with the operating principle of the superconducting nanowire single photon detector (SNSPD). .

6. Conclusion

Our group previously proposed the idea of two-dimensional superconducting imaging method on the basis of the delay-line CB-KID method. It is characteristic of the method that works at high spatial resolution and high temporal resolution by using only the four-channel readout circuits. We emphasize that our system is extremely simpler than any other superconducting imagers. Koyama and Ishida [10] developed a theory based on the London-Maxwell theory, which gives not only the wave propagation equation but also the temperature dependence of the propagation velocity. In order to verify the theoretical validity of the Koyama-Ishida model, we carried out systematic measurements of the propagation velocity in the present work. Our results showed that the propagation velocity determined experimentally gives excellent agreement with the theoretical predictions by the London-Maxwell theory. We conclude that the Koyama-Ishida model [11] gives sound confirmation of the operating principle of the CB-KID mechanism.

Acknowledgement

This work is supported by Grant-in-Aid for Scientific Research (S) No. 23226019, and Grant-in-Aid (A) No.16H02450 from JSPS. The devices were fabricated in the clean room for analog-digital superconductivity (CRAVITY). This work is partially benefitted by the use of VDEC, the University of Tokyo with the collaboration with Cadence Corporation. This work is supported of MLF program (Proposal No. 2015A0129, No. 2015P0301).

References

[1] Rodríguez-Carvajal M J 1993 *Physica B: Phys. of Condensed Matter* **192**, 55

[2] Bellows R J, Lin M Y, Arif M, Thompson A K, and Jacobson D 1999 *Journal of The Electrochemical Society* **146**, 1099

[3] Chervenak J A, Irwin K D, Grossman N G, Martinis J M, Reintsema C D, and Huber M E 1999 *Appl. Phys. Lett.* **74**, 4043

[4] Miki S, Yamashita T, Fujiwara M, Sasaki M, and Wang Z 2010 *Optics Letters*, **35**, 2133

[5] Mazin B A, Day P K, Zmuidzinas J, Leduc H G 2002 *AIP Conf. Proc.* **1**, 309

[6] Merlo V, Salvato M, Cirillo M, Lucci M, Ottaviani I, Scherillo A, Celentano G, and Pietropaolo A 2015 *Appl. Phys. Lett.* **106**, 113502

[7] Ishida T, Nishikawa M, Fujita Y, Okayasu S, Katagiri M, Satoh K, Yotsuya T, Shimakage H, Miki S, Wang Z, Machida M, Kano T, and Kato M 2008 *J. Low Temp. Phys.* **151**, 1074

[8] Yoshioka N, Yagi I, Shishido H, Yotsuya T, Miyajima S, Fujimaki A, Miki S, Wang Z, and Ishida T 2013 *IEEE Trans. App. Supercond.* **23**, Art. ID. 2400604

[9] Shishido H, Miyajima S, Narukami Y, Oikawa K, Harada M, Oku T, Arai M, Hidaka M, Fujimaki A, and Ishida T. 2015 *Appl. Phys. Lett.* **107**, 232601

[10] Shishido H, Miki Y, Yamaguchi H, Iizawa Y, Kojima K M, Oikawa K, Harada M, Oku T, Soyama K, Miyajima S, Hidaka M, and Ishida T 2017, 17th international workshop on Low Temperature Detectors (LTD17) July 16-21, 2017, Kurume, Japan

[11] Koyama T and Ishida T, this conference, to be published

[12] Zhao Q Y, Zhu D, Calandri N, Dane A E, McCaughan A N, Bellei F, Wang H-Z, Santavicca D F and Berggren K K 2017 Nature Photonics **11**, 247-251

Superconducting devices based on coherent operation of Josephson junction arrays above 77K

Boris Chesca

Department of Physics, Loughborough University, Loughborough, LE11 3TU, United Kingdom

E-mail: B.Chesca@lboro.ac.uk

Abstract. Arrays of Josephson junctions operating coherently at 77K seem to be the ideal candidates for re-shaping the future of the electronics industry. Their advantages over their semiconducting counterparts (higher operating speed, lower power consumption/electronic noise) can be exploited in practice because of their practicality: cooling down to 77K is both cheap and easy to handle. These developments naturally fit into the increasing interest of some semiconductor technologies in lower operation temperatures. Therefore, it looks attractive to link semiconductor and 77K superconductor technologies to improve system level performance. Currently, however, in the vast majority of applications 4.2K superconducting technology is used as it provides significantly superior performances relative to devices build in the 77K technology. Recent significant improvements in the performance of superconducting devices operating as 77K, as well as, in their fabrication technologies may change all that. Several such examples will be considered here. Firstly, when coherency is achieved in large series SQUID-arrays magnetic flux sensors or voltage amplifiers can be build having record values for their output voltage and flux noise sensitivities outperforming even single SQUID-based devices operating at 4.2 K. Secondly, when coherency is achieved in parallel SQUID-arrays placed in a uniform magnetic field B, B-field tuned microwave generators can be build. Large parallel SQUID arrays were also implemented to achieve record values for current amplification at 77K, highly efficient ratchets with unidirectional magnetic vortices motion and integrated nano-magnetic sensors.

1. Introduction

Since Feynman's prediction [1] that multiple Josephson junction (JJ) devices would eventually improve performance of superconducting devices there has been considerable effort spent to achieve that. Such JJ- or SQUID (Superconducting Quantum Interference Devices) arrays are very attractive in many applications where single-JJ or single-SQUID based devices are routinely used as: magnetic sensors, low-noise amplifiers, THz receivers, analog-to-digital converters, antennas, filters, etc. Because of their practicality (cooling down to 77K is both cheap and easy to handle) and their potential to penetrate the market dominated by the semiconductor technology here my focus is exclusively on arrays operating at 77K.

There are two main challenges for a successful implementation. First challenge is the fabrication reliability of large JJ- or SQUIDs-arrays. Currently, the most reliable are the bicrystal [2] and the step-edge [3] technologies. JJ- and SQUID-arrays fabricated in the bicrystal technology showed the best results so far. However, this technology lacks the flexibility of the 2D step-edge technology. SQUIDs have been fabricated along 3 different directions in a tri-crystal configuration [4] and therefore M

IOP Conf. Series: Journal of Physics: Conf. Series **1054** (2018) 012057 doi:10.1088/1742-6596/1054/1/012057

multi-crystal technology (with M= 4, 5, ...) is potentially an interesting direction to explore in the future in order to gain a similar 2D flexibility required for the implementation of complex JJ- and SQUID-arrays architectures. Recently, significant progress has been reported in the area of step-edge junction array technology [5, 6] that offers the advantage of using low cost MgO substrates and the flexibility of implementing complex 2D large array configurations involving many tens of thousands of SQUIDs.

The second challenge is the ability to operate large JJ- or SQUIDs-arrays coherently at 77K. Recently, coherent operation has been achieved in both large series- and parallel-SQUIDs arrays leading to record values relative to single-SQUID devices for flux sensitivity [7], electromagnetic power radiation in the GHz range [8], current amplification [9] and flux-ratchet tunability [10]. To emphasize that SQUID-arrays at 77K can indeed replace single-SQUIDs operating at 4.2K in many applications, in this report a unified view is given on these four significant results.

It is important to notice that contrary to previous beliefs even a considerable spread in the properties of individual junctions will not affect the overwhelming superiority of arrays over single junctions devices. Indeed, in [7] the intrinsic spread of Josephson critical currents in large arrays has been measured and fitted with a Gaussian distribution characterized by an average standard deviation σ of about 12.2% at 77K. This is significantly smaller than at 4.2 K where σ values around 25% are typical. Large arrays with additional artificially implemented spread in the SQUID areas/inductances as large as 30% have also been fabricated [7]. The flux-coherence persisted even in this case. This is because in large arrays the gains of a global response by many junctions outweighs the losses due to differences between individual junctions.

2. Series SQUID arrays

SQUIDs being the most sensitive magnetic sensors and amplifiers are routinely used in many areas, among them: medicine (magneto- and cardio-encephalography), research in material science (as magnetometers measuring magnetic properties of materials), geology, astronomy, magnetic imaging applications, electronics, etc. At present, such devices made of low temperature superconductor (LTS) SQUIDs operating at 4.2K are being used in the vast majority of applications due to their superior flux noise performance. This is despite several significant advantages offered by high temperature superconductor (HTS) SQUIDs operating at 77K: low cost and user-friendly cooling procedures, and also potential superiority as magnetic imaging devices due to lesser thermal insulation demand, i.e., reduced separation between the sensors and the room temperature object under study. The question is then: can devices based on HTS SQUIDs and operating at 77K ever get better than their LTS counterparts based on single-SQUIDs and operating at 4.2K?

A very promising approach is to build a series SQUID array (SSA) of N non-interacting SQUIDs operating flux-coherently, because in this case [10-13] the voltage modulation ΔV linearly scales with N, the dynamic range increases as $N^{1/2}$, whereas the white flux noise $S_\Phi^{1/2}$ decreases as $1/N^{1/2}$. Consequently not only the noise properties of a SSA are superior to a single-SQUID but a much larger ΔV means their matching to room temperature readout is greatly simplified since the array impedance is N times larger than that of a single SQUID. Moreover, SSAs have the potential of also improving the bandwidth. Some of these predictions have been largely confirmed when large N (typically N is in the range 100-200) SSAs operating at 4.2K have been developed in LTS technology [10-16]. However, flux-coherency and SQUIDs non-interactivity have proved to be very challenging to achieve in large N SSAs made of HTS and operated at 77K. Indeed, it has only been achieved in relatively small N SSAs (N in the range 10-30) [17-18]. With N small, their superiority over single SQUIDs was less spectacular and they could not compete in terms of noise performance with single LTS SQUIDs operated at 4.2 K. Earlier attempts to operate large N HTS SSAs (N in the range 50-130) did not show the expected improvements in magnetic sensitivity, because the flux-coherent mode was not achieved throughout the entire array [19-20].

Recently [7], we reported on the design, fabrication and testing of a new generation of very large (N=484 and N=770) non-interacting SSAs made of $YBa_2Cu_3O_7$ (YBCO) and operating flux-coherently (see Figure 1). The SSAs were fabricated using the well-established bicrystal technology. To increase field sensitivity SQUIDs are usually connected to rectangular flux-focusers with both their dimensions much larger than the SQUID width (total width of SQUID hole and two junctions). In order not to compromise the number of SQUIDs we could integrate on a standard 10x10 mm^2 bicrystal substrate while still implementing flux-focusers for enhanced sensitivity, we developed *large area narrow flux-focusers*. Their width is identical to the SQUID's width while they are much longer. Importantly, our results showed that the larger the area of such narrow flux focusers the higher the degree of flux coherence in the operation of SSAs and SQUIDs non-interactivity within the SSAs.

Families of current-voltage characteristics were measured for various fields B and at different temperatures in the range 10 to 89 K. Voltage could be measured along the entire array or different sections of it. From such families, $V(B)$ could be constructed. Large amplitude SQUID-like oscillations with a flux quantum periodicity were observed (see inset in Figure 2). Unlike the case of single SQUIDs, $V(B)$ is amplitude modulated and suppressed to nearly zero within about 20 periods. This is well understood [5] as being a consequence of a slight variation in the periods of individual

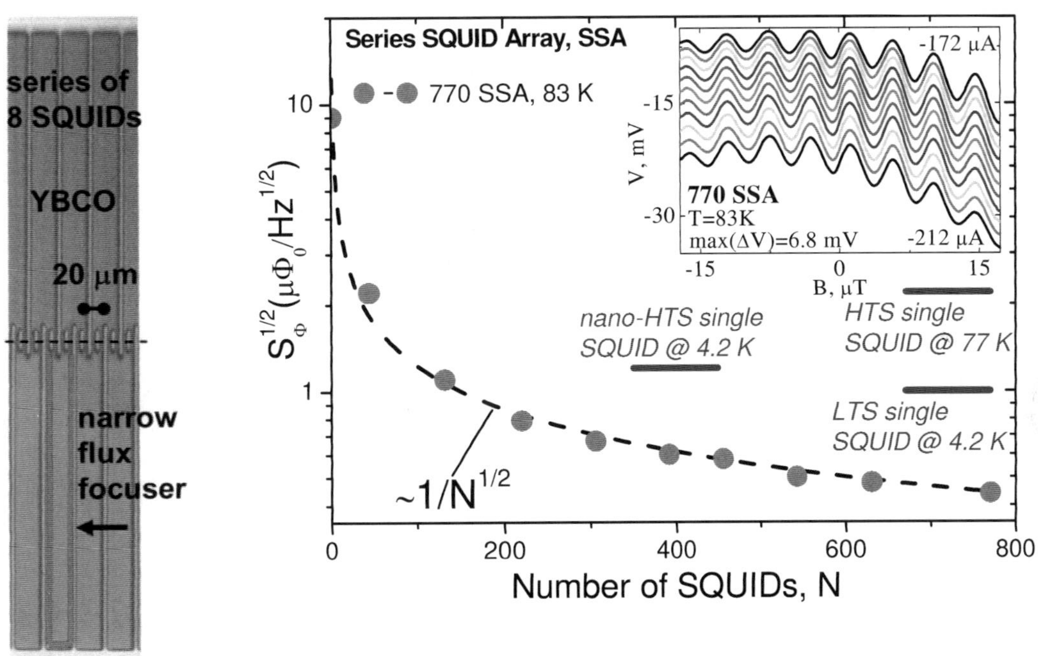

Figure 1. Left: optical micro-picture of a small part of a 770 series SQUID array showing 8 SQUIDs. The array is fabricated by optical lithography after depositing a $YBa_2Cu_3O_7$ thin film on a $SrTiO_3$ bicrystal. Each SQUID consists of two Josephson junctions connected in parallel. The junctions can be seen as 3 micron-wide narrow bridges (in brown) crossing the bicrystal boundary shown with a horizontal dotted line. Each pair of successive SQUIDs is connected in series via inductive large area narrow flux-focusers (one is highlighted in red). Right: flux density noise $S_\Phi^{1/2}$ at 1kHz versus the number of SQUIDs integrated in the 770 SQUID-array. The discontinuous line indicate the theoretical $1/N^{1/2}$ dependence. Typical flux noise levels of optimized HTS SQUIDs at 77K, nano-HTS SQUIDs at 4.2 K and LTS SQUIDs at 4.2 K are also shown as references. Inset: *V(B)* of the 770 SSA at 83K for 11 different current biases I in the range (-212,-172) μA with I changing in steps of 4μA;

SQUIDs along the array due to either variation in the SQUID loop sizes or/and non-uniform magnetic field coupling over the length of the array. This affects the coherency of the array at large applied fields. For the 770 SSA ΔV ranges from 10.8 mV at 77K to 2.2 mV at 86K. The *V(B)* at 83K is shown in the inset of Figure 2. The SSA's values of $S_\Phi^{1/2}$ in the temperature range (40-83)K were about 50 times lower than for optimized single HTS SQUIDs operating at 77K and even lower than those of single LTS SQUIDs operating at 4.2K (see Figure 2). This strongly suggests that HTS SSAs are ideal candidates to replace single LTS SQUIDs in many applications. We found [7] that to a good approximation ΔV linearly increases with N, whereas $S_\Phi^{1/2}$ decreases as $1/N^{1/2}$ suggesting that flux-coherency and SQUIDs non-interactivity were achieved.

It is important to note that here and in [7] SSA'a were investigated at the most fundamental level, regardless of particularities due to specific applications where such arrays may be used as SQUID-based instruments. Therefore the main goal was to prove their superiority over LTS single-SQUIDs as flux-to-voltage converters. Considering this the array's sensitivity is given by their flux resolution $S_\Phi^{1/2}$. On the other hand sensitivity of SQUIDs (or SQUID arrays) used as SQUID-based instruments in various specific applications is measured differently:

a) as EM/gamma/X-rays, particle detectors, gravitational waves detectors, or amplifiers is measured by its noise temperature;
b) as photon energy detectors is measured by its energy sensitivity;
c) as magnetometers is measured by its field resolution S_B;
d) as gradiometers is measured in S_B/m;

Consequently, a series SQUID array with a record $S_\Phi^{1/2}$ has the potential to operate with record sensitivity in most applications including magnetometers that measure B (provided the most advanced schemes are used for their input coil coupling, bias reversal, output read-out electronics, magnetic and electromagnetic screening, etc). It is therefore important to stress that the input coil used to measure *V(B)* in Figure 2 is neither optimally coupled, nor it is optimally designed. Indeed, in our case there is a significant mismatch between the input coil and the SQUID array loops. No effort was made to address this mismatch since our primary goal was to prove SQUID array superiority over single SQUIDs in the most fundamental way possible valid for most applications (not just as magnetometers). For that reason the *V(B)* data presented should not be used to extract the *B* field sensitivity of our SQUID array. There are many well-known schemes/techniques to significantly improve this mismatch such as single layer/multilayers flux transformers that are coupled directly or in a flip-chip configuration. Those schemes are routinely used in commercially available single HTS SQUIDs. By using those advanced schemes one can easily translate the record flux noise performance of our SQUID arrays into a record field noise performance when the SQUID-arrays are used as magnetometers. At present we consider the realization of similar coupling schemes for large serial arrays.

3. B-field tunable microwave generators

Since the discovery of the ac Josephson effect, Josephson junctions (JJ), as natural electromagnetic (EM) microwave generators, have always been attractive for applications. In practice, however, single JJs are never used because the power of the emitted EM radiation is too small. Devices consisting of multiple JJs are therefore the only realistic solution. One of the most promising approaches is when an enhanced EM radiation is produced as a result of a coherent flux flow [21] in one-dimensional (1D) arrays [22-27] placed in an externally applied magnetic field B. In this case magnetic vortices will enter the array in a form of Josephson vortices. The bias current, I, flowing across the array produces a Lorentz force, F_L, which drives the Josephson vortices unidirectionally, forming a lattice of vortices moving with a speed, v. This is accompanied by an emission of small amplitude linear waves that propagate along the array. In other words *EM* waves are excited by a (magnetic field induced) chain of propagating magnetic vortices. When the vortex spacing is commensurate with the wavelength of emitted waves, resonant modes occur. This can be viewed as the phase-locking condition of the vortex velocity and the phase velocity of one of the self-induced modes. The experimental signature of such phase locking between a train of propagating vortices and their induced *EM* radiation in a JJ array is a series of flux-flow resonances (m=1, 2, 3, 4, etc) in the current-voltage characteristics (IVCs). On a resonant current step, moving vortices couple to their induced linear waves. Further increases of the current do not lead to further linear increases in the vortices velocity, because the energy is consumed in amplifying the linear waves.

Such resonances have been observed by many groups, below 9K, in LTS JJ arrays [22-27]. The associated amplified *EM* radiation has been measured too, proving that LTS JJ arrays are suitable candidates as *B*-tunable microwave oscillators [28, 29]. After the discovery of HTS in 1986, it was anticipated that similar *B*-tunable microwave oscillators could be made of HTS JJ arrays operating at liquid nitrogen temperature. Initially the highest reported temperature for observing stable flux-flow resonances in HTS JJ arrays was 40K, achieved in $Bi_2Sr_2CaCu_2O_{8+x}$ mesa structures based on a 1D array of intrinsic JJs [30]. The first observation of multiple flux-flow resonances at 77K and above (see Figure 2) reported in [8] has been observed in asymmetric parallel arrays consisting of 440 $YBa_2Cu_3O_{7-\delta}$ JJs fabricated in bicrystal technology. The array consists (see inset in Figure 2) of 22 identical sets of 20 *JJ* of identical width (3 μm). Within each set all 20 *JJ* are connected via superconducting loops of identical width (3μm) and monotonically decreasing length from 18μm to 13μm. Recently, in my group [31], a symmetric parallel array was fabricated to investigate whether an asymmetric design is an essential condition for the observation of flux-flow resonances. The parallel array consists of 100 JJ of identical width (3 μm) connected via superconducting loops of identical size: (width, length) = (3μm x 15 μm). It was found that the flux-flow resonances were observed in

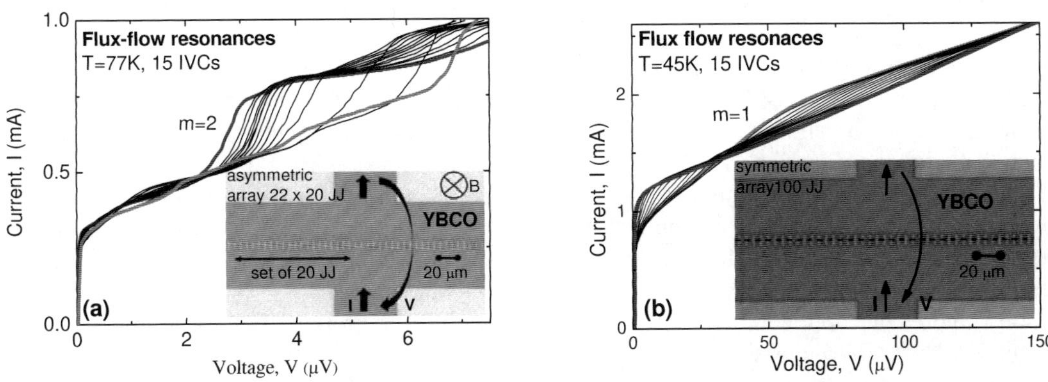

Figure 2. Families of 15 consecutive current-voltage characteristics (*IVC*s) of an asymmetric (a) and a symmetric (b) parallel arrays recorded for monotonically increasing B values. In blue (red) are the first (last) *IVC* within these sets. The position of the high-amplitude second flux-flow resonance (m=2) in (a) and of the low-amplitude first flux-flow resonance (m=1) in (b) are gradually shifted by varying B. Insets show optical micrographs of a small part of the arrays (42 JJs). The asymmetrical geometry introduces a highly preferential direction for the flux flow and strong non-linearities that ultimately leads to a significant amplification of EM.

the temperature range (30-45) K and that their amplitudes were significantly lower than for the asymmetrical design (see Figure 3). One can therefore conclude that an asymmetric design significantly enhances the coherent flow of vortices and transforms it into a very non-linear phenomenon with respect to a monotonically increasing B field. That ultimately leads to a significant amplification of the emitted EM. With our device operated at 84K on the m=1 step, one estimated a maximum power radiation $P=0.5(I\times I_{step}/I_c)^2R_N/N$ of about 20 nW at $\Phi_{ex}=\Phi_0/2$ and a B-field tunability within the frequency range (1-9) GHz. At 77K the device performance improves considerably providing a *B*-field tunable power of about 0.1 μW within the range (1.5-25) GHz. By inductively coupling multiple *M* rows of 1D JJ arrays the power of emitted *EM* radiation can be further increased. Thus, the design approach reported here shows great promise as a route to realizing high performance B-field tunable HTS microwave oscillators based on flux flow resonances.

4. Vortex-flow superconductor transistors

High temperature superconductor Josephson vortex-flow transistors (JVFT's) are the most promising candidates for high-frequency amplifying devices in all-superconducting microwave circuits [32-33]. There are two types of JVFTs: ones that are based on a single long Josephson junction (*JJ*) [34, 35] or JVFT's based on arrays of *JJs* [35-38]. A high performance JVFT for practical applications should have a high current gain, $g = \partial I_c/\partial I_{ctrl}$ (where I_c is the device critical current and I_{ctrl} is the control-gate current), a large dynamic range (the range of I_{ctrl} over which a high g can be achieved), small vortex transit times to allow high frequency operation and a relatively large transresistance, $r_m = \partial V/\partial I_{ctrl}$.

In [9] we propose a discrete JVFT based on a parallel array of 440 $YBa_2Cu_3O_{7-\delta}$ bicrystal grain boundary *JJ* that combines beneficial aspects of several previous designs. On the one hand, it has a symmetric bias configuration to ensure a large dynamic range, as well as, small vortex transit times; on the other hand it has a certain degree of inductive asymmetry implemented via an asymmetric loop configuration within the array. This asymmetry ensures the bias current produces a net inhomogeneous magnetic field along the array through self-field effects that result in a highly asymmetric $I_c(I_{ctrl})$ curve with high current gains. Of particular interest are JVFT's operating at 77K due to their low-cost cooling procedures and practicality. There have been several reports of JVFT's with current gains g as high as 6 at 30K [35], or 5.5 at 50K [37], or 14 at 70K [36], however, in all these cases $g(T)$ rapidly and monotonically decreases with temperature as it approaches 77K. Thus, the highest current gains reported so far at 77K were achieved with discrete JVFT's having an asymmetric in-line bias distribution (maximum reported values were g=3.5 [38]). Our discrete JVFT shows current gains as high as 19, a significant dynamic range of more than 20 μA at 77 K, and an unusual non-monotonic $g(T)$ dependence in the range (4.7-92)K with a maximum recorded value around 77K. The flux-flow in asymmetrical arrays is significantly altered relative to symmetrical geometries, with a highly preferential flux-flow direction leading to enhanced dynamics (sharp flux-flow resonances, EM amplification). As shown in [9] this novel behaviour can be exploited in a controllable fashion by B leading to record values for current amplification at 77K (see Figure 3). This interesting feature can find applications in developing low-noise fast first-stage current amplifiers in various electronic devices.

Another interesting application of asymmetric *JJ*-arrays is their functionality as magnetic field-tunable highly reversible vortex diodes (ratchets) as demonstrated in [10]. Thus, at 89 K the ratchet efficiency $\eta = \left|\frac{V(I)+V(-I)}{V(I)-V(-I)}\right| \times 100\%$ could be reversed from +60% to -60% with a change in B as small as 3μT. Such ratchet designs can be used to control unidirectional vortex flow vortices in superconducting devices, to remove unwanted trapped vortices in superconducting devices without the need to warm them up above their critical temperature, as well as, to build integrated nano-magnetic sensors.

5. Conclusions

Recently, several significant developments have been reported of superconducting devices such as magnetometers [7], B field-tunable microwave generators [8] and superconducting transistors [9] build based on large arrays of *JJs*- or SQUIDs-arrays. Such devices showed improved performances at 77K a temperature that can be reached with liquid nitrogen (that is cheap and user-friendly). They were fabricated using the bicrystal technology that is both sufficiently reliable and inexpensive. All these features suggest the practically and potential of this HTS 77K array technology to replace LTS single JJ or SQUIDs currently used in many applications. It is remarkable that in all these reports a proof of concept of such devices has been provided with minimal efforts invested into the optimization of their

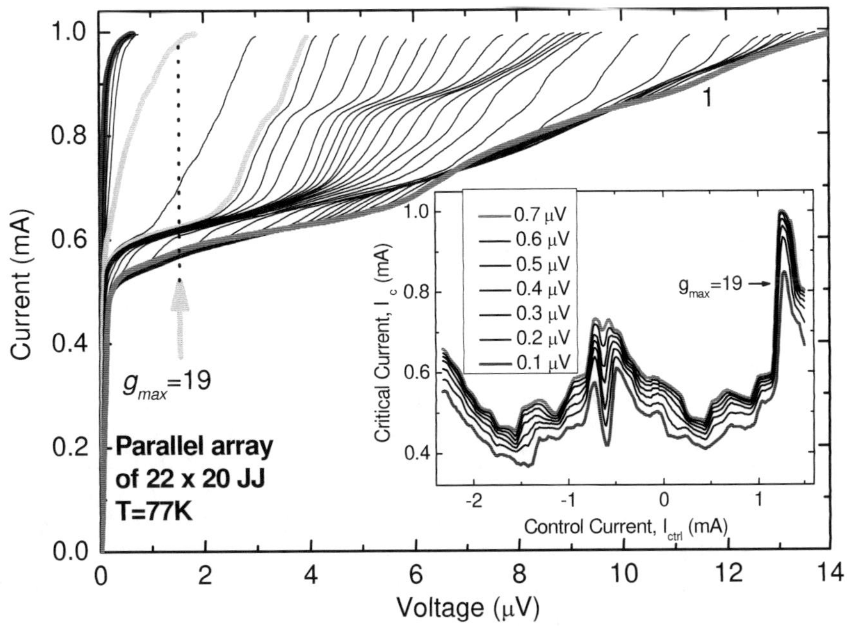

Figure 3. A set of 35 consecutive *IVC* at 77K for various values of the control current I_{ctrl} around 1.2 mA. I_{ctrl} is changed in steps of 15 μA. In blue (red) are the first (last) *IVC* within these sets. In green are the two *IV*'s for the I_{ctrl} range where a switching behaviour is observed with current gains as high as 19. Inset show highly asymmetric $I_c(I_{ctrl})$ dependencies recorded at 7 different voltages. The steepest slopes of these curves is where the transistor amplifies best.

operation. This strongly suggests that performances of such devices can be significantly improved further. This is exciting as 77K can probably be regarded as the lowest acceptable temperature that would allow superconducting electronics to challenge their semiconducting counterpart on a different, much larger, scale.

Acknowledgments

This work was supported by the Higher Education Innovation Fund (HEIF) UK and Loughborough University Enterprise under project EPG101-P4.

References

[1] Feynman R 1965 *Lectures in Physics* Volume 3 Ch. 21.
[2] Dimos D, Chaudhari P, Mannhart J, and LeGoues F. K 1988 *Phys. Rev. Lett.* **61** 219; Hilgenkamp H and Mannhart J 2002, *Rev. Mod. Phys.* **74** 485-549.
[3] Daly K. P, Dozier W. D, Burch J. F, Coons S. B, Hu R, C. E. Platt C. E and Simon R. W 1991 *Appl. Phys. Lett.* **58** 543;
[4] Schulz R. R, Chesca B, Goetz B, Schneider C.W, Schmehl A, Bielefeldt H, Hilgenkamp H, Mannhart J, and Tsuei C.C 2000 *App. Phys. Lett.* **76** 912 2000.
[5] Cho E. Y, Ma M. K, Huynh C, Pratt K, Paulson D. N, Glyantsev V. N, Dynes R. C, and Cybart S. A 2015 *Appl. Phys. Lett.***106** 252601.

[6] Du J, Lazar J.Y, Lam S. K. H, Mitchell E. E, and Foley C. P 2014 *Supercond. Sci. Technol.* 27 095005;
Mitchell E. E, Hannam K.E, Lazar J.Y, Leslie K.E, Grancea A, Keenan S.T, Lam S. K. H, and Foley C. P 2016 *Supercond. Sci. Technol.* **29** 06LT01.
[7] Chesca B, John D, Mellor C. J 2015 *Appl. Phys.Lett.* **107** 162602.
[8] Chesca B, John D, Mellor C. J 2014 *Supercond. Sci. Technol.* **27** 085015.
[9] Chesca B, John D, Kemp M, Brown J, Mellor C.J 2013 *Appl. Phys. Lett.* **103** 092601.
[10] Chesca B, John D, Pollett R, Gaifullin M, Cox J, Mellor C.J and Savel'ev S 2017 *Appl. Phys. Lett.* **111** 062602.
[11] Welty R.P and Martinis J.M 1991 *IEEE Trans. Magn.* **3** 2924.
[12] Welty R.P and Martinis J.M 1993 *IEEE Trans. Appl. Supercond.* **3** 2605.
[13] Stawiasz K.G and Ketchen M.B 1993 *IEEE Trans. Appl. Supercond.* **3** 1808.
[14] Foglietti V, Stawiasz K. G, Ketchen M.B and Koch R. H 1993 *IEEE Trans. Appl. Supercond.* **3** 3061.
[15] Li K and Hubbell S. P 1995 *IEEE Trans. Appl. Supercond.* **5** 3255.
[16] Li K and Hubbell S.P 1999 *IEEE Trans. Appl. Supercond.* **7** 3217.
[17] Li K, Hubbell S.P, Cantor R and Teepe M 1999 *IEEE Trans. Appl. Supercond.* **9** 4420.
[18] Wu S. H, Hsu M.H, Chen K.L, Chen J.C, Jeng J.T, Lai T.S, Horng H. R and Yang H. C 2006 *Supercond. Sci. Technol.* **19** S246.
[19] Lee S. G, Huh Y, Park G.S, Kim I.S, Park Y. K and Park J.C 1997 *IEEE Trans. Appl. Supercond.* **7** 3347-50.
[20] Wu C. H, Chen M. J, Chen J. C, Chen K. L, Yang H. C, Hsu M. S, Lai T.S, Tsai Y. S, Horng H. E and Chen J. H 2006 *Rev. Sci. Instrum.* **77** 033901.
[21] Eck R. E, Scalapino D.J and Taylor B. N *Phys. Rev. Lett.* **13**, 15-18 (1964).
[22] Hontsu S and J. Ishii, *J. Appl. Phys.* **63**, 2021 (1988);
[23] Ustinov A. V, Cirillo M and Malomed B. A 1993 *Phys. Rev. B* **47** 8357-60.
[24] Van der Zant H. S. J, Orlando T. P, Watanabe S and Strogatz S. H 1994 *Phys. Rev. Lett.* **74** 172-7.
[25] Duwel A. E, Watanabe S, Trias E, Orlando T.P, Van Der Zant H.S.J and Strogatz S.H 1997 *J. Appl. Phys.* **82** 4661-68.
[26] Ustinov A. V, Cirillo M, Larson B. H, Oboznov V.A, Carrelli P and Rotoli G 1995 *Phys. Rev. B* **51** 3081-91.
[27] Cirillo M, Larsen B. H, Ustinov A. V, Merlo V, Oboznov V. A, Leoni R 1993 *Phys. Lett. A* **183**, 383-9.
[28] Booi P. A. A, and Benz S. P 1994 *Appl. Phys. Lett.* **64** 2163-65.
[29] Duwel, A. E, Orlando T. P, Watanabe S, Van der Zant H. S. J 1997 *IEEE Trans. Appl. Supercond.* **7** 2897-2900.
[30] Katterwe S. O and Krasnov V. M 2011 *Phys. Rev. B* **84** 214519.
[31] Chesca B, John D and Mellor C.J 2017 unpublished.
[32] Gross R, Gerdemann R, Alff L, Bauch T, Beck A, Froehlich O.M, Koelle D and Marx A 1995 *Appl. Supercond.* **3** 443.
[33] Mannhart J 1996 *Supercond. Sci. Technol.* **9** 49.
[34] Zhang Y.M, Winkler D, Nilsson P.A and Claeson T 1995 *Appl. Phys. Lett.* **64** 1153.
[35] Berkowitz S. J, Zhang Y.M, Mallison W. H, Char K, Terzioglu E and Beasley M.R 1996 *Appl. Phys. Lett.* **69** 3257.
[36] Gerdemann R, Bauch T, Froehlich O.M, Alff L, Beck A, Koelle D and Gross R 1995 *Appl. Phys. Lett.* **67** 1010.
[37] Schuler J, Weiss S, Bauch T, Marx A, Koelle D and Gross R 2001 *Appl. Phys. Lett.* **78** 1095.
[38] Bauch T, Weis S, Haensel H, Marx A, Koelle D and Gross R 1997 *IEEE Trans. Appl. Supercond.* **7** 3605.

Magnetometer based on transfer and modulation of magnetic flux using high-critical-temperature superconductor coils

K Enpuku, M Matsuo, Y Yoshida, S Yamashita, T Sasayama and T Yoshida

Department of Electrical and Electronic Engineering, Kyushu University, Fukuoka 819-0395, Japan

Abstract. We developed a new type of magnetometer to measure low-frequency magnetic fields with an operational principle based on magnetic flux transfer and modulation. This magnetometer consists of three coils: pickup, input and readout coils. The pickup and input coils were made from high-critical-temperature superconductor (HTS) tape and were connected with very low connection resistance to form a closed loop. The magnetic flux that is collected by the pickup coil can be transferred to the input coil even at low frequencies below 1 Hz. The magnetic flux at the input coil is then detected by the readout coil using a readout scheme based on modulation of the mutual inductance M between the input and readout coils. To modulate M over time, a magnetic wire was inserted into the readout coil and a time-varying current was supplied to the wire. Using this time-varying M, the magnetic flux is converted into an amplitude-modulated voltage across the readout coil for measurements. A prototype magnetometer was fabricated for demonstration. This magnetometer can operate at low frequencies down to f = 0.5 Hz without responsivity degradation. The magnetic field noise levels were 8 and 60 pT/Hz$^{1/2}$ at 50 and 1 Hz, respectively.

1. Introduction

The induction-type magnetometer (or pickup coil) has been widely used to measure signal fields in the presence of an excitation field. This is because other sensitive sensors, such as the flux gate and giant magnetoresistive sensors, are difficult to operate correctly when they are exposed to large excitation fields. However, the main drawback of the induction-type magnetometer is that the device responsivity diminishes considerably at low frequencies because its responsivity is proportional to the frequency of the magnetic field. It is therefore important to develop a magnetometer that can measure signal fields at low frequencies with high sensitivity in the presence of the excitation field.

For this purpose, we proposed a new type of magnetometer [1]. The operation of the proposed magnetometer is based on transfer and modulation of the magnetic flux using coils made from high-critical-temperature superconductor (HTS) materials. Two schemes were used in development of this magnetometer. The first involved use of a flux-transfer scheme that is similar to the scheme used in Nb-based superconducting quantum interference device (SQUID) magnetometers [2]. A closed loop was formed using an HTS tape with very low connection resistance. This closed loop could then transfer the magnetic flux, even at low frequencies below 1 Hz. The second scheme involved development of a new readout scheme based on modulation of the coil inductance. In this case, the magnetic flux collected by the pickup coil is converted into an amplitude-modulated voltage for the readout; this scheme is similar to that of a fluxgate sensor [3], [4]. Using a prototype of this magnetometer, we experimentally confirmed the validity of its operating principle.

IOP Conf. Series: Journal of Physics: Conf. Series **1054** (2018) 012058 doi:10.1088/1742-6596/1054/1/012058

There are two possible ways to realize the proposed magnetometer based on transfer and modulation of magnetic flux. In our previous paper, we demonstrated one type of magnetometer, which was based on modulation of the coil's self-inductance [1]. In this study, we have developed the second type of magnetometer. This magnetometer configuration is composed of three coils: the pickup, input and readout coils. The mutual inductance between the input and readout coils is modulated over time to obtain the readout voltage. First, we describe the operating principle of this magnetometer. Then, we present experimental results from a prototype of the magnetometer. The magnetometer demonstrated operation at low frequencies ranging down to f = 0.5 Hz without responsivity degradation. The noise levels of the magnetometer were 8 and 60 pT/Hz$^{1/2}$ at 50 and 1 Hz, respectively.

2. Operating principle of the magnetometer

Figure 1 shows an equivalent circuit for the proposed magnetometer. This magnetometer consists of pickup, input and readout coils. The pickup and input coils are formed from HTS tape and their inductances are denoted by L_p and L_i, respectively. The two coils are connected with very low connection resistance R_c to form a closed loop. Using this closed loop, we can transfer the magnetic flux Φ_s that is collected by the pickup coil to the input coil, even at low frequencies [1]. The magnetic flux that is transferred to the input coil is then detected using the readout coil; the mutual inductance between the input and readout coils is denoted by M. A magnetic wire is inserted into the readout coil and a current I_B is supplied to the wire to modulate the value of M.

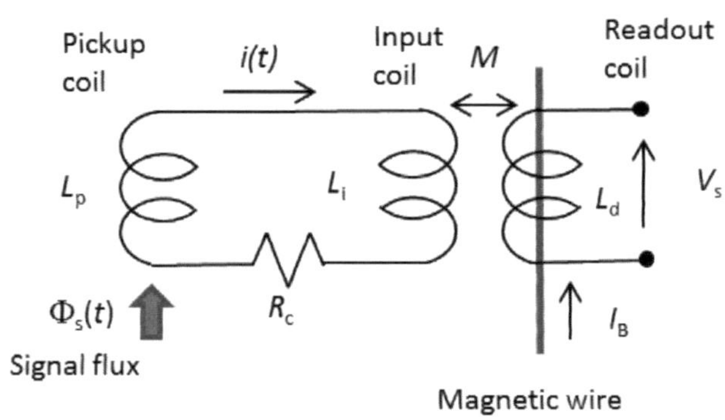

Figure 1. Equivalent circuit for the proposed magnetometer. The magnetic flux Φ_s that is collected at the pickup coil is transferred to the input coil and is then detected using the readout coil. The mutual inductance M between the input and readout coils is modulated over time by supplying a time-varying current I_B to the magnetic wire. The voltage v_s across the readout coil thus becomes an amplitude-modulated voltage with an amplitude that is proportional to Φ_s.

The operating principle of the magnetometer is described as follows. When the magnetic flux $\Phi_s(t)$ interlinks the pickup coil, a current $i(t)$ flows within the closed loop. Here, we consider the case in which the angular frequency ω_s of the signal flux satisfies the following condition [1].

$$\omega_s(L_p + L_i) >> R_c \cdot \quad (1)$$

In this case, the current i is given by

$$i(t) = \frac{1}{L_p + L_i}\Phi_s(t) \cdot \qquad (2)$$

The voltage across the readout coil, v_s, is then given by

$$v_s(t) = \frac{d}{dt}(Mi) = \frac{M}{L_p + L_i}\frac{d\Phi_s}{dt} + \Phi_s \frac{d}{dt}\left(\frac{M}{L_p + L_i}\right). \quad (3)$$

Note that the first term on the right side of the equation is given by the time-derivative of the signal flux Φ_s, and this term becomes very small at low frequencies. In contrast, the second term yields the voltage v_s if M varies with time. We use the second term in the proposed magnetometer.

The value of M can be varied over time by applying current I_B to the magnetic wire because the effective permeability of the wire is changed by I_B [3], [4]. Here, we consider the case where I_B is given by

$$I_B(t) = I_{DC} + I_{AC}\sin(\omega_m t) \quad (4)$$

where ω_m is the modulation frequency.

Under application of I_B, the mutual inductance M can be modulated over time as follows.

$$M(t) = M_0 + \Delta M(t) = M_0[1 + K\sin(\omega_m t)] \quad (5)$$

where K is a parameter that represents the degree of inductance modulation.

The second term on the right of equation (3) then gives the voltage v_s across the readout coil as

$$v_s(t) = \frac{M_0}{L_p + L_i} K\,\omega_m \cos(\omega_m t)\Phi_s(t) \quad (6)$$

where, for simplicity, we tentatively neglect the change in L_i due to I_B.

Equation (6) indicates that v_s becomes an amplitude-modulated signal with carrier frequency ω_m and an amplitude that is proportional to Φ_s. If v_s is detected using a lock-in amplifier with reference frequency ω_m, the amplifier's output, $v_o(t)$, is given by

$$v_o(t) = \frac{M_0}{L_p + L_i} K\,\omega_m \Phi_s(t) \quad (7)$$

Equation (7) shows that the voltage v_o is proportional to Φ_s but is independent of the frequency of Φ_s. Consequently, we can use the proposed magnetometer to measure magnetic fields at low frequencies without any response degradation. Based on the readout scheme used, we call the proposed magnetometer an inductance modulation scheme (IMS) magnetometer.

3. Experimental

We fabricated a prototype of the IMS magnetometer (figure 2). The pickup coil with inductance L_p and the input coil with inductance L_i were made from a (RE)BCO (rare-earth barium copper oxide) HTS tape (SF2050, SuperPower). The tape is 2 mm wide and the tape was coated with a 2-μm-thick layer of silver. The inner and outer diameters of the pickup coil were 15 mm and 35 mm, respectively. The number of turns and the inductance of this coil were N_p = 50 and L_p = 65 μH, respectively. The input coil was made from three stacked coils. The inner and outer diameters of each coil were 15 mm and 20 mm, respectively, and the number of turns in each coil was 15. The total number of turns and the inductance of the input coil were N_i = 45 and L_i = 18 μH, respectively. The pickup and input coils were soldered together. The connection resistance was estimated to be R_c = 27 μΩ, as will be shown later in the paper.

The readout coil was made from Cu wire. The diameter and the length of the coil were 5 mm and 22 mm, respectively. The number of turns and the inductance of the readout coil were N_d = 400 and L_d = 1.17 mH, respectively. A commercial amorphous magnetic wire with thickness of 35 μm and width of 500 μm (SENCY 120FC20, Aichi Steel) was inserted into the readout coil. The readout coil and the magnetic wire were placed concentrically inside the input coil, as shown in figure 2.

To modulate the mutual inductance M between the input and readout coils, a current I_B was applied to the magnetic wire. Figure 3 shows the experimentally measured change in M when DC current I_B was applied to the magnetic wire. The vertical axis represents the fractional change $\Delta M/M_0$ [see equation (5)], where M_0 = 75 μH. As shown, $\Delta M/M_0$ had a slightly complex dependence on I_B, where it increased and then decreased monotonically with increasingly positive I_B. The relative reduction was 27% at I_B = 118 mA.

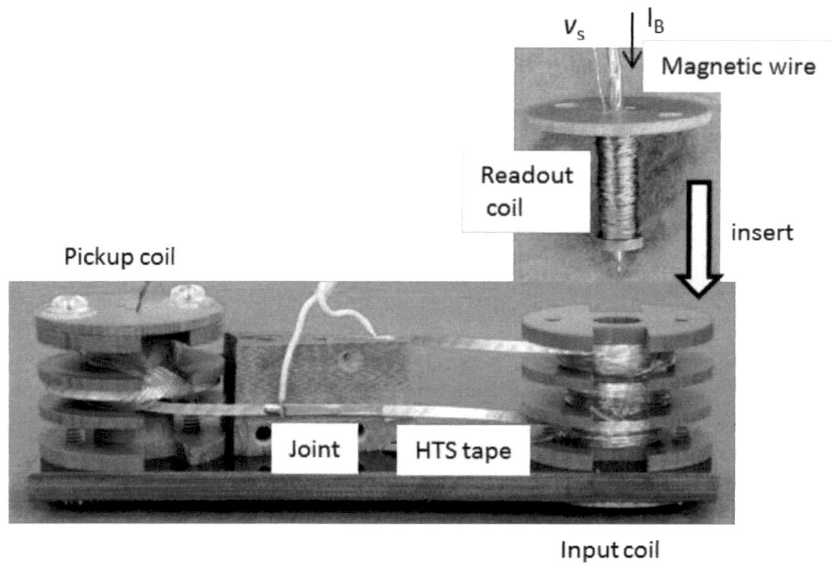

Figure 2. Photograph of the prototype magnetometer.

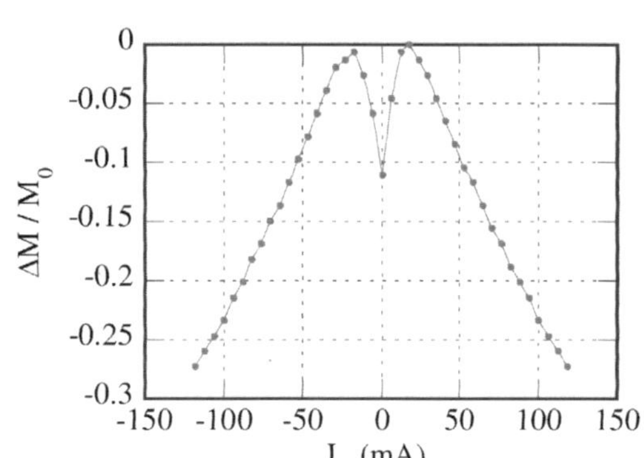

Figure 3. Relative change in inductance (ΔM/M_0) when the DC current I_B is applied to the magnetic wire.

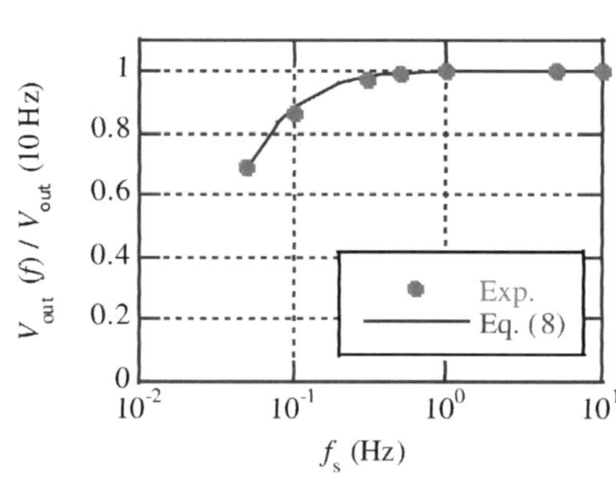

Figure 4. Frequency response of the magnetometer. The vertical axis represents the output voltage (V_{out}) of the lock-in amplifier when normalized with respect to the value at f_s = 10 Hz.

To vary M over time in the experiments, we applied the modulated current $I_B(t)$ that was given in equation (4). The experimental values in this case were I_{DC} = 55 mA, I_{AC} = 20 mA, and $f_m = \omega_m/2\pi$ = 80 kHz. In this case, the inductance $M(t)$ varies with time as shown in equation (5); an approximate value of K = 0.06 was estimated based on figure 3. Therefore, we obtain the time-varying voltage v_3 that was given in equation (6).

Figure 4 shows the magnetometer's frequency response when the frequency of the signal flux Φ_s was varied from f_s = 0.05 to 10 Hz while maintaining a fixed amplitude. During the experiment, the magnetometer was cooled to T =77 K using liquid nitrogen. The signal flux Φ_s was applied to the pickup coil with inductance L_p using a small signal coil: the mutual inductance between the pickup and signal coils was 0.7 μH. The vertical axis represents the amplitude of the output voltage (V_{out}) of the lock-in amplifier when normalized with respect to the value at f_s = 10 Hz. Note here that V_{out} was independent of f_s when f_s > 0.5 Hz. At frequencies lower than 0.5 Hz, V_{out} decreased with decreasing f_s.

Based on the equivalent circuit shown in figure 1, the frequency dependence can be determined as follows.

$$\frac{V_{out}(f)}{V_{out}(10\,\mathrm{Hz})} = \frac{2\pi f\tau}{\sqrt{1+(2\pi f\tau)^2}} \quad \text{with } \tau = \frac{L_p + L_i}{R_c} \tag{8}$$

The solid line in figure 4 was calculated using equation (8) with τ = 3 s. By substituting the values $L_p + L_i$ = 83 μH and τ = 3 s into the equation, we find that R_c = 27 μΩ. The R_c value obtained was larger than the previously reported value (R_c < 6 μΩ) [1]. Note that the value of R_c can be reduced by increasing the size of the joint area between the HTS tapes [5].

We measured the noise spectrum of the magnetometer. During the experiments, we first measured the magnetometer's responsivity using the following method. The output voltage (V_{out}) of the lock-in amplifier was measured when the amplitude of signal flux Φ_s was varied; the signal frequency was set at f_s = 3 Hz. The magnetometer output voltage, V_s, was then obtained using $V_s = V_{out}/G$, where G is the amplifier gain. The signal field B_s was calculated using the relationship $\Phi_s = B_s(\pi/4)D_{av}^2N_p$, with average diameter D_{av} = (15+35)/2 = 25 mm and N_p = 50. The B_s vs. V_s curve was obtained when B_s ranging from nT to μT was applied. In this field range, the magnetometer showed linear response, and the field-to-voltage transfer coefficient of the magnetometer was obtained from the slope of the B_s vs. V_s curve as V_s/B_s = 990 V/T. Note that the measured value showed reasonable agreement with the transfer coefficient that was calculated using equation (7); the calculated value was 924 V/T.

The noise in the magnetometer voltage signal, which is denoted by $S_V^{1/2}$, was measured inside a magnetic shielding box composed of two layers of permalloy. A signal field with frequency f_s = 3 Hz was also applied as a reference. Using the field-to-voltage transfer coefficient of V_s/B_s = 990 V/T, the voltage noise $S_V^{1/2}$ was converted into the magnetic field noise $S_B^{1/2}$. Figure 5 shows the measured spectrum of the field noise $S_B^{1/2}$; the peak shown at f = 3 Hz corresponds to the signal field. As the spectrum shows, $S_B^{1/2}$ decreased with increasing frequency f. We obtained values of $S_B^{1/2}$ = 8, 18 and 60 pT/Hz$^{1/2}$ at f = 50, 10 and 1 Hz, respectively.

The broken line in figure 5 is used to indicate the frequency dependence of the $1/f$ noise. As the spectrum shows, the measured noise displayed clear $1/f$ noise, although the origin remains unclear at present. It is therefore necessary to reduce this $1/f$ noise to greatly enhance the magnetometer's performance. For this purpose, it would be useful to develop a gradiometer because the existing magnetometer is affected quite considerably by environmental noise. While we did use the magnetic shielding box that consisted of two layers of permalloy, the environmental noise was not suppressed to a sufficient degree. It will also be necessary to determine the optimum conditions for the modulation current I_B that is supplied to the magnetic wire. This is essential because the noise generated by the magnetic wire is affected significantly by the bias condition of I_B, as previously demonstrated in the fluxgate sensor case [4], [6]-[8].

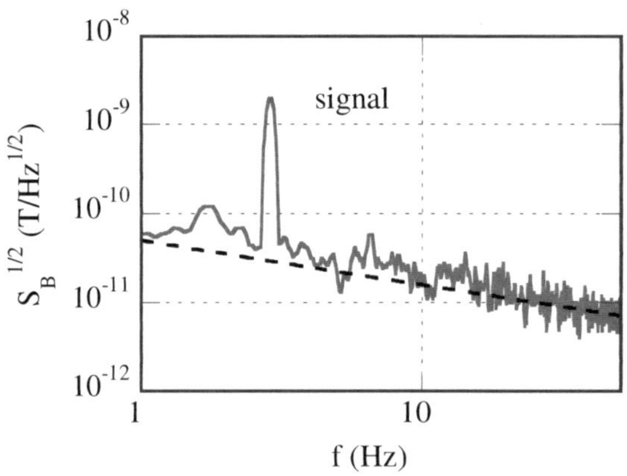

Figure 5. Magnetic field noise spectrum of the magnetometer, where the peak at $f = 3$ Hz represents the signal field. We obtained values of $S_B^{1/2} = 8$, 18 and 60 pT/Hz$^{1/2}$ at $f = 50$, 10 and 1 Hz, respectively. The broken line indicates the frequency dependence of the $1/f$ noise.

4. Conclusion

We have developed a new type of magnetometer based on transfer and modulation of magnetic flux using HTS coils. The magnetometer is expected to be able to measure signal fields at low frequencies with high sensitivity in the presence of large excitation fields. The prototype magnetometer operated at frequencies as low as $f = 0.5$ Hz without degradation of its responsivity. The magnetic field noise of the magnetometer was measured to be 8 and 60 pT/Hz$^{1/2}$ at 50 and 1 Hz, respectively. Because this noise was dominated by the $1/f$ noise, further work will be required to reduce the $1/f$ noise. It is also necessary to show operation of the magnetometer in the presence of large excitation fields.

Acknowledgments

This work was partly supported by Grant-in-Aid for Scientific Research (S) JP15H05764 from the Japan Society for the Promotion of Science.

References

[1] Enpuku K, Matsuo M, Sasayama T and Yoshida K 2017 *Appl. Phys. Express*, **10** 113101.
[2] Clarke J and Braginski A. I. 2004 *The SQUID Handbook*, Wiley-VCH Verlarg GmbH and Co. KGaA, Weinheim.
[3] Fan J, Lia P and Ripka P 2006 *J. Appl. Phys.* **99** 08B311.
[4] Sasda I and Harada S 2014 *IEEE Trans. Magn.* **50** 4007404.
[5] Zheng J, Ma H, He R, Lu Y, Song H, Cui C, Zhang Y, Bai C, Guo Y, Liu Z and Cai C 2017 *IEEE Trans. Appl. Supercond.* **27** 6603706.
[6] Koch RH and Rozen JR 2001 *Appl. Phys. Lett.* **78** 1897.
[7] Butta M and Sasada I 2013 *IEEE Trans. Magn.* **49** 62.
[8] Bazinet R, Jacas A, Confalonieri G and Vazquez M 2014 *IEEE Trans. Magn.* **50** 6500103.

IOP Conf. Series: Journal of Physics: Conf. Series **1054** (2018) 012059 doi:10.1088/1742-6596/1054/1/012059

Constructing a Vector Scanning SQUID System

Vu The Dang[1,2,*], Masaki Toji[1], Atsuki Ito[1], Yoshitsugu Ninomiya[1], Shigeyuki Miyajima[3], Ho Thanh Huy[1,2], Hiroaki Shishido[1,4], Masaru Kato[4,5], Mutsuo Hidaka[6], Masahiko Hayashi[7] and Takekazu Ishida[1,4]

[1] *Department of Physics and Electronics, Osaka Prefecture University, Sakai 599-8531, Japan*
[2] *University of Sciences, Vietnam National University HCMC, Ho Chi Minh, Viet Nam*
[3] *National Institute of Information and Communications Technology, Kobe, Hyogo 651-2492, Japan*
[4] *NanoSquare Research Institute, Osaka Prefecture University, Sakai 599-8531, Japan*
[5] *Department of Mathematical Sciences, Osaka Prefecture University, Sakai 599-8531, Japan*
[6] *National Institute of Advanced Industrial Science and Technology, Tsukuba 305-8568, Japan*
[7] *Faculty of Education and Human Studies, Akita University, Akita 010-8502, Japan*
**vuthedang@gmail.com*

Abstract We succeeded in fabricating a vector scanning SQUID sensor, which was optimized to reduce the noise and to enhance the sensitivity by choosing the screening parameter as β_L = 1. The sensors were fabricated by a superconductive foundry CRAVITY of Advanced Industrial Science and Technology (AIST) using AIST-STP2 standard process with the critical current density J_c of Josephson junctions (JJs) 320 A/cm^2. Under this condition, the critical current I_c becomes 12.5 μA in the areas of 2.2 μm x 2.2 μm of our SQUID. The sensitivity and resolution of sensors were also improved by reducing inner diameter and increasing the number of the pick-up-coil windings with employing of multi-layer technique. We designed the structure of vector pickup coils, of which the same distance is kept between each coil and the sample surface. In addition, the distance between the coils were just set at an integer multiple of a scanning step so that it becomes convenient in reconstructing a magnetic-field vector at a certain position. The fabricated SQUID sensors were operated properly, and characteristics were in good agreement with our design parameters. In order to construct the vector scanning SQUID microscope, we succeeded in connecting our sensor with a commercial flux-locked loop (FLL) readout circuit (Magnicon Inc., Model XXF-1). The spatial resolution of vector SQUID microscopy can be improved by using an *XYZ* piezo-driven scanner for the precise control of the pick-up-coil positions over the three-dimensional range of 5 mm x 5 mm x 5 mm with a step size of 10 nm at 4K. A Gifford–McMahon cryocooler with an anti-vibration mechanism will be used to conduct long-time measurements.

1. Introduction

A scanning SQUID microscope is well known to have an ultra-high-sensitive and low-noise character for measuring fields as low as 5 aT (5×10^{-18} T) [1] under noise levels as low as 3 fT/√Hz [2]. Therefore, it is always necessary for us to optimize the magnetic field measurements. Besides, SQUID devices are employed as a powerful tool in various different fields such as magnetics, electronics, biology, spintronics, security analysis, agriculture, information technology, communication, material science, and physics[3]–[8]. In the purpose of observing vortex distribution in a superconducting film, Morooka *et al.* constructed a scanning SQUID microscope with a micro DC-SQUID[9]. The cryostat consumes liquid helium to control a temperature between 3-100K and a precise scanning stage using a stepping motor. However, the magnetic field sensitivity of the sensor was limited to 4 nT/√Hz owing to the sensing area of pickup coil is 2 x 2 μm^2 and a Nb line width of 2 μm. In preceding studies, we have studied vortex distribution in a restricted superconducting plate[4], [5], [10]–[12] by using a commercial scanning SQUID microscope (SQM2000, Seiko Instruments [9]), in which a dc SQUID sensor was equipped with a 10-μm outer diameter of the pickup coil and the line width of 2 μm. Vortex images were observed clearly, and we showed that vortex distribution primarily depended on the symmetry of the geometrical pattern. However, the sensitivity and the resolution of the sensor were still restricted by a diameter of the pickup coil. For nondestructive analysis using SQUID sensor, Knauss *et al.*[13] built the MAGMA-C10 system to observe the magnetic field images in conducting failure analysis with different levels of the flip-chip packaging technology, where a small SQUID about 30 μm wide was used with the sensitivity of 20 pT. However, a single-turn pickup coil of the SQUID sensor seems not

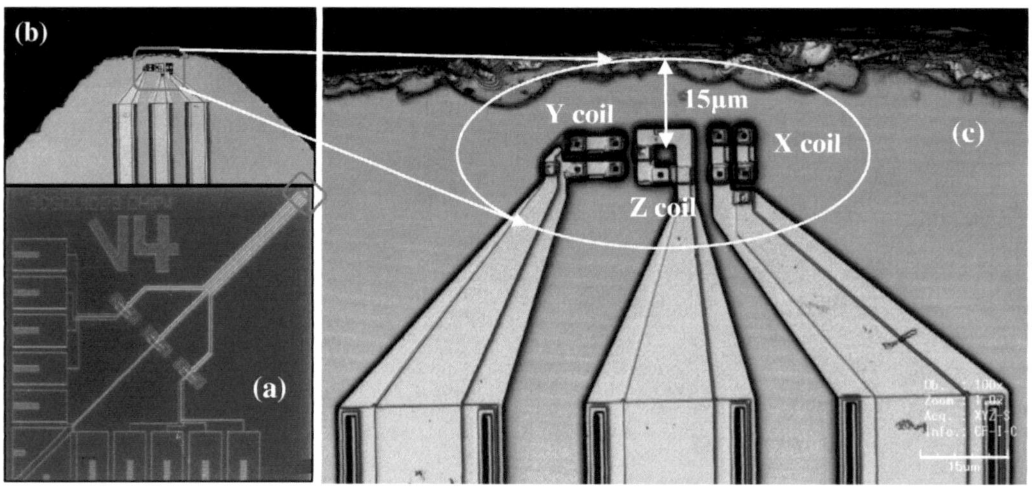

Figure 1. (a) An optical-microscope image of a vector scanning SQUID sensor. (b) A laser-microscope image of the polished corner of the vector pickup coils. (c) The details of the vector pickup loop after polishing the corner.

to have high sensitivity and fine spatial resolution enough to provide highly accurate fault analysis for small areas of the stacking layers. On the other hand, a typical scanning SQUID microscope has a single scanning pick-up coil and its SQUID readout circuit to obtain the distribution of local magnetic field component only along the *Z* direction. Therefore, a vector SQUID microscope is expected to give us an alternative tool not only for scientific research but also for various useful applications. In this study, we plan to construct vector scanning SQUID microscopy with a vector SQUID sensor fabricated by us[14], of which the sensitivity and spatial resolution are improved by reducing an inner diameter and increasing the number of pick-up-coil windings with the aid of multi-layer Nb technique. An *XYZ* piezo-driven scanner for controlling the positions of the pick-up coils is employed to provide an improved spatial resolution of magnetic field image.

2. Vector SQUID sensor

Figure 1 (a) shows a vector scanning SQUID fabricated by us [14], which was integrated on a 3 mm x 3 mm Si chip with the aid of Nb-based multi-layer technology. The vector pickup system was located at a tip of Si chip. In order to reduce the distance between vector pickup coil and the sample surface, we used a fine emery paper of roughness 5 μm to polish the tip of Si chip containing the pickup coils (see Figure 1 (b)). We succeeded in polishing the tip by the abrasion length of 70 μm to ensure a close distance between the *Z*-coil center and the end edge of the tip down to 15 μm. Therefore, the height of pickup coil from sample surface can be intentionally controlled by the *Z*-stage until the coils edge are directly touched on the sample surface and the *Z*-coil plate with a declined angle of 7 degrees (Figure 1 (b)). The geometrical configuration of our vector pickup loops was revised in this work compared to our preceding publications [14]–[16] . Figure 1 (c) is the latest version with the following features. (1) The geometry of the pickup coils was chosen to be a rectangular instead of a circle. We consider that this is preferable in achieving an enhanced spatial resolution at a much finer level than an actual coil diameter by utilizing the numerical data processing which is based on the inverse Biot-Savart law for recovering a true local magnetic field at the sample surface from observation data obtained at a finer step scan[14], [17]. Our scanning system is able to scan at a step of 10 nm at 4K by using the piezo-driven *XYZ* stages. (2) The positions of three pickup coils were optimized so as to be orthogonal with each other, and the centers of pick-up coils were carefully placed along the same single line. This design provides us a simple means how to extract the three components of a magnetic field vector at the same distance between three pick-up coils and the sample surface. The distances of the *Z*-coil center to the *X*-coil center or the *Y*-coil center were set at an integer multiple of the scanning step to calculate a magnetic field vector at the same point (x,y,z) from three channel data of the *XYZ* scanning of the SQUID microscope. (3) We built the double-winding pick-up coils in order to increase the sensitivity of SQUID sensor twice. The pickup coil system was connected directly to a SQUID circuit through a stripline with a covering superconducting Nb film. We used a gradiometer structure to couple with a SQUID washer magnetically between a feedback coil and the readout SQUID signal for canceling external electromagnetic noises. The parameters were designed in order to optimize the sensitivity and noise[18] with the screening parameter $\beta_L = 2LI_0/\Phi_0 = 1$, where I_0 is the critical current of a Josephson junction and L is the loop inductance which consists of the inductance of pickup coil, the inductance of strip-line and the inductance of two washers. We chose the AIST-STP2 standard process to fabricate our sensor. Therefore, the critical current density J_c of Josephson junctions (JJs) is 320 A/cm^2 and the critical current I_c becomes 12.5 μA in the junction area of 2.2 μm x 2.2 μm. We estimated the inductance of loop of 83 pH, the inductance of *Z* pickup coil of 20 pH for *Z*-coil with the inner size 4 μm x 4 μm and the line width of 3 μm, and the inductance of *X* or *Y* pickup coil of 10 pH of the height 2 μm. The height of *X* and *Y* coils were limited by the thickness of four-layered Nb thin films and the SiO_2 layers. Therefore, the effective areas of the *X*, *Y* pickup coils are different from that of the *Z* pick coil. We need a calibration experiment of the coil sensitivity and the numerical processing program as discussed in the preceding study[17]. Each inductance of two washers was chosen as 30 pH[18]. In order to remove the hysteretic behavior in the current–voltage curve of the SQUID sensor, two shunt resistors were connected in parallel with JJs [19], [20]. Thus, the voltage output across the SQUID was a single-valued function of an applied flux since we chose the Stewart-McCumber parameter as $\beta_c = 2\pi I_0 R_{sh} C \leq 0.7$[21], where R_{sh}, C are a shunt resistor and a self-capacitance of JJ, respectively. We calculated the C based on equation $1/C_s = 0.20 - 0.043\log_{10}J_c$, where C_s is the capacitance in an area (cm^2) [22]. Therefore, we reached the condition of $R_{sh} \leq 8\ \Omega$. We chose $R_{sh} = 7.8\ \Omega$ by taking into consideration of the standard fabrication process. The damping resistors, in parallel with washers, are chosen as 1 Ω.

3. Vector scanning SQUID system

Figure 2 shows the latest design of our vector scanning SQUID microscope, which consists of an *XYZ* piezo-driven scanner, a Gifford–McMahon cryocooler, a temperature controller, a fabricated vector SQUID sensor, and a commercial 3-CH FLL readout circuit (XFF-1 Magnicon GmbH). A vector SQUID sensor was mounted on a sensor holder (Figure 2 (c)) which was composed of a 100-μm thick cantilever (phosphor bronze), a spring plate and a twelve-pins-male connector in order to provide a soft contact between a sample and a sensor, and to make the sensor connected to the electronics system. A sample was set in face-to-face with the SQUID sensor with a declined angle of 7 degrees. The sensor holder was installed from the top of scanning SQUID system (Figure 2 (b)) and is able to slide forward for making contact with a female connector, which was soldered with 3 channel x 5 Beryllium-Copper wiring loom by following a standard prescription of connection of SQUID application with the Magnicon FLL readout circuit [23]. The flange at the top of SQUID mount is to hold SQUID sensor and to ensure a good thermal contact with the 4-K stage. The temperature of the sample is controlled by a heater and a temperature sensor. They were installed on the top flange to control the temperature by a temperature controller (Cryocon, Model 32C). The sample is mounted on a sample holder, which was

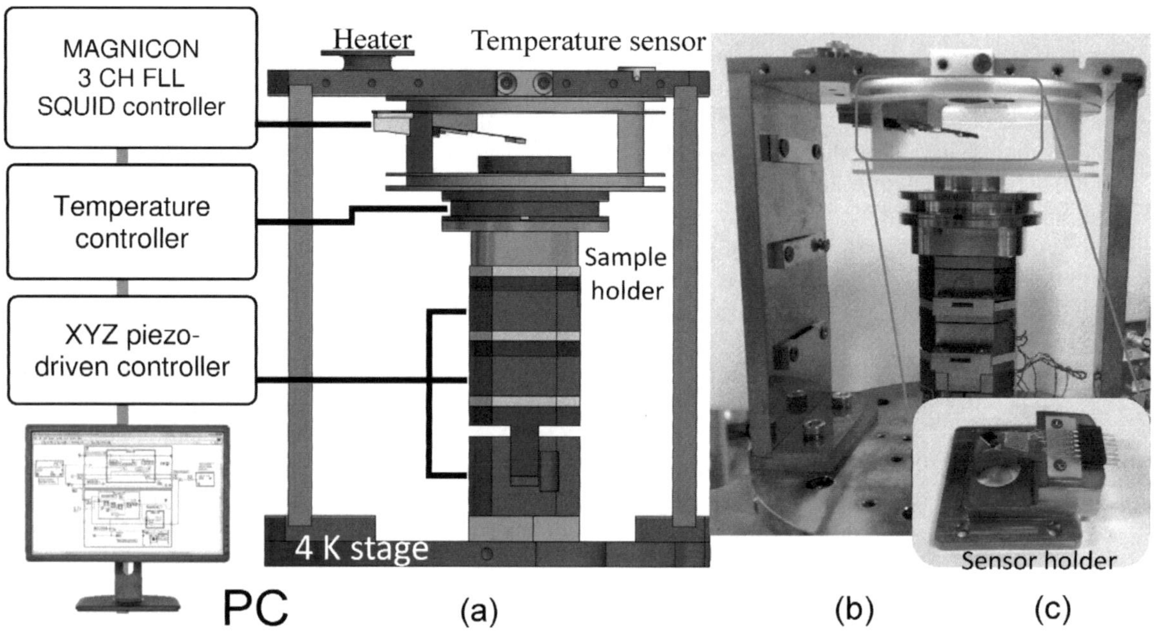

Figure 2. (a) Electronics connection to the vector-scanning microscope in the cryostat. (b) A photo of a vector scanning SQUID microscope constructed by us. (c) A vector SQUID sensor was fixed on the cantilever and the electric leads were bonded between the sensor pads and the connector terminals.

machined by oxygen-free Cu block. The sample holder was cooled down through a poor thermal conduction route to 4K stage. It took a long time for a sample stage to reach 4K in high vacuum through poor thermal conduction route, but the heat flux from the sample stage to the 4K stage can be reduced to a level less than 0.1 W and it would be enough to keep the temperature of cryostat stage at 4K within the limited cooling power of 1 W. A high resistance manganin wire of 59 Ω/m was twisted each other to wind around a copper block to provide a heater function without generating an undesired magnetic field. In order to monitor the temperature of the sample, the second temperature sensor was also installed inside of sample holder, and connected to the other channel of the temperature controller. We use an *XYZ* piezo-driven scanner for controlling the position of the sample. The sample holder is thermally linked with the 4K stage via a SUS316 stage as a poor thermal link to reduce the heat load to the refrigerator. Besides, we consider that the temperature stability of the piezo-driven scanner is important to ensure the best precision of the *XYZ* stage. The range in the *XY* plane and *Z* direction are scanned over 5 mm x 5 mm x 5 mm with a step of 10 nm at 4K which are controlled by the piezo-driven controller ANC350 motion controller (attocube systems AG). The leads from the vector SQUID sensor are fed to a vacuum connector (Lemo, EGG.3B.324.CLL) at room temperature through 15 Beryllium-Copper wiring loom, and the standard connector can be firmly coupled with the commercial Magnicon FLL readout circuit at the top of cryostat in order to keep the bandwidth 6 MHz and to reduce a disturbance from environment noises. We used a Gifford–McMahon cryocooler to decrease the temperature of the vector scanning SQUID down to 4K, which is beneficial to conduct a long-time measurement without consuming liquid helium. The commercial SQUID microscope uses liquid helium, and the measurement time is typically limited less than 24 h even if we consume 100 L of liquid helium. This is not preferable in view of conducting systematic measurements required in vortex physics under the various different conditions. We canceled an AC noise from high power compressor by setting an insulating connector in high-pressure helium line from the compressor to the GM cryocooler. The cryocooler has anti-vibration mechanism, where the main cryostat has a bellows tube and screws isolated with silicone-washers between vibrating flange and the anti-vibrating flange. Inside of the cryostat, we separated the vibration and anti-vibration sections, of which the thermal links are connected by Ag-coated high-purity soft Cu mesh wires. We coded a LabVIEW (National Instrum. Inc.) program to control all of the equipment via the USB interface or the GPIB interface.

4. Experimental results

First, we have to test the SQUID sensor designed for our vector scanning SQUID microscope. For this purpose, we set a thin-wire field coil with a diameter of 10 mm on a test package and attached in a sample holder [14]. We set the center of field coil at the tip of the vector SQUID sensor so that the field coil is able to apply the magnetic field to all the *XYZ* coils simultaneously in the cryostat for conducting measurements by the curve tracer [14]. We used a current source (Keithley Inc., Model 2400) to feed a dc current to the field coil for generating an external magnetic field to the vector SQUID sensor. The small field coil does generate inhomogeneous magnetic field, and is hence suitable for providing the three components simultaneously at the *XYZ* coils. We argue that the magnetic field from the tiny field coil has a main *Z* component and smaller *X* and *Y* components. We cooled the temperature of the vector SQUID sensor down 4K. The sensor was connected alternatively with the characteristic measurement system[14], the FLL readout circuit is used to measure the *V*-Φ characteristics and the voltage readout of sensor under an external magnetic field.

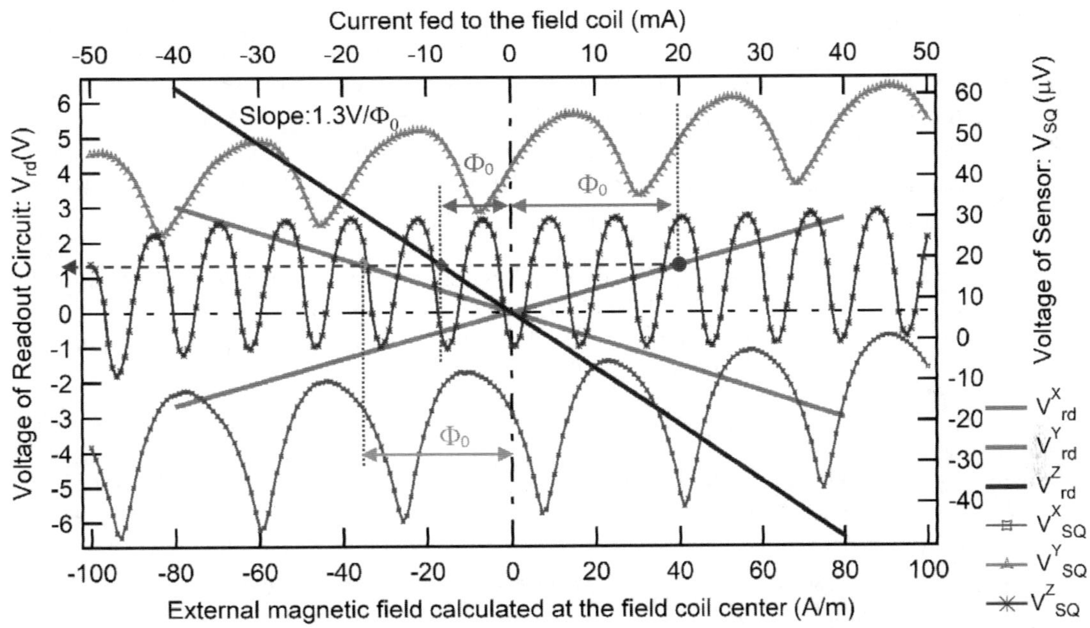

Figure 3. The comparison of the *V*- Φ characteristics with the voltage of readout circuit from Magnicon FLL readout electronics circuit, which was tuned to give V_{rd}=1.3 V against each Φ_0 external magnetic field. The horizontal scale is to refer an estimated magnetic field for the *Z* component.

Figure 3 shows a comparison of the *V*-Φ characteristics of a vector SQUID sensor which was obtained by the measurement system constructed by us, and the output voltage from the FLL readout circuit at the same external magnetic field under the bias current of 35 μA. The graph of the *V*-Φ characteristics shows that the magnetic field was measured not only by the *Z* pick-up coil but also by the *X* and *Y* coils. There, the periods of the *V*-Φ characteristics of *X, Y, Z* coils correspond to the magnetic field intensities which were estimated at the coil center of 34 A/m, 40 A/m and 16 A/m, respectively. We found that the field sensitivity against of the *Z*-magnetic field is the highest while it is lower in *X* and *Y* components of the magnetic field. While the thin-wire field coil generates an inhomogeneous magnetic field with the three components, the *Z* component is the main one in perpendicular to the *Z* pickup coil. On the other hand, the smaller areas of the X and Y pickup coil inducted a lower flux sensitivity compared to the Z coil. We can reasonably understand the different field sensitivity against measured by the *X, Y* and *Z* coils. Therefore, we confirmed the proper operation of a vector SQUID sensor to measure a vector magnetic field in the present studies. We also demonstrated that the commercial Magnicon FLL readout electronics circuit worked well to measure the three channels from the *X, Y*, and *Z* SQUID outputs. The output voltage from the readout circuit was linearly proportional to the external magnetic field and was sensitive to the polarity in the *X, Y*, and *Z* channels. According to Fig. 3, we were successful in converting from the flux-sensitive oscillation of the vector SQUID sensor to a voltage output of the Magnicon FLL readout circuit. In the Magnicon FLL readout circuit [23], we used a SQUID Viewer software to tune the conversion ratio of the FLL-output voltage per one flux quantum so that the output voltage of the readout circuit increases by 1.3 V with a one-period increase in the SQUID-loop flux. Our method is surely useful to reconstruct a vector image mapping of any magnetic fields. We believe that our vector SQUID sensor is compatible with the commercial Magnicon FLL readout circuit in converting the V-Φ characteristic to the voltage of readout circuit.

5. Summary

We designed the vector SQUID sensors and fabricated them successfully. We confirmed that our sensor is suitable for constructing a vector SQUID microscopy system. We used the measurement system built by us to investigate the fundamental characteristics of 3D SQUID sensors, systematically. We confirmed that our sensors showed a good performance, which is suitable for our novel SQUID microscope. We revealed that the voltage outputs were adequate since a commercial SQUID readout circuit were operating properly. The results were in good agreement with the profiles of the V-Φ characteristics. We are almost at the final stage in constructing a novel SQUID microscope before completing the apparatus.

Acknowledgements

This work was supported by Grant-in-Aid for Challenging Exploratory Research (No. 25600018, No.15K13979) and Grant-in-Aid for Young Scientists (B) (No. 26820130 and No. 26800192) from Japan Society for the Promotion of Science (JSPS). We thank Masaaki Maezawa for his useful discussions and suggestions how to design SQUID sensors.

References

[1] Shannon K R, 2004 *Superconductivity, the London Moment and the SQUID: Reading the Unreadable* Spencer J, Kahn B, editors. Gravity Probe B: Exploring Einstein's Universe with Gyroscopes NASA **1**, 25-27
[2] Drung D, Beyer J, Kirste A, Peters M, Ruede F, Hinnrichs C, Bundesanstalt P, and Gbr M 2007 *IEEE/CSC ESAS Eur. Supercond. NEWS FORUM*, **1**, 1–7
[3] Kirtley J R, Ketchen M B, Stawiasz K G, Sun J Z, Gallagher W J, Blanton S H, and Wind S J, 1995 *Appl. Phys. Lett.* **66**, 1138–1140
[4] Huy H T, Kato M, and Ishida T 2013 *Supercond. Sci. Technol.* **26** 065001
[5] Vu T D, Matsumoto H, Miyoshi H, Huy H T, Shishido H, Kato M, and Ishida T 2017 *Physica. C Supercond. its Appl.* **533** 44–48
[6] Fong L E, Holzer J R, McBride K K, Lima E A, Baudenbacher F, and Radparvar M 2005 *Rev. Sci. Instrum.* **76** 053703
[7] Zutic I, Fabian J, and Das Sarma S 2004 *Rev. Mod. Phys.* **76** 323–410
[8] Tanaka S, Akai T, Hatsukade Y, and Suzuki S 2009 *IEICE Trans. Electron.* **E92–C** 323–326
[9] Morooka T, Tanaka K, Odawara A, Nakayama S, Nagata A, Ikeda M, and Chinone K 2000 *Physica. C Supercond. its Appl.* **335** 157–163
[10] Huy H T, Shishido H, Hayashi M, Yotsuya T, Kato M, and Ishida T 2013 *Physica C: Supercond. its Appl.* **484** 86
[11] Huy H T, Kato M, Sato O, Hayashi M, Yotsuya T, and Ishida T 2013 *Physica C: Supercond. its Appl.* **494** 99–101

[12] Huy H T, Hayashi M, Kato M, Van Hieu N, and Ishida T 2014 *IEEE Trans. Magn.* **50** 18–21
[13] Knauss L A, Cawthorne A B, Lettsome N, Kelly S, Chatraphorn S, Fleet E F, Wellstood F C, and Vanderlinde W E 2001 *Microelectron. Reliab.* **41** 1211–1229
[14] Vu T D, Toji M, Huy H T, Miyajima S, Hiroaki S, Hidaka M, Hayashi M, and Ishida T 2017 *J. Phys. Conf. Ser.* **871** 012075
[15] Miyajima S, Okamoto T, Matsumoto H, Huy H T, Hayashi M, Maezawa M, Hidaka M, and Ishida T 2015 *IEEE Trans. Appl. Supercond.* **25** 3–6.
[16] Vu T D, Huy H T, Miyajima S, Matsumoto H, Miyoshi H, Okamoto T, Maezawa M, Hidaka M, and Ishida T 2016 *15th International Superconductive Electronics Conference, ISEC 2015* 2
[17] Hayashi M, Ebisawa H, Huy H T, and Ishida T 2012 *Appl. Phys. Lett.* **100** 182601
[18] Tesche C D and Clarke J 1977 *J. Low Temp. Phys.* **27** 301–331
[19] Stewart W C 1968 *Appl. Phys. Lett.* **12** 277–280
[20] McCumber D E 1968 *Journal of Applied Physics* **39** 3113-3118.
[21] Clarke J and Braginski A I 2004 *The SQUID Handbook* WILEY-VCH Verlag GmbH & Co. KGaA, Weinheim **1**.
[22] Maezawa M, Aoyagi M, Nakagawa H, Kurosawa I, and Takada S 1995 *Appl. Phys. Lett.* **66** 2134
[23] Manual of XXF-1 electronics, Magnicon GbR. Lemsahler Landstraße 171, D-22397 Hamburg, Germany: http://www.magnicon.com/squid-electronics/, 2013.

A Thirty-year History of Superconducting Microwave Devices and Fundamental Studies Thereof

Shigetoshi Ohshima

Yamagata University, Jonan 4-3-16, Yonezawa 992-8510,

ohshima@yz.yamagata-u.ac.jp

Abstract. Thirty years have passed since the discovery of high-temperature superconductors (HTS), and practical applications of HTS equipment and devices are moving into full-swing. From the beginning of their discovery, the use of HTS in microwave devices has been recognized as one of their most practical applications. This is because the surface resistance of HTS is about 1/1000 of that of pure copper at 1 GHz and 77 K. From early on, researchers have recognized that high performance of microwave passive devices could be realized by utilizing the characteristics of HTS. In addition, the size of HTS microwave devices is relatively small, and they can be cooled by a small cryocooler. Furthermore, we were able to establish high-quality HTS thin-film fabrication technology essential for producing high performance microwave devices in a short period of time. In view of the above, I would like to present an overview of the past thirty years of superconducting microwave devices and fundamental studies thereof.

1. Introduction

High-temperature superconductors (HTS) were discovered over thirty years ago and since then, their superconducting applications have been used in various fields. Their practical uses include bandpass filters, microwave kinetic inductance detectors (MKIDs) and transition edge sensors (TES). In this paper, we review the thirty-year history of HTS and discuss the following topics.
(1)Superconducting properties needed for high performance microwave superconducting devices
(2)Topics concerning superconducting microwave devices reported so far
(3)Practical use of superconducting bandpass filters
(4)Superconducting microwave devices expected in the future

2. Superconducting properties needed for high performance microwave superconducting devices

The most important superconducting property necessary for realizing high performance microwave devices is surface resistance (*R*s). *R*s is defined as the ratio of the electric field (E) and the magnetic field (H), R_s = E / H, when the conductor is irradiated with electromagnetic waves. Since superconductors have very low electrical resistance at frequencies below microwave frequency, their Rs is quite small compared to normal metals. The temperature dependent R_s of Al and Cu films and $YBa_2Cu_3O_y$ (YBCO) films are shown in Figure1. The R_s of YBCO thin films in the superconducting state at 1 GHz is approximately three orders smaller than that of Cu and Al thin films. This low R_s is the driving force for high performance and miniaturization of microwave superconducting devices. The R_s of superconducting thin films is small even in the magnetic field. Figure 2 shows the relationship between the applied magnetic field and the R_s of YBCO films. $R_s(90)$ and Rs(0) indicate the surface resistance when the magnetic field is applied normal and parallel to the film plane, respectively. R_s of YBCO films in a high magnetic field is quite smaller than that of pure Cu films [1-3]

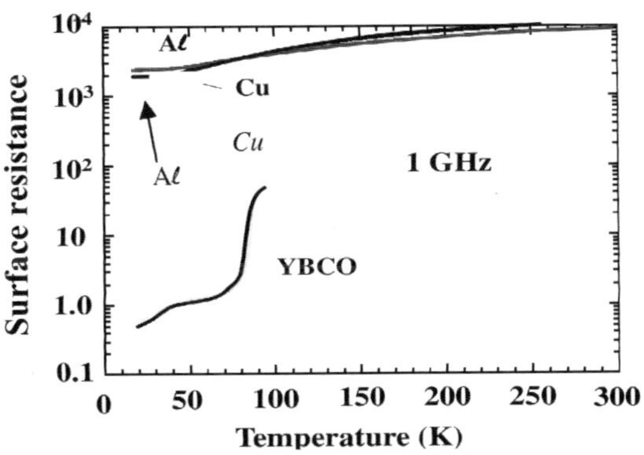

Figure 1. Temperature dependent R_s of Al, Cu and YBCO films

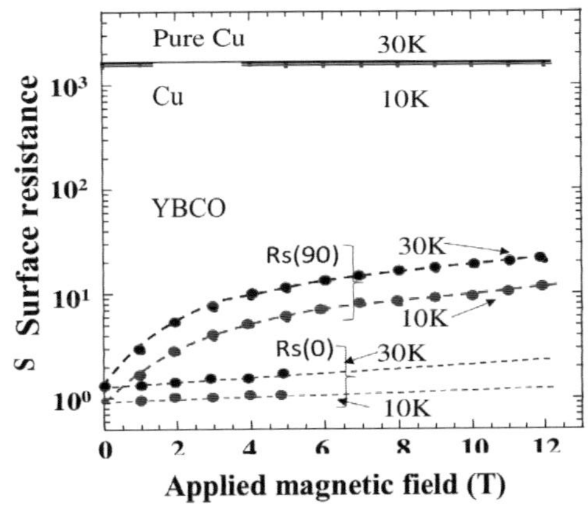

Figure 2. Applied magnetic field dependent R_s of YBCO, and Cu films

3. Topics concerning superconducting microwave devices over the past thirty years

3.1. Electrical small antenna

The first focused microwave device to be developed when HTS were discovered was an electrical small antenna. The electrical small antenna was cooled with liquid helium had a high gain and high directivity [4]. More practical applications would be possible if the electrical small antenna could be made with HTS, which could maintain a superconducting state with liquid nitrogen. Liquid nitrogen is easier to handle than liquid helium. Itoh et al. made an electrical small helical antenna with Bi-2223 wire and demonstrated that it had high gain and high directivity [5-6]. A helical antenna, however, cannot be cooled by a small refrigerator and that is a serious problem for practical use. Patch antennas can be cooled by a small refrigerator and many researchers have investigated various types of patch antennas [7-9]. Ehata et al. investigated an array square-patch antenna and mini- refrigerator for patch antennas [10]. According to their experimental results, the superiority of the superconducting antenna could be demonstrated, but it could not be put into practical use. This was due the disadvantage that superconducting antennas must be cooled to a low temperature.

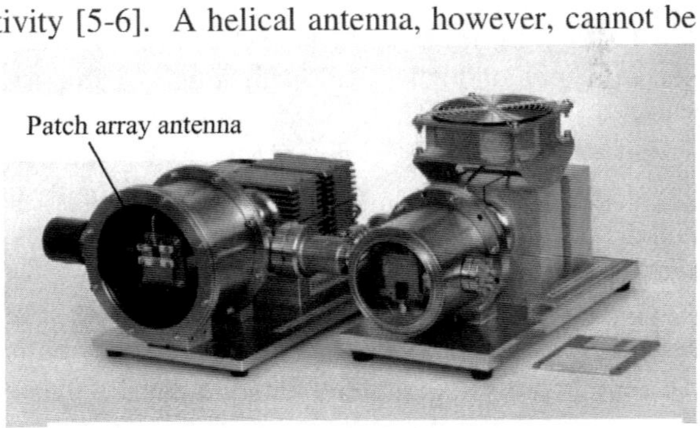

Figure 3. Picture of mini- refrigerator for superconducting patch antenna.

3.2. Superconducting bandpass filters for wireless communication

Superconducting bandpass filters have advantages compared to conventional filters, such as (1) large out-band rejection, (2) strong frequency selection and (3) low insertion loss. In the late 1990s, as the spread of wireless communication rapidly increased in the United States, deteriorating quality of communication was a big problem. This is because the usual band pass filter failed to distinguish the frequency bands of the companies. It was found that the quality of speech in mobile-phones was improved significantly by using an HTS filter and superconducting base stations have spread rapidly in the United States [11-12]. Although the use of a superconducting bandpass filters was also studied in the third generation wireless communication system IMT-2000 [13-16], unfortunately, the superconducting bandpass filter system was not adopted because of its high price

The next target of practical application for superconducting bandpass filters was software communications. Software wireless communication that can share facilities in the base station is an attractive next generation wireless communication method. It was important to develop a filter capable of shifting the resonance frequency in a short time. Several techniques for shifting the center frequency of the superconducting bandpass filters have been proposed [17-20]; however, the most practical method is to use a tuning rod. Ohshima et al. developed an automatic tuning system using dielectric rods. Figure 4 shows a picture of the apparatus. They could shift the center frequency of the bandpass filter in a very short time, less than 0.5 seconds as shown in figure 5. [21-22].

Figure 4. Automatic tuning system for HTS bandpass filter.

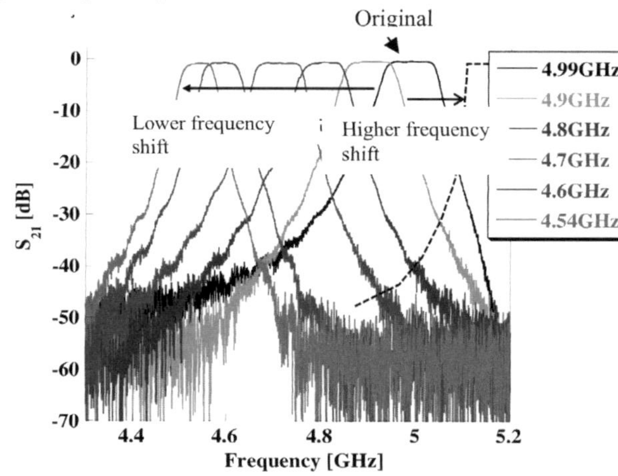

Figure 5. Center frequency shift of HTS bandpass filter using automatic tuning system shown in Fig. 4

4. Practical application of HTS bandpass filters

Several superconducting bandpass filter systems have already been put to practical use besides cellular phone base stations. Here, we show three applications: (1) reduction of city noise for radio astronomy, (2) ultra- narrow bandpass filters for weather radars. (3) ultra-wide band filters for UWB communication. UWB-communication filters are not yet practically used; however, in the future, practical use is possible. The three types of superconducting bandpass filters are introduced below.

4.1 Reduction of city noise for radio astronomy

It is necessary to receive weak radio waves coming from outer space with high sensitivity at radio astronomical observatories, which are usually installed at the top of a high mountain or in desert areas. However, it may be necessary to install such a system in the city. Kashima Space Research Center installed a radio telescope of 34m and received a weak cosmic radio wave. When third generation wireless communication became widespread, noise in the mobile phone frequency band was superimposed on weak radio waves coming from outer space and cosmic radio waves could not be observed. Therefore, it was necessary to use superconducting bandpass filters with high out-of-band cutoff characteristics. By installing superconducting bandpass filters in front of the pre-amplifier of a radio wave telescope, noise from the wireless communication base station could be removed [23-24].
A similar problem was raised in world radio telescopes installed in cities. Figure 6 shows the experimental data of the removal of city noise using a superconducting bandpass filter system [25]. Mobile noise and Wi-Fi noise could be removed by using the superconducting filter.

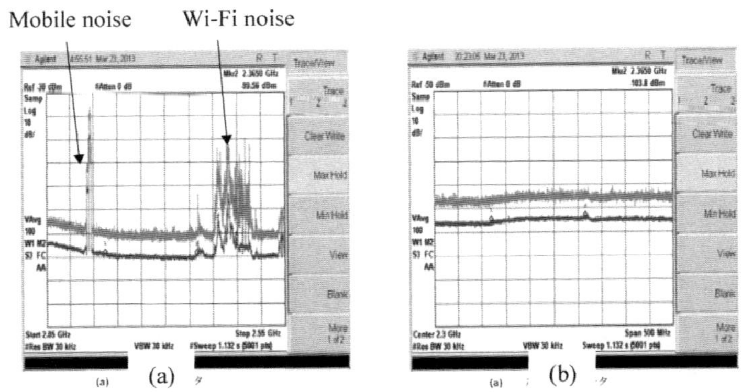

Figure 6. Reduction of city noise by superconducting bandpass filter.
(a) using a conventional filter (b) using a superconducting filter

4.2 Ultra narrow bandpass filters for a weather radar

The insertion loss and skirt characteristics of the bandpass filter depend on the R_s of the films used, but they also affect the bandwidth of the filter. As the bandwidth decreases the insertion loss increases if the R_s of the film is not small enough. In Japan, to secure the frequency of the rapidly increasing wireless LAN networks, the frequency band of weather radar have been narrowed. The problem here is the magnitude of the receiving signal. Since the power transmitted from the weather radar exceeds several dozen kW, the receiving signal may also be large; however, the power tolerance of the superconducting bandpass filter is less than 10 W. Therefore, ingenuity is necessary. Toshiba group has created the circuit as shown in figure 7 [27]. The same frequency radio waves as the transmission power do not pass through the superconducting bandpass filter and only weak signal waves with slightly shifted frequency can pass through. Through such a circuit, they could construct a highly sensitivity weather radar system.

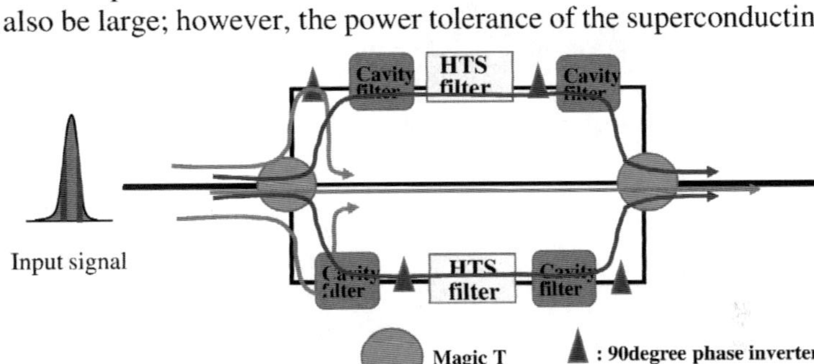

Figure 7. Circuit model of superconducting bandpass filter for weather radar

4.3 Ultra-wide band filter for UWB communication

The necessity of ultra-high speed communication has been increasing year by year and various methods are being studied. Among them, UWB communication using an ultra-wide band filter is suitable for ultra-high speed communication in a limited space and future progress is expected. Fig. 8 shows the schematic figure of the frequency band and radiation power of W-CDM, GSM and UWB communication. UWB communication has a small power transmission and wide frequency band. Superconducting bandpass filters are suitable for this type of communication and are being examined throughout the world. We also examined the superconducting bandpass filter for UWB communication. There are two frequency bands for UWB communication in Japan: 3.4 - 4.8 GHz (lower band) and 7.25 - 10.25 GHz (higher band). Fig. 9 shows the structure of a UWB superconducting filter for low band, and Fig 10 shows the results. The filter meets the specification of UWB communication [28].

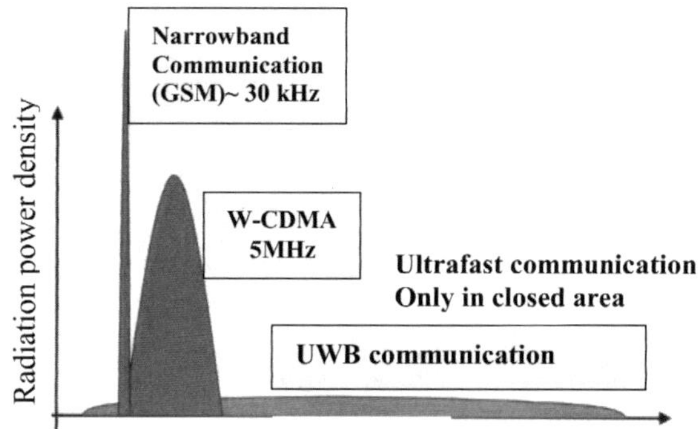

Figure 8. Schematic figure of frequency band and radiation power of W-CDM, GSM and UWB communication

Figure 9. Shape of the UWB superconducting filter

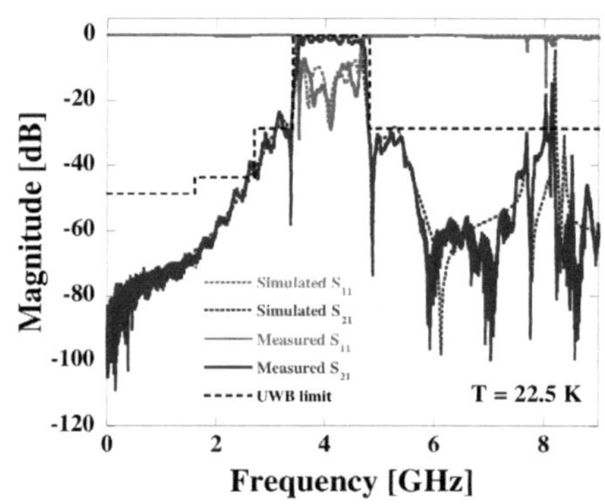

Figure 10. Frequency response of UWB superconducting filter

5. Prospective microwave superconducting devices

There are several superconducting microwave devices besides the bandpass filter. Among them, NMR pickup coils and MKIDs for terahertz wave detection are close to being put into practical application.

*5.1 NMR pickup coils*Although nuclear magnetic resonance (NMR) is fundamentally not very sensitive, improvement of sensitivity is important and is being vigorously studied in the United States [29-30] and Japan [31-32]. Equation (1) shows the S/N of an NMR system, where, Q_L, T_{eff}, T_R and η are the quality factor of the coil, effective sample temperature, pre-amplifier temperature and filling factor of coil, respectively. It is important to increase Q_L. To increase Q_L, the R_s of the pickup coil material must be small. As shown in Fig. 2, the R_s of the superconducting thin film is more than two orders of magnitude smaller than the usual metal even in a high magnetic field. Therefore, it is important to examine a superconducting pickup coil to improve NMR sensitivity. Fig. 11 shows a schematic drawing of the NMR superconducting coil. As shown in this figure, we use a 2 flats coil.

$$S/N \propto \sqrt{\frac{\eta Q_L}{T_{eff} + T_R}} \qquad (1)$$

Figure 12 shows the input power dependence on Q_L of the NMR superconducting pickup coil. The applied magnetic field is 14.1T and the resonance frequency is 800 MHz. The value of Q_L is over 15,000 with a low input power and the value decreases with increasing input power. The value of Q_L for the superconducting pickup coil is about fifty times the Q_L of the room temperature copper wire coil. Therefore, if the superconducting pickup coil is used, NMR sensitivity can be improved.

Figure 11. Schematic drawing of NMR superconducting pickup coil

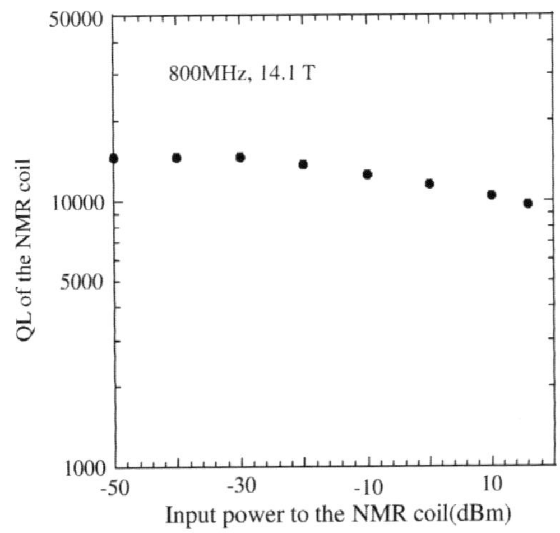

Figure 12. Input power dependence on Q_L of the NMR superconducting pickup coil

5.2 MKIDs of terahertz wave detection

In recent years, it has become important to observe the electromagnetic waves of submillimeter waves coming from outer space in order to elucidate the creation of the universe. Several methods of detecting submillimeter radio waves have been proposed, but recently MKIDs have been shown to be a useful device to detecting such waves. An MKID is an element combining an antenna for receiving a submillimeter - terahertz wave and a resonator for resonating at ~ GHz. Figure 13 shows the schematic drawing of the frequency shift of an MKID after receiving a submillimeter wave. By observing this frequency shift, it is possible to do wide-band detection from submillimeter wave to terahertz wave. The advantage of this element is that: (1) we can observe submillimeter waves to terahertz waves by evaluating the resonance in the microwave range, (2) it is easy to realize an array beyond 500 eleme Studies for the development of MKIDs are currently spreading around the world. Figure 14 shows a photograph of the MKID array reported by the Otani Group at RIKEN [33-34]. The MKIDs were fabricated of a high quality Al film prepared by the MBE method. They will be put into practical use shortly.

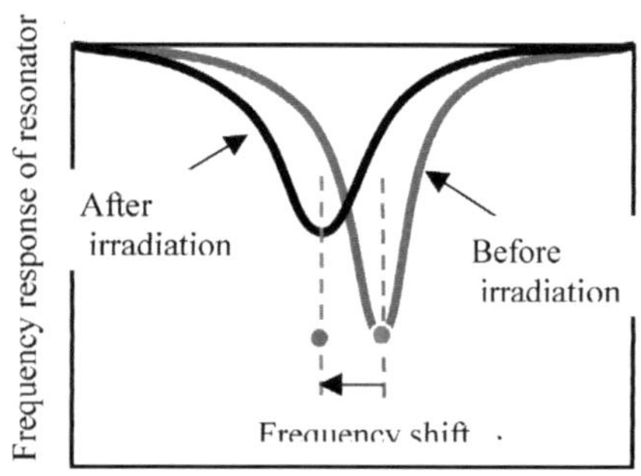

Figure 13. Frequency shift of MKID after receiving terahertz wave

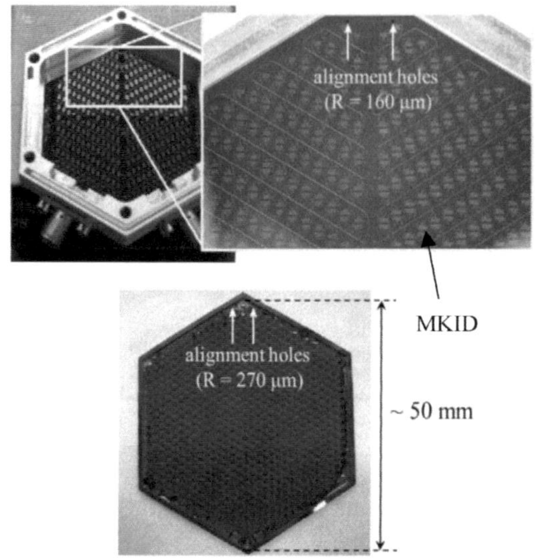

Figure 14. Photograph of MKIDs array

6. Summary

In this paper, we reviewed the thirty-year history of superconducting microwave devices, introduced practical uses of superconducting bandpass filters in recent years and discussed how superconducting NMR pickup coils and MKIDs for terahertz wave detection are expected to be put to practical use in the future. In addition, it is important to use a low R_s superconducting film to realize a high performance superconducting microwave device. It was shown that the R_s of c-axis oriented YBCO films is experimentally found to be three orders of magnitude smaller than that of thin metal films in both a non-magnetic field and a magnetic field of around 1GHz. We are expecting further applications of superconducting microwave devices in the future.

Acknowledgement

In summarizing this paper, we would like to show appreciation to the following people: Prof. B.S. Cao of Tsinghua Univ. in China, Dr. Kayano of Toshiba Corporation, Dr. Otani of RIKEN. Q of the NMR superconducting pickup coil in the magnetic field was measured by Mr. Tsuji of JEOL RESONANCE, Co, Ltd. English grammar was checked by Prof. Matthew Zisk of Yamagata University.

References

[1] Nakayama S, Honma T, Nakagawa K, Saito A and Ohshima S 2010 *Physica C* **470** 1042
[2] Nakagawa K, Honma T, Takeda K, Ono S, Kai H, Saito A, Mukaida M, Nakajima K and Ohshima S 2011 *IEEE. Trans. Applied Superconductivity*, **21** 587
[3] Honma T, Sato S, Sato K, Watanabe M, Saito A, Koike K, Kato H and Ohshima S 2013 *Pysica C*, **484** 46
[4] Adachi S, Mohri Y, Ohnuki S and Ashida K 1978 *Proc. International Symposium on Antenna and Propagation*, **A-7** 2
[5] Itoh K, Ishi O, Nagai Y, Suzuki N, Kimachi Y and Michikami O 1992 *J. Superconductivity* **5** 485
[6] Itoh K., Ishii O, Nagai Y, Suzuki N, Kimachi Y and Michikami O 1993 *IEEE Trans. Applied Superconductivity* **l.3** 2836
[7] Ohshima S, Ogasawara T and Ehata K 1997 *IEEE. Trans. Applied Superconductivity* **7** 3060
[8] Lancaster MJ, Wang HY and Hong JS 1998 *IEEE Trans. Applied Superconductivity* **8** 168
[9] Barkat O and Benghalia A 2009 *J. Infrared Millimeter and Terahertz Waves* **30** 1053

[10] Ehata K, Kusunoki M and Ohshima S 1999 *Electronics and Communications in Japan, partII-Electronics* **82** 59
[11] Simon RW, Hammond RB, Berkowitz SJ and Willemsen BA 2004 *Proc. The IEEE* **92** 1585
[12] Willemsen BA 2001 *IEEE Trans. Applied Superconductivity* **11** 60
[13] Tsuzuki G, Suzuki M and Sakakibara N 2000 *IEEE Trans. Microwave Theory and Techniques*, **48** 2519
[14] Setsune K, Enokihara A and Mizuno K 1998, *IEICE Trans. Electronics* **E81C** 1578
[15] Guo XB, Cao BS, Wei B, Zhang XP, Yin ZS, He WJ, He S, Gao LM, Zhu MH and Gao BX 2004 *Chinese J. Phys*. **42** 463
[16] Zhang XQ, Meng QD, Li F, Li CG, Li ZS, He AS, Li H and He YS 2009 *Superconductor Science Technology* **19** 394
[17] Malloy DS and Kadin AM 1997 *IEEE Trans. Applied Superconductivity* **7** 2788
[18] Miranda FA, Subramanyan G, Van Keuls FW, Romanofsky RR, Warmer JD and Mueller CH 2000 *IEEE Trans. Microwave Theory Techniques* **48** 1181
[19] Fuke H, Terashima Y, Kayano H, Yamasaki M, Aiga F and Katoh R 2001 *IEEE Trans. Applied Superconductivity* **11** 434
[20] Terashima Y, Fuke H, Aiga F, Yamasaki M, Kayano H and Kato R 2002 *Pysica C Superconductivity and Its applications* **366** 183
[21] Ohsaka M, Takeuchi S, Ono S, Lee JH, Saito A, Akasegawa A, Yamanaka K, Kurihara K and Ohshima S 2008 *Pysica C Superconductivity and Its Applications* **468** 15
[22] Saito T, Ono S, Uno M, Endo Y, Saito A, Nakajima K and Ohshima S 2011 *IEEE Trans. Applied Superconductivity* **21** 563
[23] CRL News ; 2004 **336**, (in Japanese)
http://www.nict.go.jp/publication/CRL_News/0403/frame/001_flame.html
[24] Sekido M, Kawai E and Takefuji K, "Reducton of interference signals of L-band of Kashima 34m radia telescope" CRL. News, 2014
[25] Provided data from Prof. BS Cao of Tsinghua University in China
[26] Ohshima S 2017 *J. Institute of Electronics.* **J100-C**, (in Japanese)
[27] Kawaguchi T, Shiokawa N, Nakayama K, Watanabe T, Hashimato T and Kayano H 2009 *IEICE Trans. Electronics* **E92C** 296
[28] Ishii H, Kimura T, Kobayashi N, Saito A, Ma Z, and Ohshima S 2013 *IEEE Trans. Applied Superconductivity* **23** 1500204
[29] Haward D and Hill W 1997 *IEEE Trans. Applied Superconductivity* **7** 3750
[30] Brey WW, Edison AS, Nast RE, Rocca JR, Shaha S and Withers RS 2006 *J. Magnetic Resonance* **179** 290
[31] Yamada T, Saito A, Oikawa S, Koshita K, Takahashi M, Maeda H and Ohshima S 2015 *IEEE Trans. Applied Superconductivity* **25** 1500504
[32] Koshita K, Kitajima K, Yamada T, Takahashi M, Maeda H, Saito A and Ohshima S 2016 *IEEE Trans. Applied Superconductivity* **26** 1500104
[33] Li J, Yang JP, Lin ZH, Liu D, Shi SC, Mima S, Furukawa N and Otani C 2016 *J. Low Temperature Physics* **184** 103
[34] Oguri S, Choi J, Damayanshi T, Hattori M, Hazumi M, Ishitsuka H, Karatsu K, Mima S, Minowa M, Nagasaki T, Otani C, Sekimoto Y, Tajima O, Tomita N, Yoshida M and Won E 2016 *J. Lowtemperature Physics* **184** 786

IOP Conf. Series: Journal of Physics: Conf. Series **1054** (2018) 012061 doi:10.1088/1742-6596/1054/1/012061

HTS Filter with Dielectric Rods for Tuning the Center Frequency and Trimming the Passband ripple

T Unno[1], N Sekiya[1]

[1] The Integrated Graduate School of Medicine, Engineering, and Agricultural Sciences, University of Yamanashi, Takeda 400-8511, Japan

Abstract. Tuneable filters are one of the most essential microwave components for multiband communication systems due to their attractive features. In this study, we developed high-temperature superconducting (HTS) tuneable microstrip line hair-pin resonator filter with dielectric rods for tuning the center frequency and trimming the passband ripple simultaneously. Several dielectric rods are placed above the HTS filter, and the center frequency and passband ripple are controlled by changing the distance between the dielectric rods and the resonators. Two dielectric rods are used for each resonator; one is located on the center of the resonator for tuning the center frequency and trimming the pass-band ripple, and the other is located on the open end of the resonator to tune the center frequency. The filter was designed to have a center frequency of 1.4 GHz and a bandwidth of 28 MHz by using an electromagnetic simulator based on the moment method. The HTS filter was fabricated using double-sided $YBa_2Cu_3O_y$ thin film on a 30×40×0.5 mm Al_2O_3 substrate. The dielectric rods with a dielectric constant of 39 were used. As a result, the tuning range of the filter was about 140 MHz while keeping the all most the same insertion loss with trimming. The measurement data after trimming and tuning have a good agreement with the design specifications.

1. Introduction

In the future, microwave tuneable filters will be useful for wireless communication systems. Also, High temperature superconducting (HTS) filters which have low insertion loss and sharp skirt rejection are great importance for variety of application. For wireless communication systems, HTS tuneable filters need wide-range tunability of the passband with remaining passband ripple. There are several reports on the center frequency tuneable HTS filters and trimming method for improving passband ripple of HTS filters [1]. H. Peng et al. reported the trimming method which improves the passband ripple by adjusting the resonant frequency using dielectric and metallic screws [1]. In addition, Ohshima et al. had developed a high-speed tuning system for HTS microstrip line filters using a dielectric plate for tuning the center frequency, dielectric rods for trimming the passband ripple, and conducting rods for improving the bandwidth [3]. However, the tuning system have to change three parameters such as distance between filter and the dielectric plate, the dielectric rods and the conducting rods for tuning and trimming. Thus, this mechanism is complex to tune the center frequency, trim the pass-band ripple and trim the bandwidth.

In this work, we develop a simple mechanical center frequency tuneable HTS filter. We used dielectric rods for tuning the center frequency and trimming the pass-band ripple simultaneously. To obtain large tuning range of center frequency, two dielectric rods with high dielectric constant is used for each resonator.

2. Center frequency tuning method and pass-band trimming method

We designed a 4 pole Chebyshev-type microstrip bandpass filter. The filter was constructed with stepped impedance hair-pin resonators. The center frequency f_c is set to 1.4 GHz and the bandwidth is set to 28 MHz (fractional bandwidth: 2.0%). The required coupling coefficients (k_{ij}) and external quality factors (Q_e) of the passband are $k_{12} = k_{34} = 0.019$, $k_{23} = 0.014$ and $Q_e = 43.4$. The coupling coefficients and external quality factors were calculated using design formulas reported elsewhere [8]. The filter was designed using a Sonnet EM electromagnetic simulator and a MW-STUDIO 3-dimensional simulator.

2.1. Tuning of the resonant frequency

Dielectric rods can change the resonant frequency by changing the shunt capacitance to the ground [1]. To have a great influence on the resonant frequency, the rods are placed above the resonator where the electric field is concentrated. Figure 1 shows the layout of the resonator with dielectric rods. As shown in Figure 1(b), the rods were placed above the center of the resonator (Rod A) and the edge of the resonator (Rod B). We used the Rod A with a diameter of 2.35 mm and the Rod B with a diameter of 4.0 mm. Using Dielectric rod which has low dielectric loss (tanδ = 0.00005) and a high dielectric constant (er = 39) enable to be low loss and a wide frequency tuning range. We used two rods for the resonator to increase the tuning range.

Figure 2(a) and (b) show the simulated tuneable resonant frequencies using the Rod A and Rod B independently when distances h_A and h_B shown in Fig. 1(c) were changed. As shown in Fig. 2(a), when the Rod A was used, the frequency tuning range was 54.0 MHz. As shown in Fig. 2(b), when the Rod B was used, the frequency tuning range was 79.0 MHz. The frequency tuning range using Rod B is larger than one using Rod A. This is because the large electric field is concentrated at the open-end of the resonator. To achieve a large frequency tuning range, we used the Rod A and Rod B at the same time. Figure 3 shows the simulated tuneable resonant frequencies using the Rod A and Rod B at the same time. The obtained frequency tuning range is 123 MHz.

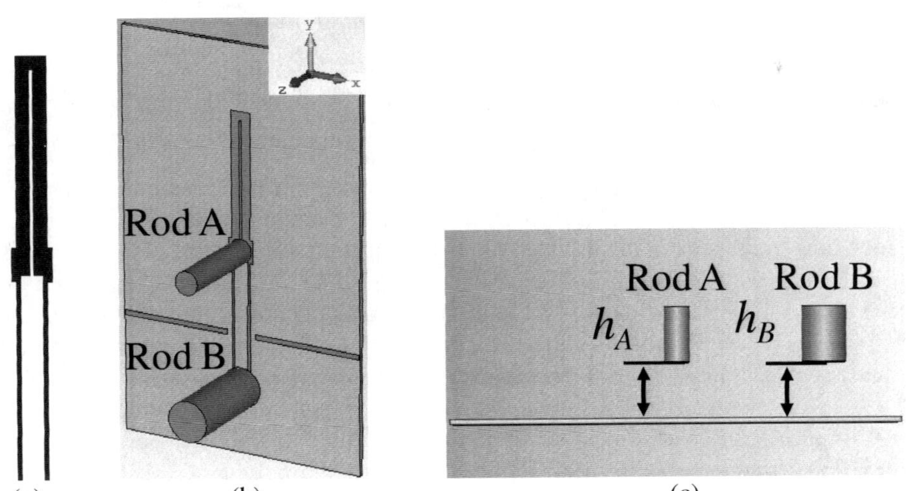

Figure 1. Layout of the resonator with the dielectric rods. (a) resonator only (b) surface (c) side

IOP Conf. Series: Journal of Physics: Conf. Series **1054** (2018) 012061 doi:10.1088/1742-6596/1054/1/012061

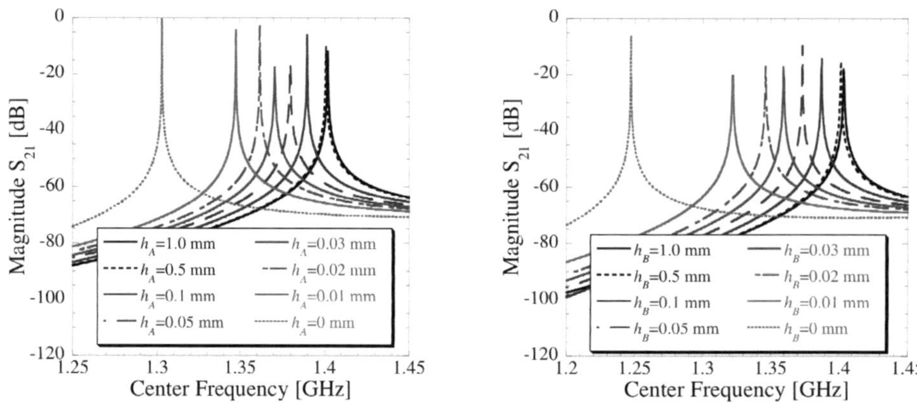

(a) (b)

Figure 2. Simulated frequency responses of the resonator with varying the gap (a) h_A (b) h_B

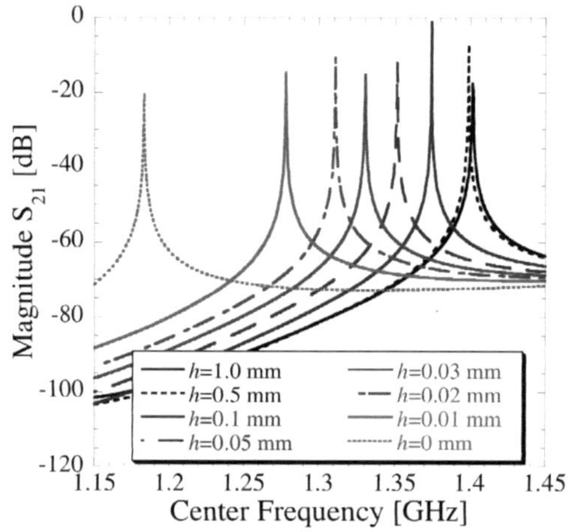

Figure 3. Simulated frequency responses of the resonator with varying the gap $h=h_A=h_B$

2.2. Simulated tuning of the center frequency and trimming of the pass-band ripple

Figure 4(a) shows the layout of the designed 4-pole bandpass filter. The filter combines stepped impedance hair-pin resonators and waveguides to adjust coupling coefficient. We previously reported that an additional waveguide placed between the comb-line coupled resonators is an effective way to adjust the total coupling coefficient [9]. Figure 4(b) shows the position of the dielectric rods which were placed above the filter. Figure 5(a) and Figure 5(b) shows simulated frequency responses of the filter before and after trimming. To trim the passband ripple, the heights of the Rod A_2 and A_3 (as shown in Figure 4(b)) were changed. It was shown the passband ripple was improved dramatically from 0.048 to 0.023 dB. Figure 5(c) shows simulated frequency responses of the filter before and after tuning center frequency and trimming passband ripple. We adjusted the height of the Rod A for both trimming passband ripple and tuning center frequency and Rod B for tuning center frequency. The maximum tuning range of the filter was approximately 145 MHz (11.0%) from 1.389 to 1.244 GHz remaining the passband ripple less than approximately 0.10 dB. Finally, based on the simulated frequency responses of the filter, we can use the dielectric rods for both tuning and trimming.

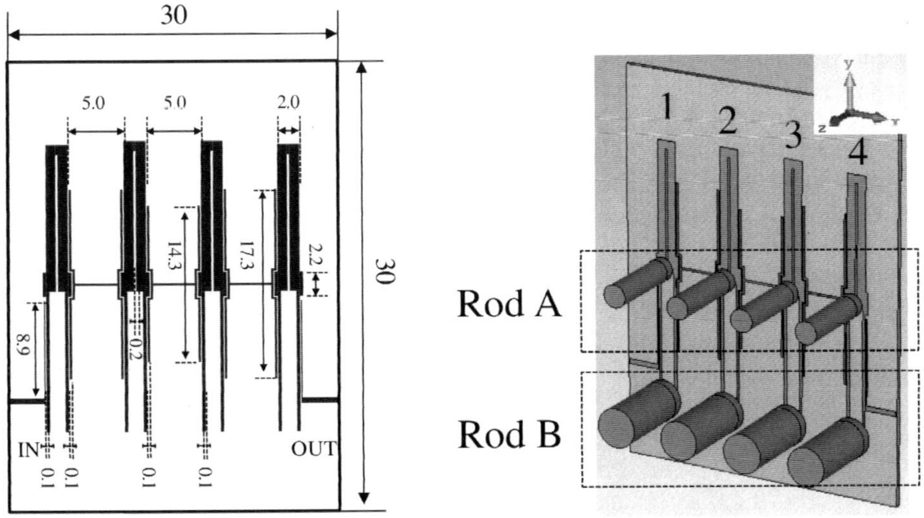

Figure 4. (a) Layout of 4-pole HTS bandpass filter. (b) Layout of 4-pole HTS bandpass filter with the rods

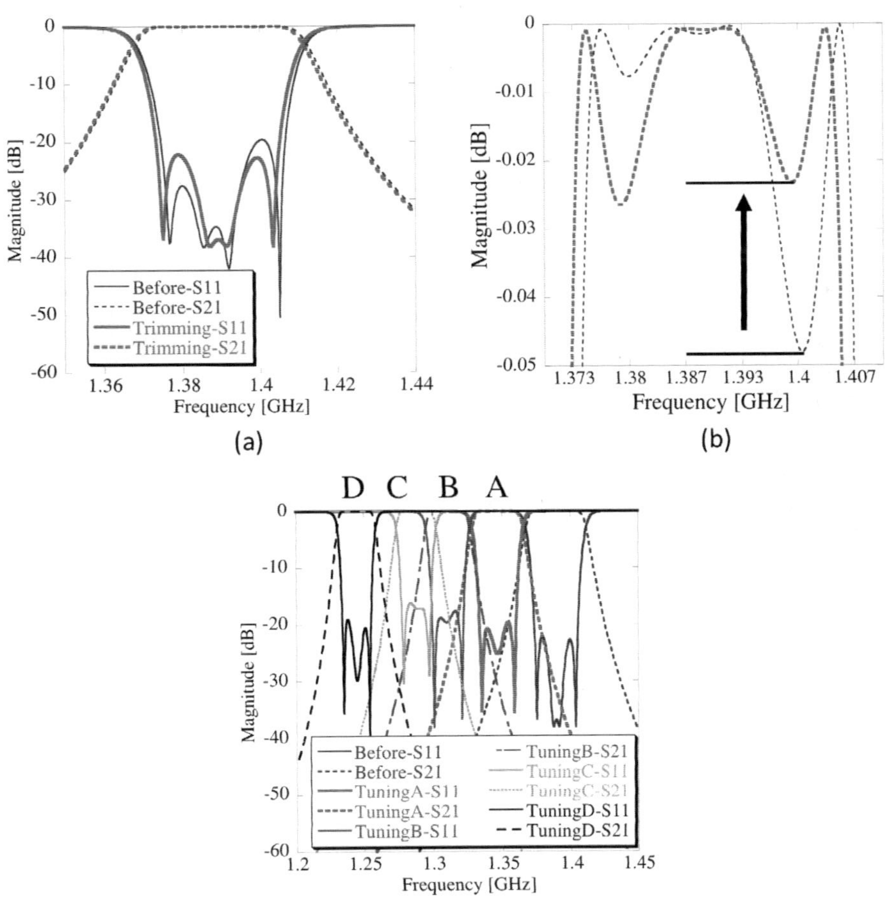

Figure 5. Simulated frequency responses of the filter with the rods. (a) before and after trimming S_{11} and S_{21}. (b) before and after trimming S_{21} (c) After both tuning and trimming.

3. Measurement of the tunable HTS filter

As illustrated in Figure 4(a), the filter was fabricated Al_2O_3 substrate with double-side-deposited 300 nm YBCO thin film. The size of the substrate is 30×40×0.5 mm and the relative dielectric constant is 9.9, respectively. The fabrication process is photolithography and ion beam etching. The filter was mounted on a metal carrier and packaged into a metal shield box, as shown in Figure 6(a). We placed the rods above the filter as shown in Figure 4(b). Figure 6(c) shows a schematic view of our tuning system. The gap h_A between resonators and dielectric rods (SG390 from Kyocera) are adjusted by mechanical screws to tune the center frequency and trim the passband ripple. The gap h_B between resonators and dielectric rods (SG390 from Kyocera) are adjusted by mechanical screws to tune the center frequency.

The fabricated HTS filter was cooled to 40 K by using a cryocooler. The measurements were carried out using a network analyser (E5071B from Agilent Technologies). Figure 6(b) shows the simulated and measured frequency responses of the filter. There was good agreement between them. Figure 7 shows the frequency responses of the filter before and after trimming passband ripple. It is found that the passband ripple was improved greatly after trimming by changing height of the Rod A_2 and A_3 (as shown in Figure 4(b)). The passband ripple was improved approximately 0.07 dB. Figure 8 shows the frequency responses of the filter before and after tuning center frequency. The maximum tuning range of the filter was approximately 140 MHz (10.57%) from 1.394 to 1.254 GHz remaining the passband ripple less than approximately 0.10 dB. The insertion loss of the passband varied from 0.08 to 0.205 dB. This very small insertion loss of the passband was due to the rods which have the small dielectric loss and the trimming the passband ripple. The 3 dB bandwidth of the filter varied from 41.1 to 25.3 MHz. This is because the coupling coefficients of the filter degreased by increasing the electric coupling of the resonator when the height between the rods and resonator was decreased. The measured center frequency, 3 dB bandwidth, insertion loss and return loss before and after tuning center frequency and trimming passband ripple are summarized in Table 1.

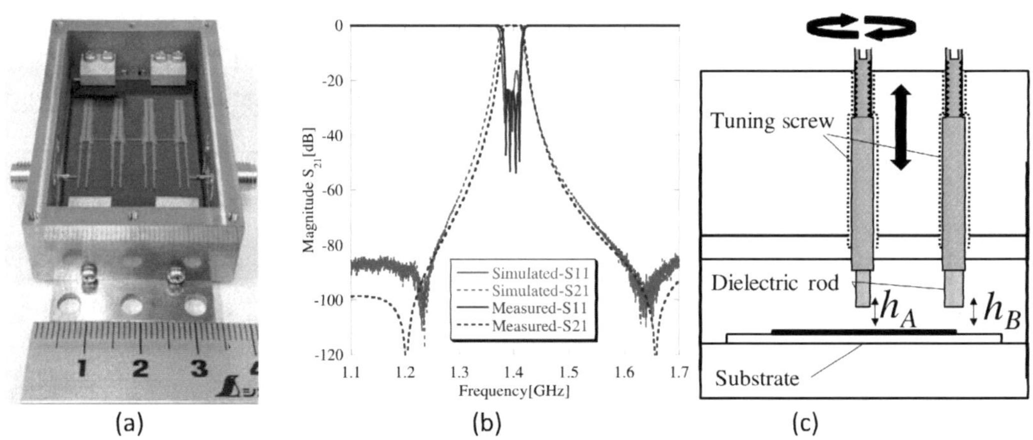

Figure 6. (a) Photograph of 4-pole bandpass filter mounted in Cu cavity. (b) Frequency response of designed and measured the filter. (c) Schematic side view of tuning system.

IOP Conf. Series: Journal of Physics: Conf. Series **1054** (2018) 012061 doi:10.1088/1742-6596/1054/1/012061

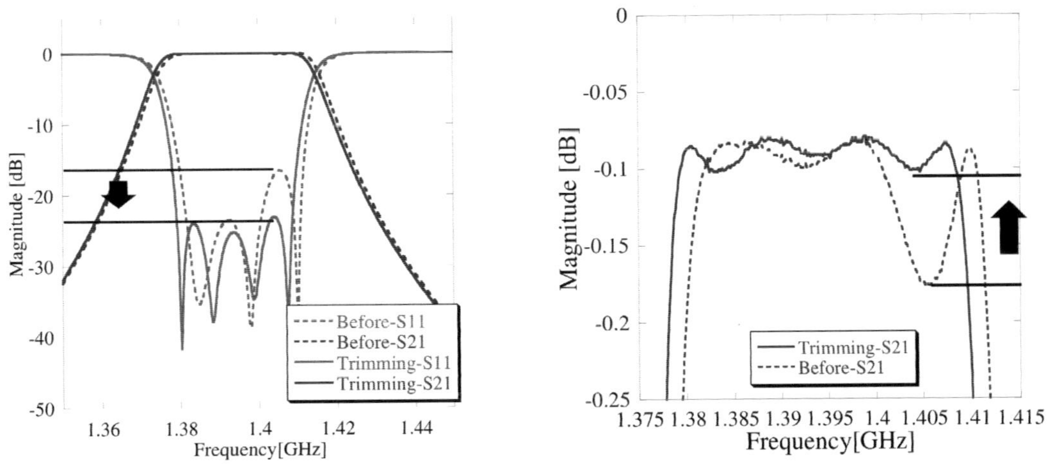

(a) (b)

Figure 7. Measured frequency responses of the filter with the rods. (a) before and after trimming S_{11} and S_{21}. (b) before and after trimming S_{21}

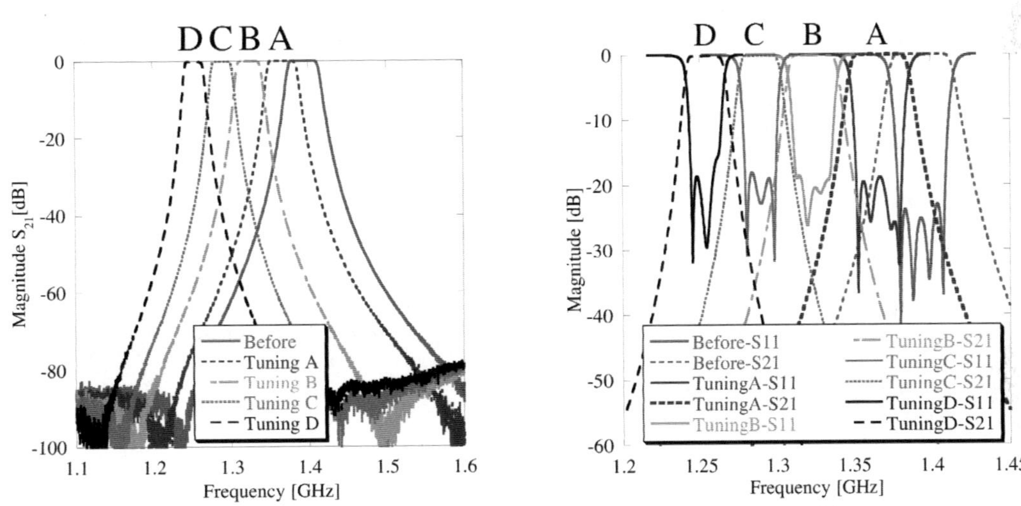

(a) (b)

Figure 8. Measured tunability of 4-pole filter before and after tuning. (a) S_{21} (b) S_{11} and S_{21}

Table 1. Measured center frequency tuning and pass-band trimming results.

	Center frequency [GHz]	-3dB Band width [MHz]	Insertion loss [dB]	Return loss [dB]
Before	1.395	41.2	-0.08	-16.5
Trimming	1.394	41.1	-0.08	-23.0
Tuning A	1.367	39.2	-0.11	-18.9
Tuning B	1.324	32.7	-0.13	-18.6
Tuning C	1.289	27.4	-0.20	-18.1
Tuning D	1.254	25.3	-0.21	-15.8

4. Conclusion

We developed a simple mechanical center frequency tuneable HTS filter. The rods are used for tuning the center frequency and trimming the passband ripple simultaneously. We designed and fabricated a 4 pole Chebyshev-type microstrip bandpass filter. From the measurement, by adjusting the gap between the resonator and the rods which were placed above the center of the resonator, the passband ripple was improved approximately 0.07 dB. The tuning range of the filter was 10.57% remaining the passband ripple less than approximately 0.10 dB by using the dielectric rods which were placed above and edge of the resonators. We obtained small variation of the insertion loss from 0.08 to 0.205 dB before and after tuning. The experimental results demonstrate the successful realization about center frequency tuning and passband ripple trimming by the dielectric rods.

Acknowledgments

The authors would like to thank Japan Steel Works, Ltd. for fabricating filter circuit.

References

[1] Peng H, Zhang X, Guo X, Mu L, Wei B, Gao B, and Cao B 2007 *Microwave and Optical Technology Letters* **49** 1565-68
[2] Aita F, Sekiya N, Ono S, Saito A, Hirano S, and Ohshima S 2006 *IEICE transactions on electronics* **89** 119-24
[3] Ohshima S, Ohsaka M, Lee J H, Takeuchi S, Kinouchi H, Ono S, and Saito A 2009 *IEEE Transactions on Applied Superconductivity* **19** 903-6
[4] Terashima Y, Fuke H, Aiga F, Yamazaki M, Kayano H, and Kato R 2002 *Physica C: Superconductivity* **366** 183-9
[5] Suo G, Guo X, Cao B, Wei B, Zhang X, Shang Z, and Zhang G 2014 *IEEE Microwave and Wireless Components Letters* **24** 170-2
[6] Peng H, Guo X, Zhang X, Wei B, Cao B, Jin S, and Lu X 2009 *IEEE Transactions on Applied Superconductivity* **19** 3683-7
[7] Sekiya N, Nakagawa Y, Saito A, and Ohshima S 2009 *Physica C: Superconductivity* **469** 1653-57.
[8] Hong J S G, and Lancaster M J 2004 *Microstrip Filters for RF/Microwave Applications* John Wiley & Sons.
[9] Sekiya N, and Sugiyama S 2015 *IEEE Transactions on Applied Superconductivity* **25** 1-5

A double-flux-quantum amplifier with a single flux-biasing line

Yuma Arai, Yoshiaki Urai, Tomoki Watanabe, Komei Higuchi, Hiroshi Shimada, and Yoshinao Mizugaki

Department of Engineering Science, The University of Electro-Communications, 1-5-1 Chofugaoka, Chofu, Tokyo, 182-8585, Japan

E-mail: y.mizugaki@uec.ac.jp

Abstract. A Double-Flux-Quantum (DFQ) amplifier can multiply its input voltage with quantum accuracy using DFQ generation at under-damped Josephson junctions. We conducted research on digital-to-analogue converters that were implemented using DFQ amplifiers for metrological applications. Conventional DFQ amplifiers were usually operated using two flux-biasing lines. In this study, we redesigned the circuit layout to realize operation using a single flux-biasing line. We designed a new layout of a 20-fold DFQ amplifier and fabricated it using a 2.5-kA/cm^2 Nb/AlO$_x$/Nb process. The measurement was performed in liquid He. The Single-Flux-Quantum (SFQ) pulse train was fed using an over-biasing method, and the input-output characteristics are acquired using an oscilloscope. When the magnetic flux bias current I_{fb1} was not supplied, the maximum input voltage V_{inmax} was 68 μV, which was 4.2 times larger than the value of 17 μV that was observed in our previous study. V_{inmax} was enhanced to 120 μV when we supplied an I_{fb1} of 1.00 mA.

1. Introduction

Several studies are being conducted to investigate the development of next-generation AC voltage standards based on applying the AC Josephson effect using several Josephson circuits including single-flux-quantum (SFQ) digital-to-analogue (D/A) converters [1–4]. One example of this is a frequency-modulation type SFQ-D/A converter [2]. This method is proposed as a means to synthesize sinusoidal 10–100 mV voltage waves at 10–100 MHz frequencies [5, 6]. However, so far, the maximum achieved output of the AC waveform generated by SFQ-D/A converters were 2.5 mV$_{pp}$ [4]. In addition to the SFQ-D/A converters, the AC programmable Josephson voltage standards (ACPJVS) and pulse-driven JVS are also candidates for next-generation AC voltage standards [5, 6]. Although ACPJVS generates large voltages that can at times be as high as 10 V, it also suffers from transient errors. Pulse-driven JVS depicts technical difficulties in case of high-frequency packaging. However, even though the output voltages of SFQ-D/A converters are small, they can negate the limitations of other JVS systems [5, 6].

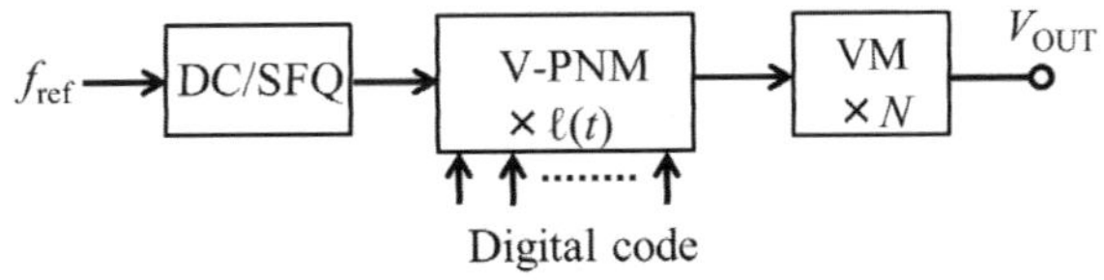

Figure 1. Block diagram of a frequency-modulation SFQ-D/A converter.

The main circuit components of a frequency-modulation SFQ-D/A converter include a DC-to-SFQ (DC/SFQ) converter, a variable pulse number multiplier (V-PNM), and a voltage multiplier (VM), as depicted in figure 1. The DC/SFQ converter varies the reference AC signal of frequency f_{ref} into an SFQ pulse train that has a repetition frequency of f_{ref}. Further, the V-PNM multiplies each SFQ pulse by $\ell(t)$ according to the digital code. In other words, the SFQ repetition frequency is modulated. Meanwhile, the VM multiplies the output voltage of $\Phi_0 \cdot f_{ref} \cdot \ell(t) \cdot N$ times. The final output voltage is expressed using the following equation:

$$V = \Phi_0 \cdot f_{ref} \cdot \ell(t) \cdot N, \qquad (1)$$

where the flux quantum $\Phi_0 = h/2e$ is a physical constant that is obtained using the Planck constant h and the elementary charge e, whereas ℓ(t) and N are integers. The output voltage determined using equation (1) can depict the quantum accuracy, which is similar to that observed in case of conventional Josephson voltage standards.

Therefore, a VM is also required to achieve quantum accuracy in case of voltage multiplication. A Double-Flux-Quantum (DFQ) amplifier is one of the VM [7]. The DFQ amplifier is a superconducting voltage multiplication circuit that utilizes DFQ generation at an under-damped Josephson junction (JJ). Since voltage is multiplied by increasing the number of flux quanta, the voltage multiplication of quantum accuracy is attainable. Moreover, the DFQ amplifier depicts an additional advantage as compared to other approaches because it operates with constant bias currents even if the voltage multiplication factor is increased by stacking the fundamental circuit cells.

As the AC Josephson effect describes, the voltage across a Josephson junction is proportional to the repetition frequency of SFQ pulses. We have fixed our target operation frequency for the DFQ amplifier to 20.0 GHz of which the corresponding Josephson voltage is 41.4 μV [8–10]. These values are high enough to be used with our recent V-PNM research where the output frequency was 16.6 GHz [11]. In addition, for stable circuit operation, the DFQ amplifier should have bias margins (operation margins for the bias current I_{sb}) of ±20% or more at 20 GHz.

In all the tests that were conducted so far, a critical-current-controlled type (I_c-type) 100-fold DFQ amplifier demonstrated accurate operation for an f_{in} value of 20.0 GHz or higher [4, 8]. As an alternative to the I_c-type, a phase-damping-controlled type (β_c-type) DFQ amplifier was also proposed as a solution [12] where β_c is the McCumber parameter defined as $2\pi I_c R^2 C/\Phi_0$[13]. Here, I_c, R, and C are the junction critical current, the resistance in parallel to the junction, and the junction capacitance, respectively. Numerical simulations predicted that the β_c-type DFQ amplifier depicted wider operating margins for f_{in} at 20.0 GHz. However, we did not confirm the correct voltage multiplication in experiments at $f_{in} \geq 20$ GHz.

In this study, we present our refinements of the circuit parameters and the physical layout of the β_c-type DFQ amplifier. Moreover, the redesigned β_c-type DFQ amplifier works using a single flux-biasing line instead of dual lines.

2. Two types of DFQ amplifiers

2.1. Operating principle of the DFQ amplifier

The DFQ amplifier comprises 2-junction loops (2JLs) that work as Josephson transmission lines, 3-junction loops (3JLs) that generate DFQs through under-damped JJs, and flux bias lines. Figures 2 and 3 depict the equivalent 3JLs circuits that were employed in I_c-type and β_c-type DFQ amplifiers, respectively. The hourglass symbols indicate the over-damped JJs (JJ.A and JJ.C in I_c-type, JJ.A in β_c-type), whereas those with crosses indicate under-damped JJs (JJ.B in I_c -type, JJ.B and JJ.C in β_c-type). By definition, an over-damped junction is where $\beta_c \leq 1$, whereas an under-damped junction is where $\beta_c >> 1$. An input SFQ pulse is fed through JJ.A to JJ.B, which causes the generation of DFQ. The DFQ generation at JJ.B allows the opposing SFQ to be induced in the 3JL. The opposite SFQ switches from

JJ.C and propagates to the subsequent 3JL (or a 2JL at the end of the DFQ amplifier). A malfunction of the DFQ amplifier indicates that the opposing SFQ flows backward through JJ.A instead of JJ.C. Therefore, it is important to determine the circuit parameters not only to make JJ.B generate the DFQ but also to make the opposite SFQs to switch to JJ.C instead of JJ.A. The difference between two DFQ amplifiers that is depicted in figures 2 and 3 is the control method that is used to control the propagation direction of opposing SFQs. Various parameters, including the β_c of the Josephson junction, are illustrated in Table 1.

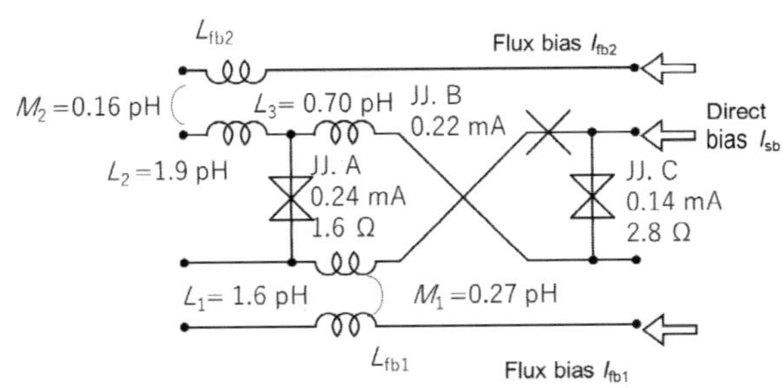

Figure 2. Equivalent circuit of a 3JL used in an I_c-type DFQ amplifier developed in the previous work.

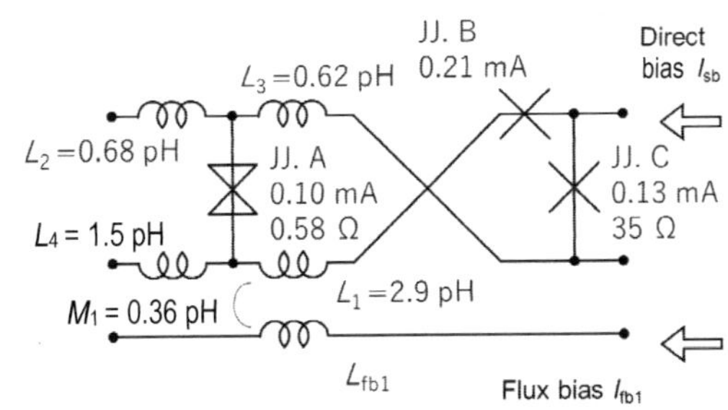

Figure 3. Equivalent circuits of a 3JL used in a β_c-type DFQ amplifier designed in this work.

Table 1. Parameters of the I_c-type and β_c -type DFQ amplifiers.

	I_c-type	β_c-type (past activity[12])	β_c-type (this research)
L_1 (pH)	1.6	2.9	2.9
L_2 (pH)	1.9	2.1	0.68
L_3 (pH)	0.7	0.62	0.62
L_4 (pH)	-	-	1.5
JJ.A I_c (mA)	0.24	0.1	0.1
JJ.A R (Ω)	1.6	0.51	0.58
JJ.A β_c	0.98	0.018	0.018
JJ.B I_c (mA)	0.22	0.21	0.21
JJ.B β_c	> 42000	> 42000	> 42000
JJ.C I_c (mA)	0.14	0.13	0.13
JJ.C R (Ω)	2.8	33	35
JJ.C β_c	1	120	135

2.2. DFQ Amplifier of the critical-current-controlled type (I_c-type)

The first proposed DFQ amplifier belonged to the I_c-type that is depicted in figure 2. It has already been implemented in frequency-modulation SFQ-D/A converters [4]. The critical current of JJ.C was fixed to be smaller than that of JJ.A to ensure that the opposing SFQ switches to JJ.C. The two magnetic flux bias lines of I_{fb1} and I_{fb2} are essential for the I_c-type DFQ amplifier. All parameters of I_c-type DFQ amplifier shown in Table 1 were used in the previous research. In this study, we did not redesign I_c-type DFQ amplifier because we wanted to focus on the designing of β_c-type DFQ amplifier.

2.3. DFQ Amplifier of the phase-damping-controlled type (β_c-type)

We searched for parameters that could expand the operational margins as compared to that of the initial I_c-type DFQ amplifier. We established that the operational margin became wider when the propagation direction control was performed based on the shunting resistor differences. As a result, the β_c-type DFQ amplifier was proposed as a means to exclude the flux-biasing lines [12]. In this study, we revised the equivalent circuit and device parameters, and the results are illustrated in figure 3. The shunting JJ.A and JJ.C resistors were tuned to make the opposing SFQ propagate through the JJ.C. Further, the 3JL parameters were determined using the SCOPE2 parameter optimization tool [14]. It should be noted that the JJ.C was tuned not to generate DFQ.

3. Layout of the β_c-type DFQ amplifier with a single magnetic-flux bias line

It should be reemphasized that we set the target values for the bias margins and maximum input frequency of the DFQ amplifier to be ±20% and 20.0 GHz, respectively and that the corresponding Josephson voltage is 41.4 μV. In our previous study, the maximum input voltage was 17.0 μV (8.2 GHz) for the β_c-type 20-fold DFQ amplifier with no magnetic-flux biasing [12].

Initially, we refined the equivalent circuit. We analyzed the cause of difference between the simulation and experimental results. We observed that, in the previous study, the inductance L_4 was not included in the equivalent circuit. Therefore, we re-optimized the circuit parameters, including L_4, using SCOPE2. The results are summarized in Table 1. In the layout design, InductEX [15] was used to perform inductance extraction of the superconducting electrodes.

In the previous study, we focused on zero flux bias operation. However, we could not demonstrate this for a 20-GHz operation. In this study, we focused on the improvement of the direct bias I_{sb} operation margin; hence, one magnetic flux bias line was implemented again. The other flux bias line, which was less important for the operation, was still excluded.

Test circuits were fabricated using a 2.5-kA/cm^2 Nb technology process that was obtained from the National Institute of Advanced Industrial Science and Technology. Figure 4 depicts the micrographs of the I_c- and the β_c-type DFQ amplifiers.

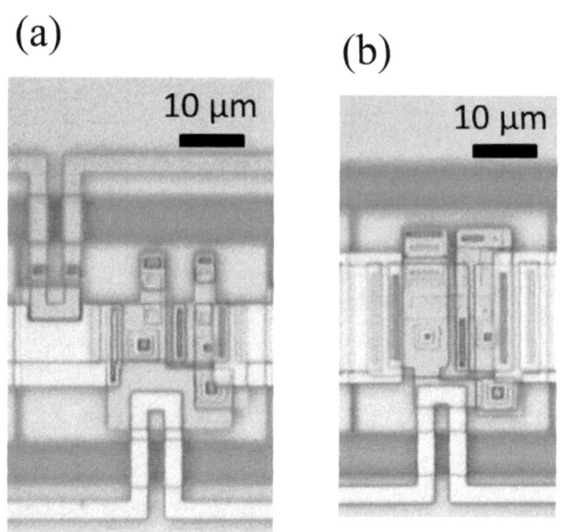

Figure 4. Photomicrographs of 3JLs
(a) I_c-type DFQ amplifier tested in the previous work.
(b) β_c-type DFQ amplifier tested in this work.

4. Operation verification of the β_c-type DFQ amplifier with a single magnetic-flux bias line

4.1. Input–output voltage characteristics

A measurement circuit diagram of a 20-fold DFQ amplifier is depicted in figure 5. The over-biasing method was adopted to feed the input SFQ pulse train where a Josephson junction in the input terminal was over-biased to continuously vary its phase. While conducting the measurements, the bias currents I_{sb}, I_{2J}, and I_{fb1} were constantly supplied. The I–O voltage characteristics were acquired with an oscilloscope using 100-fold low-noise preamplifiers.

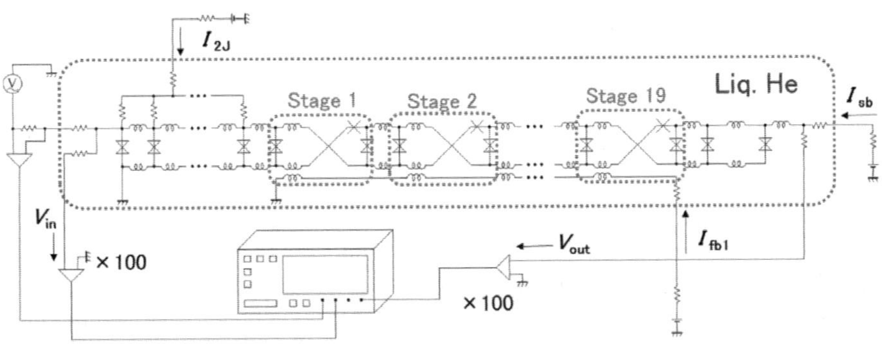

Figure 5. Measurement set-up for the I–O characteristics of a 20-fold DFQ amplifier.

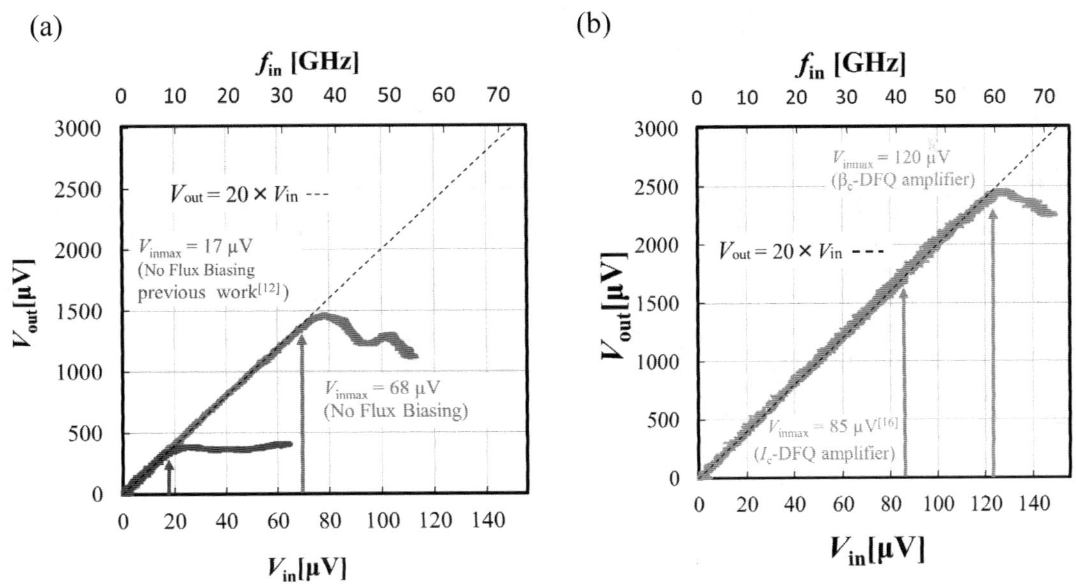

Figure 6. Relation between the V_{in} and V_{out} that were obtained in experiments. The lower and upper horizontal axes represent the input voltage V_{in} and the corresponding repetition frequency f_{in}, respectively. (a) Results for no-flux biasing; this study and the previous study for the β_c-type. (b) Results for finite flux biasing; this study and the previous study for the I_c-type. V_{inmax} of 85 μV in the previous study, where the largest value of the measured V_{in} was 89 μV, was also determined for the relative errors less than 1%.

An example of the experimental results is depicted in figure 6 (a). The dotted line represents the ideal 20-fold voltage multiplication. In this study, we defined the correct voltage multiplication for the relative errors to be within ±1%. The red points are the resultant points using zero magnetic-flux biasing (I_{fb1} = 0). The maximum input voltage (V_{inmax}) was determined to be 68 μV for which the bias conditions were I_{fb1} = 0.00 mA, I_{sb} = 0.410 mA, and I_{2J} = 1.52 mA. The corresponding repetition frequency of the SFQ pulse train, which was determined by the AC Josephson effect, was 32.7 GHz. This result was 4.2 times larger than our previous result (17.0 μV and 8.2 GHz, as depicted by the blue points in figure 6 (a)) [12]. In addition, the characteristics with finite magnetic-flux biasing (I_{fb1} = 1.00 mA) are depicted in figure 6 (b) as blue points. V_{inmax} was 120 μV of which the corresponding repetition frequency was 58.0 GHz under a biasing condition of I_{fb1} = 1.00 mA, I_{sb} = 0.379 mA, and I_{2J} = 1.52 mA. It was approximately 1.4 times larger than that of the I_c-type at 85 μV. These results are plotted as green points in figure 6 (b) [16].

4.2. Bias margins

The I–O characteristics were measured for various biasing conditions. We extracted the operable region for an input voltage exceeding 41.4 μV (20.0 GHz), which is plotted on the I_{fb1}-I_{sb} plane (as depicted in figure 7). The vertical axis is the direct bias I_{sb}, whereas the horizontal axis is the normalized magnetic-flux bias $I_{fb1}M_1/\Phi_0$. The experimental results for the upper and lower limits are plotted with red circles (●), whereas the simulation results are plotted with black squares (■).

We initially look into the numerical results. Moderately wide margins, including extremes close to ±20.8% for I_{sb}, were identified near zero magnetic-flux biasing. It indicates that this DFQ amplifier could inherently work without magnetic-flux biasing. The operational region along the horizontal axis ranges from −0.4 to +0.55.

Further, we observe the experimental results. It is obvious that the experimental operation regions that are depicted in figure 7 are quite reduced in comparison with that of the numerical results. The bias margin is ±13.6% at the maximum. There are two possible reasons for these reductions. First, it is conceivable that the layout design and/or the inductance extraction were imperfect. Specifically, for the results along the horizontal axis, it should be noted that the mutual inductance M_1 was numerically, as opposed to experimentally, determined only by InductEX [15]. The 0.36-pH value was the result obtained using InductEX, whereas 0.47 pH was a 30% larger value than that was empirically obtained in our previous experiments using similar layouts. For both the M_1 values, the operational margins along the horizontal axis were reduced by more than 30%. Second, it could be a potential reason for the presence of parameter variations in chip fabrication.

From another point of view, this study demonstrated correct voltage multiplication for an input voltage of 41.4 μV (20.0 GHz) under a zero magnetic flux-biasing condition where the I_{sb} margin was ±8.1%. This proves that the design of a β_c-type DFQ amplifier has improved.

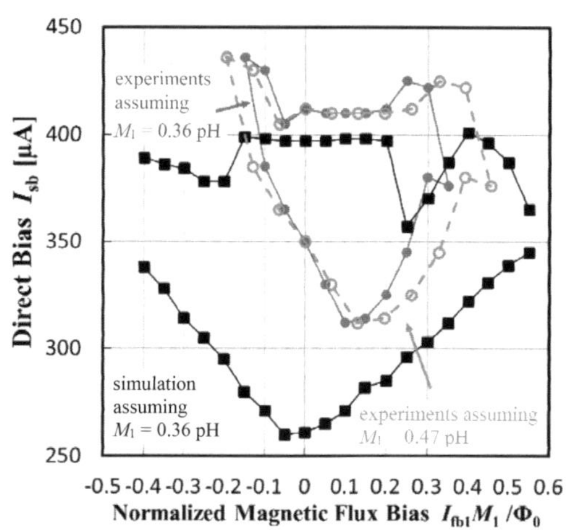

Figure 7. Bias margins for $V_{in} \geq 40.1$ μV plotted on the I_{sb}-$I_{fb1}M_1/\Phi_0$ plane.

5. Conclusion

In this study, the design and operational verification of the phase-damping-controlled-type DFQ amplifier were described. The number of magnetic-flux biasing lines that was used in our previous research was reduced from 2 to 1. It was confirmed that the maximum input voltage was 68 μV with no magnetic-flux bias, whereas the maximum input voltage with a finite magnetic-flux bias was 120 μV. This was a significant improvement over that observed in our previous result (85 μV for the I_c-type) . The maximum I_{sb} margin for an input voltage larger than 41 μV was ±13.6% in case of finite magnetic-flux biasing, whereas that for zero magnetic-flux biasing was ±8.1%.

Acknowledgments

The authors are grateful to M. Maezawa for initiating the current study. This research work was partly supported by JSPS Grant-in-Aid for Scientific Research 15K13999 and 17K04979 and by VLSI Design and Education Center (VDEC), the University of Tokyo in collaboration with Cadence Design Systems, Inc. Test circuits used in this study were fabricated using the 2.5 kA/cm^2 Nb/AlO$_x$/Nb junction technology (AIST-STP2) process in the Superconducting Clean Room (CRAVITY) at the National Institute of Advanced Industrial Science and Technology (AIST). The stable supply of liquid helium from the Coordinated Center for UEC Research Facilities is also acknowledged.

References

[1] Hamilton C A 1992 Josephson voltage standard based on single-flux-quantum voltage multiplier *IEEE Trans. Appl. Supercond.* **2** 139–42

[2] Semenov V K 1993 Digital to analog conversion based on processing of the SFQ pulses *IEEE Trans. Appl. Supercond.* **3** 2637-40

[3] Maezawa M, Hirayama F and Suzuki M 2005 Rapid single flux quantum digital-to-analog converter for ac voltage standard *Physica C* **426-31** 1674-9

[4] Mizugaki Y, Takahashi Y, Shimada H and Maezawa M 2014 9-bit superconductive single flux quantum digital-to-analog converter *Electronics Letters* **50** 1637–9

[5] Maruyama M 2008 Review on Josephson voltage standard *AIST measurement standard report* **8** 263-278 (in Japanese)

[6] Benz S P, Hamilton C A 2004 Application of the Josephson Effect to Voltage Metrology *Proceedings of the IEEE* **92** 1617-29

[7] Herr Q P 2005 Stacked double-flux-quantum output amplifier *IEEE Trans. Appl. Supercond.* **15** 259–62

[8] Sato Y, Moriya M, Shimada H, Mizugaki Y and Maezawa M 2013 Design and operation of 1000-fold voltage multiplier based on double-flux-quantum amplifier *Physics Procedia* **45** 221–4

[9] Watanabe T, Takahashi Y, Shimada H, Maezawa M and Mizugaki Y 2015 4-bit bipolar triangle voltage waveform generator using single-flux-quantum circuit *Phys. Procedia* **65** 213–6

[10] Mizugaki Y, Urai Y and Shimada H 2017 Thermally-fluctuated single-flux-quantum pulse intervals reflected in input-output characteristics of a double-flux-quantum amplifier *J. Phys. Conf. Ser.* **871** 012066

[11] Arai Y, Watanabe T, Sawada K, Shimada H and Mizugaki Y 2016 Improvement of SFQ pulse frequency in an SFQ-D/A converter (The 77th JSAP Autumn Meeting, Niigata) 15p-D61-12 Sep. 13-16 (in Japanese)

[12] Mizugaki Y and Watanabe T 2016 Design and operation of a double-flux-quantum amplifier excluding flux bias lines *IEEE Trans. Appl. Supercond.* **26** 1301104

[13] McCumber D E 1968 Effect on dc Voltage-Current Characteristics of Superconductor Weak-Link Junctions *J. phys. Appl. Phys.* **39** 3113-8

[14] Mori N, Akahori A, Sato T, Tekeuchi N, Fujimaki A and Hayakawa H 2001 A new optimization procedure for single flux quantum circuits *Physica C* **357-60** 1557-660

[15] Fourie C J, Wetzstein O, Kunert J and Meyer H G 2013 SFQ circuits with ground plane hole-assisted inductive coupling designed with InductEx *IEEE Trans. Appl. Supercond.* **23** 1300705

[16] Urai Y, Watanabe T, Takahashi Y and Mizugaki Y 2013 Mutually-coupled dc/SFQ converter tested using an on-chip pulse generator (Superconducting SFQ VLSI Workshop, Tsukuba) p 54 Nov. 21-22

IOP Conf. Series: Journal of Physics: Conf. Series **1054** (2018) 012063 doi:10.1088/1742-6596/1054/1/012063

A random-access-memory cell based on quantum flux parametron with three control lines

Hiroshi Takayama[1], **Naoki Takeuchi**[2,3], **Yuki Yamanashi**[1,2] **and Nobuyuki Yoshikawa**[1,2]

[1] Graduate School of Engineering, Yokohama National University, 79-5 Tokiwadai, Hodogaya, Yokohama 240-8501, Japan

[2] Institute of Advanced Sciences, Yokohama National University, 79-5 Tokiwadai, Hodogaya, Yokohama 240-8501, Japan

[3] PRESTO, Japan Science and Technology Agency, 4-1-8 Honcho, Kawaguchi, Saitama 332-0012, Japan

E-mail: takayama-hirosi-rd@ynu.jp

Abstract. In this study, we proposed a random-access memory (RAM) cell composed of adiabatic quantum-flux-parametron (AQFP) circuits as a component of large-scale AQFP RAMs. The memory cell is composed of a storage QFP gate and a readout AQFP gate, which is controlled by three control currents I_d, I_y and I_x. I_d is used to control an input datum. I_y is a column selection current for the write operation. I_x is a row selection current for both write and read operations. The control current margins evaluated in the simulation are larger than +/- 30%. We designed the memory cell of the size of 50 μm x 40 μm by using AIST high-speed standard process, and verified the correctness of its operation by experiments. It was shown that the amplitude of the control currents is compatible with the output currents of AQFP logic.

1. Introduction

Semiconductor integrated circuits have improved the performance by decreasing the feature size owing to the advancement of microfabrication technologies. However, an increase in heat generation from a chip due to the higher integration density becomes a crucial problem, which limits the enhancement of the performance of the present semiconductor integrated circuits.

Superconducting circuits are attracting attention as an integrated circuit technology for realizing high-end computers in the next generation because they can operate faster and consumes much less energy compared with the semiconductor circuits. Energy-efficient superconducting circuits based on rapid single flux quantum (RSFQ) circuits [1] have been proposed recently, which include eSFQ [2], ERSFQ [3], RQL [4], and LV-RSFQ [5]. We have been developing adiabatic quantum flux parametron (AQFP) circuits based on QFP circuits [6], which are extremely energy efficient superconducting circuits [7-10]. The AQFP circuit is a promising superconducting circuit for realizing ultra-low-power computing systems because the operating frequency can be 5 to 10 GHz and the bit-energy is about six orders of magnitudes lower than that of CMOS.

Large-scale random-access memories (RAMs) are necessary for realizing high-performance computing systems. There are several studies on large-scale memories using superconducting latching logic [11], partially using RSFQ logic [12], and fully using RSFQ logic up to date [13-14]. However, for the generous use of the advantage of the AQFP circuits, reduction of the energy consumption of the memories is essential. Although flip-flops based on AQFP circuits have been proposed in previous study [15-17], their circuit area is not small enough for making high-density large-scale memories. Any RAMs for large-scale memories compatible with AQFP circuits have not been realized yet.

We aim to realize an AQFP-based RAM as a cache memory whose input/output signals are compatible with those of the AQFP circuits. Even though AQFP signals can be converted to SFQ pulses, the AQFP RAM is necessary for decreasing the access time of AQFP computing systems. The AQFP circuits use bipolar signals with amplitude about several tens of microampere. For example, +50 μA of current represents logic "1" and -50 μA of current represents logic "0". Since the amplitude of control currents required for the memory cells in [11-14] is larger than 100 μA, we have to develop a memory cell drivable by the AQFP signals.

In this study, we have proposed a RAM cell composed of AQFP circuits with three control lines as a component of a large-scale AQFP RAM. The RAM cell is compatible with AQFP circuits, where the control currents are bipolar and their amplitude is the same level as that of the output current of AQFP gates. We have designed and implemented the RAM cell, whose cell size is 40 μm x 50 μm, using the AIST high-speed standard process (HSTP) [18]. The correct functions were obtained with large control current margins.

2. AQFP RAM system

An AQFP RAM system we proposed in this study is shown in figure 1. The RAM system consists of decoders, memory cells, multi-input OR gates, and a multiplexer (MUX), all of which are composed of AQFP circuits. The memory cells form a two dimensional array, where each cell is selected by three control currents, that is an x-direction current (I_x), a y-direction current (I_y) and a data input (I_d). The x-direction current (I_x) is commonly used in both write and read operations. In the write operation, a memory cell is selected by I_x and I_y and datum is written by I_d. In the read operation, memory cells in the same row are selected by I_x and the one of them is chosen by the multiplexer. The decoder of the

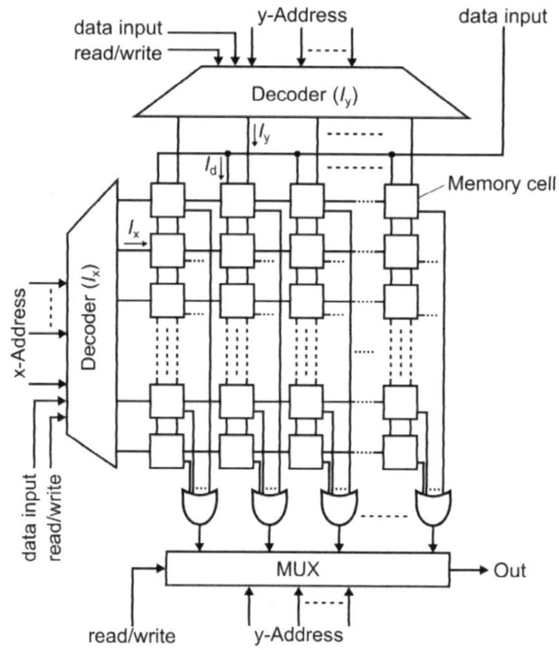

Figure 1. An AQFP RAM system proposed in this study.

AQFP RAM outputs a positive or negative signal to the selected address depending on its read/write modes and its input datum.

3. AQFP RAM cell with three control lines

Figure 2, 3, and 4 show a schematic, layout, and microphotograph of the AQFP RAM cell with three control lines proposed in this study, respectively. The 3D inductance extractor, InductEx [19], was used for extracting the value of self and mutual inductances from the layout. To obtain the compatibility with AQFP circuits, the RAM cell is composed of AQFP circuits. It consists of a storage QFP gate, which stores 1-bit information, and a read AQFP gate, which readouts the information stored in the storage gate. The storage QFP gate is biased by a DC current I_{wb}, so that the QFP has a double-well-potential and behaves as a flip-flop with bistable states. It should be noted that the nonadiabatic switching occurs essentially in the memory cell when the stored datum is rewritten. The input of the read AQFP gate is offset by a DC current I_{rb} so that the readout of the gate becomes logic "0" when the memory cell is not selected. This is satisfied by setting the condition, $I_{rb2} > 0$, because the read AQFP gate should not go to "1" state regardless of the state of the storage QFP gate when I_x is not enabled (when I_{x2} is positive in this design), where I_{rb2} and I_{x2} are the input current of the read AQFP gate caused by I_{rb} and I_x, respectively.

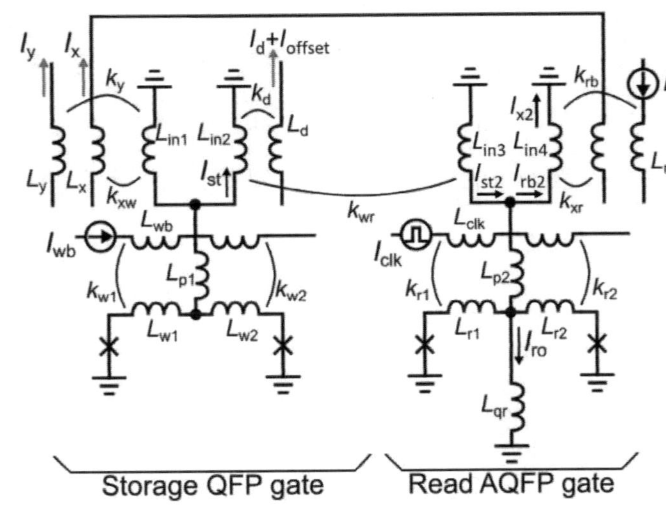

Figure 2. Schematic of an AQFP RAM cell with three control lines. The inductances are $L_d = 13.1$ pH, $L_y = 16.8$ pH, $L_x = 16.1$ pH, $L_{wb} = 12.6$ pH, $L_{in1} = 9.04$ pH, $L_{in2} = 8.41$ pH, $L_{w1} = 1.39$ pH, $L_{w2} = 1.42$ pH, $L_{p1} = 0.256$ pH, $L_{rb} = 19.9$ pH, $L_{clk} = 14.2$ pH, $L_{in3} = 30.8$ pH, $L_{in4} = 22.2$ pH, $L_{r1} = 1.36$ pH, $L_{r2} = 1.38$ pH, $L_{qr} = 8.28$ pH and $L_{p2} = 1.19$ pH. The coupling coefficients are $k_d = 0.182$, $k_y = 0.301$, $k_{xw} = 0.157$, $k_{w1} = 0.168$, $k_{w2} = 0.180$, $k_{wr} = 0.229$, $k_{xr} = 0.138$, $k_{rb} = 0.179$, $k_{r1} = 0.166$ and $k_{r2} = 0.168$ where k_{xr} , k_{w2} , k_{r2} are the coefficients between L_x and L_{in4}, between L_{w2} and L_{wb}, and between L_{clk} and L_{r2}, respectively. The parasitic coupling coefficients in the storage gate are $k_{wb,in1} = 0.00111$, $k_{wb,in2} = 0.00718$ where $k_{wb,in1}$ and $k_{wb,in2}$ are the coefficients between L_{wb} and L_{in1}, and between L_{wb} and L_{in2}, respectively. The other parasitic coupling coefficients are $k_{wb,in3} = 0.000249$, $k_{wb,in4} = 0.00399$, $k_{wb,qr} = 0.00805$, $k_{clk,in1} = 0.000813$, $k_{clk,in2} = 0.000534$, $k_{clk,in3} = 0.000793$, $k_{clk,in4} = 0.00240$, $k_{clk,qr} = 0.00397$ where $k_{a,b}$ is the coefficient between L_a and L_b. The critical currents of Josephson Junctions are all 50 μA. I_{rb2} I_{st2} and I_{x2} are the input current of the read AQFP gate caused by I_{rb} and I_{st} and I_x, respectively.

A simulation result of the AQFP RAM cell with three control lines is shown in figure 5. The RAM cell uses three control currents I_d, I_y, and I_x for write and read operations, where I_d corresponds to input datum to be written, I_y is the column selection current for the write operation and I_x is the row selection current for both write and read operations. Logic "1" is written to the RAM cell when the directions of all three control currents are positive, while logic "0" is written when all of them are negative. The information stored in the RAM cell corresponds to the direction of the current I_{st} in figure 2. The positive direction of I_{st} means logic "1" and the negative direction means logic "0". In the read operation, the RAM cells in the same row are read simultaneously when the direction of the current I_x is negative. The readout current I_{ro}, which flows synchronously with the clock current I_{clk}, is equal to the readout result of the RAM cell.

We assume that the control currents I_d, I_y and I_x are generated by AQFP circuits and that they are bipolar signals. The AQFP RAM system is driven by four phase clocks. The current I_x and the clock current I_{clk} are delayed by 90 and 180 degrees compared with the currents (I_d, I_y), respectively. Because I_x is control current for both write and read operations, the read gate must be excited by the clock current

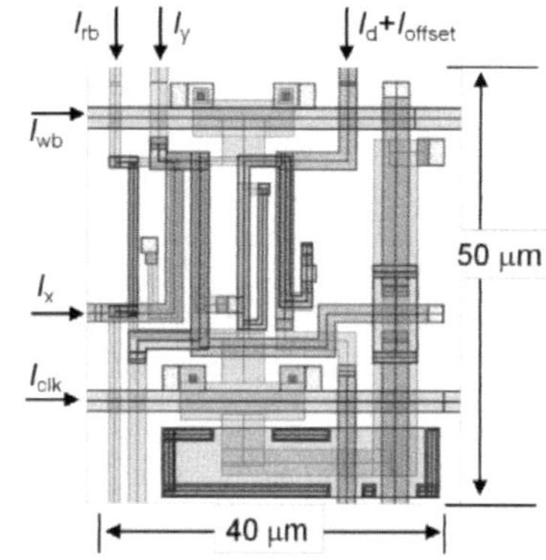

Figure 3. A layout of an AQFP RAM cell with three control lines.

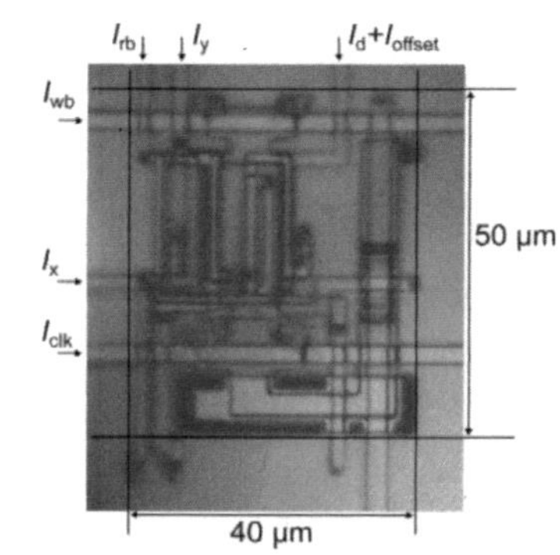

Figure 4. A microphotograph of an AQFP RAM cell with three control lines.

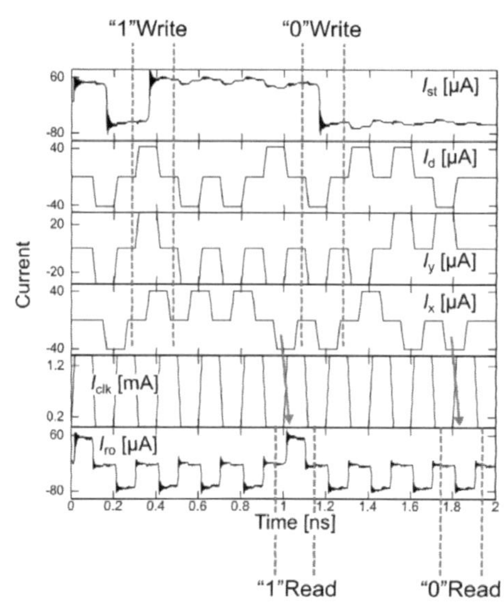

Figure 5. A simulation result of an AQFP RAM cell with three control lines.

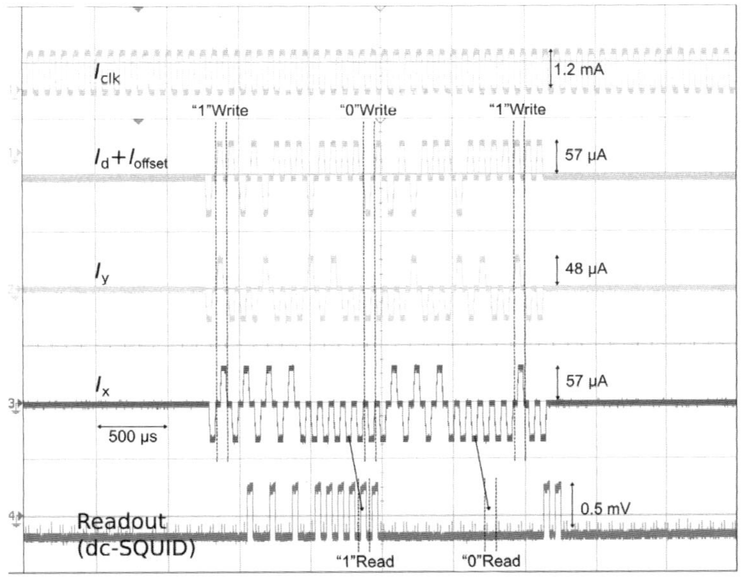

Figure 6. A measurement result of an AQFP RAM cell with three control lines.

I_{clk} while I_x is applied. The operation margin will be decreased if I_d and I_y are applied when the read gate is excited by the clock current I_{clk}.

There exists a small offset current at the input of the storage gate caused by the DC current I_{wb} and the asymmetry of the layout. The parasitic magnetic coupling between the DC current line for I_{wb} and the inductances of the storage gate causes this offset in the storage gate. To obtain correct operation, the RAM cell designed in this study requires a DC offset current I_{offset}, which is superposed on the control line for I_d. The role of I_{offset} is the cancelation of the offset current caused by I_{wb}. The DC offset current I_{offset} = 98 μA was applied in this simulation.

A measurement result of the RAM cell designed in this study is shown in figure 6. Several stages of AQFP buffers are connected to the RAM cell and the logic state of the last AQFP buffer is readout by using a dc-SQUID, which were magnetically coupled to the last buffer [7]. The result verifies that the write "1" and read "1" operation, and the write "0" and read "0" operation are successfully performed. Besides the operations for half-selection and non-selection modes are correctly obtained. It was also confirmed that the RAM cell is compatible with AQFP circuits because the amplitude of three control currents I_d, I_y and I_x are the same level as the amplitude of output currents of AQFP circuits. The DC current I_{offset} = -45 μA was applied in the measurement. The polarity of I_{offset} is different from the simulation results, probably due to the unknown parasitic couplings in the actual device structure.

Simulated and measured operation margins of the RAM cell are shown in figure 7. It can be seen from the figure that while the margin for the DC current I_{wb} is as small as +/- 7%, the margins for control currents are larger than +/- 30% in the simulation. It should be noted that I_{wb} can be applied by using the external current source, therefore its amplitude can be easily adjusted. From the figure one can also see that the measured margins for the control currents are shifted to the right side. However, we obtained the mostly reasonable measured margins because I_x has a wider operation margin than that of I_d and I_y, and the critical margin appears in I_{wb}.

The simulated energy consumption of the 1-bit AQFP RAM cell operating at 5 GHz is ranging from 8 zJ to 10 zJ depending on its read/write operation modes. As the worst case scenario, we included bit-line and word-line drivers in each memory cell, which amplify the bipolar control currents I_d, I_y, and I_x for each memory cell. Relatively large energy consumption compared with a single AQFP gate [18] is

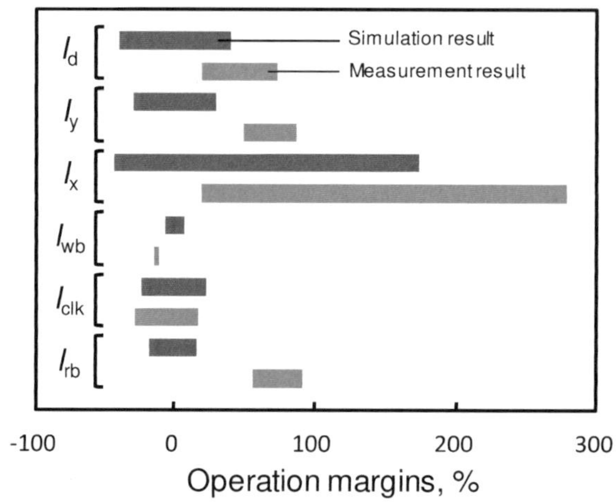

Figure 7. Operation margins of an AQFP RAM cell with three control lines. The blue indicates simulation results, and the green indicates measurement results. The figure shows the margins of the amplitude of bipolar signals I_d, I_y, I_x, the clock current I_{clk}, and the DC currents I_{wb}, I_{rb}. The margins are normalized by the amplitudes of I_d = 42 µA, I_y = 30 µA, I_x = 41 µA, I_{clk} = 1.40 mA, I_{wb} = 1.41 mA, and I_{rb} = 140 µA.

due to the non-adiabatic switching of the AQFP RAM cell caused by the back action (the gate-to-gate interaction of AQFP gates). We can estimate the average energy consumption of the 4-kbit AQFP RAM to be 36 aJ ≈ (4 kbit) x (9 zJ) per clock period at 5 GHz operation because a memory system is implemented by simply connecting the AQFP RAM cell with bit-line drivers in two dimensional structure and the most of the energy is consumed in the memory cell array.

4. Conclusion

We proposed, designed and measured an AQFP RAM cell controlled by three control lines for read and write operations. The cell size is 40 µm x 50 µm and fabricated by using the AIST high-speed standard process. We confirmed correct operations of the RAM cell by measurements and showed that the amplitude of the data write current, the row and column selection current are compatible with the output current of AQFP circuits.

Acknowledgements

The present study was supported by the Grant-in-Aid for Scientific Research (S) (No. 26220904) from the Japan Society for the Promotion of Science (JSPS). The circuits were fabricated in the Clean Room for Analog-digital superconductivity (CRAVITY) of the National Institute of Advanced Industrial Science and Technology (AIST) using the high-speed standard process (HSTP). We would like to thank Prof. C. J. Fourie for providing the 3D inductance extractor, InductEx, which was used for layout design of the circuits.

References

[1] Likharev K K and Semenov V K 1991 RSFQ logic/memory family: a new Josephson-junction technology for sub-terahertz-clock-frequency digital systems *IEEE Trans. Appl. Supercond.* **1** 3-28.

[2] Volkmann M H, Sahu A, Fourie C J and Mukhanov O A 2013 Implementation of energy efficient single flux quantum digital circuits with sub-aJ/bit operation *Supercond. Sci. Technol.* **26** 015002
[3] Kirichenko D E, Sarwana S and Kirichenko A F 2011 Zero static power dissipation biasing of RSFQ circuits *IEEE Trans. Appl. Supercond.* **21** 776-9
[4] Herr Q P, Herr A Y, Oberg O T and Ioannidis A G 2011 Ultra-low-power superconductor logic *J. Appl. Phys.* **109** 103903
[5] Tanaka M, Ito M, Kitayama A, Kouketsu T and Fujimaki A 2012 18-GHz, 4.0-aJ/bit operation of ultra-low-energy rapid single-flux-quantum shift registers *Jpn. J. Appl. Phys.* **51** 053102
[6] Hosoya M, Hioe W, Casas J, Kamikawai R, Harada Y, Wada Y, Nakane H, Suda R and Goto E 1991 Quantum flux parametron: A single quantum flux device for Josephson supercomputer *IEEE Trans. Appl. Supercond.* **1** 77-89
[7] Takeuchi N, Ozawa D, Yamanashi Y and Yoshikawa N 2013 An adiabatic quantum flux parametron as an ultra-low-power logic device *Supercond. Sci. Technol.* **26** 035010
[8] Takeuchi N, Ehara K, Inoue K, Yamanashi Y and Yoshikawa N 2013 Margin and energy dissipation of adiabatic quantum-flux-parametron logic at finite temperature *IEEE Trans. Appl. Supercond.* **23** 1700304
[9] Takeuchi N, Yamanashi Y and Yoshikawa N 2013 Simulation of sub-k_BT bit-energy operation of adiabatic quantum-flux-parametron logic with low bit-error-rate *Appl. Phys. Lett.* **103** 062602
[10] Takeuchi N, Yamanashi Y and Yoshikawa N 2015 Thermodynamic study of energy dissipation in adiabatic superconductor logic *Phys. Rev. Appl.* **4** 034007
[11] Nagasawa S, Hashimoto Y, Numata H and Tahara S 1995 A 380 ps, 9.5 mW Josephson 4-Kbit RAM operated at a high bit yield *IEEE Trans. Appl. Supercond.* **5** 2447-52
[12] Nagasawa S, Hasegawa H, Hashimoto T, Suzuki H, Miyahara K and Enomoto Y 2001 Superconducting latching/SFQ hybrid RAM *IEEE Trans. Appl. Supercond.* **11** 533-6
[13] Nagasawa S, Hinode K, Satoh. T, Kitagawa Y and Hidaka M 2006 Design of all-dc-powered high-speed single flux quantum random access memory based on a pipeline structure for memory cell arrays *Supercond. Sci. Technol.,* **19** S325
[14] Ortlepp T and Van Duzer T 2014 *IEEE Trans. Appl. Supercond.* **24** 1300307
[15] Takeuchi N, Ortlepp T, Yamanashi Y and Yoshikawa N 2014 Novel latch for adiabatic quantum-flux-parametron logic *J. Appl. Phys.* **115** 103910
[16] Tsuji N, Takeuchi N, Narama T, Ortlepp T, Yamanashi Y and Yoshikawa N 2015 Magnetically coupled quantum-flux-latch with wide operation margins *Supercond. Sci. Technol.* **28** 115013
[17] Tsuji N, Takeuchi N, Yamanashi Y, Ortlepp T and Yoshikawa N 2016 Majority gate-based feedback latches for adiabatic quantum flux parametron logic *IEICE Trans. Electron.* **E99-C** 710–6
[18] Takeuchi N, Nagasawa S, China F, Ando T, Hidaka M, Yamanashi Y and Yoshikawa N 2017 Adiabatic quantum-flux-parametron cell library designed using a $10 kAcm^{-2}$ niobium fabrication process *Supercond. Sci. Technol.* **30** 035002
[19] Fourie C J, Wetzstein O, Ortlepp T and Kunert J 2011 Three-dimensional multi-terminal superconductive integrated circuit inductance extraction *Supercond. Sci. Technol.* **24** 125015

IOP Conf. Series: Journal of Physics: Conf. Series **1054** (2018) 012064 doi:10.1088/1742-6596/1054/1/012064

Pico pulse response analysis of Josephson weak-link using time-dependent Ginzburg-Landau model

Shigeru Yoshimori

Department of Electronics and Computer Systems, Faculty of Engineering, Takushoku University, Tokyo, Japan
E-mail: yosimori@es.takushoku-u.ac.jp

Abstract The pulse wave whose width is picosecond is called the pico pulse. In this paper, the pico pulse response analysis of high Tc superconducting Josephson weak-link made of $Bi_2Sr_2Ca_2Cu_3O_{10}$ in the Terahertz (THz) region is reported. Since the photon energy of the THz radiation is greater than the energy gap of high Tc superconductor, the one dimensional time-dependent Ginzburg–Landau (TDGL) equation is used and the TDGL model as the equivalent circuit of Josephson weak-link instead of the Resistively-Shunted-Junction model is derived by use of the TDGL equation. It is found that the peak value of pulse response decreases when the pulse width decreases.

1. Introduction

The Josephson junctions are applied to ultrahigh sensitive detectors in the millimeter and Terahertz (THz) wave regions since they have the properties of high sensitivity and low noise. The THz wave is the electro-magnetic wave whose frequency is ranging from 0.1 to 10 [THz] and is going to be useful for high speed and high capacity telecommunicating, imaging, biological and chemical analysing.

Since discovery of high Tc superconductor (HTS), thin film deposition and device fabrication technology of HTS have been developed [1], [2], [3], [4]. Hua T *et al.* recently reported on the bicrystal HTS Josephson mixer's conversion efficiency at THz region [5].

In this paper, we call the pulse wave whose width is picosecond the pico pulse. Since the photon energy of the THz radiation is greater than the energy gap of HTS, we use the one dimensional time-dependent Ginzburg–Landau (TDGL) equation proposed by Gor'kov and Eliashberg and derive the TDGL model instead of the Resistively-Shunted-Junction(RSJ) model as the equivalent circuit of Josephson weak-link [6], [7], [8].

We assume that the current voltage characteristics of the bicrystal Josephson junction made of Bi-based cuprate, $Bi_2Sr_2Ca_2Cu_3O_{10}$ can be described as the Josephson weak-link whose length is L.

The pico pulse response analysis of HTS Josephson weak-link in the THz region is reported. It is found that the peak values of pulse response decreased when the pulse width decreased.

IOP Conf. Series: Journal of Physics: Conf. Series **1054** (2018) 012064 doi:10.1088/1742-6596/1054/1/012064

2. TDGL model of Josephson weak-link

Derivation of the TDGL model of Josephson weak-link was described in detail in Ref. [8].

Figure 1 is the equivalent circuit of the TDGL model, where I_d is the DC driving current, I_p the current component of the irradiated pico pulse, R(V) the nonlinear resistance of the Josephson junction, ϕ the phase difference between both electrodes, I_{J1} the current amplitude of $\sin\phi$ term well known as the Josephson current and I_{J2} the current amplitude of $\cos\phi$ term corresponding to the interference of the superconducting wave function with the quasi particle wave function.

From figure 1, we obtain the following equation.

$$I_{\mathrm{d}} + I_{\mathrm{p}}(t) = I_{\mathrm{J1}}\sin\phi + I_{\mathrm{J2}}\cos\phi + \frac{\mathrm{V}}{R(\mathrm{V})} \tag{1}$$

Here R(V) is given by

$$R(\mathrm{V}) = R_{\mathrm{n}}\left\{1 + \frac{3}{4v}\tanh\left(\frac{v}{\lambda}\right)\right\}^{-1} \tag{2}$$

where R_n is the normal resistance in the limit V $\longrightarrow$ 0 and $v = \dfrac{\mathrm{V}}{I_{\mathrm{c}}R_{\mathrm{n}}}$.

The Greek letter λ is defined as follows where τ is the relaxation time of the Cooper pair [6].

$$\lambda = \left\{\frac{4\tau}{45}\left(\frac{2e}{\hbar}I_{\mathrm{c}}R_{\mathrm{n}}\right)\left(\frac{\mathrm{L}}{\xi}\right)^2\right\}^{-1} \tag{3}$$

As can be easily understood from equation (2) in the TDGL model of the Josephson weak-link, the normal resistance of the junction is not constant but dependent on the voltage difference V. However, R(V) is constant in good approximation when the weak-link length L is short enough in comparison with the coherence length ξ. Thus, R(V) is assumed to be equal to R_n in this analysis. Moreover, I_{J1} and I_{J2} can be expressed in good approximation in the THz region as follows [8].

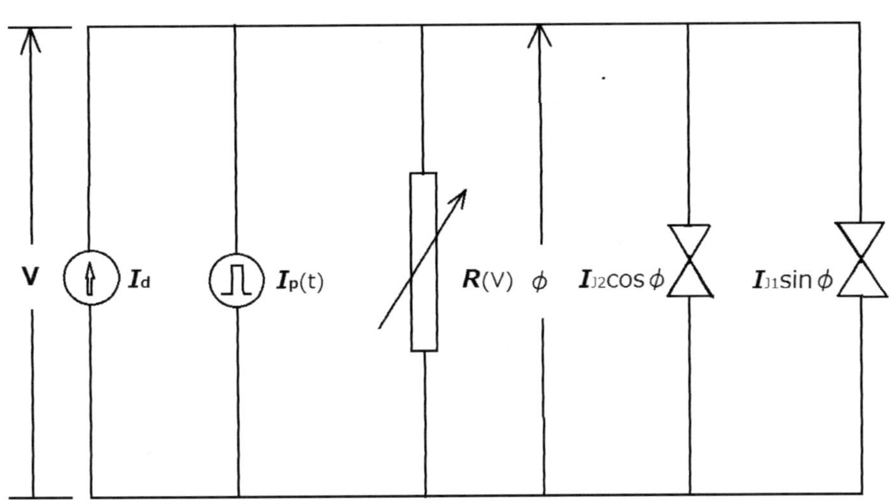

Figure 1. TDGL model for picosecond pulse response analysis of Josephson weak-link.

$$I_{J1} = I_c \tag{4}$$

$$I_{J2} = -\frac{17\tau I_c}{735}\frac{2eV}{\hbar}\left(\frac{L}{\xi}\right)^2 \tag{5}$$

where I_c is the critical current of the Josephson weak-link.

The irradiated rectangle pulse in figure 1, $I_p(t)$ is expressed as follows.

$$I_p(t) = I_{p0}\left\{\frac{1}{2} + \frac{2}{\pi}\sum_{n=0}^{\infty}\frac{\sin\left(\frac{(2n+1)\pi}{t_p}t\right)}{2n+1}\right\} \tag{6}$$

Here I_{p0} and t_p are the irradiated pulse height and pulse width, respectively.

From figure 1 with equation (4) and equation (5), the equation of motion of the phase difference ϕ is given by

$$\frac{d\phi}{dt} = \frac{i_d + i_p(t) - \sin\phi}{\alpha - \beta\cos\phi} \tag{7}$$

where i_d and $i_p(t)$ are equal to $i_d = \frac{I_d}{I_c}$ and $i_p(t) = \frac{I_p(t)}{I_c}$, respectively.

Parameters α and β are defined as follows.

$$\alpha = \frac{\hbar}{2eI_cR_n} \tag{8}$$

$$\beta = \frac{17\tau}{735}\left(\frac{L}{\xi}\right)^2 \tag{9}$$

3. Results and Discussions

The assumptions in order to analyze the pico pulse response of HTS Josephson weak-link are as follows.

(1) The critical temperature of $Bi_2Sr_2Ca_2Cu_3O_{10}$ is 110[K].

(2) The energy gaps of $Bi_2Sr_2Ca_2Cu_3O_{10}$ at 4.2[K] and 50[K] are 30[meV] and 28[meV], respectively.

(3) The normal resistance is 1.0[Ω].

(4) Values of the I_cR_n product at 4.2[K] and 50[K] are 18.5[mV] and 17.1[mV], respectively.

(5) The weak-link length is equal to the coherence length of HTS. The coherence lengths of $Bi_2Sr_2Ca_2Cu_3O_{10}$ at 4.2[K] and 50[K] are 1.02[nm] and 1.35[nm], respectively.

(6) The DC driving current normalized by the critical current, i_d is 1.1.

(7) The pulse height normalized by the critical current, $i_p = \frac{I_p}{I_c}$ is 3.0.

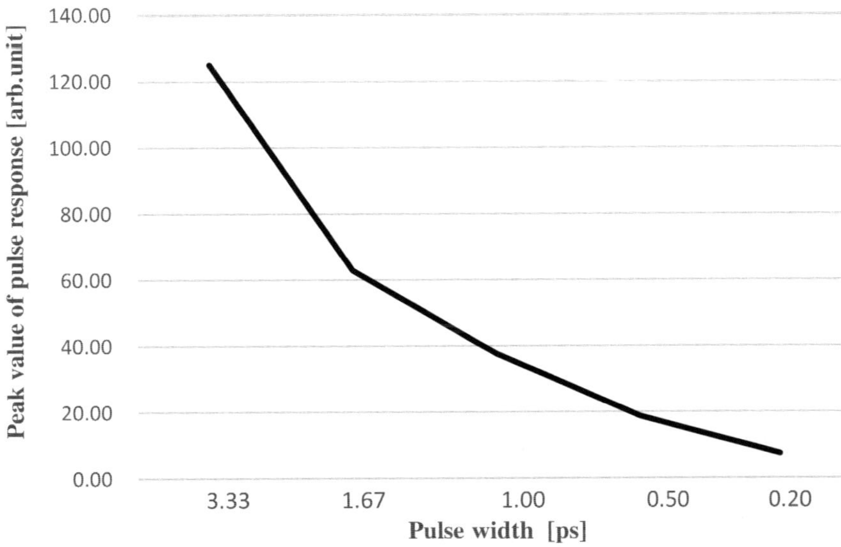

Figure 2. Peak value of pulse response as a function of pulse width as a parameter at 4.2[K].

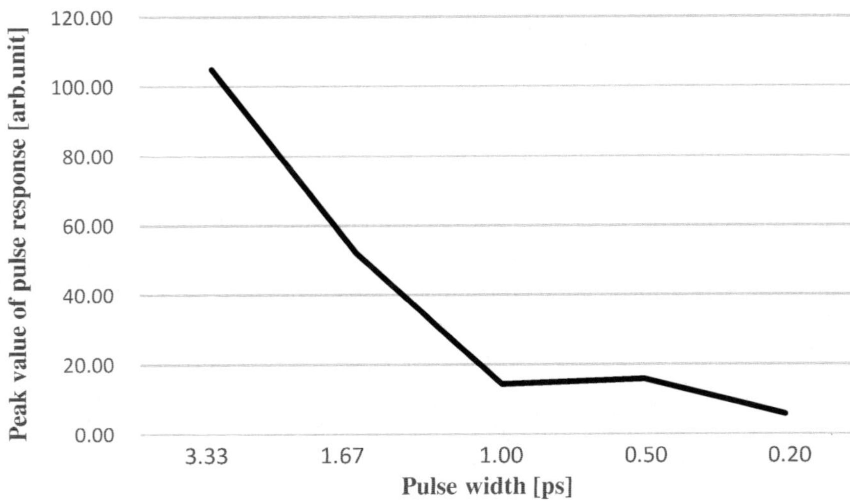

Figure 3. Peak value of pulse response as a function of pulse width as a parameter at 50[K].

We analyze the pulse width dependence of the peak value of pico pulse response and show results in .figure 2 at the liquid nitrogen temperature, 4.2[K] and figure 3 at 50[K] achieved by use of compact cryocooler.

We found from figure 2 and figure 3 that the pulse width dependence of the peak value is not dependent on the temperature and that the peak values of the pico pulse response at both temperatures decrease with decrease of pulse width.

We think that the decrease of the peak value of pulse response is due to the harmonic component of pico pulse. According to equation (6), for example, pico pulse of 0.5[ps] wide consists of the fifth harmonic component greater than 10[THz]. Influence of the fifth harmonic component is non-negligible, since the amplitude of the fifth harmonic component is greater than a tenth of the fundamental component.

The TDGL model of HTS Josephson weak-link is promising when we analyze the properties of HTS Josephson weak-link applied to developing research such as astronomical observation in the THz region.

4. Conclusions

We analysed the pico pulse response of the Josephson weak-link made of $Bi_2Sr_2Ca_2Cu_3O_{10}$ at 4.2[K] and 50[K] in the THz region using the TDGL model.

At both temperatures, we found that the peak values of pulse response decreased when the pulse width decreased.

References

[1] Amrein T *et al.* 1993 *Appl. Phys.* Lett.**63** 1978
[2] Li H Q *et al.* 1997 *Appl. Phys.* Lett.**71** 1123
[3] Klushin A M *et al.* 2002 *Physica C* **372-376** 301
[4] Matsumoto T *et al.* 2008 *Physica C* **468** 1916
[5] Hua T *et al.* 2011 *Cross Strait. Quad-Regional Radio Science and Wireless Technology Conference (IEEE)* 668
[6] Gor'kov L P and Eliashberg G M 1969 *JETP* **29** 698
[7] Jensen H H and Lindelof P E 1976 *J. Low Temp. Phys.* **23** 469
[8] Yoshimori S, Terashima T and Kawamura M 1998 *Infrared Phys. & Technol.* **39** 41

IOP Conf. Series: Journal of Physics: Conf. Series **1054** (2018) 012065 doi:10.1088/1742-6596/1054/1/012065

Replacement of NbN by $YBa_2Cu_3O_{7-\delta}$ in superconducting thin film coil in a spiral trench on a Si-wafer for compact SMESs

Y Ichiki[1], K Adachi[2], Y Suzuki[2], M Kawahara[3], A Ichiki[2], T Hioki[2], C-W Hsu[4], S Kumagai[4], M Sasaki[4], J-H Noh[5], Y Sakurahara[5], O Takai[5], H Honma[5] and T Motohiro[1,2]

[1] Graduate School of Engineering, Nagoya University, Furo-cho, Chikusa-ku, Nagoya,464-8603, Japan
[2] Green Mobility Research Institute, Institutes of Innovation for Future Society, Nagoya University, Furo-cho, Chikusa-ku, Nagoya,464-8603, Japan
[3] Kojundo Chemical Laboratory Co.,Ltd. 5-1-28, Chiyoda, Sakado-shi,Saitama, 350-0824, Japan
[4] Graduate School of Engineering, Toyota Technological Institute, Hisakata 2-12-1, Tempaku-ku, Nagoya 468-8511, Japan
[5] Materials & Surface Engineering Research Institute, Kanto-Gakuin University,1162-2, Ogikubo, Odawara, Kanagawa 250-0042, Japan

motohiro@gvm.nagoya-u.ac.jp

Abstract. To improve the energy storage volume density of our concept of a compact superconducting magnetic energy storage system, replacement of superconducting material NbN for the thin film coil in the spiral trench on a Si wafer by a high temperature superconductor $YBa_2Cu_3O_{7-\delta}$ was addressed. By using the successful result previously reported to grow a c-axis oriented film on a steel substrate, a superconducting film of Tc=45.5 K was obtained using similar buffer layers including a (100) oriented CeO_2 layer over a (100) oriented YSZ layer. The similar result reported on a $SrTiO_3$ single crystal substrate without any buffer layers indicates c-axis oriented film of Tc=90K can be obtained by further reduction of oxygen partial pressure.

1. Introduction

We have been developing superconducting thin film coils in a spiral trench on a Si wafers for compact SMESs as shown in figure1. A proof of concept has been performed using NbN thin films showing energy storage of 0.01 mJ [1]. Increasing NbN thickness by mitigating film stress [2], the stored energy increased up to 0.1 mJ [3]. Figure 2 shows a Si wafer of 101.6 mm in diameter with a latest design of the spiral trench. The 36 sided-polygonal trench in figure 1 has been changed to the 144 sided-polygonal trench. Parallel trenches are connected each other by occasional interconnections to form a bundle of 7 trenches. This bundle forms a spiral pattern throughout the wafer. In this latest design, the sevenfold increase of the critical superconducting current I can be expected while the number of the turn of the spiral: n decreases. The inductance L decreases in proportion to n. This change is advantageous since the stored electric power is in proportion to $L{\times}I^2$.

For further improvement, we moved on to replacement of NbN by $YBa_2Cu_3O_{7-\delta}$ (Y123). This replacement is to enable use of a cold source of liquid hydrogen: 20K or even of liquid nitrogen: 77K instead of use of a cryogenic refrigerator at 13K or lower in the case of NbN. Another benefit of this replacement is that the critical superconducting current density of Y123 can be about 2×10^4 A/mm^2 which is about 18 times larger than that of NbN [4]. Since the amount of magnetic energy storage is proportional to the square of the current, we may expect $18 \times 18 = 324$ times larger energy storage in the case of Y123 in comparison to that in the case of NbN. Here, it should be noted that the

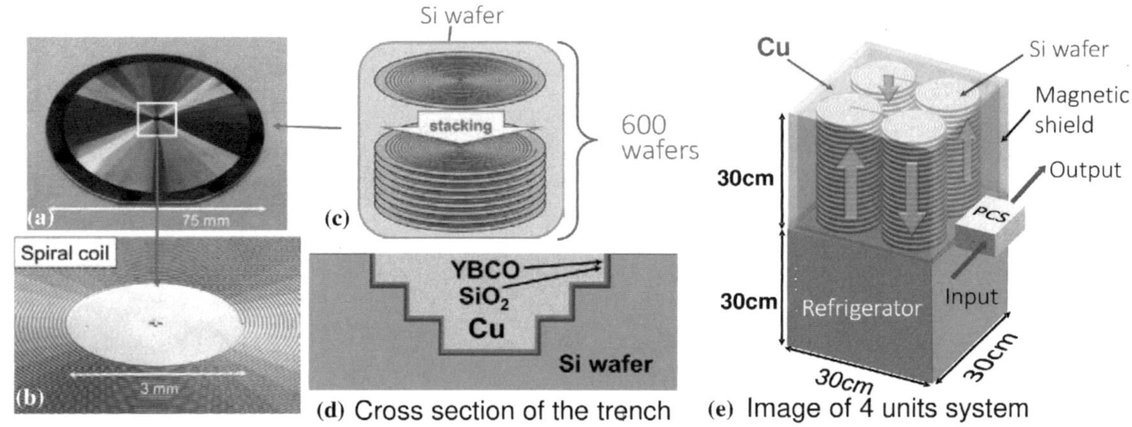

Figure 1. Superconducting coil composed of 4 coil-stack units. Each of the units is made of stacked Si wafers engraved with a superconducting thin film coil in a spiral trench formed with MEMS technology.

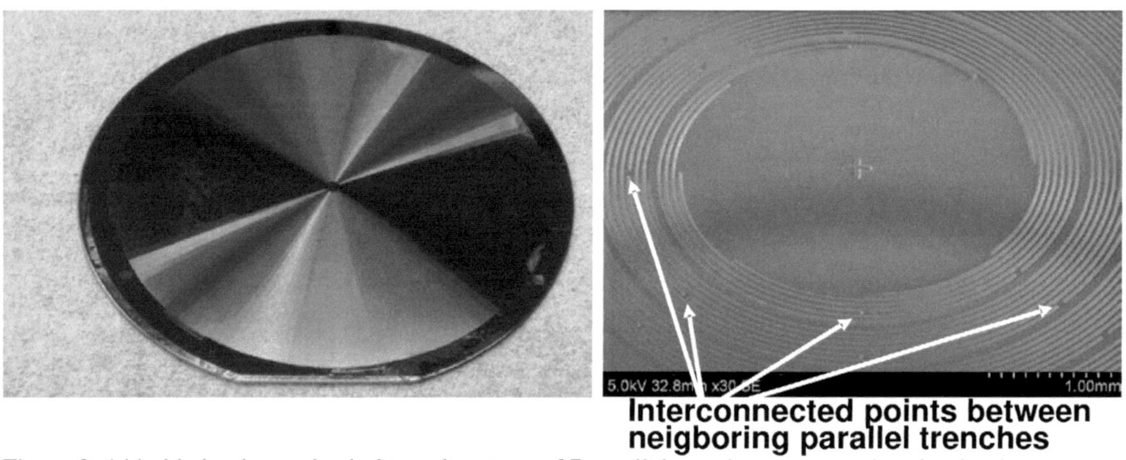

Figure 2. 144-sided polygonal spiral trench pattern of 7 parallel trenches connected each other by occasional interconnections on a Si wafer of 101.6 mm in diameter.

electromagnetic stress imposed on the superconducting coil becomes also 324 times larger. In a typical favourable design of the spiral coil in the case of NbN, the four units system shown in figure 1 together with a cryogenic refrigerator, gives the approximate estimate of the magnetic energy storage volume density as high as 1.8Wh/L and the highest electromagnetic stress near the central axis of the coil unit this is as high as 0.11 GPa which is well below the mechanical strength limit of a single crystalline Si: 4 GPa. However, these values are estimated to be 583 Wh/L and 35.6 GPa, respectively, by the replacement of NbN by Y123. This means the maximum superconducting current in the coil must be much reduced from the value derived from the critical current density. We are doing a more precise estimation at present and will be reported elsewhere. Another problem of this replacement comes from the Cu-plating process which is done in the acidic solution in the case of NbN. Since this conventional acidic solution corrodes Y123 films, we performed extensive survey of the compositions of solution and Cu-plating conditions. However, we are still on the way to find an acceptable process. Here we came across a workaround method to form a sputter-deposited thin Cu film on the Y123 film before the conventional Cu-plating. By covering the Y123 film with a sputter-deposited Cu film including the edge portion of the wafer before Cu-plating, we found the conventional Cu-plating was successfully used as shown in figure 3.

2. Experimental

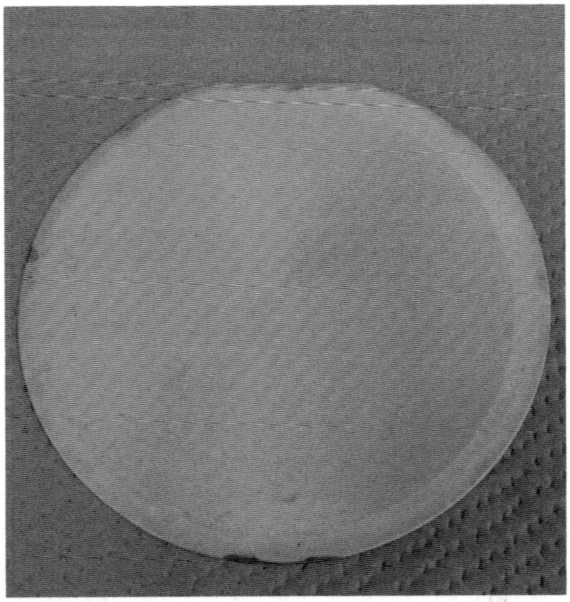

Figure 3. Homogeneously Cu-plated Y123 film on a Si substrate of 101.6 mm in diameter.

Since we utilize sufficiently matured Si-MEMS technology to form the spiral trenches as shown in figure 2 as well as the mechanical strength of the single crystal Si wafer, superconducting Y123 films must be formed on Si wafer for the present purpose. Since the superconducting current increases with the cross-sectional area of the Y123 film in the trench, Y123 films should be deposited as thick as possible. For this purpose, fluorine free metalorganic deposition (FF-MOD) by dipping process is preferable. To form Y123 films by FF-MOD, most of the works have utilized epitaxial growth on single crystal substrates, such as on $SrTiO_3$ (STO) (100) substrate [5],[6], on $LaAlO_3$ (001) substrate [7], on CeO_2-buffered R-plane sapphire (-1012) substrate [8-11], and on 40-nm-thick CeO_2-buffered yttria-stabilized zirconia (YSZ) (100) substrate [12],[13]. In most cases, acetylacetonates (Y:Ba:Cu=1:2:3) based solutions were used. According to the previous work [5], for example, c-axis oriented Y123 films show metallic temperature dependence of resistivity in the normal state and have narrow superconductive transition widths around 90K, while the films which have both c-axis and a-axis orientation show semiconductive behaviour in the normal state and show rather broad transition around 57K. Therefore, formation of c-axis oriented Y123 films (c-Y123) on Si substrate is preferable in our purpose.

On Si substrate, however, few works by MOD have been reported, but some have been reported by sputter-deposition [14], [15]. In these studies, a (100) oriented YSZ layer was firstly sputter-deposited on a Si (100) substrate succeeded by sputter-deposition of a (100) oriented CeO_2 layer (CeO_2/YSZ buffered layer) before the sputter-deposition of Y123 layer which turned out to be c-axis orientation. Therefore, we decided to form a thin Y123 layer by sputter deposition on the sputter-deposited CeO_2/YSZ buffered layer on a Si(100) wafer before the main Y123 layer formation by FF-MOD in the hope that this thin sputter-deposited Y123 layer may help the succeeding growth of c-axis Y123 film by FF-MOD.

More recently, c-Y123 layer was successfully formed on a CeO_2/YSZ buffer layer on Y_2O_3 coated steel substrates using propionates (Y:Ba:Cu=1:2.1:3.15) with additional HCl 0.1 [16-18]. In these cases, the initial precipitation of $Ba_2Cu_3O_4Cl_2$ crystallites on the CeO_2 surface was considered to be preferable for succeeding c-Y123 growth. Since this c-Y123 layer has no epitaxial relation to the steel substrate, this buffer-layer system may work also on a Si substrate for c-axis orientation of Y123.

Taking these previous results into account, we tried to form FF-MOD Y123 layer on a thin Y123 layer sputter-deposited on CeO_2/YSZ buffered Si (100) layer. However we experienced that addition of HCl into propionates solution was apt to cause unwanted sedimentations and we were not successful to make stable solutions for MOD. Therefore, we added dichloroacetic acid (DCA) into propionates (Y:Ba:Cu=1:2.1:3.15) solution in place of the HCl addition. For comparison, we also tested the commercially available acetates solution for MOD: Kojundo Chemical Lab., YBC-05(1/2/3).

3. Results

3.1. YSZ and CeO_2 sputter deposition on Si wafers

Firstly, sputter-deposition of YSZ on Si (100) wafers succeeded by CeO_2 has been performed to form CeO_2/YSZ buffer layer. Secondly, sputter-deposition of Y123 was performed on the CeO_2/YSZ buffer layer. Table 1 summarizes the process conditions of YSZ and CeO_2 sputter-deposition on Si wafers.

Table 1. Process conditions of each layer.

	Gas flow rate (sccm)	Gas pressure (Pa)	Substrate temperature (°C)	process time (h)
YSZ-deposition (500 nm)	Ar:50	5.0	575	3
CeO_2.deposition (500 nm)	Ar:50	5.0	575	3
YBCO Deposition (100nm)	Ar:2, O_2:10	24	670	3
Annealing	O_2	9.0×10^4	450	>1

Figure 4 shows X-ray diffraction patterns after YSZ deposition and after CeO_2 deposition. After YSZ deposition, a clear YSZ (200) peak can be observed. After CeO_2 deposition, a broad peak appeared overlapping the Si (200) peak which appeared the diffraction pattern after YSZ deposition. This broadness of the peak teaches this peak is not of Si (200) peak originated from the single crystal Si (100) substrate but of CeO_2 (200) peak originated from the CeO_2/YSZ buffer layer.

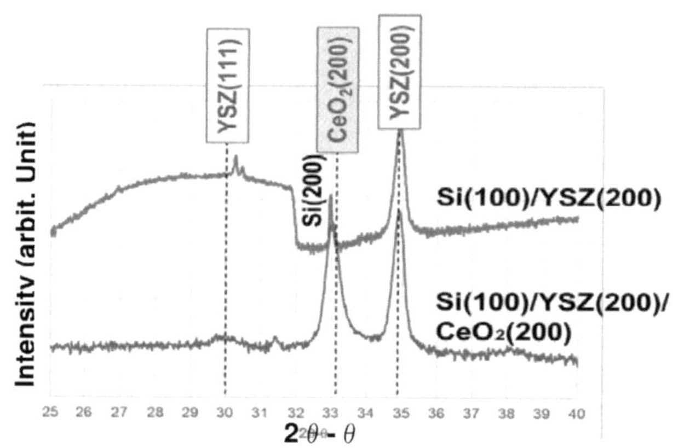

Figure 4. X-ray diffraction patterns after YSZ deposition (upper) and after CeO_2 deposition (lower).

3.2. YBCO deposition by FF-MOD

Deposition of Y123 was performed by FF-MOD using the acetate solution and the propionate solution at various annealing temperatures such as 713 °C, 733 °C, 753 °C ...etc. and O_2 partial pressure of 10 Pa (N_2 balance) for 75 min. At higher temperature such as 830 °C, formation of remarkable $BaCeO_3$ was observed. Figure 5 displays the schematic cross-sectional structures of the three samples. X-ray diffraction peaks of Y123 are scarcely observed in sputter-deposited Y123. Figure 6 shows X-ray diffraction patterns of FF-MOD samples (a) by acetate solution, (b) propionate solution with DCA and (c) propionate

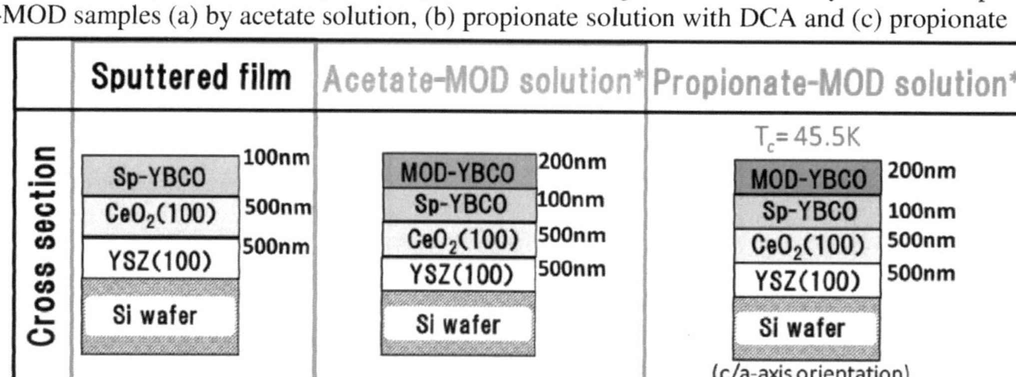

Figure 5. Schematic cross-sectional structures of the samples.

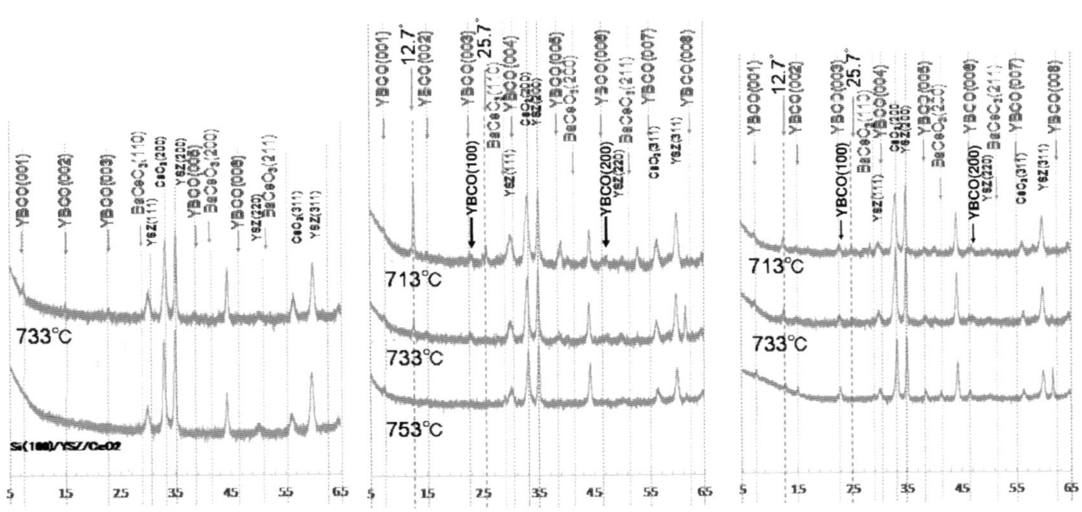

(a) FF-MOD by acetate solution **(b) FF-MOD by propionate solution with DCA** **(c) FF-MOD by propionate solution without DCA**

Figure 6. X-ray diffraction patterns of FF-MOD samples (a) by acetate solution, (b) propionate solution with DCA and (c) propionate solution without DCA.

solution without DCA. Diffraction peaks of Y123 are evidently observed in (a) FF-MOD by acetate solution. In the case of FF-MOD by propionate solution with DCA(b), c-Y123 and a-Y123 peaks can be clearly observed. The peaks at the diffraction angles of 12.7° and 25.7° corresponds to $Ba_2Cu_3O_4Cl_2$ crystallites as described in the previous reports [16-18]. However, the similar peaks were also observed in the diffraction patterns of the sample formed without DCA as shown in figure 6(c). These peaks cannot be of $Ba_2Cu_3O_4Cl_2$ but must be oxide origin $Ba_2Cu_3O_{6-x}$. Figure 7 shows temperature dependence of resistivity in the case of (b) measured by 4 probes in a cryogenic refrigerator. As temperature decreased, the resistivity increased showing semiconductor-like behaviour. At 45.5 K, superconductive transition was clearly observed. The samples (a) did not show superconductive transition.

4. Discussion

The diffraction patterns shown in figures 6(b) and (c) tell existence of the mixture of c-Y123 and a-Y123. In the reference [5] which is study of Y123 deposition by FF-MOD on STO (100) substrate, the similar combination of the X-ray diffraction pattern in figure 5 (c) and the temperature dependence of resistivity in figure 7 was shown. This fact indicates a metallic temperature dependence of resistivity and superconducting transition around 90 K caused by pure c-Y123 may be obtained with decreasing O_2 partial pressure. As for another possibility, we may get c-Y123 for more enhanced crystallization at higher annealing temperature. However, with increase in the annealing temperature, x-ray diffraction peaks of $BaCeO_3$ appeared as shown in figure 6(b) and (c), indicating that annealing temperature should be reduced. Although further experiments are indispensable, but a possibility of the fact that the sputter-deposited CeO_2/YSZ buffer layer on Si (100) substrate with

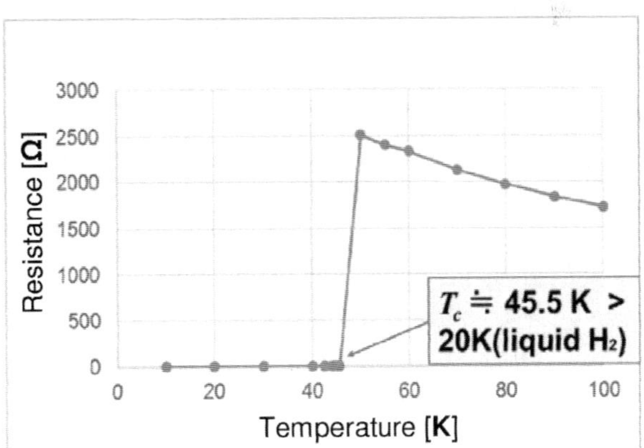

Figure 7. Temperature dependence of resistivity of the sample (c).

additional thin sputter-deposited Y123 film might have played the similar role on c-Y123 layer formation at relatively lower temperature might be brought about for further inspection.

5. Conclusion

By using the propionate FF-MOD solution with addition of DCA, a mixture of c-Y123 and a-Y123 film which shows superconducting transition at 45.5 K was obtained on CeO_2/YSZ buffered Si (100) wafer. These results were in good agreement with previously reported ones for FF-MOD on STO (100) substrate. From this previous literature, further reduction of O_2 partial pressure may lead to pure c-Y123 film formation. This results also teaches the propionate FF-MOD solution which was successful to form c-Y123 film on the steel substrate also works on CeO_2/YSZ buffered Si (100) substrate. By further adjustment of the experimental condition, we expect to deposit c-Y123 film on the Si wafers with spiral trenches as shown in figure 2 for a significant improvement of our compact SMES.

Acknowledgement

This work is partly supported by NEDO Advanced Research Program for Energy and Environmental Technology (16101979-0)

References

[1] Sugimoto N, Iguchi N, Kusano Y, Fukano T, Hioki T, Ichiki A, Bessho T and Motohiro T, 2017, *Supercond. Sci. Technol.* **30**, 015014(11pp)
[2] Suzuki Y, Iguchi N, Adachi K, Hioki T, Ichiki A, Hsu C-W, Kumagai S, Sasaki M and Motohiro T, 2017, *IOP Conf. Series: Journal of Physics: Conf. Series* **871**, 012071(8pp)
[3] Suzuki N, Iguchi N, Adachi K, Ichiki V, Hioki T, Hsu C-W, Sato R, Kumagai S, Sasaki M, Noh J-H, Sakurahara Y, Okabe K, Takai O, Honma H, Watanabe H, Sakoda H, Sasagawa H, Doy H, Zhou S, Hori H, Nishikawa S, Nozaki T, Sugimoto N and Motohiro T, 2017, *IOP Conf. Series: Journal of Physics: Conf. Series* **897**, 012019(7pp)
[4] Sugimoto N and Motohiro T 2012 *Vacuum* **93** 13-21
[5] Manabe T, Kondo W, Mizuta S and Kumagai T, 1994, *J. Mater. Res.* **9**, 858-65
[6] Vermeir P, Feys J, Schaubroek J, Verbeken K, Lommens P and Driessche I V, 2012, *Materials Research Bulletin* **47**, 4376-82
[7] Manabe T, Kondo W, Yamaguchi I, Sohma M, Tsuchiya T, Tsukada K, Mizuta S and Kumagai T, 2005, *Physica C* **417**, 98-102
[8] Manabe T, Sohma M, Yamaguchi I, Kondo W, Tsukada K, Mizuta S and Kumagai T, 2004, *Physica C* **412-414**, 896-9
[9] Manabe T, Sohma M, Yamaguchi I, Kondo W, Tsukada K, Kamiya K, Mizuta S and Kumagai T, 2005, *IEEE Trans. on Appl. Superconductivity* **15**, 2923-6
[10] Yamaguchi I, Manabe T, Sohma M, Tsukada K, Kondo W, Kamiya K, Mizuta S, and Kumagai T, 2006, *IOP Conf. Series: Journal of Physics: Conf. Series* **43**, 369-72
[11] Manabe T, Yamaguchi I, Tsukada K, Sohma M, Kondo W, Kamiya K and Kumagai T, 2007, *IEEE Trans. on Appl. Superconductivity* **17**, 3491-4
[12] Tsukada K, Furuse M, Sohma M, Manabe T, Yamaguchi I, Kondo W, Fuchino S and Kumagai T, 2005, *Jpn. J. Appl. Phys.* **44**, 4914-8
[13] Yamaguchi I, Tsukada K, Kondo W, Sohma I. V. M, Kamiya K, Kumagai T and Manabe T, 2007, *IEEE Trans. on Appl. Superconductivity* **17**, 3495-8
[14] Rombola G, Ballarini V, Chiodoni A, Gozzelino L, Mezzetti E, Minetti B, Pirri C F and Tresso E, 2005, *Int'l J. Mod. Phys.* B **19**, 4605-17
[15] Botta D, Camerlingo C, Chiodoni A, Fabbi F, Geraldo R, Ghigo G, Gozzelino L, Laviano F, Minetti B, Pirri C F, Rombola G, Tallarida G, Tresso E and Mezzetti E, 2005, *Eur. Phys. J.* B **48**, 359-65

[16] Motoki T, Shimoyama J, Yamamoto A, Ogino H, Kishio K, Honda G and Nagaishi T, 2014, *Supercond. Sci. Technol.* **27**, 095017(7pp)
[17] Motoki T, Shimoyama J, Ogino H, Kishio K, Roh J, Tohei T, Ikuhara Y, Horii S, Doi T, Honda G and Nagaishi T, 2016, *Supercond. Sci. Technol.* **29**, 015006(7pp)
[18] Motoki T, Ikeda S, Honda G, Nagaishi T, Nakamura S and Shimoyama J, 2017, *Appl. Phys. Express* **10**, 023102(4pp)

IOP Conf. Series: Journal of Physics: Conf. Series **1054** (2018) 012066 doi:10.1088/1742-6596/1054/1/012066

R&D of applied superconductivity by a small business: experiences and future perspective

D Pooke, V Chamritski, M Christian, M Fee, T Huang

HTS-110,1B Quadrant Drive
Lower Hutt 5010, New Zealand

E-mail: d.pooke@hts-110.com

Abstract. The increasing commercial availability of High-T_c Superconducting (HTS) wire in the decade following the discovery of this amazing class of materials opened the door to a range of unique application opportunities. The early international focus was on the power industry with a promise of a significant transformation in efficiency and supply security; however the realities of funding large and technically demanding prototypes and the lengthy timelines involved in realistic commercialisation pathways, meant that for a small business nearer-term opportunities would need to be explored. From a background in materials research and wire development activities, HTS-110 was established over 13 years ago to design and manufacture HTS magnets for a wide range of scientific and industrial applications; over the last decade significant progress has been made in the commercialisation of HTS magnets, both by HTS-110 and other manufacturers, paralleling the improvements in wire performance and quality. In this paper we review the developments of HTS magnet technology and applications leading to a range of niche application areas, from sample environments for synchrotron and neutron beamlines and other materials analysis applications, through to developments for the demanding realm of magnetic resonance, all of which leverage the benefits of high current density relative to copper and a relatively high operating temperature compared to the low-temperature superconductors. New future application areas promise to extend these developments and greatly expand the role of HTS magnets in our industrial society.

1. Background

Following the discovery of the High-T_c superconducting materials, New Zealand, like many countries around the globe, invested heavily in research and development of these remarkable superconductors since it was easily argued that this represented an opportunity to radically alter the generation, distribution and use of energy around the world. The leap in efficiency and security would drive widespread deployment, associated with rapidly falling wire prices due to the large scale of manufacture that would be required to meet this market demand. For almost two decades then, there ensued a programme first of materials discovery and characterisation followed rapidly with the development efforts targeting wire production in order to enable the promised revolution in the electricity industry. On the basis of materials and processes that had been identified by the team in New Zealand, a patent licensing agreement was reached with what was then the pre-eminent producer of first generation ("BSCCO") wire in the world. From this relationship a core body of local knowledge developed regarding the properties and performance of this wire, extending to the understanding that it seemed unlikely that wire manufacturing itself would be a viable option in the New Zealand environment. However establishing a manufacturing business to build devices using the

IOP Conf. Series: Journal of Physics: Conf. Series **1054** (2018) 012066 doi:10.1088/1742-6596/1054/1/012066

wire seemed attractive; the initial target would be for industrial, scientific and medical applications rather than directly for the expected large markets in the electricity sector. This suited the type and scale of industry in New Zealand, and did not preclude partnerships in the electricity industry for components as the demand arose. Importantly it would also demonstrate a commercial outcome from the public funding of HTS research, even if extensive growth of such a business would likely rest on other industries to drive volume manufacturing of the wire.

On this basis, HTS-110 was formed in 2004. Initially with just three employees and no firm customers such was in the interest in the HTS machine development that we soon were manufacturing coils for several organisations, extending then into magnet assemblies. We benefitted from early adopters in the move to fully integrated magnet systems, and with partners in meeting the needs of niche industrial users. Sitting behind the magnet development were research and development programmes, first to establish robust coil manufacturing technologies for the HTS wire, then continual efforts to extract higher performance for the magnet systems. Given the size and resources of the company these developments were necessarily targeted at near-term commercial outcomes rather that very future-focussed efforts. An underpinning HTS research and development team continued within the government-funded research environment, that team reformed as the Robinson Research Institute and through which HTS-110 has also benefitted from the wider R&D activity.

2. Technology development for coils and components

The basic information required for magnet design centred on wire performance which was a rapidly moving target in the early years. This incorporated knowledge of the anisotropic field dependence of critical current as a function of temperature, the development of reliable design tools for electromagnetic and thermal integration, plus the more practically focussed matters underpinning a reliable coil manufacturing process such as turn-to-turn insulation selection and resin selection for vacuum-pressure impregnation. Ultimately these selections would need to be tailored to both the application requirements (for example turn-to-turn or coil-to-ground dielectric strength), or even in responding to the properties of the particular wire chosen. For example, in the manufacture of the first REBCO wire coils it was quickly discovered that a re-evaluation coil manufacturing and handling processes was needed due to coil performance failures; these resulted from damage to the wire during thermal contraction on cool-down of the coil. Resin selection was an important factor and HTS-110 successfully demonstrated paraffin wax as an interim alternative while other resins were identified.

2.1. Non-contact I_c wire qualification

Especially in the early years of HTS wire manufacture, wire performance was usually very inconsistent on a length basis and given the high cost it was necessary to provide a quality assurance measure that would enable reliable coil performance. A non-contact I_c measurement system was developed to give high-resolution long-length I_c data on incoming wire.

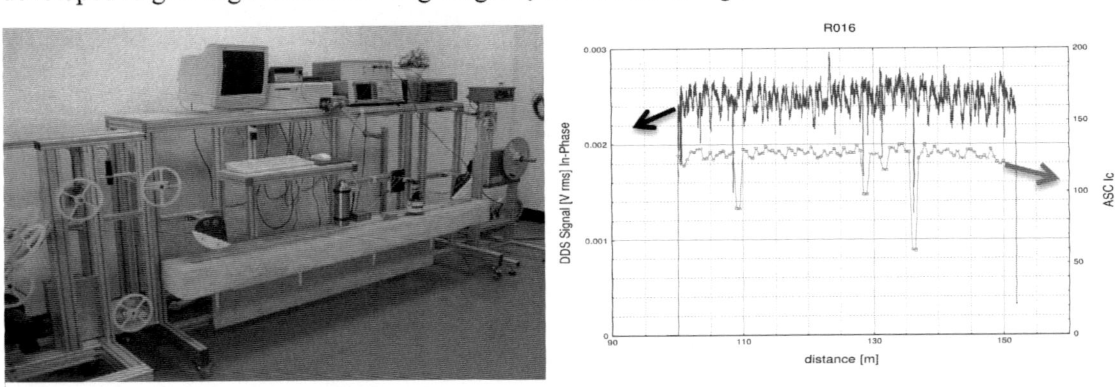

Figure 1. In-line non-contact I_c measurement system for wire qualification (left) with graph (right) showing correspondence between the induced signal and transport data measured on the same sample.

IOP Conf. Series: Journal of Physics: Conf. Series **1054** (2018) 012066 doi:10.1088/1742-6596/1054/1/012066

This was based on a small AC magnet inducing local supercurrents in the wire the effect of which could be detected through a differential Hall sensor setup operated with a lock-in detector. Calibration against wire with known transport current data provided a system with excellent correlation to the transport data yet able to give rapid scanning of wire at 77 K (figure 1). This became an essential tool in the manufacturing process.

2.2. *Wire characterisation I_c (B, ϑ, T)*

Capability development towards routine characterisation of transport properties of HTS wires ultimately led to the development, though our research partners in Robinson Research Institute, of an automated test instrument capable of I_c(B, ϑ,T) measurements to approximately 1 kA on HTS samples over the temperature range of 15-100K and with fields of 0-8 T applied at any desired angle to the tape plane [1]. The magnetic field is generated by an HTS 8 T dipole magnet constructed by HTS-110 and is integrated with a sample cryogenic system based on forced helium gas flow through the sample chamber, employing a separate cryocooler to that used by the magnet. This system, named "SuperCurrent" has been of particular value in characterising the transport properties of second-generation wire types.

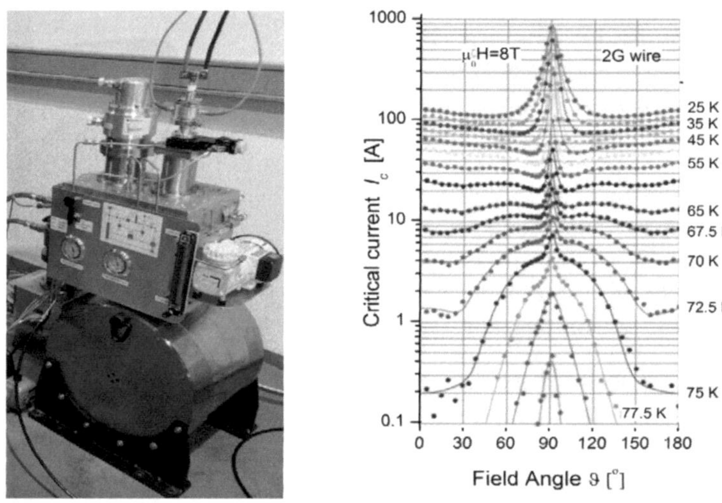

Figure 2. The "SuperCurrent" wire critical current measurement system showing the 8 T magnet and sample cryogenic manifold (left) and example data from a REBCO sample at 8 T applied field.

3. Magnets

3.1. *The first large HTS magnets*

The first large-scale physics magnet built with HTS wire was a ion-beam steering magnet installed in 1997 into a radio-carbon dating facility at the Institute of Geological and Nuclear Sciences in New Zealand [2]. This magnet was important in demonstrating the robustness of HTS magnets and the longevity of the wire and coil performance over many years of operation (which had not previously been demonstrated). Ultimately the magnet was decommissioned due to failure of the compressor for the cryocooler. This magnet led on to the development of a general-purpose H-frame 3 T laboratory electromagnet [3] with HTS coils operating at a temperature of around 30K, cooled by a single stage GM cryocooler; this system is still employed today for some transport measurements on HTS samples.

IOP Conf. Series: Journal of Physics: Conf. Series **1054** (2018) 012066 doi:10.1088/1742-6596/1054/1/012066

3.2. Higher-field – magnets for synchrotron x-ray and neutron scattering sample environments
We were fortunate in having "early adopter" push for high fields in a compact envelope for the Hahn Meitner Institut at their high resolution x-ray diffraction and resonant magnetic scattering beamline MAGS at the Berlin Electron Synchrotron facility BESSY [4]. This resulted in a 5.4 T split-pair magnet with system mass sufficiently low that it could be mounted in a commercial goniometer and was able to be rotated by 90° with respect to the beam axis and scattering plane. The magnet was cooled with a pulse-tube refrigerator set at an angle of 45° to the vertical; this allowed operation of the magnet in the two orientations without suffering from the large cooling performance drop associated with pulse-tube coolers when tipped towards the horizontal orientation. An aluminium cryostat helped to ensure an overall low system mass, again for compatibility with the goniometer. This design has been replicated for other synchrotron facilities.

At about the same time a much larger 5 T split pair magnet was designed and manufactured for the Australian Nuclear Science and Technology Organisation [4]; this unit had large optical access angles, was designed to be compatible with the user's existing sample cryo-furnace, and was also to sit on a rotating stage but with limited range of motion. 12km of HTS wire was used and when running, almost 60 tonnes of internal compressive forces had to be managed, requiring a carefully designed combination of titanium, carbon fibre and other composite components. Whereas the x-ray diffraction magnet was protected with a cold-diode bank the neutron scattering magnet, with a stored energy of order 0.5 MJ, required a room-temperature energy dump circuit to be designed; this unit provided a rapid switched-resistance discharge of the magnet, with the switching control powered by the magnet stored energy itself. Following these two magnets the management of stored energy has remained with room-temperature dump resistors, either actively engaged or always connected. More recently the quench protection heaters have also been employed successfully by other groups [5].

Since these two magnets, several more systems have been designed for neutron and synchrotron beamline sample environments, many of which has been of modest field strength (2-3 T), but offering compact physical envelopes combined with large optical and sample access. An interesting possibility with the soft-iron yoke is the control of zero-field points in the magnetic circuit, which is of particular concern for experiments with polarised neutrons [6]. Magnets for UHV applications have also been delivered, either with the UHV chamber embedded inside the HTS magnet, or with the HTS magnet mounted inside a UHV chamber [7,8]. In the latter case the magnet vacuum system is kept separate from the UHV system, and in both cases attention had to be paid to maintaining the HTS coils at resin-compatible temperatures during vacuum "bake-out".

One of the largest magnets to date, with 8.6 T in a room temperature bore of 110 mm, was manufactured for LARIAT II (Large Area Rapid Imaging and Analysis Tool, Synchrotron Research Inc., USA), used for imaging and chemical analysis with soft x-rays. Imaging is achieved with a CCD detector housed in a separate HTS detector magnet, the strength of which determines the image

Figure 3. Examples of sample environment magnets for synchrotron and neutron endstations. Left: a 5 T split-pair magnet for resonant magnetic scattering and high-resolution X-ray diffraction. Middle: The pair of magnets for the LARIAT II imaging spectromter. Right: a 3 T magnet for polarized neutron scattering.

magnification [9]. Another novel system is a compact cylindrical-bodied magnet (diameter 150 mm) of modified H-frame design offering in-plane field which can be rotated around a fixed sample, providing a mechanical vector field control with in-plane optical access of 180° for all field angles and an out-of-plane opening angle of over 8° [10].

4. Fast ramping magnets, beam scanning

Responding to a need for rapid high-throughput characterisation of magnetic materials, for example in the data-storage industry or in the development of permanent magnets, a series of fast ramping magnets has been developed. This has entailed the careful characterisation of AC losses in the HTS coil systems at ramp rates which are low compared to loss studies performed near utility distribution frequencies (50/60 Hz) or higher. In this case, the in-cycle peak losses may be more important than average losses and may be observed as in-cycle coil-pack temperature deviations. It is difficult to calculate the thermal load in a complex magnet geometry as it depends on multiple factors including field magnitude, orientation, tape I_c, operating current, tape cross section, and other factors. The loss is also difficult to minimise mechanically since there are currently limited commercial options for reducing the aspect ratio of the HTS tapes. Further potential solutions may conflict – for example: higher I_c wire will operate safely at higher temperature so will be less sensitive to local heating caused by AC loss, but coils built with high I_c tape will have higher losses and coil pack temperature rises. Nevertheless, good progress has been made, with the latest generation of magnets designed for measurement via MOKE (Magneto-optical Kerr Effect) able to perform 7 T four-quadrant ramp cycles in 60 s on a continuous basis (figure 4).

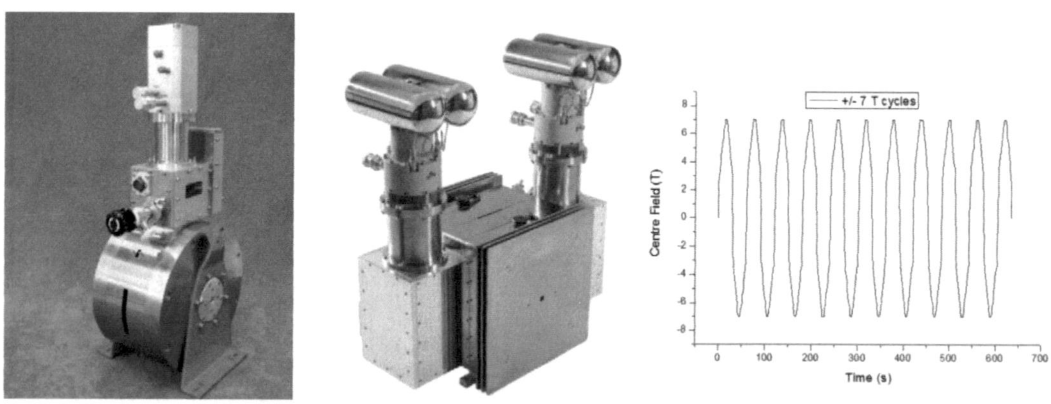

Figure 4. Split-coil (dipole) magnets for high throuh-put MOKE measurements; the optical access is through the magnet poles (iron), while the sample is inserted in the slots between the split coil pairs. Left: 5 T model, Middle: 7 T model (capable of 8 T at slower ramp rate), Right: Ramp cycles to 7 T with 62 s four-quadrant cycletime; non-linearity is due to the iron circuit.

5. Magnetic resonance

A significant research and development effort at HTS-110 has been directed towards magnetic resonance applications, particularly NMR for chemical analysis in industrial applications. While much international effort focusses on extending the available realm of NMR fields above the limit of LTS conductors (for NMR above 1 GHz, which is of interest especially for the analysis of large organic molecules), the effort at HTS-110 has been to leverage the mechanical robustness of HTS magnets, and the absence of liquid cryogens, to enable chemical spectroscopy in industrial locations. The first NMR magnet was a small 2 T split pair magnet designed initially for petrochemical applications; this has progressed through 4.7 T solenoids (200 MHz NMR) to, most recently, the first 400 MHz instrument to be installed directly in a fume hood of a chemistry laboratory for reaction monitoring in

IOP Conf. Series: Journal of Physics: Conf. Series **1054** (2018) 012066 doi:10.1088/1742-6596/1054/1/012066

the pharmaceutical industry (figure 5). These magnets are all actively driven with high-stability power supplies of sufficient quality for standard NMR field-locking systems to operate.
The field homogeneity requirements, which approach the part-per-billion level for high-resolution NMR, cannot be addressed with superconducting shims as in LTS magnets due to the lack of persistent joint technology for the BSCCO wire employed. Instead a passive ferromagnetic shimming approach, similar to that of MRI, has been adapted which brings field uniformity over the sample volume to the level of 1 ppm scale, then conventional room-temperature shims are employed for the finer scale field correction.

With the wide HTS conductors, screening currents induced during the ramp to field can have a significant impact on the field uniformity and also on the temporal stability, with the combined effect of screening currents being to reduce the central field in a solenoid. If ramped directly to field then long-term drift is observed yielding an increasing field, and of drift strength that would be problematic both in mapping and NMR operation of the magnet; moreover the field uniformity may change with time. Understanding and mitigation of screening currents is an on-going research activity internationally [11,12], but a *sufficient* control strategy has been to employ multi-stage ramp cycles which depress the effect of screening currents.

In addition to NMR systems, MRI magnets using both BSCCO and REBCO conductors have been successfully demonstrated [13].

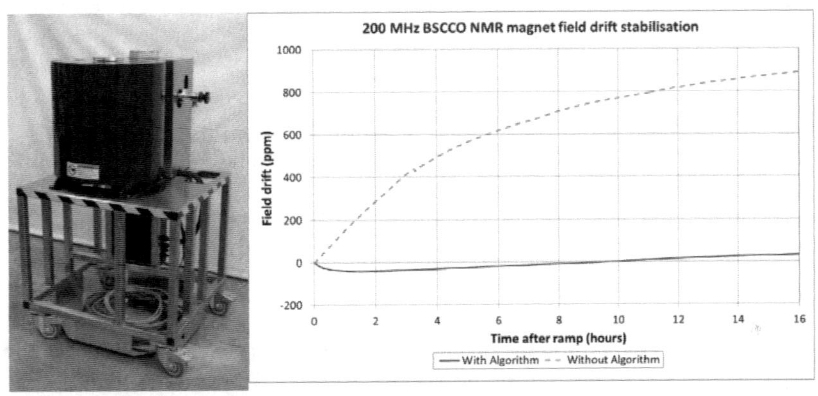

Figure 5. Left: A 400 MHz HTS NMR magnet with standard 54 mm room temperature bore. Right: comparison of magnet drift suppression with a multi-stage ramp cycle.

6. The future

Wire cost, engineering current density, anisotropy, mechanical strength, all remain issues requiring further development for wider uptake of HTS technology. Reducing overall facility requirements is also desirable for magnet systems, which then require improvements in efficiency and reliability of power supplies and crycoolers. As much as higher field strengths are a research focus in many national laboratories, widespread practical applications do not necessarily need the generation of such high fields but nevertheless would benefit from any improvements in conductor and coil manufacturing technologies. Of equal importance is the development of architectures better suited to AC-type applications, even in the realm of magnets for materials analysis where high-throughput is demanded. These modified architectures may also make a useful contribution to the limitation and control of magnetisation currents which is important for both particle accelerators and other applications requiring high field uniformity such as NMR. Beyond these wire developments, successful uptake in industry requires extensive system-wide integration and optimisation. With that, the industry is moving forward; several commercial niche areas for HTS magnets now exist (though requiring limited volumes of wire) and application areas are developing that promise to lead to significantly increased wire manufacturing volume, such as compact fusion systems.

7. Conclusions

In parallel with improvements in conductor performance, a wide range of HTS magnets have been produced of increasing performance and sophistication. As a small company, HTS-110 has concentrated on development of magnets for commercial applications and this has necessarily focused R&D effort into near-term issues. Out of this, niche scientific and industrial application using HTS conductors are a commercial reality. One relatively long-running development investment for the company has been in the area of magnetic resonance, culminating the in the production of a 400 MHz NMR system with performance comparable to conventional LTS systems yet able to be sited directly in the users' laboratory. Such systems demonstrate the advantages accruing to the high operation temperature, namely stability, mechanical robustness and a compact cryogenic envelope, in addition to the absence of liquid cryogens. The application horizons are expanding with further wire performance improvements and with the associated system-level integration necessary for commercial applications.

References

[1] Strickland N, Hoffmann C and Wimbush S, 2014 A 1 kA class cryogen-free critical current characterization system for superconducting coated conductors *Rev. Sci. Instrum.* **85** 113907

[2] Pooke D, Tallon J, Buckley R, Todd G, Kalsi S, Szczepanowski S, Sidi-Yekhlef A, Snitchler G, Picard H, Schwall R, MacKinnon B and Neale R Cryo-integrated HTS ion-beam switching magnet 1998 *Proc. 15th Int. Conf. on Magnet Technology (MT-15)*, ed L Lin, G Shen and L Yan (Science Press, Beijing) pp 789-92

[3] Pooke D, Chamritski V, Gibson S, Staines M, Fee M and Buckley R 2004 A versatile laboratory electromagnet with HTS coils *IEEE Trans. Appl. Superconductivity* **14** 1202-5

[4] Pooke D, Chamritski V, Fee M, Gibson S , King T, Tallon J, Meissner M, Fayerherm R, Olsen R, Kennedy S and Robinson R 2009 HTS 5 tesla synchrotron and neutron beamline magnets *IEEE Trans. Appl. Superconductivity* **19** 1372-5

[5] Noyes P, Markiewicz W, Voran A, Sheppard W, Pickard K, Jarvis J, Weijers H and Gavrilin A 2012 *IEEE Trans. Appl. Superconductivity* **22** 4704204

[6] Huang T, Pooke D and Chamritski V 2017 A passively shielded HTS magnet for polarized neutron scattering, *Magnet Technology-25* Amsterdam, The Netherlands Mon-Af-Po1

[7] Huang T, Gao X, Pooke D, Chamritski V, Briggs N, Christian M, Gibson S, Mitchell J, Miles M and deFeijter J 2012 A commercial HTS dipole magnet for X-ray Magnetic Circular Dichroism (XMCD) experiments *IEEE Trans. Appl. Superconductivity* **22** 4203504

[8] Lazić Z, Chamritski V, Pooke D, Valvidares S, Pellegrin E, Ferrer S, Granados X and Obradors X 2013 Integrating UHV (Ultra High Vacuum) and HTS (High Temperature Superconducting) magnets for X-ray synchrotron based experiments *J. Phys. Conf. Ser.* **425** 102003

[9] Fischer D, Jaye C, Scammon K, Sobol P and Principe E 2010 A novel large area imaging NEXAFS spectrometer for combinatorial chemical and structural analysis *Microscopy and Microanalysis* **16** 380-1

[10] Huang T, Pooke D, Chamritski V and McCrohon M 2016 Compact Rotating In-Plane HTS Magnet for X-Ray Scattering Sample Environment *IEEE Trans. Appl. Superconductivity* **26** 4500304

[11] Hwang Y, Jang J, Ahn M, Park Y and Lee S 2016 Feasibility study for reduction of the screening current induced field in a 2G high temperature superconducting coil *Supercond. Sci. Technol.* **29** 105008

[12] Okabe Y, Honda T and Kajikawa K 2017 Numerical reproduction of screening-current-induced fields in HTS tape windings using finite element method *J. Phys.: Conf. Series* **871** 012047

[13] Parkinson B 2016 Design considerations and experimental results for MRI systems using HTS magnets *Supercond. Sci. Technol.* **30** 014009

IOP Conf. Series: Journal of Physics: Conf. Series **1054** (2018) 012067 doi:10.1088/1742-6596/1054/1/012067

Effect of electromagnetic force on a quad-pancake coil wound with a Nb_3Sn CIC conductor

T Obana[1], K Takahata[1], S Hamaguchi[1], H Chikaraishi[1], S Imagawa[1], T Mito[1], H Murakami[2], K Natsume[2] and K Kizu[2]

[1]National Institute for Fusion Science, 322-6 Oroshi, Toki, Gifu, 509-5292, Japan
[2]National Institutes for Quantum and Radiological Science and Technology, 801-1 Mukoyama, Naka, Ibaraki, 311-0193, Japan

obana.tetsuhiro@nifs.ac.jp

Abstract. In the performance verification tests of the JT-60SA central solenoid model coil, the hydraulic characteristics of the model coil improved during the coil energization. As a cause of this phenomenon, it was suggested that a new cooling channel is made by the cable deformation due to the electro-magnetic (EM) force. To investigate the new cooling channel, the EM structural coupling analysis was conducted. The analytical results indicated that new cooling channels in the coil winding occur, and the cooling channels become larger as the position of the cable moves away from the center of the coil winding.

1. Introduction

Performance verification tests of the JT-60SA central solenoid (CS) model coil were conducted to verify the coil manufacturing process and fabrication jigs [1]. The test results indicated that the model coil met the design requirements, and the hydraulic characteristics of the model coil improved during coil energization. The occurrence of a new cooling channel in the coil winding is considered as a cause of the improvement. To investigate the new cooling channel, electromagnetic (EM) structural coupling analysis for the model coil was carried out. In this paper, the coupling analysis and the compressive tests for the analysis are described.

2. Model coil

A model coil is wound with a Nb_3Sn cable-in-conduit (CIC) conductor. As shown in Figure 1, the configuration of the model coil is a quad-pancake [1]. The inner and outer diameters are 1.3 m and 2 m, respectively. The number of turns is 40. Regarding the CIC conductor, the specifications of the conductor are as follows: the number of Nb_3Sn strands is 216, the number of Cu wires is 108, void fraction is 34 %, jacket size is 27.9 mm×27.9 mm, and cabling diameter is 21 mm [2].

3. Compressive tests of a Nb_3Sn cable

For the structural analysis of the model coil, overall elastic modulus [3] of a Nb_3Sn cable was measured. The measurement was conducted using the universal testing machine (AG-10kNX plus) under room temperature. Figure 2 shows the schematic view of the experimental apparatus. A Nb_3Sn cable removed from the CIC conductor was set on the U-shaped jig made from SUS316L. Then, the cable was covered with the cap which is made of a part of the conduit removed from the CIC conductor. Finally, the T-shaped jig made from SUS316L was set on the cap. In the measurement, the cable deformation was measured in the vertical direction while compacting the cable by the universal testing machine.

Figure 3 shows the relation between the compressive force per unit length and the displacement of the cable. The cable is composed of many voids, so that the displacement changes with very little compressive force. In the displacement over 1.2 mm, however, the compressive force increases drastically with the increase of the displacement. This phenomenon will occur because the voids in the cable decrease with the compressive force, and the stiffness of the cable increases.

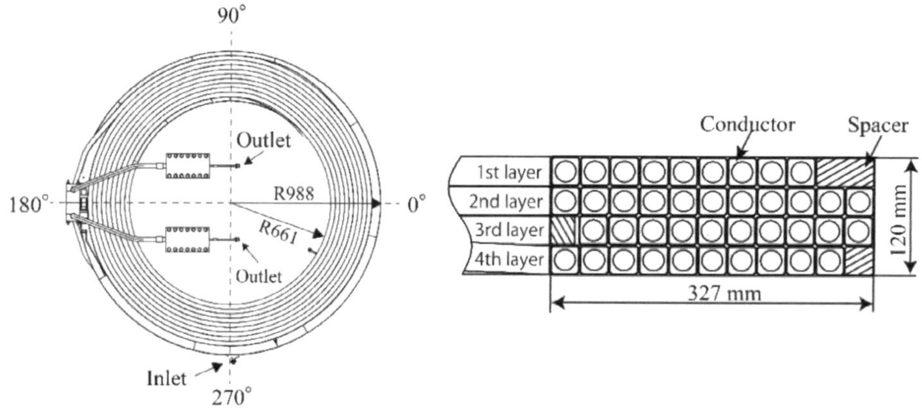

(a) (b)

Figure 1. Configuration of the CS model coil. (a) Top view (b) Cross-sectional view of the coil at 0°.

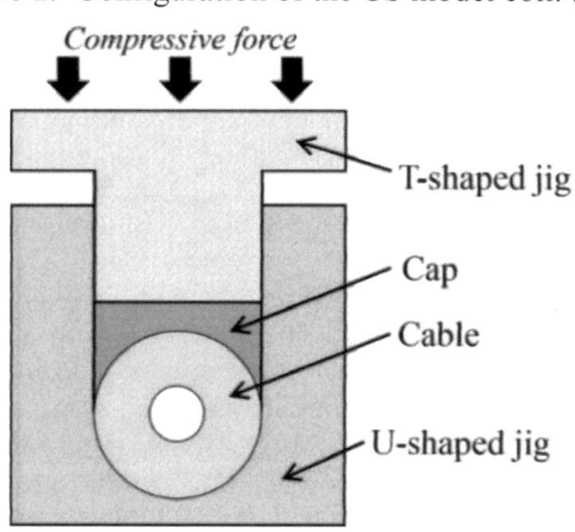

Figure 2. Schematic view of the apparatus for the compressive test.

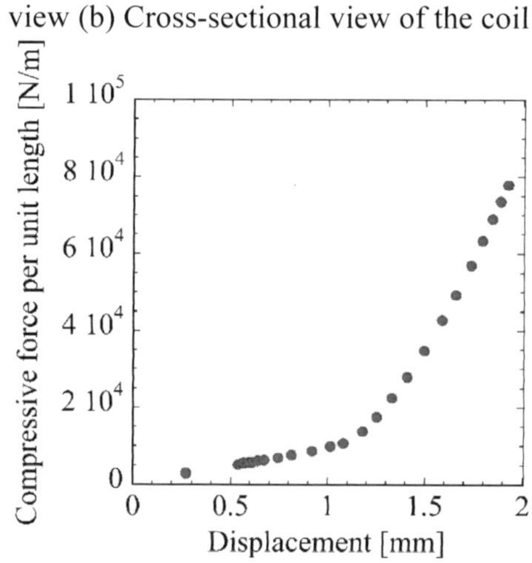

Figure 3. Result of the compressive test.

4. EM structural coupling analysis of the coil

To understand the deformation of the coil configuration, the EM structural coupling analysis was conducted using ANSYS, which is a commercial software product.

4.1. Model and boundary condition

Based on the configuration of the coil winding, the models for the coupling analysis were developed. Figure 4 shows the model for EM analysis of the coil. The model is a two-dimensional model, the configuration of which is an axial and vertical symmetry. The components of the model are as follows: cable, structural material, vacuum area, and infinite boundary area. The cable was assumed to be a conductor, the cross-section of which is a circle without a central cooling channel and twisted strands. The structural material was treated as an integrated conduit of the CIC conductor in the coil winding.

The permeability of each component was assumed to be the permeability in vacuum. In terms of the boundary conditions for the EM analysis, the natural condition was used in r-axis, the fixed boundary condition was used in z-axis, and an infinite boundary condition was used in other boundaries as illustrated in Figure 4 (a).

Figure 4 (b) shows the model consisting of the cable and structural material for the structural analysis. The model is the same as the EM analysis model removing the vacuum and infinite boundary area as illustrated in Figure 4 (a). In the structural analysis, a contact element was used between the cable and the structural material to simulate precisely the cable deformation due to the EM force. Regarding material property, the Young's modulus of 200 GPa and the Poisson's ratio of 0.3 are used in the structural material. Additionally, in the cable used in the model, the Poisson's ratio was 0.3 and the Young's modulus was 0.4 GPa which was the overall elastic modulus derived from the compressive test results of a twisted cable removing a conduit from the CIC conductor. In terms of the boundary conditions, the bottom surface of the material structure was fixed vertically.

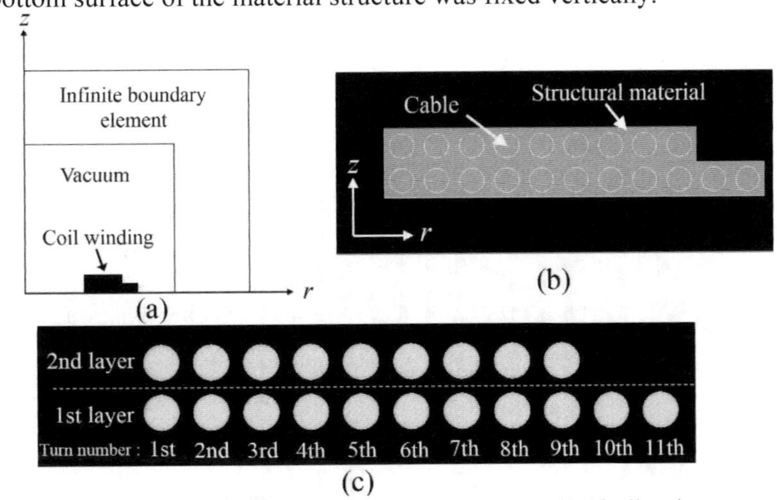

Figure 4. (a) Schematic view of the EM analysis model. The coil winding is composed of cables and structural material. (b) Coil winding in the EM analysis model. (c) Position of the cable in the coil winding.

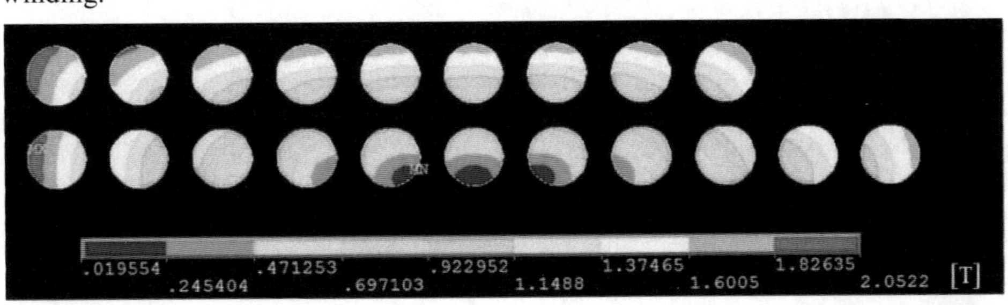

Figure 5. The distribution of the magnetic flux density at the cable of the coil winding.

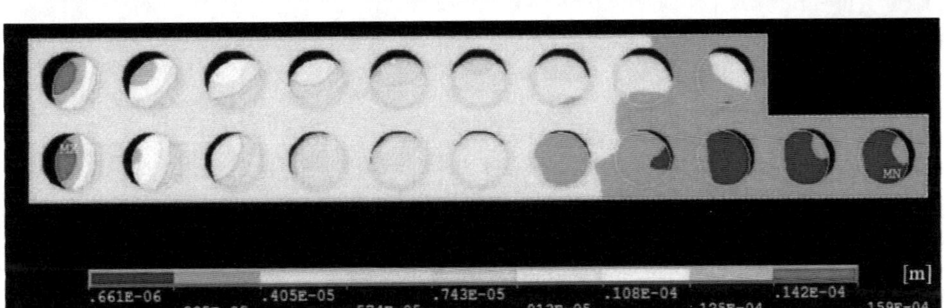

Figure 6. Deformation in the coil winding due to the EM force. The displacement at ×500 magnification is shown.

4.2. Results

The magnetic field analysis of the model coil was conducted where the coil current was 25 kA. The magnetic flux density in the cable is shown in Figure 5. The magnetic flux density increases as the position moves away from the middle of the coil winding. At the coil inner side, the magnetic flux density becomes maximum.

After the magnetic field analysis, the structural analysis was conducted with the data of the EM force derived from the magnetic field analysis described above. Figure 6 shows the result of the structural analysis when the coil current is 25 kA. The cable deformation, which means an occurrence of a new cooling channel, becomes larger as the position of the cable moves away from the center of the coil winding. At the coil inner side, the cable deformation is maximized.

To evaluate the cable deformation at each cable position, the cable deformation was calculated, assuming that there is no deformation in the structural material. Figures 4 (c) and 7 show the position of the cable in the coil winding and normalized space caused by the cable deformation at each cable position, respectively. The space in the second layer is larger than that in the first layer, and the space in the middle turn number tends to be small. The result indicates that the EM force creates non-uniform space distribution in the coil winding.

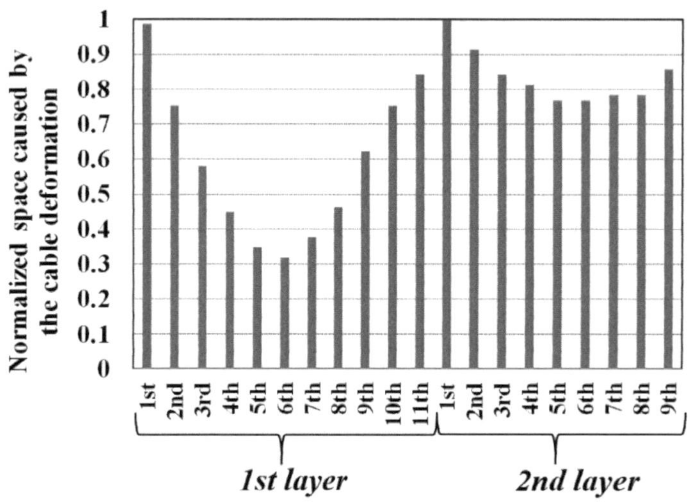

Figure 7. Normalized space caused by the deformation at each cable position. The space at each cable position is normalized by the space at the first turn in the second layer, which is the largest space.

5. Conclusion

To understand the deformation of the JT-60SA CS model coil during the coil energization, the EM structural coupling analysis for the model coil was conducted. The results of the analysis indicated that new cooling channels occur by the deformation of the cable in the coil winding, and the cable deformation becomes larger as the position of the cable moves away from the center of the coil winding.

Acknowledgment

The authors thank Dr. H. Tamura of NIFS for his technical support.

References

[1] Obana T, Murakami H, Takahata K, Hamaguchi S, Chikaraishi H, Mito T, Imagawa S, Kizu K, Natsume K and Yoshida K 2015 *Physica C*, **518** 96-100.

[2] Obana T, Takahata K, Hamaguchi S, Kizu K, Murakami H, Chikaraishi H, Noguchi H, Kobuchi T, Moriuchi S, Imagawa S, Mito T, Tsuchiya K, Natsume K, Yoshida K, Nomoto K, and Kim T 2016 *Cryogenics*, **73** 25-41.

[3] Nijhuis A, Noordman N, Kate H, Mitchell N and Bruzzone P 1999 *IEEE Trans. Appl. Supercond*, **9** 165-168.

Protection System for Normal Transitions in a Single-phase Bi2223 Full Superconducting Transformer by the Active Power Method under Flowing Currents of Various Frequencies

N Nanato, T Ono, T Adachi, T Yamanishi

Graduate School of Natural Science and Technology, Okayama University,
3-1-1, Tsushima Naka, Kita-ku, Okayama, 700-8530, Japan

nanato@okayama-u.ac.jp

Abstract. The authors have developed a large AC current supply with a single-phase Bi2223 full superconducting transformer. The supply is very small and light and can output currents of various values and frequencies. In order to operate the supply safely, a protection system for normal transitions in the transformer is essential. The authors have proposed the system based on the active power method, which detects the normal transitions as dissipated active power in the normal area. Previous studies have shown that the protection system can work well for the transformer in transporting constant current of 60 Hz. The supply is used under transporting currents of various values and frequencies. AC loss and iron loss in the transformer are increased by increasing the value and frequency and then they are detected as active power in a superconducting state. Therefore these losses cause incorrect recognition of the normal transitions. In this paper, the authors propose the system which can detect correctly the normal transitions regardless of the losses. The experimental results show its usefulness for the transformer.

1. Introduction

In recent years, high temperature superconducting (HTS) power applications have been developed [1]. The authors have developed a small and light power source with a single-phase Bi2223 HTS transformer [2]. The power source can supply AC currents of various values and frequencies to measure current conduction characteristics of an HTS wire. For safe operation, it is essential to protect the transformer from excessive heating in normal transitions. The authors have proposed a protection system based on the active power method as a detection method of the normal transitions. This method detects the normal transitions by measuring dissipated active power in normal areas [2-7]. Previous studies have shown that the protection system can work well for the transformer in transporting constant current of 60 Hz [2]. However, the system does not work well in transporting currents of various values and frequencies because AC loss and iron loss in the transformer are increased by increasing the value and frequency. These losses are detected as active power in a superconducting state. Therefore they cause incorrect recognition of the normal transitions. In this paper, the authors propose an improved system which can detect correctly the normal transitions regardless of the losses. Experimental results show the proposed method is useful for detection of the normal transitions and protection of the HTS transformer under flowing currents of various values and frequencies.

IOP Conf. Series: Journal of Physics: Conf. Series **1054** (2018) 012068 doi:10.1088/1742-6596/1054/1/012068

2. Active power method

2.1. Iron loss

Specifications and configuration of the developed single-phase Bi2223 HTS transformer is shown in Table 1 and Figure 1, respectively. Its primary and secondary coils are Bi2223 HTS coils with cylindrical and solenoidal shapes. Both coils are co-axially arranged and have an iron core. Only the coils are cooled in liquid nitrogen and the iron core is always in room temperature air.

Figure 2 shows an electric circuit diagram of the developed power source with the HTS transformer, and Figure 3 shows an equivalent circuit of Figure 2 converted to its primary side. In Figures 2 and 3,

v_{sc1} : primary voltage, v_{sc2} : secondary voltage, i_1 : primary current, i_2 : secondary current,
r_1 : resistance of the primary coil after the normal transitions,
r_2 : resistance of the secondary coil after the normal transitions,
L_1 : self-inductance of the primary coil, L_2 : self-inductance of the secondary coil,
M : mutual inductance,
v_{cc1} : voltage of a cancel coil, m_1 : mutual inductance of a cancel coil,
a : turn ratio ($= N_1/N_2, N_1$ and $N_2$are numbers of turns of the primary and secondary coils, respectively),
i_0 : excitation current, g_0 : excitation conductance, b_0: excitation susceptance,
Z : load impedance.

A variable current and variable frequency power supply with a low rated current is connected to the primary side and then a large current is outputted from the secondary side.

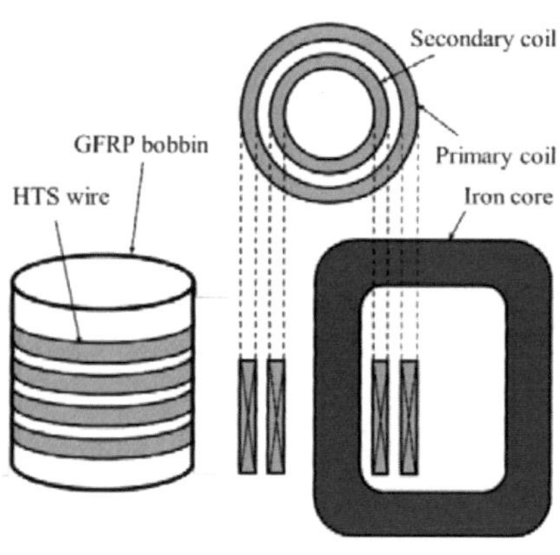

Figure 1. Configuration of the HTS transformer.

Table 1. Specifications of the Bi2223 HTS transformer

Transformer		
Phase		Single
Primary rated voltage / current		100 Vpeak /25 A_{peak}
Secondary rated voltage / current		2.5 Vpeak /1000 A_{peak}
	Primary coil	Secondary coil
Inner diameter	100 mm	83 mm
Outer Diameter	110 mm	89 mm
Height	95 mm	45 mm
Number of Turns	120	3
Inductance	509 mH	0.325 mH
I_C	53 A (a single tape)	1520 A (a bundle of 8 tapes)

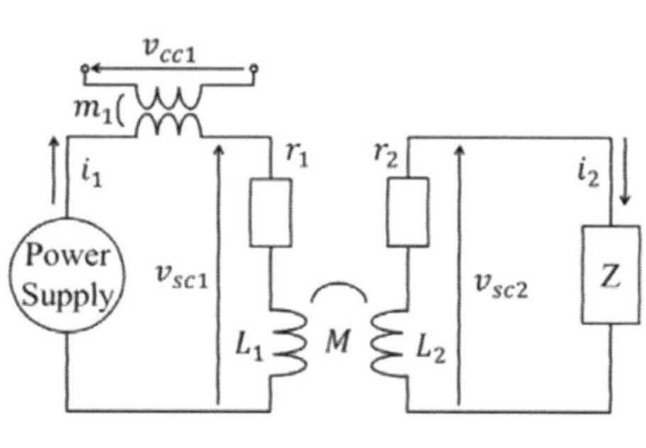

Figure 2. An electric circuit diagram of the developed power source.

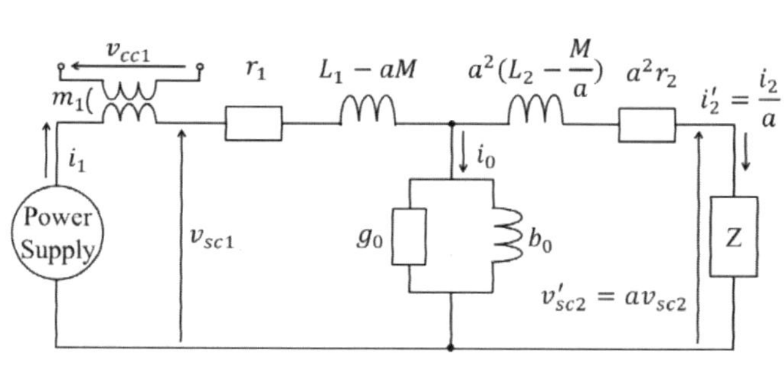

Figure 3. An equivalent circuit of the transformer converted to the primary side.

At first the authors propose the active power method which detects the normal transitions by measuring dissipated active power regardless of the iron loss.

In Figure 3, potential deference Δv between v_{SC1} and v'_{SC2} is expressed as follows.

$$\Delta v = v_{\mathrm{SC1}} - v'_{\mathrm{SC2}} = r_1 i_1 + (L_1 - aM)\frac{di_1}{dt} + a^2 r_2 i'_2 + a^2\left(L_2 - \frac{M}{a}\right)\frac{di'_2}{dt}. \tag{1}$$

The excitation current i_o is very smaller than i'_2 and then i_1 is approximately derived as follows.

$$i_1 = i_0 + i'_2 \fallingdotseq i'_2 = \frac{i_2}{a}. \tag{2}$$

Equation (1) is transformed by equation (2) as follows.

$$\begin{aligned}\Delta v &= r_1 i_1 + a r_2 i_2 + \left\{(L_1 - aM) + a^2\left(L_2 - \frac{M}{a}\right)\right\}\frac{di_1}{dt} \\ &= r_1 i_1 + a r_2 i_2 + (L_1 + a^2 L_2 - 2aM)\frac{di_1}{dt}.\end{aligned} \tag{3}$$

The cancel coil is an air-core transformer and its output voltage is shown as follows.

$$v_{\mathrm{CC1}} = m_1 \frac{di_1}{dt}. \tag{4}$$

Then a resistive voltage is derived as follows.

$$k = \frac{L_1 + a^2 L_2 - 2aM}{m_1}, \tag{5}$$

$$\Delta v' = \Delta v - k v_{\mathrm{CC1}} = r_1 i_1 + a r_2 i_2. \tag{6}$$

$\Delta v'$ is only the resistive voltage generated the normal transitions. This voltage includes no voltage due to the excitation conductance g_0, therefore the normal transitions can be detected by measuring this voltage signal regardless of the iron loss. In order to remove electromagnetic noise in the voltage signal, $\Delta v'$ is converted to an instantaneous active power P and the P is filtered resulting P_f as follows.

$$P = \Delta v' i_1 = \Delta v' i'_2 = r_1 i_1^2 + r_2 i_2^2, \tag{7}$$

$$P_f = \frac{P}{1 + sT}. \tag{8}$$

Equation (8) is expressed by Laplace transform and T is a time constant of the filter.

2.2. AC loss

P_f has no iron loss however it has AC loss signal as the active power signal (r_1 and r_2 include equivalent resistance of the AC loss). Then the AC loss P_L is calculated and subtracted from equation (8) as follows.

$$P' = P_f - P_L. \tag{9}$$

In order to calculate the AC loss, it is necessary to consider the magnetic field distribution because the AC loss generated in the superconducting wire depends on the magnetic field. The primary and secondary coils are cylindrical and solenoidal shapes, therefore the magnetic field distribution of each coil can be obtained by summing up the magnetic field created by a single circular current loop. The AC loss generated in each turn of the coil can be calculated by using the calculated magnetic flux density. B_p is defined as the maximum value of the magnetic field vertically applied to the turn and f is defined as frequency of the transport current. Hysteresis loss in the AC loss is proportional to $B_p f$

when the strength of the magnetic field is equal to or greater than a certain value B_s, and is proportional to $B_p^3 f$ when it is less than B_s, and the coupling loss in the AC loss is proportional to $B_p^2 f^2$. Then the AC loss generated in the entire coil can be obtained as follows [8].

$$P_L = A_{31}\left(\sum_{k=1}^{m} lB_{pk}^3\right)f + A_{11}\left(\sum_{k=m+1}^{N} lB_{pk}\right)f + A_{22}\left(\sum_{k=1}^{N} lB_{pk}^2\right)f^2. \tag{10}$$

l : Length of one turn of the coil,
B_{pk} : The kth lowest applied magnetic field in all turns,
N : Number of turns,
m : Number of turns in less magnetic field than B_s.

The coefficients $\boldsymbol{A_{31}}$, $\boldsymbol{A_{11}}$ and $\boldsymbol{A_{22}}$ are calculated by least squares method using actual measurement value of the AC loss of each coil. Equation (10) calculates the AC losses in the primary and secondary coils, $\boldsymbol{P_{L1}}$ and $\boldsymbol{P_{L2}}$, respectively, and $\boldsymbol{P_L} = \boldsymbol{P_{L1}} + \boldsymbol{P_{L2}}$.

It was obtained that B_s=0.12 mT, A_{31}=60000, A_{11}=0 and A_{22}=0.01 in the primary coil, and B_s=0.029 T, A_{31}=5000, A_{11}=5.8 and A_{22}=0.1 in the secondary coil.

Note that the frequency is calculated using the voltage of the cancel coil and the primary current i_1 as follows.

$$f = \frac{V_{cc1}}{2\pi m_1 I_1}. \tag{11}$$

Capital letters mean root mean square values.

Equation (9) can detect the normal transitions without incorrect detection caused by the iron loss and AC loss.

When $\boldsymbol{P'}$ reaches a specified threshold $\boldsymbol{P_{th}}$ [9], the normal transitions are recognized. The method can detect the normal transitions in the primary and/or secondary coils.

3. Energization tests for a Bi2223 HTS transformer

P_f and P' were measured under flowing the secondary currents of various values and frequencies. Figures 4 and 5 show experimental results of P_f and P', respectively.

Figure 4 shows that P_f increased as the value and/or frequency of the current increased because of the AC loss. Whereas P' in Figure 5 was supressed under about 10 W because the AC loss was removed. It is supposed that the difference under about 10 W occurred because the calculated B_{pk} had some difference from the actual magnetic field which depended on the magnetic characteristics of the

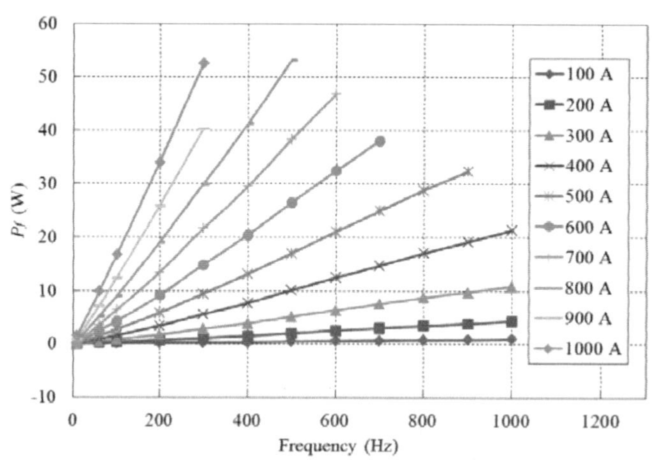

Figure 4. P_f in transporting currents with various values and frequencies.

actual iron core. However P' signal is enough small for detecting the normal transitions in the transformer as mentioned at the next chapter.

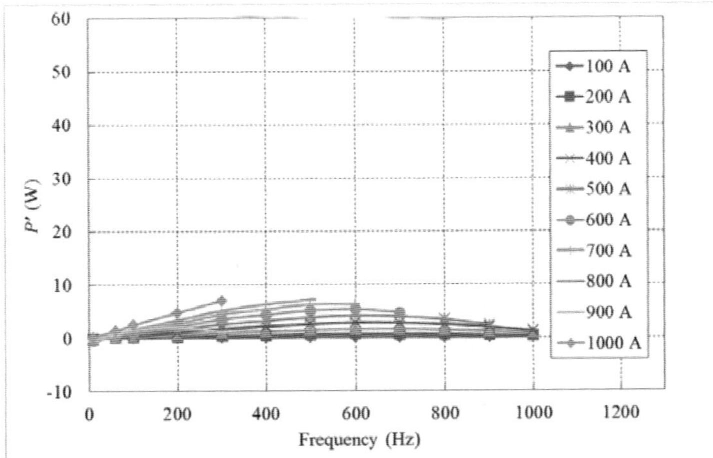

Figure 5. P' in transporting currents with various values and frequencies.

4. Protection tests for the Bi2223 HTS transformer

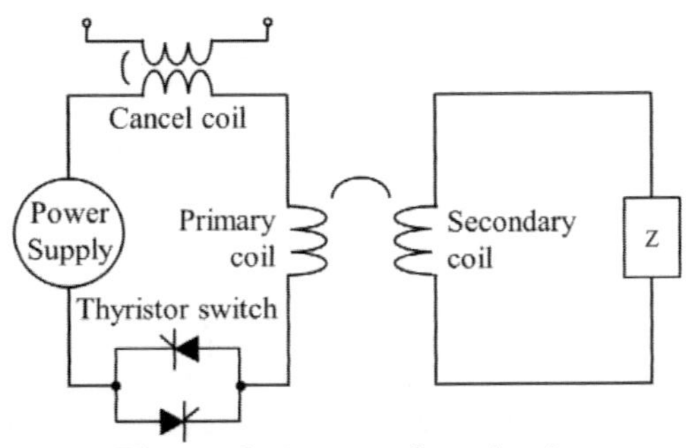

Figure 6. A protection circuit.

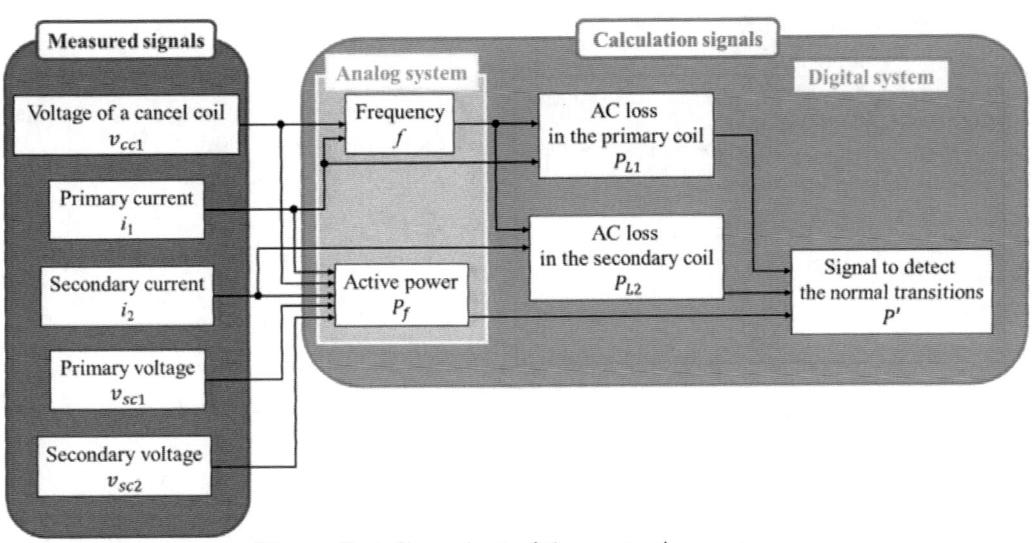

Figure 7. A flow chart of the protection system.

Protection tests for the normal transitions in the transformer was carried out to confirm effectiveness of the proposed method. The protection circuit is shown in Figure 6. Z in Figure 6 was a resistor of 0.3 mΩ. AC current of 25 A_{peak}, 60 Hz and 300 Hz was supplied to the primary coil and then secondary coil outputted a current of 1 kA_{peak}, 60 Hz and 300 Hz. The normal transition was occurred by a heater mounted on a surface of the secondary coil. When P' reaches a specified threshold P_{th}, the thyristor switch turns off and the primary and secondary currents are shut off. Then joule heating in the normal zone is suppressed and its temperature rise is also suppressed.

Figure 7 shows a flowchart of the protection system. Firstly, v_{sc1}, v_{sc2}, i_1, i_2 and v_{cc1} are measured. Secondly, f and P_f are calculated by an analog system, and P_{L1} and P_{L2} are calculated by a digital system. The analog system deals with the all measured signals which are AC signals for correct calculations. On the other hand, the digital system, which is a highly customizable and flexible system, is for DC signals. Finally, a signal P' to detect the normal transitions is calculated.

The test results for 60 Hz and 300 Hz are shown in Figures 8 and 9, respectively. A specified threshold P_{th} has to be set at 50 W for both tests [9]. In Figure 8, (a), (b) and (c) show respectively P_f, P' and temperature of the normal area. When P' reached at P_{th} =50, the transport currents were shut off and the temperature was suppressed less than 170 K which was under permitted temperature of the coil [10]. In this test, the normal transitions can be detected by P_f as well as P' because it was less than P_{th} = 50 in the superconducting states until about 18 s.

As well as Figure 8, Figure 9(a), (b) and (c) show respectively P_f, P' and temperature of the normal area, and when P' reached at P_{th} =50, the transport currents were shut off and the temperature was suppressed less than 170 K. However in this case, the normal transitions cannot be detected by P_f because it was greater than P_{th} = 50 in the superconducting states until about 3 s. These results show that P' is useful for detecting the normal transitions in the HTS transformer under flowing currents of various values and frequencies.

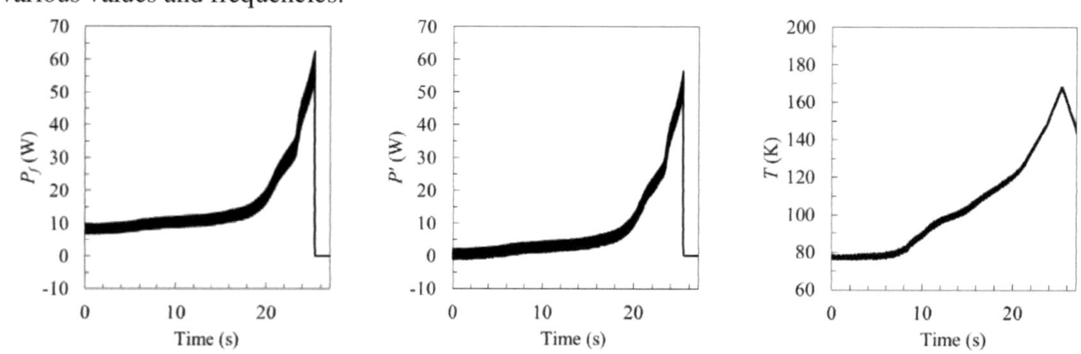

(a) P_f Signal. (b) P' Signal. (c) Temperature in the normal area.

Figure 8. Test results at 1 kA, 60 Hz current.

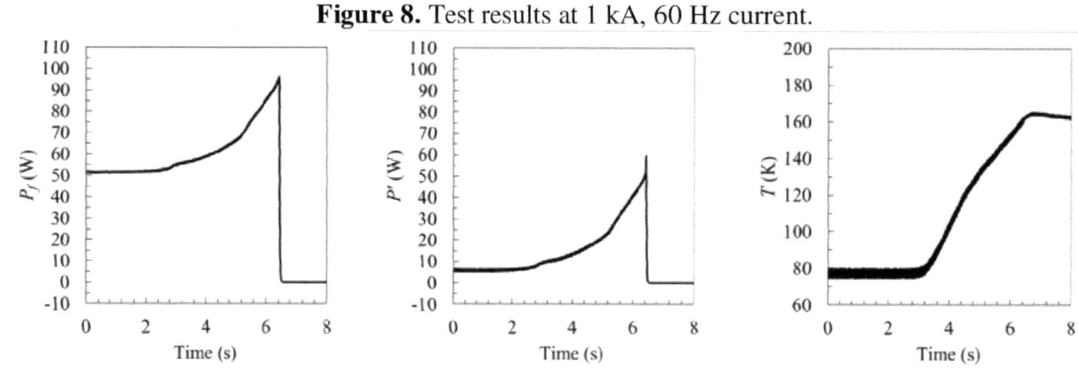

(a) P_f Signal. (b) P' Signal. (c) Temperature in the normal area.

Figure 9. Test results at 1 kA, 300 Hz current.

5. Conclusions

The authors proposed the improved active power method as a detection method of the normal transitions regardless of the iron loss and AC loss for the Bi2223 HTS transformer. As experimental results for the HTS transformer, it was confirmed that the system could work well even in cases of transporting currents of various values and frequencies.

Acknowledgements

This work was supported by Electric Technology Research Foundation of Chugoku.

References

[1] Shiohara Y, et, al. 2013 Future prospects of high Tc superconductors-coated conductors and their applications *Physica C* **484** 1
[2] Nanato N, Kobayashi Y 2014 Quench Detection and Protection for High Temperature Superconducting Transformers by Using the Active Power Method *Physics Procedia* **58** 264
[3] Nanato N, Nakamura K 2004 Quench detection method without a central voltage tap by calculating active power *Cryogenics* **44** 1
[4] Takeuchi K, et al. 2008 Fundamental studies for the application of quench protection systems based on an active power method for cryocooled LTS coils *Cryogenics* **48** 148
[5] Inoue T, Nanato N, Sasaki K, Murase S 2010 Quench protection system for two-layer superconducting coils by an active power method *Physica C* **470** 1870
[6] Nanato N, et al. 2012 Quench Detection/Protection of a Cryocooled NbTi Superconducting Magnet by using an Active Power Method *Physics Procedia* **27** 428
[7] Nanato N, et al. 2012 Quench protection tests of a cryocooler cooled 6 T NbTi superconducting magnet by an active power method *Cryogenics* **52** 321
[8] Akita S, Ishikawa T, Tanaka T 1991 AC Losses in Superconducting Coils *T. IEE Japan* **111**-B 402 (in Japanese)
[9] Nanato N, Asai W and Murase S 2012 Study on criterion for quench detection/protection of superconducting magnet based on active power method *Physics Procedia* **27** 416
[10] Iwasa Y 2009 Case Studies in Superconducting Magnets, *Design and Operational Issues Second Edition, Springer* 468

IOP Conf. Series: Journal of Physics: Conf. Series **1054** (2018) 012069 doi:10.1088/1742-6596/1054/1/012069

High resolution locating of normal transitions in a high temperature superconducting coil by capacitor type voltage terminals

N Nanato, H Kumagai

Graduate School of Natural Science and Technology, Okayama University,
3-1-1, Tsushima Naka, Kita-ku, Okayama, 700-8530, Japan

nanato@oakayama-u.ac.jp

Abstract. It is important to locate positions of normal transitions in a high temperature superconducting coil for identifying its design and fabrication weakness. The authors have proposed capacitor type voltage terminals as a contactless method to measure voltages of the coil windings through insulation of the windings. This method was useful for locating the positions of the normal transitions. Higher resolution of the location is achieved by attaching many terminals to the coil. Then shorter electro-conductive sheets of capacitor type voltage terminals are needed. However, the shorter sheets make the measured voltage signals small and the signals are supposed to be incorrectly measured. In this paper, the authors propose a method to achieve high resolution locating by long electro-conductive sheets of capacitor type voltage terminals. Numerical values of the measured voltage by the terminals depend on the positions of the normal transitions and therefore analysis of the values can achieve high resolution locating without increasing the number of the terminals. Through experimental results for a Bi2223 HTS coil, it was confirmed that the proposed method could achieve the high resolution locating.

1. Introduction

Locating normal transitions in high temperature superconducting (HTS) coils is important to discuss thermal stability of the coils. The authors have presented capacitor type voltage terminals as a contactless method to measure voltages of the coil windings [1-3]. References [2,3] show that the method is useful for locating the positions of the normal transitions. It is needed in order to make resolution of the location much higher that many voltage terminals are mounted on the windings, and then distance between adjacent two terminals are shorten. However, it makes the measurement system complex and makes measured voltage values small and then it is supposed to cause incorrect measurement. In this paper, the authors propose a method to achieve high resolution locating by using less voltage terminals based on the capacitor terminal method. Numerical values of the measured voltage by the terminals depend on the positions of the normal transitions and therefore analysis of the values can achieve high resolution locating without increasing the number of the terminals. Experimental results for a Bi2223 HTS coil showed that the proposed method could achieve the high resolution locating.

2. Principle of location method

Figure 1 shows an outline of the contactless voltage measurement method [1]. A pair of electrically conductive sheets is attached on the surface of an HTS coil and is connected with an outside capacitor. Then a capacitor is formed at the each sheet and a voltage of the coil winding is measured as a voltage of the outside capacitor based on voltage dividing by the three capacitors. C_1 and C_2 are capacitances at the two sheets, and C_S and v_C are respectively a capacitance and a voltage of the outside capacitor.

Figure 2 shows two states at a certain time after occurrence of the normal transitions as a 1D model on distribution of the normal zones in the HTS wire. When the normal zone occurs at (A) in Figure 2 which is between two sheets, relationship between the voltage of the HTS wire v_Z and the voltage of the outside capacitor v_C is shown as follows [2].

$$v_Z = \frac{C_1C_2 + C_1C_S + C_2C_S}{C_1C_2} v_C = \alpha v_C, \tag{1}$$

where

$$\alpha = \frac{C_1C_2 + C_1C_S + C_2C_S}{C_1C_2}. \tag{2}$$

Whereas the normal zone occurs at (B) in Figure 2 which is under the sheet, a following equation is obtained [2].

$$\alpha v_C = v_Z - \left\{\frac{L - l - l_S}{L} r(l) + \frac{1}{L}\int_0^l r(x)\, dx\right\} i. \tag{3}$$

Here, l_S is the length between the left end of the electro-conductive sheet (point "a") and the left end of the normal zone, l is the length of the normal zone, $L - l - l_S$ is the length between the right end of the normal zone and the right end of the sheet, L is the length of the sheet, $r(x)$ is the resistance of the

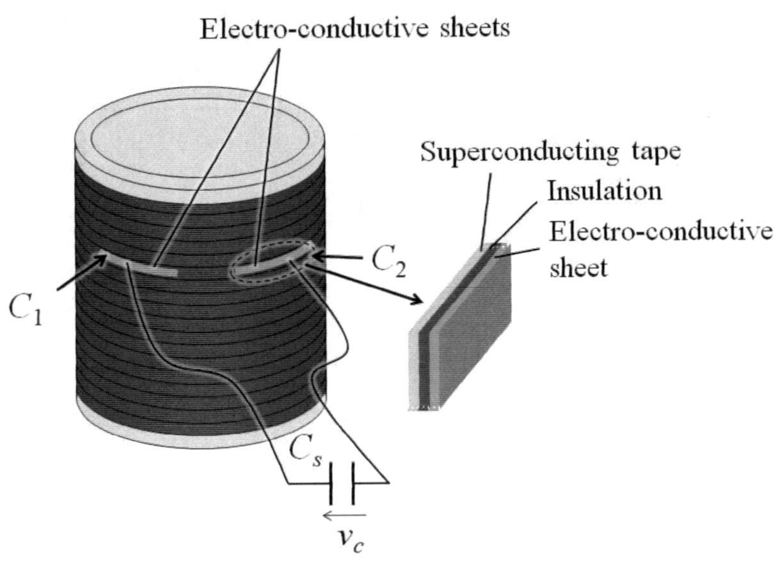

Figure 1. Outline of the contactless voltage measurement method.

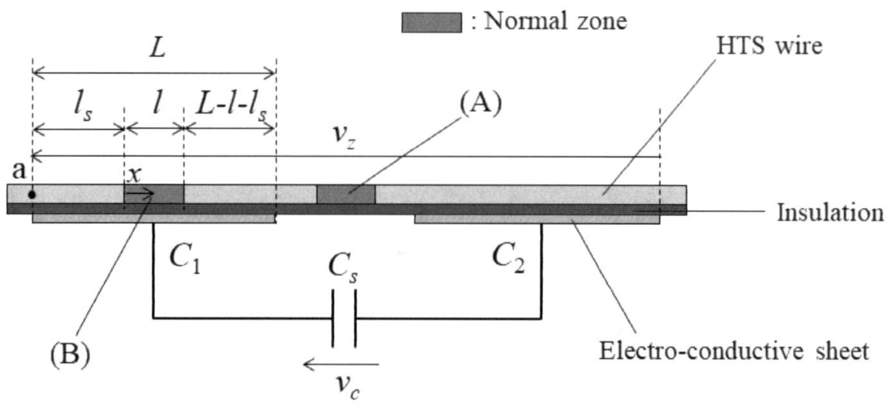

Figure 2. 1D model on distribution of the normal zones in the HTS wire.

normal zone at x ($x = 0$ at the left end of the normal zone) and i is a transport current through the HTS wire.

From the equations (1) and (3), the closer the position of the normal transition is the area between two sheets, the smaller the second term of the equation (3) becomes, and then the ratio of $\alpha v_C / v_Z$ increases. In case of (A), $\alpha v_C / v_Z$ becomes 100 % and in case of (B) $\alpha v_C / v_Z$ becomes smaller than 100 % with dependence of the position of the normal transition. Therefore locating normal transitions can be achieved by measuring the ratio of $\alpha v_C / v_Z$. Based on this method, high resolution locating can be achieved by less electro-conductive sheets of capacitor type voltage terminals. Moreover less sheets can actualize long sheets to make measured voltages larger for more correct measurement.

3. Experiments for a Bi2223 HTS coil

3.1. Experimental setup

Experiments on the locating of the normal transitions were carried out for a Bi2223 HTS coil. Figure 3 and Table 1 show an outline of a test Bi2223 HTS coil and its specifications, respectively. AC current of 60 Hz and 150 A_{peak} was supplied to the HTS coil which was cooled in liquid nitrogen. The normal transitions were caused by four heaters mounted on the positions I to IV of the 2nd turn of the HTS coil shown in Figure 4. Voltage terminals except the capacitor terminal were soldered to the windings to confirm the positions of the normal transitions and measured voltages by the terminals were v_1 to v_{11}. Distances between adjacent two terminals were 30 mm except 60 mm for v_6. Table 2 shows specifications of the capacitors.

The other capacitor terminal was attached on the 7th turn of the HTS coil. A voltage of the 2nd turn v_{C1} had an inductive voltage and therefore it was cancelled by subtracting a voltage of the 7th turn v_{C2} from v_{C1} as follows.

$$v_{Cd} = v_{C1} - k_C v_{C2} \,. \tag{4}$$

k_C was a ratio of an inductive voltage in v_{C1} to that in v_{C2} in superconducting state. v_{Cd} had some electromagnetic noise and therefore it was transformed into an active power and filtered shown as following equations to locate the positions of the normal transitions more precisely.

$$P_C = v_{Cd} i \,, \tag{5}$$

$$P_C' = \frac{P_C}{1 + sT} \,. \tag{6}$$

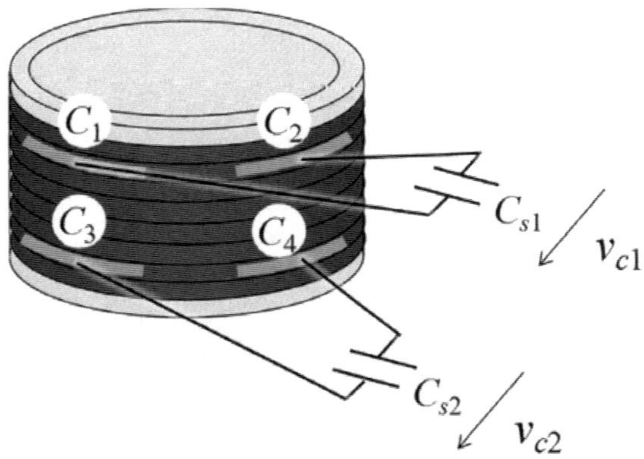

Figure 3. Outline of a test Bi2223 HTS coil.

Table 1. Specifications of HTS coil

Inncr diameter (mm)	140
Outer diameter (mm)	150
Height (mm)	95
Number of turns	7.5
I_c (at 77 K, self-field) (A)	189

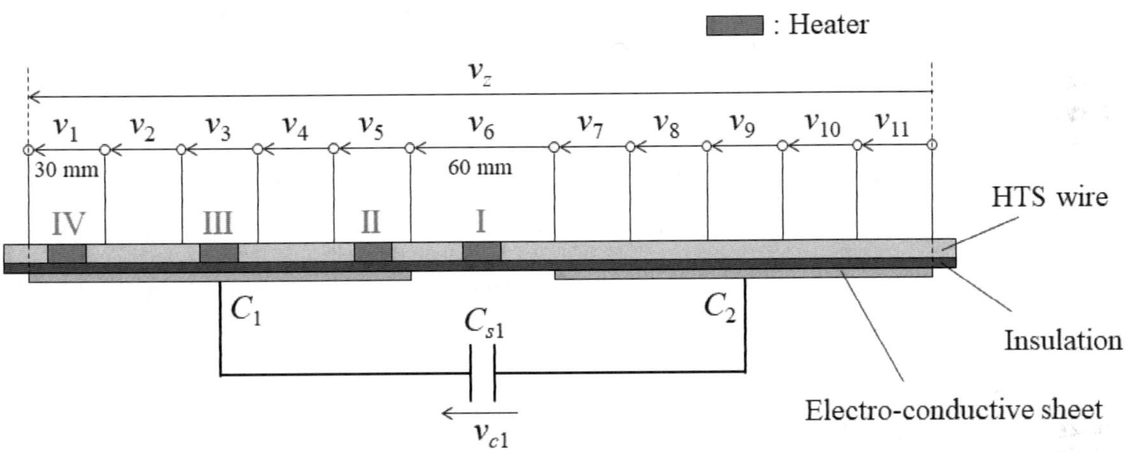

Figure 4. Attached voltages terminals and heaters.

Table 2. Specifications of capacitors

Outside capacitors : C_{S1} and C_{S2} (nF)	100
Electro-conductive sheet capacitors : C_1 (nF)	0.231
: C_2 (nF)	0.279
: C_3 (nF)	0.276
: C_4 (nF)	0.267
Length of the sheets (mm)	150
Width of the sheets (mm)	4.2
Distance between adjacent two sheets (mm)	60

Equation (6) is expressed by Laplace transform and T is a time constant of the filter.

As well as the inductive voltage in v_C, that in v_Z is removed resulting in v_{Zd}.

As the results, the ratio of $\alpha v_{Cd}/v_{Zd}$ was shown as a following equation.

$$\frac{\alpha v_{Cd}}{v_{Zd}} = \frac{\alpha v_{Cd} i}{v_{Zd} i} = \frac{\alpha P_C}{v_{Zd} i} = \frac{\alpha P'_C}{(v_{Zd} i)'}. \tag{7}$$

$(v_{Zd} i)'$ means filtered $v_{Zd} i$.

3.2. Experimental results

Figures 5-8 show experimental results for the occurrence of the normal transitions at I-IV, respectively. (a) and (b) in each figure were shown by root mean square values. (a) in each figure shows the voltages of the areas with the normal zones. It was found that the each normal transition occurred at the position of the each heater and the normal zone propagated into its adjacent areas. (b) in each figure shows v_{Zd} and (c) shows active power signals P'_c.

By using the v_{Zd}s and the P'_cs, the each ratio of $\alpha v_{Cd}/v_{Zd}$ was calculated for the normal zone size of 30 mm in each case. For example, in case of II, following calculations were carried out.

- Experimental value of equation (7)
  - Normal zone size became 30 mm at 15.5 s when v_4 and v_6 started to increase as shown in Figure 6 (a).
  - Based on the equation (7), experimental value of $\alpha v_{Cd}/v_{Zd}$ was calculated from the v_{Zd} and P'_c which were values at 15.5 s in Figures 6 (b) and (c), respectively. $i = 120/\sqrt{2}$.
  - The experimental value was normalized by $\alpha v_{Cd}/v_{Zd}$ measured in case of I in order to supress measurement errors.
- Theoretical value of equation (3)
  - Equation (3) was transformed into following one.

$$\frac{\alpha v_{Cd}}{v_{Zd}} = 1 - \frac{1}{v_{Zd}} \left\{ \frac{L - l - l_s}{L} r(l) - \frac{1}{L} \int_0^l r(x) dx \right\} i. \tag{8}$$

  - $r(x)$ was assumed to have a liner resistivity for x and then the equation (8) could be derived as

$$\frac{\alpha v_{Cd}}{v_{Zd}} = \frac{l + 2l_s}{2L}. \tag{9}$$

  - In case of II, the theoretical value was obtained for l = 30 mm and l_s = 120 mm.

Table 3 shows the calculation results. Experimental results were in good agreement with the theoretical values. The results show that the proposed method can locate the position of the normal zone for the normal zone size of 30 mm. However, it was found from other experimental results that it was hard to locate the position of shorter normal zone than 30 mm. In case of IV, the value of $\alpha v_{Cd}/v_{Zd}$ becomes very small value according to equation (9) for l_s = 0. Therefore P'_c became a low S/N ratio

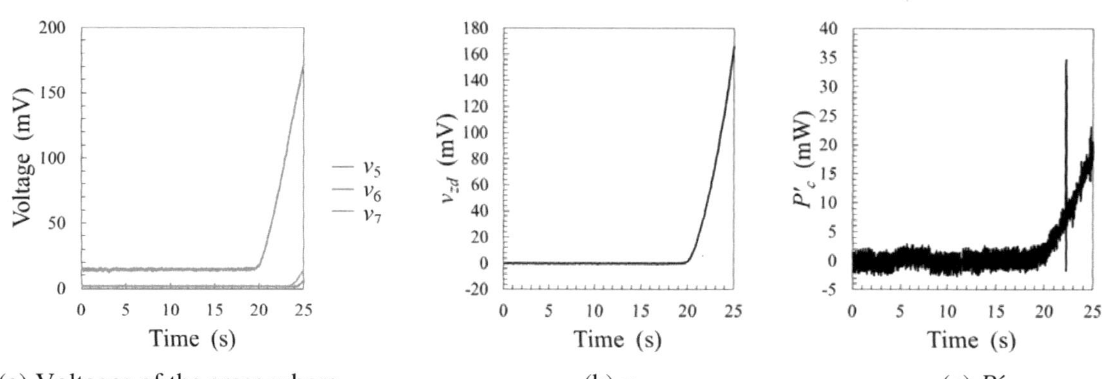

(a) Voltages of the areas where the normal transition occurred (v_5, v_6, v_7) (b) v_{zd} (c) P'_c

Figure 5. Experimental results for the normal transition at I.

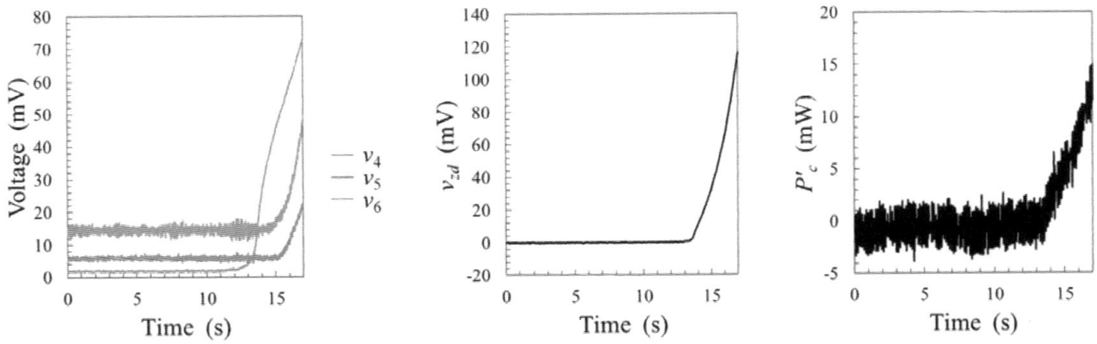

(a) Voltages of the areas with the normal zones (v_4, v_5, v_6)

(b) v_{Zd}

(c) P_c'

Figure 6. Experimental results for the normal transition at II.

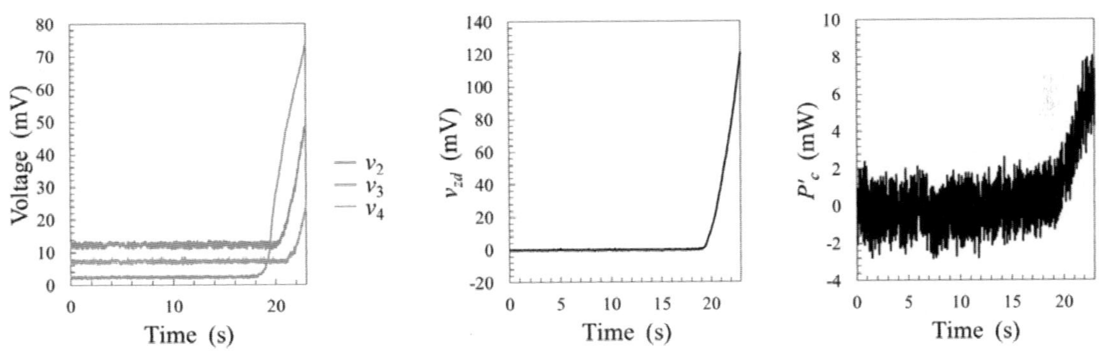

(a) Voltages of the areas with the normal zones (v_2, v_3, v_4)

(b) v_{Zd}

(c) P_c'

Figure 7. Experimental results for the normal transition at III.

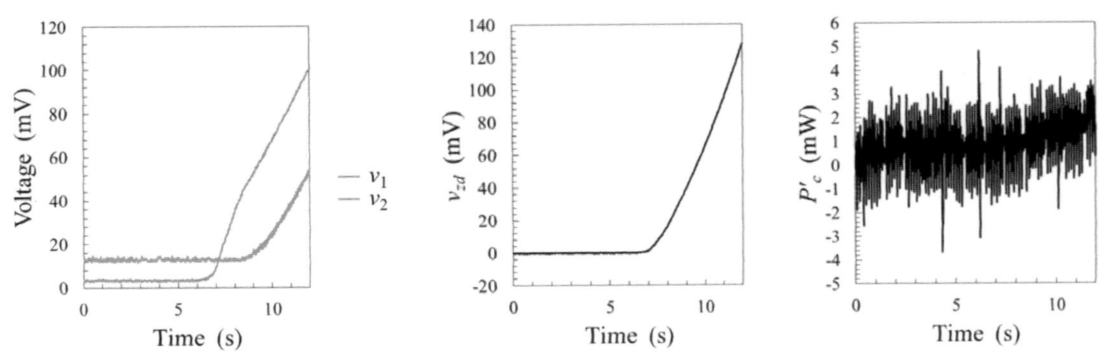

(a) Voltages of the areas with the normal zones (v_1, v_2)

(b) v_{Zd}

(c) P_c'

Figure 8. Experimental results for the normal transition at IV.

Table 3. Measured and theoretical result of each $\alpha v_{cd}/v_{zd}$.

Position of the normal transition	Experimental value $\frac{\alpha v_{Cd}}{v_{Zd}}$ (%)	Theoretical value $\frac{\alpha v_{Cd}}{v_{Zd}}$ (%)
I (between the two sheets)	100	100
II ($l = 30$ mm, $L - l - l_s = 0$ mm, $l_s = 120$ mm)	93.6	90
III ($l = 30$ mm, $L - l - l_s = 60$ mm, $l_s = 60$ mm)	50.9	50
IV ($l = 30$ mm, $L - l - l_s = 120$ mm, $l_s = 0$ mm)	9.5	10

signal for the shorter normal zone in case of IV and it was hard to be measured correctly. The authors will study on locating for the shorter normal zone especially for case IV hereafter.

In these experiments, many voltage terminals were soldered on the superconducting wire for verifying the principle of the proposed method, however, only two soldered terminals on the both ends of the HTS coil are needed for v_Z in actual use. Then the inductive voltage in v_Z is removed by a cancel coil [4].

4. Conclusions

In this paper, the authors proposed a method to achieve high resolution locating of the normal transitions in the HTS coil by less and long electro-conductive sheets of capacitor type voltage terminals. The proposed method achieves the location by calculating a ratio of a voltage of a capacitor voltage terminal to a voltage of the HTS coil wire. Experimental results for a Bi2223 HTS coil showed the proposed method could locate the position of the normal zone of 30 mm. The authors will study on higher resolution method.

Acknowledgements

This work was supported by Electric Technology Research Foundation of Chugoku.

References

[1] Nanato N and Nishiyama K 2014 Non-destructive Detection of Normal Transitions in High Temperature Superconducting Coil *Physics Procedia* **58** 260

[2] Nanato N and Nishiyama K 2015 Locating of normal transitions in a Bi2223 high temperature superconducting coil by non-contact voltage measurement method *Cryogenics* **72** 53

[3] Nanato N, Okura K, Kumagai H and Aoyama 2017 Locating of normal transitions in a Bi2223 high temperature superconducting coil by using capacitor type voltage terminals and the active power method *Journal of Physics Conference Series* **871** 012084

[4] Nanato N and Kobayashi Y 2014 Quench Detection and Protection for High Temperature Superconducting Transformers by Using the Active Power Method *Physics Procedia* **58** 264

IOP Conf. Series: Journal of Physics: Conf. Series **1054** (2018) 012070 doi:10.1088/1742-6596/1054/1/012070

Study on a Magnetic Flux Detection Coil for Detection of Normal Transitions in a Hybrid Single-phase Bi2223 Superconducting Transformer by the Active Power Method

N Nanato, S Tanaka, S Tenkumo

Graduate School of Natural Science and Technology, Okayama University, 3-1-1 Tsushima Naka, Kita-ku, Okayama, 700-8530, Japan

nanato@okayama-u.ac.jp

Abstract. The authors have been developing a small and light power source with a hybrid single-phase Bi2223 high temperature superconducting (HTS) transformer. The transformer consists of a primary copper coil and a secondary Bi2223 superconducting coil and can output current of 500 A. For safe operation of the transformer, it is important to detect normal transitions in the transformer and to protect it from excessive heating in the normal area. The authors have presented a protection system based on the active power method and magnetic flux detection coil. However, in the conventional method, the active power does not become zero in a superconducting state due to AC loss and therefore incorrect recognition may be caused. In this paper, the authors propose a method to reduce the loss signals by configuration and mounting position of the magnetic flux detection coil. Through experimental results for a hybrid single-phase Bi2223 superconducting transformer, the authors show that the proposed method can detect the normal transitions more accurately than the conventional one.

1. Introduction

The authors have been developing a hybrid single-phase Bi2223 HTS transformer [1] used in the small and light AC current source [1-3]. The transformer consists of a primary copper coil and a secondary Bi2223 HTS coil and the transformer can output current of 500 A. For safe operation of the transformer, it is necessary to detect normal transitions in the secondary coil to protect from excessive heating in the normal area. The authors have presented a protection system based on the active power method and a magnetic flux detection coil attached on the inside of the secondary coil [1]. In this protection system, the normal transitions are detected by measuring active power dissipated in the only secondary coil. However, in the conventional method, the active power does not become zero in a superconducting state due to AC loss and therefore the conventional detection method has a possibility of incorrect detection of the normal transitions.

The authors propose a method to reduce the loss signals by configuration and mounting position of the magnetic flux detection coil. The magnetic flux detection coil has only one turn and is mounted on the center of the outside of the secondary coil to measure the magnetic flux which causes much AC loss. In this paper, the authors show that the proposed method can detect the normal transitions more accurately than the conventional one through experimental results for a hybrid single-phase Bi2223 HTS transformer.

2. Protection tests for a hybrid single-phase Bi2223 HTS transformer

2.1. Configuration and mounting position of a magnetic flux detection coil

In the conventional detection method of the normal transitions, the active power dissipated in the only secondary coil can be calculated by the magnetic flux detection coil which is mounted according to the secondary coil on the inside of the bobbin [1]. Mounting position of the magnetic flux detection coil and an equivalent circuit converted to the primary side of the hybrid single-phase Bi2223 HTS transformer are shown in Figures 1 and 2, respectively. In Figure 2,

v_1 : primary voltage, i_1 : primary current, r_1 : primary resistance,
x_1 : leakage primary reactance, i_0 : excitation current, g_0 : excitation conductance,
b_0 : excitation susceptance, i_2 : secondary current, x_2 : leakage secondary reactance,
r_2 : secondary resistance generated after the normal transitions,
v_{SC2} : secondary voltage, Z : secondary load impedance,
a : turn ratio of the primary coil and secondary coil,
a' : turn ratio of the primary coil and magnetic flux detection coil,
v_f : voltage of the magnetic flux detection coil.

A secondary resistive voltage is detected by difference between $a'v_f$ and av_{sc2} and active power dissipated in the secondary coil is calculated as follows.

$$P_2 = \left(a'v_f - av_{sc2}\right)\frac{i_2}{a} = (a^2 r_2 \frac{i_2}{a})\frac{i_2}{a} = r_2 {i_2}^2. \quad (1)$$

$$P'_2 = \frac{P_2}{1+sT}. \quad (2)$$

In order to reduce an electromagnetic noise, P_2 is filtered resulting in P'_2, which is expressed by Laplace transform and T is a time constant of the filter.

In the conventional method, P'_2 has some AC loss signals which cause false detection of the normal transitions. Therefore the authors propose an improved magnetic flux detection coil with only one turn which is mounted on the center of the outside of the secondary coil as shown in Figure 3. The magnetic flux detection coil can detect most of the flux applied to HTS wire vertically. Much of AC loss is hysteresis loss caused by time varying of the vertical magnetic flux. As the result, the one turn coil can measure much of AC loss in the secondary coil and therefore $a'v_f$ has more AC loss signals than that of the conventional detection coil. Then $a'v_f$ becomes close to av_{sc2} in equation (1) and the AC loss signals in P'_2 is reduced.

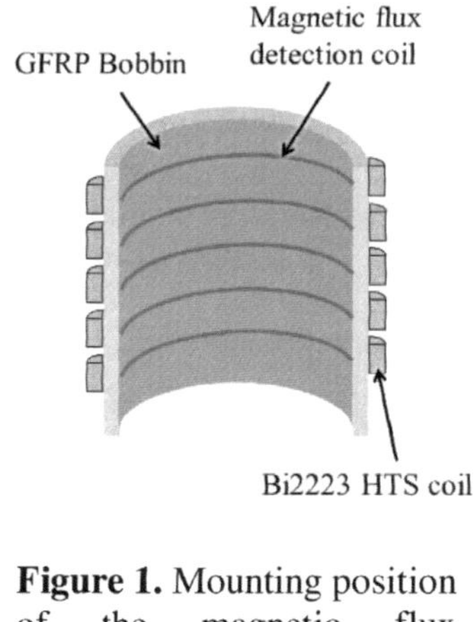

Figure 1. Mounting position of the magnetic flux detection coil.

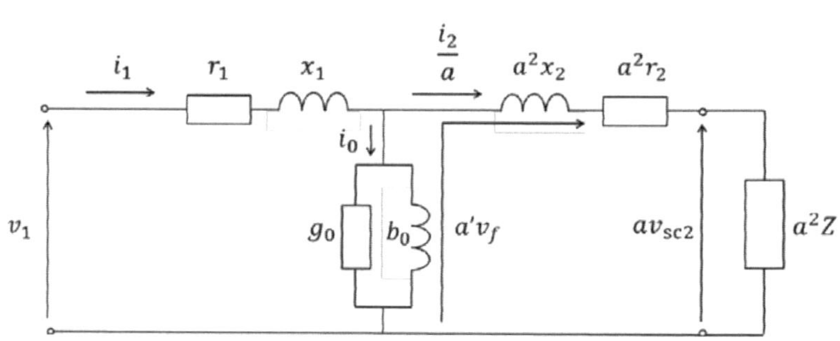

Figure 2. An equivalent circuit converted to the primary side of the hybrid single-phase Bi2223 HTS transformer.

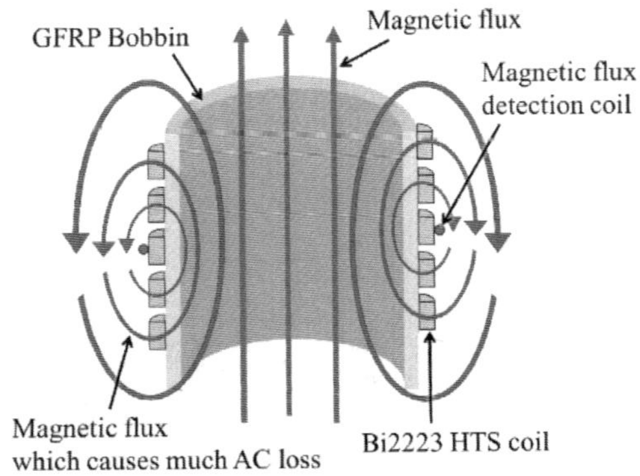

Figure 3. Magnetic flux detection coil with one turn mounted on the center of the secondary coil

2.2. Comparison of the conventional magnetic flux detection coil and the improved one
In order to verify the proposed method, P'_2s were measured as a function of transport current from 100 to 500 A_{peak} for a hybrid single-phase Bi2223 HTS transformer. Configuration and specifications of the HTS transformer are shown in Figure 4 and Table 1, respectively [1]. Experimental results are shown in Figure 5. P'_2 measured by the improved method was under about 1/3 of that measured by the conventional one. The results show that the proposed method is effective for reducing the AC loss signals.

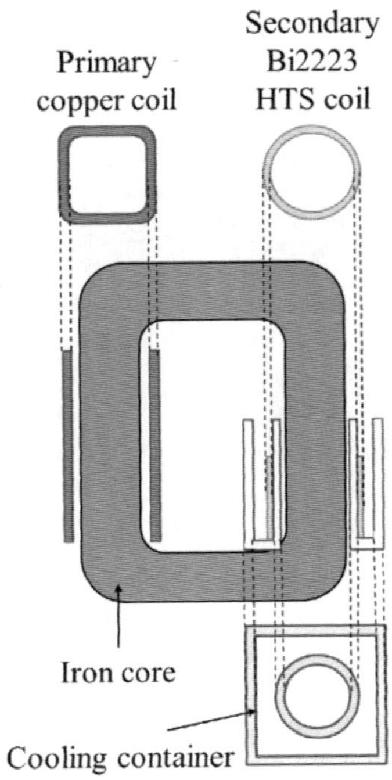

Figure 4. Configuration of the hybrid single-phase Bi2223 HTS transformer.

Table 1. Specifications of a hybrid single-phase Bi2223 HTS transformer.

	Primary coil (Copper)	Secondary coil (Bi-2223)
	63 mm (One side length)	100 mm (Inner diameter) 106 mm (Outer diameter)
Height	230 mm	65 mm
Number of Turns	120	5
I_C (at 77 K, self-field)	-	760 A

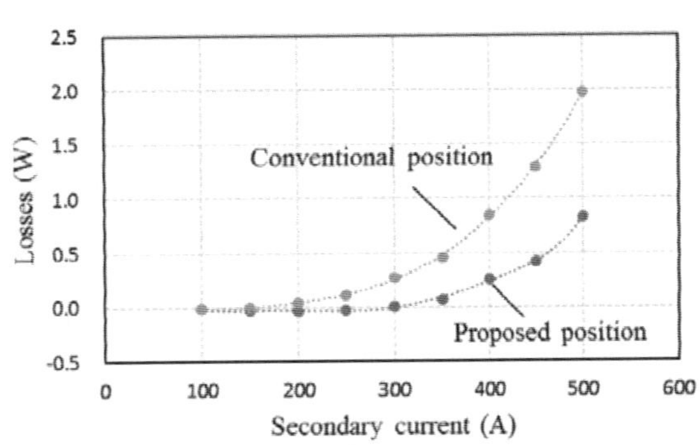

Figure 5. P'_2 as a function of transport current.

2.3. Detection and protection tests of a hybrid HTS transformer

Detection and protection tests for the HTS transformer were carried out by the proposed method. A protection circuit is shown in Figure 6. P'_2 reaches a specified threshold P_{th}, the thyristor switches are turned off and transport current is shut off.

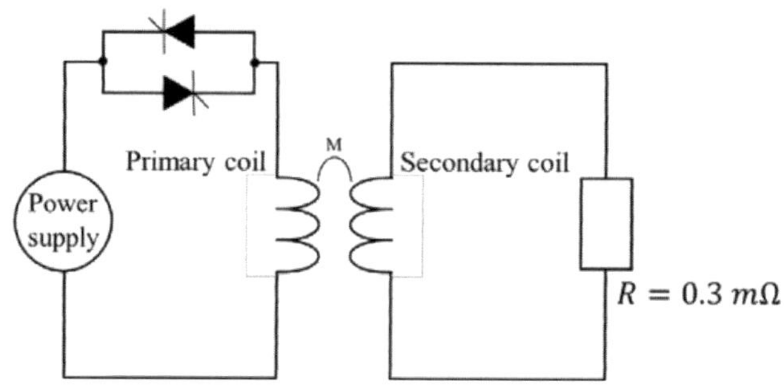

Figure 6. A protection circuit.

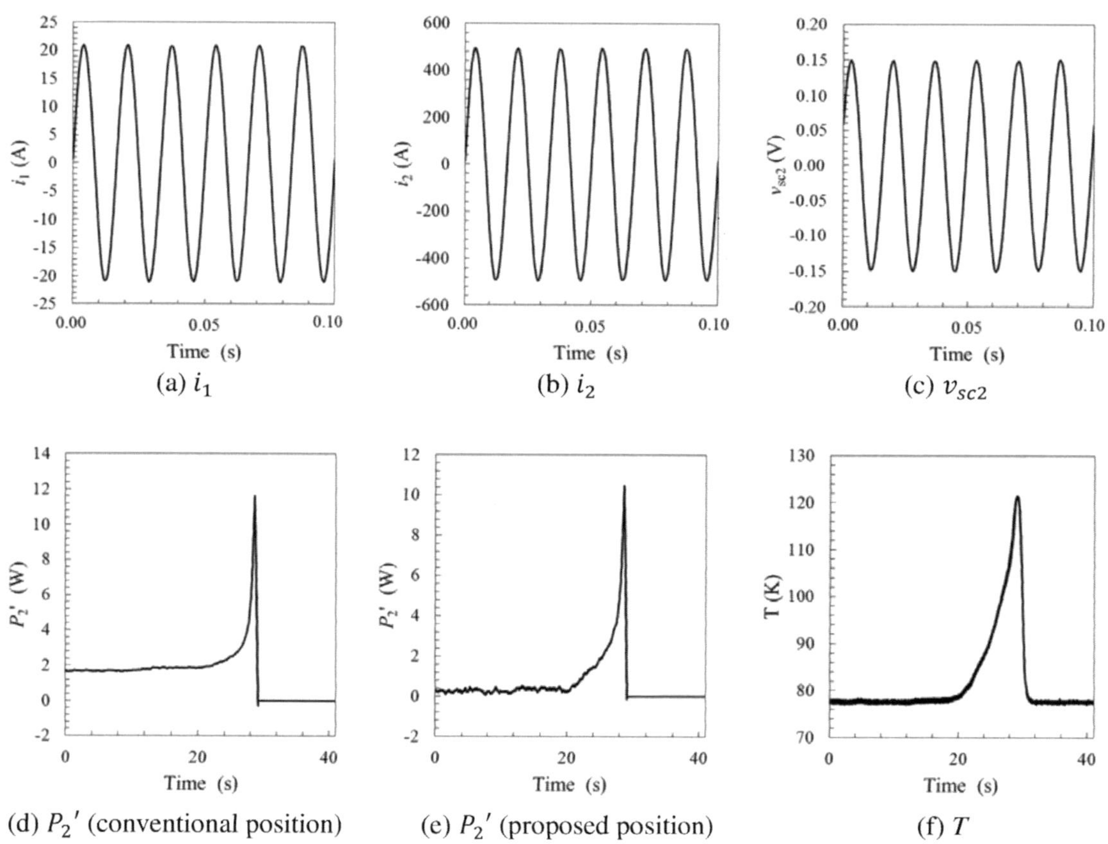

Figure 7. Test results for the hybrid single-phase Bi2223 HTS transformer.

The normal transitions were occurred by a heater mounted on the uppermost turn of the secondary coil under transporting the secondary current of 500 A_{peak} and 60 Hz. The heater has width of 12 mm and height of 4.2 mm. Figure 7 shows the experimental results. Figures 7 (a)-(c) show expanded waveforms for time axes in a superconducting state. Figures 7 (a) and (b) show the primary current of 21 A_{peak}, 60 Hz and the secondary current of 500 A_{peak}, 60 Hz, respectively. Figure 7 (c) shows the voltage across the secondary coil which equals to the voltage of the connected resistor. Figures 7 (d) and (e) show the active power signal P'_2 in the secondary coil before and after the normal transition for the conventional method and the proposed one, respectively. In a superconducting state until about 20 s, P'_2 measured by the proposed method was smaller than that by the conventional one. This shows that the proposed method can reduce the loss signals more than the conventional one as mentioned above and therefore the normal transitions can be detected with a higher SN ratio than that of the conventional method [1]. The thyristor switches were turned off when P'_2 measured by the proposed method reached a specified threshold P_{th}=10 W [4]. Figure 7 (f) shows temperature of the normal zone near the heater on the secondary coil. The maximum temperature of the secondary coil was suppressed to 122 K which was lower than permissive temperature [5].

The test results show that the proposed method can detect the normal transitions more correctly than the conventional one.

3. Conclusion

The authors have proposed the active power method using a magnetic flux detection coil as a detection method of the normal transitions. In this paper, they proposed a method to reduce loss signals included in the active power in a superconducting state by configuration and mounting position of the magnetic flux detection coil. This method made SN ratio of the detection signal higher than that of the conventional one. Experimental results verified the improvement of the method for hybrid HTS transformer. The authors will study on more correct detection.

Acknowledgements

This work was supported by Electric Technology Research Foundation of Chugoku.

References

[1] Nanato N, Nakamura S and Tanaka S 2017 Detection of normal transitions in a hybrid single-phase Bi2223 high temperature superconducting transformer by using the active power method and a magnetic flux detection coil *Journal of Physics: Conference Series* **871** 012085

[2] Nanato N and Kobayashi Y 2014 Quench Detection and Protection for High Temperature Superconducting Transformers by Using the Active Power Method *Physics Procedia* **58** 264

[3] Nanato N, Kishi N, Tanaka Y and Kondo M 2017 Basic study for a large AC current supply with a single-phase air-core Bi2223 high temperature superconducting transformer *Journal of Physics: Conference Series* **871** 012101

[4] Nanato N, Asai W and Murase S 2012 *Physics Procedia* **27** 416

[5] Iwasa Y 2009 Case Studies in Superconducting Magnets, *Design and Operational Issues Second Edition, Springer* 468

IOP Conf. Series: Journal of Physics: Conf. Series **1054** (2018) 012071 doi:10.1088/1742-6596/1054/1/012071

Design of a high temperature superconducting magnet for a single silicon crystal growth system

V Q Dao [1], C K Lee [2], J H Choi [2], M Park [1] and I K Yu [1]

[1] Changwon National University, Changwon, GN, 51140, Korea
[2] Supercoil Co., Ltd, Changwon, GN, 51140, Korea

yuik@cwnu.ac.kr

Abstract. This paper presents a study on the design of a high-temperature superconducting (HTS) magnet for a Czochralski single silicon-crystal growth system by evaluating the temperature and flow distributions of silicon melt at the cusp magnetic field. A two-dimensional finite element method (FEM) simulation model was built to determine the effects of the magnetic field on the temperature and flow distributions in the silicon melt. The characteristics of the HTS magnet were analyzed using a three-dimensional FEM model. The HTS magnet was designed using 2G HTS wire and the magnet was validated through FEM simulation. The simulation results showed that the melt convection was significantly suppressed by the Lorentz force, and that the temperature distribution was uniform in the silicon melt under the cusp magnetic field. The shape of the HTS magnet was determined as a magnet ring with a magnetic flux density of 0.35 T at the center of the crucible bottom. The fundamental design specifications and the data obtained from this study can be applied to the development of a real silicon-crystal growth system.

1. Introduction

Czochralski (Cz) technology is widely used as a single silicon-crystal growth method, in which a crucible is used to hold the melt from which a crystal is grown. To improve the quality of the crystals, the static magnetic fields of the external magnet around the crucible are used, which are well known for suppressing of the melt convection and the temperature fluctuations when the magnetic field strength is 0.1 – 0.5 T [1]–[4]. There are three types of magnetic field used in the Cz method: horizontal magnetic field, vertical magnetic field and cusp magnetic field. The cusp magnetic field, in which the free surface of the melt is centered between two opposite fields generated by two magnets, has the advantages of both horizontal and vertical magnetic fields [3]–[6]. However, to increase crucible size and crystal diameter, the application of a static magnetic field of sufficient strength requires large coil systems that consume a substantial amount of electric power [2]. Thus, the superconducting magnet technique is an attractive solution for reducing the dimensions and energy consumption of crystal growth systems.

Nowadays, low-temperature superconducting magnets (NbTi) for single silicon-crystal growth have already been commercialized and widely used. However, high-temperature superconducting (HTS) magnets will operate efficiently and reliably on magnet quench due to their high critical temperature and high capacity, which can lead to a large temperature margin [7]-[8].

In this paper, a high-temperature superconducting (HTS) magnet was proposed for a 300 mm silicon-crystal growth system, and the temperature and flow distributions of a silicon melt within the cusp magnetic field were analyzed. Based on the physical parameters of the Cz crystal growth system, a two-dimensional (2D) axisymmetric finite element method (FEM) simulation model was built. The velocity field and temperature distribution in the silicon melt were analyzed for two cases: one with no magnetic field and one with cusp magnetic field. Then, the magnetic field strength and shape were determined to design the HTS magnets using 2G HTS wire with an expected operating temperature of 30 K or less. The metal was insulated using stainless steel tape to provide quench protection and to improve thermal conduction.

A characteristic analysis of the magnet was conducted using a three-dimensional (3D) FEM simulation. It was found that melt convection was significantly suppressed by the Lorentz force and that the temperature gradient of the silicon melt was significantly reduced under the cusp magnetic field. The magnetic flux density at the center of the crucible bottom was 0.35 T. The operating current of the magnet was 344 A with four double pancake coils (DPCs). The total wire length was 11.5 km. The fundamental design specifications and the data obtained from this study can be applied to the development of a real silicon-crystal growth system.

2. Magnetic fields in a Cz silicon-crystal growth system

2.1. Configuration of a 300 mm single silicon-crystal growth system

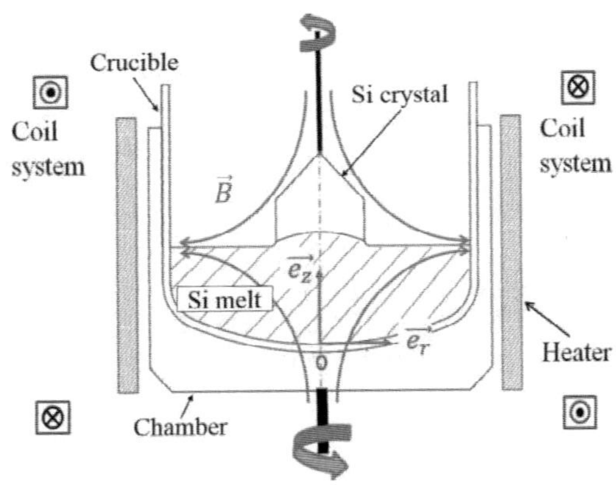

Figure 1. Configuration of a Cz silicon-crystal system

In an industrial Cz single silicon growth process, a crystal is grown from molten silicon contained in a silica crucible as shown in figure 1. The melt flow in a crucible is determined based on the material properties of the silicon melt, as shown in table 1. Table 2 shows the parameters of the Cz single silicon-crystal growth system, which were chosen to model the melt flow with a crucible having a diameter of 900 mm. The crystal has a diameter of 300 mm. The silicon melt has relatively high electric conductivity, allowing the melt flow in the crucible to be affected by electromagnetic fields [9]-[11].

Table 1. Important properties of silicon melt

Parameters	Values
Mass density, ρ	2,580 kg/m^3
Dynamic viscosity, η	0.00086 kg/ms
Specific heat, C_p	1,000 J/(kg. K)
Thermal conductivity, k	67 W/(mK)
Electric conductivity, σ	1.2× 10^6 S/m
Volumetric expansion rate, α	0.00014 1/K
Melting temperature, T_0	1,685 K

Table 2. Parameters of the Cz single silicon-crystal growth system

Parameters	Values
Crucible diameter, D_{cruc}	900 mm
Crystal diameter, D_{crys}	300 mm
Melt height, H	300 mm
Crucible rotation rate, Ω_{cruc}	5 rpm
Crystal rotation rate, Ω_{crys}	15 rpm
Melt mass, m	300 kg

2.2. Magnetic field effects on the silicon melt

Because the silicon melt has high conductivity, the magnetic field has a considerable influence on melt convection through the Lorentz force, $\overrightarrow{F_s}$, which is described as

$$\overrightarrow{F_s} = \vec{J} \times \vec{B}, \tag{1}$$

where $\vec{J}$ is the induced current density determined by

$$\vec{J} = \sigma(\vec{v} \times \vec{B}), \tag{2}$$

where, $\vec{v}$ is the melt velocity field and $\vec{B}$ is the magnetic induction. The Lorentz force opposes to the direction of the melt flow.

Normal axisymmetric flow is driven by buoyancy, electromagnetic force and heat transfer inside the crucible wall. Assuming that the fluid is incompressible and that the Boussinesq approximation is valid, the time-averaged momentum equation is described as in [2], [12]- [14],

$$\rho(\vec{v}.\nabla)\vec{v} = -\nabla p + \eta \nabla^2 \vec{v} - \rho \vec{g} \alpha (T - T_0) + \overrightarrow{F_s}, \tag{3}$$

where p is the pressure, ρ is the reference density, η is the dynamic viscosity, α is the coefficient of volume expansion, $\vec{g}$ is the gravitational acceleration, and T is the temperature of the silicon melt. Mass conservation and energy transport equations with $\nabla \vec{v} = 0$ are

$$\rho C_p (\vec{v}.\nabla) T = \nabla.(k \nabla T), \tag{4}$$

where C_p is the specific heat, and k is the thermal conductivity.

In this paper, the simulation results for a cusp magnetic field revealed a coil configuration in which the vertical component of the magnetic field equaled zero at the melt free surface. The distribution of the field inside the melt was very inhomogeneous. For simplicity, the field was modeled with solenoidal analytical expressions and the cusp magnetic field as a linear field [2] given as

$$\vec{B} = B/R(r\overrightarrow{e_r} + 2(H - z)\overrightarrow{e_z}), \tag{5}$$

where R and H are the radius and height of the melt in the crucible, respectively. The corresponding unit vectors $\overrightarrow{e_r}$ and $\overrightarrow{e_z}$ are described in figure 1.

3. Modelling of silicon melt flow with a cusp magnetic field

3.1. 2D FEM simulation model

Based on the parameters of the Cz crystal growth system given in tables 1 and 2, a 2D axisymmetric FEM simulation model was built as shown in figure 2. The temperature and flow distributions in the silicon melt were analyzed without magnetic field and with a cusp magnetic field. The crystal rotation rate was 15 rpm, and the crucible counter-rotation rate was 5 rpm. The crystal diameter was 300 mm, and the diameter of the crucible inside wall was 900 mm. The melt height was 280 mm. The temperature of the crucible wall heated by AC induction heater was 1,685 K, and the temperature of the crystal surface in contact with the liquid silicon was 1,400 K. The operating conditions of this analysis were recognized as the most appropriate for the Cz crystal growth process [1]-[3].

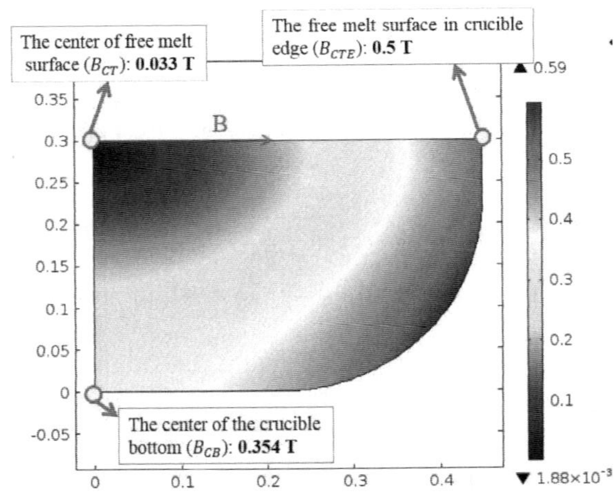

Figure 2. FEM simulation model and the magnetic flux density results for a cusp magnetic field

The shape, size, and the location of the HTS magnet were determined by the magnetic flux density at points at the center of the free melt surface, B_{CT}; the free melt surface in the crucible wall, B_{CTE}; and the centre of the crucible bottom, B_{CB}, respectively.

3.2. FEM analysis results and discussions

The comparisons of the velocity field and temperature distributions in the crucible without a magnetic field and with a cusp magnetic field in the 2D FEM simulation results are shown in figures 3 and 4, respectively.

For the cusp magnetic field, the melt convection in the crucible wall and the upward melt flow under the crystal were suppressed, while mixing of the melt under the crystal remained strong because the silica dissolution in the crucible wall was reduced. The maximum velocity of the melt flow under the contact surface between the crystal and the free melt surface was 0.158 m/s. Compared to the simulation with no magnetic field case, the temperature distribution in the melt was more homogeneous and the temperature gradient was significantly reduced.

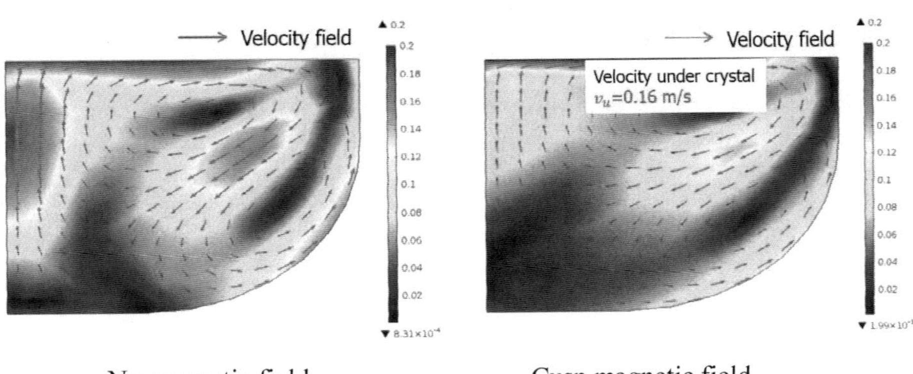

Figure 3. Velocity field of the silicon melt in the FEM simulations

IOP Conf. Series: Journal of Physics: Conf. Series **1054** (2018) 012071 doi:10.1088/1742-6596/1054/1/012071

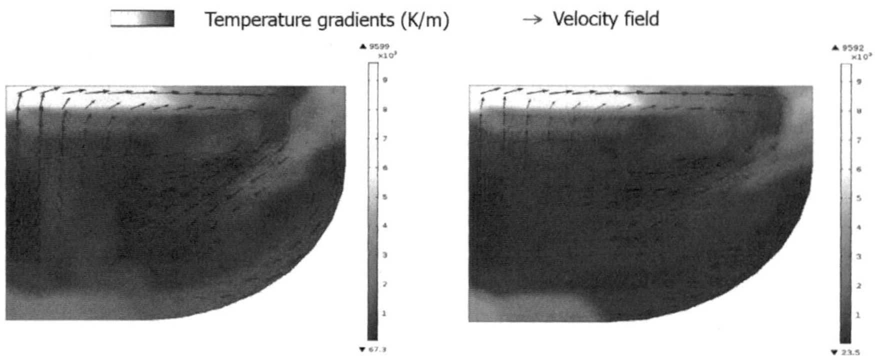

No magnetic field Cusp magnetic field

Figure 4. Temperature gradient of the silicon melt in the FEM simulations

The change in the maximum velocity of the melt flow due to the magnetic flux density at the center of the crucible bottom is shown in figure 5. The figure shows that suitable value for magnetic field strength was determined, and the minimum velocity of the melt flow was achieved at the magnetic flux density of 0.35 T at the center of the crucible bottom.

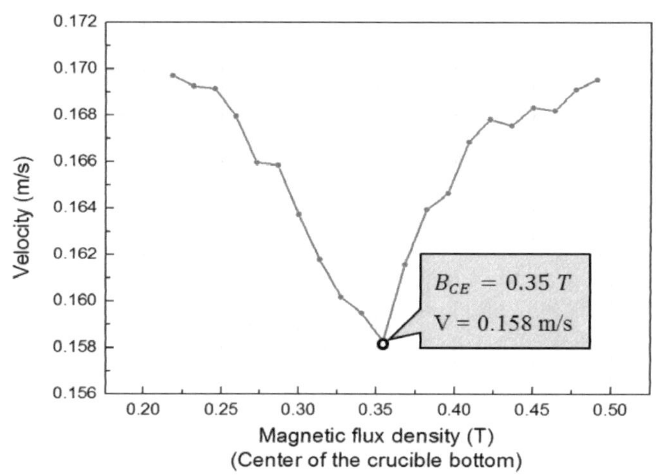

Figure 5. The velocity of the silicon melt at the center of the crucible bottom based on magnetic flux density

4. Design of an HTS magnet for a 300 mm single Si crystal growth system

4.1. Determination of specifications of the HTS magnet

Figure 6 shows the design process for the HTS magnet for a Cz 300 mm single silicon-crystal growth system.

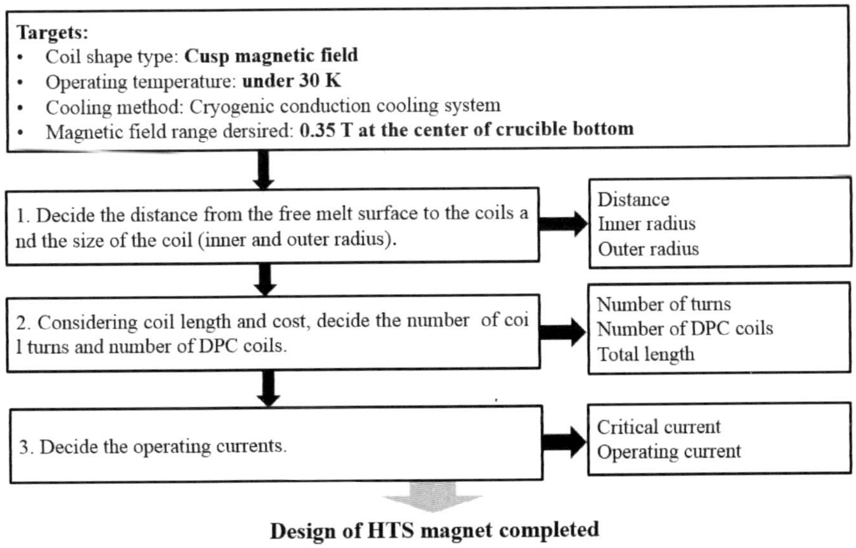

Figure 6. Design flow chart for the HTS magnet

First, the design targets of the HTS magnet were determined. Here, the authors designed the magnet based on the cusp field configuration and selected a ring-shaped coil. The magnet was cooled below 30 K using a cryogenic conduction cooling method, and the target magnetic flux density was 0.35 T at the center of the crucible bottom. Second, the distance from the free melt to the magnet and the size of the magnet rings (inner and outer radius) were decided. Third, the number of turns and number of DPCs were determined. The coil length and cost were considered. Finally, the critical current of the HTS magnet was estimated and the operating current was chosen. The characteristic analysis of the magnets was conducted using FEM simulation.

To estimate the critical current of the HTS magnet, a 3D FEM simulation model was designed. The perpendicular magnetic flux density at an operating current of 1 A was 0.00797 T. The 2G YBCO HTS wires used in the design were manufactured by SuNam company, and were 12 mm wide and 0.22 mm thick. The critical current was 600 A at a temperature of 77 K. The perpendicular magnetic flux density of the magnet was compared to the critical current characteristics curve as shown in figure 7 to determine the critical current of the HTS magnet [15].

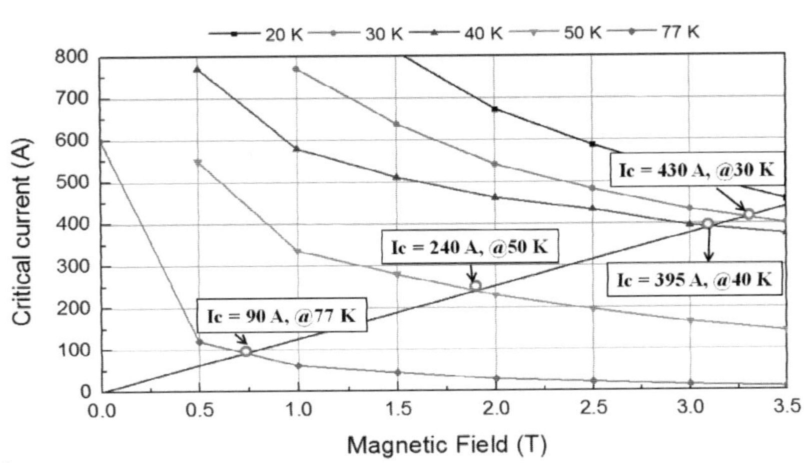

Figure 7. Critical current and field properties of the SuNam conductor (YBCO; $I_c = 600$ A at 77 K)

IOP Conf. Series: Journal of Physics: Conf. Series **1054** (2018) 012071 doi:10.1088/1742-6596/1054/1/012071

At 30 K, the critical current was 430 A, and the target operating current was 344 A which is 80 % of the critical current. Figure 8 shows the configuration of the HTS magnet in the Cz 300 mm single silicon-crystal growth system. The ring-shaped DPCs were applied to the magnet, and four DPCs made up the magnet rings. The inner and outer radii of the HTS magnet were 810 mm and 900 mm, respectively; the distance from the free melt surface to the coils was 400 mm, and the number of turns of one single package coil (SPC) was 500. Table 3 provides detailed specifications for the HTS magnet.

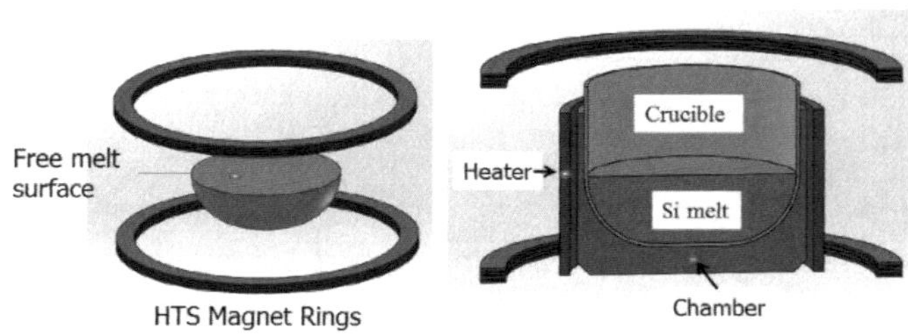

Figure 8. Configuration of the HTS magnets

Table 3. Specifications of the HTS magnet

Parameters	Values
Distance from the free melt surface to coils	400 mm
Inner radius	800 mm
Outer radius	910 mm
Number of DPCs	4ea
Number of turns of the SPC	500 turns
Superconducting wire	YBCO
Thickness of HTS tape	0.22 mm
Width of HTS tape	12 mm
Magnetic field direction	Cusp magnetic field
Magnetic field strength (center of crucible bottom)	0.35 T
Critical current (I_c)	430 A
Operating current ($I_{op}= 0.8 *I_c$)	344 A

4.2. Confirmation of the design results

A 3D FEM simulation model was built to analyze the characteristics of an HTS magnet and to confirm the design results as shown in figure 9. The magnetic flux density at the top of the crucible edge and the center of the crucible bottom were 0.3 T and 0.35 T, respectively. The results of the magnetic flux density satisfied the initial goal of the study.

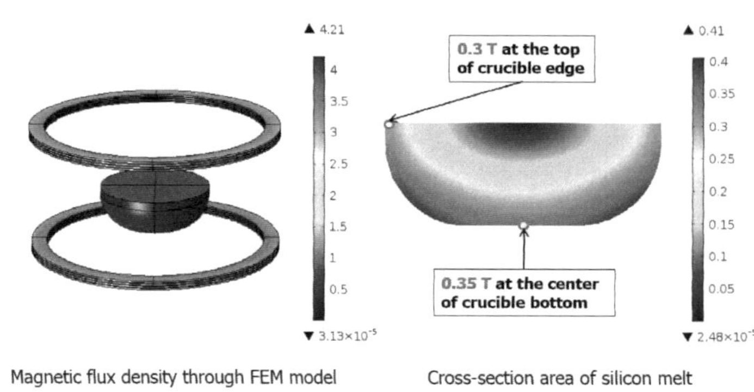

Figure 9. Confirmation of the desired magnetic field for the HTS magnet

5. Conclusion

This paper presents the results of designing an HTS magnet for a single silicon-crystal growth system. The authors analyzed the effects of static magnetic fields on silicon melting in a single silicon-crystal growth system. Magnetic field strength and shape were determined for an HTS magnet designs, and the characteristics of the designed magnet were analyzed using FEM simulations to determine its specifications. A ring-shaped DPC was applied to the HTS magnet, and the operating current of the HTS magnet with four DPCs was 344 A. The number of turns of one SPC was 500 with a total wire length of 11.5 km. The melt convection was significantly suppressed by the Lorentz force, and the temperature distribution was uniform in the silicon melt under the cusp magnetic field. The magnetic flux density of 0.35 T at the center of the crucible bottom was achieved. The basic design specifications and data from this study can be effectively applied to the development of real silicon-crystal growth systems.

Acknowledgement

This research is financially supported by Changwon National University in 2017–2018.

References

[1] Ammon V Application of magnetic fields in industrial growth of silicon single crystal *Magnetohydrodynamics* **42** no 4 pp 427–444

[2] Galindo V 2002 Crystal growth melt flow control by means of magnetic fields *Energy Conversion and Management* **43** no 3 pp 309–316

[3] Hirata H, Hoshikawa K 1989 Silicon crystal growth in a cusp magnetic field *Journal of Crystal Growth* **96** no 4 pp 747–755

[4] Kalaev V V 2007 Combined effect of DC magnetic fields and free surface stresses on the melt flow and crystallization front formation during 400mm diameter Si Cz crystal growth *Journal of Crystal Growth* **303** no 1 pp 203–210

[5] Enger S, Gräbner O, Müller G, Breuer M, Durst F 2001 Comparison of measurements and numerical simulations of melt convection in Czochralski crystal growth of silicon *Journal of Crystal Growth* **230** no 1 pp 135–142

[6] Liu X, Liu L, Li Z, Wang Y 2012 Effects of cusp-shaped magnetic field on melt convection and oxygen transport in an industrial CZ-Si crystal growth *Journal of Crystal Growth* **354** no 1 pp 101-108

[7] De Potter A, Fischer (of Bruker ASC), 2010 Improments of Si-Crystal growth processes for PV applying strong magnetic fields *InterPV* pp 82-85

[8] Narlikar A V 2004 *High Temperature Superconductivity 1: Materials* p 326

[9] Muiznieks A, Krause A, Nacke B 2007 Convective phenomena in large melts including magnetic fields *J. Cryst. Growth* **303** pp 211–220

[10] Weichaob H, Ding L and Ni Z 2015 Influence of thermal and flow distribution for the Czochralski growth process of Si under external magnetic field *The 27th Chinese Control and Decision Conference* pp 3314-18.

[11] Vegad M, Bhatt N M 2014 Review of some Aspects of Single Crystal Growth Using Czochralski Crystal Growth Technique *In Procedia Technology* **14** pp 438–446

[12] Series R W, Hurle D T J 1991 The use of magnetic fields in semiconductor crystal growth, In Journal of Crystal Growth **113** no 1–2 pp 305–328

[13] Hirata H, Hoshikawa K 1989 Silicon crystal growth in a cusp magnetic field *Journal of Crystal Growth* **96** no. 4 pp 747–755

[14] Daggolu P, Ryu J W, Galyukov A, Kondratyev A 2016 Analysis of the effect of symmetric/asymmetric CUSP magnetic fields on melt/crystal interface during Czochralski silicon growth *Journal of Crystal Growth* **452** pp 22–26

[15] Moon S H, 2014 HTS wire development and industrialization at SuNAM. *https://indico.cern.ch/event/308828/contributions/1680708/attachments/589810/811813/20140521_SuNAM_at_WAMH_DESY_vf.pdf*

IOP Conf. Series: Journal of Physics: Conf. Series **1054** (2018) 012072 doi:10.1088/1742-6596/1054/1/012072

Comparison of Trapped Field Characteristic of Bulk Magnet System Using Various Type Refrigerators

K Yokoyama[1], A Katsuki[2], A Miura[2] and T Oka[3]

[1] Division of Electrical and Electronic Engineering, Ashikaga University, 268-1 Omae-cho, Ashikaga, Tochigi 326-8558, Japan
[2] Department of Electrical and Electronic Engineering, Graduate School of Engineering, Ashikaga University, 268-1 Omae-cho, Ashikaga, Tochigi 326-8558, Japan
[3] Materials Science and Engineering, College of Engineering, Shibaura Institute of Technology, 3-7-5 Toyosu, Koto-Ward, Tokyo, 135-8548 Japan

Abstract. We developed several type superconducting bulk magnets with the goal of their industrial application, and study to improve a magnetic field activated by pulsed field magnetization (PFM). When considering a practical use of bulk magnet, it is important problem to choose a suitable refrigerator to cool a bulk superconductor. This paper investigated trapped field characteristics when using several type refrigerators; one is a Stirling refrigerator with an ultimate temperature of 40 K at no-load condition and cooling capacity of 11 W at 77 K. The other is a dual-stage GM type refrigerator with the cooling ability of 13 K and 5 W at 20 K. Now, we developed a new bulk magnet system using a dual-stage GM refrigerator with the cooling ability of 12 K and 12 W at 20 K aiming to improve a trapped field by removing heat generated after a pulse application quickly. Using a GdBCO bulk with dimensions of 60 mm in diameter and 20 mm thick, magnetizing tests were carried out and the relationship between trapped field performance and cooling ability was investigated. Even the bulk magnet using Stirling refrigerator with relatively low cooling capacity could trap a high magnetic field of 3.0 T on the bulk surface and maximum total flux of 2.0 mWb. The trapped magnetic field of new bulk magnet system was improved greatly at low temperature; the maximum trapped field of 3.6 T and the maximum total flux of 2.9 mWb were achieved.

1. Introduction

The application of REBCO bulk as magnetic field generator has attracted much attention because the high magnetic field over 2 T can be trapped with a compact system and low running cost, and accordingly, various industrial applications are considered [1-5]. From this background, many advanced studies about material fabrication [6,7] and magnetizing method [8-11] are progressing. However the relationship between trapped field performance and a cooling ability has received little attention, and thus, it's necessary to discuss the selection of refrigerator when considering a practical use of a bulk magnet system.

We study to improve of trapped field of REBCO bulk magnet by exciting with pulsed field magnetization (PFM) aiming at an industrial application. In our previous research, the several bulk magnet systems using a dual-stage GM refrigerator with an air-cooled compressor [12] and a Stirling refrigerator without compressor [13] were developed based on the idea that the system was simplified by not using a chiller. On the other hand, a cooling ability may not always be enough. Now, we developed a new bulk magnet system using a dual-stage GM refrigerator with a water-cooled compressor aiming to enhance a trapped field by improving a cooling ability. The main specifications of the above magnet systems were listed in Table 1. The bulk system using a new refrigerator (referred to as "12WGM") had more than 2 times of cooling ability of the bulk system using the existing GM refrigerator (referred to as "5WGM"). When magnetic flux penetrates into the sample, the bulk always generates heat due to pinning loss and viscous loss. Therefore, it is necessary to make the bulk trap magnetic flux while considering the relationship between heat generation and cooling. In this paper, we pay attention to the difference of cooling ability on different types of refrigerators. Then, the magnetizing characteristics are investigated by applying a single pulsed field to a GdBCO bulk 60 mm in diameter and 20 mm thick while varying the amplitude of applied fields and temperature in each bulk magnet system.

Table 1. Specifications of the bulk magnet systems

Label	12WGM	5WGM	Stirling
Photo			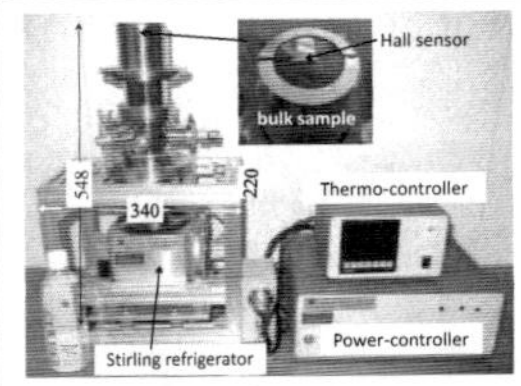
Refrigerator	Dual-stage GM (RM20,ULVAC)	Dual-stage GM (RS271,AISIN)	Stirling (CryoTel CT, UNPOWER)
Lowest temperature (no load)	12 K	13 K	40 K
Cooling capacity	12 W@20 K (2nd stage) 35 W@77 K (1st stage)	5 W@20 K (2nd stage) 19 W@80 K (1st stage)	11 W@77 K
Power consumption	5.0 kW / 3ϕ 200 V	1.6 kW / 3ϕ 200 V	160 W / 1ϕ 100 V
Compressor	Water-cooled	Air-cooled	Unnecessary

2. Experimental

A GdBCO bulk 60 mm in diameter and 20 mm thick with the high trapped field performance of 2 T at 77 K (Nippon Steel & Sumitomo Metal Corporation), which was reinforced mechanically with a stainless steel ring 2 mm thick (SUS316L), was set to the sample stage connected to a cold head of the refrigerator using a sample holder made of a stainless (SUS304). A Hall sensor (BHT-921, F.W.BELL) was adhered at the center of bulk surface with Kapton tape to measure the time variation of the magnetic flux during magnetization, and a Cernox temperature sensor (CX-1030-SD-HT-1.4L/J, LakeShore) was also adhered near the Hall sensor with the GE7031 varnish to measure the time variation of temperature. After a vacuum vessel was covered and evacuated by a diffusion pump, the bulk was cooled to 20, 30, 40, and 50 K. A single pulsed field was applied while changing the amplitudes from 3.1 to 7.0 T for each temperature, where the rising time and fall time of the pulse were 10 ms and 100 ms, respectively. During magnetization, the magnetic flux density and temperature were measured in an interval of 100 μs and 100 ms, respectively. After magnetization, the trapped field map at 7 mm above the bulk surface was measured using the three-dimensional Hall sensor (BH-703, F.W.BELL) with a 2-mm pitch and the total magnetic flux was calculated using the two-dimensional distribution.

3. Results and discussion

Figure 1 shows magnetic field density at the center of bulk surface, B_z, as a function of applied field, B_{app}, for 12WGM, 5WGM and Stirling refrigerators at 20 K and 50 K, where there is the result of Stirling refrigerator only at 50 K. At 20 K, a maximum value of B_z was reached at B_{app}=5.4 T in both 12WGM and 5WGM, and the maximum values were 3.6 T and 2.6 T, respectively. The maximum trapped field of 12WGM was approximately 1.4 times as high as that of 5WGM, indicating that a cooling ability has a strong influence on trapped field property. In higher applied fields, the B_z value was decreased due to flux flow caused by heat generation for both systems. At 50 K, the peak of B_z shifted toward lower applied field in 12WGM and 5WGM because J_c was decreased with increasing temperature, resulting that magnetic flux was able to penetrate the bulk easily. Then, the maximum values were 2.8 T for B_{app}=4.6 T in 12WGM and 2.4 T for B_{app}=3.9 T in 5WGM. On the other hand, the B_z value of Stirling system was increased with applied field, reaching a peak of 3.0 T for B_{app}=6.2 T. This means that even the bulk magnet using Stirling refrigerator with relatively low cooling capacity can trap a high magnetic field though the peak value was smaller than that of the refrigerator with high cooling capacity.

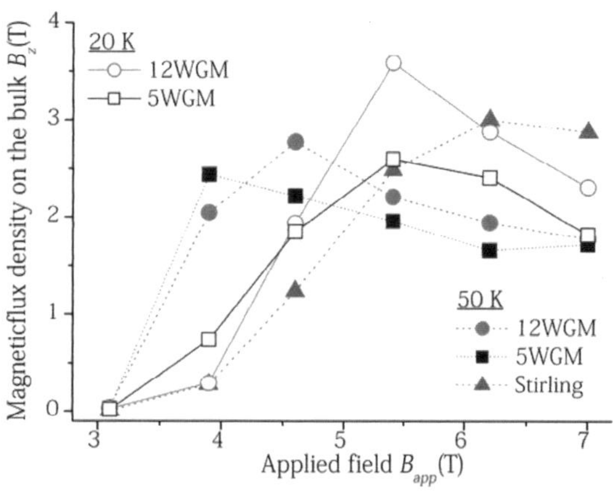

Figure 1. Applied field dependence of magnetic flux density at the center of bulk surface at 20 K and 50 K

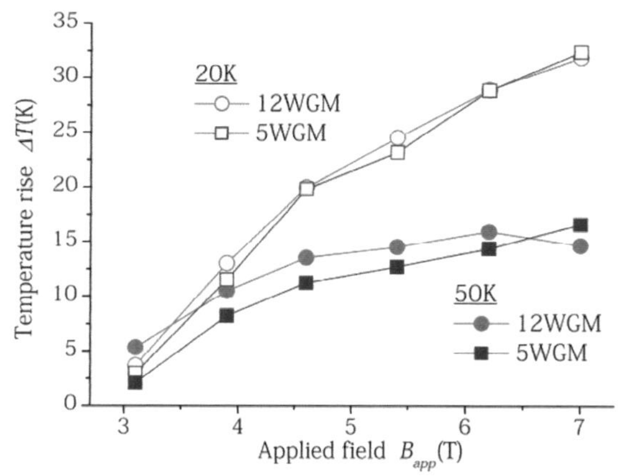

Figure 2. Comparison of maximum temperature rise on the bulk surface at 20 K and 50 K

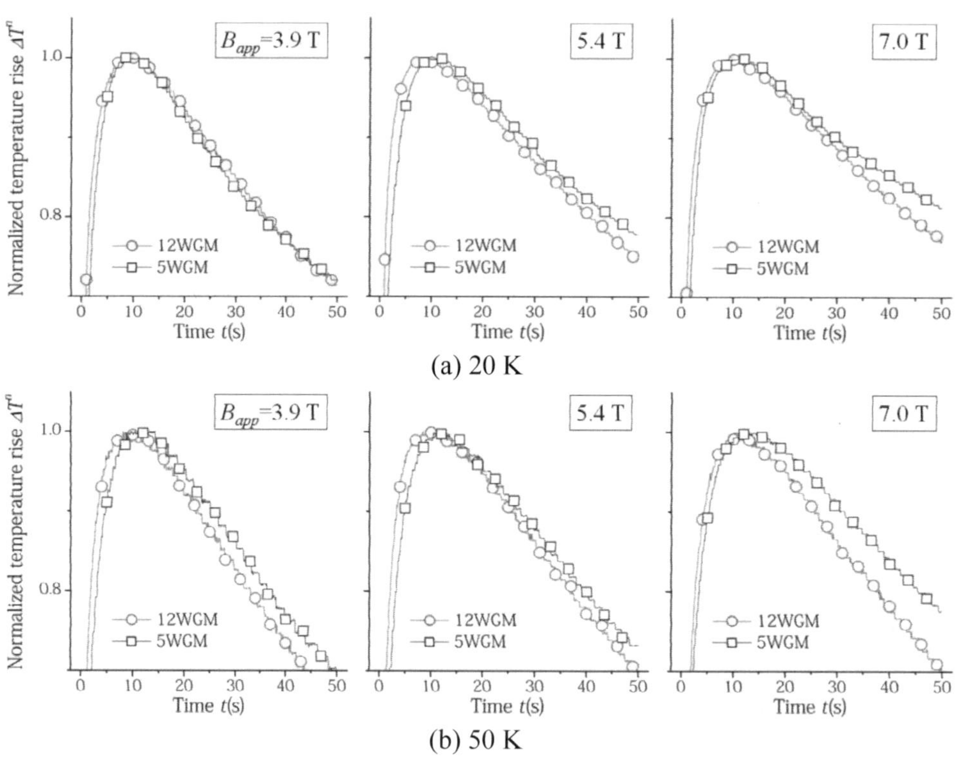

Figure 3. Time variations of normalized temperature rise by the peak value at 20 K and 50 K

Figure 2 compares the maximum temperature rise between both GM systems at 20 K and 50 K. Since a critical current density is increased with a decrease in temperature, the magnetic shield becomes high. Then, heat generation caused by pinning loss is increased at lower temperature, and therefore, the temperature rise at 20 K was higher than that at 50 K. Also, all data increased with applied fields. At 20 K, both curves were almost same, and the maximum ΔT was 31.9 K in 12WGM and 32.4 K in 5WGM for B_{app}=7.0 T. At 50 K, on the other hand, the ΔT value of 12WGM was about 1.2 times higher than that of 5WGM, and the maximum values were 16.0 K and 16.6 K, respectively. The difference of trapped field shown in Figure 1 cannot be explained in terms of the amplitude of temperature rise shown in Figure 2. Then, the measured time variation of temperature was normalized by the peak value as illustrated in Figure 3. The decrease of ΔT after reaching a peak was faster in 12WGM as compared with 5WGM in all data, indicating that the quick removal of heat by high cooling ability of a refrigerator leads to improvement of the trapped field. On the other hand, the rise time of 12WGM was slightly faster than that of 5WGM. However, the rise time of applied field was not changed because the same magnetizing equipment was used in both magnetizing experiments. We will consider the reason for a difference in rise time of temperature in detail.

Figure 4 shows the trapped field distributions at 20 K and 50 K. At 20 K, the magnetic flux did not penetrate the center of bulk for B_{app}=3.9 T, however the distribution became concentric for B_{app}=5.4 T and the area of high trapped field over 1.4 T appeared widely. For B_{app}=7.0 T, a decrease of trapped field in 5WGM was larger than that of 12WGM. At 50 K, the distribution was concentric even for low applied field of 3.9 T in both GM system, and the amplitude of magnetic flux was reduced with increasing an applied field. In Stirling system, on the other hand, the magnetic flux did not reach the center of bulk for B_{app}=3.9 T, while the area of high trapped filed over 1.4 T for B_{app}=7.0 T as almost same as that for B_{app}=5.4 T. Using these distribution maps, the total magnetic flux, Φ, and the maximum magnetic flux density on the pole surface, B_z, were calculated, where the maximum B_z was not always the value at the center of magnetic pole. Figure 4 compares the total magnetic flux, Φ (bar

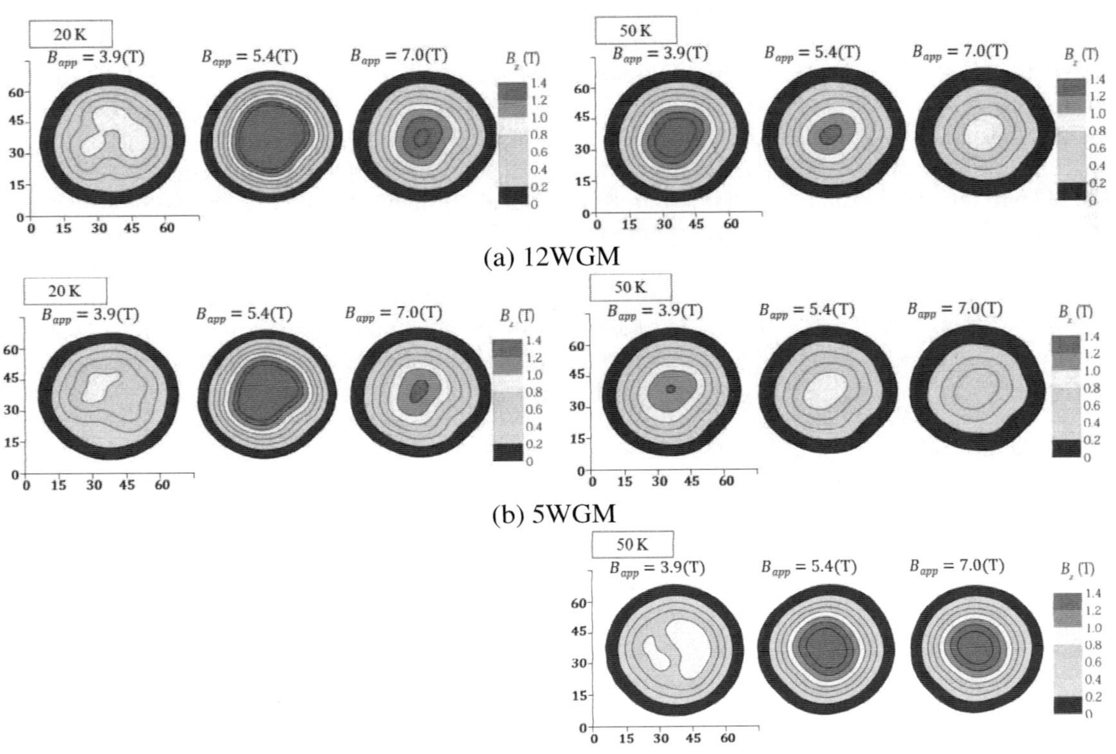

Figure 4. Trapped field distributions in 12WGM, 5WGM and Stirling systems at 20 K and 50 K.

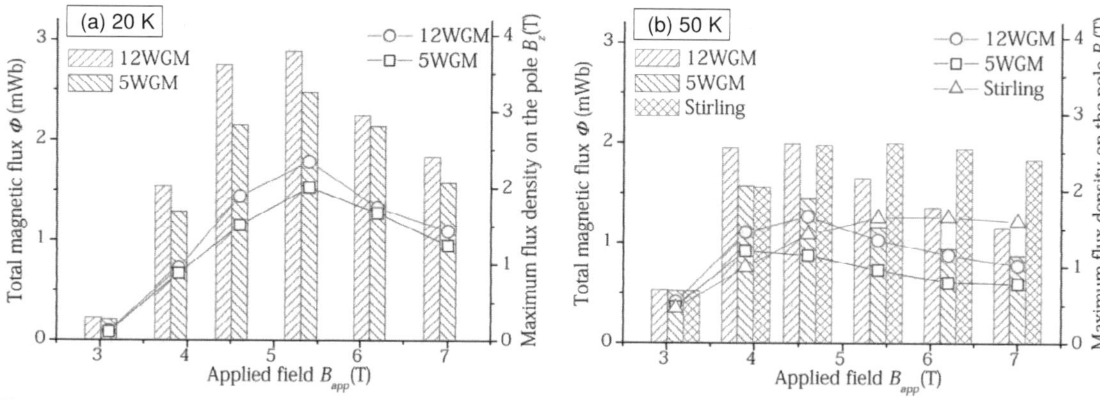

Figure 5. Comparisons of total magnetic flux (bar graph, left axis) and maximum flux density on the pole surface (line graph, right axis) as a function of applied field at 20 K and 50 K.

graph, left axis), and the maximum trapped flux density, B_z (line graph, right axis), as functions of the applied field at 20 K and 50 K. At 20 K, the maximum Φ and B_z were 2.9 mWb and 2.3 T for B_{app}=5.4 T in 12WGM, and 2.5 mWb and 2.0 T in 5WGM. At 50 K, they were 2.0 mWb and 1.7 T for B_{app}=4.6 T in 12WGM, and 1.6 mWb and 1.2 T for B_{app}=3.9 T in 5WGM. In Stirling system, on the other hand, they were 2.0 mWb for B_{app}=5.4 T and 1.7 T for B_{app}=6.2 T, respectively. When comparing between 12WGM system and Stirling system, the former can trap a higher magnetic field at lower applied field. These results suggest that the magnetizing system can be downsized and this is a major advantage for a practical use of bulk magnet system.

4. Conclusions

We investigated the influence of the cooling ability to the trapped field performance using several type bulk magnet systems using different refrigerators; a new dual-stage GM refrigerator with the cooling capacity of 12 W at 20 K, an existing dual-stage GM refrigerator with5 W at 20 K and 13 K and a Stirling refrigerator with 11 W at 77K and 40 K. The time variations of magnetic flux density and temperature on the bulk surface and magnetic field distributions on the pole surface were measured by applying a single pulsed field was applied while varying an amplitude and temperature using a GdBCO bulk 60 mm in diameter and 20 mm thick. The experimental results indicated the followings; even the bulk magnet using Stirling refrigerator with relatively low cooling capacity could trap a high magnetic field of 3.0 T on the bulk surface and maximum total flux of 2.0 mWb. Moreover, the trapped magnetic field of 12WGM was improved greatly at low temperature; the maximum trapped field of 3.6 T and the maximum total flux of 2.9 mWb were achieved. We will make the data which is the guideline to select a refrigerator for industrial application of bulk magnets by performing the more detailed magnetizing test in the above bulk magnet systems.

Acknowledgments

This work was supported by JSPS KAKENHI Grant Number JP15K05951.

References

[1] Ohsaki H, Sekino M, Suzuki T and Terao Y 2009 *2009 Int. Conf. Clean Electrical Power* 479
[2] Felder B, Miki M, Zigang D, Tsuzuki K, Shinohara N, Izumi M and Hayakawa H 2011 *IEEE Trans. Appl. Supercond.* **21** 2213
[3] Nakamura T, Tamada D, Yanagi Y, Itoh Y, Nemoto T, Utumi H and Kose K 2015 *J. Magn. Reson.* **259** 68
[4] Nakagawa K, Mishima F, Akiyama Y and Nishijima S 2012 *IEEE Trans. Appl. Supercond.* **22** 4903804
[5] Oka T, Takayanagi Y, Machida S, Ichiju K, Fukui S, Ogawa J, Sato T, Ooizumi M, Tsujimura M and Yokoyama K 2016 *IEEE Trans. Appl. Supercond.* **26** 3700204

[6] Nariki S, Teshima H and Morita M 2016 *Supercond. Sci. Technol.* **29** 034002
[7] Durrell J H, Dennis A R, Jaroszynski J, Ainslie M D, Palmer K G B, Shi Y, Campbell A M, Hull J, Strasik M, Hellstrom E and Cardwell D A 2014 *Supercond. Sci. Technol.* **27** 082001
[8] Fujishiro H, Tateiwa T, Fujiwara A, Oka T and Hayashi H 2006 *Physica C* **445-448** 334
[9] Zhou D, Shi Y, Zhao W, Dennis A R, Beck M, Ainslie M D, Palmer K G B, Cardwell D A and Durrell J H 2017 *Supercond. Sci. Technol.* **27** 6800704
[10] Ainslie M D, Fujishiro H, Mochizuki H, Takahashi K, Shi Y, Namburi D K, Zou J, Zhou D, Dennis A R and Cardwell D A 2016 *Supercond. Sci. Technol.* **29** 074003
[11] Zhou D, Ainslie M D, Shi Y, Dennis A R, Huang K, Hull J R, Cardwell D A and Durrell J H 2017 *Appl. Phys. Lett.* **110** 062601
[12] Yokoyama K, Oka T, Kondo N and Hosaka S 2013 *Physica C* **484** 343
[13] Yokoyama K, Kulawansha M P E S, Yuanding Z and Oka T 2016 *IEEE Trans. Appl. Supercond.* **26** 4601204

Design of 23kV 50MVA class HTS Cable in South Korea

Jin Bae Na[1], Heo Gyung Sung[1], Chang Yeol Choi[1], YoongSeo Jang[1], and Yang Hun

1. LS Cable&System, LS Tower, 127, LS-ro, Dongan-gu, Anyang-si, Gyeonggi-do 14119, Republic of Korea

Abstract. The world first commercial project for the superconducting applications is already starting in South Korea. The first step of the project named SSS (Superconducting Smart platform Station in South Korea) is to operate AC 23kV 50MVA class HTS cable system in the power grid of Korea Electric Power Corporation (KEPCO) in order to increase the power capacity and realize the stable operation. The above HTS cable system will connect two substations between ShinGal and HeungDuk. Total length of the HTS cable is approximately over 1km including 2 sets of the normal joint box and 2 sets of the termination which are the out-door type accessories. Before performing the type test, various preliminary tests with the short cores were performed to confirm design of the HTS cable. To meet the design goals, the HTS layers of the core consist of the two conducting and one shield layer. The characteristics of the HTS wire is the non-magnetic 2nd generation wire. It can reduce the magnetization loss caused by itself and adjacent the HTS wires. The measured AC loss is around 0.3 W/m at 1,260 Arms and increasing trend of AC loss by operating current is the well fitted of simulation results. In case of short-circuit test, experimental temperature is approximately 77K and increased maximum temperature is under 92.8 K which is enough below the boiling temperature of LN2 at gage pressure 6.7bar.(g). A 70-meter-long the HTS cable including unit the test samples and an 85-meter-long LN2 return pipe will manufacture and installed for the type test at KEPCO Power Training Center in Gochang, South Korea. The type test procedure is following the Cigre TB538 and KEPCO standard those are including the 20 cycles load test, voltage tests such as AC withstand and impulse, repeated cooling process three times as prequalification quality test and visual inspection after disassembling. In the third quarter of 2018, LS Cable&System will supply and install AC 23kV HTS cable system in South Korea.

1. Introduction

SUPERCONDUCTING APPLICATIONS using the high temperature superconductor (HTS) such as cable, the rotating machines and the superconducting fault current limiter are applicable in the real power grid with the stable operation [1-2]. The superconducting applications such as the HTS cable, the motor and the SFCL for the power grid are state-of-the-art technology. Especially, the HTS cables can transfer the large capacity power source with low losses, and are expected to be not only environmental applications but also application with electromagnetic interference free performance which means no influences of adjacent other cables. It has a compact size compared with the conventional power apparatus. However, these apparatus such as HTS cable have to consider easy connection with the existing applications in the power grid. Thus, the various R&D projects were performed and the demonstration projects were being promoted in the world.

Since 2001, Korea Electric Power Corporation (KEPCO), LS Cable&System (LS C&S) and various institutes have been cooperating national R&D program that was Development of Advanced Power system by Applied Superconductivity technologies (DAPAS) funded by the Ministry of Education, Science and Technology in South Korea. During the 10 years of DAPAS program, the basic design concept of distribution and transmission level HTS cable system was established and manufactured the HTS cable for type test to verify design factor for applying power gird. The HTS cable even distribution level can have bulk power capacity like conventional cable in transmission level which means it can replace existing conventional transmission cable and save cost of the civil work for installation of extra cable in the power grid. Thus, KEPCO and LS C&S installed 23kV 50MVA class HTS cable in the Icheon substation. This HTS cable system was approximately 500 meters inculding 2 sets of the outdoor termination, 1 set of the normal joint-box and 1 set of decompression type as cryo-cooler. Type of this HTS cable was 3 Phase-in-One cryostat type and the unmanned operation at the Incheon substation was conducted 18 months among 20 months. The next step was increasing the transmission capacity as 23kV 120MVA. Installation route of next project was approximate 100 meters. These demonstration schedules were called the Green Superconducting Electric Power Network at Icheon Substation (GENI). Furthermore, the demonstration of AC 154kV 600MVA HTS cable system and DC 80kV 500MW HTS cable system in Jeju Island was performed to investigate the stability and the operation condition in the real power grid. To verify performance and reliability of the HTS cables, our research group performed design of the HTS cable system based on several FEM simulations such as the calculation of current distribution according to the various pitch values of each HTS layer, the calculation of AC loss, and the thermal contraction force during the cool-down and the warm-up. Based on above design process, the HTS cables with short length were manufactured and measured preliminary test according to TB538 and KEPCO's standard. From the above the preliminary tests and the demonstration in Jeju site and Ichecho substation, LS C&S developed and demonstrated the world first HVDC HTS cable, which was bipolar type of DC ±80kV 500MW class and total length was 500 meters, and the highest voltage and the longest length cable as AC 154kV 600MVA class HTS cable in Jeju site. In case of AC 154kV HTS cable in Jeju site, the one phase cable length was over the 1km and experienced various installation conditions that are through the pipe duct and ground over and over again of the 1km. This system consists of the three phases of 154kV, the six outdoor type terminations, the six normal joints and the one set of the cooling system (10kW@69K). It was energized in 2016. Especially, the terminations of those DC and AC HTS cable were applied the fixed-type termination. It can withstand thermal contraction force during the cool-down and the warm-up. It means that the thermal contraction force can be absorbed by the installation method using the snake shape in a straight section of the cable installation route in the power grid.

Based on the experience and the results from DAPAS program, GENI program and Jeju demonstrations, the world first commercial HTS cable project has launched by KEPCO in 2016 that is named SSS (Superconducting Smart platform Station in South Korea) project. The main goal of SSS project is established commercialization stage and making test-bed of the smart platform in South Korea. 3-Phase in One Cryostat type of the HTS cable will manufacture 1,035-meter-long to transfer power between ShinGal and HeungDuk substations and return pipe will manufacture 1,040-meter-long for circulation of LN2. Table 1 and table 2 represent system requirements of AC 23kV 50MVA class HTS cable system in the power grid. The first stage of ShinGal project is setting up cryogenic system which is using Reverse Turbo Brayton type cryo-cooler made by Taiyo Nippon Sanso (TNSC) in Japan. Next step is performing the type test of AC 23kV 50MVA class HTS cable with 2 set of out-door type termination and 1 set of normal joint box. Procedure of type test is complied with TB538, which was published by Cigre Working Group, and KEPCO's standard. The voltage level and power capacity of this project is 23kV 50MVA class and various types of HTS wires will apply. The cryo-cooler was already installed in HeungDuk substation. Now LS C&S is preparing type test at KEPCO's Power Testing (PT) center. The installation of the HTS cable will complete in the second quarter of 2018.

Figure 1 shows a diagram of the HTS cable system of the ShinGal project. Dot line explains of flowing direction of LN2 in the HTS cable system including return pipe. Accessories for the return pipe also need. Configuration of the return pipe is similar to the double metal sheath of the HTS cable. The 1st metal sheath is making area for flowing LN2 and the 2nd metal sheath is making another wall to make vacuum status. Thus, the double metal sheath is blocking the heat invasion between inside and outside of the HTS cable. The multi insulation layer (MLI) also applied to reduce light penetration. Finally, thermal spacer also added to avoid and minimize contact area of the 2nd metal.

Table 1. The Power Grid Conditions by KEPCO.

Item	Unit	Requirement
Nominal Voltage	kV	AC 23
Power Capacity	MVA	50
Rated Current	A	1,260
HTS Cable Length	m	1,035
Return Pipe Length	m	1,040
Ground	-	Sheath
BIL	kV	150
Short Circuit Current	kA/sec	25/0.5

Table 2. Installation Status and Operation Conditions.

Item	Unit	Requirement
Installation Condition	-	Duct / Tunnel / Ground
Cable Construction	-	3-Phase in One Cryostat
No. of Termination / Joint box	set	2 / 2
Pressure Range in Normal Term	bar(g)	7.60 ~ 6.70
Temperature Range in Normal Term	K	69.0 ~ 74.8
Instantaneous Max. Temperature during the Fault Accident	K	92.8

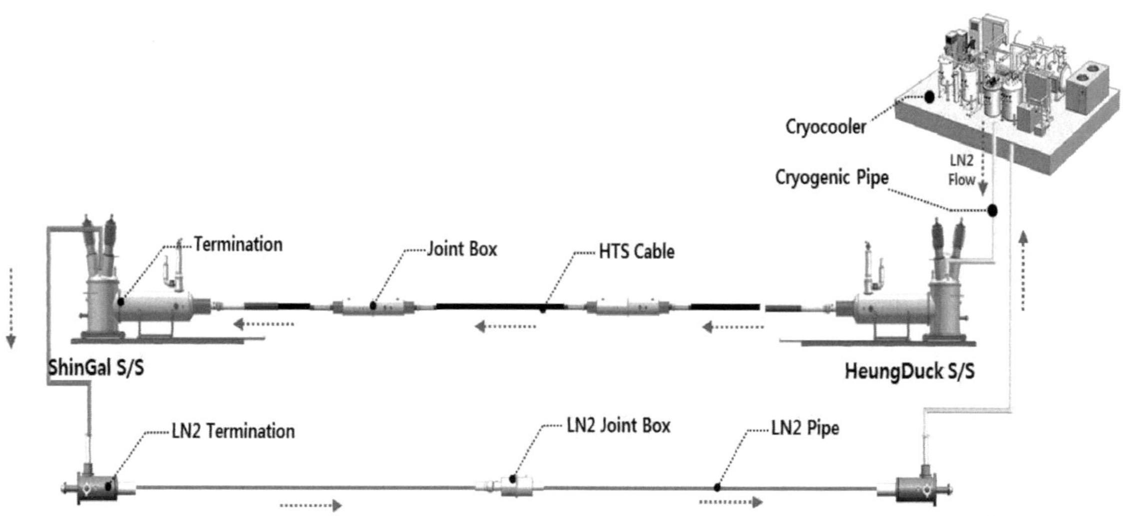

Figure 1. Diagram of the ShinGal Project.

2. Preliminary Test

Table 1 shows requirements of the power grid. Firstly, LS C&S was doing the engineering design of the whole system such as a pressure drop and the difference temperature between the inlet and the outlet of the HTS cable based on the installation status. Table 2 shows engineering design of AC 23kV 50MVA HTS cable system between ShinGal and HeungDuk substations. To make compact size and competitive price, the core configuration such as a former size, number of the HTS wires, the pitch of the HTS wire layers and the PPLP insulation thickness was designed based on the preliminary test results and productivity conditions.

The former made of thin and of round type copper stranded wires avoids skin effect during the operation. Applicable magnitude of current in each HTS wire layer could be calculated by outlet temperature of the HTS cable because the critical current of the HTS wire should be changed by temperature of LN_2. Furthermore, current distribution between the HTS layers with non-magnetic subtract such as hastelloy and stainless steel must be investigated. For increasing mechanical strength of the HTS wire, lamination material such as copper and brass could be adapted to withstand any stress during the manufacturing process and installation of the HTS cable. The double aluminium cryostats with corrugated type were also adapted to make LN_2 channel and to make vacuum wall. After manufacturing short length of the HTS cable, it was cut to investigate and its results compared with the previous design concept.

The inner diameter of duct pipe in the power grid is 200mm. Thus, the maximum diameter of the HTS cable diameter should be limited under approximate 153mm. Then the HTS cable could be installed without extra construction of tunnel and duct.

2.1. Manufactured Sample Cable

To perform the preliminary tests, the sample cables were manufactured based on the previous design value. Two kinds of HTS wires were used in these sample cables for investigating its feasibility and performance during the manufacturing and test. Total number of conducting layer and shield layer is 2 layers and 1 layer, respectively. The calculated AC loss of shield layer was generally much less compared with conducting layer. It means that the HTS wire with the high critical current should be arranged in the conducting layer for the reducing AC loss if other design factors were perfectly same. The short core of Core #1 and Core #3 has similar critical current of the conducting layer except the shield layer. The different amount of critical current at shield layer between the Core #1 and the Core #3 is around 1,000 A. However, core configuration such as pitch, wire number, insulation thickness and layer number is perfectly same among three cores. Table 3 shows the critical current of each core and measured temperature is 77.3 K [4]. In case of Core # 3, the HTS wires of conducting and shield layer was made by SuNAM and AMSC.

Table 3. The Critical Current of Each Core.

Item	Conducting Layer	Shield Layer	Wire Manufacture
Core #1	5,623.0 A	3,712.0 A	SuNAM
Core #2	3,439.0 A	2,731.0 A	AMSC
Core #3	5,608.0 A	2,718.0 A	Conducting Layer : SuNAM Shield Layer : AMSC

2.2. The Bend Test

The diameter for the bending test follows Cigre TB538 which is international recommendation for the AC HTS cable [5]. However, bending diameter at Cigre TB538 is too larger than applicable steel drum diameter. Table 4 shows diameter of the bend test where small d is the nominal diameter of the core and capital D is the nominal overall diameter of cryostat of the HTS cable. Three short cores were bent around a test cylinder which was simulated of steel drum. The test condition is ambient temperature and at least one complete turn and unwound without axial rotation. The test core was rotated through 180 degrees and bending process repeated 3 times in each bending diameter. Before preforming the bending test, we measured the critical current at 77K and self-field. After finishing the bend test, the critical current was also measured again and its results compared with previous results. To follow Cigre TB538, degradation of the critical current should be under 5%, definitely. Figure 2 shows the photo of the wooden frame and the short core bend for the test. Also, Figure 3 shows the experimental results before and after the bending test among three cores. It means that no degradation of the critical current at 77.3 K temperature was found. These results could be applied to determine steel drum diameter for manufacturing and delivery of AC 23kV HTS cable.

Table 4. The Diameter of Bending Test.

Item	Unit	Value
Single Core	-	25(d+D) + 5%
Three Core	-	20(d+D) + 5%

Figure 2. The Wooden Frame for the Bending Test.

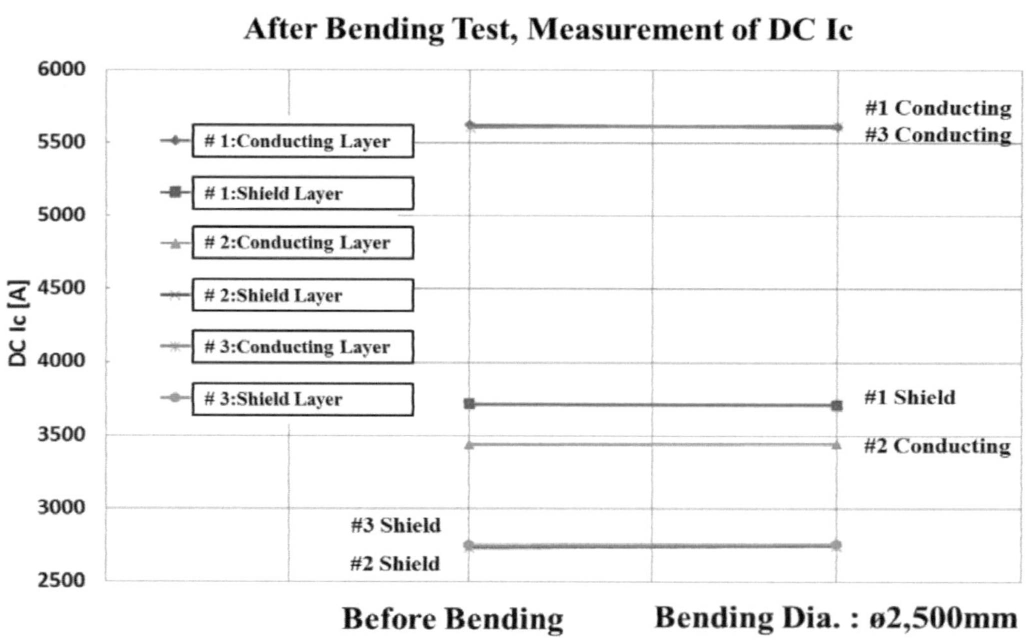

Figure 3. Before and After Bending Test, Measurement of DC Ic.

2.3. FEM Simulation for AC Loss Calculation

One of FEM simulation results in conducting layer was shown in Figure 4. This result shows an influence of the magnetic field to itself and adjacent the HTS wires. The magnetic flux distortion at edge was occurred because of gap distance between adjacent HTS wires. As the gap distance decreases, the magnetic field becomes parallel magnetic field. These characteristics were occurred in every superconducting layer. Thus, beneficial way of reducing AC loss should be made small gap distance because magnetic field at the edge of the HTS wire should be made parallel magnetic flux as possible as [3]. Dot circle in Figure 4 well explains above phenomenon. Figure 4 shows that distortion of magnetic field by the gap distance. Furthermore, this characteristic reduces the critical current of the HTS wires because this arrangement of the HTS wires is generally occurring leakage magnetic field in each layer. During the superconducting operation, AC loss in the HTS cable will be generated due to the magnetic field in radial direction which is a perpendicular field of the REBCO layer. Moreover, AC loss will be changed by current distribution of each layer. Those two simulation processes which were calculated AC loss and current distribution were repeated to satisfy cooling capacity of cryo-cooler and engineering design goals.

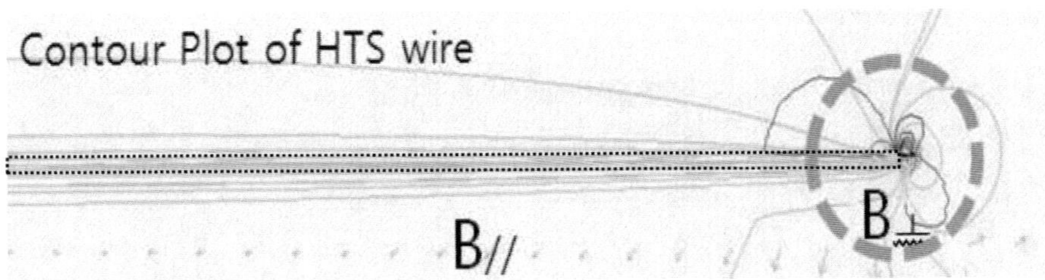

Figure 4. Magnetic Calculation of HTS wire for AC Loss.

2.4. Measurement of AC Loss

Based on above design process, three short cores manufactured and experiments AC loss measurement as Figure 5. To measure AC loss, we used an electrical method with cancelation coil. Actually, voltage and current are 90 degrees shifted in AC. The voltage signal from the HTS cable was measured by cancelling the inductive voltage component. To make same phase between voltage and current, the cancellation coil could be used and its configuration could be controlled such as number of coil turns and diameter of coil. Thus, phase shift was controlled by a cancelation coil [6].

Critical current of conducting layer at Core #2 is the lowest of among three short cores. Thus, measurement AC loss of Core #2 is the highest compared with the other cores. AC loss of the Core #1 and Core #3 were similar and measured value was approximately 0.18 W/m/phase at rated current. These results mean that the HTS wire with higher critical current should be arranged at conducting layer to reduce AC loss.

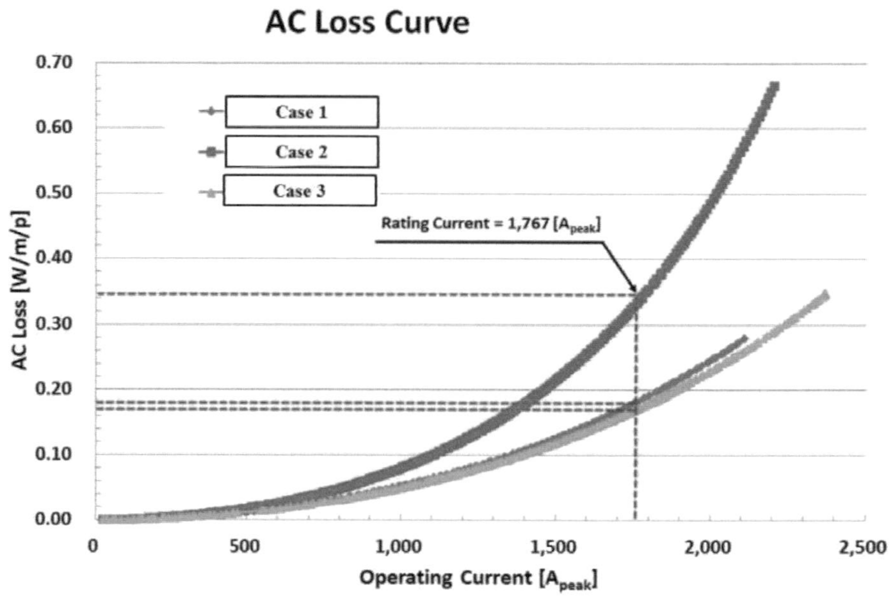

Figure 5. Measurement of AC Loss at 77.3K.

2.5. Short Circuit Test

The fault current will be bypassing to copper during the fault condition. The fault level of distribution in South Korea is 25 kA for 0.5 second which means the HTS cable should be withstood the maximum fault level. Preparation of the circuit test is shown in Figure 6 and the fault current is going through instant three phases and end of three phases are connection as common. This test method is the most sever experimental condition of circuit test of Core #1. Several thermal couples are attached at former surface to measure the rising temperature during the fault current loading. Figure 7 shows experimental results. The increased temperature of the HTS cable during fault condition is 18 K. These results are compared with boiling temperature of LN_2 at pressure status of outlet of the HTS cable. Estimated pressure at the outlet of the cable is 6.7 bar(g) and boiling temperature at this pressure of LN2 is approximately 98.5 K. Thus, the maximum raising temperature of the HTS cable during the fault is under 92.8 K. The HTS cable will be able to withstand the its fault condition because the increased maximum temperature was less than boiling point temperature of LN_2 at pressure of the outlet of the HTS cable.

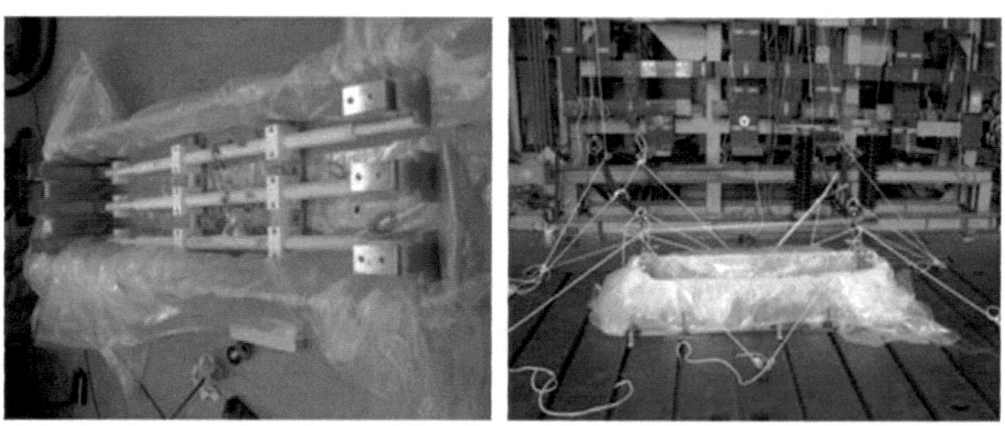

(a) The Test Set-up (b) The Experiments View

Figure 6. The Short Circuit Test of the HTS Cable.

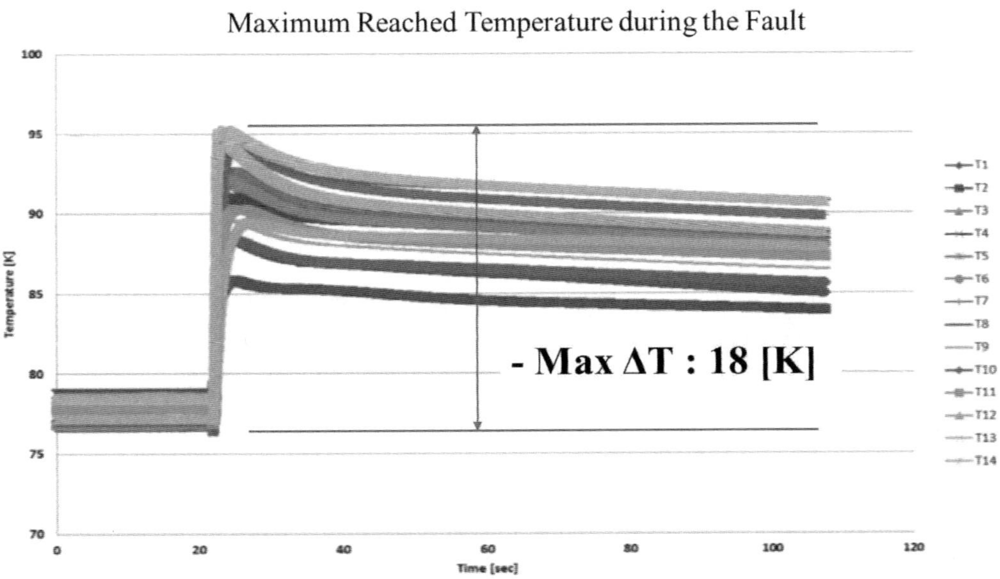

Figure 7. The Raising Temperature during the Fault.

3. Type Test and Prequalification Quality Test of 23kV 50MVA HTS Cable

For investigating AC 23kV 50MVA HTS cable, the bending test, AC loss, DC Ic and short circuit tests with short sample cable were conducted based on Cigre TB538 and KEPCO standard [5]. For performing the type test, sequence of the whole test is well explained in Table 4. The bending test will be also conducted as the routine test before the HTS cable shipment. Generally, gaseous nitrogen could be used the pressure test because it can easy flow and apply pressure. The one cycle of load cycle test spends 24 hours and the HTS cable will experience 20 cycles. During the test, rated current for 8 hours per cycle was applied the HTS cable. After finish the type test, 2 times repeated cool-down and warm-up as thermal cycle are going to progress as prequalification quality (PQ) test due to induce huge thermal shock. The HTS cable will be experienced mechanical stress during the thermal cycles. From the previous test, it found that moved HTS cable by thermal contraction force was saturated within 3 cycles of the thermal cycle. Thus, total number of repeated thermal cycle is 3 cycles as PQ test during the whole test. Figure 8 shows the schematic sketch for the type test of AC 23kV 50MVA class HTS cable system including the cable, the return pipe and the accessories.

Table 5. Specifications of Type Test.

Item		Value
Bending Test	3 times	20(d+D) + 5% = 3,613 mm + 5%
Pressure Test	30 min.	13.2 bar(g) (Gaseous Nitrogen)
Pressure Drop Test	-	Cable ≤2.5 mbar/m
Load Cycle	2.0Uo 8.0hr	27kV / Current : 1,760 Apeak
	2.0Uo 16hr	27kV / No Load : Only Voltage
Impulse Voltage	Positive	10 times : + 150 kV
	Negative	10 times : - 150 kV
AC Voltage	2.5Uo	33 kV
Thermal Cycle	2 times	Repeated Cool-down and Warm-up

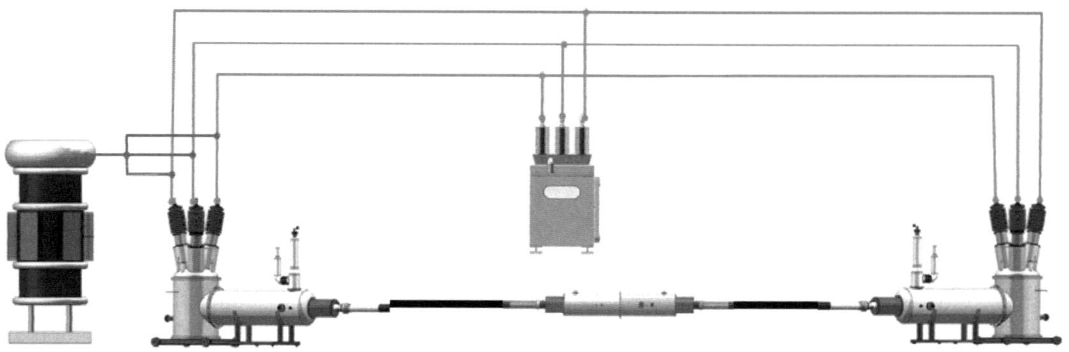

Figure 8. Schematic Sketch for Type Tess.

4. Type Test and Prequalification Quality Test of 23kV 50MVA HTS Cable

The preliminary tests and FEM simulation were investigated to verify design of the HTS cable. LS C&S was make three kinds of short cores according to critical current at 77.3 K. The experimental test of Measurement value of AC loss was under 0.6 W/m at rated current. Also, no degradation of critical current by bending test at diameter of 2,500 mm was found. The increased maximum temperature during the fault condition was approximately 18 K which means that pressure can suppres LN2 to avoid bubbling at transient current condition. Based on above experimental and design results, LS C&S plan to make AC 23kV 50MVA class HTS cable and return pipe for circulation of LN2. The return pipe has been already manufactured, and the manufacturing HTS cable will be finished until second quarter of next year. The type test and PQ test will be finished until third quarter of next year.

References

[1] S. Akita, "Superconducting technologies in electric power application", 2004 IDW (International DAPAS Workshop)

[2] J. Cho , J. Bae and H. Kim "Development and testing of 30 m HTS power transmission cable", IEEE Trans. Appl. Supercon., vol. 15, no. 2, pp.1719 -1722 2005

[3] N. Amemiya, M. Nakahata, "Numerical study on AC loss characteristics of superconducting power transmission cables comparing coated conductors with magnetic substrate," Physica C, vol. 463-465, pp 775-780, 2007

[4] KS C 6111-6, Measurement method for the DC critical current of superconducting power cables, 2009

[5] Cigre TB538 " Testing of Superconducting Cable Systems", Sept. 2013

[6] Jae-Ho Kim, Chul Han Kim, Gopal Iyyani, Jozef Kvitkovic, and Sastry Pamidi, "Transport AC Loss Measurements in Superconducting Coils",IEEE Trans, Appl. Supercon. Vol. 21, no. 3, pp.3269-3272 2011

IOP Conf. Series: Journal of Physics: Conf. Series **1054** (2018) 012074 doi:10.1088/1742-6596/1054/1/012074

Installation Design of 23kV 50MVA class HTS Cable in South Korea

Young-jin Won[1], Dong Cheol Koo[1], Joong-gu Jeon[1], Jae Hwan Song[1], Yong Hee Won[2], Young Woong Kim[2], and Jin Bae Na[2]

1. Korea Electric Power Corporation, 140, Teogyero, Jung-gu, Seoul 04629, Republic of Korea
2. LS Cable&System, LS Tower, 127, LS-ro, Dongan-gu, Anyang-si, Gyeonggi-do 14119, Republic of Korea

Abstract. As an initial step of this innovation, in September of 2016, KEPCO (Korea Power Electric Corporation) has launched the world first commercial superconducting cable project, named **SSS** (**S**uperconducting **S**mart platform **S**tation in South Korea) project, which will be not only the first commercial trial in the grid but also a test bed for Smart Superconducting Platform in South Korea. The main target design of this project is installation of 23kV 50MVA class HTS cable a system in the power grid. Type of this HTS cable is 3 phases in One Cryostat which is a remarkable way to transfer large power with low voltage and no electric magnetic field that means prevention of heating of other cores of phases and two metal sheaths of the cable. Total length of HTS cable between ShinGal and HeungDuk substations is slightly over 1km and it has 2 sets of normal joint box and 2 sets of termination. In this HTS cable system, a mechanical stress during the cool-down and warm-up is a major issue for increasing stability to set-up the HTS cable system and to connect existing power lines. To solve this issue of SSS project, we developed a simulation program which used Equivalent Solving Method for reducing and predicting thermal contraction force of the HTS cable under the mechanical stress. Also, mechanical strengths of metal sheaths, that made of aluminium material and the former that was stranded of copper wires was measured by the tensile tests. Coefficient of elasticity and thermal expansion according to various temperatures was calculated by the experimental data. And then, the analytical method was applied to verify installation method by approximate 40m length HTS cable that has various installation conditions such as vertical shape, often called snake method, at straight route in installation path. From the above procedures, we find that mechanical strength of each component of the HTS cable can calculate according to installation route in the power grid.

1. Introduction

Superconducting applications such as a HTS cable, an HTS motor and SFCL for power grid is state-of-the-art technology [1]. Especially, HTS cables can transfer large electrical power source with low losses and are expected to be not only environmental applications but also electromagnetic interference free performance which means no influence adjusting other cables. It has a compact size compared with conventional power apparatus. However, these apparatuses are also considering that are being well-fit and connection of existing applications in the power grid. Thus, various R&D projects were performed and demonstrations were being promoted in the world.

Figure 1 shows the case study for the replacement of conventional cable and the installation method of an additional HTS cable in a tunnel in the grid. This way can make more stable power grid and save cost for transferring power source than existing only installation conventional cables because one circuit of HTS cables can flow over 5 times power higher than one circuit of a high voltage conventional cable and utilize empty duct line in tunnel. Thus, HTS cable is value-creating product and powerful solution to be able to change existing power apparatus.

Since 2001, Korea Power Electric Cooperation (KEPCO), LS Cable&System (LS C&S) and various institutes have been cooperated national R&D program that was Development of Advanced Power system by Applied Superconductivity technologies (DAPAS) funded by the Ministry of Education, Science and Technology in South Korea. During the 10 years of the DAPAS program, a basic design concept of distribution and transmission level HTS cable system was established. The HTS cables were manufactured for the short length cable tests and the type test to verify our design factor for applying power gird. The HTS cable in the distribution grid of 23kV class line voltage in South Korea can transfer the bulk power capacity like conventional cable in transmission level which means the HTS cable can replace the existing conventional transmission cable and save cost of civil work to install extra cable in the power grid. Thus, KEPCO and LS C&S installed 23kV 50MVA class HTS cable in the Icheon substation and operated for 20 months. Total length and configuration of this HTS cable are approximately 500m with 2 sets of outdoor termination, 1 set of normal joint-box and the 3 Phase-in-One cryostat type, respectively. During 20 months, unmanned operation in the power grid was for 18 months. The next step of this project is increasing power transmission capacity as 23kV 120MVA. In case of this HTS cable, the installation route is approximately 100 meters. The cable includes 2 sets of out-door termination and type of cooling system is decompression unit which is compact type by putting all components in the container. These demonstration schedules and projects are called Green Superconducting Electric Power Network at the Icheon Substation (GENI). Based on experience and results from DAPAS and GENI program, KEPCO made the world first commercial HTS cable project for Superconducting Smart Platform in South Korea. Construction of the cooling system including a turbo brayton type croyo-cooler and cryogenic pipe was already started to install in HeungDuk substation. Power capacity of this project is 23kV 50MVA class and HTS wires are 2nd generation type. 3 Phase-in-One type of HTS cable of 1,035 meters will be manufactured to transfer electrical power between ShinGal and HeungDuk substations.

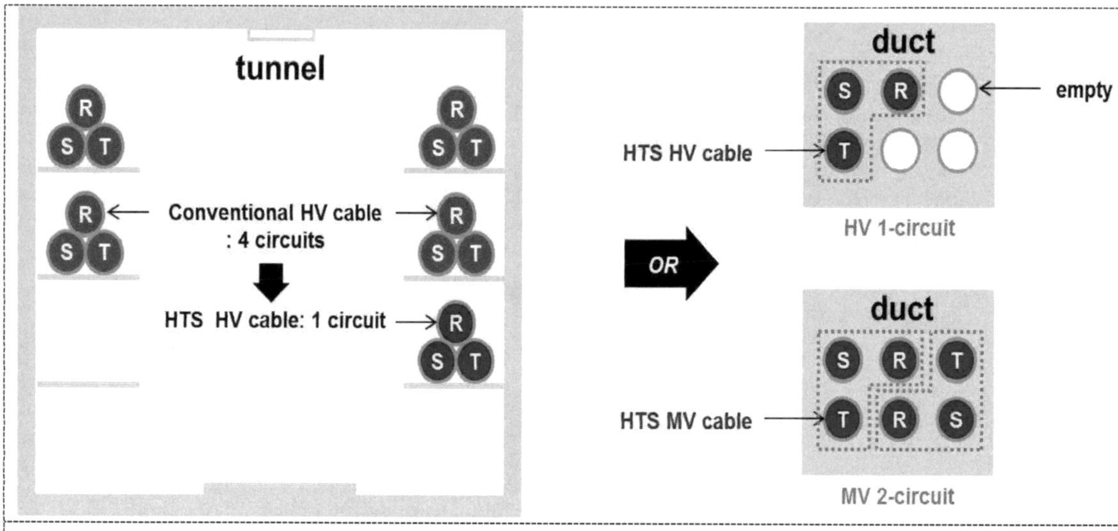

Figure 1. HTS cable in the tunnel.

2. Preliminary Test

Firstly, to gain the mechanical properties of aluminium metal sheath and the copper stranding former, tensile tests were performed at room-temperature (RT) and 77.3 K cryogenic condition. These experimental results should be applied basic input data for investigating FEM simulation program that estimated thermal contraction forces during cool-down and warm-up. This is first step to be increasing stability and accuracy of the simulation program. Finally, approximately 50 meter cable was installed to confirm simulation program.

2.1. Tensile Test for Metal Sheath

The mechanical property of an HTS cable can be changed cross-section of the area of each component such as former, 1st and 2nd metal sheaths. However, thermal contraction force of the 2nd metal sheath is very small because temperature difference between inside and outside is almost zero. The measured tension tests of values of former and the 1st metal sheath in the straight line are 2.01 tons and 1.58 tons, respectively. The length of the testing sample of the former is 1,420 mm at RT. The ratio of the shrinking length of former by cool-down process until 77.3 K is 0.29% compared the original length of the former. The saturated time until 77.3 K of 1st metal sheath is only 15 minutes. Fig. 2 shows the experimental configuration for the tension to gain the mechanical properties of the former and metal sheath at RT and 77.3 K. These experimental results were applied to adjust input data of FEM simulation for increasing accuracy.

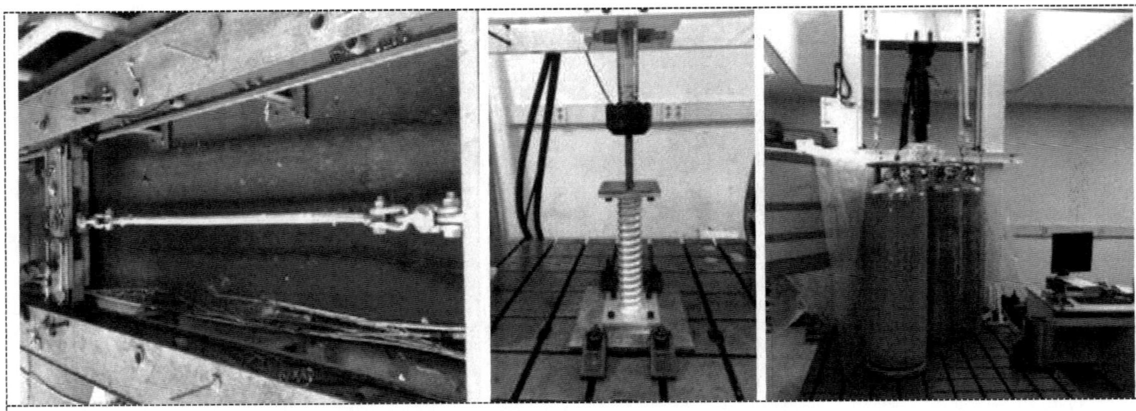

(a) Tensile Test of the Former at RT (b) Tensile Test of the Sheath (c) Test Equipment

Figure 2. Tensile Tests at RT and 77.3K.

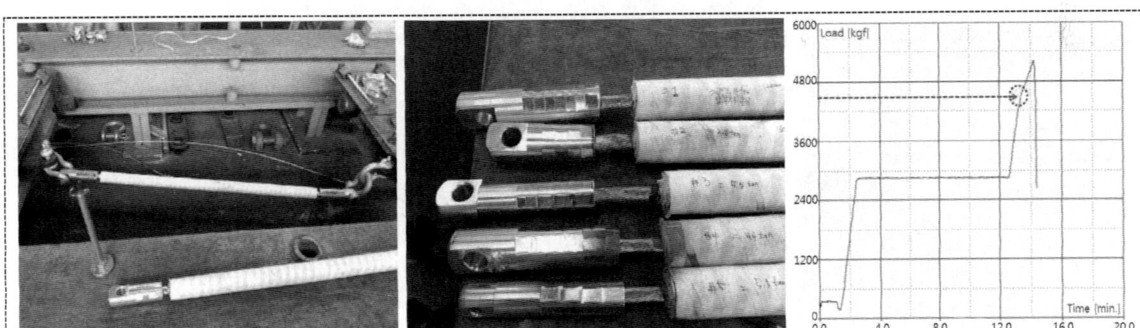

(a) Test Set-up (b) After Tensile Test (c) The Test Result

Figure 3. Tensile Test for Pulling Eye at RT.

2.2. Tensile Test for Pulling Eye

To connect the HTS cable to the winch, pulling eye is needed. Configuration of the pulling eye consists of hook for connecting winch and pressing area for connecting the copper former. In general, installation tension is approximately less than 2.0 tons. Thus, tension strength of the pulling eye should be performed according to the inner diameter of various hooks and the length of pressing area. Fig. 3

shows experimental set-up and the results of the after tensile tests of the pulling eye with the hook of 30.0 mm inner diameter and the pressing length of 25.5 mm, respectively. Before the tensile test as Fig. 3(a), the pulling eye was so close the core. As Fig. 3(b) shown, the former, made of the round shape of the copper strand, was stretched by the force of the longitudinal direction. The yield strength of the experimental case is approximately 4.5 tons and breaking strength is approximately 5.2 tons. Furthermore, the 1st and 2nd metal sheath was welded to pulling eye after manufacturing the HTS cable. In general, after the welding process, the yield strength will be increased more than only pressing with the former.

2.3. Electromagnetic Force during the Transient Status

Electromagnetic force will be generated in this 1st metal sheath during the fault condition because the HTS wires at the shield layer are experienced to quench status. The pressure condition of the HTS cable is 8.5 bars and the electromagnetic force of vertical direction is 18.51 kg/cm, respectively. Table 1 shows the specifications of the aluminium metal sheath for the FEM simulation. Based on the above status, possible thickness of aluminium metal sheath should be calculated by FEM simulation to confirm the structural feasibility during the fault. The input values, which were gained by the tension tests, for simulation of aluminium sheath strength are modulus of elasticity, yield stress and tensile stress. Each value at RT is 70 GPa, 43 MPa and 72 MPa, respectively. Generally, these properties are increasing at cryogenic condition. Thus, the measurement at RT is the worst condition. Based on FEM simulation of effective stress according to various thicknesses of metal sheath, suitable aluminium thickness of 1st metal sheath is 3.0 mm then safety margin against for electromagnetic force is over 1.2 times during the fault.

Table 1. The Specifications of the 1st Aluminium Metal Sheath.

Item	Value
Inner Diameter	83.3 mm
Wave Height	4.4 mm
Pitch	28.0 mm
Average Thickness	3.0 mm
Outer Diameter	98.1 mm

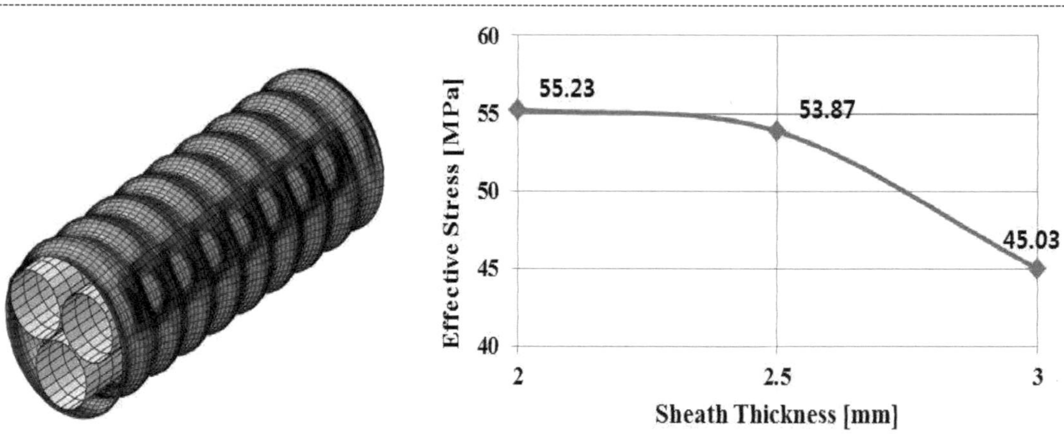

(a) Modelling for FEM simulation (b) The FEM simulation result

Figure 4. Effective Stress according to Aluminium Sheath Thickness.

3. Installation Design for 23kV 50MVA HTS Cable

The total length of an installation route is approximately 1,035 meter, including 2 sets of the normal joint box, 2 sets of outdoor termination and the return pipe for the LN_2 circulation are prepared to install the HTS cable system into the power grid. Table 2 shows the installation route of 23kV 50MVA class HTS cable and return pipe. During performing the installation process, there are five bending areas including one pass-manhole (P/M) as an off-set area and two manholes for normal join box (N/J). Diameter of each bending point is less than 11.0 meters, which means that the HTS cable isn't receive any damage form the installation. For installing the 1,040 meter HTS cable and return pipe, the steel drum would be set on the under roller (U/R) at the ShinGal substation and N/J #1 or N/J #2. Then it is pulled into the duct and tunnel from the ground section by a winch and scaffold. Figure 5 shows that installation route of 23kV 50MVA class HTS cable between HeungDuk and ShinGal substations. Initial R means bending area where the HTS cable would be moving to the center point of the bending area owing to the thermal contraction force. The pulling eye for connecting the winch should be the pressed of copper former and welding the 1st and the 2nd metal sheaths, respectively.

Generally, installation method was divided to 3 cases by considering route condition. Table 3 shows case studies of installation method. The case #1 is a fixed termination type which is the best way to connect the existing power cable in the power grid. However, the whole thermal contraction force should be absorbed by the HTS cable, the joint box, and the cleat. The case #2 is a free sliding method which means the termination is freely moving in the longitudinal direction of the thermal contraction force. There is limitation to move the termination at the substation and also check the moving effect of reducing the fore. The case #3 is an applicable method by considering route condition. For applying the snake method, the substation has enough to straight line to set-up steel frame for making wave shape of the HTS cable. As increasing the cleat stroke, the thermal contraction force could be absorbed by the cleat. It has limitation to absorb whole force. Thus, the case #3 also considers sliding the termination but it constrains the moving length of the termination.

Table 2. Installation Route of 23kV 50MVA HTS Cable and Return Pipe.

Item	Value
Installation Route	1,035 m
Normal Joint Box (NJ)	2 sets
Termination	2 sets
Manhole / Pass-Manhole (P/M)	two / one
Cable Type	3 Phase-in-One Cryostat
Return Pipe for LN_2	Yes

Table 3. Case Studies of Installation Method.

Case Study		Contents
1	Fixed Termination	Severe Condition
2	Free slide of Termination	Checking Sliding Effect at Termination
3	Free Slide + Up to Cleat Stroke Up	Reduce Reaction Force by the Thermal Contraction

3.1. Tensile Test for Pulling Eye

A metal structure should be considered to withstand the HTS cable weight, including LN_2 and the thermal contraction force during the cool-down and the warm-up. The angle support is considered to static force in normal operation and transient force during the fault. The hanger structure is also considered to weight per unit area against the weight of HTS cable. Fig. 6 shows the metal structure in the tunnel, which consists of the angle support, hanger and bolt.

3.2. Installation Design for HTS Cable

For the installation 1,035 meter HTS cable and return pipe for LN_2 circulation, total number of the steel drum is three for cable and two for return pipe. To follow local regulation, maximum diameter of HTS cable is less than 153.0 mm because the inner diameter of corrugated hard polyethylene (ELP) pipe is 200.0 mm. The main role of the ELP pipe is protecting cable in duct and the power cable is installed in the ELP pipe. After the installation, DC or AC voltage test shall be performed to check damage of the overs-heath of the HTS cable during the installation. Advanced verifications for stable installation are following below:

Table 3 shows specifications and check points of the bobbin test and mini cable test to keep the stability of the HTS cable before the installation. The rod type metal material can be applied the continuity test to check duct status then the mini cable is passing the available duct. The mini cable has similar or same diameter compared with actual the HTS cable to install in the duct.

- Advanced verifications for the installation of the HTS cable
  1) The Bobbin Test and The Mini Cable (Piece) Test
     : For checking duct condition by the bobbin
     : The feasibility test for installation with same diameter of the HTS cable piece
  2) Communication with Main Operator
     : The main operator control the whole system by control box and order to drive the winch
  3) Pre-Installed cleat and metal structure at the ground and the tunnel
  4) Checking the tension by a tension meter during the installation

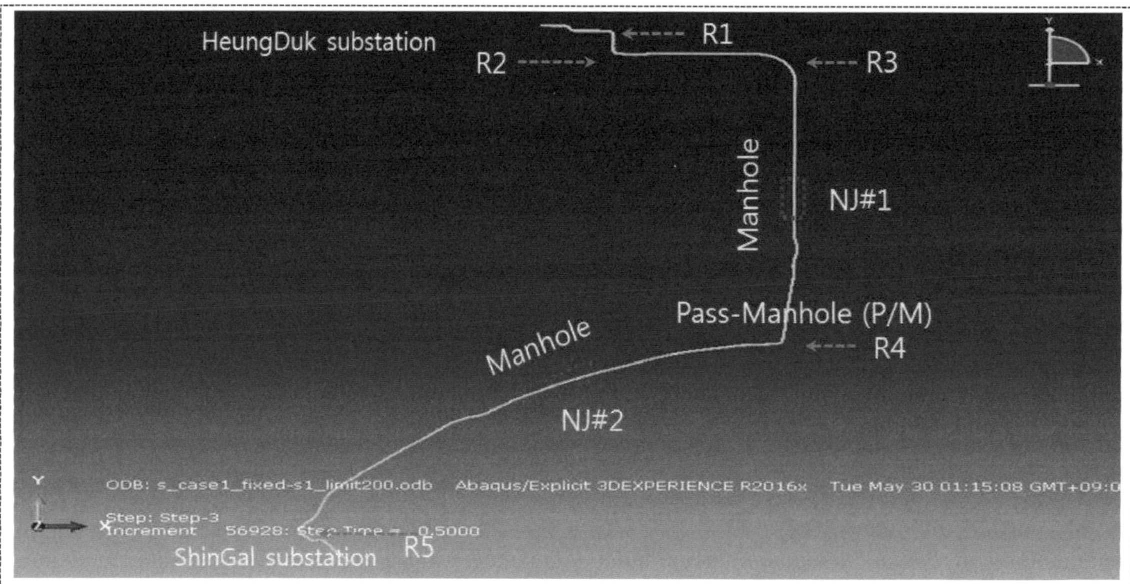

Figure 5. Installation Path of the HTS Cable and the Return Pipe for the circulation of LN_2.

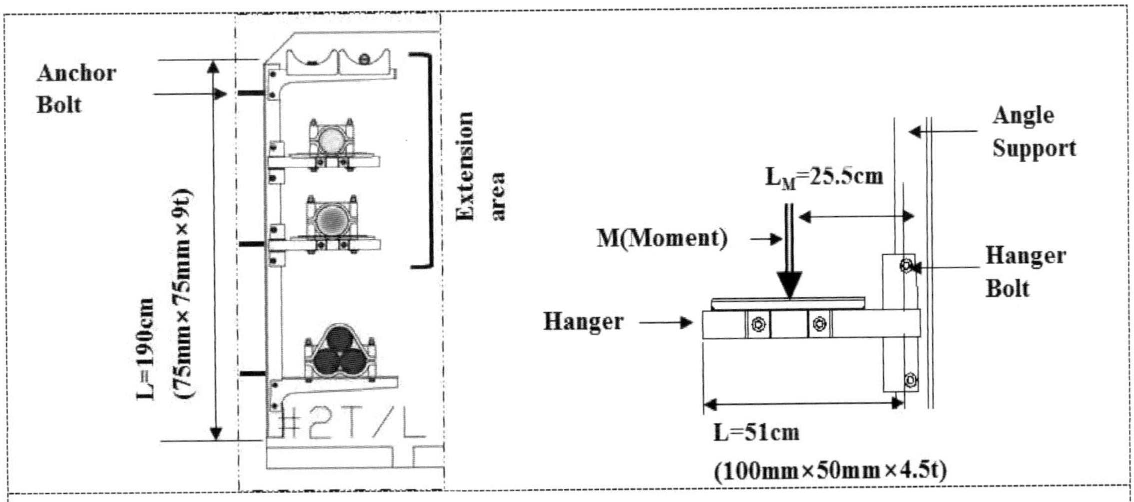

(a) The Angle Support (b) The Hanger and The Bolt

Figure 6. The Metal Structure in Tunnel.

- Case Studies for Installation
  1) Case #1, Fixed Termination
     : Increasing thermal contraction force to 450 kgf until R3 bending point
     : However, average force approximate 180 kgf/m^2
  2) Case #2, Sliding Termination
     : Movement length of the termination is less than 150 mm
     : Increasing the force of sliding cleat at bending point of R2 and R3
  3) Case #3, Sliding Termination and Increasing Cleat Stroke
     : In the case of the free sliding at P/M, thermal contraction force is less than 300kgf
     : The termination movement is less than 300 mm, the contact force approximate 500kgf/m^2

Item	Value
Short Length and Dia. of The Bobbin	600.0 mm and ø164.0 mm
Specifications	No crush and dent
Check Point	Every duct
Short Length and Dia. of Mini Cable	6.0 m and ø153.0 mm
Specifications	No twist and damage
Check Point	Pass the bobbin test

Based on the case studies from the FEM simulation, the suitable installation method can be chosen for reducing the thermal contraction force during the cool-down and the warm-up. The case #3 with the snake method in straight section near the each substation area, increasing the cleat stroke, and the limitation moving the length of the termination is applicable way to be stable and reducing the thermal contraction force of the HTS cable.

4. Discussion and Summary

For increasing stability and safety of the HTS cable during the installation, various preliminary tests were conducted to gain the mechanical strength of each component of HTS cable such as the copper former, the 1st and the 2nd metal sheath. From the experimental results, the FEM simulation program made to calculate the electromagnetic force according to various thickness of the aluminium metal sheath and choose the installation method between HeungDuk and ShinGal substations. The snake installation method is best way to absorb thermal contraction force during the cool-down and the warm-up. However, the straight section near the both substations is not enough to reduce the thermal force. In this case, the sliding termination and increasing cleat stroke can be applied the method by basing on the FEM simulation results.

References

[1] Ryu Kang-Sik, Jo Young-Sik and Park Minwon 2006 Overiew of the development of the advanced power system by the applied superconductivity technologies program in Korea *Supercond. Sci. Technol.* **19** (3) S102–S108

IOP Conf. Series: Journal of Physics: Conf. Series **1054** (2018) 012075 doi:10.1088/1742-6596/1054/1/012075

Development of Tri-axial superconducting cable system for type test

Tasuku Kitamura[1], Kazuhisa Adachi[1], Hideo Sugane[1], Tatsuhisa Nakanishi[1], Yuji Aoki[1], Nobuhiro Mido[1], Masataka Iwakuma[2] and Takayo Hasegawa[1]

[1] SWCC SHOWA CABLE SYSTEMS CO., LTD, No. 4-1-1, Minami-hashimoto, Chuo-ku, Sagamihara-city, Kanagawa, 252-0253 JAPAN

[2] KYUSYU UNIVERSITY, 744, Motooka, Nishi-ku, Fukuoka-city, Fukuoka, 819-0395, JAPAN

e-mail: t.kitamura311@cs.swcc.co.jp

Abstract. We developed a tri-axial superconducting cable system with the rated voltage of 22kV and operating current of 3 kA at liquid nitrogen temperature. We designed the superconducting cable and the termination that was cooled subcooled liquid nitrogen. We used in-house YBCO tapes and a winding machine to prepare the conductor that the I_c -value of 4,200 A or more per phase at liquid nitrogen temperature. The outer diameters of the conductor and the cryostat were approximately 50 mm and 130 mm, respectively. The length, the diameter and weight of each termination were approximately 4,000 mm, 450 mm and 600 kg. The thickness of the insulation layer, which was made from made from PPL paper, was determined by estimation of minimum break down electrical field from Weibull distribution analysis of results, which were obtained by break down tests of the model cables. Regarding the cooling condition of the cooling system, the liquid nitrogen temperatures of the inlet and the outlet of cable system was designed to be 65 K and 70 K, respectively. We manufactured a 25 m tri-axial superconducting cable, and cut out it to get 2 m samples for the I_c test and voltage test at liquid nitrogen temperature. The critical current of each phase was higher than 4,500 A. There was no break-down of each insulation phase at 52 kV for 30 minutes the voltage test. We also manufactured a pair of termination connection for system test and no break-down occurred each insulation phase at 52 kV for 30 minutes. We performed type test of cable system including the superconducting cable and the termination, based on "Recommendations for Testing of Superconducting cables (CIGRE Technical Brochures 538)" [1].

1. Introduction

The superconducting cables have been developed in Europe, USA, and Asia. The AC loss and heat invasion loss are a very important issue to be solved in superconducting cables. Tri-axial superconducting cables are superior in AC loss and the heat invasion properties [2]. A 11 kV-3 kA tri-axial super-conducting cable was developed and tested in Ohio state in USA [3]. A 13.2 kV-3 kA one was also tested at Essen in Germany [4]. In this paper, 22 kV- 3kA tri-axial superconducting cable system is developed as the busber between a generator and a step-up transformer.

2.Development of Tri-axial superconducting cable system for type test

2.1 Construction of Tri-axial Superconducting Cable

YBCO tape is manufactured by Trifluoroacetic acid - Metal Organic Deposition (TFA-MOD) method. The YBCO tape has 0.2 mm thick and 4 mm in width.3). The construction of the tri-axial superconducting cable is shown in Fig.1. The cable has inner cold pipe made of stainless steel in center of the cable section. And it has 3 phase layer superconducting layer, insulation layer and outer of inner cold pipe. The AC loss of the cable is little because the cable is axial construction. The returnable pipe is not needed because the cable has inner and outer cold layer, and be able to apply GO/RETURN cooling.

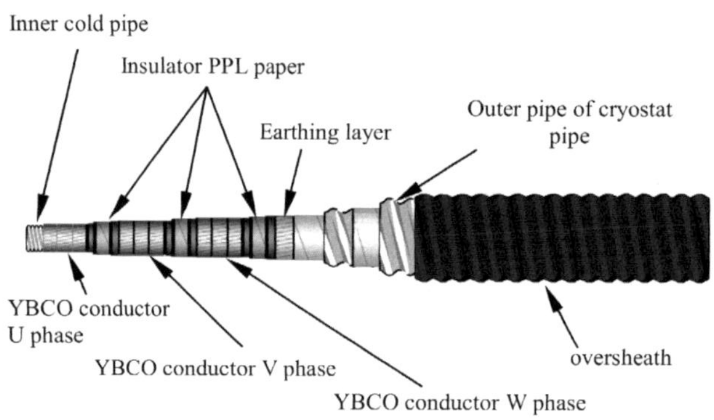

Fig. 1 Schematic illustration of the superconducting cable prototype

TABLE. 1 CONSTRUCTION OF TRI-AXIAL SUPERCONDUCTING CABLE PROTOTYPE

ITEM	DIAMETER (mm)
Inner cooling SUS corrugated tube	24
U phase / YBCO superconducting tapes	25
U phase / PPL paper insulation	32
V phase / YBCO superconducting tapes	34
V phase / PPL paper insulation	41
W phase / YBCO superconducting tapes	42
W phase / PPL paper insulation	49
Earthling phase Cu tapes	52
Inner aluminium corrugated tube	83
Multi layered insulation	85
Outer aluminium corrugated tube	121
Oversheath	131

2.2 Fundamental test of cable insulation material

Cable insulation is used with PPL paper (Polypropylene laminated paper). At first, we manufactured model a cable which has insulation 1.2mm thickness. And we examined AC voltage breakdown test and lighting impulse breakdown test, calculated E_L with Weibull plot. The Weibull plots are shown in Fig. 2. Fig. 2 shows that the minimum break down voltage of AC breakdown is 36.9 mm/kV and the minimum breakdown voltage of lightning impulse is 87.4 kV/mm. Using these results we designed that the insulation thickness of each phase was 2.0mm.

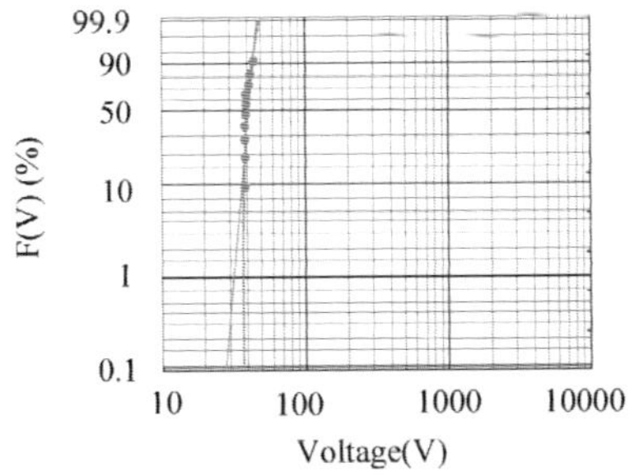

(a) AC breakdown test
E_L=36.9V/mm

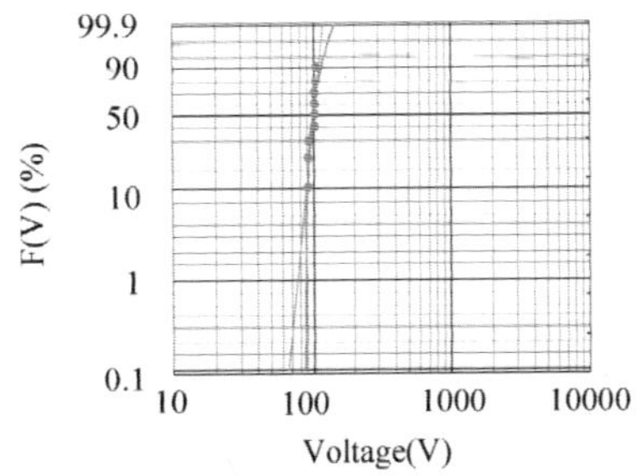

(b)lightning impulse breakdown test
E_L =87.4kV/mm

Fig.2 Weibull Plot data of PPL paper in LN2

2.3 Short circuit test for superconducting cable

The completed tri-axial superconducting cable has no normal conductor for protection against the excessive fault excess current. Here, a new protection method was studied out. A XLPE by-pass cable was connected to the tri-axial superconducting cable in parallel for each phase. The cross-sectional area of the XLPE cupper cable was 250 mm^2. The three phases of the XLPE cable were located in a room temperature space. The parallel connection of the XLPE copper cable with an insufficient current capacity against the excessive fault current which amounts to several tens of kg produces a tremendous function as the after-mentioned. It was predicted by numerical simulation in advance to the short circuit test. The block diagram of the test circuit of the short circuit test for the tri-axial superconducting cable is shown in Fig. 3. The arrangement of the tri-axial cable connected with the XLPE copper cables. To observe the transport current in each current path, CTs were arranged both on the superconducting cable and the XLPE copper cable.

The short circuit test was carried out by using a 200 MVA short circuit generator. One side of the terminals of the tri-axial cable was connected to the generator for each phase and the other side of the terminals was shorted with each other. The observed voltage and the current waveforms of W-phase are shown in Fig. 4. We can see that the total short circuit current decayed slowly during the short circuit. The current of superconducting cable was rapidly reduced because the current of HTS cable transferred to the XLPE cable. We can see that the current waveform of the superconducting cable is non-linear against the applied voltage. It suggests that the superconducting cable did not shift to a normal state and the temperature of every phase of superconducting cable did not exceed the critical temperature, T_c of YBCO. However, since the current exceeded critical current(B, T) and the flux-flow resistance was generated the transport current in the superconducting cable was decreased. We also conducted the short circuit test on other phases. We obtained the results of V and U phase. We also find similar tendency of V and U phase, compared to the results of W phase.

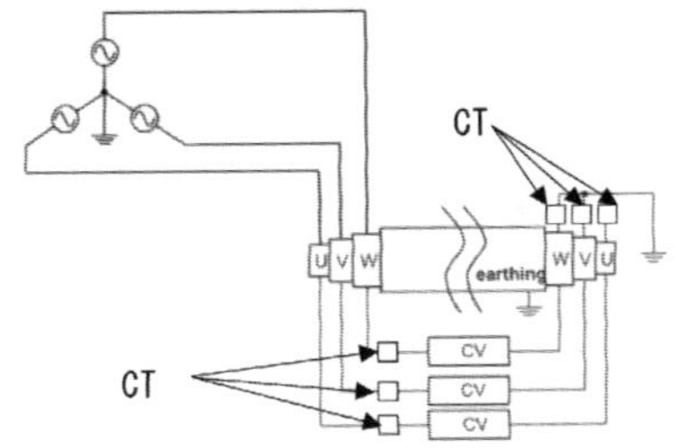

Fig.3 Block diagram of the circuit of the short circuit test of the tri-axial superconducting cable

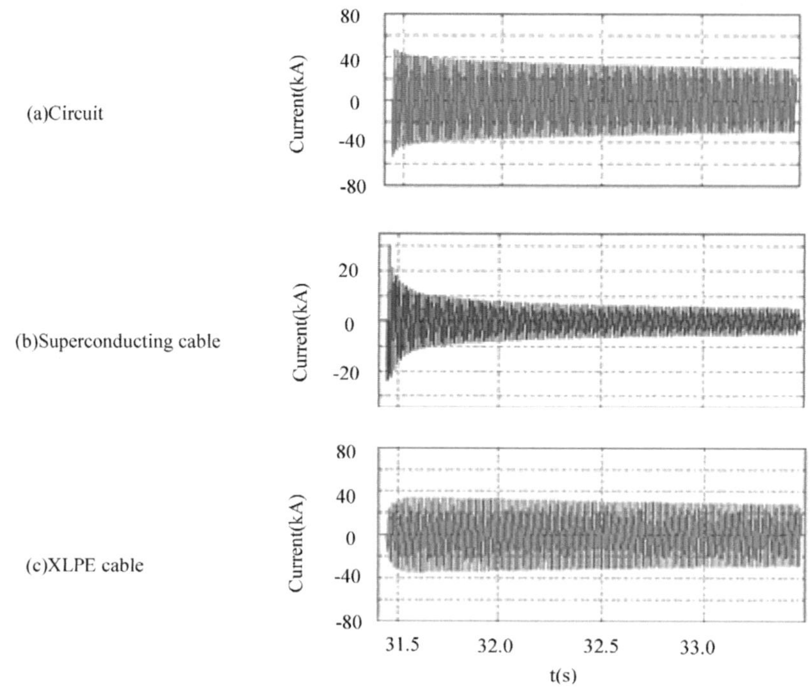

Fig.4 Three phase short circuit current waveform (W-Phase)

2.4 Construction of Tri-axial Superconducting cable Termination

We designed superconducting bus termination is designed. Each conducting layers are connected to outer termination with current lead. Each insulation layers are electric field control with stress relief cone by paper. Fig.5 shows the view of the performance test of the model sample of bus termination.

Fig. 5 A view of termintation model sample insulation test

IOP Conf. Series: Journal of Physics: Conf. Series **1054** (2018) 012075 doi:10.1088/1742-6596/1054/1/012075

3.Type test of Tri-axial superconducting cable system

We performed type test of cable system including superconducting cable, termination connection and cooling system based on "Recommendations for Testing of Superconducting cables (CIGRE Technical Brochures 538)"

3.1 Cable bending test and electric characteristic test of superconducting cable

The manufactured cable was tested by the cable bending test described in "3.0 Type Tests on Cable Systems" of CIGLE TB 538. We conducted the bending test the 20m length cable. We bent the cable around a mandrel with specified outer diameter. The minimum diameter of the mandrel is 4,675 mm from the test condition of TB538. We used a cable drum with 3,200 mm outer diameter as the mandrel. Our test condition was severe, compare to the specified condition CIGLE TB 538. We bent the cable around cable drum using the winding machine in our factory. At first, the 20 m cable was wound around the cable drum. Then the cable was drawn out from the drum. And the cable was rotated 180 degrees. Then the cable was winded around the drum again. After that, the cable was draw out from the drum. This process was a one cycle of winding. We repeated this process three times. Fig. 6 shows the view of the bending test at our factory.

Fig. 6 A view of the bending test

Based on "3.0 Type Tests on Cable" in CIGLE TB 538, we measured DC critical current of the cable after the bending test. We cut the 2 m length conductor from the 20 m length. We attach the 6 current terminals to the edge of each conductor. Fig. 7 shows a schematic draw of the critical current measurement of the W-Phase conductor. We measured the DC critical at the liquid nitrogen temperature. We used a DC current source and a DC voltage meter. We also recorded the temperature of liquid nitrogen in the measuring container. The criteria of the critical current is 1 μ V/cm.

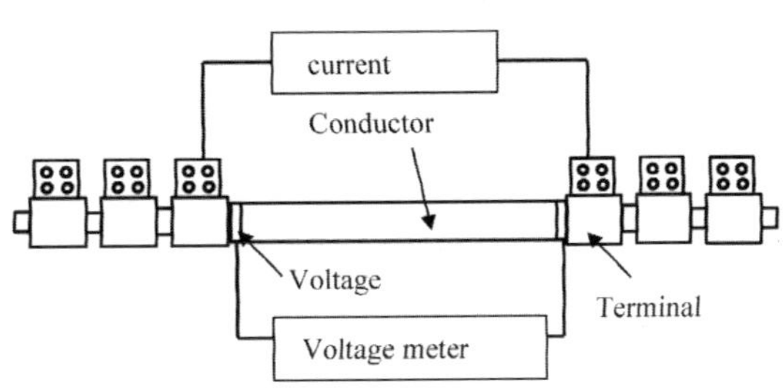

Fig. 7 A schematic of DC critical current measurement

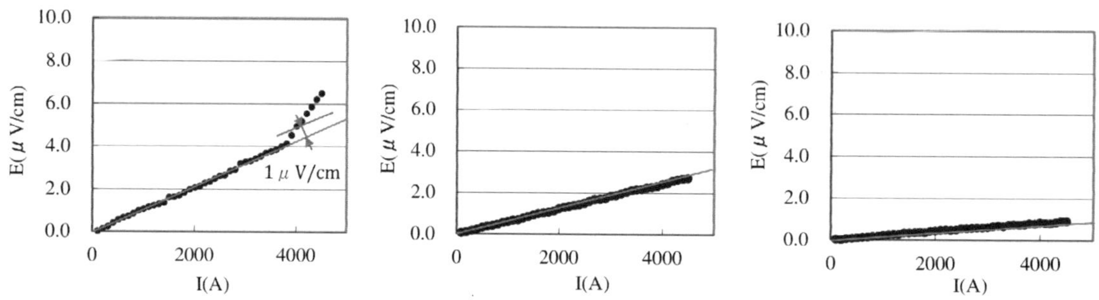

(a)U-phase I-E curve (b)V-phase I-E curve (c)W-phase I-E curve

Fig. 8 I-E curve of U,V,W-phase conductor

DC critical current of the U-phase is 3,800 A. I_c of V-phase is larger than 4,300 A. I_c of W-phase is also 4,300 A. The requirement of critical current should be more than 95% of 3,000 A. It is 2,700 A. Those measurement values are larger than the requirement. We confirmed the DC critical currents of the cable after the bending test larger than recommended value in TB 538.

3.2 Assembling of superconducting cable system for type test

By using the 18m superconducting cable after the bending test and two terminations, we constructed the cable system for type test. Fig. 9 shows a schematic draw of the cable circuit for the type test.

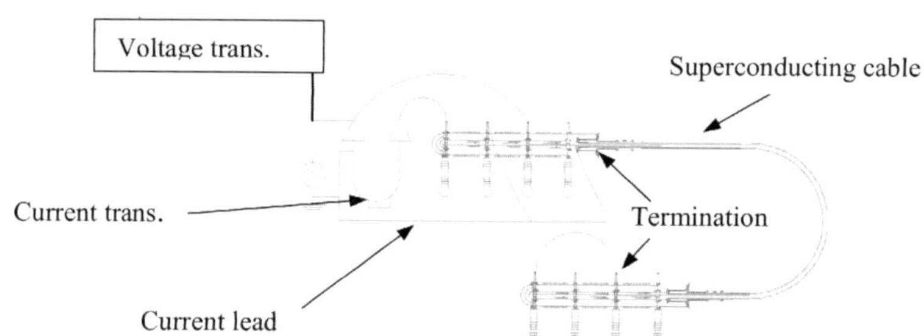

Fig. 9 A schematic draw of cable circuit for type test

The terminations were connected to the both ends of cable. The superconducting cable was bent by 3m radius. The voltage lead from the voltage transformer was connected to the U and V phase by lead cable terminals. The terminal of the U phase and the shield phase were connected together and connected to the ground. We also decided to connect lead cable with the cross sectional area of 5,000 mm^2 to the U and W phase. We also selected the cross section of lead cable with AC 3,000 A capability at room temperature. When we conduct heat cycle test, subcooled liquid nitrogen flowed and was circulated into the inside of the inner corrugate and terminals. The temperature range was 70-73 K. Fig. 10 shows a view of the type test cable system.

Fig. 10 A view of type test cable system.

After construction of the type test cable system, we started to introduce liquid nitrogen into the system. After a 72 hours cooling procedure, the superconducting cable and two terminations were cooled to the specified temperature. The temperature of inlet was 70 K, and the temperature of the outlet was 73 K with circulating subcooled liquid nitrogen. Then we moved next test procedure.

3.3 Type test of the cable system

After cooling process of superconducting cable system, we conducted type test.

3.3.1 Voltage loading test

We conducted Voltage loading test. TABLE 2 shows voltage loading test condition and results.

TABLE. 2 TEST CONDITION AND RESULTS OF VOLTAGE LOADING TEST

ITEM	CONTENT	RESULT
Voltage loading test	42kV loading, 3,000A current flow continuous 8hours on / 16 hours off 20times	Passed

We confirmed that the voltage loading test was passed.

4. Conclusion

We developed a tri-axial superconducting cable system for the type test with the rated voltage of 22 kV and the operating current of 3 kA at liquid nitrogen temperature. We designed superconducting cable, termination with using subcooled liquid nitrogen. We studied fundamental characteristics of superconducting cable and terminal. Using those test results, we defined the detail of the specification of the cable system by manufacturing 20 m superconducting cable and two terminals. We constructed the superconducting system for the type test. We achieved that the loading test was passed.

Acknowledgement

This paper is based on the results obtained from a project subsidized by the New Energy and Industrial Technology, Development Organization (NEDO).

IOP Conf. Series: Journal of Physics: Conf. Series **1054** (2018) 012075 doi:10.1088/1742-6596/1054/1/012075

References

[1] "Recommendations for Testing of Superconducting cables", CIGRE Technical Brochures 538, 2013

[2] Kazuhisa Adachi, Hideo Sugane, Tianlong Wang, Hiroki Ohnishi, Shigeki Sano, Kei Shiohara, Tasuku Kitamura, Nobuhiro Mido, Tsutomu Koizumi, Takayo Hasegawa, Masayuki Konno, and Masataka Iwakuma, "Development of 22 kV HTS Triaxial Superconducting Bus" 2017 IEEE TRANSACTIONS ON APPLIED SUPERCONDUCTIVITY, VOL. 27, NO. 4, JUNE

[3] Mark Stemmle, Frank Merschel, Mathias Noe, Achim Hobl, "AmpaCity－ Advanced superconducting medium voltage system for urban areapower supply" 2014 IEEE PES T&D Conference and Exposition 10.1109/TDC.2014.6863566

[4] J. A. Demko, I. Sauers, D. R. James, M. J. Gouge, D. Lindsay, M.Roden, J. Tolbert, D. Willen, C. Traeholt, C. T. Nielsen, "Triaxial HTS Cable for the AEP Bixby Project" IEEE Transactions on Applied Superconductivity, 2007, Volume: 17, Issue: 2,10.1109/TASC.2007.897842 IEEE,

Heat leak and pressure drop measurements of the 1000 m class superconducting DC power transmission system in Ishikari

Hirofumi Watanabe[1], Yury V Ivanov[1], Noriko Chikumoto[1], Satarou Yamaguchi[1], Kotaro Ishiyama[2], Zenji Oishi[2], Michihiko Watanabe[3] and Takato Masuda[3]

[1]Chubu University, Kasugai, Aichi 487-8501, Japan
[2]Chiyoda Corporation, Yokohama, Kanagawa, 220-8765, Japan
[3]Sumitomo Electric Industries, Ltd., Osaka, Osaka, 554-0024, Japan

E-mail: h_watanabe@isc.chubu.ac.jp

Abstract. The second cooling test of the 1000 m class superconducting DC power transmission system constructed in Ishikari, Japan, was performed in 2016. In this cooling test, the heat leaks to the low temperature parts and the pressure drops of the liquid nitrogen circulation were measured for each section of the 1000 m system. The heat leaks were measured at different outer pipe temperatures of the cryogenic pipe constituting the transmission line. The total heat leaks of the cable system, composed by the cryogenic pipe and the terminals, were 1.746 and 2.091 kW at the outer pipe temperatures of -2.4 and 17.4 °C, respectively, which were consistent with those measured in the first cooling test. The heat leak of the cryogenic pipe increased by 30-40 % by the increase of the outer pipe temperature from -2.4 and 17.4 °C. The pressure drop of the cable system was 42.80 kPa and that of all the system, including the cooling system, was 98.4 kPa at the flow rate of 36.03 L/min. The degradation of the heat leak and pressure drop characteristics were not observed between the first and second cooling tests. Based on the present results, the heat leak, the temperature rise and the pressure drop of 10 km systems were estimated. From this estimation it can be said that the cooling and the circulation of 10 km systems are possible by the present level of technology.

1. Introduction

500 m class and 1000 m class superconducting DC power transmission systems were constructed in Ishikari, Japan [1, 2, 3]. This project, we call, Ishikari project. The 500 m system connects a photovoltaic power plant to an internet data center and the transmission of electricity was successfully performed in 2015. The 1000 m system was constructed to obtain data for future longer transmission systems. In the winter of 2015, the first cooling test was performed and the tests about the cooling and the circulation, such as a heat leak measurement and a pressure drop measurement, were mainly performed. The results of these measurements were reported in the previous paper [4]. The second cooling test was performed from July to December in 2016. In this test, the heat leak and the pressure drop were measured again to test changes in these properties between these periods.

The heat leak of the cryogenic pipe constituting a transmission line affects the heat load to the cooling system, in particular, in long distance transmission lines. The heat leak data at the different outer pipe temperatures, that is, the different environmental conditions, are required to design superconducting transmission systems. In the first cooling test, the heat leak in each section of the 1000 m system was measured at the outer pipe temperature of the cryogenic pipe of -3.0 °C [4]. In the second cooling test the heat leak was measured at -2.4 °C together with at 17.4 °C to see the difference of the heat leak at different outer pipe temperatures.

For a section of the transmission line of the 1000 m system in Ishikari project, the cryogenic pipe with a radiation shield was adopted [5]. The heat leak to the inner pipe which surrounded by the radiation shield was measured to be 0.034±0.033 W/m in the first cooling test [4]. This was quite small, though the relative uncertainty of the measurement was nearly 100 % and the accuracy is not enough to design longer transmission lines over several tens of km. This result was estimated from the temperature rise of the circulating liquid nitrogen and the accuracy of the measurement was almost limited by the uncertainties of the temperature sensors due to the small temperature rise of the liquid nitrogen. To obtain higher accuracy, a low flow rate experiment was performed to increase the temperature rise along the flow channel.

The estimation of the pressure drop property of the liquid nitrogen circulation is quite important to design cooling and circulation systems in superconducting power transmission systems, because higher discharge pressure of the pump will be required as the increase of the pressure drop, which makes the capacity of the pump lager. In addition to this, because the output power of the pump can be a heat load to the cryocooler through the friction loss of the liquid nitrogen, that is, the pressure drop and the loss of the pump itself, it can be larger than the heat leak of the cryogenic pipe in the case of long transmission lines. Therefore, data about the pressure drop are required for the design of future longer transmission systems. From this point of view, the pressure drop of each section of the 1000 m system was measured again in the second cooling test.

2. 1000 m system in Ishikari project

The details of the 1000 m system in Ishikari project have already been explained elsewhere [3, 4]. Only the details connecting to the present paper will be explained. The cryogenic pipe constituting a transmission line is extended into outdoors in a U-shaped configuration from the buildings where terminals and a cooling system are installed. The length of the transmission line is approximately 1000 m. The cable has two joins, at which the cryogenic pipe is sectioned. The section 1 is a straight part with a length of 475 m, the section 2 has a curved part whose radius is 7 m with a length of 137 m and the section 3 is a straight part with a length of 379 m. Different types of cryogenic pipes are used for each section. In the section 1, a cryogenic pipe with a radiation shield is used, while in the section 2 and 3 a cryogenic pipe without a radiation shield is used. In both the designs, two inner pipes are installed in a single outer pipe for the circulation. Smooth pipes are used for the inner pipes to reduce a pressure drop. The inner pipe in which the cable core is installed is called a "cable pipe" and the other pipe to return back the liquid nitrogen for circulation is called a "return pipe". Therefore, the distance of the circulation is approximately 2000 m, though the length of the transmission line is approximately 1000 m. In the cryogenic pipe with a radiation shield, the cable pipe is surrounded by a radiation shield to shield the heat from the outer pipe at the room temperature. The radiation shield is thermally attached to the return pipe. Therefore, almost all the incident heat on the radiation shield will be removed by the liquid nitrogen flowed in the return pipe. It is expected that the heat leak to the cable pipe and the temperature rise along it will be significantly reduced, provided the cryogenic pipe with a radiation shield is used.

Figure 1 shows a piping and instrumentation diagram of the 1000 m system. In this figure the ranges of the cooling system and the cable system termed in this paper are shown. The 1000 m system equips two Turbo-Brayton cryocoolers (TB1 and TB2) with a nominal cooling power over 2 kW at 66 K and two Stirling cryocoolers (ST1 and ST2) with a nominal cooling power over 1 kW at 77 K. Centrifugal pumps (CP1 and CP3) and a positive displacement pump (CP2) are installed. CP1 and CP2 are in the cooling system and situated at the terminal A side and CP3 is situated at the terminal B side, which is 1000 m away from the terminal A. To measure the flow rate of liquid nitrogen, two

Coriolis type flowmeters are equipped. The reservoir has a volume of 2 m^2, which has a function to regulate the base pressure of the system. Temperature sensors (labelled with T) and pressure gauges (labelled with P) are equipped to monitor and to estimate the conditions of the 1000 m system. The flow channel of the liquid nitrogen can be switched by valve settings. In the second cooling test, ST1 and ST2 were not used and only one or two Turbo-Brayton cryocoolers were used depending on the types of experiments. During the heat leak and pressure drop experiments only CP1 was used, while CP2 and CP3 were cut off from the flow channel by the valve setting. On the other hand, during a low flow rate experiment, only CP2 was used.

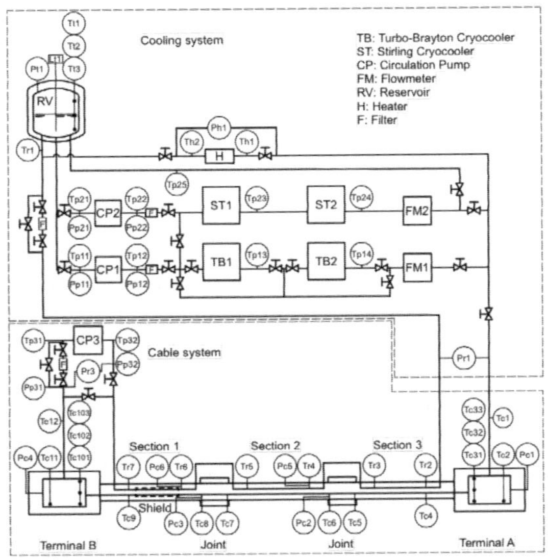

Figure 1. Piping and instrumentation diagram of the 1000 m system.

3. Heat leak measurements

Figure 2 shows an example of the liquid nitrogen temperatures during the heat leak measurement together with the flow rate of the liquid nitrogen and the outer pipe temperatures of the cryogenic pipe. The legend in the top figure is explained in figure 1. Tu1-Tu4 in the bottom figure were the outer pipe temperatures measured in the outdoors at a few meters away from the building where Terminal A is installed. The temperature of the liquid nitrogen rises as it flows from Tc1 to Tr1 for the heat leak of each section. Tc1 is just downstream of the cryocoolers and nearly constant due to the controlled outlet temperature of the Turbo-Bryton cryocoolers. The others vary depending on the flow rate and the outer pipe temperatures.

The heat leak was calculated with the temperature rise, the flow rate and the specific heat capacity of the liquid nitrogen. The temperature rise was estimated from the temperature difference in each section measured with respect to the flow rate [6]. The time required for the travel of the liquid nitrogen in each section was considered. The temperature at the outlet of a section was compared with that at the inlet measured at the time before required for the travel. The data which stayed for an enough long period of time in the range of the outer pipe temperature of -2.4±2.7 °C were used for the estimation in the case of the example in figure 2. The regions used for the estimation are shown in figure 2 with the red circles. The results of the heat leaks of the cryogenic pipe in this study are summarized in table 1 for the outer pipe temperatures of -2.4±2.7 and 17.4±2.6 °C. The temperature ranges of liquid nitrogen were 76.4±2.1 and 71.7±2.5 K for the outer pipe temperatures of -2.4±2.7 and 17.4±2.6 °C, respectively. The vacuum degree between the inner and outer pipes during these measurements was less than 0.005 Pa. The results obtained in the first cooling test at the outer pipe temperatures of -3.0±3.5 °C and at the liquid nitrogen temperature of 76.8±2.0 K [4] are summarized in table 2 for a reference. The results for the different sections of the 1000 m system are summarized in table 3. The numbers in the parentheses are the values of experimental uncertainties referred to the corresponding last digits of the results.

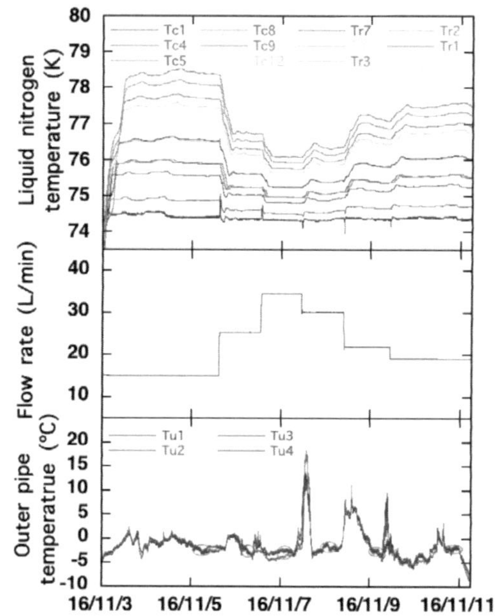

Figure 2. Liquid nitrogen temperatures during the heat leak measurement together with the flow rate and the outer pipe temperatures.

Table 1. Heat leaks per unit length of each section at different outer pipe temperatures. The numbers in the parentheses are the values of experimental uncertainties referred to the corresponding last digits of the results.

	Outer pipe -2.4±2.7 °C			Outer pipe 17.4±2.6 °C		
Section	Cable (W/m)	Return (W/m)	Total (W/m)	Cable (W/m)	Return (W/m)	Total (W/m)
1	0.033(35)	0.832(36)	0.865(50)	0.024(34)	1.204(38)	1.228(51)
2	0.92(13)	0.51(14)	1.42(19)	1.34(13)	0.63(17)	1.96(21)
3	0.792(45)	0.431(46)	1.222(65)	1.021(45)	0.535(63)	1.556(77)

Table 2. Heat leaks per unit length measured in the first cooling test [4]. The numbers in the parentheses are the values of experimental uncertainties referred to the corresponding last digits of the results.

	Outer pipe -3.0±3.5 °C		
Section	Cable (W/m)	Return (W/m)	Total (W/m)
1	0.034(33)	0.851(33)	0.886(47)
2	0.95(12)	0.52(13)	1.47(17)
3	0.818(42)	0.462(42)	1.280(59)

At the outer pipe temperature of -2.4 °C the heat leak is down to 0.865±0.050 W/m per two inner pipes in section 1. At this temperature the heat leak to the cable pipe was measured to be 0.033±0.035 W/m and quite small in comparison with the other sections. This is because almost all the heat leak to the cable pipe was received with the radiation shield and removed with the liquid nitrogen in the return pipe. Between the first and second cooling tests, the heat leaks were measured at the similar outer pipe temperature of -3.0 and -2.4 °C. The results of these measurements agree well with each other within the range of the uncertainties. In the second cooling test, heat leaks were measured at different outer pipe temperatures. As expected, the heat leak was increased in all the sections of the cryogenic pipe. The heat leak increases approximately 30-40 % by the increase of the outer pipe temperature of 20 °C.

The heat leaks of the terminals are approximately 230 W as shown in table 3. These values were obtained during no load condition, though they contain a portion of the conductive heat through current leads. Because the temperature in the buildings where the terminals were installed was air conditioned, the heat leaks of the terminals are almost independent on the outer pipe temperatures. The total heat leaks of the cable system not including the cooling system were from 1.690±0.023 to 2.091±0.022 kW depending on the range of the outer pipe temperatures. The results of the present study were consistent with those obtained in the first cooling test. The heat leak property was not changed between the two cooling tests.

Table 3. Heat leaks of each component of the 1000 m system. The results at -3.0 °C were obtained in the first cooling test [4]. The numbers in the parentheses are the values of experimental uncertainties referred to the corresponding last digits of the results.

Outer pipe (°C)	Terminal A (kW)	Terminal B (kW)	Cable pipe (kW)	Return pipe (kW)	Total (kW)
-3.0	0.199(17)	0.233(16)	0.461(16)	0.676(16)	1.690(23)
-2.4	0.225(18)	0.248(17)	0.459(17)	0.658(18)	1.746(20)
17.4	0.226(18)	0.249(17)	0.600(17)	0.901(20)	2.091(22)

4. Pressure drop measurements

Pressure drop was measured with pressure gauges and a differential pressure gauge installed along the flow channel shown in figure 1. The total pressure drop of the 1000 m system is the difference of the pressures as Pp12-Pp11. The pressure drop of the cable system was measured with a differential pressure gauge of Pr1. The pressure drops of the cable and the return pipes are Pc1-Pc4 and Pp32-(Pc1-Pr1), respectively. Since these gauges are installed at different heights, the obtained values were corrected by them.

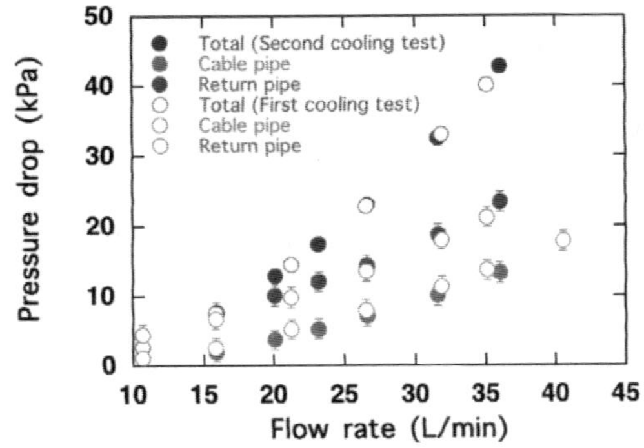

Figure 3. The pressure drop of the 1000 m system measured in the first [4] and second cooling test.

Figure 3 shows the results of the pressure drop measurements in the second cooling test together with those measured in the first cooling test [4]. The measurements were performed during the temperature measurement shown in figure 2. The average temperature measured at Tc1 was 74.3 K, while those at Tr1 were ranged from 76.1 to 78.4 K depending on the flow rate. Since the average temperature measured at Tc1 in the first cooling test was 74.9 K, the experiments were performed in the similar temperature ranges between the first and second cooling tests. The pressure drop increases as the increase of the flow rate almost proportional to the second power of the flow rate. The pressure drops of the cable pipe and the return pipe were 13.3 and 23.4 kPa, respectively, at the flow rate of 36.03 L/min. The pressure drop of the cable system itself was 42.80 kPa at this flow rate, which contained the pressure drop in the adjunctive piping and the terminals. For the additional pressure drop of 55.6 kPa by the cooling system, the total pressure drop of the 1000 m system was 98.4 kPa at 36.03 L/min. These results agree well with those obtained in the first cooling test as shown in the figure. The degradation of the pressure drop characteristics was not observed between the first and second cooling tests.

5. Low flow rate experiment

A low flow rate experiment was performed at the flow rate around 4 L/min with the 1000 m system to simulate a temperature variation of the liquid nitrogen at 40 L/min in 10 km systems. The result is shown in figure 4. The positive displacement pump (CP2) was used to reduce the flow rate and the actual flow rate during the measurement was 3.9 L/min. The spikes before 16/11/21, in particular, in Tr7 originated from the thermal oscillations near CP3 for forgetting to close a valve near it. After the valve was closed, the spikes disappeared. Two days after the flow rate was reduced to 3.9 L/min, the temperature variation reached a steady state. At this state the temperature variation varied with the outer pipe temperature. Stable circulation was confirmed even in this low flow rate condition at 3.9 L/min.

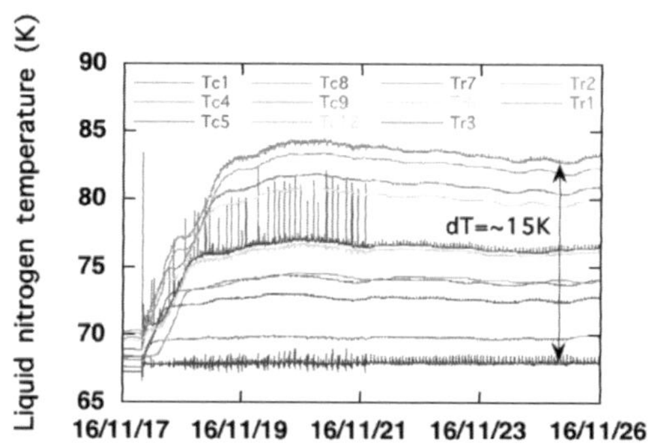

Figure 4. Temperature variations in the low flow rate experiment. The flow late was 3.9 L/min.

The heat leaks of the cable pipe surrounded with a radiation shield were 0.034±0.033, 0.033±0.035 and 0.024±0.034 W/m for the outer pipe temperatures of -3.0, -2.4 and 17.4 °C, respectively, as shown in table 1 and table 2. These values are in the range of uncertainties and, thus, consistent with each other. The relative uncertainties are around 100 % and the accuracies of the results are not enough for further estimations to design longer transmission lines. As already mentioned in Section 1, this is mainly attributed to the uncertainties of the temperature sensors used in the measurements. To overcome this problem, the flow rate was reduced to expand the temperature rise along the cable pipe. The temperature differences were measured between Tc8 and Tc9 with the flow rates from 9.9 to 3.9 L/min in addition to the flow rates over 15 L/min. The temperature rise was estimated with the same method explained in Section 3 [6]. The vacuum degree between the inner and outer pipes was less than 0.01 Pa. The obtained heat leak was 0.0634±0.0059 W/m. The relative uncertainty is reduced to less than 10 %. This result is also consistent with the other results if the ranges of the uncertainties are considered.

6. Estimation for longer transmission lines

The heat leak, the temperature rise, and the pressure drop of 10 km transmission lines are calculated with the present results. The results are shown in table 4. The use of the cryogenic pipe with a radiation shield in Ishikari project and the flow rate of 20 L/min are supposed. In the calculation, the heat leak at the outer pipe temperature of 17.4 °C shown in table 1, 1.204 W/m, is used for the return pipe and that obtained by the low flow rate experiment, 0.0634 W/m, is used for the cable pipe. The heat leak of the terminals is the sum of the heat leaks of the terminals at the outer pipe temperature of 17.4 °C in the table 3. The specific volume and the specific heat capacity of the liquid nitrogen at 77 K are used for convenience. The pressure drop of the terminals is calculated by subtracting the pressure drops of the cable pipe and the return pipe from that of the cable system shown in figure 3.

The heat leak to the cable pipe is calculated to be 0.634 kW. Consequently, the temperature rise along the cable pipe, which affects the current characteristics of the cable core, is 1.14 K. This value is quite small, if the transmission distance is considered. The pressure drop of the cable pipe is calculated to be 40.5 kPa and that of the return pipe is 78.5 kPa. The total pressure drop of the cable system is 120.2 kPa. If the pumps with a similar capacity with those used in Ishikari project are installed at each side of the terminals, the circulation will be possible. Therefore, it can be said that the cooling and the circulation of 10 km systems are possible by the present level of technology.

Table 4. Extrapolated values for 10 km systems with the present results at the flow rate of 20 L/min.

	Heat leak (kW)	Temperature rise (K)	Pressure drop (kPa)
Cable pipe	0.634	1.14	40.5
Return pipe	12.04	21.61	78.5
Terminals	0.475	0.852	1.2
Total	13.15	23.60	120.2

7. Conclusion

The second cooling test of the 1000 m system in Ishikari project was performed. The heat leaks were measured at the different outer pipe temperatures of the cryogenic pipe. The total heat leaks of the cable system were 1.746 and 2.091 kW at the outer pipe temperatures of -2.4 and 17.4 °C, respectively. The heat leaks measured in this study were consistent with those measured in the first cooling test. The heat leak of the cryogenic pipe increased by 30-40 % by the increase of the outer pipe temperature from -2.4 and 17.4 °C. The pressure drop of the cable system was 42.80 kPa and that of all the system, including the cooling system, was 98.4 kPa at the flow rate of 36.03 L/min. The degradation of the heat leak and pressure drop characteristics were not observed between the first and second cooling tests. Based on the present measurements, the heat leak, the temperature rise and the pressure drop of 10 km systems were estimated. From the estimation it can be said that the cooling and the circulation of 10 km systems are possible by the present level of technology.

Acknowledgement
This work was supported in part by the Japanese Ministry of Economy, Trade and Industry (METI) and New Energy and Industrial Technology Development Organization (NEDO).

References
[1] Yamaguchi S, Koshizuka H, Hayashi K and Sawamura T 2015 *IEEE Trans. Appl. Supercond.* **25** 5402504

[2] Yamaguchi S, Ivanov Y, Watanabe H, Chikumoto N, Koshiduka H, Hayashi K and Sawamura T 2016 *Physics Procedia* **81** 182-6
[3] Chikumoto N, Watanabe H, Ivanov V Y, Takano H, Yamaguchi S, Koshizuka H, Hayashi K and Sawamura T 2016 *IEEE Trans. Appl. Supercond.* **26** 5402204
[4] Watanabe H *et al.* 2017 *IEEE Trans. Appl. Supercond.* **27** 5400205
[5] Watanabe H, Ivanov V Y, Hamabe M, Chikumoto N, Takano H and Yamaguchi S 2015 *Physics Procedia* **67** 239-44
[6] Watanabe H, Sugino M, Sun J, Ivanov Y, Hamabe M, Kawahara T and Yamaguchi S 2011 *Physica* C **471** 1304-7

Fluid characteristic of liquid nitrogen flowing in HTS cable

O Maruyama[1] and T Mimura[1]

[1] Tokyo Electric Power Company Holdings, Yokohama, Japan
E-mail:maruyama.osm@tepco.co.jp

Abstract. For long distance operation of High-temperature superconducting (HTS) cables, evaluations of heat transfer and fluid flow dynamics of liquid nitrogen (LN_2) flowing in the HTS cable is important. However, the LN_2 flow is complicated when the cable core is housed in a spiral corrugate cryostat-pipe and positioned at an eccentric position to the center of the cryostat-pipe. In this study, the effect of a configuration of a spiral corrugate cryostat-pipe on the pressure drop was evaluated using computer simulation. As a result, it was confirmed that the calculated pressure drop in the spiral corrugate pipe is 17 times larger than that in the straight pipe due to the shrinking and expanding of the LN_2 flow at the concave domain. This shrinking and expanding of LN_2 flow was assumed to be the flow in a gradual contraction pipe and a gradual expansion pipe respectively, so that the pressure drop was calculated by a theoretical arithmetic analysis. For verification of this calculation, the calculated value was compared with the measured value of a HTS cable system, which was constructed in NEDO project. As a result, these values were agreed well each other and it indicated that the pressure drop in the spiral corrugate pipe could be evaluated by this calculation.

1. Introduction

The HTS cables have significant merits of compactness and larger power transmission capacity. In the future, these cables are expected to be installed in a power grid as replacement of existing power cables. For operation of the HTS cables in the power grid, it is necessary that the HTS coated conductors in the long distance cable are maintained at the low temperature stably and efficiently by liquid nitrogen (LN_2). It is necessary that the HTS cable can be bended at winding on a drum or installating on varying condition so that configuration of a cryostat-pipe of the HTS cable is corrugated. However, the LN_2 flow is complicated when the cable core is housed in the spiral corrugated-cryostat-pipe and positioned at an eccentric position to the center of the cryostat-pipe. Therefore, evaluations of heat transfer and fluid flow dynamics of LN_2 flowing in the HTS cable are important for the HTS cable system design including both the cooling system and the circulating pump.

2. Previous work and Objective

For simulation of the LN_2 flow in the HTS cable, referring to the 275 kV/3 kA high-voltage cable (HV cable, shown in figure 1) which has been developed in the M-PACC project [1]-[3], the pressure drop and temperature rise were calculated by theoretical analysis for the case of the simple cable configuration which is assumed that the cable core was arranged in the centerline position of the straight cryostat-pipe [4]. In this analysis, the dependences of volumetric flow of the LN_2 on the pressure drop and the temperature rise were evaluated. As a result, it was confirmed that the temperature rise decreased and the pressure drop increased significantly with increasing the volumetric flow of LN_2. However, the LN_2 flow pattern is complicated such as the case of eccentric arrangement of the cable core. A computer simulation analysis is effective in evaluating these characteristics with

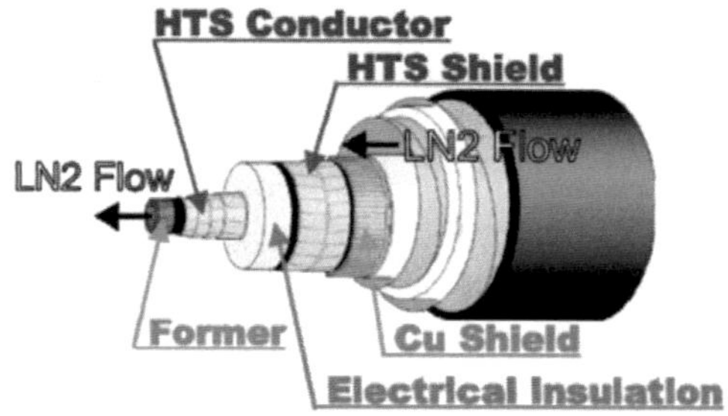

Figure 1. Structure of the HV cable.

the complicated LN_2 flow so that the accuracy of the computer simulation model was evaluated by comparison of the theoretical arithmetic analysis model for the simplified cable configuration [5]. As a result, the values of the computer simulation analysis showed a good agreement with that of the theoretical arithmetic analysis. Then, using this computer simulation model, the effects of the eccentric arrangement on the pressure drops and the temperature distribution of the cable core housed in the straight pipe were simulated [5]. As a result, the pressure drop of the eccentric arrangement became to about two-third as compared with that of the center configuration and no significant temperature difference of the HTS conductor of each configuration was confirmed due to the existence of a Cu outermost shielding layer of the cable which has high thermal conductivity.

In this study, the effect of a configuration of a spiral corrugate cryostat-pipe on the pressure drop was evaluated using this computer simulation. Furthermore, the calculated value was compared with the measured values of the HTS cable system, which was constructed in NEDO project [6].

3. Design of calculation model

The design specifications of referring the HV cable for calculation model are listed in Table 1 [2], [5]. In this HV cable design, LN_2 flows through both the inner pipe and the annulus as shown in figure 1. It is assumed that the pressure gradients for the inner and the outer (annulus) regions in the pipe are assumed to be equal. Total flow rate and the temperature of LN_2 at the cable inlet are assumed to be 20 liter/min and 65 K, respectively. The physical properties of LN_2 using in these calculations are listed in Table 2 [7]. In this model, the cable cross section is modeled as shown in figure 2 and two types of heat sources are considered; one is the heat intrusion from outer wall of the cryostat-pipe, and the other is the AC losses and the dielectric loss of the cable core. The heat intrusion is defined as 1 W/m

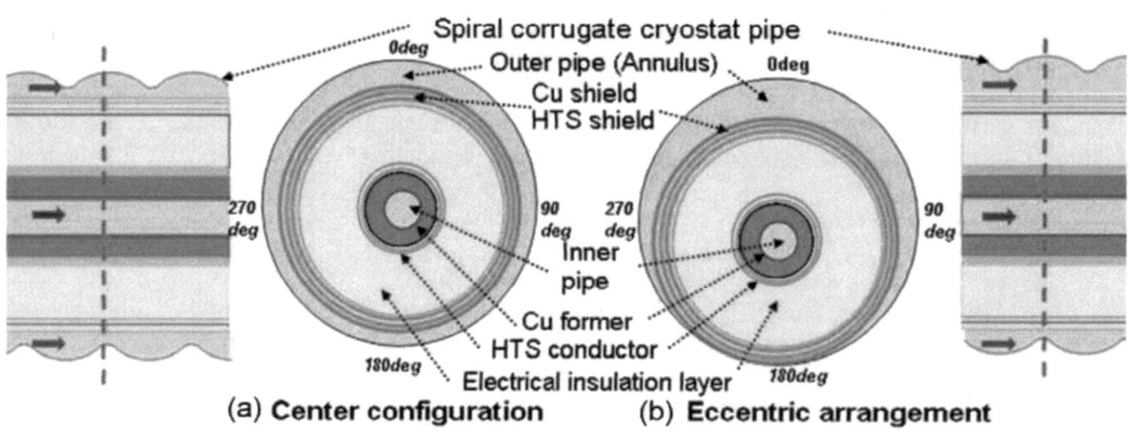

Figure 2. Calculation model for effect of corrugate pipe. (a) Center configuration (b) Eccentrici arrangement.

Table 1. HV Cable specifications

Cable component (Layer)	Architecture and materials	Outer diameter (mm)	Thermal conductivity (W/m·K)
Inner pipe	LN_2 flow	14	0.18
Former	Cu stranded	30.6	646
HTS conductor	2 layers YBCO tape	34	646
Electrical insulation	PPLP 22 mm thick	79.4	0.14
HTS shield	1 layer YBCO tape	80	646
Cu shield	3 layers Cu tape	85	646
Annulus (Outer pipe)	LN_2 flow	98.5	0.18

Table 2. Physical properties of LN_2

Items	Value
Specific heat C_p	2000 J/kg·K
Density ρ	860 kg/m^3
Viscosity μ	210 μPa·s
Thermal conductivity λ	0.18 W/m·K

[8]. The dielectric loss from the electrical insulation layer is 0.6 W/m, AC losses of the HTS conductor layer and the HTS shield layer are 0.13 and 0.07 W/m, respectively [2], [3]. The heat generation in each layer is assumed to be uniform. The pressure drop with the condition described above was calculated by the computer simulation analysis using the FLUENT, which is computational fluid dynamics simulation software, with coupling the equation of continuity, the Navier-Stokes equations and the advection thermal diffusion as described in (1), (2) and (3), respectively. The Finite volume method was applied and solved by the Semi-Implicit Pressure Linked Equations (SIMPLE) method. Furthermore, the *k-ε* model is used for the turbulence model in this simulation. The HTS cable was modelled three-dimensionally in 3.3 meters as shown in figure 3 (a). Figure 3 (b) shows a mesh cut of cross section at the modelled HTS cable. For long distance (axial) direction analysis, this simulation was calculated by periodic boundary condition.

$$\frac{\partial \rho}{\partial t}+\frac{\partial(\rho u_i)}{\partial x_i}=0 \tag{1}$$

$$\frac{\partial(\rho u_j)}{\partial t}+\frac{\partial(\rho u_i u_j)}{\partial x_i}=-\frac{\partial p}{\partial x_j}+\frac{\partial}{\partial x_i}\left[(\mu+\mu_t)\frac{\partial u_j}{\partial x_i}\right]+f_g \tag{2}$$

$$\rho C_p\frac{\partial T}{\partial t}+\rho C_p u_j\frac{\partial T}{\partial x_i}=\lambda\frac{\partial^2 T}{\partial x_j \partial x_j} \tag{3}$$

where u is vector of flow velocity, μ is viscosity λ is thermal conductivity and f_g is mean buoyancy. With this calculation model, the pressure drop of LN_2 in the steady-state conditions are provided as following sections.

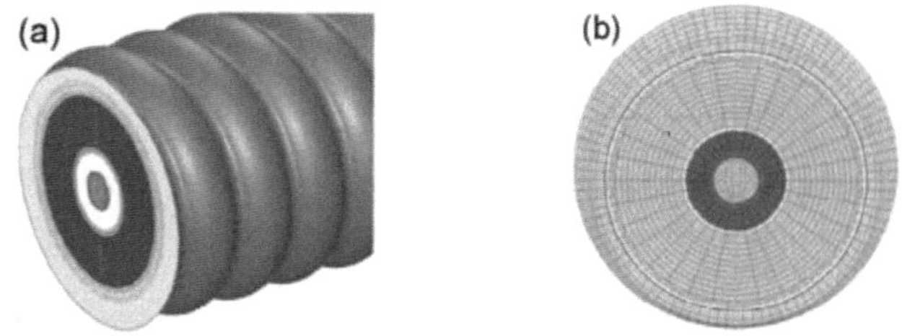

Figure 3. (a) 3-D modelling HTS cable (b) Mesh cut of HTS cable cross-section

4. Results of calculation and discussions

4.1. LN_2 fluid flow and pressure drop

The calculated pressure drop of the spiral corrugate configuration is 177 Pa/m and its value is 17 times larger than that of the straight pipe configuration, whose diameter assumed to be the equal to the hydraulic diameter of the spiral corrugate pipe. Even if the diameter of the straight pipe is assumed to be the equal to the inner diameter of spiral corrugate pipe, the value of the spiral corrugate configuration is 3 times larger than that of the straight pipe. The reason of this pressure drop increasing was discussed by simulated fluid flow in the pipe, as followings.

The simulated result of the LN_2 fluid flow in the corrugated pipe of the central core configuration is shown in figure 4. In this figure, it was confirmed that the simulated streamlines are divided into two different main flows. One is a spiral fluid flow, which is shown in the blue lines, in a convex domain in a corrugate pipe. Another is the slow spiral fluid flow, which is the shown in the yellow lines, through the long distance (axial) direction in a corrugate pipe. Figures 5 and 6 show the contour maps of LN_2 velocities of the circumferential flow and axial flow. As shown in figure 5, the spiral flow along the groove of the corrugation is confirmed. However, its velocity is about one-third of the axial flow which is shown in figure 6, so its influence on the pressure drop might be small.

As shown in figure 6, the stagnant flow region in a groove of the corrugation is confirmed. It indicates that the axial flow hardly flow into this groove. In this figure, at the concave domain by the corrugation of the pipe, the velocity is higher than that of the convex domain. This is due to the shrinking of LN_2 flow at concave domain, as shown in figure 7. Figure 8 shows the contour map of the LN_2 pressure. In this figure, it was confirmed that the pressure drop occurred this concave domain approximately 5 Pa by 1 pitch. Accordingly, it was confirmed that the pressure drop in the spiral corrugate pipe increased significantly at the concave domain of the corrugation due to the shrinking and expanding of the flowing LN_2.

In the case of the eccentric arrangement of the cable core, the calculated pressure drop is 143 Pa/m. The calculated pressure drop of the eccentric arrangement became to about four-fifth as compared with that of the center configuration. Accordingly, it was confirmed that the eccentric arrangement is effective for reduction of the pressure drop in the case of the spiral corrugate pipe as well as the straight pipe case [5].

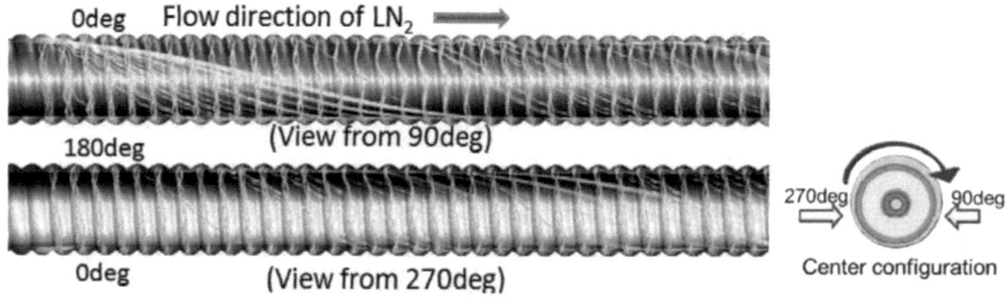

Figure 4. Streamline of LN_2 flowing in corrugated pipe.

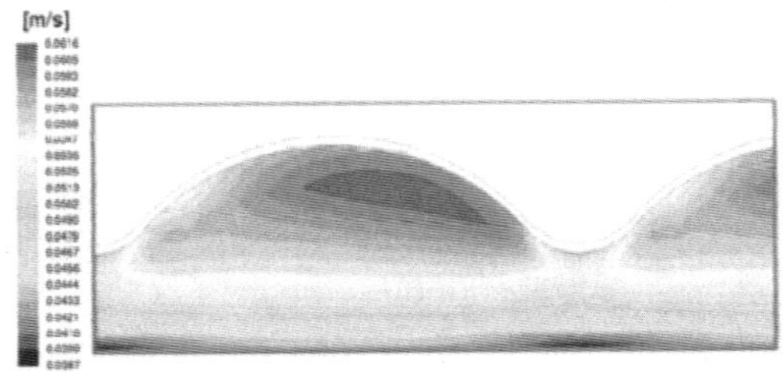

Figure 5. Contour map of LN_2 velocity of the circumferential flow.

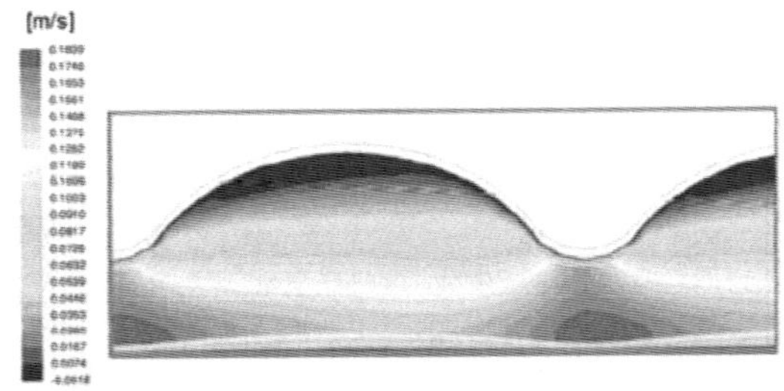

Figure 6. Contour map of LN_2 velocity of the axial flow.

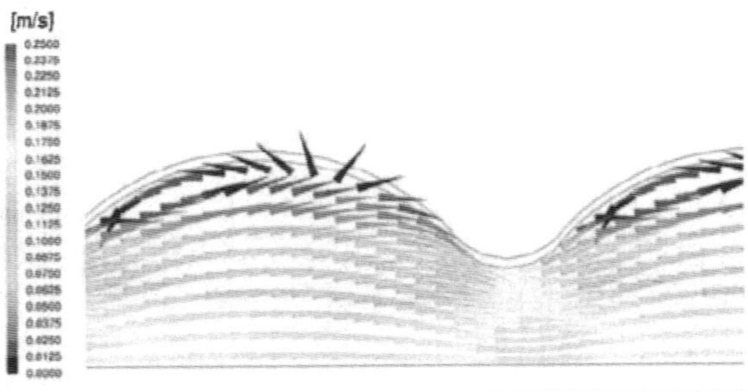

Figure 7. Vector of LN_2 velocity of the axial flow.

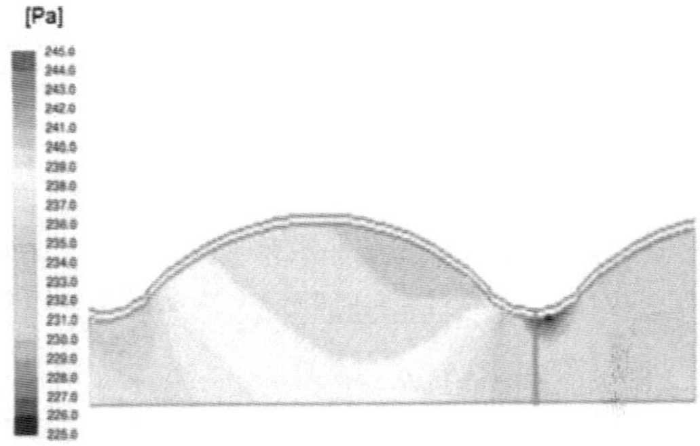

Figure 8. Contour map of LN_2 pressure.

4.2. Theoretical arithmetic analysis

As described in section 4.1, the pressure of LN_2 mainly reduced by the shrinking and expanding of the LN_2 flow path at the concave domains of this pipe. So, this flow was assumed to be consist of the two flows which are flowing in a gradual contraction pipe and a gradual expansion pipe. These pressure drops in each pipe were evaluated individually by theoretical arithmetic analysis, as followings.

In case of the gradual expansion pipe, the velocity of LN_2 decreases and the pressure increases generally. The energy of layers at the boundary is so small that the increased pressure may stop the flow there or even reverse it. As a result, eddies form and the flow separates from the pipe wall. The greater the angle of divergence of an expanding pipe the more pronounced these phenomena and the greater the turbulence loss. Therefore, the pressure drop of the LN_2 per unit length ($\Delta P/\Delta L$) can be given by (4) and (5) [9], as followings.

$$\frac{\Delta p}{\Delta L} = \frac{\xi}{s} \times \frac{1}{2} \rho v^2 \tag{4}$$

$$\xi = \left(0.011\theta^{1.22}\right)\left(1 - \frac{A_1}{A_2}\right)^2 \tag{5}$$

where s is pitch of the groove of the corrugation per unit length, v is velocity at outlet domain of the expansion pipe, ξ is loss coefficient, θ is the divergence angle of pipe and A is the cross-section area of LN_2 path. The subscript numbers 1 and 2 indicate intake and outlet of the expansion pipe. These are assumed to be the concave and convex domain of the corrugated pipe, respectively and the calculation model is shown in figure 9. As described above, the pressure drop values with varying the volumetric flow were calculated.

In case of the gradual contraction pipe, the velocity of LN_2 increases and pressure drops generally.

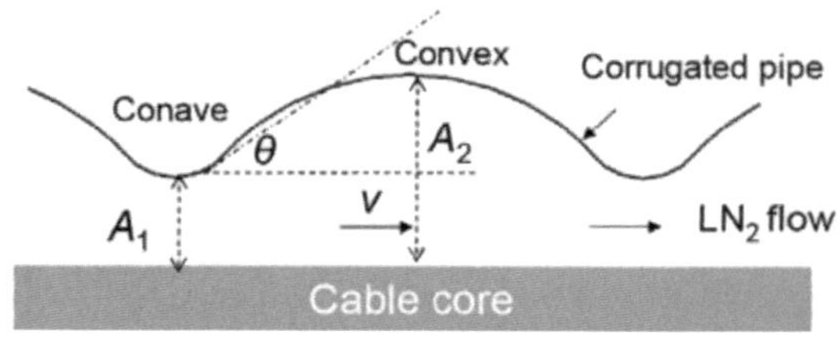

Figure 9. Calculation model for theoretical arithmetic analysis

The flow is from higher to lower pressure, which eliminates the causes of turbulence and separation.

Accordingly, the loss in a gradual contraction are due to friction only. The pressure drop caused by friction in the fluid boundary layer on pipe wall is calculated using friction factor, which is function of Reynolds number. However, as shown in figure 6, the axial flow hardly flow into the groove of pipe corrugation so that the domain of the friction by pipe wall is decreased signicantly. Accordingly, the loss caused by the pipe friction was ignored in this study.

Furthermore, in the case of the eccentric arrangement of the cable core, the pressure drop should be calculated with considering an empirical formula depends on eccentricity ratio which is the position of the cable core moved from the center to the bottom of the cryostat pipe [5], [10].

4.3. Comparison with measured value of HTS cable system

For verification of this calculation method described in section 4.2, the calculated value was compared with the measured value of a 20 meter long HTS cable system shown in figure 10, which was constructed by Furukawa electric at Shenyang in China in the NEDO project. The specifications of HTS cable in this system are different from that of table 1, as shown in table 3.

The calculated and measured values with varying the volumetric flow are shown in figure 11. As a result, these values were agreed well each other with varying the volumetric flow. It indicated that the pressure drop in the spiral corrugate pipe could be evaluated by these equations.

Table 3. Cable specifications for evaluating fluid flow.

Cable component	Architecture	Outer diameter (mm)
Inner pipe	LN_2 flow	14
Cable core	Dummy core	79
Annulus (Outer pipe)	LN_2 flow	104
Core position	Eccentric arrangement	-

Figure 10. 20-meter long HTS cable demonstration system at Shenyang in China.

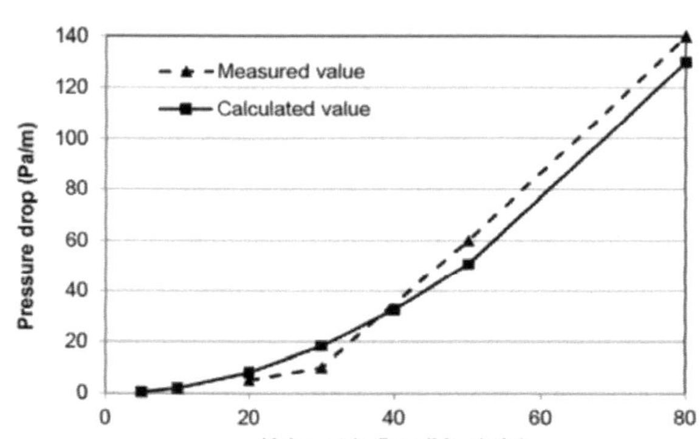

Figure 11. Pressure drop of LN_2 with varying the volumetric flow.

5. Conclusion

The effect of a configuration of a spiral corrugate cryostat-pipe on the pressure drop was evaluated using computer simulation. As a result, it was confirmed that the calculated pressure drop in the spiral corrugate pipe is 17 times larger than that in the straight pipe due to the shrinking and expanding of the LN_2 flow at the concave domain. This shrinking and expanding of LN_2 flow was assumed to be the flow in a gradual contraction pipe and a gradual expansion pipe respectively, so that the pressure drop was calculated by evaluation equations theoretically. For verification of this calculation, the calculated value was compared with the measured value of a HTS cable system, which was constructed in NEDO project. As a result, these values were agreed well each other and it indicated that the pressure drop in the spiral corrugate pipe could be evaluated by this calculation.

Acknowledgments

This work was supported by the New Energy and Industrial Technology Development Organization (NEDO).

I would like to thank to Furukawa electric for their kind supply of the HTS cable specifications and experimental data. Furthermore, I would like to thank to Yuh Shiohara (Honorary Director General of International Superconductivity Technology Center - Superconductivity Research Laboratory) and Teruo Izumi (National Institute of Advanced Industrial Science and Technology) for their kindly support and instructing specially.

References

[1] Shiohara Y, Yoshizumi M, Takagi. Y and Izumi. T, 2013 *Physica C* **484**, 1-5
[2] Mukoyama S, Yagi M, Mitsuhashi T, Nomura T, Ten J, Nakayama R, et al., 2013 *IEEE Trans. Appl. Supercond.* **23**(3) 5402804
[3] Maruyama O, Ohkuma T, Masuda T, Ashibe Y, Mukoyama S, Yagi M, et al., 2013 *IEEE Trans. Appl. Supercond.* **23**(3) 5401405
[4] Maruyama O, Ohkuma T, Izumi T, 2013 *Physics Procedia* **45** 285–288
[5] Maruyama O, Ohkuma T, Izumi T, Shiohara Y, 2014 *IEEE Trans. Appl. Supercond.* **25**(3) 5401505
[6] Mukoyama S, 2015, *ISTEC Superconductivity Web21*
[7] Yagi M, Mukoyama S, Noboru I, Osamu S, et al., 2005 *Furukawa Review*, **28** 53-60
[8] Cryogenics and Superconductivity society of *Japan, The Superconductivity and cryogenics Handbook*, Ohmsha, 1993, 1052-1088
[9] JSME, *JSME Mechanical Engineer' s Handbook*, 1988, A4-72
[10] I.E.Idelchik, *"Handbook of Hydraulic Resitance 3rd Edition"*, 1996 *Begell house, Inc.*, pp 121-122

Three-Dimensional Thermal Analysis of an SFCL REBCO Coil Immersed in Liquid Nitrogen

Kezhen Qian[1], Toshiki Shiratani[2], Yutaka Terao[2] and Hiroyuki Ohsaki[2]

[1] Graduate School of Engineering, The University of Tokyo, Tokyo 113-8654, Japan
[2] Graduate School of Frontier Sciences, The University of Tokyo, Kashiwa 277-8561, Japan

E-mail: qiankezhen@ohsaki.k.u-tokyo.ac.jp

Abstract. A resistive type superconducting fault current limiter (SFCL) using REBCO tapes has shown its advantages in power applications. However, the hotspot problem, which can be caused by the uniformity of critical current density J_C over long-length of REBCO tapes, is still a threat to the SFCL safety. In this paper, we have studied the limiting performance and transient temperature rise distribution of a nonintersecting type non-inductive solenoidal SFCL REBCO coil using three-dimensional electromagnetic and thermal coupled FEM analysis. Firstly, we modelled a local degradation of J_C and studied its influence on the transient temperature rise distribution during the over current. The results indicated that a narrow and severe J_C degradation leads to the concentration of current and heat generation in the normal J_C area, which is opposite to the case of a relatively wider and slighter J_C degradation area. Furthermore, REBCO tapes with different thickness of substrate and Ag over-layer were analysed to study the influence on limiting performance and temperature rise. A thicker substrate leaded to a slower temperature rise because of its higher heat capacity, and the Ag over-layer of 4 μm showed better limiting performance and thermal stability because of its better thermal conductivity as well as uniform temperature rise and SN transition.

1. Introduction

A resistive type superconducting fault current limiter (SFCL) using REBCO tapes has shown its advantages in limiting fault currents and improving the reliability of a power system for its compactness and rapid increase in resistance [1]–[4]. However, critical current density (J_C) uniformity over long-length of REBCO tapes may lead to a local overheating and burning of the tapes, which are called as the hotspot phenomenon. In the previous research, two types of non-inductance solenoidal coils for resistive type SFCL were studied on their current limiting properties and recovery characteristics [5]. In addition, the electrical and thermal behaviours of several types of REBCO tapes during the fault current limitation were analysed and evaluated with a simulation model, consequently, tapes with no stabilizer were considered more suitable for constructing SFCLs [6]. In reference [7], a 2D J_C distribution map of such REBCO tape with no stabilizer was measured using the scanning hall probe microscopy (SHPM).

In this paper, we have studied the thermal characteristics of an REBCO coil for SFCL use during the fault current limitation with the electromagnetic and thermal coupled FEM analysis. The nonintersecting type non-inductive solenoidal winding [5] consisting of the stabilizer free REBCO tape was assumed for the REBCO coil, which concept is shown in figure 1. This coil consists of an inner and an outer winding that are wounded bifilarly and connected in series to reduce the inductance of the coil. A local

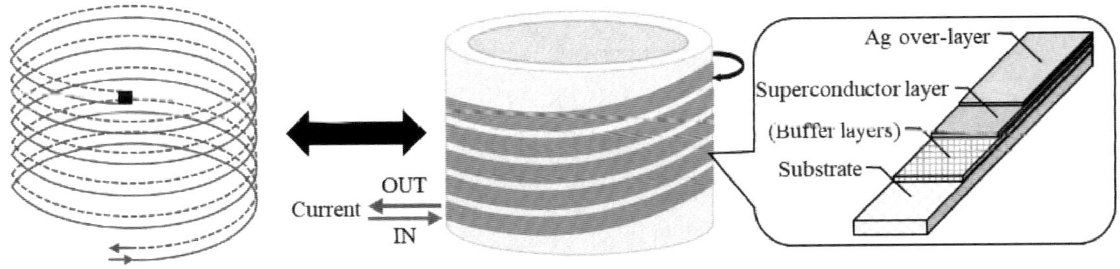

Figure 1. Structure and schematic diagram of nonintersecting type non-inductive winding REBCO coil for resistive type superconducting fault current limiter (SFCL).

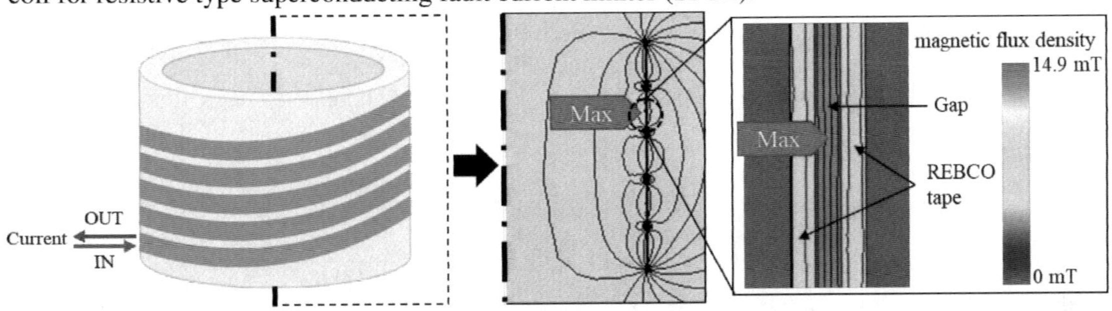

Figure 2. Cross-sectional view of the magnetic field distribution for the nonintersecting type non-inductive winding REBCO coil with a current of 100 A. Local magnetic field distribution is enlarged and indicates that the magnetic flux is not cancelled out between the outer and inner winding where the flux direction is almost parallel to the surface of REBCO tape.

degradation of J_C was modelled and the influence of degradation area on the temperature rise distribution during the over current was analysed. Moreover, REBCO tapes with different thickness of substrate and Ag over-layer were analysed to study the influence on limiting performance and temperature rise, then more appropriate thickness was suggested.

2. Numerical Analysis Method and Conditions

2.1. Numerical Analysis Method

Electromagnetic and thermal coupled analysis was utilized for the numerical analysis of SFCL REBCO coil, because the operating characteristics of REBCO tapes are strongly affected by the temperature rise. In the electromagnetic analysis, the finite element method based on current vector potentials was used [8]. The governing equation of current vector potential $\boldsymbol{T}$ defined by $\boldsymbol{J} = \nabla\times\boldsymbol{T}$ ($\boldsymbol{J}$: current density) is given by the following equation:

$$\nabla \times (\rho\nabla \times \boldsymbol{T}) = -\frac{\partial \boldsymbol{B}}{\partial t} \tag{1}$$

where ρ is the electric resistivity, and $\boldsymbol{B}$ is the magnetic flux density. A thin-plate approximation was employed for computing the current distribution in the REBCO tapes considering its high length-thickness ratio.

A nonlinear E-J relation (E: electric field) based on the power law (2) was used for modelling the electromagnetic property of superconductors [8]. Temperature and magnetic flux density dependence of the critical current density J_C was expressed by (3) [9, 10], where T is the temperature and B is the component of magnetic flux density parallel to the surface of REBCO tape. The perpendicular magnetic field component is negligible because of the bifilar winding (figure 2). The magnetic flux density dependence was considered because the magnetic flux between the outer and inner winding cannot be

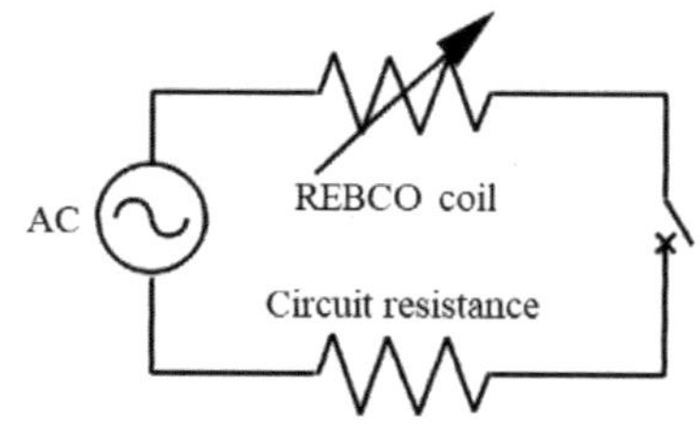

Figure 3. Schematic diagram of over-current simulation circuit.

cancelled out, even though the non-inductive solenoidal winding was employed for the REBCO coil in this paper. Furthermore, flux density B was calculated according to the calculated circuit current and the linear relation between the current and the magnetic flux density because of the coil's nonmagnetic property. Additionally, the linear relation was obtained from the static electromagnetic analysis using JMAG (figure 2).

$$E = E_C \left(\frac{J}{J_C}\right)^n \tag{2}$$

$$J_C(T) = J_{C0}[1-(T/T_C)^2]^{\frac{3}{2}} \times \frac{1}{(1+B/B_C)^{0.6}} \tag{3}$$

The governing equation of the three-dimensional thermal analysis was given by:

$$\rho c \frac{\partial T}{\partial t} = \left[\frac{\partial}{\partial x}\left(\kappa \frac{\partial T}{\partial x}\right) + \frac{\partial}{\partial y}\left(\kappa \frac{\partial T}{\partial y}\right) + \frac{\partial}{\partial z}\left(\kappa \frac{\partial T}{\partial z}\right)\right] + Q \tag{4}$$

where κ is the thermal conductivity, T is the temperature, ρ is the mass density, c is the specific heat, and Q is the Joule heat. Additionally, the temperature dependence of κ, ρ, and c were considered in the analysis [11]–[14]. The analysis region of the REBCO tapes coinciding with substrate, cooling characteristics of liquid nitrogen [8], and heat sources originated in Joule loss in the Ag and superconductor layer were given by the boundary conditions.

2.2. Analysis Conditions

The REBCO coil used in the numerical analysis is shown in figure 1, and the schematic diagram of over-current simulation circuit is shown in figure 3. The specifications of the simulation circuit and the REBCO coil are summarized in table 1 and table 2.

Table 1. Specifications of simulation circuit and REBCO coil.

Power-supply voltage	500 V
Circuit resistance	0.05 Ω
Coil diameter	31.8 cm
Number of turns for both windings	5
Number of tapes per turn	1
Coil pitch	1.35 cm
Over current duration	100 ms

Table 2. Specifications of the REBCO tape used in the coil.

REBCO tape	SuperPower SF12100
Ag over-layer thickness	2 μm
Substrate thickness	100 μm
I_C (J_{C0}) @77 K, self-field	468 A (3.9×10^{10} A/m^2)
E_C	1 μV/cm
T_C	90 K
B_C	118.2 mT

We explain the experiment results [7] and simulation results in the next chapter. In *3.1*, a comparison is conducted between over-current experiment results [7] and simulation results, which are based on the same specifications: an over-current of 2000 A and one meter SF12100 tape. Meanwhile, the simulation specifications in *3.2* and *3.3* are shown in table 1. In *3.2*, with the same macroscopic J_C distribution in

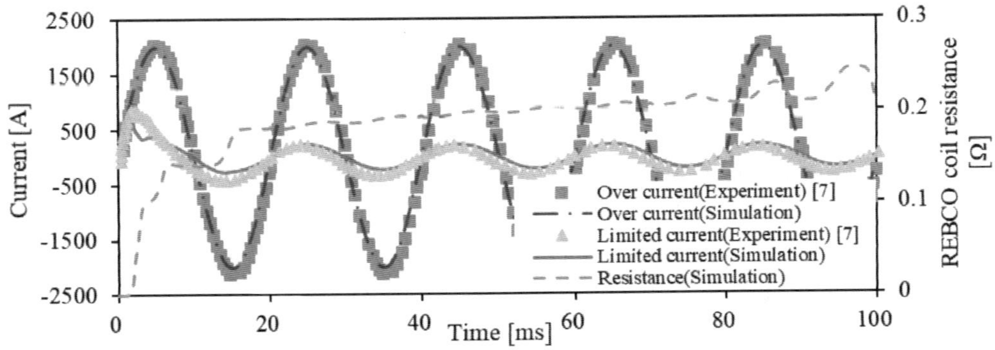

Figure 4. Over-current simulation results of 1 m SF12100 tape, which show good agreement with the experiment results [7].

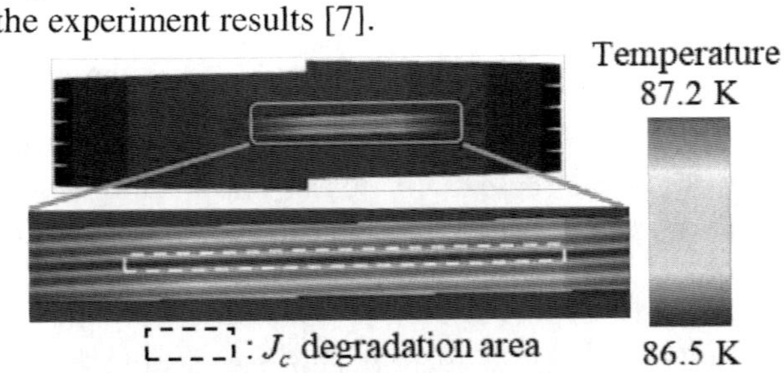

Figure 5. Temperature distribution of the REBCO coil with a J_C degradation (degrading to 25% of the rated J_C in 63 mm×2.4 mm) at the rising phase of current t = 0.725 ms.

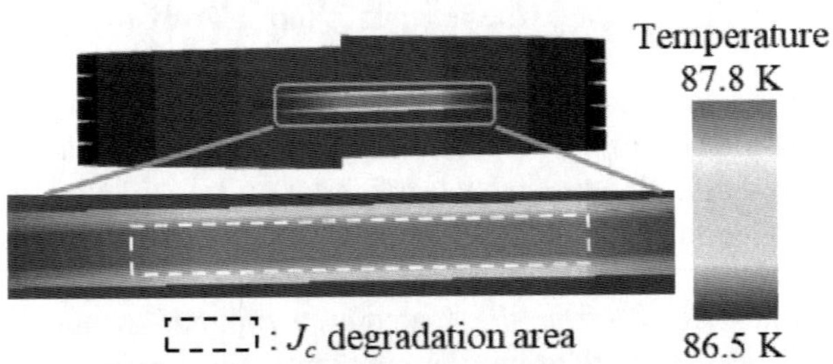

Figure 6. Temperature distribution of the REBCO coil with a J_C degradation (degrading to 75% of the rated J_C in 63 mm×7.2 mm) at the rising phase of current t = 0.725 ms.

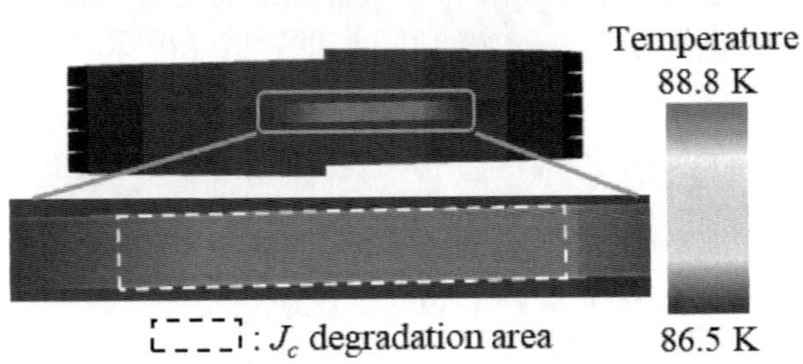

Figure 7. Temperature distribution of the REBCO coil with a J_C degradation (degrading to 85% of the rated J_C in 63 mm×12 mm) at the rising phase of current t = 0.725 ms.

longitudinal direction, 3 cases with different J_C degradation area in width direction are analysed to study the influence on the temperature rise. In 3.3, several cases with different thickness of substrate and Ag over-layer are analysed to study the influence on limiting performance and temperature rise.

3. Numerical Analysis

3.1. Comparison with experimental results

To examine the accuracy of the numerical analysis model, an over-current simulation was conducted with an over-current of 2000 A and one meter SF12100 tape. The simulation results (figure 4) show good agreement with the experimental results [7]. In addition, a local degradation of J_C (85% of the rated J_C), which is the same as the experimental tape, was also modelled in the numerical analysis.

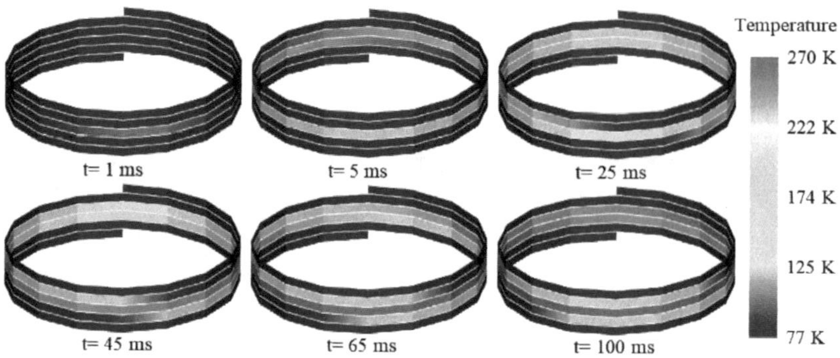

Figure 8. Development of temperature rise of the REBCO coil in 100 ms with a J_C degrading to 25% of the rated J_C in 63 mm×2.4 mm.

3.2. Influence of J_C degradation area

To study the influence of different J_C degradation area in width direction on the temperature rise distribution during the over current, 3 cases of thermal analysis were carried out in the assumptions that the macroscopic J_C distribution in longitudinal direction is the same in all 3 cases but the widths of J_C degradation area are different (figure 5–7). According to the analysis results, the difference mainly appears during the rising phase of current. As in the case of narrow and severe J_C degradation (degrading to 25% of the rated J_C in one-fifth of the tape width, figure 5), the temperature rise in the J_C degradation area is lower than that in the normal J_C area, which indicates that the current bypasses the J_C degradation area and flows through the normal J_C area. Therefore, in this case, the hotspot is more likely to occur in the normal J_C area.

On the other hand, in the case of wide and slight J_C degradation (degrading to 75% of the rated J_C in three-fifth of the tape width, figure 6), the temperature rise in the J_C degradation area is higher than that in the normal J_C area. Because the current cannot completely bypass the J_C degradation area, meanwhile, the current through the J_C degradation area generates more Joule heat. In this case, a hotspot is more likely to occur in the J_C degradation area.

With the development of temperature rise and heat conduction, the heated area expands from the J_C degradation area (figure 8). However, different widths of J_C degradation area show little influence on the temperature rise along longitudinal direction or the peak temperature of the tape during the over current.

3.3. Influence of different thickness of REBCO tape layers

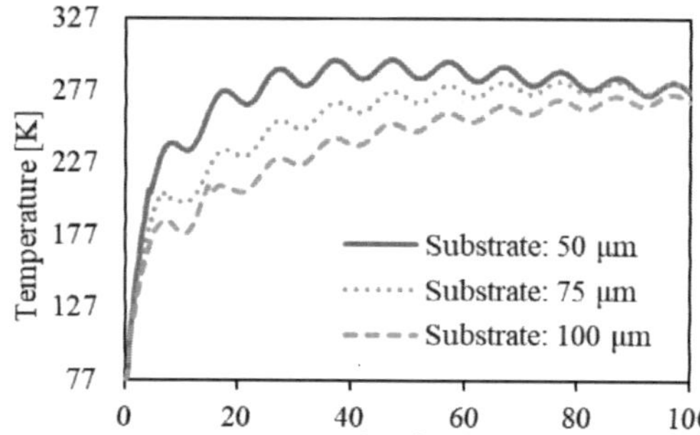

Figure 9. Maximum temperature of the REBCO coils with different substrate thickness.

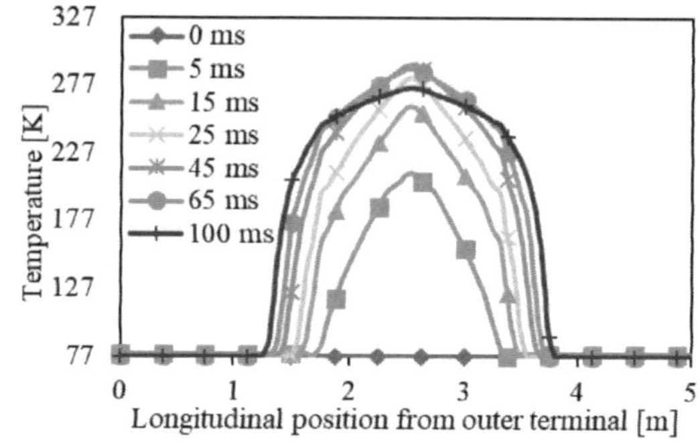

Figure 10. Temperature distribution along the REBCO tape with 50 μm substrate in longitudinal direction, which starts from the outer terminal of the outer winding, at different times.

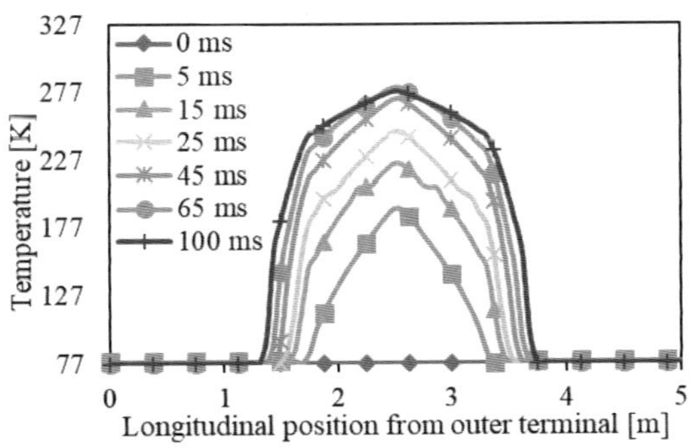

Figure 11. Temperature distribution along the REBCO tape with 75 μm substrate in longitudinal direction, which starts from the outer terminal of the outer winding, at different times.

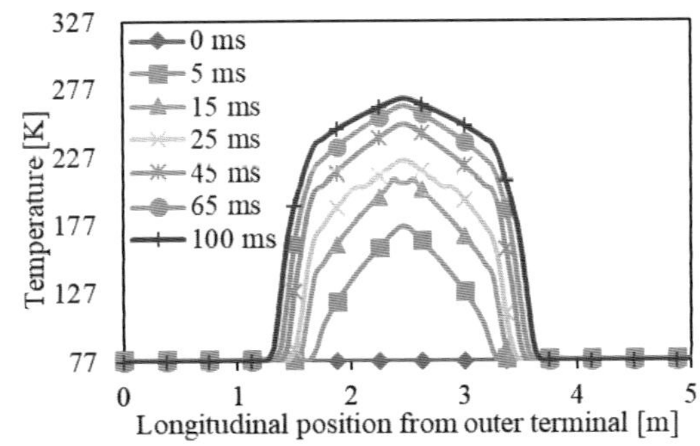

Figure 12. Temperature distribution along the REBCO tape with 100 μm substrate in longitudinal direction, which starts from the outer terminal of the outer winding, at different times.

3.3.1 Influence of substrate thickness. Three cases with substrate thickness of 50 μm, 75 μm, and 100 μm were analysed to study the influence on current limiting performance and temperature rise. The maximum temperature of the REBCO coil during the over current is shown in figure 9. The temperature distribution along the REBCO tape of outer winding in longitudinal direction at different times, which is 5 meters long and from the outer terminal to the connection of two windings, is shown in figure 10–12. Thinner substrate leads to a higher temperature rise because of its lower heat capacity, yet, the thickness of substrate has less influence on the limited current. Therefore, considering the heat capacity and thermal stability as well as mechanical strength, a relatively thicker substrate is favourable.

3.3.2 Influence of Ag over-layer thickness. Six cases with Ag over-layer thickness of 2 μm, 3 μm, 4 μm, 6 μm, 8 μm, and 10 μm were analysed to study the influence on current limiting performance and temperature rise. According to the results, the temperature distribution with an Ag over-layer of 2 μm is quite different from other cases (figure 13–16). In figure 13, a significant hotspot occurs at the J_C degradation area because of the poor thermal conductivity and high resistance of a thin Ag over-layer. As in the cases with Ag over-layer of 3 μm or higher, the hotspot phenomenon and the maximum temperature (figure 19) are both suppressed with the thicker Ag over-layer, meanwhile, a nearly uniform temperature rise along the REBCO tape in longitudinal direction is observed. Therefore, the limited currents rise even lower with the Ag over-layer of 3 μm and 4 μm that have higher overall resistances (figure 17, 18). In comparison with the case of 2 μm, a REBCO tape with 4 μm Ag over-layer shows

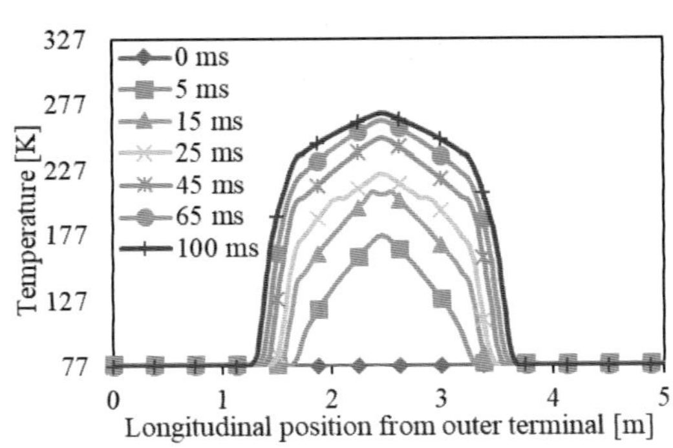

Figure 13. Temperature distribution along the REBCO tape with 2 μm Ag over-layer in longitudinal direction, which starts from the outer terminal of the outer winding, at different times.

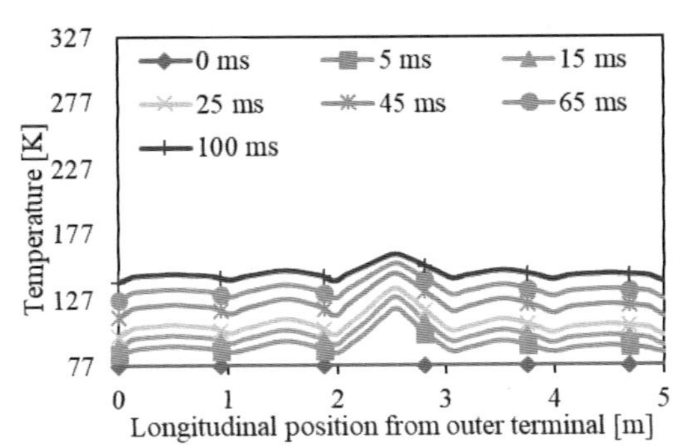

Figure 14. Temperature distribution along the REBCO tape with 3 μm Ag over-layer in longitudinal direction, which starts from the outer terminal of the outer winding, at different times.

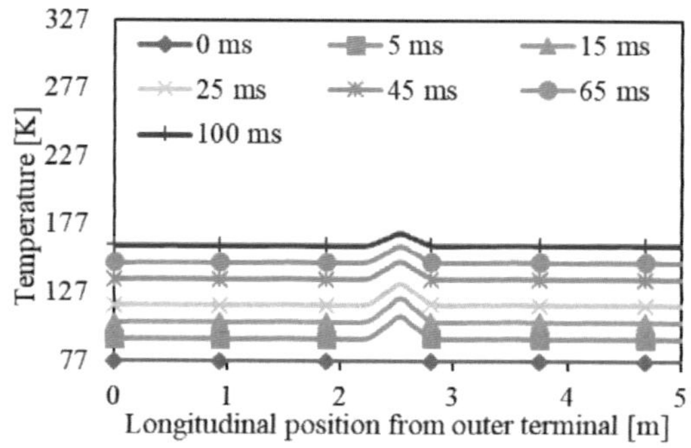

Figure 15. Temperature distribution along the REBCO tape with 4 μm Ag over-layer in longitudinal direction, which starts from the outer terminal of the outer winding, at different times.

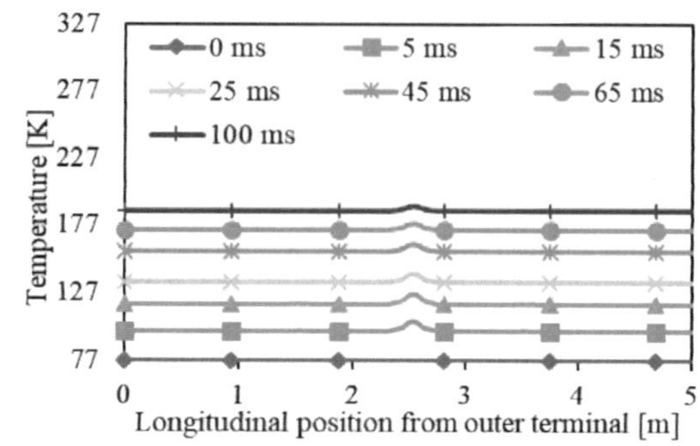

Figure 16. Temperature distribution along the REBCO tape with 6 μm Ag over-layer in longitudinal direction, which starts from the outer terminal of the outer winding, at different times.

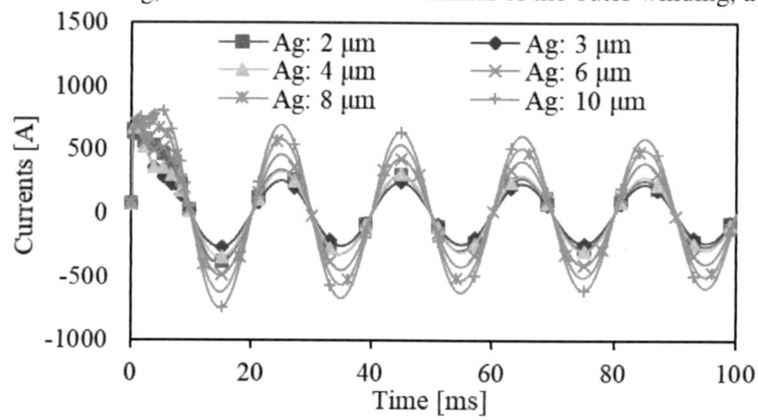

Figure 17. Limited currents of REBCO coil with different Ag over-layer thickness.

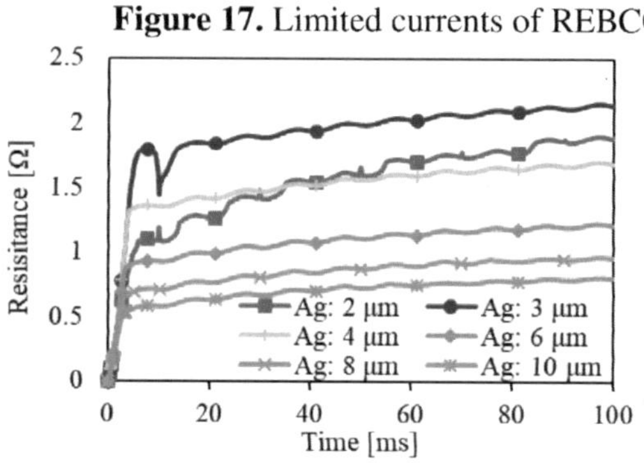

Figure 18. Resistance of the REBCO coil with different Ag over-layer thickness.

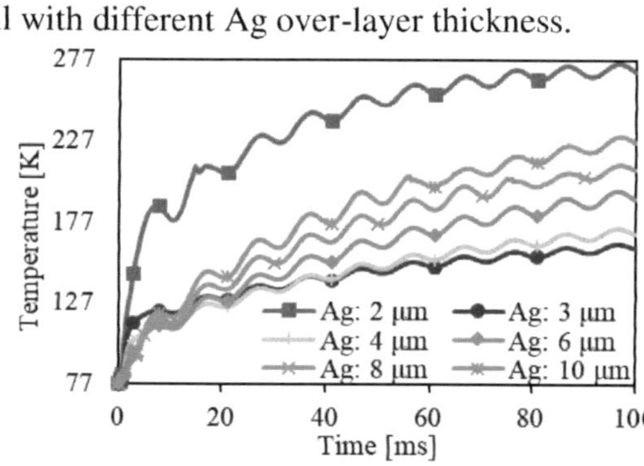

Figure 19. Maximum temperature of the REBCO coil with different Ag over-layer thickness.

better limiting performance and thermal stability in this analysis.

4. Conclusion

In this paper, we have built a three-dimensional electromagnetic and thermal coupled FEM analysis model. Using this model, a nonintersecting type non-inductive solenoidal REBCO coil for SFCL use was modelled, and its thermal characteristics during the over current were analysed with different J_C degradation area and thickness of REBCO tape layers. The development of temperature rise distribution and heat conduction was investigated, and indicated that a narrow and severe J_C degradation leads to the concentration of current and heat generation in the normal J_C area, which is opposite to the case of a relatively wider and slighter J_C degradation area. Then, several cases with different thickness of substrate and Ag over-layer were analysed. A thicker substrate leaded to a slower temperature rise because of its higher heat capacity, and the Ag over-layer of 4 μm showed better limiting performance and thermal stability in this analysis because of its better thermal conductivity as well as uniform temperature rise and SN transition.

References

[1] Morandi A 2013 State of the art of superconducting fault current limiters and their application to the electric power system *Physica C: Superconductivity* **484** 242–7

[2] Ruiz H S, Zhang X and Coombs T A 2015 Resistive-Type Superconducting Fault Current Limiters: Concepts, Materials, and Numerical Modeling *IEEE Transactions on Applied Superconductivity* **25** 1–5

[3] Neumueller H W *et al.* 2009 Development of Resistive Fault Current Limiters Based on YBCO Coated Conductors *IEEE Transactions on Applied Superconductivity* **19** 1950–5

[4] Noe M and Steurer M 2007 High-temperature superconductor fault current limiters: concepts, applications, and development status *Supercond. Sci. Technol.* **20** R15

[5] Liang F, Yuan W, Zhu J, Zhang M, Venuturumilli S, Li J, Patel J, Zhang G and Zhang Z 2017 Experimental Test of Two Types of Non-Inductive Solenoidal Coils for Superconducting Fault Current Limiters Use *IEEE Transactions on Applied Superconductivity* **27** 1–5

[6] Majka M, Kozak J and Kozak S 2017 HTS Tapes Selection for Superconducting Current Limiters *IEEE Transactions on Applied Superconductivity* **27** 1–5

[7] Kar S, Li X-F, Selvamanickam V and Rao V V 2017 Current distribution mapping in insulated (Gd,Y)BCO based stabilizer-free coated conductor after AC over-current test for R-SFCL application *IOP Conf. Ser.: Mater. Sci. Eng.* **171** 012118

[8] Ohsaki H, Sekino M and Nonaka S 2009 Characteristics of Resistive Fault Current Limiting Elements Using YBCO Superconducting Thin Film With Meander-Shaped Metal Layer *IEEE Transactions on Applied Superconductivity* **19** 1818–22

[9] Matsushita T, Wada H, Kiss T, Inoue M, Iijima Y, Kakimoto K, Saitoh T and Shiohara Y 2002 Critical current properties in superconducting Y-123 tapes *Physica C: Superconductivity* **378–381** 1102–7

[10] Zermeño V M R and Grilli F 2014 3D modeling and simulation of 2G HTS stacks and coils *Supercond. Sci. Technol.* **27** 044025

[11] Lu J, Choi E S and Zhou H D 2008 Physical properties of Hastelloy® C-276™ at cryogenic temperatures *Journal of Applied Physics* **103** 064908

[12] Fujishiro H, Ikebe M, Naito T, Noto K, Kohayashi S and Yoshizawa S 1994 Anisotropic Thermal Diffusivity and Conductivity of YBCO(123) and YBCO(211) Mixed Crystals *Jpn. J. Appl. Phys.* **33** 4965–70

[13] Zhang M, Matsuda K and Coombs T A 2012 New application of temperature-dependent modelling of high temperature superconductors: Quench propagation and pulse magnetization *Journal of Applied Physics* **112** 043912

[14] Smith D R and Fickett F R 1995 Low-Temperature Properties of Silver *J Res Natl Inst Stand Technol* **100** 119–71

IOP Conf. Series: Journal of Physics: Conf. Series **1054** (2018) 012079 doi:10.1088/1742-6596/1054/1/012079

Design and performance analysis of a 1500 A, 400 mH superconducting DC reactor coil using 2G multi-ply HTS wire

J I Lee[1], T K Kim[1], M Park[1] and I K Yu[1]

1. Changwon National University, Changwon, GN, 51140, Korea

E-mail: yuik@cwnu.ac.kr

Abstract. This paper deals with the design and performance analysis of a superconducting DC reactor using multi-ply high temperature superconducting (HTS) wire for a high-current DC reactor. To design the high-current DC reactor, 2G multi-ply HTS wire (SuNAM Co., Ltd.) was used to achieve a high ampacity. The critical current was measured according to the bending diameter in order to confirm the bending characteristics of the 2-ply and 3-ply wires. The HTS coils for a 1500 A, 400 mH HTS DC reactor were designed based on the characteristics of multi-ply HTS wire. Electromagnetic analysis of the toroid-type DC reactor was performed with the finite element method. HTS coils were wound with a D-shaped double pancake type bobbin. The critical current of the double pancake coil (DPC) modules was measured in a liquid nitrogen vessel. The performance test results were compared with the analysis results. Based on the critical current of DPC measured at 77 K, it was confirmed that 1500 A could be flowed at the operating temperature of 20 K. These results will be effectively utilized for the design and fabrication of an HTS DC reactor that can be used in real power systems.

1. Introduction

A DC reactor is a power system application device that reduces the current ripple and harmonics generated by thyristors during AC-DC conversion. It is usually used in high-voltage direct current (HVDC) systems. Conventional DC reactors are manufactured with metal conductors. However, there are some disadvantages to using metal conductors. First, a metal conductor has resistance, which inevitably causes electrical loss. As the current increases, the power loss also increases. Second, when DC reactors are made with metal conductors, the volume and weight become very large, which limits the installation space and insulation structure. These disadvantages can be overcome by using a superconducting wire that has zero resistance and a high current density [1]-[2]. However, a high temperature superconducting (HTS) DC reactor cannot be used in real kA-class systems. This is because the HVDC system current is higher than the critical current of commercial HTS wires. In addition, when HTS coils are manufactured, the critical current is decreased due to the magnetic field, and it is impossible to use in real HVDC systems with operating currents that exceed 1000 A. The authors propose an HTS coil-based DC reactor that can be applied to actual systems. The design target is based on high inductance compared to copper-conductor DC reactors used in conventional HVDC systems with a current capacity of 1500 A.

In this paper, a toroid-type HTS magnet for a DC reactor system was designed using 2G multi-ply HTS wires, which were made by soldering two or three overlapping HTS wires. The toroid-type magnet is suitable for a DC reactor, which should be compact with a large energy capacity because the inductance has been increased. As the inductance increases, the magnetic flux density per current also increases. However, in the toroid-type coil, the perpendicular magnetic field that has a greater influence on the critical current of the HTS coil is reduced. The toroid-type coil can make a high operating current and low magnetization loss [3]-[9]. The target inductance and current level of the toroid-type HTS DC reactor were 400 mH and 1500 A, respectively. Compared to the characteristics of commercial copper reactors, the target capacity is set to a similar capacity and the inductance value is doubled. The magnetic fields and the inductances of the magnet using 2-ply and 3-ply HTS wires were analysed by the finite element method (FEM) program. Two types of double pancake coil (DPC) were fabricated using 2-ply and 3-ply HTS wires and experimented to obtain practical data. These fundamental data will be usefully applied to fabricate the entire system of a toroid-type HTS DC reactor.

2. Characteristics of multi-ply HTS wires

2.1. Specifications of multi-ply HTS wires

Multi-ply HTS wire is suggested to increase the critical current of conventional single-ply HTS wire. Multi-ply HTS wire could increase the critical current by stacking single-ply HTS wires several times. Thus, multi-ply HTS wires were fabricated from 12 mm 600 A at 77 K wire from SuNAM Co., Ltd. Figures 1 and 2 show the critical currents of single-ply HTS wire at 77 K and 20 K, respectively. The maximum output current of the power supply used for the measurement was 2000 A, so we could not measure the higher critical current.

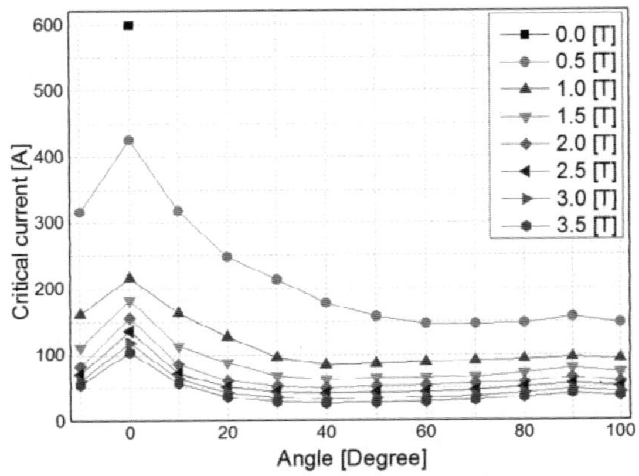

Figure 1. Critical current of single-ply HTS wire at 77 K

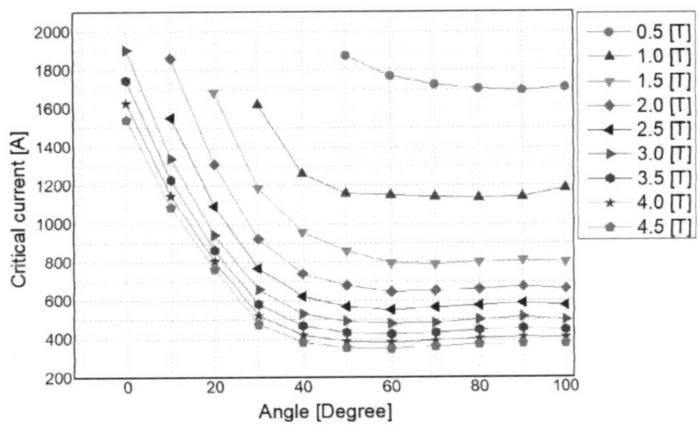

Figure 2. Critical current of single-ply HTS wire at 20 K

The HTS wire was stacked to 2-ply or 3-ply and then soldered using In-Sn. The average thicknesses of the fabricated 2-ply and 3-ply HTS wires were 0.34 mm and 0.48 mm, respectively. These thicknesses include stacked HTS wire, kapton tape for insulation, and solder for the HTS wire stacking. Figure 3 shows the critical current of 2-ply, 3-ply, and single-ply HTS wires at 77 K. The reference voltage for the critical current is 1 uV / cm. The critical currents of single-ply, 2-ply, and 3-ply were measured as 614 A, 1166 A, and 1713A, respectively. The 2-ply wire has a 95% critical current of twice the critical current of a single-ply wire, and the 3-ply wire has a 93% critical current of three times the critical current of a single-ply wire.

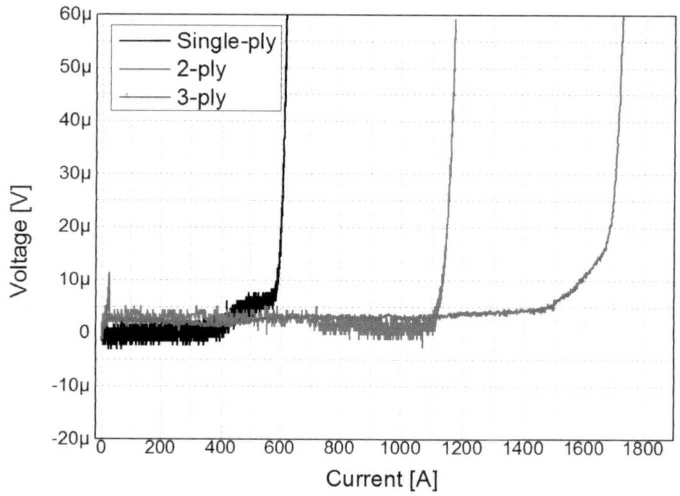

Figure 3. Critical currents of single-ply, 2-ply and 3-ply HTS wires at 77 K

2.2. Bending test of multi-ply HTS wires

2G HTS wire is fabricated in tape form according to the characteristics of the material. Therefore, the degradation of critical current due to bending occurs. The manufacturer of the HTS wire provides the information for the minimum double bending diameter of a single-ply HTS wire. The minimum double bending diameter of the single-ply HTS wire must be greater than 30 mm. However, in the case of the multi-ply HTS wire, it is necessary to measure the critical current according to the bending diameter because the ampacity characteristics of 2-ply or 3-ply wires were changed. Diameters from 70 mm to 40 mm were measured every 10 mm in a liquid nitrogen temperature. The HTS wire was bent using a constant weight as shown in figure 4.

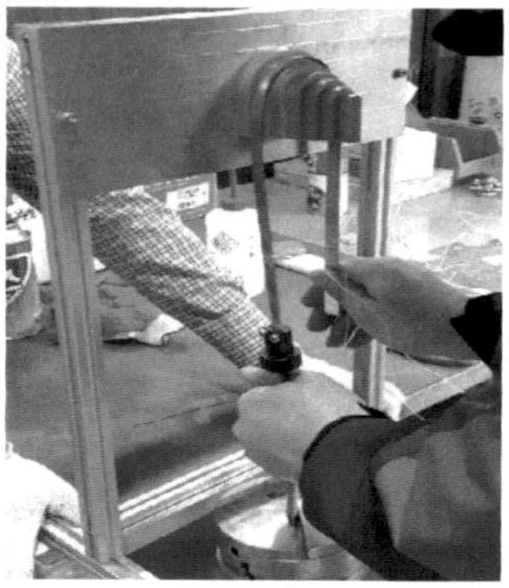

Figure 4. Jig for the HTS wire bending

The bent wire was straightened again, and then the critical current was measured in liquid nitrogen. Figure 5 shows the experimental results at the point of degradation of the critical current. The critical currents of the 2-ply and 3-ply HTS wires were decreased in the bending diameter ranges of 50 ~ 60 mm and 60 ~ 70 mm, respectively. In the figure, the voltage of 10 uV is due to the high current ramp rate of the power supply and is not a factor in the results.

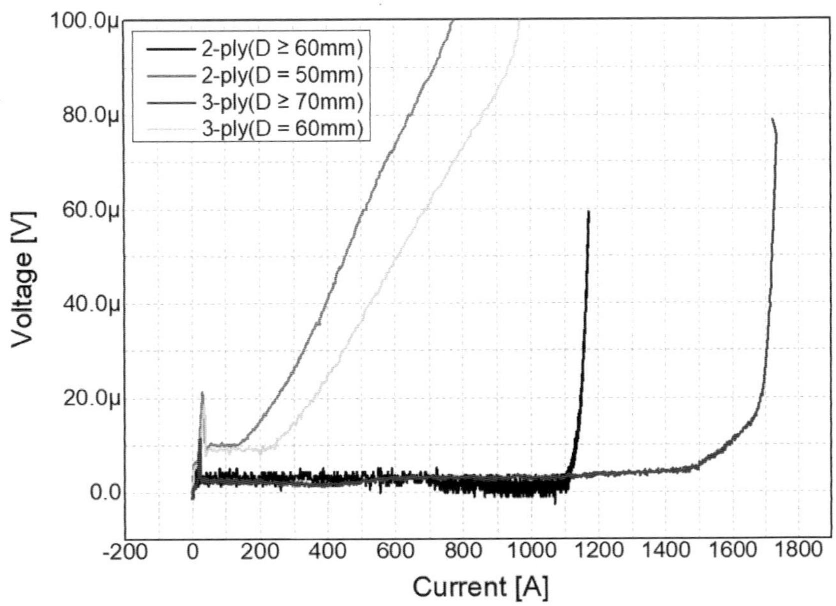

Figure 5. Measurement of the critical current while considering bending diameter

3. Design of the 1500 A, 400 mH superconducting toroid-type DC reactor coil

3.1. Design of the toroid-type DC reactor coil

Figure 6 shows the detailed structure of a D-shaped HTS coil and a toroid-type DC reactor magnet. The DPC was joined at the inner straight portion of the D-shaped coil with a 5 mm gap between the two single pancake coils. The distance from the center of the DPC to the center of the reactor was 295 mm, and each DPC was arranged at an angle of 12 degrees. In figure 6, the parameter "N" is the thickness of the HTS wire multiplied by the number of turns. N was determined through analysis by the design goals of the 1500 A and 400 mH.

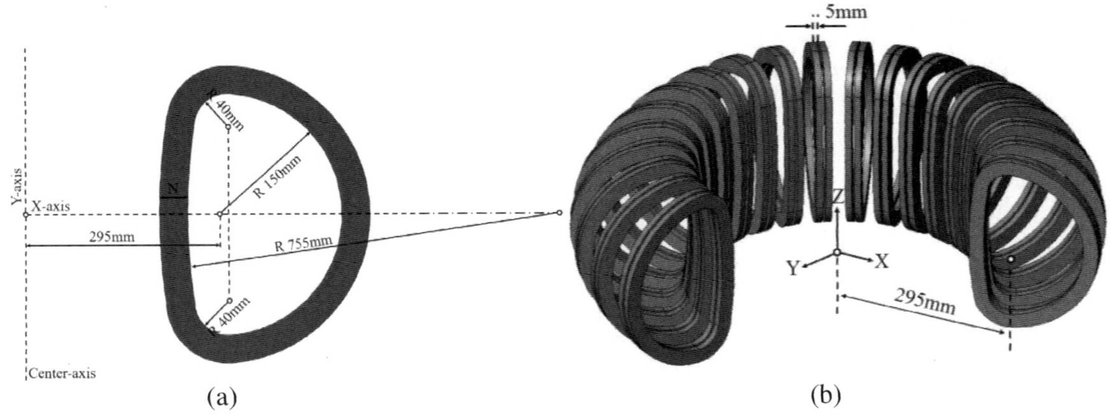

Figure 6. Detailed structure of the HTS DC reactor magnet: (a) a D-shaped HTS coil, (b) a toroid-type DC reactor magnet.

3.2. FEM simulation results of the DC reactor coil

Figure 7 shows the results of the magnetic field distribution analysis, which was conducted on one-half of the DPC with a selected boundary condition [10].

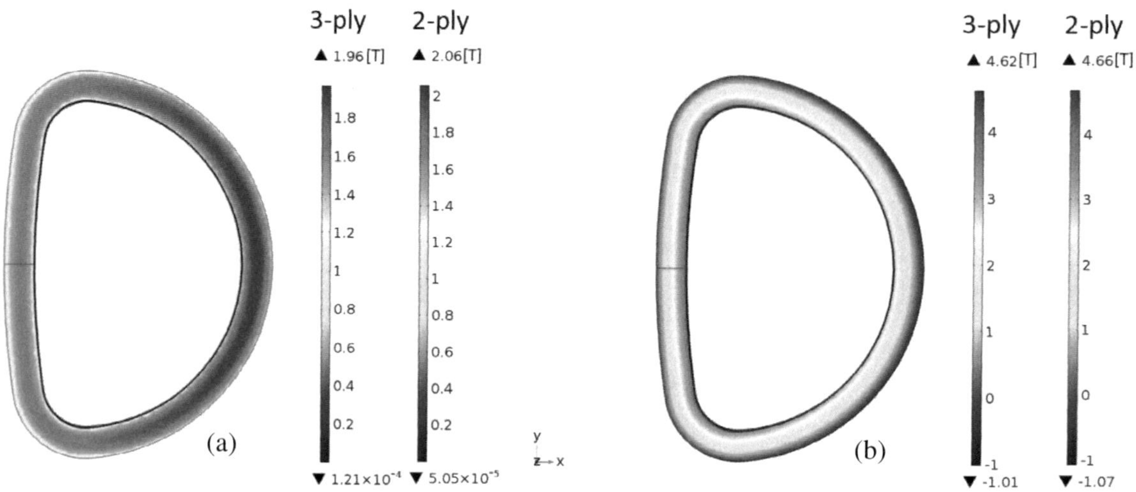

Figure 7. Electromagnetic analysis of the DC reactor coil: (a) the perpendicular magnetic field, (b) the parallel magnetic field

The values were applied to all of the DPCs in order to determine the overall parameters of the magnets. When the operating current was assumed to be 1500 A, the maximum perpendicular flux density and the maximum parallel flux density of the 2-ply HTS DC reactor magnet were 2.06 T and 4.66 T, respectively; in the case of the 3-ply DC reactor magnet, they were 1.96 T and 4.62 T, respectively. The calculation results of 2-ply and 3-ply inductances were 401.66 mH and 405.85 mH, respectively. The detailed specifications of the 1500 A, 400 mH toroid-type HTS DC reactor magnets are shown in Table 1.

Table 1. Specifications of the toroid-type multi-ply HTS DC reactor magnet.

Items	2-ply	3-ply
Wire width	12.2 mm	12.2 mm
Wire thickness	0.34 mm	0.48 mm
Co-winding SUS-tape thickness	0.1 mm	-
Number of turns	58 turns	58 turns
Reactor inductance	401.66 mH	405.85 mH
Perpendicular magnetic field at 1500 A	2.06 T	1.96 T
Parallel magnetic field at 1500 A	4.66 T	4.62 T
Calculated HTS DC reactor critical current at 20 K	2280 A	3380 A

4. Fabrication and performance test of a DPC module

DPC module bobbins for the critical current measurement of multi-ply HTS coils were produced. Figure 8 shows the detailed structure of a DPC module coil in a multi-ply HTS DC reactor. The 2-ply and 3-ply HTS wires were wound into the DPC modules for 58 turns.

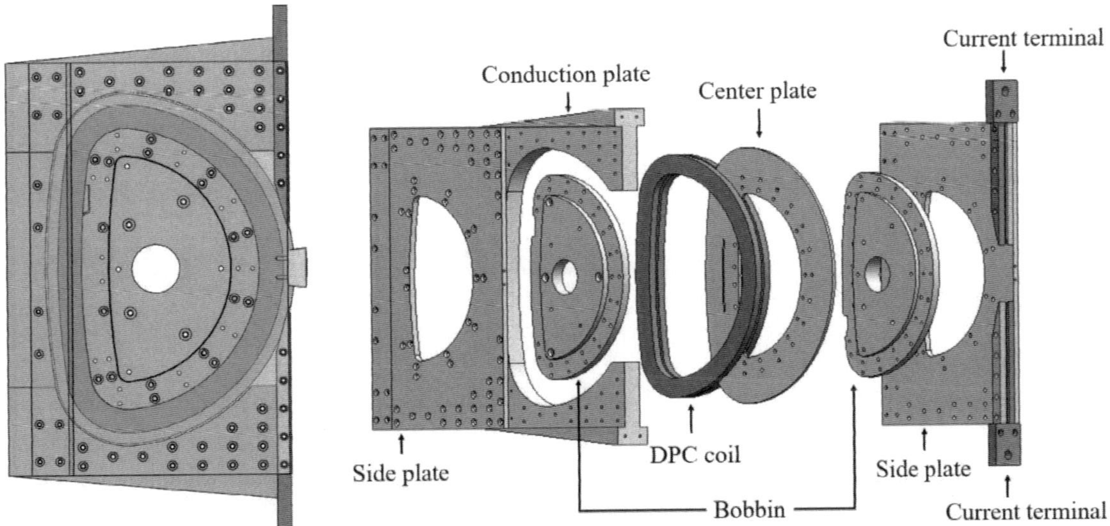

(a) (b)

Figure 8. 3D modeling of a fabricated DPC: (a) the assembled DPC module, (b) the parts of the DPC module

The calculated critical current of the 2-ply DPC module was 315 A, and the inductance was 5.62 mH. In the case of the 3-ply DPC module, the critical current was 420 A, and the inductance was 5.59 mH. The critical current was calculated by considering the intensity of the magnetic field and the angle dependency [11]. The critical currents of the 2-ply and 3-ply DPC modules were measured at 77 K. Figure 9 shows the actual DPC using 2-ply and 3-ply HTS wires. Voltage signal lines for the critical current measurement were attached near the current terminals.

(a) (b)

Figure 9. DPC module coil with signal line connection for voltage measurement (red circle): (a) 2-ply DPC, (b) 3-ply DPC

Figure 10 shows the critical current measurement results of DPC modules at 77 K. The experimental results show that the 2-ply and 3-ply DPC modules' critical currents were 322 A and 421 A, respectively. The experiment was performed at a current ramp rate of 1 A / s. As a result, as the voltage is biased to about 5.7 mV, the inductance is 5.7 mH due to V = L (di / dt).

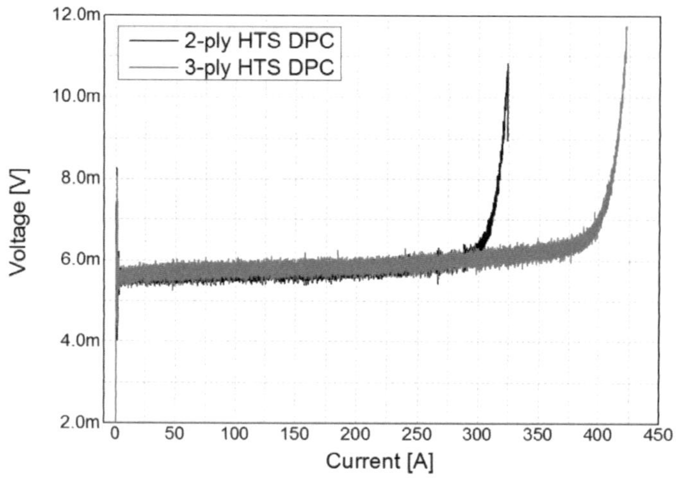

Figure 10. The critical current measurement results of DPC modules

5. Conclusions

In this study, a 1500 A, 400 mH D-shaped toroid-type HTS DC reactor magnet with multi-ply HTS wire was designed. The characteristics of the multi-ply HTS wire and the reduction of the critical current according to the bending diameter were presented. The multi-ply HTS wire included 2-ply and 3-ply HTS wires. The critical current degradation of the multi-ply HTS wire was less than 7%. The operating current and inductance of the designed DC reactor were concluded to be 1500 A, 401.66 mH (2-ply), and 405.85 mH (3-ply). The critical current of the 2-ply HTS DC reactor, when considering the intensity and angle of the magnetic field at 20 K, was 2280 A; it had a margin of about 34% at the operating current of 1500 A. In the case of the 3-ply HTS DC reactor, 3380 A had a margin of about 56%. Considering that the critical current margin of the superconducting applications is about 30% on average, it can be confirmed that the target value of this paper can be achieved sufficiently with the 2-ply HTS wire. In the case of an HVDC system, which needs a higher current capacity, 3-ply (or more) HTS wire can be used. Through the experiments, the performance characteristics (inductance and critical current) of a DPC module for the toroid-type HTS DC reactor magnet were tested in a liquid nitrogen temperature. The fundamental data will be applied to an actual-scale HTS DC reactor system on a real HVDC system.

Acknowledgment

This research was supported by Korea Electric Power corporation [grant number: R16XA01]

References

[1] Kim K, Kim J G, Jung H, Kim S, Lee S, Park M and Yu I K 2013 Design of a 400 mH 400 A toroid-type HTS DC reactor magnet *IEEE Trans. Appl. Supercond.* **23** no.3 p 4601104
[2] Kim K, Go B S, Park H C, Kim S K, Kim S, Lee S, Oh Y, Park M and Yu, I. K 2015 Development of toroid-type HTS DC reactor series for HVDC system *Physica C: Supercond. Appl.* **518** pp 159-65
[3] Schoenung S M, Meier W R, and Schafer W J, 1991 A comparson of largescale toroidal and solenoidal SMES systems *IEEE Trans. Magn* **27** pp 2324–8
[4] Higashikawa K, Nakamura T, Shikimachi K, Hirano N, Nagaya S, Kiss T and Inoue M 2007 Conceptual design of HTS coil for SMES using YBCO coated conductor *IEEE Trans. Appl. Supercond.***17** no.2 pp 1990–3

[5] Higashikawa K, Nakamura T, Sugano M, Shikimachi K, Hirano N and Nagaya S 2008 Performance improvement of YBCO coil for highfield HTS-SMES based on homogenized distribution of magnetically mechanically influenced critical current *IEEE Trans. Appl. Supercond.* **18** no.2 pp 758–61
[6] Lee S, Kim W S, Park S H, Lee J W, Park C, Bae J H, Seoung K C, Lee H, Lee J H, Choi K and Hahn S 2010 Design of HTS modular magnets for a 2.5 MJ toroidal SMES : ReBCO vs BSCCO *IEEE Trans. Appl. Supercond.* **20** no.3 pp 1324–8
[7] Lieurance D W 1997 Global cost optimization of 1-10 MWh toroidal SMES *IEEE Trans. Appl. Supercond.* pp 14–7
[8] Kwak S Y, Park M J, Kim W S, Hahn S Y, Lee S W, Lee J K, Choi K D, Han J H, Bae J H, Kim S K, Kim K D, Kim H J, Seong K C, Jung H K and Hahn S Y, 2008 The Optimal Design of 600 kJ SMES Magnet Based on Stress and Magnetic Field Analysis *IEEE Trans. Appl. Supercond.* **18** no.2 pp 713–6
[9] Park M J, Kwak S Y, Kim W S, Lee S W, Hahn S Y, Lee J K, Han J H, Choi K D, Jung H K, Seong K C and Hahn S Y 2007 Stress analysis of HTS magnet for a 600 kJ SMES *IEEE Trans. Appl. Supercond.* **17** no.2 pp 1994–7
[10] Lee K, Sim K, Lee C and Kim S 2015 Estimation of the critical current of race-track HTS magnet considering angular dependency *Progress in Supercond. Cryogen.* 17 no.3 pp 47-50
[11] Kang M, Koo M, Lee H, and Cha G 2009 Current Estimation of an HTS BSCCO Magnet Having Multiple Power Sources Based on the Field Dependent E–J Relation *IEEE Trans. Appl. Supercond.* **19** no.3 pp 1257-61

Design and Demonstration of a Double-pancake Coil for SMES using MgB_2 Multi-strand cable

T Yagai[1], S Mizuno[1], T Okubo[1], S Mizuochi[1], M Kamibayashi[1], M Jinbo[1], T Takao[1], Y Makida[2], T Shintomi[2], N Hirano[3], T Komagome[4], K Tsukada[4], T Onji[5], Y Arai[5], M Tomita[5], D Miyagi[6], M Tsuda[6], and T Hamajima[4]

[1] Faculty of Science and Technology, Sophia University, 7-1, Kioi, Chiyoda, Tokyo, Japan 102-8554.
[2] High Energy Acceleration Research Organization, 1-1, Oho, Tsukuba, Ibaraki, Japan, 305-0801.
[3] Chubu Electric Power, 1, Higashi-Shinmachi, Higashi-ku, Nagoya, Aichi, Japan, 461-8680.
[4] MAYEKAWA MFG. Co., Ltd, 3-14-15, Botan,Koto-ku, Tokyo, Japan, 135-8482.
[5] Railway Technocal Research Institute, 2-8-38, Hikari-cho, Kokubunji, Tokyo, Japan. 185-8540.
[6] Faculty of Engineering, Tohoku University, 6-6-05, Aramaki, Aoba, Sendai, Miyagi, Japan. 980-8579

tsuyoshi-yagai@sophia.ac.jp

Abstract. MgB_2 round wires are now commercially available with applicable critical current density in magnetic fields up to 5 T. One of the promising applications using MgB_2 is a superconducting magnetic energy storage (SMES) coil. In our project, multi-strand cables with 600 A nominal current are designed for double-pancake (DP) coils with 400 mm inner diameter. The coil production processes in our project are wind-and-react (W&R) and react-and-wind (R&W) methods, in which a Rutherford-type cable consisting of eight (W&R) or ten wires (R&W) with specific twist pitch is reacted and then used to wind the DP coils, to investigate the suitable manufacturing process for MgB_2 coils. The final goal is to fabricate five-stacked DP coils, made up of four DP coils by the R&W method and one DP by the W&R method with a total number of 512 turns, forming a 30 kJ SMES coil and to demonstrate compensation of voltage fluctuation on the BUS line of 48 V DC in a micro grid consisting of renewable energy source and liquid hydrogen supply system in our project. The first DP coils made by the two manufacturing processes are constructed and tested with regard to their properties such as critical current density in liquid helium to keep uniform temperature distribution throughout the coils.

1. Introduction

MgB_2 round wires attract the interests of researchers from the point of view of applying them to large-scale power devices like superconducting magnetic energy storage (SMES). For compensation of power variations produced by photovoltaic cells and wind turbines, our group has developed a new DC-microgrid system (48 V), in which the voltage variations due to fluctuating weather conditions would be well controlled by using a combination of a small SMES magnet made of Bi2223, a fuel cell, and a water electrolysis device [1][2]. The system has worked successfully, however, manufacturing process of SMES coils consisting of cost-effective wire should be pursued for installation in the real power grid system.

The MgB_2 production process resembles Nb_3Sn production, which needs heat treatment, and the strain sensitive characteristics of superconducting properties. Unlike the Nb_3Sn strand, MgB_2 wire has bending strain sensitivity even before the heat treatment. This might cause unexpected deterioration during conductor and coil fabrication process in any production method. Although the W&R process seems reasonable for the construction of these coils, the R&W process deserves our attention in terms of simple and easy fabricating features and turn-to-turn insulation, where applied strain to the wires is acceptable. Although many research groups have started the investigation of characteristics of cable using MgB_2 wires, the current capacities are not suitable for large-scale application or coil winding [3-7]. Our aim in this study is to design double-pancake (DP) coils for the SMES using a conductor with large current capacity by twisting the MgB_2 wires with allowable strain level. First, we performed a design study of a Rutherford-type cable for fabricating large-scale DP coils based on the differential geometry and space curve theory to optimize the cable parameters such as twist pitch and cable toughness. Second, the DP coils were produced by introducing both R&W and W&R methods to investigate the feasibility of the processes for large-scale coil production. Finally, we will present the first demonstration results of large-scale DPs using multi-stranded conductors which satisfy the applied bending strain criterion of strain-sensitive MgB_2 strands, at liquid Helium temperature.

2. Specifications of MgB_2 conductors and coils

2.1. Conductor specifications

Schematics of strand and conductor designed for DP coil is the Rutherford-type cable, as shown in figure 1, and their specifications are shown in table 1. The strands selected for making DPs are provided by Hyper Tech Inc., named NM strand. The diameter of the strand is almost identical 0.83 mm for 30-NM and 0.84 mm for 24-NM. The number in the strand name represents the number of filaments. The Cu ratio of strands are 12 % and 20 % for 30-NM and 24-NM, respectively. The permissible bending strain which is specified by the manufacturer, is 4 % and 0.23 %, before and after heat treatment, respectively.

Table 1. Specifications of strands and conductors.

Items		Conductor for DP-1 (W&R)	Conductor for DP-2 (R&W)
Strand	Type	30-NM	24-NM
	Diameter [mm]	0.83	0.84
	Heat Treatment Status	Un react	Unreact
	Num. of filaments	30	24
	Jc [A/mm^2] @20K, 2T	1780	
	Fill Factor	0.2	0.11
	Cu ratio [%]	12	20
	Permissible strain (before / after heat treatment)	4 / 0.23 %	
Conductor	Num. of Strands	8+4CuZn	10
	Former	Cu	NA
	Dimension [mm]	5.06×2.86	4.35×1.43
	Insulation	S-glass	kapton

In this research, we introduced a Rutherford-type, rectangular cross-section conductor for fabricating the DP coils. The rated current is determined to be 600 A for development of large-scale MgB_2 conductor for actual coil winding. To realize the current capacity, we considered not only the number of strands but also the thermal stability in case of quench event. The conductor using 30-NM, which has low Cu ratio, has Cu mandrel and four CuZn strands in it. On the other hand, the conductor using 24-NM has no metals in the conductor design with enough stability using 10 strands. The rated transport current is 600 A, which is almost half of the maximum current capacity estimated by the temperature limitation under perfect adiabatic conditions and yields 1.76 T maximum magnetic flux density for a net stored energy of 30 kJ by stacking two types of MgB_2 coils.

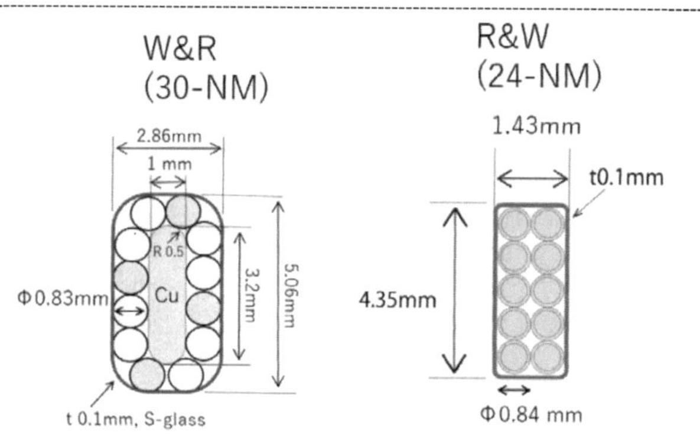

Figure 1. Schematic view of conductor cross sections both for W&R and R&W coils

2.2. Specification of double-pancake coils

The schematic view of the cross section of the stacked DP coils is shown in figure 2. The items listed in table 2 are the specifications of the DP-1, DP-2, and SMES coil. In figure 2, there are two types of coils, using different types of conductors. One is a DP by the conductor with NM30 while the other is DP coils by the conductor using NM24. The difference between the two DP coils is not only in the conductors used but also the production method. The DP-1 coil using NM30 conductor is fabricated by using the W&R method, while DP-2 is by using the R&W method. The reason for introducing two different methods is to demonstrate which one is the best as a production method for large-scale coil fabrication of future SMES coils using MgB_2. At this moment, the R&W method has some advantages against the W&R method with regard to forming inter-turn insulation, impregnation, preventing heat treatment of terminal and easier dimension control. Therefore, we determined the ratio of number of DPs, one W&R coil (DP-1) and four R&W coils (DP-2,3,4,5). The whole coil would have an inner diameter of 400 mm and an outer diameter of 606 mm, with the height at around 55 mm.

Table 2. Specifications of double-pancake coil

Items (DP)	DP-1	DP-2	SMES Coil (5-DP coils)
Inner diameter[mm]		400	
Outer diameter[mm]		606	
Height[mm]	11.1	11	55
Inductance[mH]	3.9	9.1	170
Rated current[A]		600	
Num. of turns	72	110	512
Maximum B field[T]	0.4	0.61	1.76
B field at the center[T]	0.11	0.17	0.77
Stored energy [kJ]	0.7	1.6	30

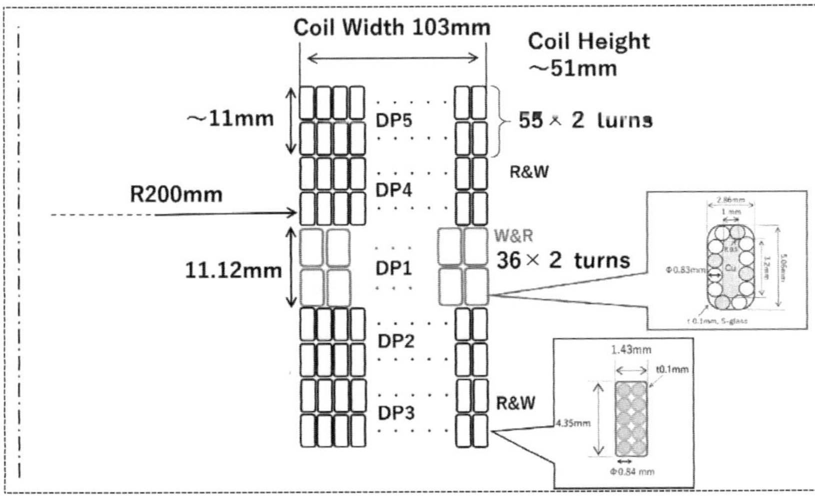

Figure 2. Schematic of cross section of 30 kJ SMES coil. Two types of DPs will be used to confirm the feasibility of both R&W and W&R production methods

3. Theoretical Background

3.1. Governing equations of space curve

Because the conductors are fabricated by twisting and assembling the strands, resultant strand traces in the coil form complex space curves. To calculate the traces, we introduce the governing equations of the space curve. First, the given position on a strand section is expressed as the following vector, as shown in figure 3, that is,

$$\boldsymbol{r}(t) = x(t)\boldsymbol{i} + y(t)\boldsymbol{j} + z(t)\boldsymbol{k} \tag{1}$$

where t is a parameter $\boldsymbol{i}, \boldsymbol{j}, \boldsymbol{k}$ are unit vectors in the x, y, z directions, respectively. The tangent vector $\boldsymbol{t}$ is obtained by differentiating the equation (1) with respect to parameter t,

$$\boldsymbol{t} = \frac{d\boldsymbol{r}}{dt} = \frac{dx}{dt}\boldsymbol{i} + \frac{dy}{dt}\boldsymbol{j} + \frac{dz}{dt}\boldsymbol{k} \tag{2}$$

$$s = \int_a^b \left|\frac{d\boldsymbol{r}}{dt}\right| dt = \int_a^b \sqrt{\left(\frac{dx}{dt}\right)^2 + \left(\frac{dy}{dt}\right)^2 + \left(\frac{dz}{dt}\right)^2}\, dt \tag{3}$$

The curvature vector $\boldsymbol{R}$, which is a key parameter for estimating bending strain, will be obtained by differentiating the tangent vector in equation (2) with respect to the arc length in equation (3), resulting in,

$$\boldsymbol{R} = \frac{(\boldsymbol{t} + d\boldsymbol{t}) - \boldsymbol{t}}{ds} = \frac{d^2x}{ds^2}\boldsymbol{i} + \frac{d^2y}{ds^2}\boldsymbol{j} + \frac{d^2z}{ds^2}\boldsymbol{k} \tag{4}$$

To evaluate the bending strain, we need the curvature radius, which will be calculated by taking the absolute value of the curvature vector $\boldsymbol{R}$ in equation (4). The inverse of the radius is expressed as follows [8],

$$\frac{1}{\rho}=\sqrt{\left(\frac{d^2x}{ds^2}\right)^2+\left(\frac{d^2y}{ds^2}\right)^2+\left(\frac{d^2z}{ds^2}\right)^2} \tag{5}$$

By using these equations, the bending strain ε applied to the strands in conductor and coil windings is represented as

$$\varepsilon=\frac{r_{fil}}{\rho}\times 100\ [\%] \tag{6}$$

where r_{fil} is the radius of filament region. Once the strand traces are calculated, the strain distributions along strand lengths are obtained.

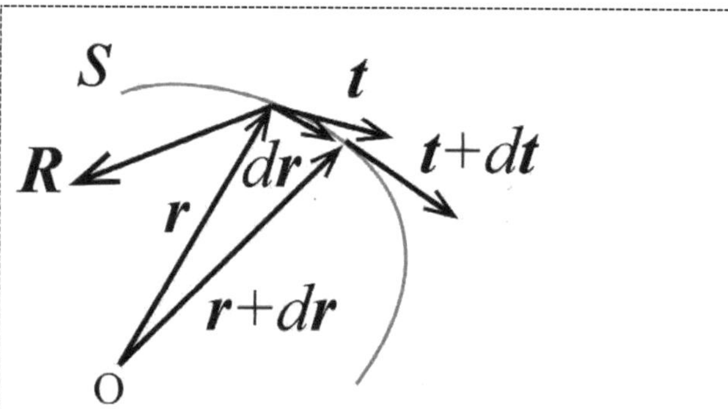

Figure 3. Definition of three-dimensional vectors for deriving space curve equations

4. Analysis of strain dependence on the twist pitch

When we take a twist pitch, the strand traces are obtained. The differential geometry gives us the local bending radius distribution along strand axis, then the strain distribution is estimated. By taking maximum value of the strand, one twist pitch design gives one representative strain value.

Figure 4 (a) shows the result of strain calculation for the Rutherford-type cable using the 30-NM strand.

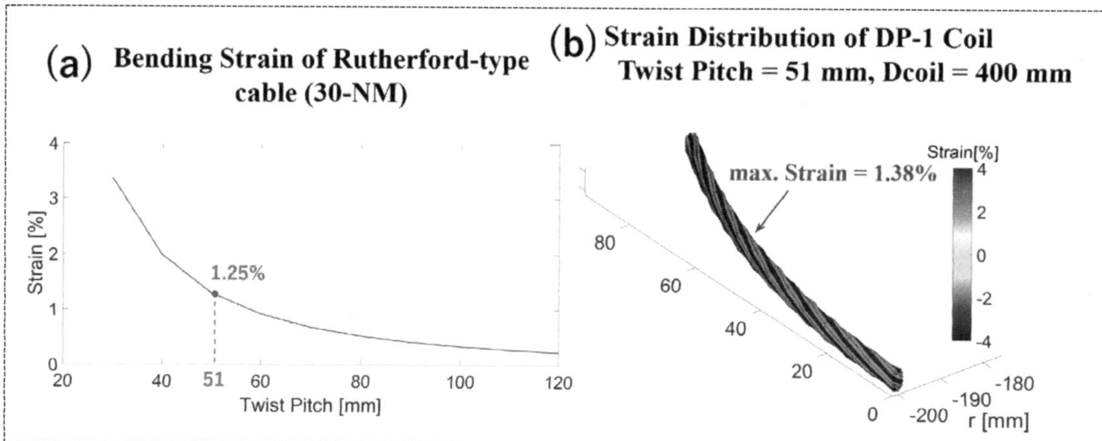

Figure 4. (a) Strain Calculation vs. twist pitch for 30-NM Rutherford conductor and (b) strain distribution in DP-1 coil in case of 51 mm twist pitch.

The range of twist pitch for strain calculation is from 30 mm to 120 mm. As for the Rutherford-type cable for W&R DP-1 coil, the strain is well below the permissible strain 4 % which is specified by the manufacturer. With regard to the twist pitch, any pitch is applicable for conductor and DP coil fabrication. However, the twisting machine is not able to select the continuous pitch length because the pitch is determined by the combination of gears in the machine. Besides, the conductor toughness is also an important parameter to make the winding process easier. Finally, the 51 mm twist pitch was selected for the DP-1 conductor.

Figure 4 (b) shows a part of the calculated strain distribution of DP-1. The maximum strain appears at the edge of the conductor, while the strain is still lower at flat surface of the conductor. The value is 1.38 %, slightly higher than the maximum strain in the straight conductor. Once the coil is wound, the strain remains constant because DP-1 introduces the W&R method. The rated current of 600 A is expected to be realized in DP-1 coil due to enough margin of bending strain.

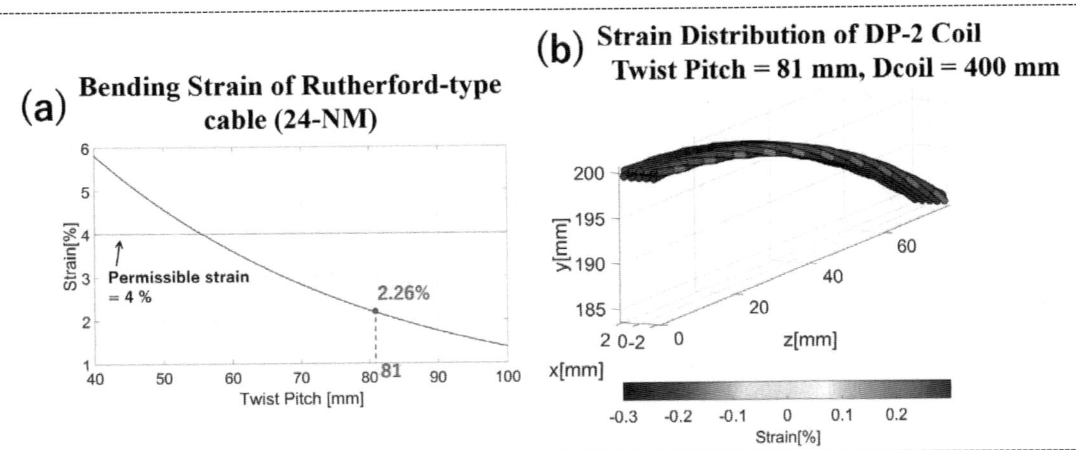

Figure 5. (a) Strain Calculation vs. twist pitch for 24-NM Rutherford cable and (b) strain distribution in DP-2 R&W coil in case of 82 mm twist pitch.

Figure 5(a) shows the result of strain calculation in Rutherford-type cable without mandrel using 10-24-NM strands. The range of strain calculation is performed from 40 mm to 100 mm. Unlike the 30-NM Rutherford conductor using 30-NM with Cu mandrel, the thin dimension due to the mandrel-free design causes higher strain around the edge of the conductor. Therefore, a longer twist pitch should be selected for enough strain margin. Therefore, the 82 mm pitch was selected but with reduced thickness compared to the original thickness of 1.68 mm for getting better tightness.

Figure 5(b) shows a part of strain distribution in DP-2 R&W coil. After conductor fabrication, the Rutherford-type cable is wound on a specialized bobbin with 800 mm for heat treatment. In the treatment process, the strain in the conductor is initialized and the permissible strain becomes 0.23 %. After winding, the maximum strain appears at the outer flat side of the winding, which shows 0.17 % strain still below the permissible strain. As a brief summary of the design activity, the analysis ensures that the coils may be operated without any problem.

5. DP coil test at liquid helium temperature

Figure 6 shows the schematic diagram of settings of DP-1 and DP-2 tests. The inset is the picture after finishing the setup. The DPs are series-connected to feed the same transport current and intentionally exposed to a more severe magnetic field and electromagnetic force conditions. The power supply has enough capacity to feed larger current than the rated one, up to 700 A. The coil voltages are measured by a nano-volt meter, to prevent the quench event by sending a signal to the circuit breaker. To verify the field generation, we put the hole probes at the center of the coil and 165 mm radial position, which detects almost maximum field of the coil. The DPs are immersed into liquid helium at 4.2 K without any external field. The coil current shows that the DP coils are in good operation, without enormous voltage occurrence. The maximum B field at the 165 mm in radial position is 0.54 T, which is identical to 0.53 T calculated by FEM using COMSOL Multiphysics software. This is the first demonstration of large-scale DP coil operation using conductors consisting of multi-stranded configuration. As the next step, we will prepare for the LH2 indirect cooling experiment in which the operating temperature is in the range of 20.2 K to 25 K, and also DP-3,4,5 coil production.

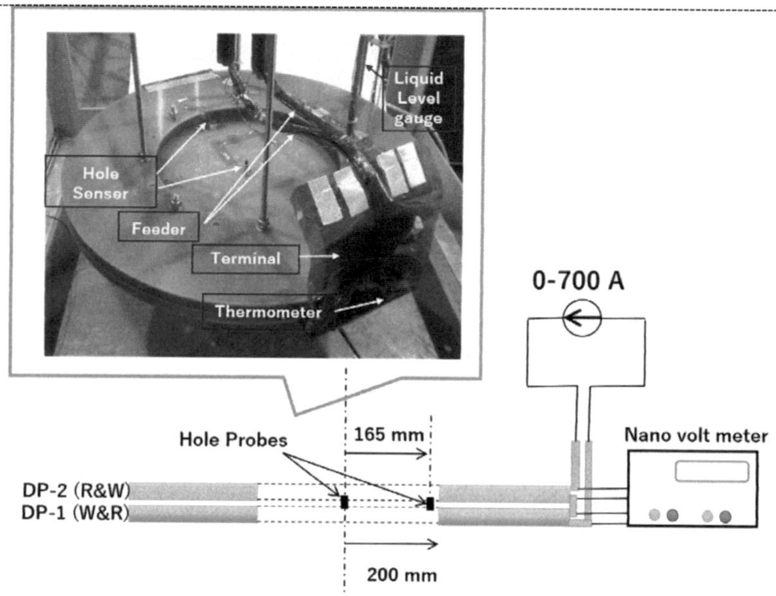

Figure 6. Experimental setup of series-connected DP-1 and DP-2 at liquid helium temperature. The external field is not applied in this demonstration.

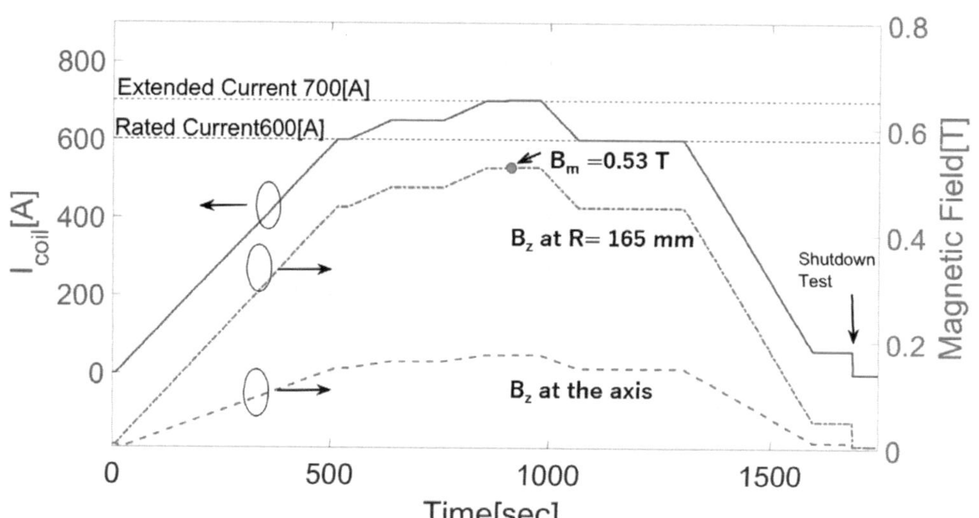

Figure 7. Experimental result of continuous current feed to the coils up to 700 A. The ramp up and down of the transport current was successfully performed without any fault.

6. Conclusion

We carried out the design study of the coil for large-scale, 30 kJ SMES for investigating the feasibility of commercially available MgB_2 strands. The conductors forming large-scale coils are designed based on the Rutherford-type cable geometry with specified twist pitches in which the applied strains are well below the permissible strains of strain-sensitive MgB_2 strands. The first DP coils fabricated by introducing both the R&W and W&R methods showed excellent performance at liquid helium temperature, proving the validity of the design parameters of the conductors. This is the first and only demonstration of a large-scale DP coil operation using multi-stranded MgB_2 conductors all over the world. In the next phase, the remaining three DP coils will be produced in the same way as the process used for DP-2 production and liquid hydrogen indirect-cooling operation to demonstrate the compensation of power fluctuation in a SMES device.

Acknowledgments

This work was supported by JST ALCA Grant Number JPMJAL1002, Japan.

References

[1] Hamajima T, Amata H, Iwasaki T, Atomura N, Tsuda M, Miyagi D, Shintomi T, Makida Y, Takao T, Munakata K and Kajiwara M, 2012 *IEEE Trans. Appl. Supercond.*, **22** 5701704.
[2] Miyagi D, Sato R, Ishida N, Sato Y, Tsuda M, Hamajima T, Shintomi T, Makida Y, Takao T and Iwaki K. 2015 *IEEE Trans. Appl. Supercond.*, **25** 5700305.
[3] Musenih R, Greco M, Razeti M and Tavilla G, 2007 *Supercond. Sci. Technol.* **20** 235.
[4] Kovac P, Hušek I, Melišek M and Holúbek T, 2008 *Supercond. Sci. Technol.* **21** 125003.
[5] Holúbek T, Schlachter I S and Goldacker W, 2009 *Supercond. Sci. Technol.* **22** 055011.
[6] Holúbek I, Kováč P, Melišek T and Kopera L, 2009, *Cryogenis,* **49** 366.
[7] Kováč P, Hušek P, Melišek T, Kopera. L and Reissner M, 2010, *Supercond. Sci. Technol.* **23** 025014.
[8] G. Farin, *Curves and Surfaces for CAGD 5th edition*, 2002 (Morgan Kaufmann)
[9] Kopera L, Kováč P, Kulich M, Melišek T, Rindfleisch M, Yue J and Hušek I, 2017, *Supercond. Sci. Technol.* **30** 015002.

Superconducting rotating machines: A review of the past 30 years and future perspectives

T Nitta

The University of Tokyo, Tokyo 113-8656, Japan (emeritus professor)

Abstract. Research and development of superconducting rotating machines has a history spanning greater than 30 years. Earlier, superconducting homo-polar DC machines that used low temperature superconductors may have found the first practical applications in superconducting rotating machines; however, it cannot be said with certainty, given their use in secretive war ships. Superconducting synchronous generators cooled by liquid helium are well developed, particularly by the Super-GM project in Japan. The paper discusses fundamental subjects in the design, manufacture, operation, control, and performance in power systems. The development of oxide superconducting wires and bulks, has found applications in superconducting electric motors (mainly synchronous ones), the structures of which are very similar to the generators. The paper also discusses the power system problems addressed by superconducting equipment, which have arisen due to the increase in the number of renewable sources of energy. The paper assesses the solutions offered by superconductivity to the above mentioned challenges.

1. Introduction

Research and development on applied superconductivity to electric machines began in the 1960's after researchers developed superconducting wires for direct current with stable performance.

In general, electrical machinery can be broadly classified into transformers and rotating machines. Alternating current flows in a transformer. On the other hand, direct current is used for field windings of DC machines and synchronous machines.

At the time, large capacity DC machines were required for ship propulsion, given that the technologies of power electronics were under development.

DC machines can be broadly categorized into homo-polar and hetero-polar machines. The hetero-polar machines, which are widely used for motion control such as in electric cars, have limitation for larger capacity due to problems on commutation. On the other hand, homo-polar machines require higher magnetic induction without the presence of magnetic material. This requirement led to the development and application of superconducting magnets.

A demand for large capacity synchronous generators, whose capacities at that time were less than those of steam turbines, let to the use of superconducting field windings.

In the early days, metal superconductors cooled by liquid helium were used to develop superconducting rotating electric machines such as DC homo-polar machines and synchronous generators.

IOP Conf. Series: Journal of Physics: Conf. Series **1054** (2018) 012081 doi:10.1088/1742-6596/1054/1/012081

At present, induction motors are being developed through the use of superconductor cooling systems that utilize the resistive states of superconductors. The items for obtaining the summaries of developing superconducting rotating machines are presented.

2. Rotating electric machine by use of low temperature superconductor [1]

This section introduces homo-polar DC machines and synchronous generators, whose application of DC superconducting magnets cooled by liquid helium produced successful results.

2.1 Homo-polar machines

The need for large capacity DC machines for electric ship propulsion systems led to the development of superconducting homo-polar DC machines. At present, several electric propulsion ships are operated by induction motors with power electronics devices. However, the superconducting homo-polar machines are considered because of their characteristics from an electrical machinery perspective.

2.1.1 Structure

Figure 1 and 2 illustrate two types of superconducting DC homo-polar machines, that is, disc type and drum type, respectively.

In the disc type homo-polar machine, the induced voltages are applied to the armature disc using superconducting magnets. The schematic of the armature disc is illustrated in Figure 1 (right inset). The use of superconducting magnet provides higher magnetic flux density than conventional magnets.

In rotating electrical machines, armature reaction phenomenon is an important problem. In particular, in homo-polar machines, the magnetic flux due to the armature current near the armature can be cancelled by the armature current flowing through the reaction torque disc. Such machines developed to overcome armature reaction have exhibited good performance. However, the disc type structure can lead to problems associated with mechanical vibration.

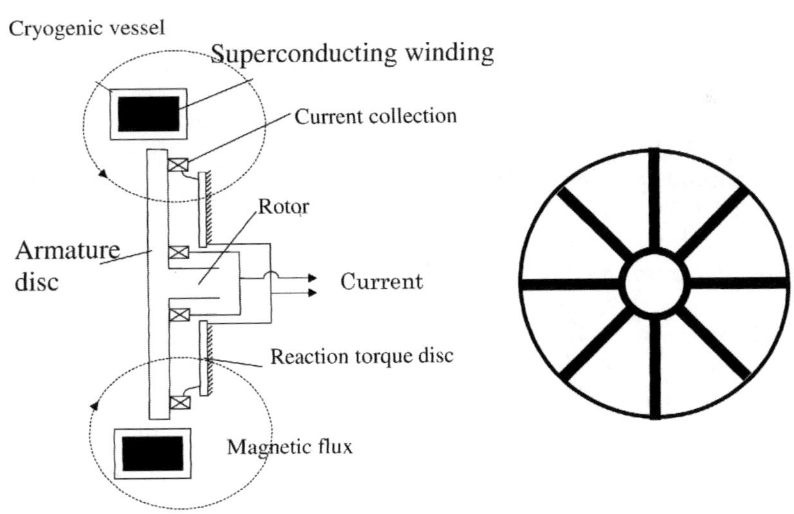

Figure 1 Disc type homo-polar DC machine
Left: cross section Right: armature disc

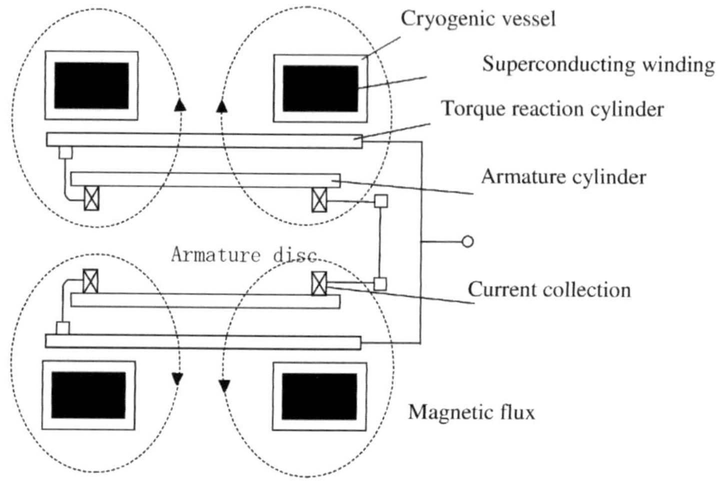

Figure 2 Drum type Homo-polar DC machine

2.1.2 Properties

The homo-polar DC machine behaves as current source. A critical aspect is the performance in terms of armature current collection. Significant research efforts have been focussed on developing brushes for current collection; however, an optimal solution has not been found as yet.

These machines have been used for thee electrical propulsion of war ships. Then, there are not so many papers opened.

In the early 1980's, a production line of the disc type-machines was observed at the International Research and Development facility in Newcastle upon Tyne, UK.

However, in present day, large capacity DC machines are much less desired. Instead, large capacity machines that facilitate speed control are being developed through the use of induction motors with power electronic technologies. Even to this day, there are several electric propulsion ships, such as MS Queen Elizabeth. However, the newer applications of these machines will not be developed unless characteristics, such as current source and armature reaction, are not addressed.

2.2 Synchronous generator [2][3]

Synchronous generators were developed by improving cooling methods for field and armature windings. However, conventional cooling methods did not suffice for the cooling requirements of large capacity machines. In such cases, the application of superconducting technology could realize better machines from not only a capacity point of view but also others. This potential was further explored by electric and power system engineers. Subsequently, research and development of this technology was actively pursued in the 1970's in USA, USSR, Germany, France, and so on. However, these research projects have been pretty much terminated. The national project in Japan (also known as Super-GM) (1988–2004) may in fact be the last such research project. The developments in the design, manufacturing, and operation have matured. The following structure and the properties are based mainly on the research and development of Super-GM.

2.2.1 Structure

In the field of rotating metal superconducting winding cooled by liquid Helium and the armature of conventional winding without magnetic materials, the fundamental structure has been defined as shown in Fig. 3

Liquid He is supplied through the transfer coupling to the rotor. The evaporating He from the rotor cools the torque tube and the radiation shield (cold damper). Subsequently, the gas escapes through the transfer coupling. The rotor comprises space to accumulate He, winding retaining cylinder, vacuum space, radiation shield, vacuum space, and room temperature (warm) damper. The stator consists of an air gap armature winding and a magnetic shield.

Helium transfer coupling is achieved through the use of a fluid magnetic seal. Thus, liquid He can be supplied stably and efficiently from the stationary part to the rotating parts. Thus, the He gas evaporating in the rotor is transferred from the rotating part to the stationary part through the coupling. Let us consider the behaviour of liquid and gaseous helium in the rotor. Inside the rotor, the liquid He flows to the outer part due to the density gradient and gaseous He flows towards the inner part due to the centrifugal force. This phenomenon is referred to as the "thermal siphon effect". Thus, the pressure of gaseous He in the inner part of rotor is reduced, thereby enabling the supply of liquid He to the inside of the rotor. This effect is called the "self-pumping effect". The discovery of above effects have led to the simplification of the interior structure of the rotor.

On the other hand, the torque tube shrinks due to the difference between room temperature and cryogenic temperature. The resultant shrinkage exerts stress on the tube. To address this issue, two shrink relaxation methods have been developed—the double bearing method and the flexible support method.

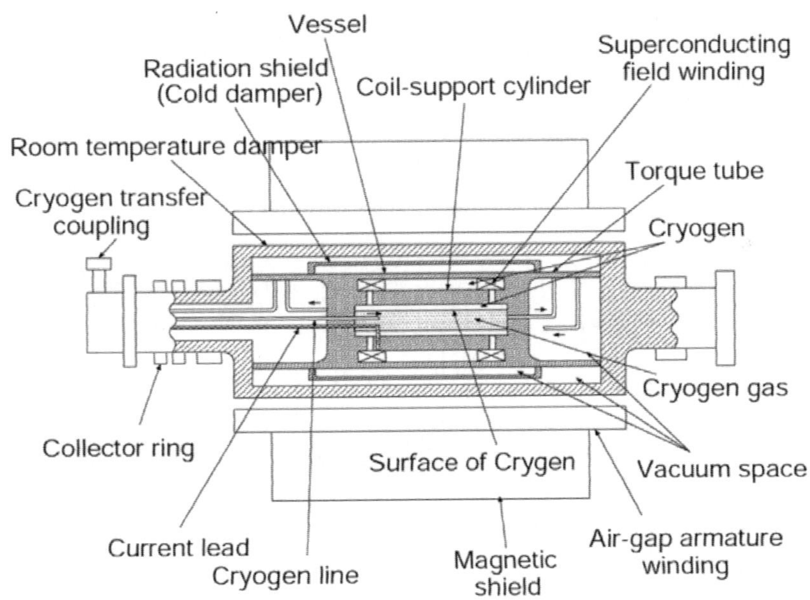

Figure 3 Cross section of superconducting generator cooled by liquid He

As illustrated in Figure 3, the superconducting field winding is affixed on the cylinder of the nonmagnetic material using slots. The shape is of the saddle is similar to that of a conventional machine. The rotor has two shields, referred to as called cold damper and room temperature damper (warm damper). The cold damper, which is affixed in the vacuum space between room temperature and cryogenic temperature, has two functions. First, it acts as a radiation shield. In addition, it also shields the field winding from the magnetic flux generated by armature current. It is cooled by the evaporated

He gas in the rotor. The room temperature damper is affixed to the surface of the rotor. The damper plays the same role as a damper winding in a conventional machine.

The armature winding consists of copper winding. The winding is affixed onto the cylinder of non-magnetic material by using slots. This reduces the synchronous impedance.

Field winding and armature one are directly in magnetic flux, then, stand for the magnetic strength. The mechanical support system of the winding is critical to the functioning of the machine.

2.2.2 Properties

As mentioned earlier, the superconducting synchronous generator provides has several advantages over conventional generators, such as
larger capacity, Higher efficiency, particularly for partial loads, Improved power system stability, Larger capacity in reactive power supply, and smaller size and lighter weight. In addition, the superconducting synchronous generator provides larger capacity during imbalanced load operations, including a larger capacity to absorb harmonic currents and improved voltage regulation. It also exhibits higher terminal voltage and an improved ability to handle sub-synchronous resonance in power systems. Moreover, the simulations are in better agreement with experimental results.

..

Table 1 Characteristics of superconducting synchronous generator developed by Super-GM.

	Slow-response-excitation type (0.1 p.u./s)		Quick-response-excitation-type (1.0 p.u./s)
	A	B	
Specifications			
Capacity (MVA)	83	83	73
Voltage (kV)	10	10	10
Current (A)	4,792	4,792	4,215
Power factor	0.9	0.9	0.9
Rotation speed (m^{-1})	3600	3600	3,600
Synchronous reactance (p.u.)	0.35	0.35	0.45
Field current Rated (A)	3,000	3,000	3,200
Maximum (A)	3,600	3,600	4,500
Rotor structure Room temperature damper Cold damper/Radiation shield Thermal shrink relaxation	Single layer type Three-layer type Double bearing	Squirrel cage type Single layer type Flexible disc	Three-layer type Single layer type Flexible support
Property superconductor	High stability	High current density	Low AC loss
Stator structure	Air-gap winding Double transposed conductor Water cooled		

IOP Conf. Series: Journal of Physics: Conf. Series **1054** (2018) 012081 doi:10.1088/1742-6596/1054/1/012081

Given that the synchronous reactance is very small, it was pointed out that the short circuit current is relatively large. However, the large current at short circuit occurs only at the instance of the short circuit occurrence. At that instance, the generator reactance is transient or sub-transient, the value of which is almost identical to that in conventional generators.

Moreover, it has been pointed out that the larger time constant of the field winding of superconducting generators may negatively impact power systems. However, the role of time constant in field windings of conventional generators is that of a cold damper in a superconducting generator. Under fault conditions in power systems, a larger time constant prevents a significant reduction in the field current. The characteristic is rather suited for power systems.

In addition to the above, a full superconducting generator with both superconducting field winding and armature winding was examined; however, such a generator does not have the above mentioned features. This may be attributed to the poor characteristics of the AC superconductor developed at that time.

2.2.3 Concluding remarks

As part of the Super-GM project (1988- 2004), long-term operation [4] with refrigerator, and synchronous condenser operation were successfully carried out using 70-MW class superconducting generators.[5] In these developments, three types of rotors and a stator were manufactured (Refer to Table 1). The small generator was used to carry out power operations in real power systems.[6] These generators do not face issues in practical implementation; however, there is scope for further improvements. It should be noted that a synchronous condenser operation was successfully carried out using a high temperature superconductor in a real power system.

The characteristics of superconducting synchronous generators can be obtained by the structures. Conversely the structure may be determined by characteristics requested

3. Present and future scope for development of superconducting rotating machines [7]

Now that superconductors that can operate in at rather high temperature have been discovered, the applications for rotating electric machines are increasing. These include, synchronous generators and motors, induction machines, and new concept machines. In use of superconductors, wire, bulk, thin film and so on have been examined. Research studies are also investigating several types of refrigerants, including liquid He, liquid N_2, liquid H_2, gaseous He, and gaseous Ne. Therefore, summarizing all these advances is beyond the scope of this paper. Before considering the summaries, the present state of electrical rotating machines must be considered. It is known that more than 50% of electric energy is consumed by motors. This percentage is expected to increase in the future. Therefore, more efficient motors must be developed, for example, by using permanent magnets as well as achieving elegant control through power electronics technologies. Therefore, the pressing question then becomes that of "can superconducting rotating machines provide solutions to these problems"? The answer might actually be "yes". It is well known that the nature of electric power supply is rapidly changing, with an increase in renewable energy sources. This poses new challenges to existing power systems such as power system stability, voltage stability, and harmonic current. Can rotating superconducting machines provide solutions to the newly developing challenges? The answer might actually be "yes".

Synchronous machines, induction machines, and DC machines have found wide and diverse applications. Large capacity DC machines were replaced by AC machines (synchronous and induction machines). The machines whose capacities are less than 100 W are used widely in electric and electronic devices. On the other hand, synchronous generators are applied in power stations. Similarly, induction motors works have found widespread applications. Small motors have almost been replaced by DC motors. Similarly, large motors have been replaced by synchronous motors owing to their higher efficiency and power factor, particularly in large power applications owing to features such as, simplistic structure, low cost, and robustness. Similarly, synchronous machines have been adopted for large power applications.

From the above, it is evident that summarizing the research and developments is exhaustive.

In this paper, we have summarized the scope and applications for both generators and motors. The cryogenic temperature in rotors and stator have been discussed, both for , wire and bulk. In addition, synchronous motors and induction motors have been discussed in terms of capacity and type of magnets for both AC or DC. Moreover, the cooling method have been discussed including pool and forced conduction. Similarly, the discussion has encompassed cryogen, for liquids, gas or others, and their application purpose, including in air crafts, ships, and wind powers.

At present, several feasibility studies have been conducted on the application of superconductivity to rotating machines like in the earlier phase of development. These developments are important. However, it may need to determine the goal of the studies. What characteristics are good for practical uses?

For example, generators, except those used in the generation of wind power, constant speed generators. These are suitable for the cooling systems of the cryogenic rotor.

As regards motor applications, changes in the motor speed are critical challenges, (as in the case of wind power generators), particularly with cryogenic rotors, where controlling the temperature can prove to be quite challenging. Therefore, AC homo-polar generators with cryogenic stator can find potential application in wind power generation.

On the other hand, induction motors that use superconducting squirrel-cage rotors are attractive given their nonlinear resistance characteristics and the high temperature superconductors. The subject is in completion for conventional ones. Therefore, current research and developments in super conducting rotating machines are beginning to attract the attentions of electrical and power system engineers.

4. Conclusions

The paper sought to describe the developments in the application of superconductivity to rotating electric machines. The paper also comments on their scope for future development. Thus, taking into account the changing needs of society and the developments in the features of rotating superconducting machines, such machines continue to be developed. Conversely, the developments of those machine has the potential to lead or change the world.

References

[1] R Bumby, "Superconducting rotating electrical machines", Cladendon Press Oxford, 1983

[2] T Ageta et al., "Recent Development Progress of 70MW Class Superconducting Generator", CIGRE Meeting 2000

[3] T Nitta, "Progress of technology for superconducting generator", Cryogenics, 42, 2002, pp. 151-155

[4] M Shibuya, et al., "Recent Development Progress of 70 MW Class Superconducting Generator", 18th ICEC 2000.

[5] Y Imai et al., "Field test a 70 MW class superconducting generator I the mode of a synchronous condenser", Electrical Engineering in Japan. Vol.141, pp. 17-24, November 2002. Wiley

[6] T Kishida et al., "A Series of Studies on Superconducting Generator with High Response Excitation," Proceedings 16th ICEC/ICEM,P.909-912, 1997.

[7] Abstracts of CEC 2017 (https://indico.cern.ch/event/578092/timetable/?view=standard), MT25 2017 (http://mt-25.org/onewebmedia/MT25%20Book-of-Abstracts-Presentations.pdf), and EUCAS 2017 (https://www.symporg-registrations.com/symporg/frontend/reg/tDailyAgenda.csp?pageID=49376&ef_sel_menu=1322&eventID=112&mode=.

IOP Conf. Series: Journal of Physics: Conf. Series **1054** (2018) 012082 doi:10.1088/1742-6596/1054/1/012082

Optimal Design of a Superconducting Motor for Electric-drive Aeropropulsion Based on Finite-Element Analysis and Genetic Algorithm

Weilu Kong[1], Yutaka Terao[2] and Hiroyuki Ohsaki[2]

[1]Graduate School of Engineering, the University of Tokyo, Japan
[2]Graduate School of Frontier Sciences, the University of Tokyo, Japan

E-mail: ohsaki@k.u-tokyo.ac.jp

Abstract. Electric-drive Aeropropulsion is one of the key technologies in design of future aircrafts because the electric motor drive has high efficiency and the this would reduce operational cost and pollutant emission. This paper discusses the optimization of electromagnetic design of superconducting motors. Finite element analysis, multi-objective genetic algorithm and analytical equations are applied in the optimization. Power density, power loss including AC loss of superconducting materials, and torque ripple are considered in the objective function. The results of optimization suggest that the application of superconducting motor is feasible for electric-drive aeropropulsion with power density higher than 16.4 kW/kg.

1. Electric-drive Aeropropulsion

A global reduction of the emissions of pollutants into the atmosphere is needed and a large part of the emissions come from transportation vehicles, including airplanes due to the nature of combustion-based propulsion system. With the application of Electric-drive Aeropropulsion (EA), operational cost, pollutant emission, and noise can be reduced largely due to the high efficiency of electric motor.

The overall cost of a passenger airline is assumed to reduce by 40% with EA [1]. Moreover, EA meets the following requirements: 1. Reduction of operational cost, 2. Low emissions of pollutants, and 3. Low noise.

However, conventional machines exhibit output power densities D_p typically lower than 1 kW/kg and have reached an optimum in terms of performance [3], leaving little room for improvement.

On the contrary, superconducting rotating machines are destined to play a key role as the enabling technology for the EA because it is supposed to achieve D_p comparable to turbine engines in excess of 10-20kW/kg [2]. So superconducting motor smaller and lighter than a gas turbine can be potentially built, which makes EA possible.

2. Design of Superconducting Motors

2.1. Introduction to Superconducting Motor

The concept of electrical ducted fan has been proposed that takes advantage of the high bypass ratio of engine and locally replace the engine core with a SuperConducting Motor (SCM) [2].

As Fig. 1 shows, a system configuration of EA consisting of turbine, generator AC/DC converter, DC Bus, DC/AC inverter, SCMs and ducted fans is considered. For that, we design SCMs used in this system. Energy source can be replaced with fuel cell when energy density of cell increases to be on a pair with a that of current energy source.

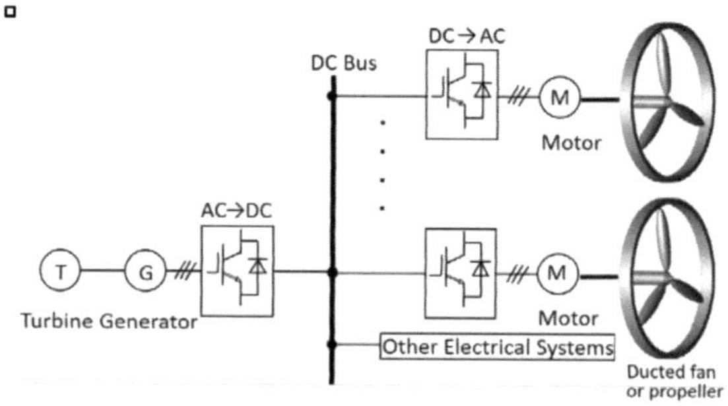

Figure 1. Example of EA System Configuration

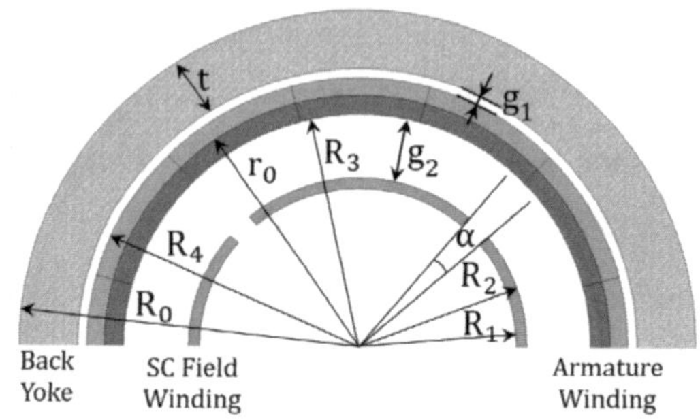

Figure 2. Structure of SCMs

Fig. 2 shows the structure of SCMs we considered. YBCO field coils in the rotor produce magnetic field in copper armature windings for Partially SuperConducting Motors (PSCMs) or in MgB2 armature windings for Fully SuperConducting Motors (FSCMs) in the stator. The motor has an air-core (i.e., non-magnetic) construction in the rotor and air gap windings in the stator. Liquid hydrogen (LH2) as the fuel, is also considered as coolant to cool down superconducting materials.

Table 1 lists the necessary variables for design and optimization.

Table 1. Parameters for SCMs

Parameters	Symbols
Inside radius of field winding (m)	R_1
Outside radius of field winding (m)	R_2
Inside radius of armature winding (m)	R_3
Average radius of armature winding (m)	r_0
Outside radius of armature winding (m)	R_4
Outside radius of back yoke (m)	R_0
Thickness of back yoke (m)	t
Winding pole angle (*rad*)	α
Effective length (m)	L_{ef}
Mechanical gap (m)	g_1
Air gap (m)	g_2

2.2. *Materials for SCMs*

We choose High-Temperature Superconductor (HTS) YBCO wire, as the field coils because YBCO wire has high critical current density even among high external magnetic field. Copper or MgB_2 superconducting wire shown in Fig. 3, are chosen as armature winding due to the potential of reduction of AC loss due to multi-filament structure of MgB_2 wire, and back yoke uses low high-frequency iron-loss steel sheet. The operation conditions of the materials are shown in Table 2. The J_{ave} is the average current density considered for the initial design. In optimization, J_{ave} of copper depends on the area of armature winding and J_{ave} of MgB_2 depends on the area of armature winding and external magnetic flux density.

Table 2. Materials and Operation Conditions

Parts	Materials	Operation Temperature (K)	J_{ave} (A/mm^2)
Field winding	YBCO	20	177
Armature winding I	Copper	293	7.5
Armature winding II	MgB_2	20	48
Back yoke	20JNEH1200	293	/

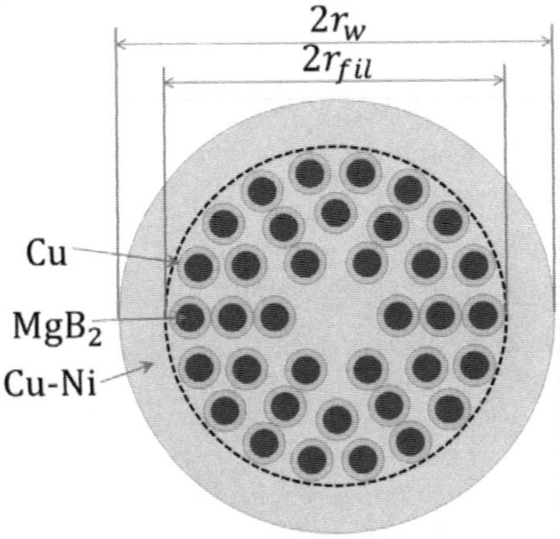

Figure 3. Cross Section of MgB_2 Wire

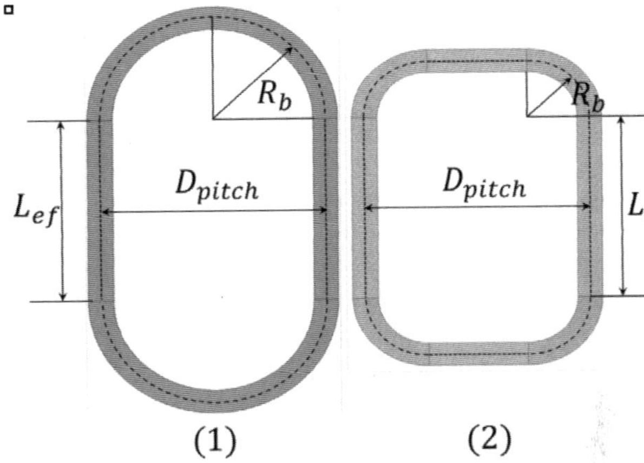

Figure 4. (1). Field Coil Racetrack Structure; (2). Armature Winding Structure

2.3. *Differences between PSCM and FSCM*

For PSCM, the superconducting material YBCO wire is for DC application, so the Joule loss of copper armature winding: W_{Cu}, loss due to current leads of field coils: W_{Lead} and Iron loss of back yoke: W_{Fe} are considered as loss. Because only field winding of PSCM is superconducting material, cryogenic chamber is set outside the rotor, in which LH_2 flows to cool down the field winding.

For FSCM, loss due to current leads of both field coils and armature windings W_{Lead}, 100 times AC loss of MgB2 at 20K for AC application: $100 \times W_{AC,20K}$, and W_{Fe} are considered as loss. Because both of field winding and armature winding of FSCM are superconducting materials, cryogenic chamber is set outside the rotor, and outside the stator, respectively. LH_2 flows to cool down the field winding and armature windings in the 2 chambers and back yoke is isolated to avoid cooling back yoke.

The W_{Lead} are considered as a resistance 0.02Ω for YBCO field coils and MgB_2 armature windings [7]. Because of Joule loss, average current densities: J_{ave} of copper armature winding are limited to 7.5 (A/mm^2) (with liquid cooling in ducts) respectively. However, J_{ave} of MgB_2 wire can achieve 48 (A/mm^2) that creates higher interlinkage flux, which realizes higher D_p.

2.4. *Specifications of SCMs*

The SCMs feasible for EA should meet the following requirements: 1. Power: 1 to 30 MW, 2. Revolving Speed: 2000 to 12000 rpm, 3. Number of pole pairs: 2 to 4. One near-term challenge of National Aeronautics and Space Administration (NASA) is to design a 4MW FSCM with $D_p \geq 16.4$ kW/kg and efficiency $\geq 99\%$ [5].

The target of the motor output power is 3.0, 4.0 and 5.0 MW class in this paper. If the jet engines in airplane for around 180-passenger class are replaced with electrical motors, the overall output capacity should be 44 MW. Therefore, a distributed EA needs 15, 11 and 9 motors, respectively. The phase voltage of the motors is set to be 1000 V that is among the range between ± 4.5 kVDC at the DC bus in Fig. 1 [4]. Common specifications for different SCM cases are shown in Table 3. The operation for YBCO and MgB_2 is 20 K, and we plan to use liquid hydrogen for the coolant.

Table 3. Specifications of SCMs

Parameters	PSCM	FSCM
Output Power P (MW)	3.0, 4.0, 5.0	
Line Voltage U_L (V)	1732	
Line Current I_L (A)	1177, 1569, 1961	
Revolving Speed N (rpm)	5000	
Number of pole pairs p	2	
Wires of Armature Windings	Copper	MgB_2

2.5. Analytical Design and Equations

The field coil structure is assumed as racetrack structure and armature winding structure is assumed as similar structure in AC loss and D_p calculation, as shown in Fig. 4. The average length of the winding: L_{ave} is expressed as following:

$$L_{ave} = 2L_{ef} + 2\left(D_{pitch} - 2R_b\right) + 2\pi R_b \tag{1}$$

where D_{pitch}: winding pitch and R_b: bending radius, respectively. We assumed R_b of YBCO as $D_{pitch}/2$, MgB_2 windings as 60 mm [6], and that of Cu as 20 mm.

Radial components B_ρ and tangential field components B_θ of magnetic flux density of 4-pole motor in the region at angular location θ and radius ρ of the field winding are given by (2)-(3) and those of armature winding are given by (4)-(5), respectively [7], as a function of ρ and θ. Fig. 2 shows necessary variables for SCM analytical design and optimization.

From field coils:

$$B_\rho(\rho, n) = \frac{\mu_0 \cdot J_f \cdot 2 \cdot \cos(n \cdot \alpha)}{\pi \cdot n \cdot (2n+2)} \cdot \left(R_2^{2+2n} - R_1^{2+2n}\right) \cdot \left\{\rho^{-2n-1} + \frac{\rho^{2n-1}}{(R_0-t)^{4n}}\right\} \cdot \cos(2n\theta) \tag{2}$$

$$B_\theta(\rho, n) = \frac{\mu_0 \cdot J_f \cdot 2 \cdot \cos(n \cdot \alpha)}{\pi \cdot n \cdot (2n+2)} \cdot \left(R_2^{2+2n} - R_1^{2+2n}\right) \cdot \left\{\rho^{-2n-1} - \frac{\rho^{2n-1}}{(R_0-t)^{4n}}\right\} \cdot \sin(2n\theta) \tag{3}$$

From armature windings:

$$B_\rho(\rho, n) = \begin{cases} \frac{\mu_0 \cdot J_a \cdot 2 \cdot \cos(n \cdot \alpha)}{\pi \cdot n} \cdot \left[\ln\left(\frac{R_4}{R_3}\right) + \left\{\frac{R_4^{2+2n} - R_3^{2+2n}}{(2n+2) \cdot (R_0-t)^{4n}}\right\}\right] \cdot \rho^{2n-1} \cdot \cos(2n\theta), & \text{if } R_2 \le \rho \le R_3 \\ \frac{\mu_0 \cdot J_a \cdot 2 \cdot \cos(n \cdot \alpha)}{\pi \cdot n \cdot (2+2n)} \cdot \left\{ \begin{array}{c} 1 - \left(\frac{R_3}{\rho}\right)^{2n+2} + (2n+2) \cdot \ln\left(\frac{R_4}{\rho}\right) \\ + \frac{R_4^{2+2n} - R_3^{2+2n}}{(R_0-t)^{4n}} \cdot \rho^{2n-2} \end{array} \right\} \cdot \rho \cdot \cos(2n\theta), & \text{if } R_3 \le \rho \le R_4 \end{cases} \tag{4}$$

$$B_\theta(\rho, n) = \begin{cases} \frac{-\mu_0 \cdot J_a \cdot 2 \cdot \cos(n \cdot \alpha)}{\pi \cdot n} \cdot \left[\ln\left(\frac{R_4}{R_3}\right) + \left\{\frac{R_4^{2+2n} - R_3^{2+2n}}{(2n+2) \cdot (R_0-t)^{4n}}\right\}\right] \cdot \rho^{2n-1} \cdot \sin(2n\theta), & \text{if } R_2 \le \rho \le R_3 \\ \frac{\mu_0 \cdot J_a \cdot 2 \cdot \cos(n \cdot \alpha)}{\pi \cdot n \cdot (2+2n)} \cdot \left[\begin{array}{c} \left\{-1 + \left(\frac{R_3}{\rho}\right)^{2n+2} - (2n+2) \cdot \ln\left(\frac{R_4}{\rho}\right)\right\} \\ - \frac{R_4^{2+2n} - R_3^{2+2n}}{(R_0-t)^{4n}} \cdot \rho^{2n-2} \end{array} \right] \cdot \rho \cdot \sin(2n\theta), & \text{if } R_3 \le \rho \le R_4 \end{cases} \tag{5}$$

where ρ: radius, n: odd number, J_f: field coil current density, J_a: armature winding current density, and θ: angular location, respectively.

Equation (2)-(5) are used in analytical design and AC loss calculation. Fig. 5 shows airgap magnetic field distribution of one case. Both of the radial component and tangential component have good agreement.

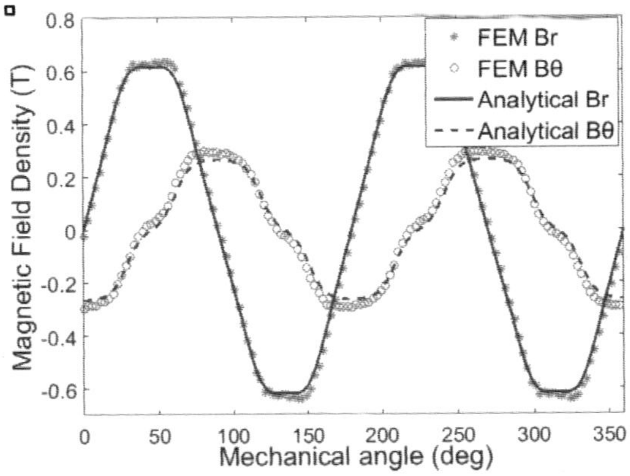

Figure 5. Comparison between Result of Analytical Equation and FEM

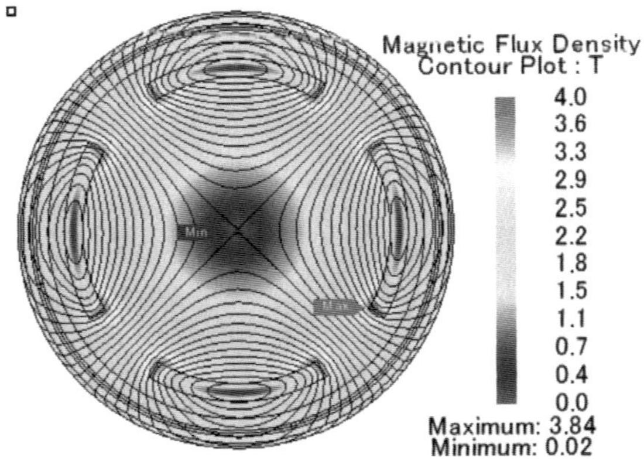

Figure 6. Magnetic Field Distribution of Optimized 5.0 MW PSCM

3. Optimization on SCMs

In the case of SCM for EA, the output power density D_p, electrical loss W_{loss} and torque ripple T_r are the main concerns. To acquire a model with balance of D_p, W_{loss} and T_r, we optimized analytical design with FEM and Multi-Objective Genetic Algorithm (MOGA). FEM analysis is carried out with J-MAG ®.

3.1. AC Loss Calculation

When considering the calculation of AC loss of FSCM, we consider the hysteresis loss: W_h and coupling loss: W_c. The W_h is expressed as following [8]:

$$W_h = L_a \cdot n_f \cdot \pi \cdot R_0^2 \cdot q_h \tag{6}$$

$$q_h = \frac{16}{3\pi} \cdot f \cdot B_{max} \cdot r_{one} \cdot J_c \tag{7}$$

$$L_a = m \cdot p \cdot L_{ave} \cdot n_{para} \tag{8}$$

where L_a: MgB_2 wire length, n_f: number of MgB_2 filaments, r_{one}: filament radius, q_h: hysteresis loss per unit volume, f: frequency, B_{max}: maximal magnetic flux density on armature winding, J_c: critical current of MgB_2 wire, m: number of phase and n_{para}: number of parallel connected MgB_2 wires, respectively.

The W_c is expressed as following [9]:

$$W_c = L_a \cdot \pi \cdot r_w^2 \cdot q_c \tag{9}$$

$$q_c = \left(\frac{r_{fil}}{r_w}\right)^2 \cdot \frac{(2\pi f)^2 \cdot \tau_{se}}{1+(2\pi f \tau_{se})^2} \cdot \frac{B_{max}^2}{\mu_0} \tag{10}$$

$$\tau_{se} = \frac{1}{2} \cdot \left(\frac{L_p}{2\pi}\right)^2 \cdot \mu_0 \cdot \sigma_{eff} \tag{11}$$

where r_w: MgB_2 wire radius, r_{fil}: radius of filament region, τ_{se}: time constant, L_p: twist pitch and σ_{eff}: transverse conductivity, respectively. The characteristics of MgB_2 wires are based on samples [9] and commercial products [10], and the properties are listed in Table 4.

Table 4. Properties of MgB_2 Wires

r_w	r_{fil}	r_{one}	L_p	n_f
0.255 mm	0.193 mm	1.646×10^{-2} mm	9 mm	36
Copper Component	Fill Factor of MgB_2		J_c (20 K, 2.0 T)	σ_{eff}
0.15	0.15		1780 A/mm^2	1.36×10^6 S/m

3.2. *Motor Mass Calculation*
In the calculation of D_p, the mass densities of copper, MgB_2 wire, YBCO wire and 20JNEH1200 are 8960 kg/m^3, 7995 kg/m^3, 8940 kg/m^3 and 7650 kg/m^3, respectively. The mass of other parts including bearing and supporting parts are considered as 25% of the field coils, armature windings and back yoke [11].

3.3. *Objective of Optimization*
The objective of optimization is to optimize the SCM design to be with $D_p \geq 16.4$ kW/kg, $W_{loss} \leq 1\%$ for PSCM or $W_{loss} \leq 10\%$ for FSCM, and $T_r \leq 4\%$.

3.4. *Cost Function and Bound Constraints*
Set the variable vector $\vec{X}$

$$\vec{X} = \left[R_0, t, \left(\frac{R_4 - R_3}{2}\right), (R_3 - R_2), L_{ef}, \alpha\right] \tag{12}$$

The cost function Z, consists of D_p, W_{loss}, and T_r and we adjust the importance of each parameter by weighted sum approaches, thus:

$$Z(\vec{X}) = w_1 \cdot \frac{1}{D_p(\vec{X})} + w_2 \cdot W_{loss}(\vec{X}) + w_3 \cdot Tr(\vec{X}) \tag{13}$$

where w_1, w_2, and w_3 are the weights of D_p, W_{loss}, and T_r, which denote the importance of the parameters, respectively. We set the ratio of weights $w_1 : w_2 : w_3$ to be 1:1:1, which means the importance of the three properties are considered the same.

Thus, the optimization is to minimize $Z(\vec{X})$ subject to bound constraints:

$$X_c^l \leq X_c \leq X_c^u \quad (c = 1, 2, \ldots, 6) \tag{14}$$

where X_c^l and X_c^u arfe the lower and upper bounds of a design variable, respectively.

We set the upper bounds of R_0 and L_{ef} as around half of the current turbofan engine IAE V2500, and the other bounds referring to initial design and our previous optimization result as following:

$$\overrightarrow{X_c^l} = [200\text{ mm}, 5\text{ mm}, T_a\ (\text{mm}), 4\text{ mm}, 400\text{ mm}, 10^\circ] \tag{15}$$

$$\overrightarrow{X_c^u} = [400\text{ mm}, 30\text{ mm}, T_a\ (\text{mm}), 50\text{ mm}, 1500\text{ mm}, 40^\circ] \tag{16}$$

where T_a is the thickness of one-layer armature winding that varies with I_L and J_{ave}. While g_1 and g_2 are set to be constant, 10 mm and 40 mm, respectively [12].

4. Results and Discussion
Thus, with 50-generation MOGA, we got optimal results of PSCM and FSCM as shown in Table 5. According to the result of PSCM and FSCM, we designed motors with high D_p, among which, 5.0

MW PSCM, 4.0 MW FSCM and 5.0 MW FSCM surpass 16.4 kW/kg. And, all of the PSCMs meet the requirement of low W_{loss}, and even 5.0 MW FSCM, which has the highest loss, almost meets the requirement of 10 %. So, the optimization improves the properties with a good balance.

The FSCMs have higher D_p than PSCMs of the same output power, while much higher loss considering cryogenic loss and loss from the current lead connecting the terminal of MgB_2 armature winding.

For PSCM and FSCM, the D_p increases as P increases because the thickness of field winding increases, which causes a higher airgap magnetic flux density B_g. And, as P increases, the W_{loss} of PSCM decreases while that of FSCM increases because the W_{Cu} changes little, W_{Lead} remains the same and W_{Fe} decreases much when the volume of back yoke decreases, and because the P increases, the loss per output power decreases. However, the L_a of MgB2 increases with P and since the W_{AC} accounts for the vast majority of loss of FSCM, the W_{loss} increases as P raises.

Among the variables, the thickness of back yoke, t and the thickness of field coils, thus R_2-R_1, has the most effect on the D_p and W_{loss} because, when we keep the L_{ef} constant, the increase of t will increase the volume of back yoke, which cause the increase of W_{Fe} and mass simultaneously.

Table 5. Optimal Result of SCMs

w_1: w_2: w_3	1:1:1					
Type	PSCM			FSCM		
P (MW)	3.0	4.0	5.0	3.0	4.0	5.0
D_p (kW/kg)	5.42	13.92	17.12	5.85	17.08	21.21
W_{loss}%	0.54	0.36	0.33	6.96	7.71	10.40
T_r%	1.10	0.69	0.75	0.65	0.35	0.58
W_{AC}@20 K (W)	/	/	/	1697.47	2527.62	4338.04
B_g (T)	1.61	1.83	2.14	1.34	1.48	1.84
R_1 (mm)	166.41	158.72	211.29	272.54	200.90	151.12
R_2 (mm)	185.86	181.71	236.31	286.07	217.40	175.51
R_3 (mm)	225.86	221.71	276.31	326.07	257.40	215.51
R_4 (mm)	237.20	231.30	284.47	332.61	265.49	221.96
R_0 (mm)	271.02	246.60	301.09	366.21	281.03	238.07
t (mm)	23.82	5.30	6.62	23.59	5.54	6.11
α (deg)	25.57	24.90	27.81	26.22	25.89	27.45
L_{ef} (mm)	982.48	1048.75	671.45	743.37	1062.59	1051.00

Among the 6 types of optimized motor, the 5.0 MW PSCM meets the requirements best, so the magnetic field distribution of it is shown in Fig. 6, which illustrates the high airgap magnetic field density is the key reason of high D_p of PSCM. The external field of field coils is below 4 T, which ensures the application and high current density of YBCO wires. Change of 5 MW PSCMs D_p and W_{loss} with t and thickness of field winding are shown on Fig. 7 and Fig. 8, which help us understand the trend of the change of the two key properties.

5. Conclusion

We have carried out electromagnetic design of 5000-rpm PSCM and FSCM with output power 3.0 MW, 4.0 MW and 5.0 MW for EA. Using real samples of superconducting wires, we calculated AC loss and realized low AC loss $\leq 1\%$. With FEM and MOGA, we optimized design and got model that minimized cost function. And, we succeeded to optimize PSCM that meet near-term challenge for EA with $D_p = 17.12$ kW/kg, $W_{loss} = 0.33\%$ and $T_r = 0.75\%$. We optimized FSCM to higher D_p and around 10% W_{loss}, which is low to P but is still hard to take the heat.

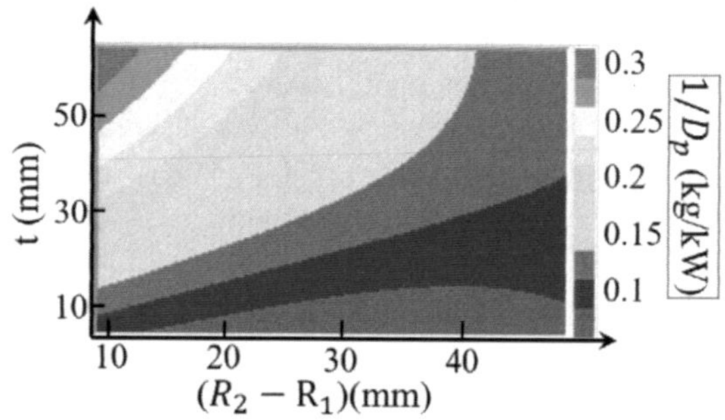

Figure 7. D_p of 5 MW PSCMs as a Function of *t* and Thickness of Field Winding

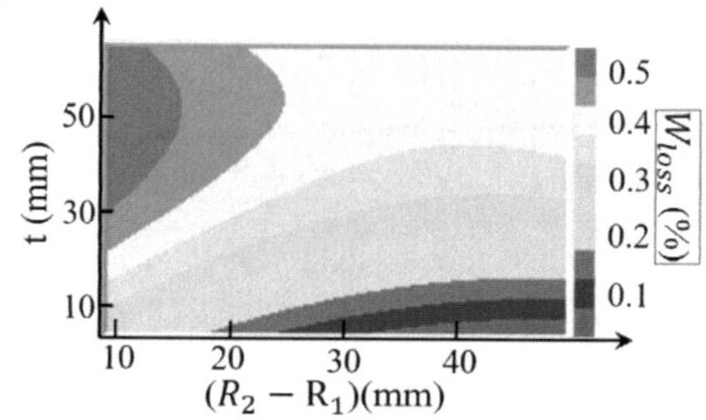

Figure 8. W_{loss} of 5 MW PSCMs as a Function of *t* and Thickness of Field Winding

With optimization, we understand the dependence of variables on the properties we need to optimize, and we testified the effectiveness of MOGA and weighted sum approaches in the optimization of SCMs for EA. As the next step, we plan to adjust the weights to get higher D_p for PSCM and lower W_{loss} for FSCM.

References

[1] Nishizawa A and Kobayashi H 2014 *Journal of IEE. Japan.* **134** 84
[2] Luongo C A, Masson P J, Nam T, Mavris D, Kim H D, Brown G V, Waters M and Hall D 2009 *IEEE Trans. Appl. Supercond.* **19** 1055
[3] Masson P J, Brown G V, Soban D S and Luongo C A 2007 *Supercond. Sci. Technol.* **20** 748
[4] Jones C E, Norman P J, Galloway S J, Armstrong M J and Bollman A M 2016 *IEEE Trans. Appl. Supercond.* **26** 1
[5] Jankovsky A 2016 *Electric & Hybrid Aerospace Technology Symposium* 2016
[6] Tanaka H, Kodama M, Ichiki Y, kusunoki T, Kotaki H, Suzuki T, Nishi K and Okamoto K 2017 *IEEE Trans. Appl. Supercond.* **27** 1
[7] Kalsi S S 2011 *Applications of High Temperature Superconductors to Electric Power Equipment* (New Jersy: John Wiley & Sons)
[8] Carr W J, Jr. 2011 *AC Loss and Macroscopic Theory of Superconductors Second Edition* (London: Taylor & Francis)
[9] Magnusson N, Lindau S, Taxt H and Runde M 2014 *Supercond. Sci. Technol.* **27** 105003
[10] Hypertech, 2017 *MgB_2 and Nb_3Sn Brochure Rev 8*
[11] Woodson H H, Stekly Z J J and Halas E 1966 *IEEE Trans. Power Appara. Syst.* **85** 264
[12] Jang H-M, Muta I, Hoshino T, Nakamura T, Kim S-W, Sohn M-H, Kwon Y-K and Ryu K-S 2003 *IEEE Trans. Appl. Supercond.* **13** 2197

Design and analysis of air-core superconducting generator for wind power applications

H W Cho[1], M Feddersen[2], K S Haran[3]

[1]Department of Electrical, Electronics and Communication Eng. Edu., Chungnam National University, Daejeon 34134, Korea
[2]Zunum Aero Inc., Champaign, IL 61821, U.S.A.
[3]Department of Electrical and Computer Engineering, University of Illinois at Urbana-Champaign, Urbana, IL 61820, U.S.A.

E-mail: hwcho@cnu.ac.kr, fedders2@illinois.edu, kharan@illinois.edu

Abstract. In this paper, the design and analysis of the air-core superconducting (SC) generators are carried out. Air-core SC generators have the advantage of reducing the gross weight of the system because they do not have a high-weight iron core that constitutes a magnetic circuit, and can be a good design option with high power density. The magnetic field acting on the outside of the air-core SC generator can be shielded by using an active shielding SC coil. In this study, we compared the weight and output power of an air-core SC generator with an active shielding SC coil to that of conventional SC generators, and studied design conditions for application to wind power generation systems. The results show that the total weight of the generator with actively shielded SC generator has lower weight compare with conventional magnetically shielded SC generator.

1. Introduction

In the movement towards reducing CO_2 emissions and our dependence on fossil fuels, wind power is considered one of the main technologies in delivering renewable energy to the world. The European Wind Energy Association has predicted a need for 300 GW of installed wind power by 2030, with 120 GW of that expected offshore [1]. As of Jan 2016, there is 131 GW of onshore and 11 GW of offshore installed wind capacity in the EU [2]. This relatively low offshore capacity can be improved by higher power turbines, especially as wind turbine power ratings have been increasing in recent years with sizes on the order of 10-20 MW on the horizon. As wind turbine power needs increase further, generator technology must similarly improve to enable high-power wind generator development. One such technology is the inclusion of superconducting (SC) wire in generator designs to produce greater magnetic field strength with less material [3]. SC wind generator designs are typically direct drive, low speed, and high torque, and due to very high air-gap magnetic flux densities from the SC coils (3-5X compared to conventional designs) much of the flux-carrying steel within the machine can be eliminated, resulting in a more lightweight and compact turbine.

There are many feasibility and design studies with regard to SC generators for wind power applications. [4] and [5] presented 10 MW and 2 MW class SC wind turbine generators and demonstrated their finite element analysis (FEA) results and test methodologies. Their successful studies and experiments are shown the generator design and potential challenges of SC generators for mega-watt scale wind power applications. [6] presented a feasible electromagnetic designs of SC generators for a 10 MW direct-drive wind turbine. The topology comparison of SC generators with more non-magnetic cores and with more iron core were described for low weight system.

With the introduction of a technology such as SC coils into wind generator design, new spaces open up for novel topologies and implementations. One such topology is the actively-shielded air-core SC machine described in previous work [7]. Originally designed for aircraft applications, this machine has not been analysed at larger scales such as wind generators. Furthermore, the potential for such a topology in offshore direct-drive applications has not yet been explored. In this paper, a first-pass at wind generator sizing is performed for the actively shielded air-core SC topology. The aim is to reduce the design weight for further optimization and to give some intuition as to the relative state of SC technology in potential wind generator applications.

2. Actively shielded air-core SC generator

2.1. Concept of actively shielded air-core SC machine

The actively-shielded air-core SC design is an inside-out synchronous machine, operating similar to a permanent magnet generator. SC dc field coils replace the permanent magnets, but they are held stationary in the outer portion of the machine to simplify the superconductor cooling scheme. The 3-phase copper armature coils are part of the rotating inner portion of the machine and are connected to an outside 3-phase system through slip rings. To reduce the magnetic field propagation outside the machine, active SC shielding coils are placed just beyond the dc field coils. The design of these coils has been optimized to minimize the outside magnetic field while maximizing the air-gap flux density within the machine; this tradeoff was achieved through a multi-objective electromagnetic optimization in prior work [8]. Figure 1 shows the main electrical components of the actively-shielded air-core machine. In figure 2, the architecture of actively shielded air-core superconducting machine model shows an example machine utilizing the air-core design with a cryocooler module, a torque tube formed around the SC coils, and slip rings for the rotating armature windings.

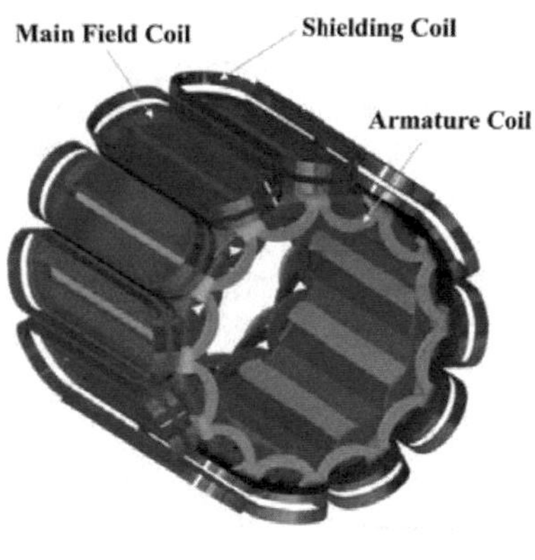

Figure 1. Main electrical components of the actively-shielded air-core SC machine.

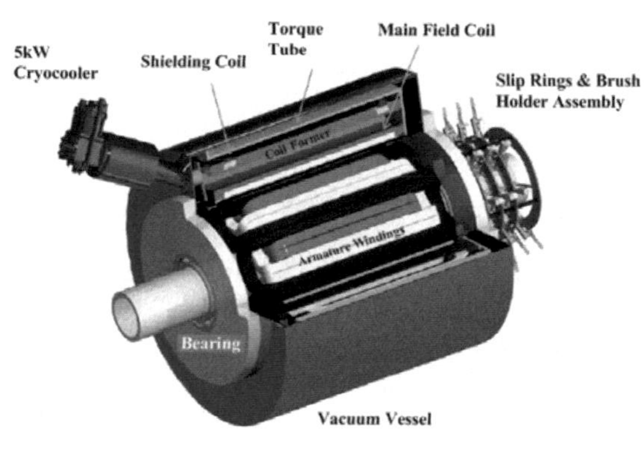

Figure 2. Architecture of actively-shielded air-core SC machine.

2.2. Air-core SC generator for wind power applications

An reference SC wind power generator concept design was used as a starting point for sizing the air-core wind generators. In this reference design, low temperature superconducting magnets

IOP Conf. Series: Journal of Physics: Conf. Series **1054** (2018) 012083 doi:10.1088/1742-6596/1054/1/012083

were placed in the rotor, with copper windings around iron teeth in the stator [9]. Some key aspects in this reference design which are absent from the proposed air-core design include iron present in the stator, NbTi used as the SC wire material, and the SC coils placed in the rotor. In the air-core design there is very little flux-carrying iron in the machine, Nb_3Sn is used as the SC wire material, and the SC coils are held in the stator. It is expected that Nb_3Sn will allow for stronger magnetic flux density with less wire, and the lack of iron will drastically reduce the machines weight. The SC generator design constraints are listed in Table 1.

Table 1. SC generator design constraints.

Item		Value
Rated power	(MW)	10
Rated speed	(rpm)	10
Rated voltage	(kV)	3.3
Outer diameter	(mm)	4800
Armature copper depth	(mm)	127
Armature current density	(A/mm^2)	3.7
Ampere-turns per pole	(kAT/pole)	906.4
Pole count		36

2.3. Sizing procedure for wind power applications

To allow for consistent comparisons between the reference and the air-core designs (with magnetic core and with active shielding SC coils), certain parameters were fixed during the sizing process. These include the rated speed, armature current density, and SC field coil amp-turns per pole as listed in Table 1. Additionally, the parameters of the SC coils were unchanged since the coils had been previously optimized to minimize outside magnetic field while maximizing air-gap flux density [7]. The diameter was limited below 4.8 m and the armature depth was limited below 127 mm (5 in), both based on the reference design [9]. These quantities were varied with the intent of minimizing the machines SC wire length, since SC wire is the most expensive component in the design, and SC wire is priced based on length and current-carrying capacity. Copper weight and SC wire weight were also examined since they contribute to material cost and overall machine weight. In offshore wind applications, the diameter of the generator module is not particularly restricted when compared to land based shipping limitations. For this reason, it is not uncommon to have diameters in excess of 4 m for offshore wind generators. Results were obtained through 2-D electromagnetic simulations of the generator cross section. For each value of diameter and armature depth, the active length needed for the desired power rating was calculated via simulation. Quantities for copper weight, SC weight, and SC wire length were then calculated based on the geometry of the machine and windings as well as known material properties. A fill factor of 0.5 was assumed for both copper and SC wires.

3. Design results and discussion

3.1. Flux density distribution

An 3D CAD model of a 10 MW class actively-shielded air-core SC generator for wind power application is shown in Fig. 3. The SC generator with magnetic shielding core (Air-core design 1) and the SC generator with actively-shielding coils (Air-core design 2) were designed to compare the performance and volume of the air-core type SC generator. As described above, the two-dimensional shape of the machine's armature, physical airgap, and SC coils is the same, and the outermost diameter is fixed as 4.8 m.

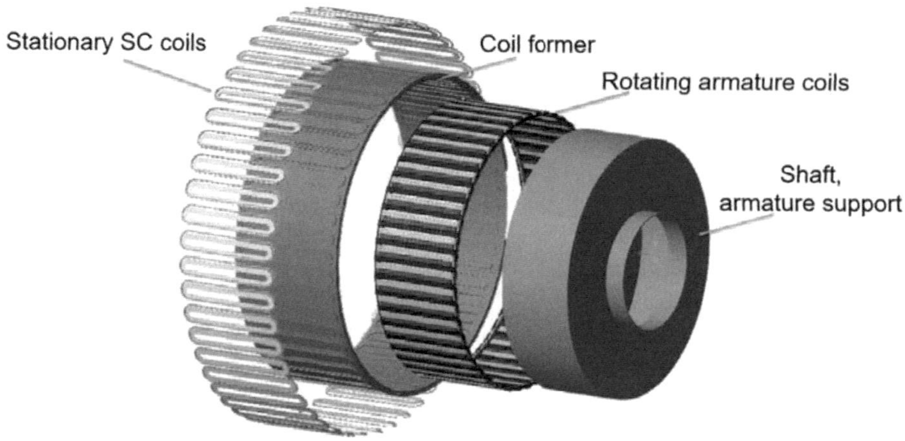

Figure 3. 10MW class actively-shielded SC generator for wind power applications.

Figures 4 and 5 show the magnetic flux density distribution inside the SC generator with the magnetic shielding core. It can be seen that the magnetic shielding core and active shielding SC coils installed at the outermost part of the machine is properly shielding the high magnetic flux by the SC main field coil. In air-core design 1, Magnetic shielding core was applied with ferromagnetic stainless steel-SS400, and it was assumed that it had ideally one-piece shape.

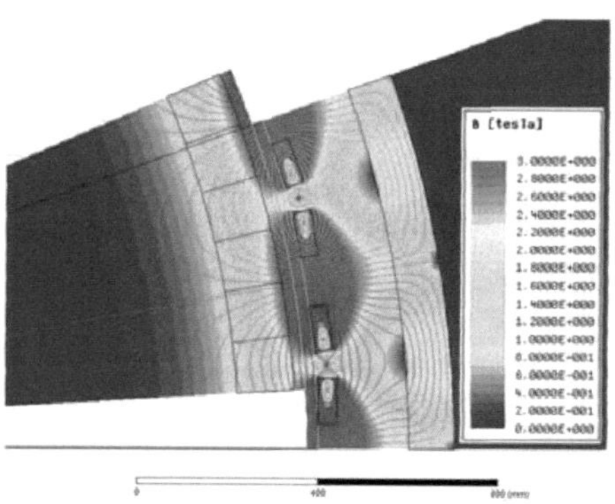

Figure 4. Magnetic field distribution of magnetic core shielded SC generator.

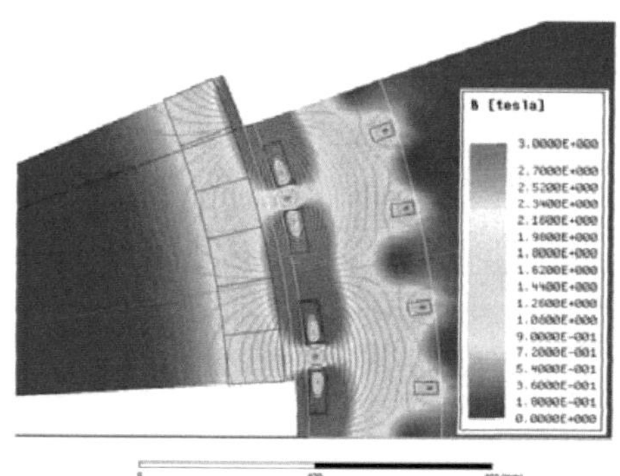

Figure 5. Magnetic field distribution of actively-shielded SC generator.

Figure 6 shows the flux density distribution in the airgap of the designed air-core sc generators. In air-core design 1, the maximum value of airgap flux density is about 3.6T, and in air-core design 2 it is about 3.4T. It can be seen that the airgap flux density of the air-core design 2 is about 5% smaller in the identical 2D shape.

Figure 7 shows the flux density analysis results to confirm the magnetic flux shielding performance of the generators at the boundary of 5.8 m in diameter. The maximum radial

IOP Conf. Series: Journal of Physics: Conf. Series **1054** (2018) 012083 doi:10.1088/1742-6596/1054/1/012083

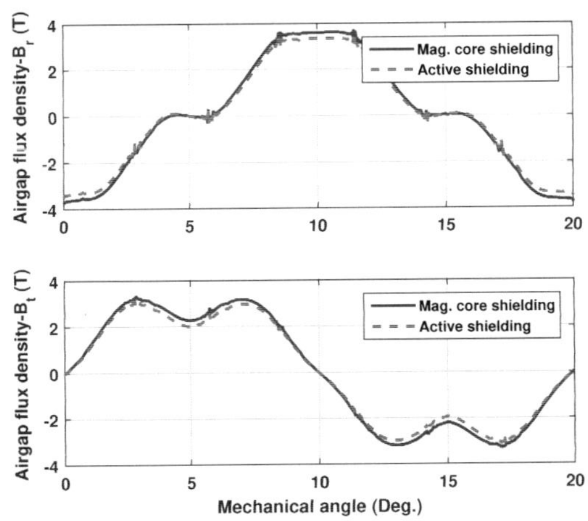

Figure 6. Comparison of airgap flux density distribution (at the center of electrical gap).

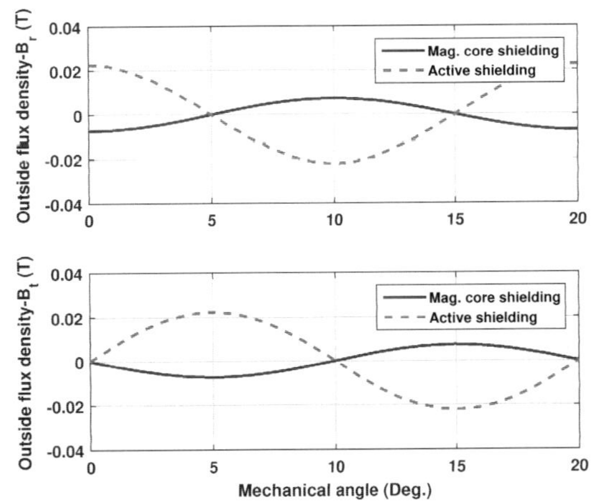

Figure 7. Comparison of outside flux density distribution (at 5.3 m-diameter).

flux density is about 0.007 T in the air-core design 1 and about 0.022 T in the air-core design 2. It can be seen that the magnetic flux shielding performance at the set boundary is about three times better in the air-core design 1 case.

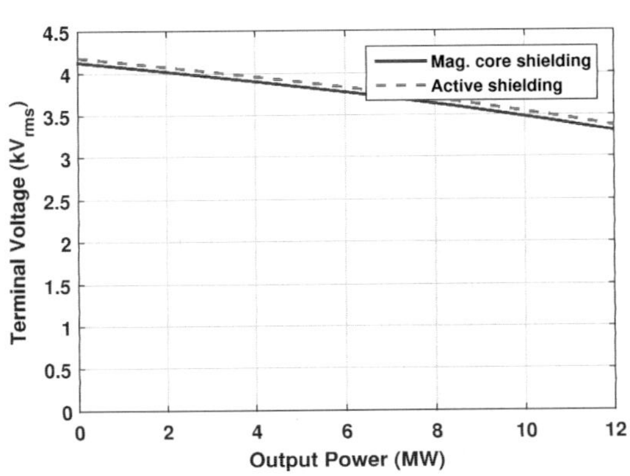

Figure 8. Output power vs. terminal voltage characteristics.

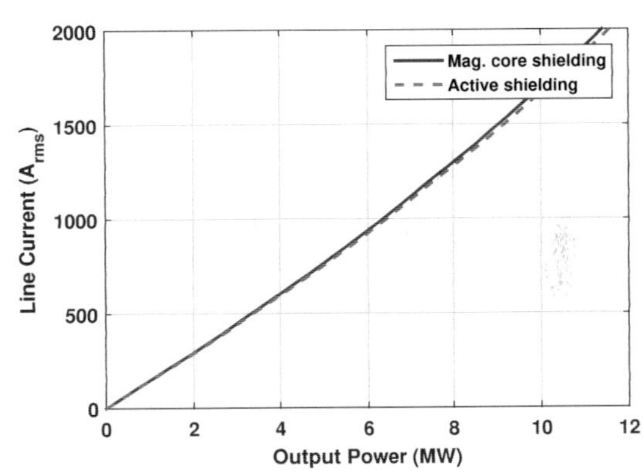

Figure 9. Output power vs. line current characteristics.

3.2. Output characteristics and weight

As machine diameter increases, the amount of required copper and SC wire decreases. This decrease follows an inverse relation in accordance with the machine's constant D^2L factor; as the diameter increases, the active length must decrease exponentially to maintain the same power. Additionally, since the geometry of the SC coils was held constant, an increase in machine diameter requires an increase in the pole count. This adds to the SC wire weight and length, and becomes more pronounced for larger diameters as the pole count increases and the active length decreases. Increasing the armature depth allows for more torque- producing material within the machine, which decreases the active length for a given diameter. In theory, there would come a point where the additional copper does not add significantly to the torque of the machine, and instead only increases the weight. This did not appear in the results, since that turning point is at a value beyond the current range.

Figures 8 and 9 show the output power vs. terminal voltage characteristics of two generators designed in this paper. Both generators have a terminal voltage of about 3.3-3.5 kV at the same operating point with an output of about 10 MW.

Table 2. Design parameters of reference and magnetic-core/actively-shielded 10MW SC generator.

Item		Reference	Air-core design 1	Air-core design 2
Shield type			Magnetic Core	Active shielding
Active length	(m)	1.88	2	2.2
Rated current	(A)	1750	1800	1800
Armature current density	(A/mm^2)	3.6	3.7	3.7
SC coil material		NbTi	Nb_3Sn	Nb_3Sn
SC coil total length	(km)	720	800	1066
SC coil total weight	(kg)	3840	4266	5685
Armature copper weight	(kg)	9270	6000	6000
Shield core weight	(kg)	-	45377	0
Total	(ton)	-	55.64	11.68

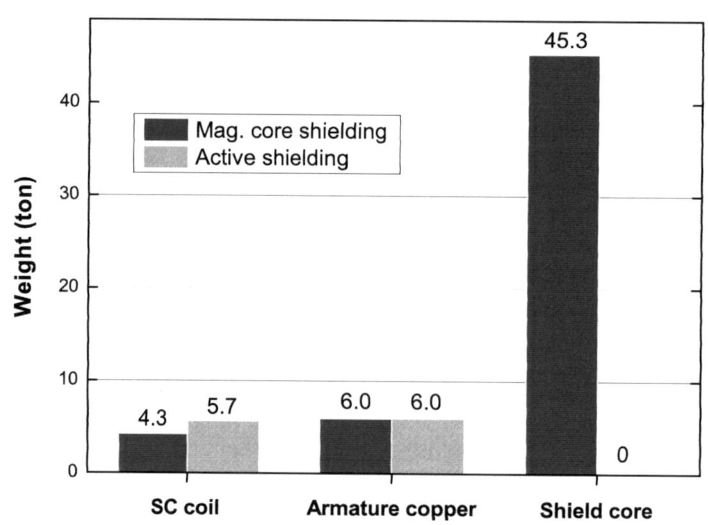

Figure 10. Comparison of weight for electrical components.

Design results are presented in Table 2 for a comparison of possible generators utilizing the air-core superconducting topology. Design parameters of reference and magnetic core shielding and actively-shielded SC generator are listed. As shown in the table, the magnetic core installed for the shielding of the magnetic flux emitted to the outside in the air-core design 1 has a weight of about 45ton, which is more than 81.6% of the total electrical component weight in the generator. Figure 10 shows the comparison of weight for electrical components of magnetic core shielding and active shielding SC machines. This result also confirms that the total weight of the generator with the active shielding SC coil is about 21% of the generator weight with the magnetic shielding core when the two designed generators have the same output.

4. Conclusion

In this paper, 10MW class SC generator for offshore wind power application was designed and tested for its characteristics. The outside diameter of the SC generator was limited to 4.8 m by referenced design and an inside-out synchronous generator with a 3-phase armature rotating inside was designed. In the air-core design 2 with active shielding SC coils, the weight reduction effect of 79% or more was confirmed when comparing two devices having the same output. It is concluded that engineering trade-off studies are needed to reduce the cost by increasing the SC coil quantity and decreasing the weight of the whole machine.

In the immediate future, more constraints will be added to the optimization study. These include the weight of support structures, the cost of SC wire, and the addition of cooling for the SC wires. Additionally, optimization of the SC coil geometry would expand the design space by decoupling the pole count from the diameter. An analysis of thermal limits within the machine would also prove useful, as it could potentially allow for higher current densities within the armature and shorter active lengths. A more exhaustive multi-objective approach is the most reasonable next step to implement these improvements.

Acknowledgments

This work was supported by the research fund of Chungnam National University.

References

[1] Abrahamsen A B, Mijatovic N, Seiler E, Trholt C, Norgard P, Pedersen N, Andersen N and Ostergard J 2010 *Supercond Sci. Technol.* **23** 034019
[2] Wind in power 2015 European statistics EWEA
[3] Abrahamsen A B, Jensen B B, Seiler E, Mijatovic N, Rodriguez-Zermeno V M, Andersen N H and Ostergard J 2011 *Physica C: Superconductivity* **471** 1464
[4] Song X, Liu D, Polinder H, Mijatovic N, Holboll J, and Jensen B B 2017 *IEEE Trans. Appl. Supercon.* **27** 5201505
[5] Song X, Mijatovic N, Kellers J, Buhrer C, Rebsdorf A V, Hansen J, Christensen M, Krause J, Putz H, Wiezoreck J, and Holboll J 2017 *IEEE Trans. Appl. Supercon.* **27** 5201205
[6] Liu D, Polinder H, Abrahamsen A B, Ferreira J A 2017 *IEEE Trans. Appl. Supercon.* **27** 5202007
[7] Haran K S, Loder D C, Deppen T O and Zheng L 2016 *IEEE Trans. Appl. Supercond* **26** 5202508
[8] Loder D C, Haran K S 2015 *IEEE Electric Machines and Drives Conference* 1709
[9] Fair R et al 2012 *Superconductivity for large-scale wind turbines* (Online Electronic Materials)

IOP Conf. Series: Journal of Physics: Conf. Series **1054** (2018) 012084 doi:10.1088/1742-6596/1054/1/012084

Structural design and heat load analysis of a flux pump-based HTS module coil for a large-scale wind power generator

O Tuvdensuren[1], H J Sung[1], B S Go[1], T T Le[1], M Park[1] and I K Yu[1]

[1] Changwon National University, Changwon, GN, 51140, Korea

yuik@cwnu.ac.kr

Abstract. Superconducting generators typically require a power supply, current lead (CL) and slip ring to deliver DC current to a high-temperature superconducting (HTS) coil, which causes a conduction heat load. On the other hand, a flux pump (FP) is possible to supply DC current to the HTS coils of the generator without the heat transfer loads. This paper deals with a structural design and heat load analysis of an FP-based HTS module coil for a 12 MW wind power generator. The structures such as HTS coil bobbins, coil supports, and the connection components between the FP and the HTS coils were designed. The conduction and radiation heat loads of the FP-based HTS module coils were analysed using a 3D finite element method program. The results of the HTS module coil of the generator were compared with a conventional CL-based HTS module coil. As a result, the total heat loads of the FP-based HTS module coil were lower than the conventional CL-based HTS module coil. The structural design and heat load analysis results of the FP-based HTS module coil can effectively be utilized to develop a large-scale HTS wind power generator.

1. Introduction

In the development of offshore wind power generation, the size and weight of wind power generators are increasing, and the capacity of generators is also increasing. Many companies have tried to develop wind power generators with a larger capacity, smaller size and lighter weight. A high-temperature superconducting (HTS) generator is one potential solution for these issues. HTS generators are usually adopted for large multi-megawatts generators [1]-[3].

A large-scale HTS generator has its own set of problems, such as the need for a huge vacuum vessel and the difficulty of repairing and maintaining the HTS coils because all the HTS coils are consisted in one cryostat [4]-[6]. Normally, the HTS coils are connected in a series, if the one HTS coils fails, the other coils break down. To overcome these problems, large-scale HTS generator module coils are structurally separated with an HTS flux pump (FP) or current lead (CL) [7]-[8]. Conventional CL requires a power supply and a slip ring to transfer the DC current to the HTS coil. The CL can be a bridge between the cryogenic environment and room temperature, which causes large heat transfer loads [9]-[11].

The high cooling capacity of cooling systems is therefore important in large-scale HTS wind power generators. On the other hand, a FP can supply DC current to the HTS coils of the generator without the heat transfer loads of excitation systems [12]-[15]. However, an effective structural design and the heat load analysis considering the connection of the FP to the HTS coils are required for the HTS wind power generator.

This paper deals with a structural design and heat load analysis of an FP-based HTS module coil for a 12 MW wind power generator. The 12 MW HTS generator is composed of 30 module coils. The structures of one module coil, such as HTS coil bobbins, coil supports, and the connection components between the FP and the HTS coils, were designed. The conduction and the radiation heat loads of the FP-based and the CL-based HTS coils were analyzed using a 3D finite element methods program considering the two-stage cryo-cooler.

The results of the FP-based HTS coil of the generator were compared with a conventional CL-based HTS module coil. As a result, the total heat load of the FP-based HTS module coil was lower than that of the conventional CL-based HTS module coil. These results and structural design will be applied to a large-scale HTS wind power generator with cooling system.

2. Structural designs of the FP-based and CL-based HTS module coils

The 12 MW HTS generator consists of rotor and stator parts. The rotor components are HTS field coils, a cryostat and supporting structure, and a rotor body. The stator components are comprised of stator teeth, copper stator coils, and a gear to rotate the FP exciter. The generator has 30 HTS modules. Table 1 shows the specifications of the 12 MW HTS generator [3].

Table 1. Specifications of the 12 MW HTS generator

Items	Values
Rated output power	12 MW
Rated line to line voltage	6.6 kV
Rated armature current	1.07 kA
Rated wind speed	11.4 m/s
Rotational speed	8 rpm
Rated torque	15.04 MNm
Number of poles	30 ea
Operating temperature	20 K

All the HTS field coils are structurally separated and not connected series as shown in figure 1(a). The HTS module consists of HTS field coils, coil bobbins, bobbin supports, a heat exchanger, a CL or FP, a cryostat, and a cryo-cooler. The HTS coils are covered by bobbin supports to protect the HTS coils from a high mechanical force. Heat exchangers are used to transfer heat between the HTS coils and the cold head of the cryo-cooler. The coil supports secures the HTS coils and blocks the heat transfer from the cryostat.

The FP exciters are located in each module and classified as rotating and stationary parts as shown in figure 1(b). The rotating parts have permanent magnets, an iron-shaft, an iron disk, and an FP exciter gear at room temperature. These parts are rotated by interlocking gear teeth of the generator and FP exciter. The stationary parts include an HTS stator wire, an iron-disk, an iron shaft, and an iron-ring at a low temperature. The stationary parts are located in the cryostat of the module to connect the HTS stator wire directly to the HTS field coil.

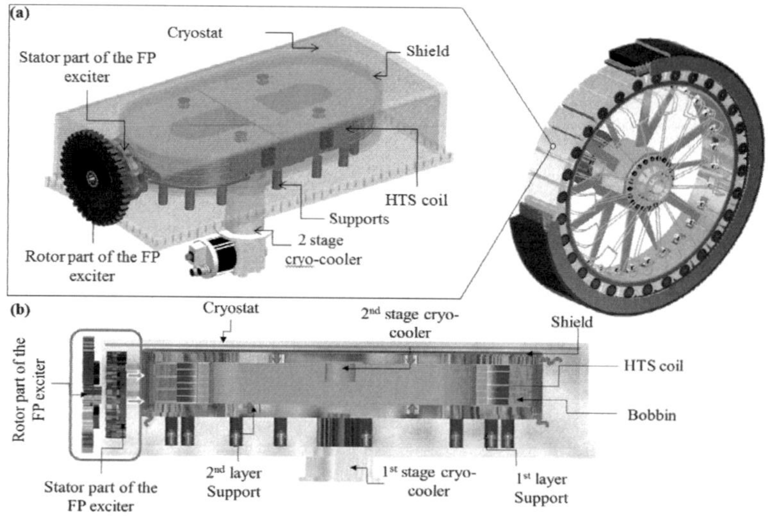

Figure 1. Configuration of the FP-based HTS module coil

The CL-based HTS module coil consists of an HTS field coil, coil bobbins, bobbin supports, a heat exchanger, current leads, a cryostat, and a cryo-cooler. Figure 2 shows the structure design of the CL-based HTS module coil.

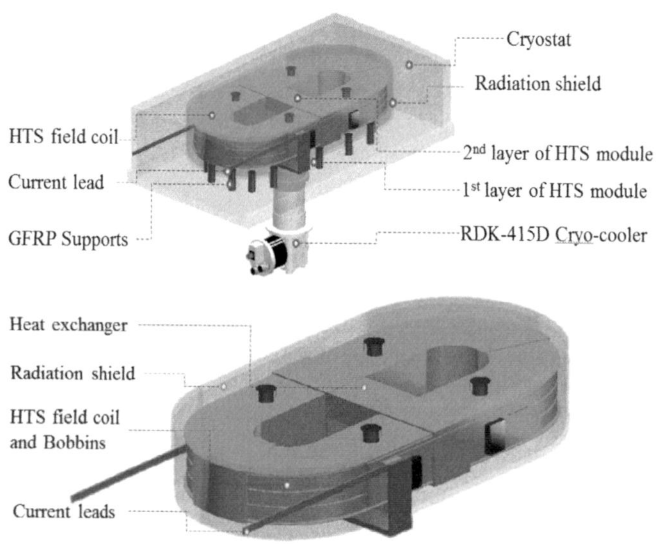

Figure 2. Configuration of the CL-based HTS module coil

The CLs for transferring current from room temperature to the cryogenic environment. To design the CLs, the important factors are the thermal conductivity of materials and the calculating shapes (length, cross sectional area and cooling surface area). The minimum heat load and ideal length of the CL can be calculated by equations (1) and (2).

$$Q_{op} = I\sqrt{2\int_{T_L}^{T_H} \rho(T)k(T)dt} \quad (1)$$

$$\frac{L}{A} = \frac{1}{I}\int_{T_L}^{T_H} \frac{k(T)}{\sqrt{2\int_{T_L}^{T_H} \rho(T)k(T)dt}} dT \quad (2)$$

The thermal conductivity $k(T)$ and the electric resistivity $\rho(T)$ are inversely related, according to Wiedemann- Franz law (3)

$$k(\mathrm{T})\rho(T) = L_0 T \quad (3)$$

Here, the Lorentz number is $L_0 = 2.45 \times 10^{-8}\, W\Omega K^{-2}$. Hence, equations (1) and (2) can be rewritten as:

$$Q_{op} = I\sqrt{L_0(T_H^2 - T_L^2)} \quad (4)$$

$$\frac{L}{A} = \frac{1}{I}\int_{T_L}^{T_H} \frac{k(T)}{\sqrt{L_0(T_H^2 - T)}} dT \quad (5)$$

where, L is the length of CL, A is the cross sectional area, and I is the carrying current, T_H and T_L represent the temperature of the cold and warm ends, $k(T)$ is the thermal conductivity, and $\rho(T)$ is the electric resistivity. The current leads are made up of a normal conducting material at the room temperature end, so copper or brass is used generally. Figure 3 shows the configuration of the CL for the HTS module coil. The current leads including copper current terminals and the cooling block; current feedthroughs were designed using a 3D CAD program.

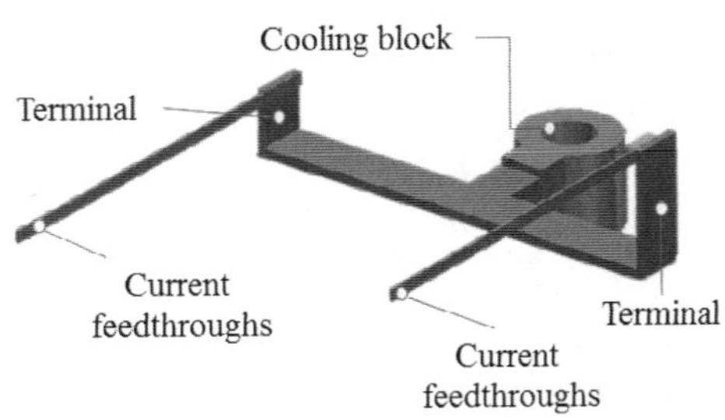

Figure 3. Configuration of the current leads

3. Heat loss analysis of the HTS module coil

Excitation loss: The sum of iron and joule heat losses of the HTS FP exciter is the excitation loss. The total resistance of the exciter and operating current were 79 μΩ and 352 A, respectively. Therefore, the joule heat loss of the exciter was 10 W, and the iron loss of the stator part was 0.2 W. The total excitation loss was 10.2 W in the generator module. The design and analysis of the HTS FP-based exciter were discussed in [5].

Eddy current loss: The 12 MW HTS generator adopts three-phase armature coils, which are conventional copper coils at room temperature. They induce eddy currents to the HTS module. The eddy current loss of the HTS module is generated in the coil bobbin, bobbin supports, and heat exchanger. The total eddy current loss in one HTS module coil of the 12 MW HTS generator was 15 W which was analyzed using an FEM program and discussed in [5].

Conduction and radiation losses in the HTS module coil: The supports were located in both the 1st and 2nd layers of HTS module. The conduction heat losses of supports are transferred from ambient temperature to the shield and from shield temperature to coil temperature. Based on the capacity map of a two-stage GM cryo-cooler, in the second stage, to achieve the temperature under 20 K at the HTS coils, the total heat load should be less than 27.5 W. In the first stage, the total heat loss should be less than 52 W to maintain the temperature of the shield less than 65 K using the two stage GM cryo-cooler in the HTS module.

To reduce the radiation loss of the coil, a radiation shield was used. The radiation loss between the shield and cryostat is estimated as:

$$\epsilon\sigma(T_c^4 - T_r^4) \tag{6}$$

where, ϵ is the effective total thermal emissivity of the shield, σ is the Stefan Boltzmann constant, and T_c and T_r are the cryostat temperature and thermal shield temperature.

4. Heat load analysis and simulation results

4.1. Heat load analysis of the CL-based HTS module coil

The thermal characteristics of the CLs were analyzed using a 3D FEM program. For the simulation, the operating current was 352 A. The temperature distribution analysis results of the CL-based HTS module coil is shown in figure 4. The temperature distribution in the HTS coils is shown in figure 4 (a). In the 2nd layer, the cold head temperature was 19 K, as shown in figure 4 (b). The temperature distribution in the supports of the CL-based HTS module coil is shown in figure 4 (c) and (d), respectively.

As a result, the temperature of the HTS coil was 19.8 K. In the 1st layer, the conduction heat load from supports was 14.3 W, and the radiation heat load was 4.9 W, the conduction heat load from the current leads was 31.8 W. The total heat load of the 1st layer was 51 W.

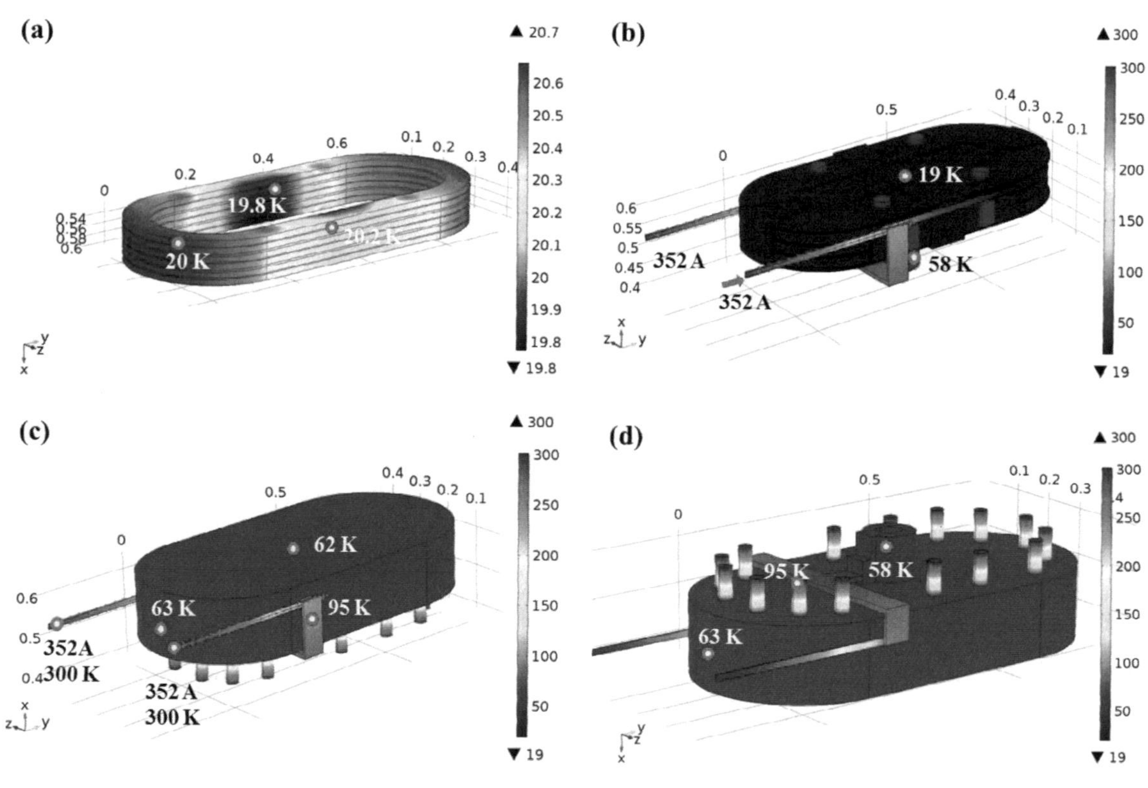

Figure 4. Temperature distributions in the HTS coil and supports of the CL-based HTS module

IOP Conf. Series: Journal of Physics: Conf. Series **1054** (2018) 012084 doi:10.1088/1742-6596/1054/1/012084

In the case of the 2nd layer, the temperature of the shield was 63 W. The total heat load of the 2nd layer was 27.3 W. Table 2 shows the simulation results of the total heat load of the CL-based HTS module coil.

Table 2. Results of the heat load analysis in the CL-based HTS module coil

	Items	Values
1st layer	Conduction heat load from the current lead	31.8 W
	Conduction heat load from the supports	14.3 W
	Radiation heat load	4.9 W
	Cooling power of the 1st layer	52 W
	Total heat load of the 1st layer	51 W
2nd layer	Eddy current heat load	15 W
	Conduction heat load from the 2nd supports	12.3 W
	Cooling power of the 2nd layer	27.5 W
	Total heat load of the 2nd layer	27.3 W

4.2. Heat load analysis of the FP-based HTS module coil

Figure 5 shows the temperature distributions analysis results of the FP-based HTS module coil. The temperature distribution in the HTS coils is shown in figure 5 (a). Figure 5 (b) and (c) shows the temperature distributions of supports in both layers of the FP-based HTS module coil. Figure 5 (d) shows the temperature distribution in the cooling block of the FP-based HTS module coil.

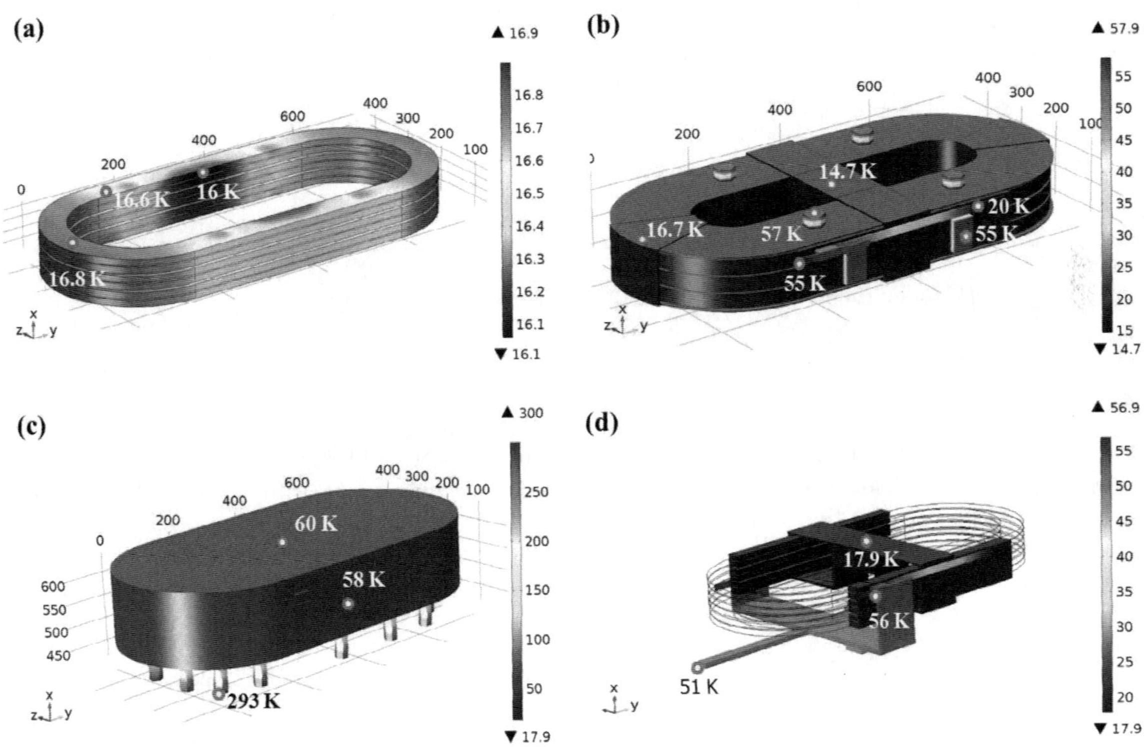

Figure 5. Temperature distributions in the HTS coil and supports of the FP-based HTS module

As a result, the temperature of the HTS coil was 19.3 K. In the 1st layer, the conduction heat load from supports was 15.5 W, and the excitation load was 10.2 W; the radiation heat load was 4.7 W. The total heat load of the 1st layer was 30.4 W. In the 2nd layer, the total heat load was 27 W. The temperature of the shield was 58 K. The eddy current load was 15 W, and the conduction heat load from supports in this layer was 12 W. Table 3 shows the simulation results of the total heat load of the FP-based HTS module coil.

Table 3. Results of the heat load analysis in the FP-based HTS module coil

	Items	Values
1st layer	Excitation load	10.2 W
	Conduction heat load from the supports	15.5 W
	Radiation heat load	4.7 W
	Cooling power of the 1st layer	52 W
	Total heat load of the 1st layer	30.4 W
2nd layer	Eddy current load	15 W
	Conduction heat load from the 2nd supports	12 W
	Cooling power of the 2nd layer	27.5 W
	Total heat load of the 2nd layer	27 W

Table 4 shows the comparative analysis results of the heat load between the CL and FP. The total heat loads of CL and FP were 78.3 W and 57.4 W, respectively.

Table 4. Comparative results of the heat load between the CL and FP

		Heat load	
	Items	Flux pump	Current lead
1st layer	Excitation load/Conduction heat load from the current lead	10.2 W	31.8 W
	Conduction heat load from the supports	15.5 W	14.3 W
	Radiation heat load	4.7 W	4.9 W
	Total heat load of the 1st layer	30.4 W	51 W
2nd layer	Eddy current load	15 W	15 W
	Conduction heat load from the 2nd supports	12 W	12.3 W
	Total heat load of the 2nd layer	27 W	27.3 W
	Total heat load	57.4 W	78.3 W

5. Conclusions

This paper discusses the structural design and heat load analysis of an FP-based HTS module coil for a 12 MW wind power generator. The conduction and radiation heat loads of the FP-based HTS coils were analyzed using a 3D finite element method program considering a two-stage cryo-cooler. The design procedure was performed at an operating temperature of 20 K, and the overall heat loss at each layer of the HTS module coil was less than the cooling capacity of each stage of the cryo-cooler. The results of the FP-based HTS coil of the generator were compared with the conventional CL-based HTS module coil. As a result, the temperature of the HTS coil of the FP-based HTS module was 19.3 K, and the total heat load was 57.4 W. The temperature of the HTS coil of the CL-based HTS module was 19.8 K and the total heat load was 78.3 W. The temperature of the FP-based HTS module coil was lower than that of the CL-based HTS module coil because the total heat load of the FP-based HTS module coil was smaller than that of the conventional CL-based HTS module coil. In conclusion, the FP-based HTS module was evaluated to be more advantageous in terms of better cooling capacity and thermal stability margins. The structural design and heat load analysis results of the FP-based HTS module coil can be effectively utilized in the development of large-scale HTS wind power generators.

Acknowledgments
This research was supported by Changwon National University in 2017-2018.

References

[1] Young G P, Woo S L, Jeyull L, Seunghyun S, Young J H, Yoon D C and Tae K K 2016 Operational characteristics of HTS coils with flux diverters in semipersistent mode under alternating magnetic field *IEEE Trans. App. Supercond.* **26** no 4 p 5204105

[2] Xinkai Z, Ming Ch, Peng H, Gan Zh and Xinch Wei 2017 Analysis of dual stator HTS reluctance rotor brushless doubly fed wind generator *Electrical Machines and systems(ICEMS) International Conference* Australia pp 1-4

[3] Sung H J, Badcock R A, Go B S, Park M, Yu I K and Jiang Z 2016 Desing of a 12 MW HTS Wind Power Generator Including a Flux Pump Exciter *IEEE Trans. App. Supercond.* **26** no 3 p 5206205

[4] Sung H J, Badcock R A, Jiang Z, Choi J, Park M and Yu I K 2016 Design and heat load analysis of a 12 MW HTS wind power generator module employing a brushless HTS exciter *IEEE Trans. App. Supercond.* **26** no 4 p 5205404

[5] Sung H J, Go B S, Jiang Z, Park M and Yu I K 2016 Heat loss analysis-based design of a 12 MW wind power generator module having an HTS flux pump exciter *Physica C: Supercon. and its appl.* **530** pp 133-137

[6] Tuvdensuren O, Dao V Q, Le T T, Go B S, Park M and Yu I K 2017 Design of a current lead for a large scale superconducting wind power generator *Korean institue of electrical engineers 48th conference (KIEE)* pp 1065-66

[7] Dao V Q, Kim T, Sung H J, Choi J, Park M and Yu I K 2017 Thermal optimization of current leads for HTS DC reactor Magnet *International symposium on innovation in information technology and application (ISIITA) international conference 2017* pp 17-21

[8] Go B S, Sung H J, Park M and Yu I K 2017 Structural design of a Module coil for a 12 MW class HTS generator for wind turbine *IEEE Trans. App. Supercond.* **27** no 4 p 5202405

[9] Sung H J, Park M, Go B S and Yu I K 2016 A Study on the required performance of a 2G HTS wire for HTS wire for HTS wind power generators *Supercond. Sci. Technol.* **29** p 9

[10] Le T T, Sung H J, Tuvdensuren O, Park M and Yu I K 2017 Thermal design of a superconducting module for a 12 MW wind power generator *Korean institue of electrical engineers 48th conference (KIEE)* pp 1100-01

[11] Jiang Z, Hamilton K, Amemiya N, Badcock R A and Bumby C W 2014 Dynamic resistance of a high-Tc superconducting flux pump *Appl. Phys. Lett.* **105** no 11 p 115008

[12] Haran K, Mantak H, Edward Ch and Devdatta K 2014 Direct drive HTS wind generator design for commercial applications *International conference on electrical machines (ICEM)* pp 491-495

[13] Go B S, Sung H J, Choi J, Park M and Yu I K 2016 Modularization of the HTS Field Coils for a Large-Scale Superconducting Generator *IEEE Trans. App. Supercond.* **26** no 3 p 5202905

[14] Shafaie R, Kalantar M 2013 Design of a 10 MW class wind turbine HTS synghronous generator with optomized field winding *IEEE Trans. App. Supercond.* **23** no 4 p 5202307

[15] Chris W B, Sinhoi P, Andres E P, Zhenan J, James G S, Sung H J 2017 Frequency Dependent Behavior of a Dynamo-Type HTS Flux pump *IEEE Trans. App. Supercond.* **27** no 4 p 5200705

Numerical Simulation on Behaviour of Magnetic Beads in Magnetic Filter for Medical Protein Screening System using High Gradient Magnetic Separation

M Kubota, Y Mori, S B Kim and H Ueda

Graduate School of Natural Science and Technology, Okayama University, 3-1-1, Tsushima Naka, Kita-ku, Okayama 700-8530, Japan.

E-mail address: hirsohi.ueda@okayama-u.ac.jp

Abstract. Recently, Antibody drugs are attracting attention in the medical field. Antibody using immune function has high efficacy and side effects are relatively few, so it is expected to become the mainstream of future medicine. Indispensable technologies for development of antibody drugs are continuous, large amount and high-speed separation of medical protein. Therefore, we have proposed superconducting high gradient magnetic separation (HGMS) system for medical protein. If HGMS system using superconducting magnet is realized, processing time of separation can be greatly shortened. In the previous studies, it was possible to separate 200 nm magnetic beads suspended in pure water, with the result that the capture ratio was 97.8% and the collection ratio was 94.1%. However, depending on the type of the suspension, the clogging of the magnetic filter and the phenomenon that the magnetic nano-beads are not trapped have been confirmed. Therefore, in this study, the magnetic force of the magnetic filter is evaluated by electromagnetic fluid and particle tracing analysis considering the behaviour of the magnetic beads two-dimensionally in the magnetic filter.

1. Introduction

The antibody is a type of immune, which mainly present in blood and body fluids, recognizing and destroying viruses invaded the body as antigens. Antibody drugs are expected to have less side effects, achieve high and long efficacy in the body, thus reducing the number of dossing drugs. Therefore, the development of new antibody drugs has been actively carried out all over the worldwide. The global market scale is estimated to grow 10 times in the next 10 years [1]. On the other hand, as a disadvantage, it can be mentioned that the current antibody drugs are very expensive. Regarding this problem, continuous, large amount and high-speed separation by improvement of producing technology leads to reduction of production cost in the industry.

We have been conducting research on development of high gradient magnetic separation (HGMS) system using magnetic beads for antibody screening [2-6]. In this system, by applying a high magnetic field to the magnetic filter by the superconducting magnet, a high gradient of magnetic field and magnetic force is generated around the magnetic wire. In previous studies, the prototype HGMS system using superconducting magnets was constructed [4-6]. The dimensions of the system cabinet were 517 mm wide × 617 mm deep × 1000 mm high. The cabinet must be small as it is a tabletop type. The superconducting magnet was set in the middle of the cabinet, and the room temperature bore had a 30-mm diameter. The magnetic filter was inserted in a 20-mm inner diameter tube set in the middle of the room temperature bore of the magnet. The cooling system was a Gifford-McMahon (GM) cryocooler with a cooling capacity of 6 W at 55 K and cooled via conduction. The operating temperature was 4.2 K. A resonance circuit was added to the HGMS system in order to improve the recovery of nanobeads [6]. The filter was demagnetized in the AC magnetic field by the resonance circuit to make good use of the superconducting magnet of HGMS.It was possible to separate 200-nm magnetic beads suspended in pure water, with the result that the capture ratio was 97.8% and the collection ratio was 94.1% [4]. However, depending on the type of the suspension, the clogging of the magnetic filter and the phenomenon that the magnetic nano-beads are not trapped have been confirmed. Therefore, in this study, we carried out numerical analysis to investigate the behaviour of magnetic beads around the magnetic filter and develop the magnetic filter with wire diameter, aperture, and stack interval as variables.

2. High gradient magnetic separation

At present, in the separation of medical proteins such as magnetic separation and affinity chromatography, affinity beads are mainly used as carriers. Affinity beads specifically and reversibly adsorb certain biomolecules. There are two types of affinity beads. One is nonmagnetic beads for chromatography and centrifugation and the other is magnetic beads for magnetic separation. Chromatography and centrifugal separation are difficult for continuous operation and require long time for separation, so it is unsuitable for high-speed, large-amount, and continuous separation. For beads of the same total mass, if the size is reduced to 1/10, the number of beads is 10^3 times, and the probability of attaching antibodies can be dramatically increased. However, the force caused by mass is small with downsizing the beads. On the other hand, magnetic force for even substances of small mass can be large in a high magnetic field and high gradient magnetic field. The magnetic force, F_{mag}, is the following expression when the external magnetic field is sufficiently large.

$$\boldsymbol{F}_{mag} = V_p(\boldsymbol{M}_s \cdot \boldsymbol{\nabla})H \qquad (1)$$

where V_p, and $\boldsymbol{M}_s$ are the volume, saturation magnetization of the magnetic beads, respectively. H is the magnitude of the magnetic field. Therefore, we have proposed a high gradient magnetic separation (HGMS) using a small superconducting magnet [2]. In case of using this small superconducting magnet, the magnetic field can be up to 2 - 4 T, thereby a high magnetic-field gradient and large magnetic force can be generated.

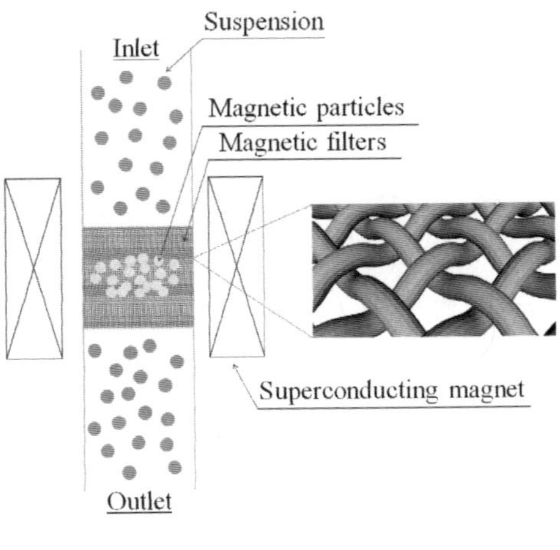

Figure 1. Schematic diagram of the superconducting HGMS system.

Figure 1 shows a schematic diagram of the superconducting HGMS system. The magnetic filter was inserted in a tube set in the middle of the room temperature bore of the magnet. HGMS systems can be divided into three major parts: pre-treatment part, magnetic separation part and post-processing part. In the pre-treatment part, the antibody in the suspension specifically attaches with an affinity substance. In the magnetic separation part, a high magnetic field is generated by the superconducting magnet and a high magnetic field gradient is generated around the magnetic filter, so the magnetic force acts on the

IOP Conf. Series: Journal of Physics: Conf. Series **1054** (2018) 012085 doi:10.1088/1742-6596/1054/1/012085

magnetic beads attaching antibodies. As a result, the magnetic beads are adsorbed by the magnetic filter. Then, the magnetic field is reduced to zero. In the post-processing part, the magnetic filter is demagnetized by a resonance circuit using a superconducting magnet in the HGMS. Next, the tube and filter are flushed with a solution, and then the nano-beads are recovered.

3. Numerical Analysis of behaviour of the magnetic beads around the magnetic filter

3.1. Analysis conditions

In order to investigate the behaviour of the magnetic beads around the magnetic filter in HGMS system using COMSOL as the parameter of wire diameter (d), aperture (A), stack interval (H), and number of stack filters as shown in Figure 2. In this study, the two-dimensional simulation was applied for the fundamental behaviour of magnetic beads. The aperture ratio, ε, is determined by the diameter and the aperture as following expression.

$$\varepsilon = (A/A + d)^2 \times 100 \quad (2)$$

where A is aperture, d is wire diameter. Table 1 lists the specifications of the magnetic filter in this simulation. The size of the magnetic filter is assumed to be commercially available from Kansai Wire Netting Co., Ltd. Currently, many of the filters for magnetic separation are S43000 because of high chemical resistance and strength. In the analysis model, a magnetic filter was arranged, and a periodic boundary condition was set for the black thick line in Figure 2.

Table 1. Specifications of magnetic line filter.

Wire diameter d (μm)	Aperture A (μm)	Aperture ratio ε (%)
30	34	28.22
30	48	37.87
30	72	49.83
40	45	28.03
40	75	42.53
40	101	51.31
50	52	25.99
50	77	36.76
50	104	45.61
60	67	27.83
60	84	34.03
60	121	44.69
70	84	29.75
70	111	37.61
70	142	44.86

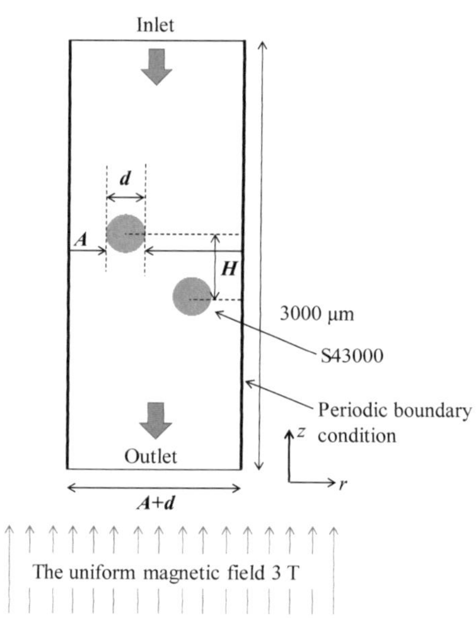

Figure 2. Schematic diagram of analysis model.

3.2. Electromagnetic field analysis

The electromagnetic field analysis was carried out in order to evaluate the magnetic force around the wire for the analysis model as shown in Figure 2. A uniform magnetic field of 3 T was applied to calculate the distribution of the magnetic force in various wire structure. Figure 3 shows the distribution of magnetic force in various magnetic wire structures. The strongest magnetic force is generated for the wire with cross section of circular, and the range of magnetic force is wider as well. From the above, it is considered that the circular cross-section magnetic wire used for magnetic separation is optimal. Figure 4 shows a vector map of the magnetic force generated around the magnetic wire. As seen in Figure 4, an attractive force is generated around the upper part of the magnetic wire. On the other hand, repulsive force is generated at the side of the magnetic wire, which is considered to greatly affect the behaviour of the magnetic beads.

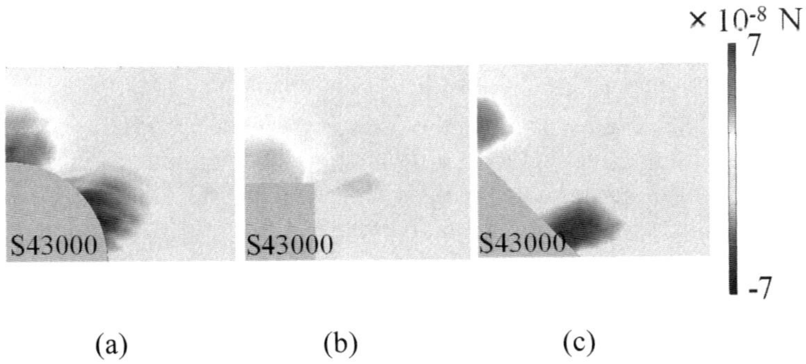

Figure 3. Distribution of magnetic force around the magnetic wire with cross section of (a) Circular shape, (b) quadrate shape, and (c) rhomboid shape.

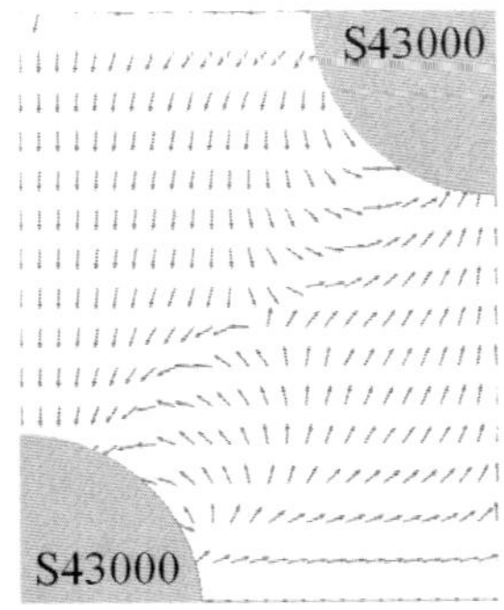

Figure 4. Vector map of magnetic force around the magnetic wire with cross sectional of circular shape.

3.3. Fluid dynamics Analysis
The fluid analysis was carried out using the model as shown in Figure 2. In this simulation, the viscosity of fluids is 1.1×10^{-3} Pa s, and fluid velocity at inlet is 5.0 mm/s, and temperature is 293.15 K. Figure 5 shows the velocity distributions of fluid for the filter with wire diameter of 30 μm and various aperture. As seen in Figure 5, because the fluid flows so as to avoid an obstacle, and the flow velocity becomes the maximum at the side of the magnetic wires in this simulation, and the flow velocity decreases in the upper and lower parts of the magnetic wire.

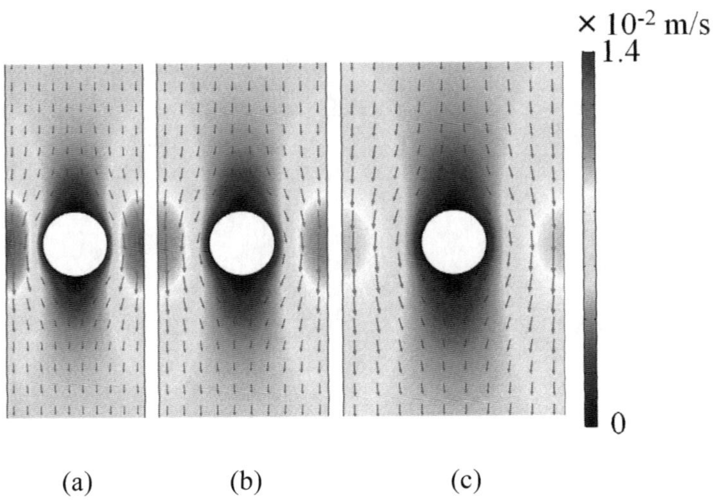

Figure 5. Distribution of flow velocity around the magnetic wire (d=30 μm) with an aperture ratio of (a) ε = 28.22%, (b) ε = 37.87%, and (c) ε = 49.83%.

3.4. Electromagnetic-fluid-particle tracing coupled analysis
Electromagnetic-fluid-particle tracing coupled analysis was conducted using the analysis model as shown in Figure 2. Analysis conditions are the same in subsections 3.2 and 3.3 applying a uniform magnetic field of 3 T and flowing water from the inlet. The mass, diameter and relative permeability of bead is 1.0×10^{-5} ng, 100 nm, and 1.05, respectively. These values supposed the beads manufactured by Tamagawa Seiki Co., Ltd with a saturation magnetization of 0.3 T. Figure 6 shows the particle trajectories when the inlet velocity of 5.0 mm/s in a single layer with a diameter of 30 μm. As seen in

Figure 6, the magnetic beads are adsorbed by the magnetic force generated on the upper part of the magnetic wire, and due to the influence of the repulsive force generated on the side of the magnetic wire, it is not adsorbed at the lower part of the magnetic wire. Figure 7 shows the change of the capture ratio for the various diameter in a single layer. When diameter is smaller, the capture ratio is higher. Furthermore, when aperture ratio (ε) is high, the capture ratio tends to decrease. However, considering the clogging of magnetic beads in previous studies, it is necessary to properly use filter size depending on the size and concentration of magnetic beads. Figure 8 shows the change in the capture ratio for the various stack interval (H) of 2-layers filter with the inlet velocity of 5.0 mm/s. It is difficult to grasp the trend since the movement of the fluid changes complicatedly depending on the stack interval and aperture. In the case of d = 30 μm, it is more efficient to select the aperture of filter rather than interval of stack in order to increase the capture ratio of beads.

Figure 9 shows the capture ratio of beads for various number of stack filters in the case of v = 5 mm/s and H = 100 μm. As seen in Figure 9, when the number of stack filters is taken as a variable under the condition of d = 70 μm, H = 100 μm, and v = 5.0 mm/s, about 80% of the magnetic beads flowing through the suspension are caught up to the 2nd layer. As a result, it is highly possibility to occurs the clogging in 1st and 2nd layers at v = 5.0 mm/s, therefore it is desirable to increase the flow velocity and the number of stack filters with the low concentration of beads in suspension.

4. Conclusion

In this study, the magnetic force of the magnetic filter is evaluated by electromagnetic fluid and particle tracing analysis considering the behaviour of the magnetic beads two-dimensionally in the magnetic filter. The filter with the diameter of 30 μm and the aperture ratio of 28.22% captures the magnetic beads most efficiently. However, in the case of 10-layers filter, the capture ratio achieves about 90% even if high aperture ratio which means low capture ratio, so in consideration of the clogging of beads, the diameter of 30 μm and aperture ratio of 49.83% is optimal. Currently, we confirmed that the capture ratio greatly fluctuates when the inflow velocity is changed. And, it can be expected to improve the treatment time of capturing beads. In order to accurately evaluate the phenomenon in a real system, we are conducting a three-dimensional electromagnetic-fluid-particle coupled analysis using a commercial magnetic mesh filter.

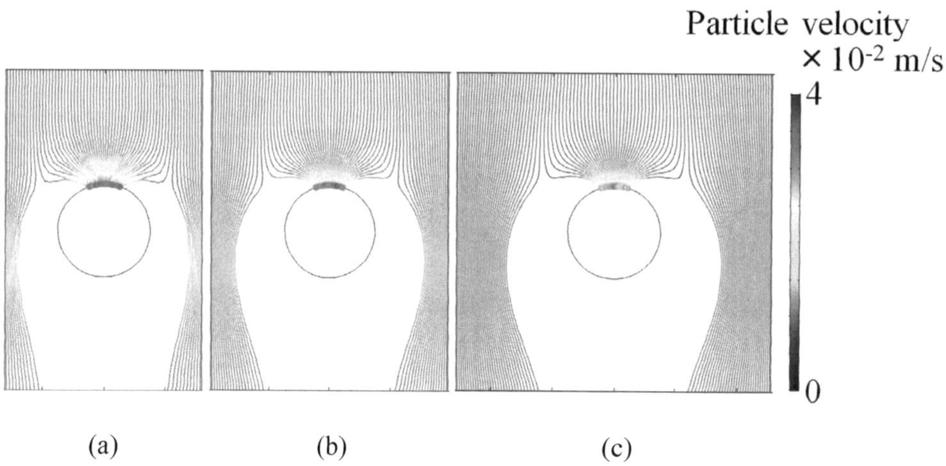

Figure 6. The trajectory of magnetic beads around the magnetic wire (d=30 μm) with an aperture ratio of (a) ε=28.22%, (b) ε=37.87%, and (c) ε=49.83%.

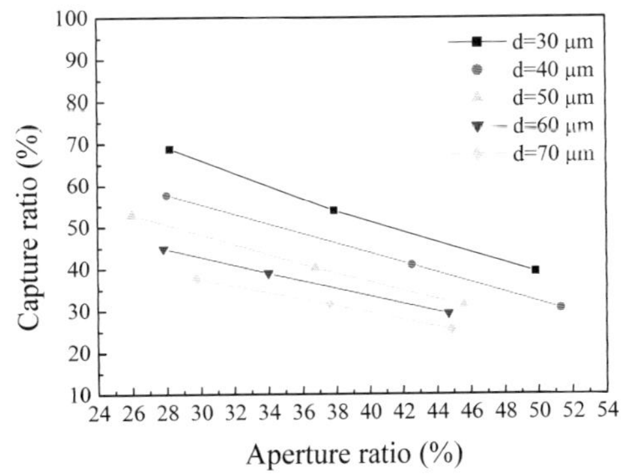

Figure 7. Capture ratio of beads for various wire diameter in single-layer filter.

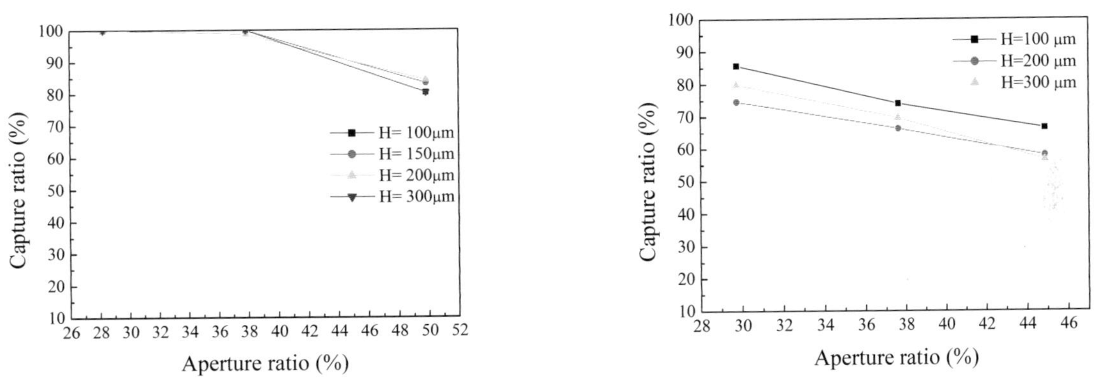

(a) $d = 30$ μm (b) $d = 70$ μm

Figure 8. Capture ratio of beads for various stack interval in 2-layers filter with $v = 5.0$ mm/s.

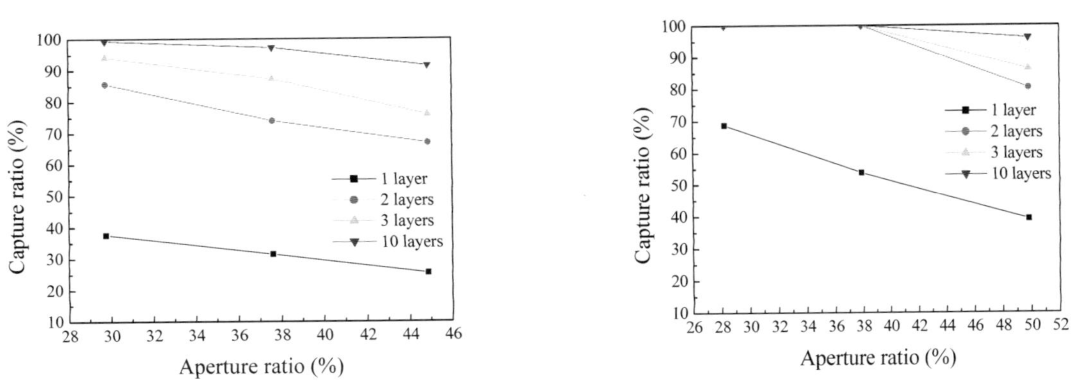

(a) $d = 30$ μm (b) $d = 70$ μm

Figure 9. Capture ratio of beads for various number of stack filters with $v = 5$ mm/s and $H = 100$ μm.

References

[1] https://www.visiongain.com/Report/1445/Next-Generation-Antibody-Therapies-Market-Forecast-2015-2025

[2] Ueda H, Agatsuma K, Kajikawa K, Furuse M, Fuchino S and Ishiyama A 2009 Design and Test of Filter of High Gradient Magnetic Separation System for Trapping Immunoglobulin in Serum *IEEE Trans. Appl. Superoncd.* **19** 2157-2161
[3] Ueda H, Agatsuma K, Fuchino S, Imura T, Furuse M, Kajikawa K, Ishiyama A, Koizumi T and Miyake S 2010 Improvement in High Gradient Magnetic Separation System for Trapping Immunoglobulin in Serum *IEEE Trans. Appl. Superoncd.* **20** 949-952
[4] Ueda H, Agatsuma K, Furuse M, Fuchino S, Kajikawa K, Kamioka Y, Iitsuka T and Nakamura S 2014 Performance of Filters in Medical Protein Separation System Using Superconducting Magnet *IEEE Trans. Appl. Supercond.* **24** 3700405
[5] Fuchino S, Furuse M, Agatsuma K, Kamioka Y, Iitsuka T, Nakamura S, Ueda H and Kajikawa K 2014 Development of Superconducting High Gradient Magnetic Separation System for Medical Protein Separation *IEEE Trans. Appl. Supercond.* **24** 4400204
[6] Kajikawa K, Ueda H, Kamioka Y, Agatsuma K, Fuchino S, Furuse M, Iitsuka T and Nakamura S 2014 Development of Degaussing System of Magnetic Filters for High Gradient Magnetic Separation to Improve Recovery Ratio of Trapped Magnetic Nanobeads *IEEE Trans. Appl. Supercond.* **24** 4400305

Magneto-Archimedes levitation of precious metals under a high magnetic field gradient

O Miura[1]*, K Yamagishi[1] and D Yamamoto[1]

[1] Department of Electrical and Electronic Engineering, Tokyo Metropolitan University, Hachioji 192-0397, Japan

e-mail: miurao@ tmu.ac.jp

Abstract. We studied the levitation properties for precious metals utilizing magneto-Archimedes effect under a high magnetic field gradient. In order to enhance the power factor of *BdB/dz* for the magnetic force in vertical direction cylindrical ferromagnetic materials were set into the room temperature bore of a 10 T superconducting solenoidal magnet. The maximum *BdB/dz* achieved the high value of −1598 T^2/m by the optimum configuration of ferromagnetic materials. The magnetic levitation properties for several kinds of precious metals such as silver, gold and platinum in manganese chloride aqueous solution as a paramagnetic liquid medium were studied. The experimental results showed that silver, gold and platinum levitated at each different height which was almost the same as its own theoretical one and could be controlled by changing configuration of ferromagnetic materials. This indicates the strong possibility of the selective separation of a target material from the solid mixture by utilizing magneto-Archimedes levitation under a high magnetic field gradient.

1. Introduction

We have been studying the selective separation of the valuable resource from the solid mixture on urban mine by utilizing magneto-Archimedes levitation [1]. The condition of stable levitation in vertical direction for materials in a liquid medium is expressed as follows,

$$B\frac{dB}{dz} = \frac{\rho_f - \rho_p}{\chi_f - \chi_p}\mu_0 g. \qquad (1)$$

Where ρ_p is the density of the material, χ_p is the magnetic volume susceptibility of the material, ρ_f is the density of the liquid medium, χ_f is the magnetic volume susceptibility of the liquid medium, g is the gravitational acceleration, μ_0 is the permeability of vacuum, B is the magnetic flux density [2, 3]. Here the most important factor of magneto-Archimedes levitation force is *BdB/dz* which is generated by the magnet in itself. However, the maximum value of *BdB/dz* remains in −434 T^2/m for even in a standard high-field 10 T superconducting solenoidal magnet (JASTEC JMTD-10T100SS). Therefore in order to levitate precious metals such as gold and platinum with the very high density and the low magnetic susceptibility, the further increase in *BdB/dz* is required.

In this study to increase the power factor, *BdB/dz* especially a magnetic field gradient some cylindrical ferromagnetic materials were set into the room temperature bore of a superconducting solenoidal magnet. The power factor significantly increased by means of the optimum configuration of ferromagnetic materials. We studied levitation properties of precious metals in this system.

2. Experimental

In order to increase the power factor *BdB/dz*, especially a magnetic field gradient we carried out an idea of arranging ferromagnetic iron cylinders in the room temperature bore of a 10 T superconducting solenoidal magnet. The axisymmetric shaped cylindrical irons stably levitated at the centre of the magnet by the strong magnetic restoring force. Figure 1 shows *z* axis distribution of *BdB/dz* at 10 T with a change in the diameter of 50 mm long iron cylinders. The maximum *BdB/dz* significantly increased from −434 T^2/m to −1598 T^2/m with a decrease in diameter of the cylinders. Moreover, it was shown that a change in diameter of the cylinders made it possible to control the rate of change of *BdB/dz* related to the resolution of separation. Figure 2 shows *z* axis distribution of *BdB/dz* at 10 T with a change in the length of 10 mm diameter iron cylinders. Maximum *BdB/dz* slightly increases with increase in length. From this result, we selected the iron cylinder with a diameter of 10 mm and a length of 50 mm generating the maximum *BdB/dz* of -1060 T^2/m in consideration of the experimental setup.

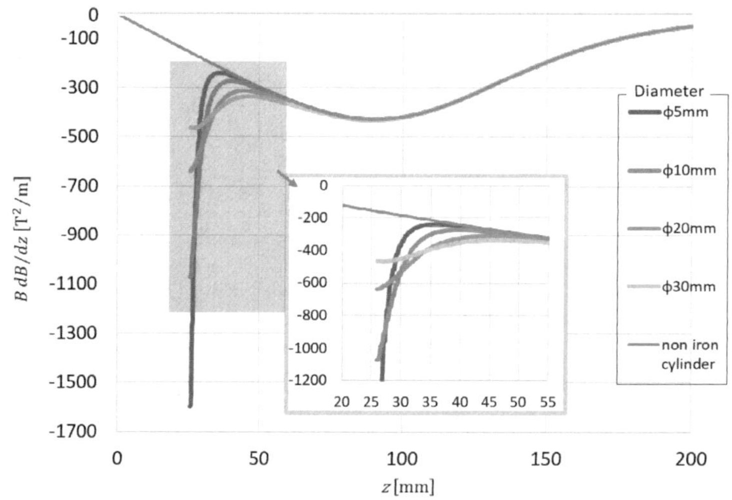

Figure 1. *z* axis distribution of *BdB/dz* at 10 T with a change in the diameter of 50 mm long iron cylinders in a radial direction of $r = 0$.

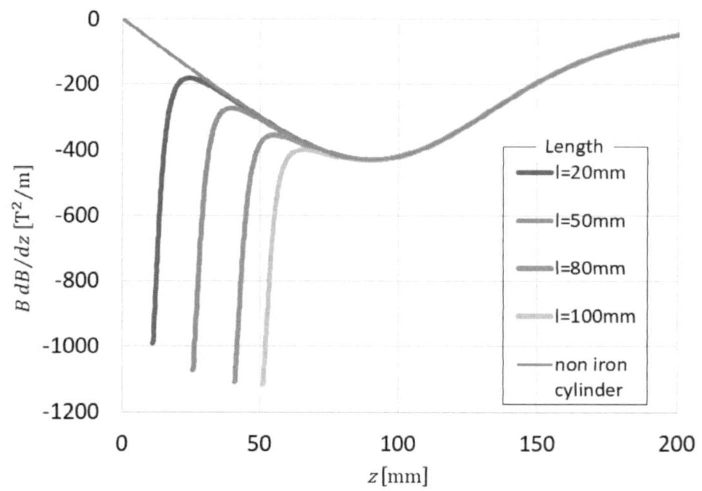

Figure 2. *z* axis distribution of *BdB/dz* at 10 T with a change in the length of 10 mm diameter iron cylinders in a radial direction of $r = 0$.

The experimental devices for magneto-Archimedes levitation studies are shown in figure 3. The iron cylinder sheathed in acrylic was set at the centre of the room temperature bore of a 10 T superconducting solenoidal magnet. $MnCl_2$ 50 wt% aqueous solution was used as a paramagnetic liquid medium. The graduated cylinder filled with the medium was put on the iron cylinder and observed by a CCD camera with a right-angle prism. We studied the magnetic levitating properties for several kinds of precious metals such as gold, silver, copper and platinum. Table 1 shows the density and the magnetic volume susceptibility measured by a SQUID magnetometer for the materials and $MnCl_2$ concentration used in the experiment.

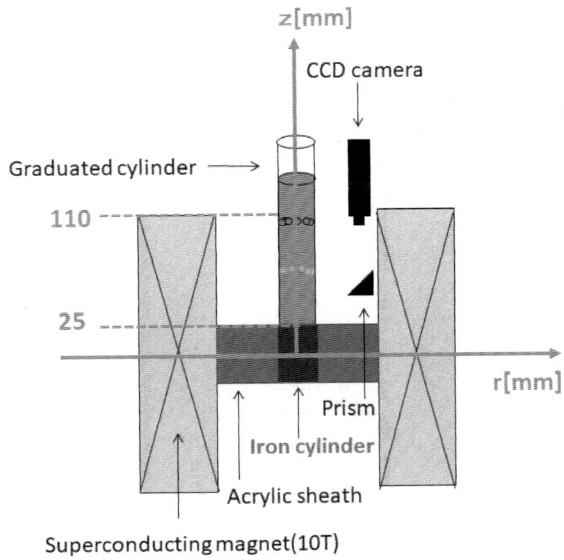

Figure 3. The experimental devices for magneto-Archimedes levitation studies.

Table1. Physical properties of the materials in the experimental condition.

Materials	Density [g/cm^3]	Magnetic susceptibility [-]
Gold	19.3	-1.70×10^{-4}
Silver	5.56	-2.91×10^{-5}
Copper	8.96	-2.25×10^{-5}
Platinum	21.5	1.05×10^{-4}
Manganese chloride aqueous solution (50 wt.%)	1.33	4.13×10^{-4}

3. Result and discussion

Figure 4 shows the magneto-Archimedes levitation for each precious metal in $MnCl_2$ aqueous solution at 10 T. It was clearly observed that all of the precious metals levitated on the iron cylinder under the high magnetic field gradient. The platinum levitated at z = 28 mm with BdB/dz of -809 T^2/m, the gold levitated at a height of z = 33 mm with BdB/dz of -380 T^2/m, the silver levitated at z = 153 mm with BdB/dz of -170 T^2/m, and the copper levitated at z = 158 mm with BdB/dz of -140 T^2/m. Table 2 shows the comparison of levitation properties between calculated values by equation (1) and experimental ones for each metal. We confirmed that the levitation height for each metal was almost the same as its own calculated value. The small differences between them for silver and copper may be caused by impurities of the metals. The levitation gap between two materials (platinum and gold, silver and copper) was 5 mm in this experiment. When considering practical use, it's necessary to expand the gap by controlling the variation of BdB/dz via change in diameter of cylinders. This indicates the strong possibility for the selective separation of a target material from the solid mixture by utilizing magneto-Archimedes levitation under a high magnetic field gradient.

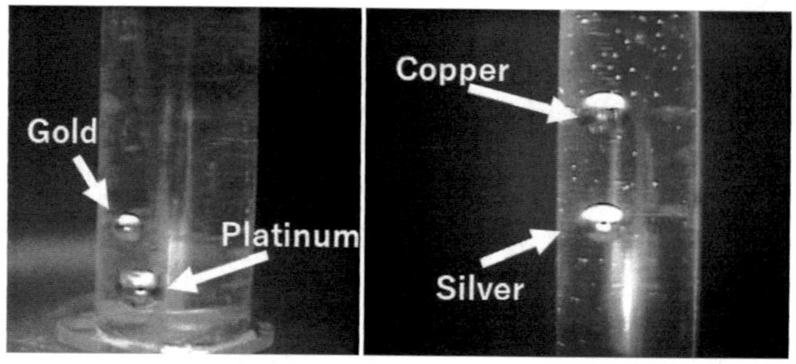

Figure 4. The magneto-Archimedes levitation for each precious metal in $MnCl_2$ solution at 10 T.

Table 2. The comparison of levitation properties between calculated values and experimental ones for each metal.

	BdB/dz [T^2/m]		Levitating height [mm]	
	Calculated value	Experimental value	Calculated value	Experimental value
Platinum	-809	-809	28	28
Gold	-380	-380	32	33
Silver	-234	-170	141	153
Copper	-215	-140	145	158

4. Conclusion

We studied the levitation properties for precious metals utilizing magneto-Archimedes effect under a high magnetic field gradient. We made it possible to generate a high magnetic field gradient by arranging ferromagnetic iron cylinders at the center of the superconducting magnet. The magnetic force factor of *BdB/dz* in vertical direction achieved the high value of −1598 T^2/m by the optimum configuration of ferromagnetic materials at 10 T. The experimental results showed that silver, copper, gold and platinum levitated at each different height which was almost the same as its own theoretical one, and could be controlled by changing the configuration of ferromagnetic materials. This indicates the strong potential for the selective separation of a target material from the solid mixture by utilizing magneto-Archimedes levitation under a high magnetic field gradient.

References

[1] Matsuura Y, Miura O 2016 *IEEE Transactions on Applied Superconductivity* **26** 4
[2] Ikezoe Y, Hirota N, Nakagawa J, Kitazawa K 1998 *Nature* **393** 749-750
[3] Hirota N, Ikezoe Y, Uetake H, Nakagawa J, Kitazawa K 2000 *Materials Transaction* **41** 976-980

IOP Conf. Series: Journal of Physics: Conf. Series **1054** (2018) 012087 doi:10.1088/1742-6596/1054/1/012087

Numerical Studies on the Dynamic Responses of Levitated High-temperature Superconductor with a Strongly Coupled Thermo-electromagnetic Model

C Q Ye[1], **T Y Gong**[1,2], **G T Ma**[1], **W J Yang**[1] and **Z Y Yang**[1]

[1] State Key Laboratory of Traction Power, Southwest Jiaotong University, Chengdu, China, 610031
[2] College of Electrical Engineering, Southwest Jiaotong University, Chengdu, China, 610031
E-mail: changing_ye@foxmail.com

Abstract. In this paper, we present a 2-D numerical model based on the H-formulation and nonlinear E-J relationship to study the dynamic responses of a bulk high-temperature superconductor(HTSC) levitated above two kinds of permanent magnetic guideway(PMG). Being different from the existing related models and results of this subject, the thermal effect was brought into account by a strong coupled model with electromagnetic and thermal. The levitation forces were calculated by finite element method and then the vertical motion of the levitated HTSC subject to external disturbance is characterized by a second-order dynamic equation which couples the electromagnetic model via the levitation force. We studied the thermal influence on the vertical force and dynamic characteristics of the HTSC levitated above two kinds of PMGs. The obtained results prove that this strongly coupled model can better simulate the dynamic response of HTSC levitation system.

1. Introduction

With the distinct merits of self-stable levitation, contactless, less friction and energy-saving, the high temperature superconductor (HTSC), such as bulk REBCO superconductors and stacked tapes of REBCO coated conductor materials, levitated above a permanent magnet (PM), is promising in the industrial applications of rail transit, magnetic bearing, and flywheel energy storage system, etc[1]-[2].

To explore the dynamic stability of HTSC levitated above PM, experiments and numerical simulations have been conducted, including the cylinder and cuboid bulks excited by PM, known as rotational superconducting magnetic bearing (SMB) and linear SMB. The nonlinear dynamic phenomena of quasi-periodic, periodic double, chaotic, bifurcation, levitation drift behaviors were discovered in rotational SMB and discussed in approximate analytical models. Comparing with the rotational SMB, published papers on the dynamics of linear SMB mostly devoted to the effects of field-cooling height, working height, off-center, derailment operation, load-weight and its respective distribution on the initial frequency, dynamic stiffness and dumping parameters of system, including that the levitator composed of multiple YBaCuO bulks[3]-[5]. The abovementioned achievements regarding the vibration, to some extent, lead to better understanding the dynamics of linear SMB. However, the thermal effect induced by vibration of magnet or flux creep and motion, should be accounted in dynamic behaviors. Hence, this paper is devoted to make a primary discussion on how the thermal influence affect on the levitation force and vertical dynamics.

This paper is organized as following, Part II introduces the linear SMB and establish the 2-D model that couples the electromagnetic and thermal fields bidirectionally, to investigate the dynamic responses of a bulk HTSC levitated. Part III presents the coupled simulation of dynamics and conclusion is provided in Part IV.

2. Theoretical modeling

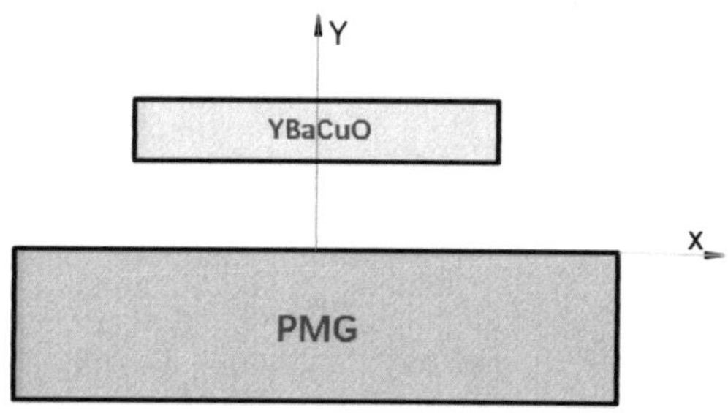

(a)

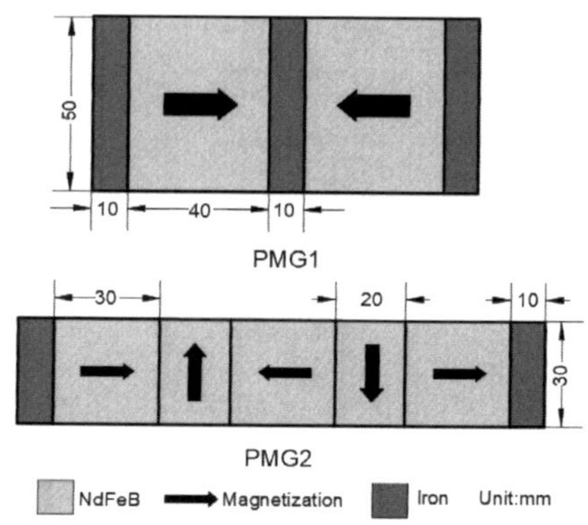

(b)

Figure 1. (a) Two-dimensional geometrical configuration of the linear superconducting magnetic bearing (SMB) composed of a bulk HTSC of YBaCuO levitated above a permanent magnetic guideway (PMG) with translational symmetry along the invisible *z*-axis, being axis-symmetrical in term of the *y*-axis of the so-called Cartesian system. (b) Two common kinds of PMG, PMG1 and PMG2(unit: mm).

The 2-D reduced structure of the linear SMB addressed in this paper, consisting of a bulk YBaCuO and a permanent magnetic guideway(PMG) that extends infinitely in the translational direction, is shown in Fig. 1, where PMG1 is a common type and PMG2 is the Halbach-derived type that concentrates most the magnetic energy to the upper side. All PMGs are made of NdFeB permanent magnets that have remanent flux density $B_r = 1$ T. The magnetic field of PMG is calculated using a 2-D magnetostatic finite element model by the mf(magnetic fields) interface in Comsol[6]. The applied bulk HTSC in this work is a melt-textured YBaCuO block. The electromagnetic governing equation for the bulk HTSC is based on the H-formulation[7]. We only consider an unidirectional coupling between the PMG model and the bulk HTSC model, through applying the superposition of external field generated by PMG and self-field on the outer boundaries of HTSC model. This hypothesis is reliable because the magnetic field generated by the supercurrent in bulk HTSC excited by the PMG does not significantly exceed the coercive field of PMs. Additionally, we calculated the magnetic field of bulk HTSC through the Biot-Savart law. Last but not the least, this model has been validated by experiments, comparing with other models by different groups[8]-[9].

2.1 Thermo-Electromagnetic modeling

The bulk HTSC is modeling by the H-formulation with the nonlinear power law of E-J relationship, the multiphysics finite element software Comsol has been successfully applied in superconductor modeling in 2D and 3D application. In this part, we will present the fundamental modeling strategy. According to Faraday's law and neglecting the displacement current, we obtain the governing partial differential equation (PDE) of HTSC in the $\boldsymbol{H}$-formulation of general form,

$$\mu\frac{\partial \boldsymbol{H}}{\partial t} + \nabla \times \boldsymbol{E} = 0. \tag{1}$$

where the electrical field $\boldsymbol{E} = \rho\boldsymbol{J} = \rho\nabla \times \boldsymbol{H}$. To satisfy the condition $\nabla \cdot \boldsymbol{H} = 0$ in finite element method discretization of the 2nd-order differential operator $\nabla \times \nabla \times$ and avoid both the ill-conditioned matrices and slowly iteration convergence in nodal finite element method, the edge elements are applied in Comsol. Considering the translational symmetry of linear SMB, the induced current and electrical field in HTSC only have the *z*-components, J_z and E_z, therefore the magnetic and electric fields can be represented by $H = [H_x, H_y, 0]$ and $E = [0,0,E_z]$ in *xy*-plane. Consequently, the governing electromagnetic equation (1) can be rewritten as

$$\mu \frac{\partial H_x}{\partial t} + \frac{\partial E_z}{\partial y} = 0, \tag{2}$$

$$\mu \frac{\partial H_y}{\partial t} - \frac{\partial E_z}{\partial x} = 0. \tag{3}$$

And the current density along the translational direction can be defined through the simplified equation (1)

$$J_z = \frac{\partial H_y}{\partial x} - \frac{\partial H_x}{\partial y}. \tag{4}$$

Equation (2) and (3) can be solved by PDE module in Comsol. The nonlinear constitute relationship in HTSC is defined by the *E*-*J* power law,

$$\rho(J_z) = \frac{E_c}{J_c} \left| \frac{J}{J_c} \right|^{n-1}, \tag{5}$$

usually, J_c is the critical current depended on the magnetic and thermal fields when E_c equals to $1\ \mu\text{V/m}$, and n is a parameter according to material used. To account the effects of magnetic and thermal fields on the critical current density $J_c(B,T)$, we apply the empirical equation[4],

$$J_c(B,T) = J_{c0} \cdot \frac{B_0}{B_0 + \sqrt{{B_x}^2 + {B_y}^2}} \cdot \frac{T_c - T}{T_c - T_0}, \tag{6}$$

where B_0 is the characteristic magnetic field based on the bulk HTSC properties, B_x and B_y represent the magnetic field components along the *x*-axis and *y*-axis directions respectively in Fig. 1, the temperature T is defined through the thermal model, and the parameters T_0 and T_c, depending on material thermal characteristic, respectively, represent the initial and critical temperatures of bulk HTSC.

Table 1. Parameters of bulk HTSC of YBaCuO used in calculations

Bulk width	24 mm
Bulk thickness	10 mm
Critical temperature, T_c	92 K [9]
Initial temperature, T_0	77 K
Critical current density at operating temperature/zero field, J_{c0}	$J_{c0} = 7.5 \times 10^7$ A/m^2 [8]
Critical electrical field, E_c	10^{-4} V/m [7]
Reference value of the magnetic induction, B_0	0.37 T [7]
n-exponent at 77 K	21 [7]
Thermal conductivity, k	6 WK^{-1}m^{-1} [10]
Heat capacity per unit volume, c	$73.75\,T^2 + 5599.78\,T + 87669.311$ J/(m^3·K)
Heat transfer coefficient, h	400 Wm^{-2}·K^{-1} [4]

We set the thermal model based on the classical governing equation for T,

$$k\left(\frac{\partial^2 T}{\partial x^2} + \frac{\partial^2 T}{\partial y^2}\right) - c\frac{\partial T}{\partial t} = -Q, \tag{7}$$

where the thermal conductivity k and heat capacity per unit volume c herein are independent of the spatial coordinates, and the source term Q, known as the Joule heat arise from the conversion of electromagnetic energy, can be calculated through

$$Q = J_z E_z. \tag{8}$$

We apply the convective boundary condition

$$k\frac{\partial T}{\partial n} + h(T - T_0) = 0, \tag{9}$$

on the bulk HTSC, which means the thermal flux exchanges between the bulk HTSC and liquid nitrogen, where h is the convective heat transfer coefficient. We summarized the values of parameters in Table 1.

2.2 Dynamics model

Once we obtain the induced supercurrent in each domain at any time instant, the levitation force of bulk superconductor is solved through Lorentz's force equation

$$F_y = \iint_S J_z B_x dxdy = \sum_{i=1}^{N} j_z^e B_x^e \, \Delta S_i^e \tag{10}$$

where the superscript e indicates an element in the mesh of bulk HTSC, S is the cross-sectional area of bulk, N are the number elements of a bulk. B_x is the combination of both the magnetic field of PMG and all rest elements.

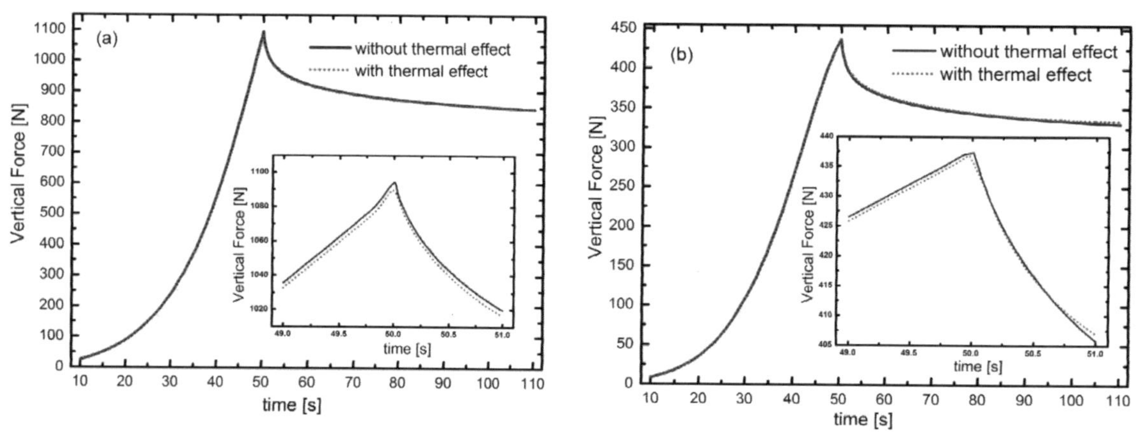

Figure 2. The levitation force acted on the bulk HTSC with the descending velocities at 1 mm/s, from 60 mm to 10 mm above the center of (a) PMG1 and (b) PMG2, when the thermal effect is taking into account and not.

We refer to the linear SMB schematically drawn in Fig. 1 to establish the electromagnetic model and dynamic equation, in which the actual 3-D problem was simplified to be a 2-D one, recalling the practical case that the PM rails usually extends infinitely in the translational direction and are fixed on the ground or an experiment platform. Generally, according to Newton's second motion law, the motion of HTSC in the vertical direction can be expressed by,

$$m\ddot{y} + F_y - mg = f_y \tag{10}$$

where m is the mass of levitator (not always equals to the mass of HTSC) and equals to 20 kg in this work, y represents the vertical displacement of levitator, therefore, $\ddot{y}$ represents the acceleration of levitator in the vertical direction, g is the gravitational acceleration (normally 9.8 N/kg), F_y is the vertical magnetic force acting in the bulk HTSC and can be solved by Lorentz's law, f_y is the

IOP Conf. Series: Journal of Physics: Conf. Series **1054** (2018) 012087 doi:10.1088/1742-6596/1054/1/012087

exciting force acting on the levitator. It is worth noting that the air resistance is such tiny that can be neglected, regarding the tiny volume of triple-seeded bulk HTSC in this work.

3. Results and discussions

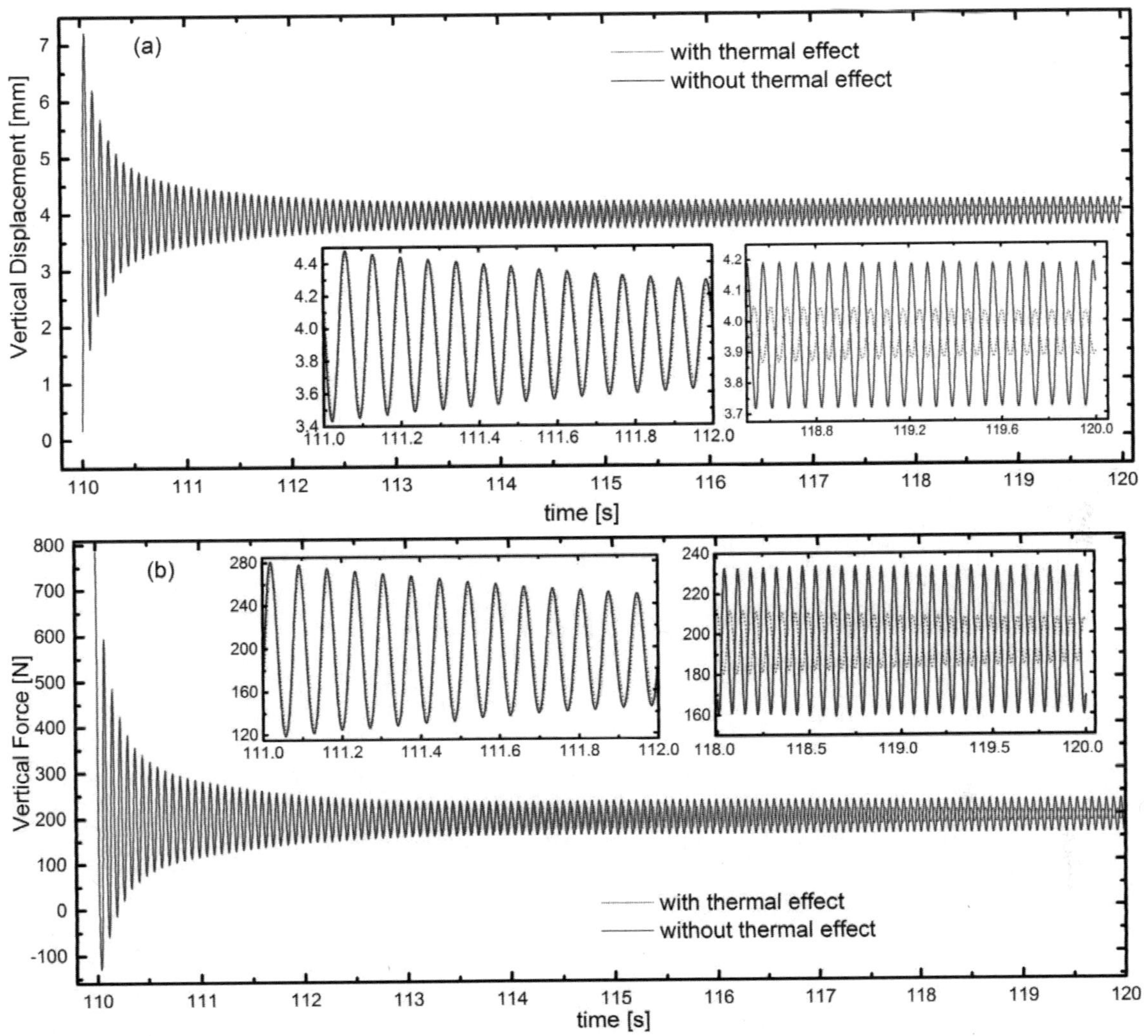

Figure 3. Dynamic responses of the bulk HTSC levitated above the center of PMG1 vary with time, (a)vertical displacement around the minimal height 10 mm, (b) vertical electromagnetic force acted on the bulk HTSC, when the thermal effect is taking into account and not.

The bulk HTSC is a melt-textured YBaCuO block that has 24 mm in width and 10 mm in thickness. To appraise the thermal effect on the dynamic responses of bulk HTSC levitated above the two kinds of PMGs, numerical simulation were carried out by starting the vertical movement at a position of 60 mm above the center of PMG where the external applied field is almost negligible at this height, and then descending to the position of 10 mm above the PMG, after the relaxation of 60 seconds, and finally vibrating for 10 seconds excited by the difference between the gravitational force and levitation force. It is noted that the descending speed of bulk HTSC is 1 mm/s to create a quasi-static state, therefore, the displacement current can be neglected. Furthermore, we set the relaxation time for 60 seconds to reduce the levitation height and oscillation center drifts during vibration that obviously displayed in Fig. 6 of Ref.[4].

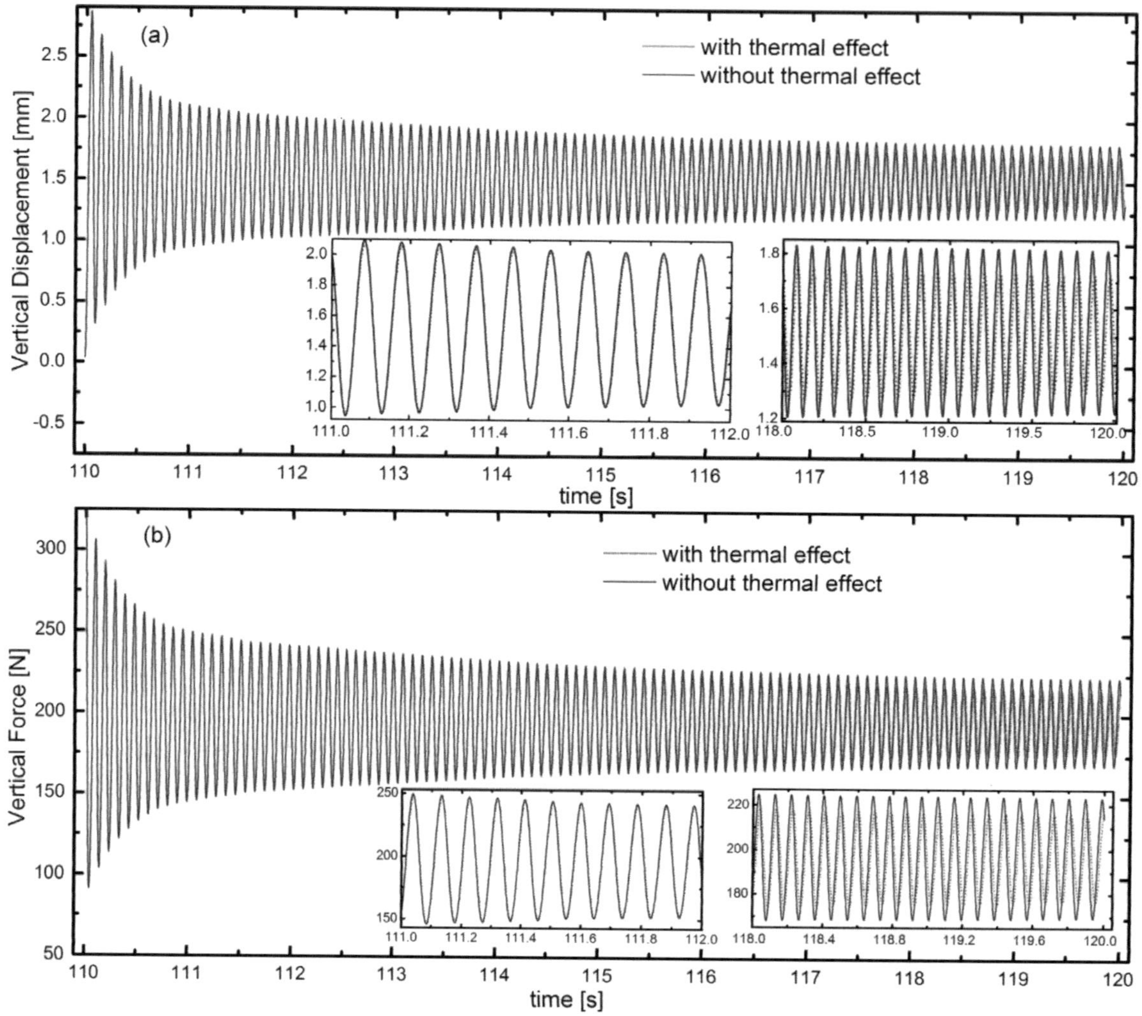

Figure 4. Dynamic responses of the bulk HTSC levitated above the center of PMG2 vary with time, (a)vertical displacement around the minimal height 10 mm, (b) vertical electromagnetic force acted on the bulk HTSC, when the thermal effect is taking into account and not.

The levitation forces acted on the bulk HTSC as a function of time from the initial height 60 mm to the minimal height 10 mm, are portrayed in Fig. 2, when the thermal effect is taking into account and not. Specially, the levitation force generated by PMG1 is almost the twice of that generated by PMG2, because the magnetic field above the center of PMG1 is much stronger than that of PMG2. Generally speaking, it is safe to conclude that, for both cases of PMG1 and PMG2, the levitation forces are almost the same, whether the thermal effect is taking into account or not, although the levitation force without accounting the thermal effect at the minimal working height is little higher than the case considering the thermal effect.

Figures 3 and 4 display the responses of displacement and levitation force of bulk HTSC levitated above the center of PMG1 and PMG2 experienced oscillations between the corresponding balance position, after relaxation for 60 seconds at the position of 10 mm, where the initial velocity is 0 mm/s. To portray the dynamic behaviors clearly, the insets in the each figure show the oscillations of displacement and vertical force from 111 to 112 s and 118 to 120 s.

Roughly speaking, both for PMG1 and PMG2, we can observed the invisible downward drift of the oscillation center both in the dynamics responses of displacement and levitation centers, whether the

IOP Conf. Series: Journal of Physics: Conf. Series **1054** (2018) 012087 doi:10.1088/1742-6596/1054/1/012087

thermal effect is taking into account or not. Comparing with the obvious downward drift of the oscillation center during the vibrations[3]-[4], the relaxation of bulk HTSC is necessary and usually applied in our experiments[5]. In the initial vibration phase, from 110 to 114 s, the oscillation behaviors, including the oscillation amplitudes, periods and phase angels, manifest almost synchronous for the cases considering thermal effect and not. As the oscillation went on from 114 to 120 s, however, the oscillation behaviors of the cases of PMG1 and PMG2, including the oscillation amplitudes and frequencies, display increasingly nonsynchronous, comparing with the models with and without thermal effects. Obviously, the oscillation amplitudes of vertical displacement and force decrease quicker, in the case of accounting the thermal effect, than that case without accounting the thermal effect, both in PMG1 and PMG2.

Specially, comparing with PMG1 in Fig. 3 and PMG2 in Fig. 4, it can be observed that the phases of PMG2 with and without thermal effects are almost corresponding from 110 to 120 s, however, the phases of PMG2 with and without thermal effects are not corresponding in the last half of oscillations. As a matter of fact, the dynamic equation of bulk HTSC levitated above the PMGs are nonlinear, as shown in Eq. (10). Furthermore, the vertical magnetic force between the bulk HTSC and PMG solved by Lorentz equation, depends not only on the displacement, velocity, time, motion condition, and motion history of bulk HTSC, but also on the magnetic field generated by PMG and physical parameters of HTSC and PMG(Fig. 5 of Ref. [3]). As a result, the superconducting levitation system demonstrates *hysteresis* characteristic and its stiffness is not a constant during the free-vibration process(see Fig. 5 of Ref. [3]). Consequently, the damping frequency and period time are not constants during the vibration(It also demonstrated in Figs. 6 and 7 of Ref. [4]). Concisely, the PMGs with different arrangement of PMs shape the magnetic field distribution, and affect the phases of vibration in Fig. 3 and Fig. 4 eventually. It can be safely concluded that the thermal effect influences the stability of SMB system and the thermal effect in SMB can be employed as a damper of practical application.

4. Conclusions

In this paper, a 2-D coupled thermo-electromagnetic modeling method for the bulk HTSC levitated above a normal and a Halbach-derived PMGs implemented in Comsol was introduced and employed with the dynamic model to study the influence of thermal effect on dynamic responses of levitated bulk HTSC. It is concluded from the numerical results that the bulk HTSC levitated above both kinds of PMG converges more quickly, when we apply the strongly coupled thermo-electromagnetic model which can better simulate the dynamic response of SMB, than that model without taking account the thermal effect. On that ground, an eddy current damper can improve the dynamic system stability. Further work should be done for better evaluate the thermal effect on the practical case of SMB.

Acknowledgements

The authors would like deeply to thank the anonymous reviewers and editors, whose critical comments considerably improved the quality of this paper. This work was supported in part by the National Natural Science Foundation of China under Grants 51475389, 51722706 and 51707164, in part by China Postdoctoral Science Foundation under Grant 2017M623055, and in part by the Sichuan Youth Science&Technology Foundation under Grant 2016JQ0003.

References

[1] Wang J, Wang S, Zeng Y, *et al.* 2002 *Physica C* **809-814** 378–381
[2] Werfel F N, Floegel-Delor U, Rothfeld R, Riedel T, Goebel B, Wippich D and Schirrmeister P 2012 *Supercond. Sci. Technol.* **25** 014007
[3] Gou X F, Zheng X J and Zhou Y H 2007 *IEEE Trans. Appl. Supercond.* **17(3)** 3795–3802
[4] Alloui L, Ben-Alia K, Bouillault F, Mimoune S M, Bernard L and Leveque J 2013 *Physica C* **487(4)** 1-10

[5] Jiang D, Ma G, Xu Y, Deng Z, Wang S and Wang J 2013 *IEEE Trans. Appl. Supercond.* **23(3)** 3600404
[6] https://www.comsol.com/
[7] Sass F, Sotelo G G, Junior R D A and Sirois F 2015 *Supercond. Sci. Technol.* **28(12)** 125012
[8] Queval L *et al.* 2016 *Trans. Appl. Supercond.* **26(3)** 3601905
[9] Ye C Q, Ma G T and Wang J S 2016 *IEEE Trans. Appl. Supercond.* **26(8)** 3603309
[10] Poole C P, Farach H A, Creswick R J and Prozorov R 2007 *Superconductivity*(Amsterdam: Elsevier)
[11] Roy F, Dutoit B, Grilli F and Sirois F 2008 *IEEE Trans. Appl. Supercond.* **18(1)** 29–35

IOP Conf. Series: Journal of Physics: Conf. Series **1054** (2018) 012088 doi:10.1088/1742-6596/1054/1/012088

Levitation stability of superconducting stators with addition of a ring-shaped magnet

Muneo Futamura[1] **and Ryo Shindo**[1]

[1]Faculty of Systems Science and Technology, Akita Prefectural University, Yurihonjo, Akita, Japan

E-mail: futamura@akita-pu.ac.jp

Abstract. High magnetic rigidity is required to optimize superconducting levitation. In this study, we propose a hybrid stator comprising a superconductor and a ring-shaped permanent magnet for improving levitation stability. We measured the oscillation characteristics of a magnet levitating above several hybrid stators with ring-shaped magnets of different thicknesses. The magnetic force between the hybrid stator and the levitating magnet was determined via semi-static displacement. By adding a ring-shaped magnet to the stator, we found that vertical attractive magnetic force and stiffness of the hybrid stator increased compared with that of a simple superconducting stator. The magnetic force showed hysteresis against the displacement of the levitating magnet. We observed that hysteresis energy loss increased with a thicker ring-shaped magnet. The oscillation frequency of the levitating magnet above the hybrid stator was higher than that of the simple superconducting stator. Faster decay due to hysteresis energy loss and small amplitude resulting from enhanced magnetic stiffness were also observed. An improvement in the stability of the levitating magnet was thus realized by the hybrid stator.

1. Introduction

Recently, energy saving has become a critical feature of mechanical systems because of an increase in energy consumption by the growing world population. Superconducting magnetic levitation technologies are deemed to be a promising approach for energy conservation [1-3]. Magnetic levitation above a bulk superconductor is caused by the penetration and pinning of the flux lines in the superconductor. The absence of physical contact between the rotor and stator in superconducting bearings effectively eliminates a major source of friction. Moreover, no active control is required for positioning the levitated magnets. The performance of applications using a superconducting levitating system depends on the stability of the levitation force. An application (e.g., superconducting bearings for energy storage flywheels) having high levitation stability can operate with high-speed rotation and store more energy. The possibility of experiencing failure will also decrease. Therefore, to achieve performance elevation in superconducting mechanical systems, improving levitation stability by increasing the magnetic stiffness of superconducting levitation is important. We propose a hybrid stator comprising a superconductor and a ring-shaped permanent magnet for high-performance bulk superconducting applications [4]. Upon addition of a ring-shaped magnet under the superconducting stator, the attractive force between the floating magnet and stator becomes greater at large distances as compared with short distances. Hence, vertical magnetic stiffness is enhanced and levitation stability is improved. In this study, we measured the levitation characteristics of a levitating magnet fixed above the proposed hybrid stator.

2. Experiment and results

2.1. Measurement preparation

A disk magnet (NdFeB magnet; diameter: 10 mm; thickness: 5 mm; mass: 2.95 g) was fixed 5 mm above a stationary part that comprised a bulk superconductor (QMG-DyBaCuO; diameter: 32 mm; thickness: 5 mm) and an additional ring-shaped magnet. The superconductor was cooled in the magnetic fields of the disk and ring magnets. For the experiment, a superconducting stator without a magnet was compared against a hybrid superconducting stator with one of the three types of ring-shaped magnets (Type A, B, and C). Table 1 shows the dimensions of the ring-shaped magnets. Each ring magnet had the same diameter but a different thickness. We measured the levitation characteristics (i.e., damping oscillation and magnetic force) of the disk magnet levitating above the superconducting stator for each condition of the stator (without magnet or with magnet type A, B, or C).

Table 1. Dimensions of the ring-shaped magnets added to the superconducting stationary part.

	Thickness	Internal diameter	External diameter
Type A	9		
Type B	13	25	30
Type C	18		

2.2. Damped oscillation

Figure 1 shows a schematic of the oscillation measurement arrangement. A disk magnet was levitated above a superconducting stator and pulled up by the magnetic field produced by a drive coil suspended above the magnet. Next, the drive field was cut, and the oscillation decay of the levitating magnet was measured using a laser displacement meter over the coil. The coil current was adjusted so that the lifting height of all oscillations was 5.9 mm.

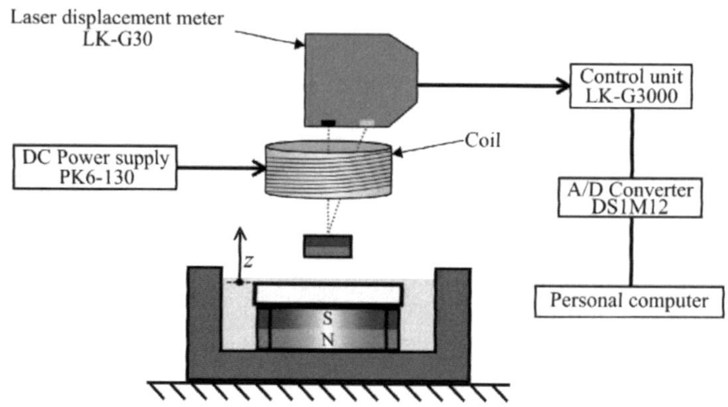

Figure 1. Schematic of the measuring apparatus for oscillation decay of the levitating magnet.

Figure 2 shows the displacement oscillation waveforms of the magnet above each stator with magnet type A, B, or C. Compared with the superconducting stator without a ring-shaped magnet, the levitating magnet above the hybrid stator oscillated and converged to a lower position because of the attractive force of the additional ring-shaped magnet. The oscillation of the hybrid stator showed a higher frequency as a function of amplitude, as shown in figure 3. The logarithmic decrement as a function of amplitude was nearly identical for all stators, as shown in figure 4. The logarithmic decrement refers to magnetic loss per lap. Given the frequency dependence of the stator (figure 3), magnetic loss per second is larger (i.e., oscillation attenuates more quickly) for a hybrid stator with a thicker magnet. The magnetic force acting on the magnet was measured to examine the loss per reciprocation in detail.

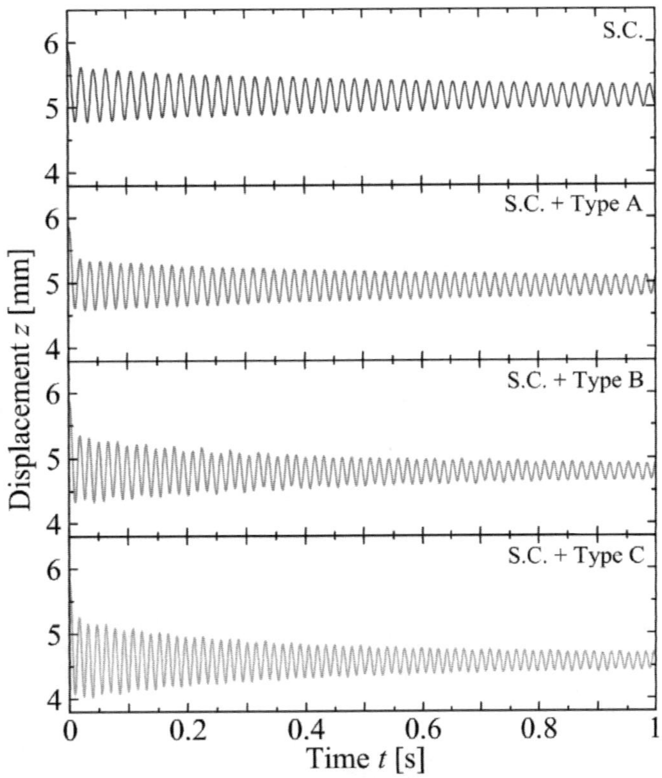

Figure 2. Measured damping waveforms of various stators.

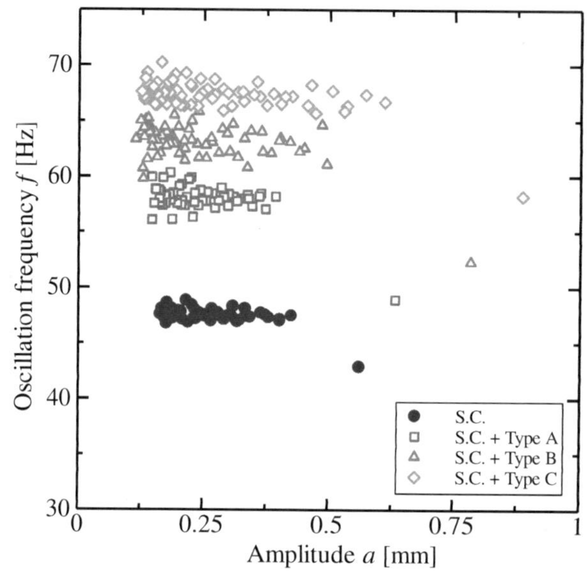

Figure 3. Oscillation frequency as a function of amplitude.

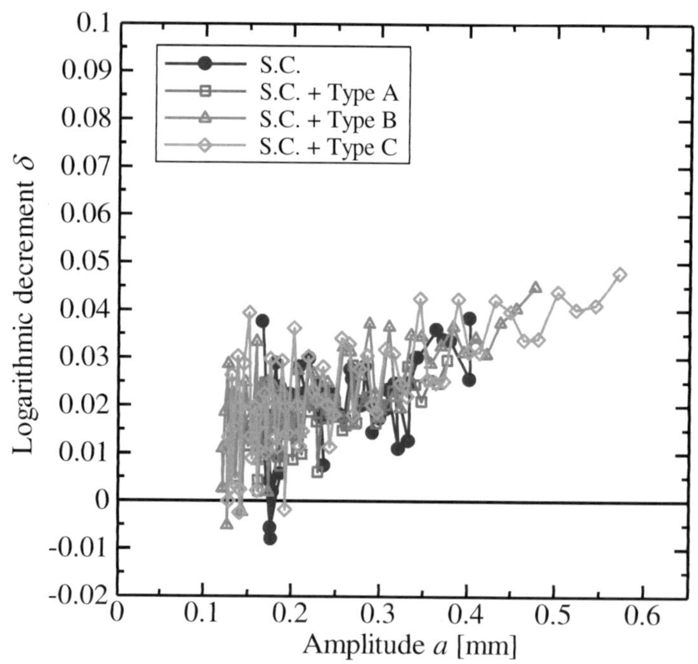

Figure 4. Amplitude dependence of logarithmic decrement.

2.3. Vertical magnetic force

Figure 5 shows a schematic of the measuring apparatus. The vertical magnetic force between the levitating magnet and the superconducting stator was measured using digital force gauges. The magnet moved quasi-statically within the same range as the respective oscillations, as shown in figure 2.

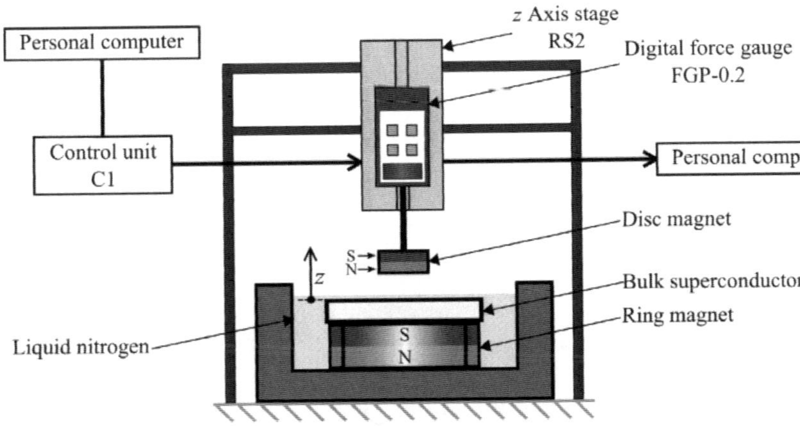

Figure 5. Measuring apparatus for measuring the vertical magnetic force of the magnet and superconducting stator.

The magnetic force showed hysteresis against the displacement of the magnet, as shown in figure 6. The area of the hysteresis loop was equivalent to the magnetic loss due to the depinning of the flux lines from the pins in the superconductor. Oscillation damping was due to the energy loss that was induced by the motion of fluxes in the superconductor. Generally, loss due to magnetic flux drag is large for large magnetic fields [5]. As shown in table 2, magnetic loss was comparatively greater in the thicker magnet than in the thinner magnet.

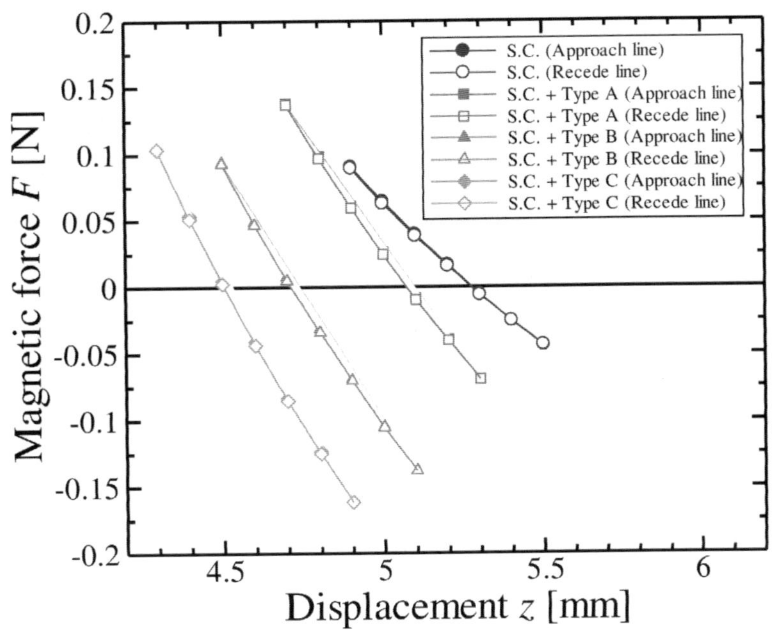

Figure 6. Vertical magnetic force of the magnet and superconducting stator. Hysteresis loop for magnetic force and displacement of the magnet for a given period.

IOP Conf. Series: Journal of Physics: Conf. Series **1054** (2018) 012088 doi:10.1088/1742-6596/1054/1/012088

Table 2. Energy loss per lap

Stator type	Energy loss [N mm]
S.C.	4.58×10^{-4}
S.C. + Type A	7.75×10^{-4}
S.C. + Type B	10.42×10^{-4}
S.C. + Type C	10.83×10^{-4}

Figure 7 shows the magnetic stiffness determined from the average magnetic force on the approach and recede lines. Upon addition of the ring-shaped magnet to the stator, magnetic stiffness increased, which subsequently increased the oscillation frequency, as shown in figure 3.

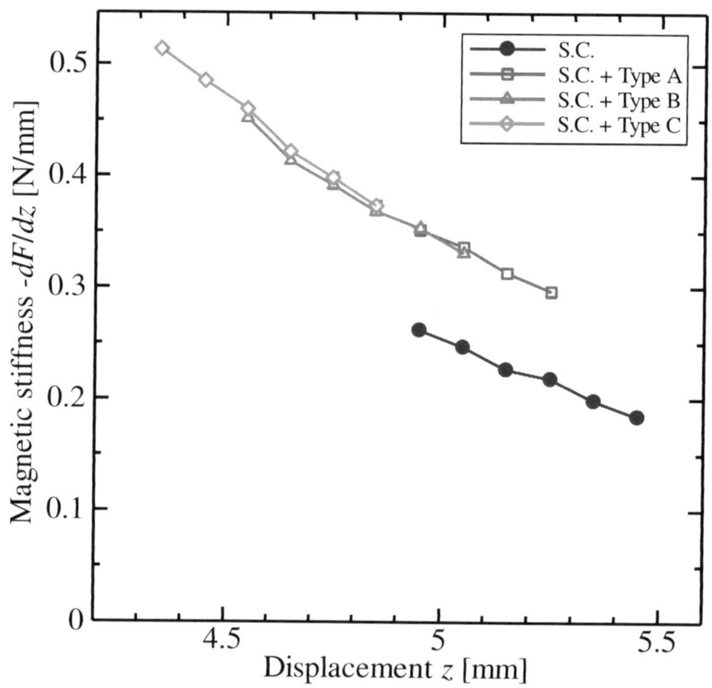

Figure 7. Vertical magnetic stiffness of the levitating magnet for each stator type. Magnetic stiffness was improved for all hybrid stators.

3. Discussion

As shown in figure 3, a higher oscillation frequency was obtained for the thicker ring-shaped magnet because the spring constant (i.e., magnetic stiffness) increased when acting on the levitated magnet owing to the low levitating position. The oscillation frequency is determined by

$$f = \frac{\pi}{2}\sqrt{\frac{k}{m}} \tag{1}$$

Table 3 shows the magnetic stiffness at the convergent position of each oscillation in figure 2 and the frequency calculated using (1). The calculated frequency (table 3) and measured oscillation frequency (figure 3) are nearly identical. Therefore, we assume that the spring constant value is reasonable.

The magnetic force acting on the levitating magnet is defined as follows:

$$F = MV\frac{dB}{dz}, \tag{2}$$

where M is residual magnetization, V is volume of the levitating magnet, and dB/dz is the magnetic field gradient around the levitating magnet. Hence, the magnetic stiffness (dF/dz) is proportional to the second derivative of the magnetic field. Using Coulomb's law, the magnetic field on the center axis of the ring-shaped magnet is determined as follows:

$$B = \frac{B_r}{2}\left\{\left[1+\left(\frac{D/2}{z+T}\right)^2\right]^{-0.5} - \left[1+\left(\frac{D/2}{z}\right)^2\right]^{-0.5} - \left[1+\left(\frac{d/2}{z+T}\right)^2\right]^{-0.5} + \left[1+\left(\frac{d/2}{z}\right)^2\right]^{-0.5}\right\}, \tag{3}$$

where, B_r is the remanent magnetization, T is the thickness, and D and d are the outer and inner diameters of the ring-shaped magnet, respectively. Figure 8 (a) shows the calculated magnetic field gradient of each ring-shaped magnet. A magnetic force acts in order to move the magnet toward the point (Type A: z = 5.8, B: z = 4.9, and C: z = 4.0) where the magnetic field gradient crosses the horizontal axis. The magnetic force of the ring-shaped magnet and the superconducting pinning magnetic force, which keep the magnet at the initial pinning point (z = 5), are combined to improve levitation stability.

Figure 8 (b) shows the second-order gradient of the magnetic field , indicating small differences in the values of all ring-shaped magnets. This second-order gradient of the ring-shaped magnet contributes to the magnetic stiffness of the superconducting flux pinning. Consequently, the magnetic stiffness of the hybrid stators is greatly improved compared to a simple superconducting stator without a ring-shaped magnet, as shown in figure 7.

Table 3. Magnetic stiffness at the convergent position and calculated oscillation frequency.

Stator type	Convergent position [mm]	Magnetic stiffness [N/mm]	Oscillation frequency [Hz]
S.C.	5.2	0.22	43.8
S.C. + Type A	5.0	0.34	54.3
S.C. + Type B	4.8	0.38	57.1
S.C. + Type C	4.6	0.44	61.5

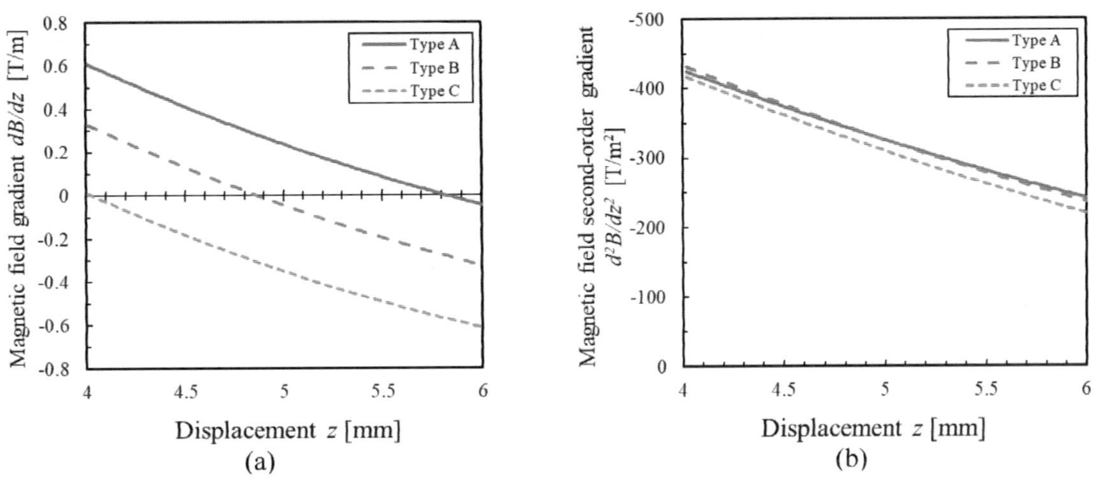

Figure 8. (a) Magnetic field gradient and **(b)** Magnetic field second-order gradient calculated from (3).

4. Conclusion

We measured the levitation characteristics of a levitating magnet fixed above a proposed hybrid superconducting stator that included a ring-shaped magnet. Damping oscillation and magnetic force of the levitating magnet over the hybrid stator were measured. The results showed increased magnetic stiffness, which subsequently produced high-frequency oscillation. Magnetic loss per lap in the thicker magnet was slightly greater than that in the thinner magnet. Magnetic loss per second was larger for the hybrid stator with a thicker magnet; i.e., oscillation attenuated faster with the hybrid stator. Oscillation amplitude is expected to be reduced and oscillation decay is expected to ramp up with an additional magnet.

References

[1] Werfel N F, Floegel-Delor U, Rothfeld R, Riedel T, Goebel B, Wippich D and Schirrmeister P 2012 *Supercond. Sci. Technol.* **25** 014007
[2] Yamachi N, Nishikawa T, Sakai N, Sawa K and Murakami M 2003 *Physica C* **392-396** pp 579-584
[3] Poltavets V, Kovalev K, Ilyasov R, Glazunov A, Maevsky V, Verzbitsky L, Akhmadyshev V, and Shikov A 2014 *Journal of Physics: Conference Series* **507** 032022
[4] Shindo R and Futamura M 2016 *Transactions of the JSME (in Japanese)* **82** 16-00002
[5] Irie F and Yamafuji K 1967 *J. Phys. Soc. Jpn.* **23** pp 255-268

IOP Conf. Series: Journal of Physics: Conf. Series **1054** (2018) 012089 doi:10.1088/1742-6596/1054/1/012089

Effect of vertical vibration on the stability of Side-Suspending HTS Maglev Rotating System with trimodal structure PMG

Li-Feng Zhao[1, 2], Meng-Liang Yao[1, 2], Lin-Bo Li[1, 2], Xifeng Pan[3], Da-Jin Zhou[1, 2], Jing Jiang[1, 2], Yong Zhang[1, 2] and Yong Zhao[1, 2]

[1]Key Laboratory of Magnetic Suspension Technology and Maglev Vehicle, Ministry of Education，Chengdu, 610031, China.

[2]Superconductivity and New Energy R&D Center, Southwest Jiaotong University, Chengdu 610031, China,

[3]National Engineering Laboratory for Superconducting Materials (NELSM), Western Superconducting Technologies Company, Ltd. (WST), Xi'an 710018, China.

zhaolf@home.swjtu.edu.cn

Abstract. By changing the vibration amplitude, side-suspending gap and load of side-suspending maglev system, respectively, the influence of vertical vibration on the bearing stability of the side-suspending high temperature superconducting (HTS) maglev system with trimodal structure permanent magnetic guideway (PMG) is investigated. The results show that the attenuation of the side-suspending height increases with the increase of the vibration amplitude, side-suspending gap and load， respectively. However, the overall decrease of the side-suspending height is not larger than 1.5 mm under all situations, and it will not affect the stability of the side-suspending maglev system apparently.

1. Introduction

The Meisner effect and flux pinning characteristics of HTS bulk make it possible to form a self-stabilizing system suspending over a permanent magnet without additional energy consumption [1, 2]. This characteristic has potential application prospects on flywheel energy storage [3], vacuum pipeline maglev train and so on [4, 5]. The vacuum pipeline maglev system is safe and energy-efficient because it avoids air-resistance and mechanical friction resistance. It is thus regarded as the ultimate solution for the future transportation. The future vacuum pipeline for maglev system is most likely laid on viaduct. The periodic flexure deformation of viaduct will form a low-frequency vibration source for maglev train running at high-speed. Earlier reports point out that low-frequency vibration will cause the levitation force of HTS maglev system to decrease [6, 7]. In order to study the dynamics of the maglev system running in an evacuated pipeline, a side-suspending maglev system running along a circular PMG with the diameter of 6.5 m in a vacuum pipeline has been built recently in the Superconductivity and New Energy R&D Center of Southwest Jiaotong University, China. It can reach the speed of 150km/h, and it is the fastest HTS maglev system running in the vacuum pipeline [8-10]. However, further studies show that the two unimodal structure PMGs (UMGs) of the system can be combined into a single trimodal structure PMG (TMG), which can obviously improve the running speed of side-suspending maglev system [11].

In view of the side-suspending maglev system running along the circular TMG in the evacuated pipeline, and in order to further study the effect of periodic low-frequency vibration on the safety of maglev train, this paper discusses the influence of low-frequency vertical vibration on the bearing stability of the side-suspending maglev system.

2. Experiments

The cross section diagram of the side-suspending HTS maglev rotating system with two UMGs is shown in figure 1. a. The every block with an arrow inside is the permanent magnet with the section size of 40 mm×40 mm. The arrow stands for the direction of the magnetic field for each block. The black portion represents iron with the section size of 6 mm×40 mm, the iron between two permanent magnets provides a peak of magnetic field nearby. The improved PMG structure is just to put the two UMGs together to form a single TMG, as shown in figure 1. b. As there is an additional magnetic peak formed, the TMG can provide greater centripetal levitation force when the maglev vehicle runs along the circular TMG. This improvement can obviously increase the running speed of the side-suspending maglev rotating system without changing the system structure greatly.

Due to the installation error of the circular TMG in the vacuum pipeline, the side-suspended maglev system will be subject to low-frequency vibrations in both vertical and radial directions when it's running along the TMG. The effect of the radial vibration on the centripetal levitation force for the system is equivalent to the influence of the alternating field on the levitation force for the traditional maglev mode of a HTS bulk suspending over a permanent magnet, and it has been much discussed [12]. However, for the side-suspending system in this paper, the effect of vertical vibration on the carrying capacity of the side-suspending system perhaps decreases the stability of the system. Therefore, we design the vibration test device, as shown in Fig. 2, to evaluate the effect of low-frequency vibration in the vertical direction on the carrying stability of the side-suspending maglev system. In Fig. 2, the section size of TMG is the same as that in Fig. 2. b with the length of 350 mm. The cryostat's outer size is 35 mm×150 mm×90 mm.

The test device is consisted of a vibrator, a TMG fixed with the vibrator, 6 HTS bulks fixed in a cryostat and two laser displacement sensors. Magnetic components of B_x and B_z at the height of 10mm over the TMG surface are presented in figure 3.

The experiment is designed to evaluate the dynamic stability of the side-suspending system under vibration with different amplitude, side-suspension gap and load of the vehicle. The experimental conditions are set as follows, respectively. a) The effect of vibration amplitude: the side-suspending gap was set as 10 mm, and the vibration amplitude of the vibration source was set as 1, 2, 3, 4 mm, respectively. b) The effect of different side-suspending gap: the vibration amplitude was set as 3mm, and the side-suspending gap was set as 5, 8, 10, 13, 15 mm, respectively. c) The effect of different maglev loads: the vibration amplitude was set as 3 mm, and the side-suspending gap was10 mm, the extra load for the maglev vehicle was 0, 125, 250, 375 and 500g, respectively. After the experimental conditions had been set up, the liquid nitrogen was fulfilled into the cryostat to make the HTS bulks achieve superconducting state and keep 10 minutes, then laser displacement sensors were used to record the primary side-suspending height of cryostat (H_0) and displacement variations of the TMG and cryostat in the vertical direction caused by vibration, respectively.

When the maglev vehicle was accelerated from standstill to cruising speed along the circular TMG, HTS bulks experienced the alternating magnetic field caused by the installation error of TMG and the frequency of it increased gradually. Therefore, in this experiment, the vibration frequency was set to increase from 0 to 20 Hz in 5 minutes and further keep the frequency of 20 Hz for 5 minutes, and then the final side-suspending height of cryostat (Hv) could be obtained when the vibration device was closed. Thus H_0-H_v represents the attenuation of side-suspending height.

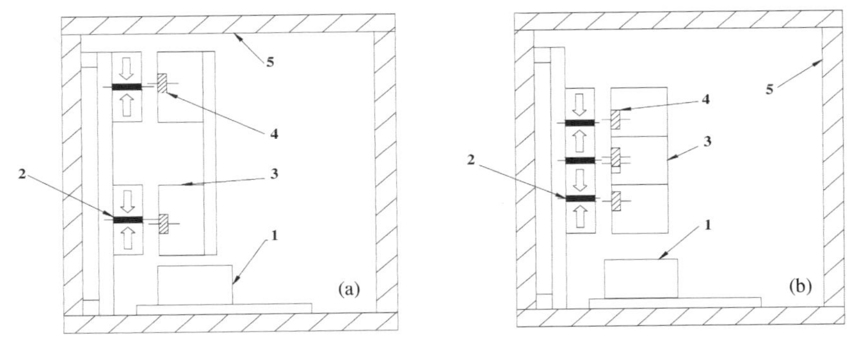

Figure 1. Cross-sectional structure of the side-suspended maglev system in the evacuated tube. (a) Established maglev system possessing two parallel UMGs; (b) The maglev system with one TMG; (1)Linear motor. (2) Iron. (3) Liquid nitrogen container. (4) HTS bulk. (5) Vacuum tube.

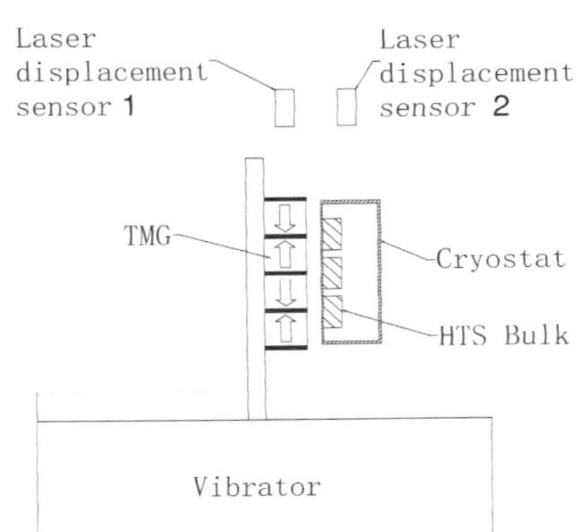

Figure 2. Schematic diagram of the vibration test device.
The cross-section size of TMG is 40mm× 190mm, and the length of it is 350 mm.
The outer size of cryostat is 35 mm× 150 mm× 90 mm.

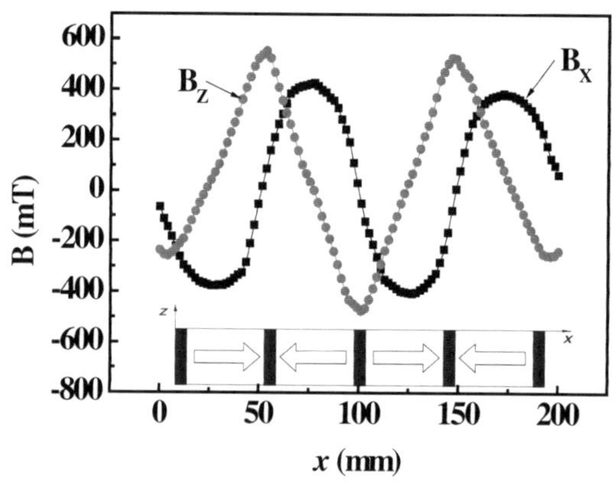

Figure 3. Simulation of B_x and B_z at the height of 10mm over the TMG surface.

3. Results and discussion

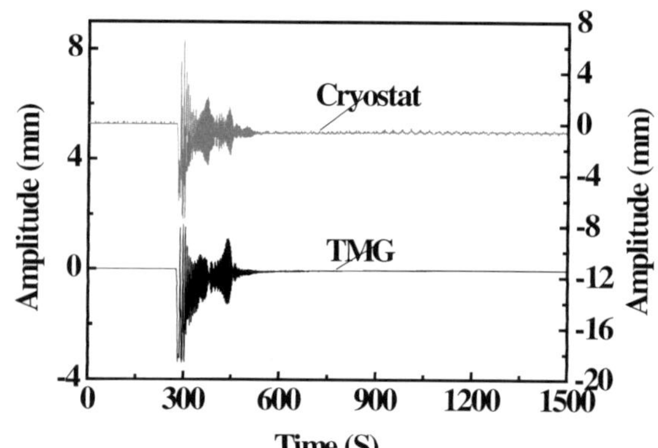

Figure 4. The time-domain vibration response of the cryostat with the TMG at the gap of 10 mm and the vibrator source amplitude of 2 mm.

A typical time-domain vibration response of the cryostat with the TMG at the gap of 10 mm and the vibrator source amplitude of 2 mm is shown in figure 4. In figure 4, when the vibrator starts to work, frequency of it is increased from 0 to 20 Hz in 300 s. In order to observe the attenuation of amplitude of cryostat with time under the frequency of 20 Hz, it has been kept for 15 minutes. With the increase of frequency, the amplitudes of cryostat and TMG decrease rapidly. This should be accord with the situation of the side-suspending maglev running along the TMG, for the reason that higher speed means higher frequency, and Mass inertia of maglev vehicle leads to smaller amplitude at the same time. Furthermore we can also figure out in figure 4 that the side-suspending height decrease during the process, and it does not decrease with time any more when it is kept at the frequency of 20 Hz. Abruptly increased amplitudes for both cryostat and TMG observed with the increase of frequency are perhaps associated with their eigenfrequencies, respectively. Furthermore, we notice that the slight inclination phenomenon for cryostat can be observed with the decrease of side-suspending height, i.e., the top of the cryostat is slightly away from the TMG, and the bottom of it is close to the TMG to a small extent. Thus we keep the laser displacement sensor 2 to aim at the middle of top of the cryostat to diminish the effect of it on the measurement.

a) The effect of vibration amplitude on the stability of the side-suspending maglev system.
The attenuation of the side-suspending height of the maglev vehicle with the vibration amplitude is shown in figure 5. The results in figure 5 indicate that the attenuation of side-suspension height increases with the augment of vibration amplitude. As we know that superconductor bulk captures some flux when it is field cooled to achieve superconducting state. The periodical vertical vibration for the side-suspending maglev system shown in figure 2 will cause the relative position between HTS bulks and the TMG to be changed periodically. According to the magnetic components of B_x and B_z at the height of 10 mm over TMG surface, as shown in figure. 3, it will cause the bulks to experience an alternating magnetic field. Some earlier results state that the guidance force is elastic with the lateral displacement in a certain extent [13]. However, in this experiment, with the increase of amplitude, the attenuation of side-suspending height of the maglev system increases continuously. This indicates that the vibration process leads to the AC loss due to HTS bulks' experiencing the alternating magnetic field. Earlier results noted that the alternating magnetic field leads to an irreversible attenuation of the guidance force [14]. Ogawa et al. claimed that alternating magnetic field can lead to the temperature of HTS bulk to increase, and thus reduce the J_c and trapped magnetic flux [15-17]. Furthermore, the guidance force declines with the decrease of the captured magnetic flux [13]. Therefore, in our experiment, it is observed that, with the increase in amplitude, the AC loss increases and further causes the attenuation of side-suspending height to increase.

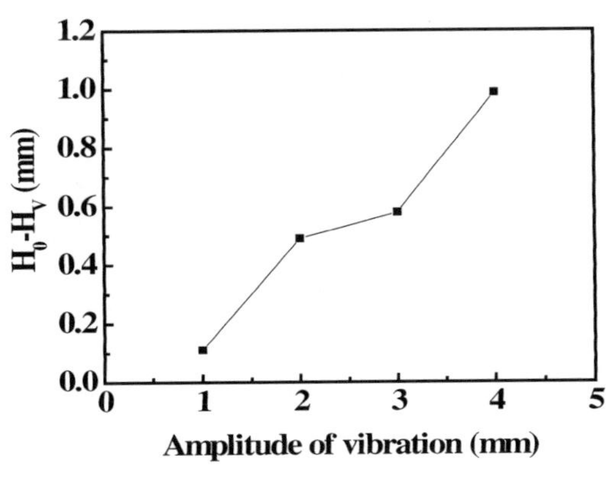

Figure 5. The attenuation of side-suspending height with the vibration amplitude of 1, 2, 3 and 4 mm at the gap of 10 mm.

b) The effect of different side-suspending gap on the stability of the side-suspending maglev system.
The experimental results are shown in figure 6. It is shown that the vibration will cause the attenuation of the side-suspending height to increase with the increase of the side-suspending gap. For the traditional maglev mode, as the guidance force increases with the increase of the captured flux [13], the stiffness of the guidance force increases with the decrease of the suspending gap

[18-20]. Therefore, when the side-suspending gap is smaller, magnetic flux captured by bulks is larger and so does the side-suspending stiffness, the bulks only need to drop a short distance to make up for the attenuation of guidance force caused by AC loss. For larger side-suspending gap, the side-suspending stiffness is smaller, for the AC loss induced by vibration, a larger distance is required for the side-suspending maglev to drop to make up for its AC loss. Therefore, it can be seen that, with the increase of the side-suspending gap, the vibration will result in the increase of the attenuation of the side-suspending height.

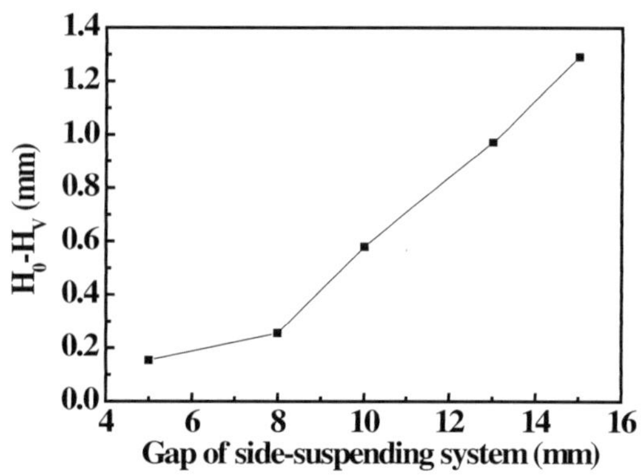

Figure 6. The attenuation ofside-suspension height with vibration amplitude of 3 mm at the gap of 5, 8, 10, 13 and 15 mm

c) The effect of the load carried by the maglev system on the stability of the side-suspending maglev system.

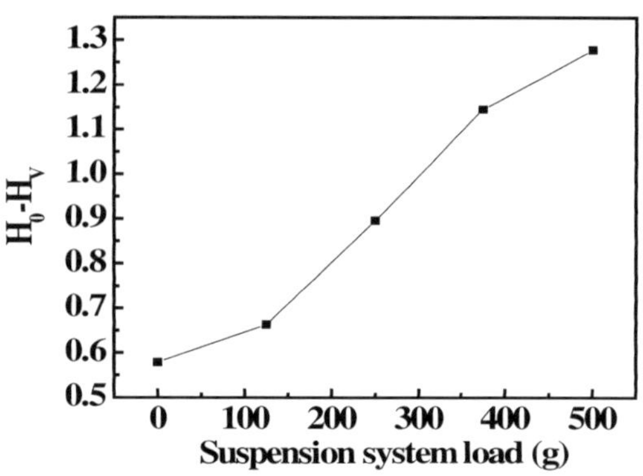

Figure 7. The attenuation of side-suspension height with the load of 0, 125, 250, 350 and 500g at the vibration amplitude of 3 mm and the gap of 10 mm

In addition, we also measured the effect of vertical vibration on the side-suspending height under different load. The results indicate that the attenuation of side-suspending height increases with the increase of load, as shown in figure 7. It perhaps means that larger load leads to larger AC loss of the bulks, and thus the larger attenuation of the side-suspending height. Moreover, under the same maglev stiffness, the maglev system with larger load needs to descend a larger distance to regain the side-suspending carrying capacity to balance its total weight. Therefore, we observe that under vibration, the attenuation of side-suspending height increases with the augment of load.

It is necessary to be remarked that under the vibration, the side-suspending system carrying capacity will not continue to decay obviously with the increase of vibration time, as shown in figure 4. Furthermore, decrease of side-suspending height seems mainly to occur when the vibrator start to work with the largest amplitude. Similar effect of AC magnetic field on levitation force was experimentally investigated by Smolyak et. al [21]. It is pointed out that at the beginning of the application of AC magnetic field, the levitation force attenuates rapidly, and then tends to a constant value gradually. In addition, we note that although the vibration leads the side-suspending height to decrease for the system, the overall decay of side-suspending height is not more than 1.5 mm. Thus slight decrease in side-suspending height will not result in a significant decrease in centripetal levitation force [8].

4. Summary

The vertical vibration will cause the HTS bulks to undergo alternating magnetic field and cause the attenuation of side-suspending height to increase due to the AC loss. Increasing the side-suspending gap, vibration amplitude and the load of the maglev system will further cause the attenuation of side-suspending height to increase, respectively. However, as the overall decrease of side-suspending height is smaller than 1.5 mm, this will not have a greater impact on the centripetal levitation force and the stability of the side-suspending maglev system. Moreover, as we know that decreasing the working temperature of HTS may increase its performance and diminish AC loss apparently, the HTS components should work at a temperature much lower than 77K for the future application of HTS maglev system.

Acknowledgments

In this paper, the authors wish to thank the fund for the project support, including the Program of International S&T Cooperation (Grant No. 2013DFA51050), the National Nature Science Foundation of China (grant No. 51271155,51377138), the Fundamental Research Funds for the Central Universities under No. 2682016ZDPY10, Foundation of Key Laboratory of Magnetic Suspension Technology and Maglev Vehicle, Ministry of Education.

References

[1] Hull J R 2000 *Supercond. Sci. Technol.* **13** R1-R15

[2] Werfel F N, Floegel-Delor U, Rothfeld R, Riedel T, Goebel B, Wippich D and Schirrmeister P 2012 *Supercond. Sci. Technol.* **25** 014007

[3] Mukoyama S, Nakao K, Sakamoto H, Matsuoka T, Nagashima K, Ogata M, Yamashita T, Miyazaki Y, Miyazaki K, Maeda T and Shimizu H 2017 *IEEE Trans. Appl. Supercond.* **27** 3600804

[4] Raghunathan R S, Kim H D and Setoguchi T 2002 *Progr. Aerosp. Sci.*, **38** 469

[5] Ma G T 2013 *IEEE Trans. Appl. Supercond.*, **23** 3024

[6] Tsuda M, Susa T, Ohyama T, Ishiyama A and Kohayashi S 2000 *IEEE Trans. Appl. Supercond.* **10** 894

[7] Ohyama T, Shimizu H, Tsuda M and Ishiyama A 2001 *IEEE Trans. Appl.Supercond.*, **11** 1988

[8] Liang Gang, Zhao Lifeng, Yang Jinlei, Ma Jiaqing, Zhang Yong and Zhao Yong 2015 *IEEE Trans. Appl.Supercond.*, **25** 3601406

[9] Zhou Dajin, Cui Chenyu, Zhao Lifeng, Zhang Yong and Zhao Yong 2017 *Supercond. Sci. Technol.* **30** 25019

[10] Zhou Dajin, Cui Chenyu, Zhao Lifeng, Zhang Yong and Zhao Yong 2017 *IEEE Trans. Appl.Supercond.* **27** 3600107

[11] Zhao Lifeng, Du Yidong, Pan Xifeng, Jiang Jing, Zhang Yong and Zhao Yong. *Rare Metal Mat. Eng.* In press.

[12] Yoshida Y, Uesaka M, Miya K 1994 *IEEE Trans. Magnetics* **30** 3503

[13] Wang X R, Song H H, Ren Z Y, Zhu M, Wang J S, Wang S Y and Wang X Z 2003 *Physica C*, **386** 536

[14] Zhang L C, Wang J S, Wang S Y and He Q Y 2007 *Physica C* **459** 43-46

[15] Ogawa J, Iwamoto M, Tsukamoto O, Murakami M and Tomita M 2002 *Physica C* **372** 1754

[16] Ogawa J, Iwamoto M and Yamagishi K 2003 *Physica C* **386** 26

[17] Yamagishi K, Ogawa J, Tsukamoto O, Murakami M and Tomita M 2003 *Physics C* **392-396** 659

[18] Johansen T H, Mestl H, and Bratsberg H 1994 *J. Appl. Phys.***75** 1667

[19] Johansen T H, Yang Z J, Bratsberg H, Helgesen G and Skjeltorp A T 1991 *Appl. Phys. Lett.*, **58** 179

[20] Johansen T H, Bratsberg H, Yang Z J, Guo S J and Loberg B 1991 *J. Appl. Phys.*, **70** 7496

[21] Smolyak B M, Ermakov G V and Chubraeva L I 2007 *Supercond. Sci. Technol.* **20** 406

Institute of Physics
Dirac House, Temple Back
Bristol BS1 6BE UK

ISSN: 1742-6588
ISBN 978-1-5108-6825-0

9 781510 868250

CURRAN ASSOCIATES INC.
proceedings.com